Lecture Notes in Computer Science 15071

Founding Editors

Gerhard Goos
Juris Hartmanis

The series Lecture Notes in Computer Science (LNCS), including its subseries Lecture Notes in Artificial Intelligence (LNAI) and Lecture Notes in Bioinformatics (LNBI), has established itself as a medium for the publication of new developments in computer science and information technology research, teaching, and education.

LNCS enjoys close cooperation with the computer science R & D community, the series counts many renowned academics among its volume editors and paper authors, and collaborates with prestigious societies. Its mission is to serve this international community by providing an invaluable service, mainly focused on the publication of conference and workshop proceedings and postproceedings. LNCS commenced publication in 1973.

Aleš Leonardis · Elisa Ricci · Stefan Roth ·
Olga Russakovsky · Torsten Sattler · Gül Varol
Editors

Computer Vision – ECCV 2024

18th European Conference
Milan, Italy, September 29–October 4, 2024
Proceedings, Part XIII

 Springer

Editors
Aleš Leonardis 🔟
University of Birmingham
Birmingham, UK

Elisa Ricci 🔟
University of Trento
Trento, Italy

Stefan Roth 🔟
Technical University of Darmstadt
Darmstadt, Germany

Olga Russakovsky 🔟
Princeton University
Princeton, NJ, USA

Torsten Sattler 🔟
Czech Technical University in Prague
Prague, Czech Republic

Gül Varol 🔟
École des Ponts ParisTech
Marne-la-Vallée, France

ISSN 0302-9743   ISSN 1611-3349 (electronic)
Lecture Notes in Computer Science
ISBN 978-3-031-72623-1   ISBN 978-3-031-72624-8 (eBook)
https://doi.org/10.1007/978-3-031-72624-8

This Springer imprint is published by the registered company Springer Nature Switzerland AG
The registered company address is: Gewerbestrasse 11, 6330 Cham, Switzerland

If disposing of this product, please recycle the paper.

Foreword

Welcome to the proceedings of the European Conference on Computer Vision (ECCV) 2024, the 18th edition of the conference, which has been held biennially since its founding in France in 1990. We are holding the event in Italy for the third time, in a convention centre where we can accommodate nearly 7000 in-person attendees.

The journey to ECCV 2024 started at CVPR 2019, where the general chairs met at a sketchy bar in Long Beach to discuss ideas on how to organize an amazing future ECCV. After that, Covid came and changed conferences probably forever, but we maintained our excitement to organize an awesome in-person conference in 2024, particularly having seen the wonderful event that was ECCV 2022 in Tel Aviv. And here we are, 5 years later, delighted that we finally get to welcome all of you to Milan!

First and foremost, ECCV is about the exchange of scientific ideas, and hence the most important organizational task is the selection of the programme. To our Program Chairs, Aleš Leonardis, Elisa Ricci, Gül Varol, Olga Russakovsky, Stefan Roth, and Torsten Sattler, our entire community owe a huge debt of gratitude. As you will read in their Preface to the proceedings, they dealt with over 8,500 submissions, and their diligence, thoroughness, and sheer effort has been inspirational. From the early stages of designing the call for papers, through recruiting and selecting area chairs and reviewers, to ensuring every paper has a fair and thorough assessment, their professionalism, commitment, and attention to detail has been exemplary. From all the choices that we made to organize ECCV, choosing this amazing team of Program Chairs has been, without a doubt, our very best decision.

The Program Chairs were advised and supported by an Ethics Review Committee, Chloé Bakalar, Kate Saenko, Remi Denton, and Yisong Yue, who provided invaluable input on papers where reviewers or Area Chairs had raised ethical concerns.

The Publication Chairs, Jovita Lukasik, Michael Möller, François Brémond, and Mahmoud Ali, did great work in assembling the camera-ready papers into these Springer volumes, dealing with numerous issues that arose promptly and professionally.

With the ever-expanding importance of workshops, tutorials, and demos at our major conferences comes a corresponding increase in the challenge of organizing over 70 workshops and 9 tutorials attached to the main conference. The Workshop and Tutorial Chairs, Alessio Del Bue, Jordi Pont-Tuset, Cristian Canton, and Tatiana Tommasi, and Demo Chairs, Hyung Chang and Marco Cristani, worked tirelessly to coordinate the very varied demands and structures of these different events, yielding an extremely rich auxiliary program which will enhance the conference experience for everyone who attends.

The Industry Chairs, Shaogang Gong and Cees Snoek, working with Limor Urfaly and Lior Gelfand, managed to attract numerous companies and organizations from all around the world to the industrial Expo, making ECCV 2024 an international event that not only showcases excellent foundational research, but also presents how such research can be transformed in real products. It is notable that together with the big companies, several start-ups chose ECCV to present their business ideas and activities.

As usual in the recent main conference, we also organized Speed Mentoring and Doctoral Consortium events. While the latter is an institutional occasion allowing senior PhDs to show their work, the former is especially important to answer to the many doubts and uncertainties young scholars have while undertaking a career in Computer Vision and Artificial Intelligence. We warmly thank our Social Activities Chairs, Raffaella Lanzarotti, Simone Bianco and Giovanni Farinella, and the Doctoral Consortium Chairs, Cigdem Beyan and Or Litany, for their dedication to organize such events, so important for our young colleagues, as well as all the mentors who were available to share their experience.

As the conference becomes larger and larger, the overall budget increases to a level where uncertainties in estimates of attendance can result in significant losses, perhaps enough to significantly damage the prospects of running the conference in the future. This risk imposed the need to require authors to commit to full conference registrations, with the undesirable consequent possibility to cause hardship to authors, particularly student authors, from organizations where funding attendance is difficult. To mitigate this effect, we allocated travel grants to students who might otherwise have had difficulty in attending. Our Diversity Chairs, David Fouhey and Rita Cucchiara, put tremendous effort into a wonderfully careful and thoughtful programme, balancing diversity in all its forms with a changing budget landscape, allowing us to support the participation of many people who might not otherwise have been able to attend.

This year's conference also marks the move to a unified, multi-year conference website, based on that used for other leading computer vision and AI conferences. This was an effort that we decided to take on to make the ECCV website more familiar to the community and support the organization of the future ECCV editions. Our thanks go to Lee Campbell of Eventhosts who managed the entire design and development of the new website and provided responsive support and updates throughout the conference organization timeline.

Our Finance Chairs, Gérard Medioni and Nicole Finn, provided invaluable advice and assistance on all aspects of the financial planning of the conference, in conjunction with the professional conference organizers AIM Group, where Lavinia Ricci led a fantastic team that managed thousands of details from space planning to food to registrations. In particular, Elder Bromley and Mara Carletti were a tremendous support to the entire conference organization. The conference centre, Mico DMC, also provided an excellent service, showing considerable flexibility in adapting the venue and processes to the particular demands of an academic-focused conference.

Finally, we cannot forget to thank our Social Media Chair, Kosta Derpanis, supported by Abby Stylianou and Jia-Bin Huang, who, with their presence on social media, spread ECCV news and responded to small and big questions in a timely manner, helping the community to promptly receive information and fix their problems.

October 2024

Andrew Fitzgibbon
Laura Leal-Taixé
Vittorio Murino

Preface

ECCV 2024 saw a remarkable 48% increase in submissions, with 8585 valid papers submitted, compared to 5804 in ECCV 2022, 5150 in 2020, and 2439 in 2018. Of these, 2387 papers (27.9%) were accepted for publication, with 200 (2.3% overall) selected for oral presentations.

A total of 173 complete submissions were desk-rejected for various reasons. Common issues included revealing author identities in the paper or supplementary material (for example: by including the author names, referring to specific grants in the acknowledgments, including links to GitHub accounts with visible author information, etc.), exceeding the page limit or otherwise significantly altering the submission template, or posting an arXiv preprint explicitly mentioning the work as an ECCV 2024 submission. The desk-rejected submissions also include 14 papers identified as dual submissions to other concurrent conferences, such as ICML, NeurIPS, and MICCAI. Seven papers were rejected among the accepted papers due to plagiarism issues, after the Program Chairs conducted a thorough plagiarism check with the iThenticate software followed by manual inspection.

The double-blind review process was managed using the CMT system, ensuring anonymity between authors and Area Chairs/reviewers. Each paper received at least three reviews, amounting to over 26,000 reviews in total. To maintain the quality and the fairness of the reviews, we recruited 469 Area Chairs (ACs) and more than 8200 reviewers. Of the recruited reviewers, 7293 ultimately participated.

Area Chairs were chosen for their technical expertise and reputation; many of them had served in similar roles at other top conferences. We also recruited additional Area Chairs through an open call for self-nominations. Among the selected ACs, around 18% were women. Geographical distribution was taken into account during the selection process, with 31% of ACs working in Europe, 38% in the Americas, 27% in Asia, and 4% in the rest of the world. Despite our best efforts to consider diversity factors in the selection process, the gender diversity and geographic diversity of ACs is still below what we were hoping for.

To ensure a smooth decision-making process, we grouped the ACs into triplets. Upon accepting their role, ACs were requested to enter into the CMT system domain conflicts and subject areas, as well as information about their DBLP, TPMS (Toronto Paper Matching System), and OpenReview profiles. These were used to determine whether different subject areas were sufficiently covered, to detect conflicts, and to automatically form AC triplets. An AC triplet was a group of three ACs who worked together throughout the process, which helped to facilitate discussion, calibrate the decision-making process, and provide support for first-time ACs. AC triplets were manually refined by Program Chairs to ensure no triplet contained more than two major time zones to enable easier coordination for virtual AC meetings. Paper assignments were made such that no AC within the triplet was conflicted with any of the papers. Each AC was responsible for overseeing 18 papers on average.

In each triplet, we identified a Lead AC who was an experienced member of the computer vision community willing to take on additional administrative duties. The primary responsibilities of Lead ACs included coordinating the AC triplet to ensure deadlines were met and offering guidance to less experienced ACs. They also served as the main contact person with the Program Chairs and helped in handling papers if an original AC was unable to complete their tasks.

Reviewers were invited from reviewer pools of previous conferences and also selected from the pool of authors. We asked experienced ACs to recommend more potential reviewers. We further recruited reviewers through self-nominations, where researchers volunteered to review for ECCV 2024 via an open call. The self-nominations were then filtered, selecting those reviewers with at least two papers published in related conferences such as ECCV, ICCV, CVPR, ICLR, NeurIPS, etc.

Similarly to ACs, reviewers were also asked to update their CMT, Toronto Paper Matching System (TPMS), and OpenReview profiles. Reviewers had the option to classify themselves as either senior researchers or junior researchers and students, with the number of assigned papers varying accordingly (a quota of 6 for senior researchers with 3 or more times reviewing experience, and a quota of 4 otherwise). On average, each reviewer was assigned about 4 papers.

Conflicts of interest among authors, ACs, and reviewers were managed automatically via conflict domains and co-authorship information from DBLP. Paper assignments to ACs and reviewers were determined using an algorithm that combined subject-area affinity scores from CMT, TPMS, and OpenReview, along with additional criteria such as ensuring that every paper had at least one non-student reviewer. For AC assignments, we additionally incorporated a newly built affinity score based on embedding similarity with the papers on the Google Scholar profile of the ACs. Once the assignments were made, ACs and reviewers were asked to promptly report any undetected conflicts of interest or other concerns that would preclude them from handling the assigned papers to ensure immediate reassignment, if necessary.

Given the widespread use of Large Language Models (LLMs), ECCV 2024, like previous conferences, implemented a special policy regulating their use. Authors were permitted to use any tools, including LLMs, in preparing their papers, but they remained fully responsible for any misrepresentation, factual inaccuracies, or plagiarism. Reviewers were also allowed to use LLMs to refine the wording of their reviews; however, they were held accountable for the accuracy of their content and were strictly prohibited from inputting submissions into an LLM. One challenge we ran into was that tools that detect LLM-generated content were not able to distinguish between text polished vs. text written from scratch by an LLM. Thus, it became quite difficult to enforce the policy in practice.

Overall, the challenges encountered during the review process were consistent with those faced at previous computer vision conferences. For instance, a small number of reviewers ultimately did not submit their reviews or provided brief, uninformative reviews, necessitating last-minute reassignment to emergency reviewers. As the community expands and the number of submissions rises, securing enough qualified reviewers and performing a quality check of the received reviews becomes increasingly difficult.

The review process included a rebuttal phase where authors were given a week to address any concerns raised by the reviewers. Each response was limited to a single PDF page using a predefined template. Rebuttals with formatting or anonymity violations were removed by the Program Chairs. ACs then facilitated discussions with reviewers to evaluate the merits of each submission. While the goal was to reach a consensus, the final decision was left to the ACs in the triplet to ensure fairness. As is common, the ACs within each triplet were tasked with organizing meetings to discuss papers, especially borderline cases. The entire process was conducted online, with no in-person meetings. For each paper, a primary and a secondary AC were assigned from the AC triplet. The primary AC was responsible for entering the decisions and meta-reviews into the CMT system, which were then reviewed and approved by the secondary ACs. Besides making accept/reject decisions, the ACs were also required to provide recommendations regarding oral/poster presentations and award nominations.

Reviewers and ACs were asked to flag papers with potential ethical concerns. An ethics review committee, consisting of four leading researchers from both industry and academia, was appointed. The role of the committee was to evaluate papers escalated by the AC for further investigation. In total, 31 papers were flagged to the committee. The committee requested that the authors of 15 papers address specific concerns in their camera-ready submissions. The authors all complied with the requests.

Following the recent IEEE decision to ban the use of the photograph of Lena Forsén, we automatically identified all submissions that included this image. Although the ban did not officially apply to ECCV, we nevertheless shared the context with the authors of the affected papers and suggested that they remove the photograph in the camera-ready submission. All authors complied.

To help the ACs to track the overall progress of the reviewing and decision-making processes, we created AC triplet activity dashboards for each AC triplet (via Google sheets), which were updated on a regular basis. During the review process, the activity dashboard enabled ACs to track the progress of incoming reviews for each paper and promptly identify papers that necessitated attention, e.g., papers potentially requiring the invitation of emergency reviewers, or papers with suspiciously short reviews. During the decision-making process, the dashboard was also used by the Program Chairs to identify papers with incomplete review information, e.g., for which reviewers did not enter their final recommendation, the AC did not enter their final decision or meta-review, the secondary AC did not confirm the recommendation, etc.

At the end of the decision-making progress, the Program Chairs carefully examined the very small number of cases where ACs overruled a unanimous reviewer recommendation for acceptance or rejection. In such instances, it was crucial for the primary and secondary ACs to explain in detail the motivation behind the decision. To ensure a fair evaluation, an additional Area Chair, not originally part of the decision-making triplet, was appointed to provide an independent opinion and advise the Program Chairs in determining whether to uphold or reverse the overturn.

Inevitably, after decisions were released, some authors were dissatisfied. The authors were given the possibility to appeal the final decision in case of procedural issues encountered during the review process by filling out a form to directly contact the Program Chairs. Valid reasons for appealing a decision included policy errors (e.g., reviewers or ACs enforcing a non-existent policy), clerical errors (e.g., the meta-review clearly indicated an intention to accept a paper, but it was mistakenly rejected), and significant misunderstandings by the reviewers or ACs. In total, we received 59 appeals and upheld 6 of them. One was due to a clear policy error by an AC, and the paper decision changed from reject to accept following a successful appeal. The other 5 papers were already conditionally accepted – but the AC/reviewers wrongly flagged them as papers for which the associated dataset was required to be released prior to final acceptance; this condition was removed following the appeal.

After the camera-ready deadline, additional checks were conducted, particularly for papers contributing a new dataset and for those flagged for potential ethical concerns. Specifically, during the submission phase, authors were required to specify if the primary contribution of their papers was the release of a new dataset, and for these papers, a valid dataset link had to be provided by the camera-ready deadline. A total of 435 submissions included a dataset as part of their contribution, and all complied with the requirement to release the dataset by the camera-ready deadline.

We sincerely thank all the ACs and reviewers for their invaluable contributions to the review process. Their dedication and expertise were crucial in maintaining the high standards of the conference. We would also like to thank our Technical Program Chair, Sascha Hornauer, who did tremendous work behind the scenes, and to the entire CMT team for their prompt support with any technical issues.

October 2024

Aleš Leonardis
Elisa Ricci
Stefan Roth
Olga Russakovsky
Torsten Sattler
Gül Varol

Organization

General Chairs

Andrew Fitzgibbon   Graphcore, UK
Laura Leal-Taixé   NVIDIA, Italy
Vittorio Murino   University of Verona & University of Genoa, Italy

Program Chairs

Aleš Leonardis   University of Birmingham, UK
Elisa Ricci   University of Trento, Italy
Stefan Roth   Technical University of Darmstadt, Germany
Olga Russakovsky   Princeton University, USA
Torsten Sattler   Czech Technical University in Prague, Czech Republic
Gül Varol   Ecole des Ponts Paris Tech, France

Technical Program Chair

Sascha Hornauer   Mines Paris – PSL, France

Industrial Liaison Chairs

Cees Snoek   University of Amsterdam, Netherlands
Shaogang Gong   Queen Mary University of London, UK

Publication Chairs

Francois Bremond   Inria, France
Mahmoud Ali   Inria, France
Jovita Lukasik   University of Siegen, Germany
Michael Moeller   University of Siegen, Germany

Poster Chairs

Aljoša Ošep   Carnegie Mellon University, USA
Zuzana Kukelova   Czech Technical University in Prague,
                              Czech Republic

Diversity Chairs

David Fouhey   New York University, USA
Rita Cucchiara   Università di Modena e Reggio Emilia, Italy

Local Chairs

Raffaella Lanzarotti   Università degli Studi di Milano, Italy
Simone Bianco   University of Milano-Bicocca, Italy

Conference Ombuds

Georgia Gkioxari   California Institute of Technology, USA
Greg Mori   Borealis AI & Simon Fraser University, Canada

Ethics Review Committee

Chloé Bakalar   Meta, USA
Kate Saenko   Boston University, USA
Remi Denton   Google Research, USA
Yisong Yue   Caltech, Asari AI & Latitude AI, USA

Workshop and Tutorial Chairs

Alessio Del Bue   Istituto Italiano di Tecnologia, Italy
Cristian Canton   Meta AI, USA
Jordi Pont-Tuset   Google DeepMind, Switzerland
Tatiana Tommasi   Politecnico di Torino, Italy

Finance Chairs

Gerard Medioni   Amazon, USA
Nicole Finn   c to c events, USA

Demo Chairs

Hyung Chang   University of Birmingham, UK
Marco Cristani   University of Verona, Italy

Publicity and Social Media Chairs

Konstantinos Derpanis   York University & Samsung AI Centre Toronto,
                                          Canada
Jia-Bin Huang   University of Maryland College Park, USA
Abby Stylianou   Saint Louis University, USA

Social Activities Chairs

Giovanni Maria Farinella   University of Catania, Italy
Raffaella Lanzarotti   Università degli Studi di Milano, Italy
Simone Bianco   University of Milano-Bicocca, Italy

Doctoral Consortium Chairs

Cigdem Beyan   University of Verona, Italy
Or Litany   NVIDIA & Technion, USA

Web Developer

Lee Campbell   Eventhosts, USA

Area Chairs

Ehsan Adeli	Stanford University, USA
Aishwarya Agrawal	University of Montreal & Mila & DeepMind, Canada
Naveed Akhtar	University of Western Australia, Australia
Yagiz Aksoy	Simon Fraser University, Canada
Karteek Alahari	Inria, France
Jose Alvarez	NVIDIA, USA
Djamila Aouada	Interdisciplinary Centre for Security, Reliability and Trust, University of Luxembourg, Luxembourg
Andre Araujo	Google, Brazil
Pablo Arbelaez	Universidad de los Andes, Colombia
Iro Armeni	Stanford University, USA
Yuki Asano	University of Amsterdam, Netherlands
Karl Åström	Lund University, Sweden
Mathieu Aubry	Ecole des Ponts ParisTech, France
Shai Bagon	Weizmann Institute of Science, Israel
Song Bai	ByteDance, Singapore
Xiang Bai	Huazhong University of Science and Technology, China
Yang Bai	Tencent, China
Lamberto Ballan	University of Padova, Italy
Linchao Bao	University of Birmingham, UK
Ronen Basri	Weizmann Institute of Science, Israel
Sara Beery	Massachusetts Institute of Technology, USA
Vasileios Belagiannis	University of Erlangen–Nuremberg, Germany
Serge Belongie	University of Copenhagen, Denmark
Sagie Benaim	Hebrew University of Jerusalem, Israel
Rodrigo Benenson	Google, Switzerland
Cigdem Beyan	University of Bergamo, Italy
Bharat Bhatnagar	Meta Reality Labs Research, Switzerland
Tolga Birdal	Imperial College London, UK
Matthew Blaschko	KU Leuven, Belgium
Federica Bogo	Meta Reality Labs Research, Switzerland
Timo Bolkart	Google, Switzerland
Katherine Bouman	California Institute of Technology, USA
Lubomir Bourdev	Primepoint, USA
Edmond Boyer	Inria, France
Yuri Boykov	University of Waterloo, Canada
Eric Brachmann	Niantic, Germany

Wieland Brendel	University of Tübingen, Germany
Gabriel Brostow	University College London, UK
Michael Brown	York University, Canada
Andrés Bruhn	University of Stuttgart, Germany
Andrei Bursuc	valeo.ai, France
Benjamin Busam	Technical University of Munich, Germany
Holger Caesar	TU Delft, Netherlands
Jianfei Cai	Monash University, Australia
Simone Calderara	University of Modena and Reggio Emilia, Italy
Octavia Camps	Northeastern University, Boston, USA
Joao Carreira	DeepMind, UK
Silvia Cascianelli	University of Modena and Reggio Emilia, Italy
Ayan Chakrabarti	Google Research, USA
Tat-Jen Cham	Nanyang Technological University, Singapore
Antoni Chan	City University of Hong Kong, China
Manmohan Chandraker	University of California, San Diego, USA
Xiaojun Chang	University of Technology Sydney, Australia
Devendra Singh Chaplot	Mistral AI, USA
Liang-Chieh Chen	TikTok, USA
Long Chen	Hong Kong University of Science and Technology, China
Mei Chen	Microsoft, USA
Shizhe Chen	Inria, France
Shuo Chen	RIKEN, Japan
Xilin Chen	Institute of Computing Technology, Chinese Academy of Sciences, China
Anoop Cherian	Mitsubishi Electric Research Labs, USA
Tat-Jun Chin	University of Adelaide, Australia
Minsu Cho	Pohang University of Science and Technology, Korea
Hisham Cholakkal	Mohamed bin Zayed University of Artificial Intelligence, United Arab Emirates
Yung-Yu Chuang	National Taiwan University, Taiwan
Ondrej Chum	Czech Technical University in Prague, Czech Republic
Marcella Cornia	University of Modena and Reggio Emilia, Italy
Marco Cristani	University of Verona, Italy
Cristian Canton	Meta AI, USA
Zhaopeng Cui	Zhejiang University, China
Angela Dai	Technical University of Munich, Germany
Jifeng Dai	Tsinghua University, China
Yuchao Dai	Northwestern Polytechnical University, China

Dima Damen	University of Bristol, UK
Kostas Daniilidis	University of Pennsylvania, USA
Antitza Dantcheva	Inria, France
Raoul de Charette	Inria, France
Shalini De Mello	NVIDIA Research, USA
Tali Dekel	Weizmann Institute of Science, Israel
Ilke Demir	Intel Corporation, USA
Joachim Denzler	Friedrich Schiller University Jena, Germany
Konstantinos Derpanis	York University, Canada
Jose Dolz	École de technologie supérieure Montreal, Canada
Jiangxin Dong	Nanjing University of Science and Technology, China
Hazel Doughty	Leiden University, Netherlands
Iddo Drori	Boston University and Columbia University, USA
Enrique Dunn	Stevens Institute of Technology, USA
Thibaut Durand	Borealis AI, Canada
Sayna Ebrahimi	Google, USA
Mohamed Elhoseiny	King Abdullah University of Science and Technology, Saudi Arabia
Bin Fan	University of Science and Technology Beijing, China
Yi Fang	New York University, USA
Giovanni Maria Farinella	University of Catania, Italy
Ryan Farrell	Brigham Young University, USA
Alireza Fathi	Google, USA
Christoph Feichtenhofer	Meta & FAIR, USA
Rogerio Feris	MIT-IBM Watson AI Lab & IBM Research, USA
Basura Fernando	Agency for Science, Technology and Research (A*STAR), Singapore
David Fouhey	New York University, USA
Katerina Fragkiadaki	Carnegie Mellon University, USA
Friedrich Fraundorfer	Graz University of Technology, Austria
Oren Freifeld	Ben-Gurion University, Israel
Huan Fu	University of Sydney, China
Yasutaka Furukawa	Simon Fraser University, Canada
Andrea Fusiello	University of Udine, Italy
Fabio Galasso	Sapienza University, Italy
Jürgen Gall	University of Bonn, Germany
Orazio Gallo	NVIDIA Research, USA
Chuang Gan	MIT-IBM Watson AI Lab, USA
Lianli Gao	University of Electronic Science and Technology of China, China

Shenghua Gao	University of Hong Kong, China
Sourav Garg	University of Adelaide, Australia
Efstratios Gavves	University of Amsterdam, Netherlands
Peter Gehler	Zalando, Germany
Theo Gevers	University of Amsterdam, Netherlands
Golnaz Ghiasi	Google DeepMind, USA
Rohit Girdhar	Carnegie Mellon University, USA
Xavier Giro-i-Nieto	Universitat Politecnica de Catalunya, Spain
Ross Girshick	Allen Institute for Artificial Intelligence, USA
Zan Gojcic	NVIDIA, Switzerland
Vladislav Golyanik	Max Planck Institute for Informatics, Germany
Stephen Gould	Australian National University, Australia
Liangyan Gui	University of Illinois Urbana-Champaign, USA
Fatma Guney	Koc University, Turkey
Yulan Guo	Sun Yat-sen University, China
Yunhui Guo	University of Texas at Dallas, USA
Mohit Gupta	University of Wisconsin-Madison, USA
Minh Ha Quang	RIKEN Center for Advanced Intelligence Project, Japan
Bumsub Ham	Yonsei University, Korea
Bohyung Han	Seoul National University, Korea
Kai Han	University of Hong Kong, China
Xiaoguang Han	Shenzhen Research Institute of Big Data & Chinese University of Hong Kong (Shenzhen), China
Tatsuya Harada	University of Tokyo & RIKEN, Japan
Tal Hassner	Weir AI, USA
Xuming He	ShanghaiTech University, China
Felix Heide	Princeton & Algolux, USA
Anthony Hoogs	Kitware, USA
Timothy Hospedales	Edinburgh University, UK
Mahdi Hosseini	Concordia University, Canada
Peng Hu	College of Computer Science, Sichuan University, China
Gang Hua	Wormpex AI Research, USA
Jia-Bin Huang	University of Maryland College Park, USA
Qixing Huang	University of Texas at Austin, USA
Sharon Xiaolei Huang	Pennsylvania State University, USA
Junhwa Hur	Google, USA
Nazli Ikizler-Cinbis	Hacettepe University, Turkey
Eddy Ilg	Saarland University, Germany
Ahmet Iscen	Google, France

Nathan Jacobs   Washington University in St. Louis, USA
Varun Jampani   Stability AI, USA
C. V. Jawahar   International Institute of Information Technology
                               - Hyderabad, India
Laszlo Jeni   Carnegie Mellon University, USA
Jiaya Jia   Chinese University of Hong Kong, China
Qin Jin   Renmin University of China, China
Jungseock Joo   NVIDIA & University of California, Los
                               Angeles, USA
Fredrik Kahl   Chalmers, Sweden
Yannis Kalantidis   NAVER LABS Europe, France
Vicky Kalogeiton   Ecole Polytechnique, IP Paris, France
Evangelos Kalogerakis   University of Massachusetts Amherst, USA
Hiroshi Kawasaki   Kyushu University, Japan
Margret Keuper   University of Mannheim & Max Planck Institute
                               for Informatics, Germany
Salman Khan   Mohamed bin Zayed University of Artificial
                               Intelligence, United Arab Emirates
Anna Khoreva   Bosch Center for Artificial Intelligence, Germany
Gunhee Kim   Seoul National University, Korea
Junmo Kim   Korea Advanced Institute of Science and
                               Technology, Korea
Min H. Kim   Korea Advanced Institute of Science and
                               Technology, Korea
Seon Joo Kim   Yonsei University, Korea
Tae-Kyun Kim   Korea Advanced Institute of Science and
                               Technology & Imperial College London, Korea
Benjamin Kimia   Brown University, USA
Hedvig Kjellström   KTH Royal Institute of Technology, Sweden
Laurent Kneip   ShanghaiTech University, China
A. Sophia Koepke   University of Tübingen, Germany
Iasonas Kokkinos   University College London, UK
Wai-Kin Adams Kong   Nanyang Technological University, Singapore
Piotr Koniusz   Data61/CSIRO & Australian National University,
                               Australia
Jana Kosecka   George Mason University, USA
Adriana Kovashka   University of Pittsburgh, USA
Hilde Kuehne   University of Bonn, Germany
Zuzana Kukelova   Czech Technical University in Prague,
                               Czech Republic
Ajay Kumar   Hong Kong Polytechnic University, China
Kiriakos Kutulakos   University of Toronto, Canada

Suha Kwak   Pohang University of Science and Technology,
                                Korea
Zorah Laehner   University of Bonn, Germany
Shang-Hong Lai   National Tsing Hua University, Taiwan
Iro Laina   University of Oxford, UK
Jean-Francois Lalonde   Université Laval, Canada
Loic Landrieu   École des Ponts ParisTech, France
Diane Larlus   Naver Labs Europe, France
Viktor Larsson   Lund University, Sweden
Stéphane Lathuilière   Telecom-Paris, France
Gim Hee Lee   National University of Singapore, Singapore
Yong Jae Lee   University of Wisconsin-Madison, USA
Bastian Leibe   RWTH Aachen University, Germany
Ales Leonardis   University of Birmingham, UK
Vincent Lepetit   Ecole des Ponts ParisTech, France
Fuxin Li   Oregon State University, USA
Hongdong Li   Australian National University, Australia
Xi Li   Zhejiang University, China
Yin Li   University of Wisconsin-Madison, USA
Yu Li   International Digital Economy Academy,
                                Singapore
Xiaodan Liang   Sun Yat-sen University, China
Shengcai Liao   Inception Institute of Artificial Intelligence,
                                United Arab Emirates
Jongwoo Lim   Seoul National University, Korea
Ser-Nam Lim   University of Central Florida, USA
Stephen Lin   Microsoft Research, China
Tsung-Yi Lin   NVIDIA Research, USA
Yen-Yu Lin   National Yang Ming Chiao Tung University,
                                Taiwan
Yutian Lin   Wuhan University, China
Zhe Lin   Adobe Research, USA
Haibin Ling   Stony Brook University, USA
Or Litany   NVIDIA, USA
Jiaying Liu   Peking University, China
Jun Liu   Lancaster University, UK
Miaomiao Liu   Australian National University, Australia
Si Liu   Beihang University, China
Sifei Liu   NVIDIA, USA
Wei Liu   Tencent, China
Yanxi Liu   Pennsylvania State University, USA
Yaoyao Liu   Johns Hopkins University, USA

Yebin Liu	Tsinghua University, China
Zicheng Liu	Microsoft, USA
Ziwei Liu	Nanyang Technological University, Singapore
Ismini Lourentzou	University of Illinois, Urbana-Champaign, USA
Chen Change Loy	Nanyang Technological University, Singapore
Feng Lu	Beihang University, China
Le Lu	Alibaba Group, USA
Oisin Mac Aodha	University of Edinburgh, UK
Luca Magri	Politecnico di Milano, Italy
Michael Maire	University of Chicago, USA
Subhransu Maji	University of Massachusetts, Amherst, USA
Atsuto Maki	KTH Royal Institute of Technology, Sweden
Yasushi Makihara	Osaka University, Japan
Massimiliano Mancini	University of Trento, Italy
Kevis-Kokitsi Maninis	Google Research, Switzerland
Kenneth Marino	Google DeepMind, USA
Renaud Marlet	Ecole des Ponts ParisTech & Valeo, France
Niki Martinel	University of Udine, Italy
Jiri Matas	Czech Technical University in Pragu, Czech Republic
Yasuyuki Matsushita	Osaka University, Japan
Simone Melzi	University of Milano-Bicocca, Italy
Thomas Mensink	Google Research, Netherlands
Michele Merler	IBM Research, USA
Pascal Mettes	University of Amsterdam, Netherlands
Ajmal Mian	University of Western Australia, Australia
Krystian Mikolajczyk	Imperial College London, UK
Ishan Misra	Facebook AI Research, USA
Niloy Mitra	University College London, UK
Anurag Mittal	Indian Institute of Technology Madras, India
Davide Modolo	Amazon, USA
Davide Moltisanti	University of Bath, UK
Philippos Mordohai	Stevens Institute of Technology, USA
Francesc Moreno	Institut de Robotica i Informatica Industrial, Spain
Pietro Morerio	Istituto Italiano di Tecnologia, Italy
Greg Mori	Simon Fraser University & Borealis AI, Canada
Roozbeh Mottaghi	FAIR @ Meta, USA
Yadong Mu	Peking University, China
Vineeth N. Balasubramanian	Indian Institute of Technology, India
Hajime Nagahara	Osaka University, Japan
Vinay Namboodiri	University of Bath, UK
Liangliang Nan	Delft University of Technology, Netherlands

Michele Nappi	University of Salerno, Italy
Srinivasa Narasimhan	Carnegie Mellon University, USA
P. J. Narayanan	International Institute of Information Technology, Hyderabad, India
Nassir Navab	Technical University of Munich, Germany
Ram Nevatia	University of Southern California, USA
Natalia Neverova	Facebook AI Research, France
Juan Carlos Niebles	Salesforce & Stanford University, USA
Michael Niemeyer	Google, Germany
Simon Niklaus	Adobe Research, USA
Ko Nishino	Kyoto University, Japan
Nicoletta Noceti	MaLGa-DIBRIS & Università degli Studi di Genova, Italy
Matthew O'Toole	Carnegie Mellon University, USA
Jean-Marc Odobez	Idiap Research Institute, École polytechnique fédérale de Lausanne, Switzerland
Francesca Odone	Università di Genova, Italy
Takayuki Okatani	Tohoku University & RIKEN Center for Advanced Intelligence Project, Japan
Carl Olsson	Lund University, Sweden
Vicente Ordonez	Rice University, USA
Aljosa Osep	Technical University of Munich, Germany
Martin R. Oswald	University of Amsterdam, Netherlands
Andrew Owens	University of Michigan, USA
Tomas Pajdla	Czech Technical University in Prague, Czechia
Manohar Paluri	Meta, USA
Jinshan Pan	Nanjing University of Science and Technology, China
In Kyu Park	Inha University, Korea
Jaesik Park	Seoul National University, Korea
Despoina Paschalidou	Stanford, USA
Georgios Pavlakos	University of Texas at Austin, USA
Vladimir Pavlovic	Rutgers University, USA
Marcello Pelillo	University of Venice, Italy
David Picard	École des Ponts ParisTech, France
Hamed Pirsiavash	University of California, Davis, USA
Bryan Plummer	Boston University, USA
Thomas Pock	Graz University of Technology, Austria
Fabio Poiesi	Fondazione Bruno Kessler, Italy
Marc Pollefeys	ETH Zurich & Microsoft, Switzerland
Jean Ponce	Ecole normale supérieure - PSL, France
Gerard Pons-Moll	University of Tübingen, Germany

Laura Sevilla-Lara	University of Edinburgh, UK
Fahad Shahbaz Khan	Mohamed bin Zayed University of Artificial Intelligence, United Arab Emirates
Qi Shan	Apple, USA
Shiguang Shan	Institute of Computing Technology, Chinese Academy of Sciences, China
Viktoriia Sharmanska	University of Sussex & Imperial College London, UK
Eli Shechtman	Adobe Research, USA
Evan Shelhamer	DeepMind, UK
Lu Sheng	Beihang University, China
Boxin Shi	Peking University, China
Jianbo Shi	University of Pennsylvania, USA
Mike Zheng Shou	National University of Singapore, Singapore
Abhinav Shrivastava	University of Maryland, USA
Aliaksandr Siarohin	Snap, USA
Kaleem Siddiqi	McGill University, Canada
Leonid Sigal	University of British Columbia, Canada
Oriane Siméoni	valeo.ai, France
Krishna Kumar Singh	Adobe Research, USA
Richa Singh	Indian Institute of Technology Jodhpur, India
Yale Song	Facebook AI Research, USA
Concetto Spampinato	University of Catania, Italy
Srinath Sridhar	Brown University, USA
Bjorn Stenger	Rakuten, Japan
Abby Stylianou	Saint Louis University, USA
Akihiro Sugimoto	National Institute of Informatics, Japan
Chen Sun	Brown University, USA
Deqing Sun	Google, USA
Jian Sun	Xi'an Jiaotong University, China
Min Sun	National Tsing Hua University, Taiwan
Qianru Sun	Singapore Management University, Singapore
Yu-Wing Tai	Dartmouth College, USA
Ayellet Tal	Technion, Israel
Ping Tan	Hong Kong University of Science and Technology, China
Robby Tan	National University of Singapore, Singapore
Chi-Keung Tang	Hong Kong University of Science and Technology, China
Makarand Tapaswi	International Institute of Information Technology, Hyderabad & Wadhwani AI, India
Justus Thies	Technical University of Darmstadt, Germany

Radu Timofte	University of Wurzburg & ETH Zurich, Germany
Giorgos Tolias	Czech Technical University in Prague, Czechia
Federico Tombari	Google & Technical University of Munich, Switzerland
Tatiana Tommasi	Politecnico di Torino, Italy
James Tompkin	Brown University, USA
Matthew Trager	AWS AI Labs, USA
Anh Tran	VinAI, Vietnam
Du Tran	Google, USA
Shubham Tulsiani	Carnegie Mellon University, USA
Sergey Tulyakov	Snap, USA
Dimitrios Tzionas	University of Amsterdam, Netherlands
Maria Vakalopoulou	CentraleSupelec, France
Joost van de Weijer	Computer Vision Center, Spain
Laurens van der Maaten	Meta, USA
Jan van Gemert	Delft University of Technology, Netherlands
Sebastiano Vascon	Ca' Foscari University of Venice & European Centre for Living Technology, Italy
Nuno Vasconcelos	University of California, San Diego, USA
Mayank Vatsa	Indian Institute of Technology Jodhpur, India
Javier Vazquez-Corral	Autonomous University of Barcelona, Spain
Andrea Vedaldi	Oxford University, UK
Olga Veksler	University of Waterloo, Canada
Luisa Verdoliva	University Federico II of Naples, Italy
Rene Vidal	University of Pennsylvania, USA
Bo Wang	CtrsVision, USA
Heng Wang	TikTok, USA
Jingdong Wang	Baidu, China
Jingya Wang	ShanghaiTech University, China
Ruiping Wang	Institute of Computing Technology, Chinese Academy of Sciences, China
Shenlong Wang	University of Illinois, Urbana-Champaign, USA
Ting-Chun Wang	NVIDIA, USA
Wenguan Wang	Zhejiang University, China
Xiaolong Wang	University of California, San Diego, USA
Xiaoqian Wang	Purdue University, USA
Xiaoyu Wang	Hong Kong University of Science and Technology, USA
Xin Wang	Microsoft Research, USA
Xinchao Wang	National University of Singapore, Singapore
Yang Wang	Concordia University, Canada
Yiming Wang	Fondazione Bruno Kessler, Italy

Yu-Chiang Frank Wang	National Taiwan University, Taiwan
Yu-Xiong Wang	University of Illinois, Urbana-Champaign, USA
Zhangyang Wang	University of Texas at Austin, USA
Olivia Wiles	DeepMind, UK
Christian Wolf	Naver Labs Europe, France
Michael Wray	University of Bristol, UK
Jiajun Wu	Stanford University, USA
Jianxin Wu	Nanjing University, China
Shangzhe Wu	Stanford University, USA
Ying Wu	Northwestern University, USA
Yu Wu	Wuhan University, China
Jianwen Xie	Baidu Research, USA
Saining Xie	New York University, USA
Weidi Xie	Shanghai Jiao Tong University, China
Dan Xu	Hong Kong University of Science and Technology, China
Huijuan Xu	Pennsylvania State University, USA
Xiangyu Xu	Xi'an Jiaotong University, China
Yan Yan	Illinois Institute of Technology, USA
Jiaolong Yang	Microsoft Research, China
Kaiyu Yang	California Institute of Technology, USA
Linjie Yang	ByteDance AI Lab, USA
Ming-Hsuan Yang	University of California, Merced, USA
Yanchao Yang	University of Hong Kong, China
Yi Yang	Zhejiang University, China
Angela Yao	National University of Singapore, Singapore
Mang Ye	Wuhan University, China
Kwang Moo Yi	University of British Columbia, Canada
Xi Yin	Facebook, USA
Chang D. Yoo	Korea Advanced Institute of Science and Technology, Korea
Shaodi You	University of Amsterdam, Netherlands
Jingyi Yu	Shanghai Tech University, China
Ning Yu	Netflix Eyeline Studios, USA
Qi Yu	Rochester Institute of Technology, USA
Stella Yu	University of Michigan, USA
Yizhou Yu	University of Hong Kong, China
Junsong Yuan	University at Buffalo, State University of New York, USA
Lu Yuan	Microsoft, USA
Sangdoo Yun	NAVER AI Lab, Korea
Hongbin Zha	Peking University, China

Jing Zhang	Australian National University, Australia
Lei Zhang	Hong Kong Polytechnic University, China
Ning Zhang	Facebook AI, USA
Richard Zhang	Adobe, USA
Tianzhu Zhang	University of Science and Technology of China, China
Hengshuang Zhao	University of Hong Kong, China
Qi Zhao	University of Minnesota, USA
Sicheng Zhao	Tsinghua University, China
Liang Zheng	Australian National University, Australia
Wei-Shi Zheng	Sun Yat-sen University, China
Zhun Zhong	University of Nottingham, UK
Bolei Zhou	University of California, Los Angeles, USA
Kaiyang Zhou	Hong Kong Baptist University, China
Luping Zhou	University of Sydney, Australia
S. Kevin Zhou	University of Science and Technology of China, China
Lei Zhu	Hong Kong University of Science and Technology (Guangzhou), China
Linchao Zhu	Zhejiang University, China
Andrew Zisserman	University of Oxford, UK
Daniel Zoran	DeepMind, UK
Wangmeng Zuo	Harbin Institute of Technology, China

Technical Program Committee

Sathyanarayanan N. Aakur	Abulikemu Abuduweili	Abhinav Agarwalla
Andrea Abate	Hanno Ackermann	Shivam Aggarwal
Wim Abbeloos	Chandranath Adak	Sara Aghajanzadeh
Amos L. Abbott	Aikaterini Adam	Shashank Agnihotri
Rameen Abdal	Lukáš Adam	Harsh Agrawal
Salma Abdel Magid	Kamil Adamczewski	Antonio Agudo
Rabab Abdelfattah	Donald Adjeroh	Shai Aharon
Ahmed Abdelkader	Mohammed Adnan	Saba Ahmadi
Eslam Mohamed Abdelrahman	Mahmoud Afifi	Waqar Ahmed
	Triantafyllos Afouras	Byeongjoo Ahn
Ahmed Abdelreheem	Akshay Agarwal	Chanho Ahn
Akash Abdu Jyothi	Madhav Agarwal	Jaewoo Ahn
Kfir Aberman	Nakul Agarwal	Namhyuk Ahn
Shahira Abousamra	Sharat Agarwal	Seoyoung Ahn
Yazan Abu Farha	Shivang Agarwal	Nilesh A. Ahuja
Shady Abu-Hussein	Vatsal Agarwal	Yihao Ai

Abhishek Aich
Emanuele Aiello
Noam Aigerman
Kiyoharu Aizawa
Kenan Emir Ak
Reza Akbarian Bafghi
Sai Aparna Aketi
Derya Akkaynak
Ibrahim Alabdulmohsin
Yuval Alaluf
Md Ferdous Alam
Stephan Alaniz
Badour A. Sh AlBahar
Paul Albert
Isabela Albuquerque
Emanuel Aldea
Luca Alessandrini
Emma Alexander
Konstantinos P.
    Alexandridis
Stamatis Alexandropoulos
Saghir Alfasly
Mohsen Ali
Waqar Ali
Daniel Aliaga
Sadegh Aliakbarian
Garvita Allabadi
Thiemo Alldieck
Antonio Alliegro
Jurandy Almeida
Gal Alon
Hadi Alzayer
Ibtihel Amara
Giuseppe Amato
Sameer Ambekar
Niki Amini-Naieni
Shir Amir
Roberto Amoroso
Dongsheng An
Jie An
Junyi An
Liang An
Xiang An
Yuexuan An
Zhaochong An

Zhulin An
Saket Anand
Pavan Kumar Anasosalu
    Vasu
Codruta O. Ancuti
Cosmin Ancuti
Fernanda Andaló
Connor Anderson
Juan Andrade-Cetto
Bruno Andreis
Alexander Andreopoulos
Bjoern Andres
Shivangi Aneja
Dragomir Anguelov
Damith Kawshan
    Anhettigama
Aditya Annavajjala
Michel Antunes
Tejas Anvekar
Saeed Anwar
Evlampios Apostolidis
Srikar Appalaraju
Relja Arandjelović
Nikita Araslanov
Alexandre Araujo
Helder Araujo
Eric Arazo
Dawit Mureja Argaw
Anurag Arnab
Federica Arrigoni
Alexey Artemov
Aditya Arun
Nader Asadi
Kumar Ashutosh
Vishal Asnani
Mahmoud Assran
Amir Atapour-Abarghouei
Nikos Athanasiou
Ali Athar
ShahRukh Athar
Rowel O. Atienza
Benjamin Attal
Matan Atzmon
Nicolas Audebert
Romaric Audigier

Tristan
    Aumentado-Armstrong
Yannis Avrithis
Muhammad Awais
Md Rabiul Awal
Görkay Aydemir
Mehmet Aygün
Melika Ayoughi
Ali Ayub
Kumar Ayush
Mehdi Azabou
Basim Azam
Mohammad Farid
    Azampour
Hossein Azizpour
Vivek B. S.
Yunhao Ba
Mohammadreza Babaee
Deepika Bablani
Sudarshan Babu
Roman Bachmann
Inhwan Bae
Jeonghun Baek
Kyungjune Baek
Seungryul Baek
Piyush Nitin Bagad
Mohammadhossein Bahari
Gaetan Bahl
Sherwin Bahmani
Bing Bai
Haoyue Bai
Jiawang Bai
Jinbin Bai
Lei Bai
Qingyan Bai
Shutao Bai
Xuyang Bai
Yalong Bai
Yuanchao Bai
Yue Bai
Yunpeng Bai
Ziyi Bai
Daniele Baieri
Sungyong Baik
Ramon Baldrich

Coloma Ballester
Irene Ballester
Sonat Baltaci
Biplab Banerjee
Monami Banerjee
Sandipan Banerjee
Snehasis Banerjee
Soumya Banerjee
Sreya Banerjee
Jihwan Bang
Ankan Bansal
Nitin Bansal
Siddhant Bansal
Chong Bao
Hangbo Bao
Runxue Bao
Wentao Bao
Yiwei Bao
Zhipeng Bao
Zhongyun Bao
Amir Bar
Omer Bar Tal
Manel Baradad Jurjo
Lorenzo Baraldi
Lorenzo Baraldi
Tal Barami
Danny Barash
Daniel Barath
Adrian Barbu
Nick Barnes
Alina J. Barnett
German Barquero
Joao P. Barreto
Jonathan T. Barron
Rodrigo C. Barros
Luca Barsellotti
Myles Bartlett
Kristijan Bartol
Hritam Basak
Dina Bashkirova
Chaim Baskin
Lennart C. Bastian
Abhipsa Basu
Soumen Basu
Peyman Bateni

Anil Batra
David Bau
Zuria Bauer
Eduard Gabriel Bazavan
Federico Becattini
Nathan A. Beck
Jan Bednarik
Peter A. Beerel
Ardhendu Behera
Harkirat Singh Behl
Jens Behley
William Beksi
Soufiane Belharbi
Giovanni Bellitto
Ismail Ben Ayed
Abdessamad Ben Hamza
Yizhak Ben-Shabat
Guy Ben-Yosef
Robert Benavente
Assia Benbihi
Yasser Benigmim
Urs Bergmann
Amit H Bermano
Jesus Bermudez-Cameo
Florian Bernard
Stefano Berretti
Gabriele Berton
Petra Bevandić
Matthew Beveridge
Lalit Bhagat
Yash Bhalgat
Homanga Bharadhwaj
Skanda Bharadwaj
Ambika Saklani Bhardwaj
Goutam Bhat
Gaurav Bhatt
Neel P. Bhatt
Deblina Bhattacharjee
Anish Bhattacharya
Rajarshi Bhattacharya
Uttaran Bhattacharya
Apratim Bhattacharyya
Anand Bhattad
Binod Bhattarai
Snehal Bhayani

Brojeshwar Bhowmick
Aritra Bhowmik
Neelanjan Bhowmik
Amran Bhuiyan
Ankan Kumar Bhunia
Ayan Kumar Bhunia
Jing Bi
Qi Bi
Xiuli Bi
Michael Bi Mi
Hao Bian
Jia-Wang Bian
Shaojun Bian
Simone Bianco
Pia Bideau
Xiaoyu Bie
Roberto Bigazzi
Mario Bijelic
Bahri Batuhan Bilecen
Guillaume-Alexandre
    Bilodeau
Alexander Binder
Massimo Bini
Niccolò Bisagno
Carmen Bisogni
Anthony R. Bisulco
Sandika Biswas
Sanket Biswas
Ksenia Bittner
Kevin Blackburn-Matzen
Nathaniel Blanchard
Hunter Blanton
Emile Blettery
Hermann Blum
Deyu Bo
Janusz Bobulski
Marius Bock
Lea Bogensperger
Simion-Vlad Bogolin
Moritz Böhle
Piotr Bojanowski
Alexey Bokhovkin
Georg Bökman
Ladislau Boloni
Daniel Bolya

Lorenzo Bonicelli
Gianpaolo Bontempo
Vivek Boominathan
Adrian Bors
Damian Borth
Matteo Bortolon
Davide Boscaini
Laurie Nicholas Bose
Shirsha Bose
Mark Boss
Andrea Bottino
Larbi Boubchir
Alexandre Boulch
Terrance E. Boult
Amine Bourki
Walid Bousselham
Fadi Boutros
Nicolas C. Boutry
Thierry Bouwmans
Richard S. Bowen
Kevin W. Bowyer
Ivaylo Boyadzhiev
Aidan Boyd
Joseph Boyd
Aljaz Bozic
Behzad Bozorgtabar
Manuel Brack
Samarth Brahmbhatt
Nikolas Brasch
Cameron E. Braunstein
Toby P. Breckon
Mathieu Bredif
Romain Brégier
Carlo Bretti
Joel R. Brogan
Clifford Broni-Bediako
Andrew Brown
Ellis L. Brown
Thomas Brox
Qingwen Bu
Tong Bu
Shyamal Buch
Alvaro Budria
Ignas Budvytis
José M. Buenaposada

Kyle R. Buettner
Marcel C. Bühler
Nhat-Tan Bui
Adrian Bulat
Valay M. Bundele
Nathaniel J. Burgdorfer
Ryan D. Burgert
Evgeny Burnaev
Pietro Buzzega
Marco Buzzelli
Kwon Byung-Ki
Fabian Caba
Felipe Cadar
Martin Cadik
Bowen Cai
Changjiang Cai
Guanyu Cai
Haoming Cai
Hongrui Cai
Jiarui Cai
Jinyu Cai
Kaixin Cai
Lile Cai
Minjie Cai
Mu Cai
Pingping Cai
Rizhao Cai
Ruisi Cai
Ruojin Cai
Shen Cai
Shengqu Cai
Weidong Cai
Weiwei Cai
Wenjia Cai
Wenxiao Cai
Yancheng Cai
Yuanhao Cai
Yujun Cai
Yuxuan Cai
Zhaowei Cai
Zhipeng Cai
Zhixi Cai
Zhongang Cai
Zikui Cai
Salvatore Calcagno

Roberto Caldelli
Akin Caliskan
Lilian Calvet
Necati Cihan Camgoz
Tommaso Campari
Dylan Campbell
Guglielmo Camporese
Yigit Baran Can
Shaun Canavan
Gemma Canet Tarrés
Marco Cannici
Ang Cao
Anh-Quan Cao
Bing Cao
Bingyi Cao
Chenjie Cao
Dongliang Cao
Hu Cao
Jiahang Cao
Jiale Cao
Jie Cao
Jiezhang Cao
Jinkun Cao
Meng Cao
Mingdeng Cao
Peng Cao
Pu Cao
Qinglong Cao
Qingxing Cao
Shengcao Cao
Song Cao
Weiwei Cao
Xiangyong Cao
Xiao Cao
Xu Cao
Xu Cao
Yan-Pei Cao
Yang Cao
Yichao Cao
Ying Cao
Yuan Cao
Yuhang Cao
Yukang Cao
Yulong Cao
Yun-Hao Cao

Yunhao Cao
Yutong Cao
Zhangjie Cao
Zhiguo Cao
Zhiwen Cao
Ziang Cao
Giacomo Capitani
Luigi Capogrosso
Andrea Caraffa
Jaime S. Cardoso
Chris Careaga
Zachariah Carmichael
Giuseppe Cartella
Vincent Cartillier
Paola Cascante-Bonilla
Lucia Cascone
Pol Caselles Rico
Angela Castillo
Lluis Castrejon
Modesto
   Castrillón-Santana
Francisco M. Castro
Pedro Castro
Jacopo Cavazza
Oya Celiktutan
Luigi Celona
Jiazhong Cen
Fernando J. Cendra
Fabio Cermelli
Duygu Ceylan
Eunju Cha
Junbum Cha
Rohan Chabra
Rohan Chacko
Julia Chae
Nicolas Chahine
Junyi Chai
Lucy Chai
Tianrui Chai
Wenhao Chai
Zenghao Chai
Anish Chakrabarty
Goirik Chakrabarty
Anirban Chakraborty
Deep Chakraborty

Souradeep Chakraborty
Punarjay Chakravarty
Jacob Chalk
Loick Chambon
Chee Seng Chan
David Chan
Dorian Y. Chan
Kin Sun Chan
Matthew Chan
Adrien Chan-Hon-Tong
Sampath Chanda
Shivam Chandhok
Prashanth Chandran
Kenneth Chaney
Chirui Chang
Di Chang
Dongliang Chang
Hong Chang
Hyung Jin Chang
Ju Yong Chang
Matthew Chang
Ming-Ching Chang
MingFang Chang
Nadine Chang
Qi Chang
Seunggyu Chang
Wei-Yi Chang
Xiaojun Chang
Yakun Chang
Yi Chang
Zheng Chang
Sumohana S.
   Channappayya
Hanqing Chao
Pradyumna Chari
Korawat Charoenpitaks
Dibyadip Chatterjee
Moitreya Chatterjee
Pratik Chaudhari
Muawiz Chaudhary
Ritwick Chaudhry
Ruchika Chavhan
Sofian Chaybouti
Zhengping Che
Gullal Singh Cheema

Kunal Chelani
Rama Chellappa
Allison Y. Chen
Anpei Chen
Aozhu Chen
Bin Chen
Bin Chen
Bin Chen
Bohong Chen
Bor-Chun Chen
Bowei Chen
Brian Chen
Chang Wen Chen
Changan Chen
Changhao Chen
Changhuai Chen
Chaofan Chen
Chaofan Chen
Chaofeng Chen
Chen Chen
Cheng Chen
Chengkuan Chen
Chieh-Yun Chen
Chong Chen
Chun-Fu Richard Chen
Congliang Chen
Cuiqun Chen
Da Chen
Daoyuan Chen
Defang Chen
Dian Chen
Dian Chen
Ding-Jie Chen
Dingfan Chen
Dong Chen
Dongdong Chen
Fangyi Chen
Guang-Yong Chen
Guangyao Chen
Guangyi Chen
Guanying Chen
Guikun Chen
Guo Chen
Haibo Chen
Haiwei Chen

Hansheng Chen
Hanting Chen
Hanzhi Chen
Hao Chen
Haodong Chen
Haokun Chen
Haoran Chen
Haoxin Chen
Haoyu Chen
Haoyu Chen
Hong-You Chen
Honghua Chen
Honglin Chen
Hongming Chen
Huanran Chen
Hui Chen
Hwann-Tzong Chen
Jiacheng Chen
Jiajing Chen
Jianqiu Chen
Jiaqi Chen
Jiaxin Chen
Jiayi Chen
Jie Chen
Jie Chen
Jie Chen
Jieneng Chen
Jierun Chen
Jinghui Chen
Jingjing Chen
Jintai Chen
Jinyu Chen
Joya Chen
Jun Chen
Jun Chen
Jun Chen
Jun-Cheng Chen
Jun-Kun Chen
Junjie Chen
Kai Chen
Kai Chen
Kaifeng Chen
Kefan Chen
Kejiang Chen
Kuan-Wen Chen

Kunyao Chen
Li Chen
Liang Chen
Lianggangxu Chen
Liangyu Chen
Ling-Hao Chen
Linghao Chen
Liyan Chen
Min Chen
Min-Hung Chen
Minghao Chen
Minghui Chen
Nenglun Chen
Peihao Chen
Pengguang Chen
Ping Chen
Qi Chen
Qi Chen
Qi Chen
Qiang Chen
Qimin Chen
Qingcai Chen
Rongyu Chen
Rui Chen
Runjian Chen
Shang-Fu Chen
Shen Chen
Sherry X. Chen
Shi Chen
Shiming Chen
Shiyan Chen
Shizhe Chen
Shoufa Chen
Shu-Yu Chen
Shuai Chen
Si Chen
Siheng Chen
Simin Chen
Sixiang Chen
Sizhe Chen
Songcan Chen
Tao Chen
Tianshui Chen
Tianyi Chen
Tsai-Shien Chen

Wei Chen
Weidong Chen
Weihua Chen
Weijie Chen
Weikai Chen
Wuyang Chen
Xi Chen
Xi Chen
Xiang Chen
Xiangyu Chen
Xianyu Chen
Xiao Chen
Xiao Chen
Xiaohan Chen
Xiaokang Chen
Xiaotong Chen
Xin Chen
Xin Chen
Xin Chen
Xinghao Chen
Xingyu Chen
Xingyu Chen
Xinlei Chen
Xiuwei Chen
Xuanbai Chen
Xuanhong Chen
Xuesong Chen
Xuweiyi Chen
Xuxi Chen
Yanbei Chen
Yang Chen
Yaofo Chen
Yen-Chun Chen
Yi-Ting Chen
Yi-Wen Chen
Yilun Chen
Yinbo Chen
Yingcong Chen
Yingwen Chen
Yiting Chen
Yixin Chen
Yixin Chen
Yu Chen
Yuanhong Chen
Yuanqi Chen

Yuedong Chen
Yueting Chen
Yuhao Chen
Yuhao Chen
Yujin Chen
Yukang Chen
Yun-Chun Chen
Yunjin Chen
Yunlu Chen
Yuntao Chen
Yushuo Chen
Yuxiao Chen
Zehui Chen
Zerui Chen
Zewen Chen
Zeyuan Chen
Zeyuan Chen
Zhang Chen
Zhao-Min Chen
Zhaoliang Chen
Zhaoxi Chen
Zhaoyu Chen
Zhaozheng Chen
Zhaozheng Chen
Zhe Chen
Zhen Chen
Zhen Chen
Zheng Chen
Zhenghao Chen
Zhengyu Chen
Zhenyu Chen
Zhi Chen
Zhibo Chen
Zhimin Chen
Zhineng Chen
Zhiwei Chen
Zhixiang Chen
Zhiyang Chen
Zhiyuan Chen
Zhuangzhuang Chen
Zhuo Chen
Zhuoxiao Chen
Zihan Chen
Zijiao Chen
Ziwen Chen

Ziyang Chen
An-Chieh Cheng
Bowen Cheng
De Cheng
Erkang Cheng
Feng Cheng
Guangliang Cheng
Hao Cheng
Hao Cheng
Harry Cheng
Ho Kei Cheng
Jingchun Cheng
Jun Cheng
Ka Leong Cheng
Kelvin B. Cheng
Kevin Ho Man Cheng
Lechao Cheng
Shuo Cheng
Silin Cheng
Ta-Ying Cheng
Tianhang Cheng
Weihao Cheng
Wencan Cheng
Wensheng Cheng
Xi Cheng
Xize Cheng
Xuxin Cheng
Yen-Chi Cheng
Yi Cheng
Yihua Cheng
Yu Cheng
Zesen Cheng
Zezhou Cheng
Zhanzhan Cheng
Zhen Cheng
Zhi-Qi Cheng
Ziang Cheng
Ziheng Cheng
Soon Yau Cheong
Anoop Cherian
Daniil Cherniavskii
Ajad Chhatkuli
Cheng Chi
Haoang Chi
Hyung-gun Chi

Seunggeun Chi
Zhixiang Chi
Naoya Chiba
Julian Chibane
Aditya Chinchure
Eleni Chiou
Mia Chiquier
Chiranjeev Chiranjeev
Kashyap Chitta
Hsu-kuang Chiu
Wei-Chen Chiu
Chee Kheng Chng
Shin-Fang Chng
Donghyeon Cho
Gyusang Cho
Hanbyel Cho
Hoonhee Cho
Jae Won Cho
Jang Hyun Cho
Jungchan Cho
Junhyeong Cho
Nam Ik Cho
Seokju Cho
Seungju Cho
Suhwan Cho
Sunghyun Cho
Sungjun Cho
Sungmin Cho
Wonwoong Cho
Nathaniel E. Chodosh
Jaesung Choe
Junsuk Choe
Changwoon Choi
Chiho Choi
Ching Lam Choi
Dongyoung Choi
Dooseop Choi
Hongsuk Choi
Hyesong Choi
Jaewoong Choi
Janghoon Choi
Jin Young Choi
Jinwoo Choi
Jinyoung Choi
Jonghyun Choi

Jongwon Choi
Jooyoung Choi
Kanghyun Choi
Myungsub Choi
Seokeon Choi
Wonhyeok Choi
Bhavin Choksi
Jaegul Choo
Baptiste Chopin
Ayush Chopra
Gene Chou
Shih-Han Chou
Divya Choudhary
Siddharth Choudhary
Sunav Choudhary
Rohan Choudhury
Vasileios Choutas
Ka-Ho Chow
Pinaki Nath Chowdhury
Sanjoy Chowdhury
Sammy Christen
Fu-Jen Chu
Qi Chu
Ruihang Chu
Sanghyeok Chu
Wei-Ta Chu
Wen-Hsuan Chu
Wen-Hsuan Chu
Wei Qin Chuah
Andrei Chubarau
Ilya Chugunov
Sanghyuk Chun
Se Young Chun
Hyungjin Chung
Inseop Chung
Jaeyoung Chung
Jihoon Chung
Jiwan Chung
Joon Son Chung
Thomas A. Ciarfuglia
Umur A. Ciftci
Antonio E. Cinà
Ramazan Gokberk Cinbis
Anthony Cioppa
Javier Civera

Christopher A. Clark
James J. Clark
Ronald Clark
Brian S. Clipp
Mark Coates
Federico Cocchi
Felipe Codevilla
Cecília Coelho
Danny Cohen-Or
Forrester Cole
Julien Colin
John Collomosse
Marc Comino-Trinidad
Nicola Conci
Marcos V. Conde
Alexandru Paul
    Condurache
Runmin Cong
Tianshuo Cong
Wenyan Cong
Yuren Cong
Alessandro Conti
Andrea Conti
Enric Corona
John R. Corring
Jaime Corsetti
Jason J. Corso
Theo W. Costain
Guillaume Couairon
Robin Courant
Davide Cozzolino
Donato Crisostomi
Francesco Croce
Florinel Alin Croitoru
Ioana Croitoru
James L. Crowley
Steve Cruz
Gabriela Csurka
Alfredo Cuesta Infante
Aiyu Cui
Hainan Cui
Jiali Cui
Jiequan Cui
Justin Cui
Qiongjie Cui

Ruikai Cui
Yawen Cui
Ying Cui
Yuning Cui
Zhen Cui
Zhenyu Cui
Ziteng Cui
Benjamin J. Culpepper
Xiaodong Cun
Federico Cunico
Claudio Cusano
Sebastian Cygert
Guido Maria D'Amely di
    Melendugno
Moreno D'Incà
Feipeng Da
Mosam Dabhi
Rishabh Dabral
Konstantinos M. Dafnis
Matthew L. Daggitt
Anders B. Dahl
Bo Dai
Bo Dai
Haixing Dai
Mengyu Dai
Peng Dai
Peng Dai
Pingyang Dai
Qi Dai
Qiyu Dai
Wenrui Dai
Wenxun Dai
Zuozhuo Dai
Nicola Dall'Asen
Naser Damer
Bo Dang
Meghal Dani
Duolikun Danier
Donald G. Dansereau
Quan Dao
Son Duy Dao
Trung Tuan Dao
Mohamed Daoudi
Ahmad Darkhalil
Rangel Daroya

Abhijit Das
Abir Das
Anurag Das
Ayan Das
Nilotpal Das
Partha Das
Soumi Das
Srijan Das
Sukhendu Das
Swagatam Das
Ananyananda Dasari
Avijit Dasgupta
Gourav Datta
Achal Dave
Ishan Rajendrakumar Dave
Andrew Davison
Aram Davtyan
Axel Davy
Youssef Dawoud
Laura Daza
Francesca De Benetti
Daan de Geus
Alfredo S. De Goyeneche
Riccardo de Lutio
Maria De Marsico
Christophe De
   Vleeschouwer
Tonmoay Deb
Sayan Deb Sarkar
Dale Decatur
Thanos Delatolas
Fabien Delattre
Mauricio Delbracio
Ginger D. Delmas
Andong Deng
Bailin Deng
Bin Deng
Boyang Deng
Chaorui Deng
Cheng Deng
Congyue Deng
Jiacheng Deng
Jiajun Deng
Jiankang Deng
Jingjing Deng

Jinhong Deng
Kangle Deng
Kenan Deng
Lei Deng
Liang-Jian Deng
Shijian Deng
Weijian Deng
Xiaoming Deng
Xueqing Deng
Ye Deng
Yongjian Deng
Youming Deng
Yu Deng
Zhijie Deng
Zhongying Deng
Matteo Denitto
Mohammad Mahdi
   Derakhshani
Alakh Desai
Kevin Desai
Nishq P. Desai
Sai Vikas Desai
Jean-Emmanuel Deschaud
Frédéric Devernay
Rahul Dey
Arturo Deza
Helisa Dhamo
Amaya Dharmasiri
Ankit Dhiman
Vikas Dhiman
Yan Di
Zonglin Di
Ousmane A. Dia
Renwei Dian
Haiwen Diao
Xiaolei Diao
Gonçalo José Dias Pais
Abdallah Dib
Juan Carlos Dibene
   Simental
Sebastian Dille
Christian Diller
Mariella Dimiccoli
Anastasios Dimou
Or Dinari

Caiwen Ding
Changxing Ding
Choubo Ding
Fangqiang Ding
Guiguang Ding
Guodong Ding
Henghui Ding
Hui Ding
Jian Ding
Jian-Jiun Ding
Jianhao Ding
Li Ding
Liang Ding
Lihe Ding
Ming Ding
Runyu Ding
Shouhong Ding
Shuangrui Ding
Shuxiao Ding
Tianjiao Ding
Tianyu Ding
Wenhao Ding
Xiaohan Ding
Xinpeng Ding
Yaqing Ding
Yong Ding
Yuhang Ding
Yuqi Ding
Zheng Ding
Zhengming Ding
Zhipeng Ding
Zhuangzhuang Ding
Zihan Ding
Zihan Ding
Ziluo Ding
Anh-Dung Dinh
Markos Diomataris
Christos Diou
Alish Dipani
Alara Dirik
Ajay Divakaran
Abdelaziz Djelouah
Nemanja Djuric
Thanh-Toan Do
Tien Do

Anh-Dzung Doan
Bao Gia Doan
Khoa D. Doan
Carl Doersch
Andreea Dogaru
Nehal Doiphode
Csaba Domokos
Bowen Dong
Haoye Dong
Jiahua Dong
Jianfeng Dong
Jinpeng Dong
Junhao Dong
Junting Dong
Li Dong
Minjing Dong
Nanqing Dong
Peijie Dong
Qiaole Dong
Qihua Dong
Qiulei Dong
Runpei Dong
Shichao Dong
Shuting Dong
Sixun Dong
Siyan Dong
Tian Dong
Wei Dong
Wei Dong
Weisheng Dong
Xiaoyi Dong
Xiaoyu Dong
Xingping Dong
Yinpeng Dong
Zhenxing Dong
Jonathan C. Donnelly
Naren Doraiswamy
Gianfranco Doretto
Michael Dorkenwald
Muskan Dosi
Huanzhang Dou
Shuguang Dou
Yiming Dou
Zhiyang Frank Dou
Zi-Yi Dou

Arthur Douillard
Michail C. Doukas
Matthijs Douze
Amil Dravid
Vincent Drouard
Dawei Du
Jia-Run Du
Jinhao Du
Ruoyi Du
Weiyu Du
Xiaodan Du
Ye Du
Yingjun Du
Yong Du
Yuanqi Du
Yuntao Du
Zexing Du
Zhuoran Du
Radhika Dua
Haodong Duan
Haoran Duan
Huiyu Duan
Jiafei Duan
Jiali Duan
Peiqi Duan
Ye Duan
Yuchen Duan
Yueqi Duan
Zhihao Duan
Amanda Cardoso Duarte
Shiv Ram Dubey
Anastasia Dubrovina
Daniel Duckworth
Timothy Duff
Nicolas Dufour
Rahul Duggal
Shivam Duggal
Bardienus P. Duisterhof
Felix Dülmer
Matteo Dunnhofer
Chi Nhan Duong
Elona Dupont
Nikita Durasov
Zoran Duric
Mihai Dusmanu

Titir Dutta
Ujjal Kr Dutta
Isht Dwivedi
Sai Kumar Dwivedi
Sebastian Dziadzio
Thomas Eboli
Ramin Ebrahim Nakhli
Benjamin Eckart
Takeharu Eda
Johan Edstedt
Alexei A. Efros
Nikos Efthymiadis
Bernhard Egger
Max Ehrlich
Iván Eichhardt
Farshad Einabadi
Martin Eisemann
Peter Eisert
Mohamed El Banani
Mostafa El-Khamy
Alaa El-Nouby
Randa I. M. Elanwar
James Elder
Abdelrahman Eldesokey
Ismail Elezi
Ehsan Elhamifar
Moshe Eliasof
Sara Elkerdawy
Qing En
Ian Endres
Andreas Engelhardt
Francis Engelmann
Martin Engilberge
Kenji Enomoto
Chanho Eom
Guy Erez
Linus Ericsson
Maria C. Escobar
Marcos Escudero-Viñolo
Sedigheh Eslami
Valter Estevam
Ali Etemad
Martin Nicolas Everaert
Ivan Evtimov
Ralph Ewerth

Fevziye Irem Eyiokur
Cristobal Eyzaguirre
Gabriele Facciolo
Mohammad Fahes
Masud ANI Fahim
Fabrizio Falchi
Alex Falcon
Aoxiang Fan
Baojie Fan
Bin Fan
Chao Fan
David Fan
Haoqiang Fan
Hehe Fan
Heng Fan
Hongyi Fan
Huijie Fan
Junsong Fan
Lei Fan
Lei Fan
Lifeng Fan
Lue Fan
Mingyuan Fan
Qi Fan
Rui R. Fan
Xiran Fan
Yifei Fan
Yue Fan
Zejia Fan
Zhaoxin Fan
Zhenfeng Fan
Zhentao Fan
Zhipeng Fan
Zhiwen Fan
Zicong Fan
Sean Fanello
Chaowei Fang
Chuan Fang
Gongfan Fang
Guian Fang
Han Fang
Jiansheng Fang
Jiemin Fang
Jin Fang
Jun Fang

Kun Fang
Pengfei Fang
Rui Fang
Sheng Fang
Xiang Fang
Xiaolin Fang
Xiuwen Fang
Zhaoyuan Fang
Zhiyuan Fang
Eros Fanì
Matteo Farina
Farzan Farnia
Aiman Farooq
Azade Farshad
Mohsen Fayyaz
Arash Fayyazi
Ben Fei
Jingjing Fei
Xiaohan Fei
Zhengcong Fei
Michael Felsberg
Berthy T. Feng
Bin Feng
Brandon Y. Feng
Chao Feng
Chen Feng
Chengjian Feng
Chun-Mei Feng
Fan Feng
Guorui Feng
Hao Feng
Jianjiang Feng
Lan Feng
Lei Feng
Litong Feng
Mingtao Feng
Qianyu Feng
Ruicheng Feng
Ruili Feng
Ruoyu Feng
Ryan T. Feng
Shiwei Feng
Songhe Feng
Tuo Feng
Wei Feng

Weitao Feng
Weixi Feng
Xuelu Feng
Yanglin Feng
Yao Feng
Yue Feng
Zhanxiang Feng
Zhenhua Feng
Zunlei Feng
Clara Fernandez
Victoria Fernandez
    Abrevaya
Miguel-Ángel
    Fernández-Torres
Sira Ferradans
Claudio Ferrari
Ethan Fetaya
Mustansar Fiaz
Guénolé Fiche
Panagiotis P. Filntisis
Marco Fiorucci
Kai Fischer
Tobias Fischer
Tom Fischer
Volker Fischer
Robert B. Fisher
Alessandro Flaborea
Corneliu O. Florea
Georgios Floros
Alejandro Fontan
Tomaso Fontanini
Lin Geng Foo
Wolfgang Förstner
Niki M. Foteinopoulou
Simone Foti
Simone Foti
Victor Fragoso
Gianni Franchi
Jean-Sebastien Franco
Emanuele Frascaroli
Moti Freiman
Rafail Fridman
Sara Fridovich-Keil
Felix Friedrich
Stanislav Frolov

Andre Fu
Bin Fu
Changcheng Fu
Changqing Fu
Jianlong Fu
Jie Fu
Jingjing Fu
Keren Fu
Lan Fu
Lele Fu
Minghao Fu
Qichen Fu
Rao Fu
Taimeng Fu
Tianwen Fu
Yanwei Fu
Ying Fu
Yonggan Fu
Yun Fu
Yuqian Fu
Zehua Fu
Zhenqi Fu
Zipeng Fu
Wolfgang Fuhl
Yasuhisa Fujii
Yuki Fujimura
Katsuki Fujisawa
Kent Fujiwara
Takuya Funatomi
Christopher Funk
Antonino Furnari
Ryo Furukawa
Ryosuke Furuta
David Futschik
Akshay Gadi Patil
Siddhartha Gairola
Rinon Gal
Adrian Galdran
Silvio Galesso
Meirav Galun
Rofida Gamal
Weijie Gan
Yiming Gan
Yulu Gan
Vineet Gandhi

Kanchana Vaishnavi
    Gandikota
Aditya Ganeshan
Aditya Ganeshan
Swetava Ganguli
Sujoy Ganguly
Hanan Gani
Alireza Ganjdanesh
Harald Ganster
Roy Ganz
Angela F. Gao
Bin-Bin Gao
Changxin Gao
Chen Gao
Chongyang Gao
Daiheng Gao
Daoyi Gao
Difei Gao
Hang Gao
Hongchang Gao
Huiyu Gao
Junyu Gao
Junyu Gao
Kaifeng Gao
Katelyn Gao
Kuofeng Gao
Lin Gao
Ling Gao
Maolin Gao
Quankai Gao
Ruopeng Gao
Shanghua Gao
Shangqi Gao
Shangqian Gao
Shaobing Gao
Shenyuan Gao
Xiang Gao
Xiangjun Gao
Yang Gao
Yipeng Gao
Yuan Gao
Yue Gao
Yunhe Gao
Zhanning Gao
Zhi Gao

Zhihan Gao
Zhitong Gao
Zhongpai Gao
Ziteng Gao
Nicola Garau
Elena Garces
Noa Garcia
Nuno Cruz Garcia
Ricardo Garcia Pinel
Guillermo
    Garcia-Hernando
Saurabh Garg
Mathieu Garon
Risheek Garrepalli
Pablo Garrido
Quentin Garrido
Stefano Gasperini
Vincent Gaudilliere
Chandan Gautam
Shivam Gautam
Paul Gavrikov
Jonathan Z. Gazak
Chongjian Ge
Liming Ge
Runzhou Ge
Shiming Ge
Songwei Ge
Weifeng Ge
Wenhang Ge
Yanhao Ge
Yixiao Ge
Yunhao Ge
Yuying Ge
Zhiqi Ge
Zongyuan Ge
James Gee
Chen Geng
Daniel Geng
Xue Geng
Zhengyang Geng
Zichen Geng
Kyle Genova
Georgios Georgakis
Mariana-Iuliana
    Georgescu

Yiangos Georgiou
Markos Georgopoulos
Stamatios Georgoulis
Michal Geyer
Mina Ghadimi Atigh
Abhijay Ghildyal
Reza Ghoddoosian
Mohsen Gholami
Peyman Gholami
Enjie Ghorbel
Soumya Suvra Ghosal
Anindita Ghosh
Anurag Ghosh
Arnab Ghosh
Arthita Ghosh
Aurobrata Ghosh
Shreya Ghosh
Soham Ghosh
Sreyan Ghosh
Andrea Giachetti
Paris Giampouras
Alexander Gielisse
Andrew Gilbert
Guy Gilboa
Nate Gillman
Jhony H. Giraldo
Roger Girgis
Sharath Girish
Mario Valerio Giuffrida
Francesco Giuliari
Nikolaos Gkanatsios
Ioannis Gkioulekas
Alexi Gladstone
Derek Gloudemans
Aurele T. Gnanha
Hyojun Go
Akos Godo
Danilo D. Goede
Arushi Goel
Kratarth Goel
Rahul Goel
Shubham Goel
Vidit Goel
Tejas Gokhale
Vignesh Gokul

Aditya Golatkar
Micah Goldblum
S. Alireza Golestaneh
Gabriele Goletto
Lluis Gomez
Guillermo Gomez-Trenado
Alex Gomez-Villa
Nuno Gonçalves
Biao Gong
Boqing Gong
Chen Gong
Chengyue Gong
Dayoung Gong
Dong Gong
Jingyu Gong
Kaixiong Gong
Kehong Gong
Liyu Gong
Mingming Gong
Ran Gong
Rui Gong
Tao Gong
Xiaojin Gong
Xuan Gong
Xun Gong
Yifan Gong
Yu Gong
Yuanhao Gong
Yuning Gong
Yunye Gong
Cristina I. González
Adam Goodge
Nithin Gopalakrishnan
    Nair
Cameron Gordon
Dipam Goswami
Gaurav Goswami
Hanno Gottschalk
Chenhui Gou
Jianping Gou
Shreyank N. Gowda
Mohit Goyal
Julia Grabinski
Helmut Grabner
Patrick L. Grady

Alexandros Graikos
Eric Granger
Douglas R. Gray
Michael Alan Greenspan
Connor Greenwell
David Griffiths
Artur Grigorev
Alexey A. Gritsenko
Ivan Grubišić
Geonmo Gu
Jianyang Gu
Jiaqi Gu
Jiayuan Gu
Jinwei Gu
Peiyan Gu
Qiao Gu
Tianpei Gu
Xianfan Gu
Xiang Gu
Xiangming Gu
Xiao Gu
Xiuye Gu
Yanan Gu
Yiwen Gu
Yuchao Gu
Yuming Gu
Yun Gu
Zeqi Gu
Zhangxuan Gu
Banglei Guan
Junfeng Guan
Qingji Guan
Shanyan Guan
Tianrui Guan
Tongkun Guan
Ziqiao Guan
Ziyu Guan
Denis A. Gudovskiy
Antoine Guédon
Paul Guerrero
Ricardo Guerrero
Liangke Gui
Sadaf Gulshad
Manuel Günther
Chen Guo

Chenqi Guo
Chuan Guo
Guodong Guo
Haiyun Guo
Heng Guo
Hengtao Guo
Hongji Guo
Jia Guo
Jianwei Guo
Jianyuan Guo
Jiayi Guo
Jie Guo
Jie Guo
Jingcai Guo
Jingyuan Guo
Jinyang Guo
Lanqing Guo
Meng-Hao Guo
Mengxi China Guo
Minghao Guo
Pengfei Guo
Pengsheng Guo
Pinxue Guo
Qi Guo
Qin Guo
Qing Guo
Qingpei Guo
Saidi Guo
Shi Guo
Shuxuan Guo
Song Guo
Taian Guo
Tiantong Guo
Tianyu Guo
Wen Guo
Wenzhong Guo
Xianda Guo
Xiaobao Guo
Xiefan Guo
Yangyang Guo
Yanhui Guo
Yao Guo
Yichen Guo
Yong Guo
Yuan-Chen Guo

Yuliang Guo
Yuxiang Guo
Yuyu Guo
Zixian Guo
Ziyu Guo
Zujin Guo
Aarush Gupta
Akash Gupta
Akshita Gupta
Ankush Gupta
Anshul Gupta
Anubhav Gupta
Anup Kumar Gupta
Honey Gupta
Kunal Gupta
pravir singh gupta
Rohit Gupta
Saumya Gupta
Corina Gurau
Yeti Z. Gurbuz
Saket Gurukar
Siddharth Gururani
Vladimir Guzov
Matthew A. Gwilliam
Hyunho Ha
Ryo Hachiuma
Isma Hadji
Armin Hadzic
Daniel Haehn
Nicolai Haeni
Ronny Haensch
Meera Hahn
Oliver Hahn
Nguyen Hai
Yuval Haitman
Levente Hajder
Moayed Haji-Ali
Sina Hajimiri
Alexandros Haliassos
Peter M. Hall
Bumsub Ham
Ryuhei Hamaguchi
Abdullah J. Hamdi
Max Hamilton
Lars Hammarstrand

Hasan Abed Al Kader
    Hammoud
Beining Han
Boran Han
Dongyoon Han
Feng Han
Guangxing Han
Guangzeng Han
Hu Han
Jiaming Han
Jin Han
Jungong Han
Junlin Han
Kai Han
Kang Han
Kun Han
Ligong Han
Longfei Han
Mei Han
Mingfei Han
Muzhi Han
Sungwon Han
Tengda Han
Tengda Han
Wencheng Han
Xinran Han
Xintong Han
Yizeng Han
Yufei Han
Zhenjun Han
Zhi Han
Zhongyi Han
Zongbo Han
Zongyan Han
Ankur Handa
Tiankai Hang
Yucheng Hang
Asif Hanif
Param Hanji
Joëlle Hanna
Niklas Hanselmann
Nicklas A. Hansen
Fusheng Hao
Shaozhe Hao
Shengyu Hao

Shijie Hao
Tianxiang Hao
Yanbin Hao
Yu Hao
Zekun Hao
Takayuki Hara
Mehrtash Harandi
Sanjay Haresh
Haripriya Harikumar
Adam Harley
David M. Hart
Md Yousuf Harun
Md Rakibul Hasan
Connor Hashemi
Atsushi Hashimoto
Khurram Azeem Hashmi
Nozomi Hata
Ali Hatamizadeh
Ryuichiro Hataya
Timm Haucke
Joakim Bruslund Haurum
Stephen Hausler
Mohammad Havaei
Hideaki Hayashi
Zeeshan Hayder
Chengan He
Conghui He
Dailan He
Fan He
Fazhi He
Gaoqi He
Haoyu He
Hongliang He
Jiangpeng He
Jianzhong He
Jiawei He
Ju He
Jun-Yan He
Junfeng He
Lihuo He
Liqiang He
Nanjun He
Ruian He
Runze He
Ruozhen He

Shengfeng He
Shuting He
Songtao He
Tao He
Tianyu He
Tong He
Wei He
Xiang He
Xiangteng He
Xiangyu He
Xiaoxiao He
Xin He
Xingyi He
Xingzhe He
Xinlei He
Xinwei He
Yang He
Yang He
Yannan He
Yeting He
Yinan He
Ying He
Yisheng He
Yuhang He
Yuhang He
Zecheng He
Zewei He
Zexin He
Zhenyu He
Ziwen He
Ramya S. Hebbalaguppe
Peter Hedman
Deepti B. Hegde
Sindhu B. Hegde
Matthias Hein
Mattias Paul Heinrich
Aral Hekimoglu
Philipp Henzler
Byeongho Heo
Jae-Pil Heo
Miran Heo
Stephane Herbin
Alexander Hermans
Pedro Hermosilla Casajus
Jefferson E. Hernandez

Monica Hernandez
Charles Herrmann
Roei Herzig
Fabian Herzog
Georg Hess
Mauricio Hess-Flores
Robin Hesse
Chamin P. Hewa
    Koneputugodage
Masatoshi Hidaka
Richard E. L. Higgins
Mitchell K. Hill
Carlos Hinojosa
Tobias Hinz
Yusuke Hirota
Elad Hirsch
Shinsaku Hiura
Chih-Hui Ho
Man M. Ho
Tuan N. A. Hoang
Jennifer Hobbs
Jiun Tian Hoe
David T. Hoffmann
Derek Hoiem
Yannick Hold-Geoffroy
Lukas Höllein
Gregory I. Holste
Hiroto Honda
Cheeun Hong
Cheng-Yao Hong
Danfeng Hong
Deokki Hong
Fa-Ting Hong
Fangzhou Hong
Feng Hong
Guan Zhe Hong
Je Hyeong Hong
Ji Woo Hong
Jie Hong
Joanna Hong
Lanqing Hong
Lingyi Hong
Sangwoo Hong
Sungeun Hong
Sunghwan Hong

Susung Hong
Weixiang Hong
Xiaopeng Hong
Yan Hong
Yi Hong
Yuchen Hong
Julia Hornauer
Maxwell C. Horton
Eliahu Horwitz
Mir Rayat Imtiaz Hossain
Tonmoy Hossain
Mehdi Hosseinzadeh
Kazuhiro Hotta
Andrew Z. Hou
Fei Hou
Hao Hou
Ji Hou
Jian Hou
Jinghua Hou
Qibin Hou
Qiqi Hou
Ruibing Hou
Saihui Hou
Tingbo Hou
Xinhai Hou
Yuenan Hou
Yunzhong Hou
Zhi Hou
Lukas Hoyer
Petr Hruby
Jun-Wei Hsieh
Chih-Fan Hsu
Gee-Sern Hsu
Hung-Min Hsu
Yen-Chi Hsu
Anthony Hu
Benran Hu
Bo Hu
Dapeng Hu
Dongting Hu
Guosheng Hu
Haigen Hu
Hanjiang Hu
Hanzhe Hu
Hengtong Hu

Hezhen Hu
Jian Hu
Jian-Fang Hu
Jie Hu
Juan Hu
Junjie Hu
Kai Hu
Lianyu Hu
Lily Hu
MengShun Hu
Minghui Hu
Mu Hu
Panwen Hu
Panwen Hu
Ruizhen Hu
Shengshan Hu
Shiyu Hu
Shizhe Hu
Shu Hu
Tao Hu
Tao Hu
Vincent Tao Hu
Wenbo Hu
Xiaodan Hu
Xiaolin Hu
Xiaoling Hu
Xiaotao Hu
Xiaowei Hu
Xinting Hu
Xinting Hu
Xixi Hu
Xuefeng Hu
Yang Hu
Yaosi Hu
Yinlin Hu
Yueyu Hu
Zeyu Hu
Zhanhao Hu
zhengyu hu
Zhenzhen Hu
Zhiming Hu
Zhiming Hu
Zhongyun Hu
Zijian Hu
Andong Hua

Hang Hua
Miao Hua
Wei Hua
Mengdi Huai
Binbin Huang
Bo Huang
Buzhen Huang
Chao Huang
Chao-Tsung Huang
Chaoqin Huang
Ching-Chun Huang
Cong Huang
Danqing Huang
Di Huang
Feihu Huang
Haibin Huang
Haiyang Huang
Hao Huang
Hsin-Ping Huang
Huaibo Huang
Ian Y. Huang
Jiabo Huang
Jiahui Huang
Jiancheng Huang
Jiangyong Huang
Jiaxing Huang
Jin Huang
Jinfa Huang
Jing Huang
Jonathan Huang
Junkai Huang
Junwen Huang
Kejie Huang
Kuan-Chih Huang
Lei Huang
Libo Huang
Lin Huang
Linjiang Huang
Linyan Huang
linzhi Huang
Lun Huang
Luojie Huang
Mingzhen Huang
Qidong Huang
Qiusheng Huang

Runhui Huang
Ruqi Huang
Shaofei Huang
Sheng Huang
Sheng-Yu Huang
Shiyuan Huang
Shuaiyi Huang
Shuangping Huang
Siteng Huang
Siyu Huang
Siyuan Huang
Tao Huang
Thomas E. Huang
Tianyu Huang
Wenjian Huang
Wenke Huang
Xiaohu Huang
Xiaohua Huang
Xiaoke Huang
Xiaoshui Huang
Xiaoyang Huang
Xijie Huang
Xinyu Huang
Xuhua Huang
Yan Huang
Yaping Huang
Ye Huang
Yi Huang
Yi Huang
Yi Huang
Yi-Hua Huang
Yifei Huang
Yihao Huang
Ying Huang
Yizhan Huang
You Huang
Yue Huang
Yufei Huang
Yuge Huang
Yukun Huang
Yuyao Huang
Zaiyu Huang
Zehao Huang
Zeyi Huang
Zhe Huang

Zhewei Huang
Zhida Huang
Zhiqi Huang
Zhiyu Huang
Zhongzhan Huang
Ziling Huang
Zilong Huang
Ziqi Huang
Zixuan Huang
Ziyuan Huang
Jaesung Huh
Ka-Hei Hui
Le Hui
Tianrui Hui
Xiaofei Hui
Ahmed Imtiaz Humayun
Thomas Hummel
Jing Huo
Shuwei Huo
Yuankai Huo
Julio Hurtado
Rukhshanda Hussain
Mohamed Hussein
Noureldien Hussein
Chuong Minh Huynh
Tran Ngoc Huynh
Muhammad Huzaifa
Inwoo Hwang
Jaehui Hwang
Jenq-Neng Hwang
Sehyun Hwang
Seunghyun Hwang
Sukjun Hwang
Sunhee Hwang
Wonjun Hwang
Jeongseok Hyun
Sangeek Hyun
Ekaterina Iakovleva
Tomoki Ichikawa
A. S. M. Iftekhar
Masaaki Iiyama
Satoshi Ikehata
Sunghoon Im
Woobin Im
Nevrez Imamoglu

Abdullah Al Zubaer Imran
Tooba Imtiaz
Nakamasa Inoue
Naoto Inoue
Eldar Insafutdinov
Catalin Ionescu
Radu Tudor Ionescu
Nasar Iqbal
Umar Iqbal
Go Irie
Muhammad Zubair Irshad
Francesco Isgrò
Yasunori Ishii
Berivan Isik
Syed M. S. Islam
Mariko Isogawa
Vamsi Krishna K. Ithapu
Koichi Ito
Leonardo Iurada
Shun Iwase
Sergio Izquierdo
Sarah Jabbour
David Jacobs
David E. Jacobs
Yasamin Jafarian
Azin Jahedi
Achin Jain
Ayush Jain
Jitesh Jain
Kanishk Jain
Yash Jain
Nikita Jaipuria
Tomas Jakab
Muhammad Abdullah
    Jamal
Hadi Jamali-Rad
Stuart James
Nataraj Jammalamadaka
Donggon Jang
Sujin Jang
Young Kyun Jang
Youngjoon Jang
Steeven Janny
Paul Janson
Maximilian Jaritz

Ronnachai Jaroensri
Guillaume Jaume
Sajid Javed
Saqib Javed
Bhavin Jawade
Hirunima Jayasekara
Senthilnath Jayavelu
Guillaume Jeanneret
Pranav Jeevan
Rohit Jena
Tomas Jenicek
Simon Jenni
Hae-Gon Jeon
Jeimin Jeon
Seogkyu Jeon
Boseung Jeong
Jongheon Jeong
Yonghyun Jeong
Yoonwoo Jeong
Koteswar Rao Jerripothula
Artur Jesslen
Nikolay Jetchev
Ankit Jha
Sumit K. Jha
I-Hong Jhuo
Deyi Ji
Ge-Peng Ji
Hui Ji
Jianmin Ji
Jiayi Ji
Jingwei Ji
Naye Ji
Pengliang Ji
Shengpeng Ji
Wei Ji
Xiang Ji
Xiaopeng Ji
Xiaozhong Ji
Xinya Ji
Yu Ji
Yuanfeng Ji
Zhanghexuan Ji
Baoxiong Jia
Ding Jia
Jinrang Jia

Menglin Jia
Shan Jia
Shuai Jia
Tong Jia
Wenqi Jia
Xiaojun Jia
Xiaosong Jia
Xu Jia
Yiren Jian
Bo Jiang
Borui Jiang
Chen Jiang
Guangyuan Jiang
Hai Jiang
Haiyang Jiang
Haiyong Jiang
Hanwen Jiang
Hanxiao Jiang
Hao Jiang
Haobo Jiang
Huaizu Jiang
Huajie Jiang
Jianmin Jiang
Jianwen Jiang
Jiaxi Jiang
Jin Jiang
Jing Jiang
Kaixun Jiang
Kui Jiang
Li Jiang
Liming Jiang
Meirui Jiang
Ming Jiang
Ming Jiang
Peng Jiang
Peng-Tao Jiang
Runqing Jiang
Siyang Jiang
Tianjian Jiang
Tingting Jiang
Weisen Jiang
Wen Jiang
Xingyu Jiang
Xueying Jiang
Xuhao Jiang

Yanru Jiang
Yi Jiang
Yifan Jiang
Yingying Jiang
Yue Jiang
Yuming Jiang
Zeren Jiang
Zheheng Jiang
Zhenyu Jiang
Zhongyu Jiang
Zi-Hang Jiang
Zixuan Jiang
Ziyu Jiang
Zutao Jiang
Zutao Jiang
Jianbin Jiao
Jianbo Jiao
Licheng Jiao
Ruochen Jiao
Shuming Jiao
Wenpei Jiao
Zequn Jie
Dongkwon Jin
Gaojie Jin
Haian Jin
Haibo Jin
Kyong Hwan Jin
Lei Jin
Lianwen Jin
Linyi Jin
Peng Jin
Peng Jin
Sheng Jin
SouYoung Jin
Weiyang Jin
Xiao Jin
Xiaojie Jin
Xin Jin
Xin Jin
Yeying Jin
Yi Jin
Ying Jin
Zhenchao Jin
Chenchen Jing
Junpeng Jing

Mengmeng Jing
Taotao Jing
Jeonghee Jo
Yeonsik Jo
Younghyun Jo
Cameron A. Johnson
Faith M. Johnson
Maxwell Jones
Michael J. Jones
R. Kenny Jones
Ameya Joshi
Shantanu H. Joshi
Brendan Jou
Chen Ju
Wei Ju
Xuan Ju
Yan Ju
Felix Juefei-Xu
Florian Jug
Yohan Jun
Kim Jun-Seong
Masum Shah Junayed
Chanyong Jung
Claudio R. Jung
Hoin Jung
Hyunyoung Jung
Sangwon Jung
Steffen Jung
Sacha Jungerman
Hari Chandana K.
Prajwal K. R.
Berna Kabadayi
Anis Kacem
Anil Kag
Kumara Kahatapitiya
Bernhard Kainz
Ivana Kajic
Ioannis Kakogeorgiou
Niveditha Kalavakonda
Mahdi M. Kalayeh
Ajinkya Kale
Anmol Kalia
Sinan Kalkan
Jayateja Kalla
Tarun Kalluri

Uday Kamal
Sandesh Kamath
Chandra Kambhamettu
Meina Kan
Sai Srinivas Kancheti
Takuhiro Kaneko
Cuicui Kang
Dahyun Kang
Guoliang Kang
Hyolim Kang
Jaeyeon Kang
Juwon Kang
Li-Wei Kang
Mingon Kang
MinGuk Kang
Minjun Kang
Weitai Kang
Zhao Kang
Yash Mukund Kant
Yueying Kao
Saarthak Kapse
Aupendu Kar
Oğuzhan Fatih Kar
Ozgur Kara
Tamás Karácsony
Srikrishna Karanam
Neerav Karani
Mert Asim Karaoglu
Laurynas Karazija
Navid Kardan
Amirhossein Kardoost
Nour Karessli
Michelle Karg
Mohammad Reza Karimi
    Dastjerdi
Animesh Karnewar
Arjun M. Karpur
Shyamgopal Karthik
Korrawe Karunratanakul
Tejaswi Kasarla
Satyananda Kashyap
Yoni Kasten
Marc A. Kastner
Hirokatsu Kataoka
Isinsu Katircioglu

Kai Katsumata
Ilya Kaufman
Manuel Kaufmann
Chaitanya Kaul
Prannay Kaul
Prakhar Kaushik
Isaak Kavasidis
Ryo Kawahara
Yuki Kawana
Justin Kay
Evangelos Kazakos
Jingcheng Ke
Lei Ke
Tsung-Wei Ke
Wei Ke
Zhanghan Ke
Nikhil V. Keetha
Thomas Kehrenberg
Marilyn Keller
Rohit Keshari
Janis Keuper
Daniel Keysers
Hrant Khachatrian
Seyran Khademi
Wesley A. Khademi
Taras Khakhulin
Samir Khaki
Hasam Khalid
Umar Khalid
Amr Khalifa
Asif Hussain Khan
Faizan Farooq Khan
Muhammad Haris Khan
Naimul Khan
Qadeer Khan
Zeeshan Khan
Bishesh Khanal
Pulkit Khandelwal
Ishan Khatri
Muhammad Uzair Khattak
Vahid Reza Khazaie
Vaishnavi M. Khindkar
Rawal Khirodkar
Pirazh Khorramshahi
Sahil S. Khose

Kourosh Khoshelham
Sena Kiciroglu
Benjamin Kiefer
Kotaro Kikuchi
Mert Kilickaya
Benjamin Killeen
Beomyoung Kim
Boah Kim
Bumsoo Kim
Byeonghwi Kim
Byoungjip Kim
Changhoon Kim
Changick Kim
Chanho Kim
Chanyoung Kim
Dahun Kim
Dahyun Kim
Diana S. Kim
Dong-Jin Kim
Donggun Kim
Donghyun Kim
Dongkeun Kim
Dongwan Kim
Dongyoung Kim
Eun-Sol Kim
Eunji Kim
Euyoung Kim
Geeho Kim
Giseop Kim
Guisik Kim
Gwanghyun Kim
Hakyeong Kim
Hanjae Kim
Hanjung Kim
Heewon Kim
Howon Kim
Hyeongwoo Kim
Hyeongwoo Kim
Hyo Jin Kim
Hyung-Il Kim
Hyunwoo J. Kim
Insoo Kim
Jae Myung Kim
Jeongsol Kim
Jihwan Kim

Jinkyu Kim
Jinwoo Kim
Jongyoo Kim
Joonsoo Kim
Jung Uk Kim
Junho Kim
Junho Kim
Junsik Kim
Kangyeol Kim
Kunhee Kim
Kwang In Kim
Manjin Kim
Minchul Kim
Minji Kim
Minjung Kim
Namil Kim
Namyup Kim
Nayeong Kim
Sanghyun Kim
Seong Tae Kim
Seungbae Kim
Seungryong Kim
Seungwook Kim
Soo Ye Kim
Soohwan Kim
Sungnyun Kim
Sungyeon Kim
Sunnie S. Y. Kim
Tae Hyun Kim
Tae Hyung Kim
Taehoon Kim
Taehun Kim
Taehwan Kim
Taekyung Kim
Taeoh Kim
Taewoo Kim
Won Hwa Kim
Wonjae Kim
Woo Jae Kim
Young Min Kim
YoungBin Kim
Youngeun Kim
Youngseok Kim
Youngwook Kim
Akisato Kimura

Andreas Kirsch
Nikita Kister
Furkan Osman Kınlı
Marcus Klasson
Florian Kleber
Tzofi M. Klinghoffer
Jan P. Klopp
Florian Kluger
Hannah Kniesel
David M. Knigge
Byungsoo Ko
Dohwan Ko
Jongwoo Ko
Takumi Kobayashi
Muhammed Kocabas
Yeong Jun Koh
Kathlén Kohn
Subhadeep Koley
Nick Kolkin
Soheil Kolouri
Jacek Komorowski
Deying Kong
Fanjie Kong
Hanyang Kong
Hui Kong
Kyeongbo Kong
Lecheng Kong
Lingdong Kong
Linghe Kong
Lingshun Kong
Naejin Kong
Quan Kong
Shu Kong
Xianghao Kong
Xiangtao Kong
Xiangwei Kong
Xiaoyu Kong
Xin Kong
Youyong Kong
Yu Kong
Zhenglun Kong
Aishik Konwer
Nicholas C. Konz
Gwanhyeong Koo
Juil Koo

Julian F. P. Kooij
George Kopanas
Sanjeev J. Koppal
Bruno Korbar
Giorgos Kordopatis-Zilos
Dimitri Korsch
Adam Kortylewski
Divya Kothandaraman
Suraj Kothawade
Iuliia Kotseruba
Sasikanth Kotti
Alankar Kotwal
Shashank Kotyan
Alexandros Kouris
Petros Koutras
Rama Kovvuri
Dilip Krishnan
Praveen Krishnan
Ranganath Krishnan
Rohan M. Krishnan
Georg Krispel
Alexander Krull
Tianshu Kuai
Haowei Kuang
Zhengfei Kuang
Andrey Kuehlkamp
David Kügler
Arjan Kuijper
Anna Kukleva
Jonas Kulhanek
Peter Kulits
Akshay R. Kulkarni
Ashutosh C. Kulkarni
Kuldeep Kulkarni
Nilesh Kulkarni
Abhinav Kumar
Abhishek Kumar
Akash Kumar
Avinash Kumar
B. V. K. Vijaya Kumar
Chandan Kumar
Pulkit Kumar
Ratnesh Kumar
Sateesh Kumar
Satish Kumar

Suryansh Kumar
Yogesh Kumar
Nupur Kumari
Sudhakar Kumawat
Nilakshan
    Kunananthaseelan
Rohit Kundu
Souvik Kundu
Meng-Yu Jennifer Kuo
Weicheng Kuo
Shuhei Kurita
Yusuke Kurose
Takahiro Kushida
Uday Kusupati
Alina Kuznetsova
Jobin K. V.
Henry Kvinge
Ho Man Kwan
Hyeokjun Kweon
Donghyeon Kwon
Gihyun Kwon
Heeseung Kwon
Hyoukjun Kwon
Myung-Joon Kwon
Taein Kwon
YoungJoong Kwon
Cameron Kyle-Davidson
Christos Kyrkou
Jorma Laaksonen
Patrick Labatut
Yann Labbé
Manuel Ladron de Guevara
Florent Lafarge
Jean Lahoud
Bolin Lai
Farley Lai
Jian-Huang Lai
Shenqi Lai
Xin Lai
Yu-Kun Lai
Yung-Hsuan Lai
Zeqiang Lai
Zhengfeng Lai
Barath Lakshmanan
Rohit Lal

Rodney LaLonde
Hala Lamdouar
Meng Lan
Yushi Lan
Federico Landi
George V. Landon
Chunbo Lang
Jochen Lang
Nico Lang
Georg Langs
Raffaella Lanzarotti
Dong Lao
Yixing Lao
Yizhen Lao
Zakaria Laskar
Alexandros Lattas
Chun Pong Lau
Shlomi Laufer
Justin Lazarow
Svetlana Lazebnik
Duy Tho Le
Hieu Le
Hoang Le
Hoang Le
Thi-Thu-Huong Le
Trung Le
Trung-Nghia Le
Tung Thanh Le
Hoàng-Ân Lê
Herve Le Borgne
Guillaume Le Moing
Erik Learned-Miller
Tim Lebailly
Byeong-Uk Lee
Byung-Kwan Lee
Cheng-Han Lee
Chul Lee
Daeun Lee
Dogyoon Lee
Dong Hoon Lee
Eugene Eu Tzuan Lee
Eung-Joo Lee
Gyuseong Lee
Hsin-Ying Lee
Hwee Kuan Lee

Hyeongmin Lee
Hyungtae Lee
Jae Yong Lee
Jaeho Lee
Jaeseong Lee
Jaewon Lee
Jangho Lee
Jangwon Lee
Jihyun Lee
Jiyoung Lee
Jong-Seok Lee
Jongho Lee
Jongmin Lee
Joo-Ho Lee
Joon-Young Lee
Joonseok Lee
Jungbeom Lee
Jungho Lee
Jungwoo Lee
Junha Lee
Junhyun Lee
Junyong Lee
Kibok Lee
Kuan-Ying Lee
Kwonjoon Lee
Kwot Sin Lee
Kyungmin Lee
Kyungmoon Lee
Minhyeok Lee
Minsik Lee
Pilhyeon Lee
Saehyung Lee
Sangho Lee
Sanghyeok Lee
Sangmin Lee
Sehun Lee
Sehyung Lee
Seon-Ho Lee
Seong Hun Lee
Seongwon Lee
Seung Hyun Lee
Seung-Ik Lee
Seungho Lee
Seunghun Lee
Seungmin Lee

Seungyong Lee
Sohyun Lee
Suhyeon Lee
Sungho Lee
Sungmin Lee
Suyoung Lee
Taehyun Lee
Wooseok Lee
Yao-Chih Lee
Yi-Lun Lee
Yonghyeon Lee
Youngwan Lee
Leonidas Lefakis
Bowen Lei
Chenyang Lei
Chenyi Lei
Jiahui Lei
Na Lei
Qinqian Lei
Yinjie Lei
Thomas Leimkuehler
Abe Leite
Abdelhak Lemkhenter
Jiaxu Leng
Luziwei Leng
Zhiying Leng
Hendrik P. A. Lensch
Jan E. Lenssen
Ted Lentsch
Simon Lepage
Stefan Leutenegger
Filippo Leveni
Axel Levy
Ailin Li
Aixuan Li
Baiang Li
Baoxin Li
Bin Li
Bing Li
Bing Li
Bo Li
Bowen Li
Boying Li
Changlin Li
Changlin Li

Chao Li
Chenghong Li
Chenglin Li
Chenglong Li
Chengze Li
Chun-Guang Li
Daiqing Li
Dasong Li
Dian Li
Dong Li
Fangda Li
Feiran Li
Fenghai Li
Gen Li
Guanbin Li
Guangrui Li
Guihong Li
Guorong Li
Haifeng Li
Han Li
Hang Li
Hangyu Li
Hanhui Li
Hao Li
Hao Li
Haoang Li
Haoran Li
Haoxiang Li
Haoxin Li
He Li
Heng Li
Hengduo Li
Hongshan Li
Hongwei Bran Li
Hongxiang Li
Hongyang Li
Hongyu Li
Huafeng Li
Huan Li
Hui Li
Jiacheng Li
Jiahao Li
Jialu Li
Jiaman Li
Jiangmeng Li

Jiangtong Li
Jiangyuan Li
Jianing Li
Jianwei Li
Jianwu Li
Jiaqi Li
Jiaqi Li
Jiatong Li
Jiaxuan Li
Jiazhi Li
Jichang Li
Jie Li
Jin Li
Jinglun Li
Jingzhi Li
Jingzong Li
Jinlong Li
Jinlong Li
Jinpeng Li
Jinxing Li
Jun Li
Jun Li
Junbo Li
Juncheng Li
Junxuan Li
Junyi Li
Kai Li
Kaican Li
Kailin Li
Ke Li
Kehan Li
Keyu Li
Kun Li
Kunchang Li
Kunpeng Li
Lei Li
Lei Li
Li Li
Li Erran Li
Liang Li
Lin Li
Lincheng Li
Liulei Li
Liunian Harold Li
Lujun Li

Manyi Li
Maomao Li
Meng Li
Mengke Li
Mengtian Li
Mengtian Li
Ming Li
Ming Li
Minghan Li
Mingjie Li
Nannan Li
Nianyi Li
Peike Li
Peizhao Li
Peng Li
Pengpeng Li
Pengyu Li
Ping Li
Puhao Li
Qiang Li
Qing Li
Qingyong Li
Qiufu Li
Qizhang Li
Ren Li
Rong Li
Rongjie Li
Ru Li
Rui Li
Ruibo Li
Ruihui Li
Ruilong Li
Ruining Li
Ruixuan Li
Runze Li
Ruoteng Li
Shaohua Li
Shasha Li
Shigang Li
Shijie Li
Shikun Li
Shile Li
Shuai Li
Shuai Li
Shuang Li

Shuwei Li
Si Li
Siyao Li
Siyuan Li
Siyuan Li
Taihui Li
Tianye Li
Wanhua Li
Wanqing Li
Wei Li
Wei Li
Wei Li
Wei-Hong Li
Weihao Li
Weijia Li
Weiming Li
Wenbin Li
Wenbo Li
Wenhao Li
Wenjie Li
Wenshuo Li
Wentong Li
Wenxi Li
Xiang Li
Xiang Li
Xiang Li
Xiang Li
Xiang Li
Xiangtai Li
Xiangyang Li
Xianzhi Li
Xiao Li
Xiao Li
Xiaoguang Li
Xiaomeng Li
Xiaoming Li
Xiaoqi Li
Xiaoqiang Li
Xiaotian Li
Xiaoyu Li
Xin Li
Xin Li
Xin Li
Xinghui Li
Xingyi Li

Xingyu Li
Xinjie Li
Xinyu Li
Xiu Li
Xiujun Li
Xuan Li
Xuanlin Li
Xuelong Li
Xuelu Li
Xueqian Li
Ya-Li Li
Yanan Li
Yang Li
Yang Li
Yangyan Li
Yanjing Li
Yansheng Li
Yanwei Li
Yanyu Li
Yaohui Li
Yaowei Li
Yawei Li
Yi Li
Yi Li
Yicong Li
Yicong Li
Yifei Li
Yijin Li
Yijun Li
Yijun Li
Yikang Li
Yimeng Li
Yiming Li
Yiming Li
Yingwei Li
Yiting Li
Yixuan Li
Yize Li
Yizhuo Li
Yong Li
Yong-Lu Li
Yongjie Li
Yuanman Li
Yuanming Li
Yuelong Li

Yuexiang Li
Yuezun Li
Yuhang Li
Yuheng Li
Yulin Li
Yumeng Li
Yunfan Li
Yunheng Li
Yunqiang Li
Yunsheng Li
Yuyan Li
Yuyang Li
Zejian Li
Zekun Li
Zekun Li
Zhangheng Li
Zhangzikang Li
Zhaoshuo Li
Zhaowen Li
Zhe Li
Zhe Li
Zhen Li
Zhen Li
Zhen Li
Zheng Li
Zhengqin Li
Zhengyuan Li
Zhenyu Li
Zhichao Li
Zhihao Li
Zhihao Li
Zhiheng Li
Zhiqi Li
Zhixuan Li
Zhong Li
Zhuoling Li
Zhuowan Li
Zhuowei Li
Zhuoxiao Li
Zihan Li
Ziqiang Li
Wen Qiao Li
Dongze Lian
Long Lian
Qing Lian

Ruyi Lian
Zhouhui Lian
Jin Lianbao
Chao Liang
Chia-Kai Liang
Dingkang Liang
Feng Liang
Gongbo Liang
Hanxue Liang
Hao Liang
Hui Liang
Jiadong Liang
Jiajun Liang
Jian Liang
Jingyun Liang
Jinxiu S. Liang
Junwei Liang
Kaiqu Liang
Ke Liang
Kevin J. Liang
Luming Liang
Mingfu Liang
Pengpeng Liang
Siyuan Liang
Xiaoxiao Liang
Xinran Liang
Xiwen Liang
Yang Liang
Yixun Liang
Yongqing Liang
Youwei Liang
Yuanzhi Liang
Zhexin Liang
Zhihao Liang
Zhixuan Liang
Kang Liao
Liang Liao
Minghui Liao
Ting-Hsuan Liao
Wei Liao
Xin Liao
Yinghong Liao
Yue Liao
Zhibin Liao
Ziwei Liao

Benedetta Liberatori
Daniel J. Lichy
Maiko Lie
Qin Likun
Isaak Lim
Teck Yian Lim
Bannapol Limanond
Baijiong Lin
Beibei Lin
Cheng Lin
Chenhao Lin
Chia-Wen Lin
Chieh Hubert Lin
Chuang Lin
Chung-Ching Lin
Chunyu Lin
Ci-Siang Lin
Di Lin
Fanqing Lin
Feng Lin
Fudong Lin
Guangfeng Lin
Haotong Lin
Haozhe Lin
Hubert Lin
Hui Lin
Jason Lin
Jianxin Lin
Jiaqi Lin
Jiaying Lin
Jiehong Lin
Jierui Lin
Jintao Lin
Kai-En Lin
Ke Lin
Kevin Lin
Kevin Qinghong Lin
Kuan Heng Lin
Kun-Yu Lin
Kunyang Lin
Kwan-Yee Lin
Lijian Lin
Liqiang Lin
Liting Lin
Luojun Lin

Mingyuan Lin
Qiuxia Lin
Shaohui Lin
Shih-Yao Lin
Sihao Lin
Siyou Lin
Tiancheng Lin
Tsung-Yu Lin
Wanyu Lin
Wei Lin
Wei Lin
Wen-Yan Lin
Wenbin Lin
Xiangbo Lin
Xianhui Lin
Xiaofan Lin
Xiaofeng Lin
Xin Lin
Xudong Lin
Xue Lin
Xuxin Lin
Ya-Wei Eileen Lin
Yan-Bo Lin
Yancong Lin
Yi Lin
Yijie Lin
Yiming Lin
Yiqi Lin
Yiqun Lin
Yongliang Lin
Yu Lin
Yuanze Lin
Yuewei Lin
Zhi-Hao Lin
Zhiqiu Lin
ZhiWei Lin
Zinan Lin
Ziyi Lin
David B. Lindell
Philipp Lindenberger
Jingwang Ling
Jun Ling
Yongguo Ling
Zhan Ling
Alexander Liniger

Stefan P. Lionar
Phillip Lippe
Lahav O. Lipson
Joey Litalien
Ron Litman
Mattia Litrico
Dor Litvak
Aishan Liu
Ajian Liu
Akide L. Y. Liu
Andrew Liu
Ao Liu
Bang Liu
Benlin Liu
Bin Liu
Bin Liu
Bing Liu
Binghao Liu
Bingyuan Liu
Bo Liu
Bo Liu
Bo Liu
Boning Liu
Bowen Liu
Boxiao Liu
Chang Liu
Chang Liu
Chang Liu
Chang Liu
Chao Liu
Chengxin Liu
Chengxu Liu
Chih-Ting Liu
Chuanjian Liu
Chun-Hao Liu
Daizong Liu
Decheng Liu
Di Liu
Difan Liu
Dong Liu
Dongnan Liu
Fang Liu
Fang Liu
Fangyi Liu
Feng Liu

Fengbei Liu
Fenglin Liu
Fengqi Liu
Furui Liu
Fuxiao Liu
Haisong Liu
Han Liu
Hanwen Liu
Hanyuan Liu
Hao Liu
Haolin Liu
Haotian Liu
Haozhe Liu
Heshan Liu
Hong Liu
Hongbin Liu
Hongfu Liu
Hongyu Liu
Hsueh-Ti Derek Liu
Huidong Liu
Isabella Liu
Ji Liu
Jia-Wei Liu
Jiachen Liu
Jiaheng Liu
Jiahui Liu
Jiaming Liu
Jiancheng Liu
Jiang Liu
Jianmeng Liu
Jiashuo Liu
Jiawei Liu
Jiawei Liu
Jiayang Liu
Jiayi Liu
Jie Liu
Jie Liu
Jie Liu
Jihao Liu
Jing Liu
Jing Liu
Jing Liu
Jingyuan Liu
Jingyuan Liu
Jiuming Liu

Jiyuan Liu
Jun Liu
Kang-Jun Liu
Kangning Liu
Kenkun Liu
Kunhao Liu
Li Liu
Lijuan Liu
Lingbo Liu
Lingqiao Liu
Liu Liu
Liyang Liu
Meng Liu
Mengchen Liu
Miao Liu
Ming Liu
Minghao Liu
Minghua Liu
Mingxuan Liu
Mingyuan Liu
Nan Liu
Nian Liu
Ning Liu
Peidong Liu
Peirong Liu
Peiye Liu
Pengju Liu
Ping Liu
Qi Liu
Qiankun Liu
Qihao Liu
Qing Liu
Qingjie Liu
Richard Liu
Risheng Liu
Rui Liu
Ruicong Liu
Ruoshi Liu
Ruyu Liu
Shaohui Liu
Shaoteng Liu
Shaowei Liu
Sheng Liu
Shenglan Liu
Shikun Liu

Shilong Liu
Shuaicheng Liu
Shuaicheng Liu
Shuming Liu
Songhua Liu
Tao Liu
Tian Yu Liu
Tianci Liu
Tianshan Liu
Tongliang Liu
Tyng-Luh Liu
Wei Liu
Weifeng Liu
Weixiao Liu
Weiyu Liu
Wen Liu
Wenxi Liu
Wenyu Liu
Wenyu Liu
Wu Liu
Xian Liu
Xianglong Liu
Xianpeng Liu
Xiao Liu
Xiaohong Liu
Xiaoyu Liu
Xiaoyu Liu
Xihui Liu
Xin Liu
Xin Liu
Xinchen Liu
Xingtong Liu
Xingyu Liu
Xinhang Liu
Xinhui Liu
Xinpeng Liu
Xinwei Liu
Xinyu Liu
Xiulong Liu
Xiyao Liu
Xu Liu
Xubo Liu
Xudong Liu
Xueting Liu
Xueyi Liu

Yan Liu
Yanbin Liu
Yang Liu
Yang Liu
Yang Liu
Yang Liu
Yang Liu
Yanwei Liu
Yaojie Liu
Ye Liu
Yi Liu
Yihao Liu
Yingcheng Liu
Yingfei Liu
Yipeng Liu
Yipeng Liu
Yixin Liu
Yizhang Liu
Yong Liu
Yong Liu
Yonghuai Liu
Yongtuo Liu
Yu Liu
Yu-Lun Liu
Yu-Shen Liu
Yuan Liu
Yuang Liu
Yuanpei Liu
Yuanpeng Liu
Yuanwei Liu
Yuchen Liu
Yuchen Liu
Yuchi Liu
Yueh-Cheng Liu
Yufan Liu
Yuhao Liu
Yuliang Liu
Yun Liu
Yun Liu
Yun Liu
Yunfan Liu
Yunfei Liu
Yunze Liu
Yupei Liu
Yuqi Liu

Yuyang Liu
Yuyuan Liu
Zhaoqiang Liu
Zhe Liu
Zhe Liu
Zhen Liu
Zheng Liu
Zhenguang Liu
Zhi Liu
Zhihua Liu
Zhijian Liu
Zhili Liu
Zhuoran Liu
Ziquan Liu
Ziyi Liu
Zuxin Liu
Zuyan Liu
Josep Llados
Ling Lo
Shao-Yuan Lo
Liliana Lo Presti
Sylvain Lobry
Yaroslava Lochman
Fotios Logothetis
Suhas Lohit
Marios Loizou
Vishnu Suresh Lokhande
Cheng Long
Chengjiang Long
Fuchen Long
Guodong Long
Rujiao Long
Shangbang Long
Teng Long
Xiaoxiao Long
Zijun Long
Ivan Lopes
Vasco Lopes
Adrian Lopez-Rodriguez
Javier Lorenzo-Navarro
Yujing Lou
Brian C. Lovell
Weng Fei Low
Changsheng Lu
Chun-Shien Lu

Daohan Lu
Dongming Lu
Erika Lu
Fan Lu
Guangming Lu
Guo Lu
Hao Lu
Hao Lu
Hongtao Lu
Jiachen Lu
Jiaxin Lu
Jiwen Lu
Lewei Lu
Liying Lu
Quanfeng Lu
Shenyu Lu
Shun Lu
Tao Lu
Xiangyong Lu
Xiankai Lu
Xin Lu
Xuanchen Lu
Xuequan Lu
Yan Lu
Yang Lu
Yanye Lu
Yawen Lu
Yifan Lu
Yongchun Lu
Yongxi Lu
Yu Lu
Yu Lu
Yuzhe Lu
Zhichao Lu
Zhihe Lu
Zijia Lu
Tianyu Luan
Pauline Luc
Simon Lucey
Timo Lüddecke
Jonathon Luiten
Jovita Lukasik
Ao Luo
Cheng Luo
Chuanchen Luo

Donghao Luo
Fangzhou Luo
Gen Luo
Gongning Luo
Hongchen Luo
Jiahao Luo
Jiebo Luo
Jinqi Luo
Jinqi Luo
Jun Luo
Katie Z. Luo
Kunming Luo
Lei Luo
Mandi Luo
Mi Luo
Ruotian Luo
Sihui Luo
Tiange Luo
Wenhan Luo
Xiao Luo
Xiaotong Luo
Xiongbiao Luo
Xu Luo
Yadan Luo
Yawei Luo
Ye Luo
Yisi Luo
Yong Luo
You-Wei Luo
Yuanjing Luo
Zelun Luo
Zhengxiong Luo
Zhengyi Luo
Zhiming Luo
Zhipeng Luo
Zhongjin Luo
Zilin Luo
Ziyang Luo
Tung M. Luu
Diogo C. Luvizon
Jun Lv
Pei Lv
Yunqiu Lv
Zhaoyang Lv
Gengyu Lyu

Jiancheng Lyu
Jipeng Lyu
Junfeng Lyu
Mengyao Lyu
Mingzhi Lyu
Weimin Lyu
Xiaoyang Lyu
Xinyu Lyu
Yiwei Lyu
Youwei Lyu
Ailong Ma
Andy J. Ma
Benteng Ma
Bingpeng Ma
Chao Ma
Chuofan Ma
Cong Ma
Cuixia Ma
Fan Ma
Fangchang Ma
Fei Ma
Guozheng Ma
Haoyu Ma
Hengbo Ma
Huimin Ma
Jiahao Ma
Jianqi Ma
Jiawei Ma
Jiayi Ma
Kai Ma
Kede Ma
Lei Ma
Li Ma
Lin Ma
Liqian Ma
Lizhuang Ma
Mengmeng Ma
Ning Ma
Qianli Ma
Rui Ma
Shijie Ma
Shiqiang Ma
Shiqing Ma
Shuailei Ma
Sizhuo Ma

Tao Ma
Teli Ma
Wenxuan Ma
Wufei Ma
Xianzheng Ma
Xiaoxuan Ma
Xinyin Ma
Xinzhu Ma
Xu Ma
Yeyao Ma
Yifeng Ma
Yuexiao Ma
Yuexin Ma
Yunsheng Ma
Zhan Ma
Zhanyu Ma
Ziping Ma
Ziqiao Ma
Muhammad Maaz
Anish Madan
Neelu Madan
Spandan Madan
Sai Advaith Maddipatla
Rishi Madhok
Filippo Maggioli
Simone Magistri
Marcus Magnor
Sabarinath Mahadevan
Shweta Mahajan
Aniruddha Mahapatra
Sarthak Kumar Maharana
Behrooz Mahasseni
Upal Mahbub
Arif Mahmood
Kaleel Mahmood
Mohammed Mahmoud
Tanvir Mahmud
Jinjie Mai
Helena de Almeida Maia
Josef Maier
Shishira R. Maiya
Snehashis Majhi
Orchid Majumder
Sagnik Majumder
Ilya Makarov

Sina Malakouti
Hashmat Shadab Malik
Mateusz Malinowski
Utkarsh Mall
Srikanth Malla
Clement Mallet
Dimitrios Mallis
Abed Malti
Yunze Man
Oscar Mañas
Karttikeya Mangalam
Fabian Manhardt
Ioannis Maniadis Metaxas
Fahim Mannan
Rafal Mantiuk
Dongxing Mao
Jiageng Mao
Wei Mao
Weian Mao
Weixin Mao
Ye Mao
Yongsen Mao
Yunyao Mao
Yuxin Mao
Zhiyuan Mao
Emanuela Marasco
Matthew Marchellus
Alberto Marchisio
Diego Marcos
Alina E. Marcu
Riccardo Marin
Manuel J. Marín-Jiménez
Octave Mariotti
Dejan Markovic
Imad Eddine Marouf
Valerio Marsocci
Diego Martin Arroyo
Ricardo Martin-Brualla
Brais Martinez
Renato Martins
Damien Martins Gomes
Tetiana Martyniuk
Pierre Marza
David Masip
Carlo Masone

Timothée Masquelier
André G. Mateus
Minesh Mathew
Yusuke Matsui
Bruce A. Maxwell
Christoph Mayer
Prasanna Mayilvahanan
Amir Mazaheri
Amrita Mazumdar
Pratik Mazumder
Alessio Mazzucchelli
Amarachi B. Mbakwe
Scott McCloskey
Naga Venkata Kartheek
    Medathati
Henry Medeiros
Guofeng Mei
Haiyang Mei
Jie Mei
Jieru Mei
Kangfu Mei
Lingjie Mei
Xiaoguang Mei
Dennis Melamed
Luke Melas-Kyriazi
Iaroslav Melekhov
Yifang Men
Ricardo A. Mendoza-León
Depu Meng
Fanqing Meng
Jingke Meng
Lingchen Meng
Qier Meng
Qingjie Meng
Quan Meng
Yanda Meng
Zibo Meng
Otniel-Bogdan Mercea
Pablo Mesejo
Safa Messaoud
Nico Messikommer
Nando Metzger
Christopher Metzler
Vasileios Mezaris
Liang Mi

Zhenxing Mi
S. Mahdi H. Miangoleh
Bo Miao
Changtao Miao
Jiaxu Miao
Zichen Miao
Bjoern Michele
Christian Micheloni
Marko Mihajlovic
Zoltán Á. Milacski
Simone Milani
Leo Milecki
Roy Miles
Christen Millerdurai
Monica Millunzi
Chaerin Min
Cheol-Hui Min
Dongbo Min
Hyun-Seok Min
Jie Min
Juhong Min
Kyle Min
Yifei Min
Yuecong Min
Zhixiang Min
Matthias Minderer
Di Ming
Qi Ming
Xiang Ming
Riccardo Miotto
Aymen Mir
Pedro Miraldo
Parsa Mirdehghan
Seyed Ehsan Mirsadeghi
Muhammad Jehanzeb
    Mirza
Ashkan Mirzaei
Dmytro Mishkin
Anand Mishra
Ashish Mishra
Samarth Mishra
Shlok K. Mishra
Diganta Misra
Abhay Mittal
Gaurav Mittal

Surbhi Mittal
Trisha Mittal
Taiki Miyanishi
Daisuke Miyazaki
Hong Mo
Kaichun Mo
Sangwoo Mo
Sicheng Mo
Sicheng Mo
Zhipeng Mo
Michael Moeller
Peyman Moghadam
Hadi Mohaghegh
   Dolatabadi
Salman Mohamadi
Mirgahney H. Mohamed
Deen Dayal Mohan
Fnu Mohbat
Satyam Mohla
Tony C. W. Mok
Liliane Momeni
Pascal Monasse
Ajoy Mondal
Anindya Mondal
Mathew Monfort
Tom Monnier
Yusuke Monno
Eduardo F. Montesuma
Gyeongsik Moon
Taesup Moon
WonJun Moon
Dror Moran
Julie R. C. Mordacq
Deeptej S. More
Arthur Moreau
Davide Morelli
Luca Morelli
Pedro Morgado
Alexandre Morgand
Henrique Morimitsu
Matteo Moro
Lia Morra
Matteo Mosconi
Ali Mosleh

Sayed Mohammad
   Mostafavi Isfahani
Saman Motamed
Chong Mou
Dana Moukheiber
Pierre Moulon
Ramy A. Mounir
Théo Moutakanni
Fangzhou Mu
Jiteng Mu
Yao Mark Mu
Manasi Muglikar
Yasuhiro Mukaigawa
Amitangshu Mukherjee
Avideep Mukherjee
Prerana Mukherjee
Tanmoy Mukherjee
Anirban Mukhopadhyay
Soumik Mukhopadhyay
Yusuke Mukuta
Ravi Teja Mullapudi
Lea Müller
Norman Müller
Chaithanya Kumar
   Mummadi
Muhammad Akhtar Munir
Subrahmanyam Murala
Sanjeev Muralikrishnan
Ana C. Murillo
Nils Murrugarra-Llerena
Mohamed Adel Musallam
Damien Muselet
Josh David Myers-Dean
Byeonghu Na
Taeyoung Na
Muhammad Ferjad Naeem
Sauradip Nag
Pravin Nagar
Rajendra Nagar
Varun Nagaraja
Tushar Nagarajan
Seungjun Nah
Shu Nakamura
Gaku Nakano
Yuta Nakashima

Kiyohiro Nakayama
Mitsuru Nakazawa
Krishna Kanth Nakka
Yuesong Nan
Karthik Nandakumar
Paolo Napoletano
Syed S. Naqvi
Dinesh Reddy
   Narapureddy
Supreeth
   Narasimhaswamy
Kartik Narayan
Sriram Narayanan
Fabio Narducci
Erickson R. Nascimento
Muzammal Naseer
Kamal Nasrollahi
Lakshmanan Nataraj
Vishwesh Nath
Avisek Naug
Alexander Naumann
K. L. Navaneet
Pablo Navarrete Michelini
Shah Nawaz
Nazir Nayal
Niv Nayman
Amin Nejatbakhsh
Negar Nejatishahidin
Reyhaneh Neshatavar
Pedro C. Neto
Lukáš Neumann
Richard Newcombe
Alejandro Newell
Evonne Ng
Kam Woh Ng
Trung T. Ngo
Tuan Duc Ngo
Anh Nguyen
Anh Duy Nguyen
Cuong Cao Nguyen
Duc Anh Nguyen
Hoang Chuong Nguyen
Huy Hong Nguyen
Khai Nguyen
Khanh-Binh Nguyen

Khanh-Duy Nguyen
Khoi Nguyen
Khoi D. Nguyen
Kiet A. Nguyen
Ngoc Cuong Nguyen
Pha Nguyen
Phi Le Nguyen
Phong Ha Nguyen
Rang Nguyen
Tam V. Nguyen
Thao Nguyen
Thuan Hoang Nguyen
Toan Tien Nguyen
Trong-Tung Nguyen
Van Nguyen Nguyen
Van-Quang Nguyen
Thuong Nguyen Canh
Thien Trang Nguyen Vu
Haomiao Ni
Jiangqun Ni
Minheng Ni
Yao Ni
Zhen-Liang Ni
Zixuan Ni
Dong Nie
Hui Nie
Jiahao Nie
Lang Nie
Liqiang Nie
Qiang Nie
Ying Nie
Yinyu Nie
Yongwei Nie
Aditya Nigam
Kshitij N. Nikhal
Nick Nikzad
Jifeng Ning
Rui Ning
Xuefei Ning
Li Niu
Muyao Niu
Shuaicheng Niu
Wei Niu
Xuesong Niu
Yi Niu

Yulei Niu
Zhenxing Niu
Zhong-Han Niu
Shohei Nobuhara
Jongyoun Noh
Junhyug Noh
Nadhira Noor
Parsa Nooralinejad
Sotiris Nousias
Tiago Novello
Gal Novich
David Novotny
Slawomir Nowaczyk
Ewa M. Nowara
Evangelos Ntavelis
Valsamis Ntouskos
Leonardo Nunes
Oren Nuriel
Zhakshylyk Nurlanov
Simbarashe Nyatsanga
Lawrence O'Gorman
Anton Obukhov
Michael Oechsle
Ferda Ofli
Changjae Oh
Dongkeun Oh
Junghun Oh
Seoung Wug Oh
Youngtaek Oh
Hiroki Ohashi
Takehiko Ohkawa
Takeshi Oishi
Takahiro Okabe
Fumio Okura
Daniel Olmeda Reino
Suguru Onda
Trevine S. J. Oorloff
Michael Opitz
Roy Or-El
Jose Oramas
Jordi Orbay
Tribhuvanesh Orekondy
Evin Pınar Örnek
Alessandro Ortis
Magnus Oskarsson

Julian Ost
Daniil Ostashev
Mayu Otani
Naima Otberdout
Hatef Otroshi Shahreza
Yassine Ouali
Amine Ouasfi
Cheng Ouyang
Wanli Ouyang
Wenqi Ouyang
Xu Ouyang
Poojan B. Oza
Milind G. Padalkar
Johannes C. Paetzold
Gautam Pai
Anwesan Pal
Simone Palazzo
Avinash Paliwal
Cristina Palmero
Chengwei Pan
Fei Pan
Hao Pan
Jianhong Pan
Junting Pan
Liang Pan
Lili Pan
Linfei Pan
Liyuan Pan
Tai-Yu Pan
Xichen Pan
Xingjia Pan
Xinyu Pan
Yingwei Pan
Zhaoying Pan
Zhihong Pan
Zixuan Pan
Zizheng Pan
Rohit Pandey
Saurabh Pandey
Bo Pang
Guansong Pang
Lu Pang
Meng Pang
Tianyu Pang
Youwei Pang

Ziqi Pang
Omiros Pantazis
Juan J. Pantrigo
Hsing-Kuo Kenneth Pao
Marina Paolanti
Joao P. Papa
Samuele S. Papa
Dim P. Papadopoulos
Symeon Papadopoulos
George Papandreou
Toufiq Parag
Chethan Parameshwara
Foivos Paraperas
    Papantoniou
Shaifali Parashar
Alejandro Pardo
Jason R. Parham
Kranti K. Parida
Rishubh Parihar
Chunghyun Park
Daehee Park
Dongmin Park
Dongwon Park
Eunbyung Park
Eunhyeok Park
Eunil Park
Geon Yeong Park
Gyeong-Moon Park
Hyoungseob Park
Jae Sung Park
JaeYoo Park
Jin-Hwi Park
Jinhyung Park
Jinyoung Park
Jongwoo Park
JoonKyu Park
JungIn Park
Junheum Park
Kiru Park
Kwanyong Park
Seongsik Park
Seulki Park
Song Park
Sungho Park
Sungjune Park

Taesung Park
Yeachan Park
Gaurav Parmar
Paritosh Parmar
Maurizio Parton
Magdalini Paschali
Vito Paolo Pastore
Or Patashnik
Gaurav Patel
Maitreya Patel
Diego Patino
Suvam Patra
Viorica Patraucean
Badri Narayana Patro
Danda Pani Paudel
Angshuman Paul
Sneha Paul
Soumava Paul
Sudipta Paul
Sujoy Paul
Rémi Pautrat
Ioannis Pavlidis
Svetlana Pavlitska
Raju Pavuluri
Kim Steenstrup Pedersen
Marco Pedersoli
Adithya Pediredla
Pieter Peers
Jiju Peethambaran
Sen Pei
Wenjie Pei
Yuru Pei
Simone Alberto Peirone
Chantal Pellegrini
Latha Pemula
Abhirama Subramanyam
    V. B. Penamakuri
Adrian Penate-Sanchez
Baoyun Peng
Bo Peng
Can Peng
Cheng Peng
Chi-Han Peng
Chunlei Peng
Jie Peng

Jingliang Peng
Kebin Peng
Kunyu Peng
Liang Peng
Liangzu Peng
Pai Peng
Peixi Peng
Sida Peng
Songyou Peng
Wei Peng
Wen-Hsiao Peng
Xi Peng
Xiaojiang Peng
Yi-Xing Peng
Yuxin Peng
Zhiliang Peng
Ziqiao Peng
Matteo Pennisi
Or Perel
Gabriel Perez
Gustavo Perez
Juan C. Perez
Andres Felipe Perez
    Murcia
Eduardo Pérez-Pellitero
Neehar Peri
Skand Peri
Gabriel J. Perin
Federico Pernici
Chiara Pero
Elia Peruzzo
Marco Pesavento
Dmitry M. Petrov
Ilya A. Petrov
Mathis Petrovich
Vitali Petsiuk
Tomas Pevny
Shubham Milind Phal
Chau Pham
Hai X. Pham
Khoi Pham
Long Hoang Pham
Trong Thang Pham
Trung X. Pham
Tung Pham

Hoang Phan
Huy Phan
Minh Hieu Phan
Julien Philip
Stephen Phillips
Cheng Perng Phoo
Hao Phung
Shruti S. Phutke
Weiguo Pian
Yongri Piao
Luigi Piccinelli
A. J. Piergiovanni
Sara Pieri
Vipin Pillai
Wu Pingyu
Silvia L. Pintea
Francesco Pinto
Maura Pintor
Giovanni Pintore
Vittorio Pippi
Robinson Piramuthu
Fiora Pirri
Leonid Pishchulin
Francesca Pistilli
Francesco Pittaluga
Fabio Pizzati
Edward Pizzi
Benjamin Planche
Iuliia Pliushch
Chiara Plizzari
Ryan Po
GIovanni Poggi
Matteo Poggi
Kilian Pohl
Chandradeep Pokhariya
Ashwini Pokle
Matteo Polsinelli
Adrian Popescu
Teodora Popordanoska
Nikola Popović
Ronald Poppe
Samuele Poppi
Andrea Porfiri Dal Cin
Angelo Porrello

Pedro Porto Buarque de
    Gusmão
Rudra P. K. Poudel
Kossar Pourahmadi
    Meibodi
Hadi Pouransari
Ali Pourramezan Fard
Omid Poursaeed
Anish J. Prabhu
Mihir Prabhudesai
Aayush Prakash
Aditya Prakash
Shraman Pramanick
Mantini Pranav
B. H. Pawan Prasad
Meghshyam Prasad
Prateek Prasanna
Ekta Prashnani
Bardh Prenkaj
Derek S. Prijatelj
Véronique Prinet
Malte Prinzler
Victor Adrian Prisacariu
Federica Proietto Salanitri
Sergey Prokudin
Bill Psomas
Dongqi Pu
Mengyang Pu
Nan Pu
Shi Pu
Rita Pucci
Kuldeep Purohit
Senthil Purushwalkam
Waqas A. Qazi
Charles R. Qi
Chenyang Qi
Haozhi Qi
Jiaxin Qi
Lei Qi
Mengshi Qi
Peng Qi
Xianbiao Qi
Xiangyu Qi
Yuankai Qi
Zhangyang Qi

Guocheng Qian
Hangwei Qian
Jianing Qian
Qi Qian
Rui Qian
Shengsheng Qian
Shengyi Qian
Shenhan Qian
Wen Qian
Xuelin Qian
Yaguan Qian
Yijun Qian
Yiming Qian
Zhenxing Qian
Wenwen Qiang
Feng Qiao
Fengchun Qiao
Xiaotian Qiao
Yanyuan Qiao
Yi-Ling Qiao
Yu Qiao
Hangyu Qin
Haotong Qin
Jie Qin
Peiwu Qin
Siyang Qin
Wenda Qin
Xuebin Qin
Xugong Qin
Yang Qin
Yipeng Qin
Yongqiang Qin
Yuzhe Qin
Zequn Qin
Zeyu Qin
Zheng Qin
Zhenyue Qin
Ziheng Qin
Jiaxin Qing
Congpei Qiu
Haibo Qiu
Hang Qiu
Heqian Qiu
Jiayan Qiu
Jielin Qiu

Longtian Qiu
Mufan Qiu
Ri-Zhao Qiu
Weichao Qiu
Xuchong Qiu
Xuerui Qiu
Yuda Qiu
Yuheng Qiu
Zhongxi Qiu
Maan Qraitem
Chao Qu
Linhao Qu
Yanyun Qu
Kha Gia Quach
Ruijie Quan
Fabio Quattrini
Yvain Queau
Faisal Z. Qureshi
Rizwan Qureshi
Hamid R. Rabiee
Paolo Rabino
Ryan L. Rabinowitz
Petia Radeva
Bhaktipriya Radharapu
Krystian Radlak
Bodgan Raducanu
M. Usman Rafique
Francesco Ragusa
Sahar Rahimi Malakshan
Tanzila Rahman
Aashish Rai
Arushi Rai
Shyam Nandan Rai
Zobeir Raisi
Amit Raj
Kiran Raja
Sachin Raja
Deepu Rajan
Jathushan Rajasegaran
Gnana Praveen Rajasekhar
Ramanathan Rajendiran
Marie-Julie Rakotosaona
Gorthi Rama Krishna Sai
    Subrahmanyam

Sai Niranjan
    Ramachandran
Santhosh Kumar
    Ramakrishnan
Srikumar Ramalingam
Michaël Ramamonjisoa
Ravi Ramamoorthi
Shanmuganathan Raman
Mani Ramanagopal
Ashish Ramayee Asokan
Andrea Ramazzina
Jason Rambach
Sai Saketh Rambhatla
Sai Saketh Rambhatla
Clément Rambour
Francois Bernard Julien
    Rameau
Visvanathan Ramesh
Adín Ramírez Rivera
Haoxi Ran
Xuming Ran
Aakanksha Rana
Srinivas Rana
Kanchana N. Ranasinghe
Poorva G. Rane
Aneesh Rangnekar
Harsh Rangwani
Viresh Ranjan
Anyi Rao
Sukrut Rao
Yongming Rao
ZhiBo Rao
Carolina Raposo
Hanoona Abdul Rasheed
Amir Rasouli
Deevashwer Rathee
Christian Rathgeb
Avinash Ravichandran
Bharadwaj Ravichandran
Arijit Ray
Dripta S. Raychaudhuri
Sonia Raychaudhuri
Haziq Razali
Daniel Rebain
William T. Redman

Albert W. Reed
Aniket Rege
Christoph Reich
Christian Reimers
Simon Reiß
Konstantinos Rematas
Tal Remez
Davis Rempe
Bin Ren
Chao Ren
Chuan-Xian Ren
Dayong Ren
Dongwei Ren
Jiawei Ren
Jiaxiang Ren
Jing Ren
Mengwei Ren
Pengfei Ren
Pengzhen Ren
Qibing Ren
Shuhuai Ren
Sucheng Ren
Tianhe Ren
Weihong Ren
Wenqi Ren
Xuanchi Ren
Yanli Ren
Yihui Ren
Yixuan Ren
Yufan Ren
Zhenwen Ren
Zhihang Ren
Zhiyuan Ren
Zhongzheng Ren
Jose Restom
George Retsinas
Ambareesh Revanur
Ferdinand Rewicki
Manuel Rey Area
Md Alimoor Reza
Farnoush Rezaei Jafari
Hamed Rezazadegan
    Tavakoli
Rafael S. Rezende
Wonjong Rhee

Anthony D. Rhodes
Daniel Riccio
Alexander Richard
Christian Richardt
Luca Rigazio
Benjamin Risse
Dominik Rivoir
Luigi Riz
Mamshad Nayeem Rizve
Antonino M. Rizzo
Wes J. Robbins
Damien Robert
Jonathan Roberts
Joseph Robinson
Antonio Robles-Kelly
Mrigank Rochan
Chris Rockwell
Chris Rockwell
Ivan Rodin
Erik Rodner
Ranga Rodrigo
Andres C. Rodriguez
Cristian Rodriguez
Carlos Rodriguez-Pardo
Antonio J.
   Rodriguez-Sanchez
Barbara Roessle
Paul Roetzer
Alina Roitberg
Javier Romero
Meitar Ronen
Keran Rong
Xuejian Rong
Yu Rong
Marco Rosano
Bodo Rosenhahn
Gabriele Rosi
Candace Ross
Andreas Rössler
Giulio Rossolini
Mohammad Rostami
Edward Rosten
Daniel Roth
Karsten Roth
Mark S. Rothermel

Matthias Rottmann
Anastasios Roussos
Aniket Roy
Anirban Roy
Debaditya Roy
Shuvendu Roy
Sudipta Roy
Ahana Roy Choudhury
Amit Roy-Chowdhury
Aruni RoyChowdhury
Dávid Rozenberszki
Denys Rozumnyi
Lixiang Ru
Lingyan Ruan
Shulan Ruan
Viktor Rudnev
Daniel Rueckert
Nataniel Ruiz
Ewelina Rupnik
Evgenia Rusak
Chris Russell
Marc Rußwurm
Fiona Ryan
Dawid Damian Rymarczyk
DongHun Ryu
Sari Saba-Sadiya
Robert Sablatnig
Mohammad Sabokrou
Ragav Sachdeva
Ali Sadeghian
Arka Sadhu
Sadra Safadoust
Bardia Safaei
Ryusuke Sagawa
Avishkar Saha
Gobinda Saha
Oindrila Saha
Aditya Sahdev
Lakshmi Babu Saheer
Aadarsh Sahoo
Pritish Sahu
Aneeshan Sain
Nirat Saini
Saurabh Saini
Kuniaki Saito

Shunsuke Saito
Rahul Sajnani
Fumihiko Sakaue
Parikshit V. Sakurikar
Riccardo Salami
Soorena Salari
Mohammadreza Salehi
Leonard Salewski
Driton Salihu
Benjamin Salmon
Cristiano Saltori
Joel Saltz
Tim Salzmann
Sina Samangooei
Babak Samari
Nermin Samet
Fawaz Sammani
Leo Sampaio Ferraz
   Ribeiro
Shailaja Keyur Sampat
Alessio Sampieri
Jorge Sanchez
Pedro Sandoval-Segura
Nong Sang
Shengtian Sang
Patsorn Sangkloy
Depanshu Sani
Juan C. Sanmiguel
Hiroaki Santo
Joshua Santoso
Bikash Santra
Soubhik Sanyal
Hitesh Sapkota
Ayush Saraf
Nikolaos Sarafianos
István Sárándi
Kyle Sargent
Andranik Sargsyan
Josip Šarić
Mert Bulent Sariyildiz
Abhijit Sarkar
Anirban Sarkar
Chayan Sarkar
Michel Sarkis
Paul-Edouard Sarlin

Sara Sarto
Josua Sassen
Srikumar Sastry
Imari Sato
Takami Sato
Shin'ichi Satoh
Ravi Kumar Satzoda
Jack Saunders
Corentin Sautier
Mattia Savardi
Bogdan Savchynskyy
Mohamed Sayed
Marin Scalbert
Gianluca Scarpellini
Gerald Schaefer
Guilherme G. Schardong
David Schinagl
Phillip Schniter
Patrick Schramowski
Matthias Schubert
Peter Schüffler
Samuel Schulter
René Schuster
Klamer Schutte
Luca Scofano
Jesse Scott
Marcel Seelbach Benkner
Karthik Seemakurthy
Mattia Segù
Santi Seguí
Sinisa Segvic
Constantin Marc Seibold
Roman Seidel
Lorenzo Seidenari
Taiki Sekii
Yusuke Sekikawa
Matan Sela
Pratheba Selvaraju
Agniva Sengupta
Ahyun Seo
Jinhwan Seo
Junyoung Seo
Kwanggyoon Seo
Seonguk Seo
Seunghyeon Seo

Jinseok Seol
Hongje Seong
Ana F. Sequeira
Dario Serez
Dario Serez
David Serrano-Lozano
Pratinav Seth
Francesco Setti
Giorgos Sfikas
Mohammad Amin Shabani
Faisal Shafait
Anshul Shah
Chintan Shah
Jay Shah
Ketul Shah
Mubarak Shah
Viraj Shah
Mohamad Shahbazi
Muhammad Bilal B.
    Shaikh
Abdelrahman M. Shaker
Greg Shakhnarovich
Md Salman Shamil
Fahad Shamshad
Caifeng Shan
Dandan Shan
Hongming Shan
Xiaojun Shan
Chong Shang
Fanhua Shang
Jinghuan Shang
Lei Shang
Sifeng Shang
Wei Shang
Yuzhang Shang
Yuzhang Shang
Sukrit Shankar
Dian Shao
Mingwen Shao
Rui Shao
Ruizhi Shao
Shuai Shao
Shuwei Shao
Ron A. Shapira Weber
S. M. A. Sharif

Aashish Sharma
Avinash Sharma
Charu Sharma
Prafull Sharma
Prasen Kumar Sharma
Allam Shehata
Mark Sheinin
Sumit Shekhar
Oleksandr Shekhovtsov
Chuanfu Shen
Fei Shen
Fengyi Shen
Furao Shen
Hui-liang Shen
Jiajun Shen
Jianghao Shen
Jiangrong Shen
Jiayi Shen
Li Shen
Li-Yong Shen
Linlin Shen
Maying Shen
Qiu Shen
Qiuhong Shen
Shuai Shen
Shuhan Shen
Siqi Shen
Tianwei Shen
Tong Shen
Xiaolong Shen
Xiaoqian Shen
Yan Shen
Yanqing Shen
Yilin Shen
Ying Shen
Yiqing Shen
Yuan Shen
Yucong Shen
Yuhan Shen
Yunhang Shen
Zehong Shen
Zengming Shen
Zhijie Shen
Zhiqiang Shen
Hualian Sheng

Tao Sheng
Yichen Sheng
Zehua Sheng
Shivanand Venkanna
   Sheshappanavar
Ivaxi Sheth
Baoguang Shi
Botian Shi
Dachuan Shi
Daqian Shi
Haizhou Shi
Hengcan Shi
Jia Shi
Jing Shi
Jingang Shi
QingHongYa Shi
Ruoxi Shi
Tianyang Shi
Weishi Shi
Wu Shi
Wuxuan Shi
Xiaodan Shi
Xiaoshuang Shi
Xiaoyu Shi
Xingjian Shi
Xinyu Shi
Xuepeng Shi
Yichun Shi
Yujiao Shi
Zhenbo Shi
Zheng Shi
Zhensheng Shi
Zhenwei Shi
Zhihao Shi
Zifan Shi
Takashi Shibata
Meng-Li Shih
Yichang Shih
Dongseok Shim
Wataru Shimoda
Ilan Shimshoni
Changha Shin
Gyungin Shin
Hyungseob Shin
Inkyu Shin

Seungjoo Shin
Ukcheol Shin
Yooju Shin
Young Min Shin
Koichi Shinoda
Kaede Shiohara
Suprosanna Shit
Palaiahnakote
   Shivakumara
Sindi Shkodrani
Michal
   Shlapentokh-Rothman
Debaditya Shome
Hyounguk Shon
Sulabh Shrestha
Aman Shrivastava
Ayush Shrivastava
Gaurav Shrivastava
Aleksandar Shtedritski
Dong Wook Shu
Han Shu
Jun Shu
Xiangbo Shu
Xiujun Shu
Yang Shu
Bing Shuai
Hong-Han Shuai
Qing Shuai
Changjian Shui
Pushkar Shukla
Mustafa Shukor
Hubert P. H. Shum
Nina Shvetsova
Chenyang Si
Jianlou Si
Zilin Si
Mennatullah Siam
Sven Sickert
Désiré Sidibé
Ioannis Siglidis
Alberto Signoroni
Karan Sikka
Pedro Silva
Julio Silva-Rodríguez
Hyeonjun Sim

Jae-Young Sim
Chonghao Sima
Christian Simon
Martin Simon
Alessandro Simoni
Enis Simsar
Abhishek Singh
Apoorv Singh
Ashish Singh
Bharat Singh
Jasdeep Singh
Jaskirat Singh
Krishnakant Singh
Manish Kumar Singh
Mannat Singh
Nikhil Singh
Pravendra Singh
Rajat Vikram Singh
Simranjit Singh
Darshan Singh S.
Utkarsh Singhal
Dipika Singhania
Vasu Singla
Abhishek Kumar Sinha
Animesh Sinha
Sanjana Sinha
Saptarshi Sinha
Sudipta Sinha
Sophia A.
   Sirko-Galouchenko
Josef Sivic
Elena Sizikova
Geri Skenderi
Gregory Slabaugh
Habib Slim
Dmitriy Smirnov
James S. Smith
William Smith
Noah Snavely
Kihyuk Sohn
Bolivar E. Solarte
Mattia Soldan
Sobhan Soleymani
Samik Some
Nagabhushan Somraj

Jeany Son
Seung Woo Son
Byung Cheol Song
Chen Song
Guanglu Song
Jie Song
Jifei Song
Li Song
Liangchen Song
Lin Song
Luchuan Song
Mingli Song
Ran Song
Sibo Song
Sifan Song
Siyang Song
Weilian Song
Weinan Song
Wenfeng Song
Xiangchen Song
Xibin Song
Xinhang Song
Yafei Song
Yang Song
Yi-Yang Song
Yizhi Song
Yue Song
Zeen Song
Zhenbo Song
Zikai Song
Ekta Sood
Tomáš Souček
Rajiv Soundararajan
Albin Soutif-Cormerais
Jeremy Speth
Indro Spinelli
Jon Sporring
Manogna Sreenivas
Arvind Krishna Sridhar
Deepak Sridhar
Balaji Vasan Srinivasan
Pratul Srinivasan
Anuj Srivastava
Astitva Srivastava
Dhruv Srivastava

Koushik Srivatsan
Pierre-Luc St-Charles
Ioannis Stamos
Anastasis Stathopoulos
Colton Stearns
Jan Steinbrener
Jan-Martin O. Steitz
Sinisa Stekovic
Federico Stella
Michael Stengel
Alexandros Stergiou
Gleb Sterkin
Rainer Stiefelhagen
Noah Stier
Timo N. Stoffregen
Vladan Stojnić
Nick O. Stracke
Ombretta Strafforello
Julian Straub
Nicola Strisciuglio
Vitomir Struc
Yannick Strümpler
Joerg Stueckler
Chi Su
Hang Su
Hang Su
Kun Su
Rui Su
Shaolin Su
Sitong Su
Xingzhe Su
Xiu Su
Yao Su
Yiyang Su
Yongyi Su
Zhaoqi Su
Zhixun Su
Zhuo Su
Zhuo Su
Iago Suárez
Arulkumar Subramaniam
Sanjay Subramanian
A. Subramanyam
Swathikiran Sudhakaran
Yusuke Sugano

Masanori Suganuma
Yumin Suh
Mohammed Suhail
Xiuchao Sui
Yang Sui
Yao Sui
Heung-Il Suk
Pavel Suma
Baigui Sun
Baochen Sun
Bin Sun
Bo Sun
Changchang Sun
Che Sun
Cheng Sun
Chong Sun
Chunyi Sun
Gan Sun
Guofei Sun
Guoxing Sun
Haifeng Sun
Hanqing Sun
Haoliang Sun
He Sun
Heming Sun
Hongbin Sun
Huiming Sun
Jennifer J. Sun
Jian Sun
Jiande Sun
Jianhua Sun
Jiankai Sun
Jipeng Sun
Keqiang Sun
Lei Sun
Lichao Sun
Long Sun
Mingjie Sun
Peize Sun
Pengzhan Sun
Qiyue Sun
Shangquan Sun
Shanlin Sun
Shuyang Sun
Tao Sun

Tiancheng Sun
Wei Sun
Weiwei Sun
Weixuan Sun
Xianfang Sun
Xiaohang Sun
Xiaoshuai Sun
Xiaoxiao Sun
Ximeng Sun
Xuxiang Sun
Yanan Sun
Yasheng Sun
Yihong Sun
Ying Sun
Yixuan Sun
Yu Sun
Yuan Sun
Yuchong Sun
Zeren Sun
Zhanghao Sun
Zhaodong Sun
Zhaohui H. Sun
Zhicheng Sun
Zhicheng Sun
Haomiao Sun
Varun Sundar
Shobhita Sundaram
Minhyuk Sung
Kalyan Sunkavalli
Yucheng Suo
Indranil Sur
Saksham Suri
Naufal Suryanto
Vadim Sushko
David Suter
Roman Suvorov
Fnu Suya
Teppei Suzuki
Kunal Swami
Archana Swaminathan
Gurumurthy Swaminathan
Robin Swanson
Eran Swears
Alexander Swerdlow
Sirnam Swetha

Tabish A. Syed
Tanveer Syeda-Mahmood
Stanislaw K. Szymanowicz
Sethuraman T. V.
Calvin-Khang T. Ta
The-Anh Ta
Babak Taati
Samy Tafasca
Andrea Tagliasacchi
Haowei Tai
Yuan Tai
Francesco Taioli
Peng Taiying
Keita Takahashi
Naoya Takahashi
Jun Takamatsu
Nicolas Talabot
Hugues G. Talbot
Hossein Talebi
Davide Talon
Gary Tam
Toru Tamaki
Dipesh Tamboli
Andong Tan
Bin Tan
Cheng Tan
David Joseph New Tan
Fuwen Tan
Guang Tan
Jianchao Tan
Jing Tan
Jingru Tan
Lei Tan
Mingkui Tan
Mingxing Tan
Shuhan Tan
Shunquan Tan
Weimin Tan
Xin Tan
Zhentao Tan
Zhentao Tan
Masayuki Tanaka
Chen Tang
Chengzhou Tang
Chenwei Tang

Fan Tang
Feng Tang
Hao Tang
Haoran Tang
Jiajun Tang
Jiapeng Tang
Jiaxiang Tang
Jie Tang
Junshu Tang
Keke Tang
Luming Tang
Luyao Tang
Lv Tang
Ming Tang
Quan Tang
Shengji Tang
Sheyang Tang
Shitao Tang
Shixiang Tang
Tao Tang
Weixuan Tang
Xu Tang
Yang Tang
Yansong Tang
Yehui Tang
Yu-Ming Tang
Zheng Tang
Zhipeng Tang
Zitian Tang
Md Mehrab Tanjim
Julian Tanke
An Tao
Chaofan Tao
Chenxin Tao
Jiale Tao
Junli Tao
Keda Tao
Ming Tao
Ran Tao
Wenbing Tao
Xinhao Tao
Jean-Philippe G. Tarel
Laia Tarres
Laia Tarrés
Enzo Tartaglione

Keisuke Tateno
SaiKiran K. Tedla
Antonio Tejero-de-Pablos
Bugra Tekin
Purva Tendulkar
Minggui Teng
Ruwan Tennakoon
Andrew Beng Jin Teoh
Konstantinos Tertikas
Piotr Teterwak
Piotr Teterwak
Anh Thai
Kartik Thakral
Nupur Thakur
Sadbhawna Thakur
Balamurugan Thambiraja
Vikas Thamizharasan
Kevin Thandiackal
Sushil Thapa
Daksh Thapar
Jonas Theiner
Christian Theobalt
Spyridon Thermos
Fida Mohammad Thoker
Christopher L. Thomas
Diego Thomas
William Thong
Mamatha Thota
Mukund Varma
   Thottankara
Changyao Tian
Chunwei Tian
Jinyu Tian
Kai Tian
Lin Tian
Tai-Peng Tian
Xin Tian
Xinyu Tian
Yapeng Tian
Yu Tian
Yuan Tian
Yuesong Tian
Yunjie Tian
Yuxin Tian
Zhuotao Tian

Mert Tiftikci
Javier Tirado-Garín
Garvita Tiwari
Lokender Tiwari
Anastasia Tkach
Andrea Toaiari
Sinisa Todorovic
Pavel Tokmakov
Tri Ton
Adam Tonderski
Jinguang Tong
Peter Tong
Xin Tong
Zhan Tong
Francesco Tonini
Alessio Tonioni
Alessandro Torcinovich
Marwan Torki
Lorenzo Torresani
Fabio Tosi
Matteo Toso
Anh T. Tran
Hung Tran
Linh-Tam Tran
Minh-Triet Tran
Ngoc-Trung Tran
Phong Tran
Jonathan Tremblay
Alex Trevithick
Aditay Tripathi
Subarna Tripathi
Felix Tristram
Gabriele Trivigno
Emanuele Trucco
Prune Truong
Thanh-Dat Truong
Tomasz Trzcinski
Fu-Jen Tsai
Yu-Ju Tsai
Michael Tschannen
Tze Ho Elden Tse
Ethan Tseng
Yu-Chee Tseng
Shahar Tsiper
Hanzhang Tu

Rong-Cheng Tu
Yuanpeng Tu
Zhengzhong Tu
Zhigang Tu
Narek Tumanyan
Anil Osman Tur
Haithem Turki
Mehmet Ozgur Turkoglu
Daniyar Turmukhambetov
Victor G. Turrisi da Costa
Tinne Tuytelaars
Bartlomiej Twardowski
Radim Tylecek
Christos Tzelepis
Seiichi Uchida
Hideaki Uchiyama
Vishaal Udandarao
Mostofa Rafid Uddin
Kohei Uehara
Tatsumi Uezato
Nicolas Ugrinovic
Youngjung Uh
Norimichi Ukita
Amin Ullah
Markus Ulrich
Ardian Umam
Mesut Erhan Unal
Mathias Unberath
Devesh Upadhyay
Paul Upchurch
Shagun Uppal
Yoshitaka Ushiku
Anil Usumezbas
Yuzuko Utsumi
Roy Uziel
Anil Vadathya
Sharvaree Vadgama
Pratik Vaishnavi
Gregory Vaksman
Matias A. Valdenegro Toro
Lucas Valença
Eduardo Valle
Ernest Valveny
Laurens van der Maaten
Wouter Van Gansbeke

Nanne van Noord
Max W. F. van Spengler
Lorenzo Vaquero
Farshid Varno
Cristina Vasconcelos
Francisco Vasconcelos
Igor Vasiljevic
Florin-Alexandru
 Vasluianu
Subeesh Vasu
Arun Balajee Vasudevan
Vaibhav S. Vavilala
Kyle Vedder
Vijay Veerabadran
Ronny Xavier Velastegui
 Sandoval
Senem Velipasalar
Andreas Velten
Raviteja Vemulapalli
Deepika Vemuri
Edward Vendrow
Jonathan Ventura
Lucas Ventura
Jakob Verbeek
Dor Verbin
Eshan Verma
Manisha Verma
Monu Verma
Sahil Verma
Constantin Vertan
Eli Verwimp
Noranart Vesdapunt
Jordan J. Vice
Sara Vicente
Kavisha Vidanapathirana
Dat Viet Thanh Nguyen
Sudheendra
 Vijayanarasimhan
Sujal T. Vijayaraghavan
Deepak Vijaykeerthy
Elliot Vincent
Yael Vinker
Duc Minh Vo
Huy V. Vo
Khoa H. V. Vo

Romain Vo
Antonin Vobecky
Michele Volpi
Riccardo Volpi
Igor Vozniak
Nicholas Vretos
Vibashan V. S.
Ngoc-Son Vu
Tuan-Anh Vu
Khiem Vuong
Mårten Wadenbäck
Neal Wadhwa
Sophia J. Wagner
Muntasir Wahed
Nobuhiko Wakai
Devesh Walawalkar
Jacob Walker
Matthew Walmer
Matthew R. Walter
Bo Wan
Guancheng Wan
Jia Wan
Jin Wan
Jun Wan
Qiyang Wan
Renjie Wan
Wei Wan
Xingchen Wan
Yecong Wan
Zhexiong Wan
Ziyu Wan
Karan Wanchoo
Alex Jinpeng Wang
Angtian Wang
Baoyuan Wang
Benyou Wang
Biao Wang
Bin Wang
Bing Wang
Binghui Wang
Binglu Wang
Can Wang
Ce Wang
Changwei Wang
Chao Wang

Chaoyang Wang
Chen Wang
Chen Wang
Chen Wang
Chengrui Wang
Chien-Yao Wang
Chu Wang
Chuan Wang
Congli Wang
Dadong Wang
Di Wang
Dong Wang
Dong Wang
Dongdong Wang
Dongkai Wang
Dongqing Wang
Dongsheng Wang
X. Wang
Fan Wang
Fangfang Wang
Fangjinhua Wang
Fei Wang
Feng Wang
Feng Wang
Fu-Yun Wang
Gaoang Wang
Guangcong Wang
Guangming Wang
Guangrun Wang
Guangzhi Wang
Guanshuo Wang
Guo-Hua Wang
Guoqing Wang
Guoqing Wang
Haixin Wang
Haiyan Wang
Han Wang
Hanjing Wang
Hanyu Wang
Hao Wang
Hao Wang
Haobo Wang
Haochen Wang
Haochen Wang
Haohan Wang

Haoqi Wang
Haoran Wang
Haotao Wang
Haoxuan Wang
HaoYu Wang
Hengkang Wang
Hengli Wang
Hengyi Wang
Hesheng Wang
Hong Wang
Hongjun Wang
Hongxiao Wang
Hongyu Wang
Hongzhi Wang
Hua Wang
Huafeng Wang
Huan Wang
Huijie Wang
Huiyu Wang
Jiadong Wang
Jiahao Wang
Jiahao Wang
Jiahao Wang
Jiakai Wang
Jialiang Wang
Jiamian Wang
Jian Wang
Jiang Wang
Jiangliu Wang
Jianjia Wang
Jianyi Wang
Jianyuan Wang
Jiaqi Wang
Jiashun Wang
Jiayi Wang
Jiaze Wang
Jin Wang
Jinfeng Wang
Jingbo Wang
Jinghua Wang
Jingkang Wang
Jinglong Wang
Jinglu Wang
Jinpeng Wang
Jinqiao Wang

Jue Wang
Jun Wang
Jun Wang
Junjue Wang
Junke Wang
Junxiao Wang
Kai Wang
Kai Wang
Kai Wang
Kai Wang
Kaihong Wang
Kewei Wang
Keyan Wang
Kun Wang
Kup Wang
Lan Wang
Lanjun Wang
Le Wang
Lei Wang
Lei Wang
Lezi Wang
Liansheng Wang
Liao Wang
Lijuan Wang
Lijun Wang
Limin Wang
Lin Wang
Linwei Wang
Lishun Wang
Lixu Wang
Liyuan Wang
Lizhen Wang
Lizhi Wang
Longguang Wang
Luozhou Wang
Luting Wang
Mang Wang
Manning Wang
Mei Wang
Mengjiao Wang
Mengmeng Wang
Miaohui Wang
Min Wang
Naiyan Wang
Nannan Wang

Ning-Hsu Wang
Pei Wang
Peihao Wang
Peiqi Wang
Peng Wang
Pengfei Wang
Pengkun Wang
Pichao Wang
Pu Wang
Qi Wang
Qian Wang
Qiang Wang
Qiang Wang
Qiangchang Wang
Qianqian Wang
Qifei Wang
Qilong Wang
Qin Wang
Qing Wang
Qingzhong Wang
Qitong Wang
Qiufeng Wang
Ronggang Wang
Rui Wang
Rui Wang
Rui Wang
Ruibin Wang
Ruisheng Wang
Ruoyu Wang
Sai Wang
Sen Wang
Sen Wang
Shan Wang
Shaoru Wang
Sheng Wang
Sheng-Yu Wang
Shengze Wang
Shida Wang
Shijie Wang
Shipeng Wang
Shiping Wang
Shiyu Wang
Shizun Wang
Shuhui Wang
Shujun Wang

Shunli Wang
Shunxin Wang
Shuo Wang
Shuo Wang
Shuo Wang
Shuzhe Wang
Siqi Wang
Siwei Wang
Song Wang
Song Wang
Su Wang
Tan Wang
Tao Wang
Taoyue Wang
Teng Wang
Tengfei Wang
Tiancai Wang
Tianqi Wang
Tianyang Wang
Tianyu Wang
Tong Wang
Tsun-Hsuan Wang
Tuanfeng Wang
Tuanfeng Y. Wang
Wei Wang
Weihan Wang
Weikang Wang
Weimin Wang
Weiqiang Wang
Weixi Wang
Weiyao Wang
Weiyun Wang
Wen Wang
Wenbin Wang
Wenhao Wang
Wenjing Wang
Wenqian Wang
Wentao Wang
Wenxiao Wang
Wenxuan Wang
Wenzhe Wang
Xi Wang
Xi Wang
Xiang Wang
Xiao Wang

Xiao Wang
Xiao Wang
Xiaobing Wang
Xiaofeng Wang
Xiaohan Wang
Xiaosen Wang
Xiaosong Wang
Xiaoxing Wang
Xiaoyang Wang
Xijun Wang
Xijun Wang
Xinggang Wang
Xinghan Wang
Xinjiang Wang
Xinshao Wang
Xintong Wang
Xizi Wang
Xu Wang
Xuan Wang
Xuanhan Wang
Xue Wang
Xueping Wang
Xuyang Wang
Yali Wang
Yan Wang
Yan Wang
Yang Wang
Yangang Wang
Yangtao Wang
Yaohui Wang
Yaoming Wang
Yaxing Wang
Yaxiong Wang
Yi Wang
Yi Ru Wang
Yidong Wang
Yifan Wang
Yifeng Wang
Yifu Wang
Yikai Wang
Yilin Wang
Yilun Wang
Yin Wang
Yinggui Wang
Yingheng Wang

Yingqian Wang
Yipei Wang
Yiqun Wang
Yiran Wang
Yiwei Wang
Yixu Wang
Yizhi Wang
Yizhou Wang
Yizhou Wang
Yong Wang
Yu Wang
Yu-Shuen Wang
Yuan-Gen Wang
Yuchen Wang
Yude Wang
Yue Wang
Yuesong Wang
Yufei Wang
Yufu Wang
Yuguang Wang
Yuhan Wang
Yujia Wang
Yulin Wang
Yunke Wang
Yuting Wang
Yuxi Wang
YuXin Wang
Yuzheng Wang
Ze Wang
Zedong Wang
Zehan Wang
Zengmao Wang
Zeyu Wang
Zeyu Wang
Zhao Wang
Zhaokai Wang
Zhaowen Wang
Zhe Wang
Zhen Wang
Zhen Wang
Zhendong Wang
Zheng Wang
Zheng Wang
Zheng Wang
Zhengyi Wang

Zhennan Wang
Zhenting Wang
Zhenyi Wang
Zhenyu Wang
Zhenzhi Wang
Zhepeng Wang
Zhi Wang
Zhibo Wang
Zhihao Wang
Zhihui Wang
Zhijie Wang
Zhikang Wang
Zhixiang Wang
Zhiyong Wang
Zhongdao Wang
Zhonghao Wang
Zhouxia Wang
Zhu Wang
Zian Wang
Zifu Wang
Zihao Wang
Zijian Wang
Ziqiang Wang
Ziqin Wang
Ziqing Wang
Zirui Wang
Zirui Wang
Ziwei Wang
Ziyan Wang
Ziyang Wang
Ziyi Wang
Ziyun Wang
Frederik Warburg
Syed Talal Wasim
Daniel Watson
Jamie Watson
Ethan Weber
Silvan Weder
Jan Dirk Wegner
Chen Wei
Donglai Wei
Fangyin Wei
Fangyun Wei
Guoqiang Wei
Jia Wei

Jiacheng Wei
Kaixuan Wei
Kun Wei
Longhui Wei
Megan Wei
Mian Wei
Mingqiang Wei
Pengxu Wei
Ping Wei
Qiuhong Anna Wei
Shikui Wei
Tianyi Wei
Wei Wei
Wenqi Wei
Xian Wei
Xin Wei
Xing Wei
Xinyue Wei
Xiu-Shen Wei
Yi Wei
Yixuan Wei
Yunchao Wei
Yuxiang Wei
Yuxiang Wei
Zeming Wei
Zhipeng Wei
Zihao Wei
Zimian Wei
Jean-Baptiste Weibel
Luca Weihs
Martin Weinmann
Michael Weinmann
Bihan Wen
Bowen Wen
Chao Wen
Chenglu Wen
Chuan Wen
Congcong Wen
Jie Wen
Jing Wen
Qiang Wen
Rui Wen
Sijia Wen
Song Wen
Xiang Wen

Xin Wen
Yilin Wen
Youpeng Wen
Yuanbo Wen
Yuxin Wen
Chung-Yi Weng
Junwu Weng
Shuchen Weng
Wenming Weng
Yijia Weng
Zhenzhen Weng
Thomas Westfechtel
Christopher Johannes
    Wewer
Spencer Whitehead
Tobias Jan Wieczorek
Thaddäus Wiedemer
Julian Wiederer
Kevin Tirta Wijaya
Asiri Wijesinghe
Kimberly Wilber
Jeffrey R. Willette
Bryan M. Williams
Williem Williem
Christian Wilms
Benjamin Wilson
Richard Wilson
Felix Wimbauer
Vanessa Wirth
Scott Wisdom
Calden Wloka
Alex Wong
Chau-Wai Wong
Chi-Chong Wong
Ka Wai Wong
Kelvin Wong
Kok-Seng Wong
Kwan-Yee K. Wong
Yongkang Wong
Sangmin Woo
Simon S. Woo
Markus Worchel
Scott Workman
Marcel Worring
Safwan Wshah

Aming Wu
Bo Wu
Bojian Wu
Boxi Wu
Changguang Wu
Chaoyi Wu
Chen Henry Wu
Cheng-En Wu
Chenming Wu
Chenyan Wu
Chenyun Wu
Cho-Ying Wu
Chongruo Wu
Cong Wu
Dayan Wu
Di Wu
Dongming Wu
Fangzhao Wu
Fuxiang Wu
Gaojie Wu
Guanyao Wu
Guile Wu
Haiping Wu
Haiwei Wu
Haiyu Wu
Han Wu
Haoning Wu
Haoning Wu
Haotian Wu
Hefeng Wu
Huisi Wu
Jane Wu
Jay Zhangjie Wu
Jhih-Ciang Wu
Ji-Jia Wu
Jialian Wu
Jiaye Wu
Jimmy Wu
Jing Wu
Jing Wu
Jinjian Wu
Jiqing Wu
Jun Wu
Junfeng Wu
Junlin Wu

Junru Wu
Junyang Wu
Junyi Wu
Letian Wu
Lifang Wu
Lin Yuanbo Wu
Liwen Wu
Min Wu
Minye wu
Peng Wu
Penghao Wu
Qian Wu
Qiangqiang Wu
Qianyi Wu
Qingbo Wu
Rongliang Wu
Rui Wu
Rundi Wu
Shuang Wu
Shuzhe Wu
Tao Wu
Tao Wu
Tao Wu
Te-Lin Wu
Tianfu Wu
Tianhao Wu
Tianhao Wu
Ting-Wei Wu
Tong Wu
Tong Wu
Tsung-Han Wu
Tz-Ying Wu
Weibin Wu
Weijia Wu
Xian Wu
Xiao Wu
Xiaodong Wu
Xiaohe Wu
Xiaoqian Wu
Xiaoyang Wu
Xindi Wu
Xingjiao Wu
Xinxiao Wu
Xiuzhe Wu
Yang Wu

Yangzheng Wu
Yanze Wu
Yanzhao Wu
Yawen Wu
Yicheng Wu
Ying Nian Wu
Yingwen Wu
Yong Wu
Yuanwei Wu
Yue Wu
Yue Wu
Yuqun Wu
Yushu Wu
Yushuang Wu
Zhe Wu
Zheng Wu
Zhi-Fan Wu
Zhihao Wu
Zhijie Wu
Zhiliang Wu
Zhonghua Wu
Zijie Wu
Ziyi Wu
Zizhao Wu
Zongwei Wu
Zongyu Wu
Zongze Wu
Stefanie Wuhrer
Jamie M. Wynn
Monika Wysoczańska
Jianing Xi
Teng Xi
Bin Xia
Changqun Xia
Haifeng Xia
Jiaer Xia
Jiahao Xia
Kun Xia
Mingxuan Xia
Shihong Xia
Weihao Xia
Xiaobo Xia
Yan Xia
Ye Xia
Yifei Xia

Zhaoyang Xia
Zhihao Xia
Zhihua Xia
Zhuofan Xia
Zimin Xia
Chuhua Xian
Wenqi Xian
Yongqin Xian
Donglai Xiang
Jinhai Xiang
Liuyu Xiang
Tian-Zhu Xiang
Tiange Xiang
Wangmeng Xiang
Xiaoyu Xiang
Yuanbo Xiangli
Anqi Xiao
Aoran Xiao
Bei Xiao
Chunxia Xiao
Fanyi Xiao
Han Xiao
Jiancong Xiao
Jimin Xiao
Jing Xiao
Jing Xiao
Jun Xiao
Junbin Xiao
Junfei Xiao
Mingqing Xiao
Qingyang Xiao
Ruixuan Xiao
Taihong Xiao
Yang Xiao
Yanru Xiao
Yao Xiao
Yijun Xiao
Yuting Xiao
Zehao Xiao
Zeyu Xiao
Zihao Xiao
Binhui Xie
Chaohao Xie
Chi Xie
Christopher Xie

Chuanlong Xie
Fei Xie
Guo-Sen Xie
Haozhe Xie
Hongtao Xie
Jiahao Xie
Jiaxin Xie
Jin Xie
Jinheng Xie
Jiu-Cheng Xie
Jiyang Xie
Junyu Xie
Liuyue Xie
Ming-Kun Xie
Mingyang Xie
Qian Xie
Tingting Xie
Weicheng Xie
Xianghui Xie
Xiaohua Xie
Xudong Xie
Yichen Xie
Yiming Xie
You Xie
Yuan Xie
Yusheng Xie
Yutong Xie
Zeke Xie
Zhenda Xie
Zhenyu Xie
ZiYang Xie
Chaoyue Xing
Fuyong Xing
Jinbo Xing
Xiaoyan Xing
Xiaoying Xing
XiMing Xing
Xin Xing
Yazhou Xing
Yifan Xing
Yun Xing
Zhen Xing
Hongkai Xiong
Jingjing Xiong
Jinhui Xiong

Junwen Xiong
Peixi Xiong
Wei Xiong
Weihua Xiong
Yu Xiong
Yuanhao Xiong
Yuanjun Xiong
Yuwen Xiong
Zhexiao Xiong
Zhiwei Xiong
Yuliang Xiu
Alessio Xompero
An Xu
Angchi Xu
Baixin Xu
Bicheng Xu
Bo Xu
Chao Xu
Chenfeng Xu
Chenshu Xu
Chenxin Xu
Chi Xu
Dejia Xu
Dongli Xu
Feng Xu
Gangwei Xu
Haiming Xu
Haiyang Xu
Han Xu
Haofei Xu
Haohang Xu
Haoran Xu
Hongbin Xu
Hongmin Xu
Jiale Xu
Jianjin Xu
Jiaqi Xu
Jie Xu
Jilan Xu
Jinglin Xu
Jingyi Xu
Jun Xu
Kai Xu
Katherine Xu
Ke Xu

Kele Xu
Lan Xu
Lian Xu
Liang Xu
Linning Xu
Lumin Xu
Manjie Xu
Mengde Xu
Mengdi Xu
Mengmeng Frost Xu
Min Xu
Ming Xu
Mutian Xu
Peiran Xu
Peng Xu
Qi Xu
Qiang Xu
Qiangeng Xu
Qingshan Xu
Qingyang Xu
Qiuling Xu
Ran Xu
Renzhe Xu
Ruikang Xu
Runsen Xu
Runsheng Xu
Shichao Xu
Sirui Xu
Tongda Xu
Wanting Xu
Wei Xu
Weiwei Xu
Wenjia Xu
Wenju Xu
Wenqiang Xu
Xiang Xu
Xianghao Xu
Xiangyu Xu
Xiangyu Xu
Xiaogang Xu
Xiaohao Xu
Xin Xu
Xin Xu
Xin-Shun Xu
Xing Xu

Xinli Xu
Xinyu Xu
Xiuwei Xu
Xiyan Xu
Xudong Xu
Xuemiao Xu
Xun Xu
Yan Xu
Yan Xu
Yan Xu
Yangyang Xu
Yanwu Xu
Yating Xu
Yi Xu
Yi Xu
Yi Xu
Yihong Xu
YiKun Xu
Yinghao Xu
Yingyan Xu
Yinshuang Xu
Yiran Xu
Yixing Xu
Yongchao Xu
Yue Xu
Yufei Xu
Yunqiu Xu
Zexiang Xu
Zhan Xu
Zhe Xu
Zhengqin Xu
Zhenlin Xu
Zhiqiu Xu
Zhiyuan Xu
Zhongcong Xu
Zhuoer Xu
Zipeng Xu
Ziyue Xu
Zongyi Xu
Ziwei Xuan
Danna Xue
Fanglei Xue
Fei Xue
Feng Xue
Han Xue

Jianru Xue
Le Xue
Lixin Xue
Mingfu Xue
Nan Xue
Qinghan Xue
Shangjie Xue
Xiangyang Xue
Zihui Xue
Abhay Yadav
Amit Kumar Singh Yadav
Takuma Yagi
Tomas F Yago Vicente
I. Zeki Yalniz
Kota Yamaguchi
Shin'ya Yamaguchi
Burhaneddin Yaman
Toshihiko Yamasaki
Kohei Yamashita
Lee Juliette Yamin
Chaochao Yan
Hongyu Yan
Jiexi Yan
Kai Yan
Pei Yan
Qingan Yan
Qingsen Yan
Qingsong Yan
Rui Yan
Shaoqi Yan
Shi Yan
Siming Yan
Siming Yan
Siyuan Yan
Weilong Yan
Wending Yan
Xiangyi Yan
Xinchen Yan
Xingguang Yan
Xueting Yan
Yan Yan
Yichao Yan
Zhaoyi Yan
Zhiqiang Yan
Zhiyuan Yan

Zike Yan
Zizheng Yan
Keiji Yanai
Pinar Yanardag
Anqi Yang
Anqi Joyce Yang
Bangbang Yang
Baoyao Yang
Bin Yang
Binwei Yang
Bo Yang
Bo Yang
Boyu Yang
Changdi Yang
Chao Yang
Charig Yang
Cheng-Fu Yang
Cheng-Yen Yang
Chenhongyi Yang
Chuanguang Yang
De-Nian Yang
Dingcheng Yang
Dingkang Yang
Dong Yang
Erkun Yang
Fan Yang
Fan Yang
Fan Yang
Fan Yang
Fan Yang
Feng Yang
Fengting Yang
Fengxiang Yang
Fengyuan Yang
Fu-En Yang
Gang Yang
Gengshan Yang
Guandao Yang
Guanglei Yang
Haitao Yang
Hanqing Yang
Heran Yang
Honghui Yang
Huanrui Yang
Huiyuan Yang

Huizong Yang
Hunmin Yang
Jiange Yang
Jiaqi Yang
Jiawei Yang
Jiayu Yang
Jiazhi Yang
Jie Yang
Jie Yang
Jiewen Yang
Jihan Yang
Jing Yang
Jingkang Yang
Jinhui Yang
Jinlong Yang
Jinrong Yang
Jinyu Yang
Kaicheng Yang
Kailun Yang
Lan Yang
Le Yang
Lehan Yang
Lei Yang
Lei Yang
Lei Yang
Li Yang
Lihe Yang
Ling Yang
Lingxiao Yang
Linlin Yang
Lixin Yang
Longrong Yang
Lu Yang
Luwei Yang
Michael Ying Yang
Min Yang
Ming Yang
MingKun Yang
Mouxing Yang
Muli Yang
Peiyu Yang
Qi Yang
Qian Yang
Qiushi Yang
Ren Yang

Rui Yang
Ruihan Yang
Sejong Yang
Shan Yang
Shangrong Yang
Shiqi Yang
Shuai Yang
Shuai Yang
Shuang Yang
Shuo Yang
Shusheng Yang
Sibei Yang
Siwei Yang
Siyuan Yang
Siyuan Yang
Song Yang
Songlin Yang
Tianyu Yang
Tong Yang
Wankou Yang
Wenhan Yang
Wenhan Yang
Wenjie Yang
Wenqi Yang
William Yang
Xi Yang
Xi Yang
Xiangpeng Yang
Xiao Yang
Xiaofeng Yang
Xiaoshan Yang
Xin Jeremy Yang
Xingyi Yang
Xinlong Yang
Xitong Yang
Xiulong Yang
Xu Yang
Xuan Yang
Xue Yang
Xuelin Yang
Xun Yang
Yan Yang
Yan Yang
Yang Yang
Yaokun Yang

Yezhou Yang
Yiding Yang
Yijun Yang
Yijun Yang
Yin Yang
Yinfei Yang
Yixin Yang
Yongqi Yang
Yongqi Yang
Yue Yang
Yuewei Yang
Yuezhi Yang
Yujiu Yang
Yung-Hsu Yang
Yuwei Yang
Ze Yang
Ze Yang
Zetong Yang
Zhangsihao Yang
Zhaoyuan Yang
Zhen Yang
Zhenpei Yang
Zhibo Yang
Zhiwei Yang
Zhiwen Yang
Zhiyuan Yang
Zhuoqian Yang
Ziyan Yang
Ziyun Yang
Zongxin Yang
Zuhao Yang
Chengtang Yao
Cong Yao
Hantao Yao
Jiawen Yao
Lina Yao
Mingde Yao
Mingshuai Yao
Qingsong Yao
Shunyu Yao
Taiping Yao
Ting Yao
Xincheng Yao
Xinwei Yao
Xu Yao

Xufeng Yao
Yao Yao
Yazhou Yao
Yue Yao
Ziwei Yao
Sudhir Yarram
Rajeev Yasarla
Mohsen Yavartanoo
Botao Ye
Dengpan Ye
Fei Ye
Hanrong Ye
Jianglong Ye
Jiarong Ye
Jin Ye
Jingwen Ye
Jinwei Ye
Junjie Ye
Keren Ye
Maosheng Ye
Meng Ye
Meng Ye
Muchao Ye
Nanyang Ye
Peng Ye
Qi Ye
Qian Ye
Qinghao Ye
Qixiang Ye
Ruolin Ye
Shuquan Ye
Tian Ye
Vickie Ye
Wenqian Ye
Xinchen Ye
Yufei Ye
Moon Ye-Bin
Yousef Yeganeh
Chun-Hsiao Yeh
Raymond Yeh
Yu-Ying Yeh
Florence Yellin
Sriram Yenamandra
Tarun Yenamandra
Promod Yenigalla

Chandan Yeshwanth
Dong Yi
Hongwei Yi
Kai Yi
Ran Yi
Renjiao Yi
Xinyu Yi
Alper Yilmaz
Jonghwa Yim
Aoxiong Yin
Fei Yin
Fukun Yin
Jia-Li Yin
Ming Yin
Nan Yin
Ruihong Yin
Tianwei Yin
Wenzhe Yin
Xiaoqi Yin
Yingda Yin
Yu Yin
Yufeng Yin
Zhenfei Yin
Xianghua Ying
Xiaowen Ying
Naoto Yokoya
Chen YongCan
ByungIn Yoo
Innfarn Yoo
Jinsu Yoo
Sungjoo Yoo
Hee Suk Yoon
Jae Shin Yoon
Jihun Yoon
Sangwoong Yoon
Sejong Yoon
Sung Whan Yoon
Sung-Hoon Yoon
Sunjae Yoon
Youngho Yoon
Youngseok Yoon
Youngseok Yoon
Yuichi Yoshida
Ryota Yoshihashi
Yusuke Yoshiyasu

Chenyu You
Haoran You
Haoxuan You
Shan You
Yang You
Yingxuan You
Yurong You
Chan-Hyun Youn
Kim Youwang
Nikolaos-Antonios
    Ypsilantis
Baosheng Yu
Bei Yu
Bruce X. B. Yu
Chaohui Yu
Chunlin Yu
Cunjun Yu
Dahai Yu
En Yu
En Yu
Fenggen Yu
Gang Yu
Haibao Yu
Hanchao Yu
Hang Yu
Hao Yu
Hao Yu
Haojun Yu
Heng Yu
Hong-Xing Yu
Houjian Yu
Jianhui Yu
Jiashuo Yu
Jing Yu
Jiwen Yu
Jiyang Yu
Kaicheng Yu
Lei Yu
Lidong Yu
Lijun Yu
Mulin Yu
Peilin Yu
Qian Yu
Qihang Yu
Qing Yu

Rui Yu
Ruixuan Yu
Runpeng Yu
Shaozuo Yu
Shuzhi Yu
Sihyun Yu
Tan Yu
Tao Yu
Tianjiao Yu
Wei Yu
Weihao Yu
Wenwen Yu
Xi Yu
Xiaohan Yu
Xin Yu
Xin Yu
Xuehui Yu
Yingchen Yu
Yongsheng Yu
Yunlong Yu
Zehao Yu
Zhaofei Yu
Zhengdi Yu
Zhengdi Yu
Zhixuan Yu
Zhongzhi Yu
Zhuoran Yu
Zitong Yu
Chun Yuan
Chunfeng Yuan
Hangjie Yuan
Haobo Yuan
Jiakang Yuan
Jiangbo Yuan
Liangzhe Yuan
Maoxun Yuan
Shanxin Yuan
Shengming Yuan
Shuai Yuan
Shuaihang Yuan
Wentao Yuan
Xiaoding Yuan
Xiaoyun Yuan
Xin Yuan
Xin Yuan

Yixuan Yuan
Yu-Jie Yuan
Yuan Yuan
Yuan Yuan
Yuhui Yuan
Zheng Yuan
Zhuoning Yuan
Mehmet Kerim Yücel
Dongxu Yue
Haixiao Yue
Kaiyu Yue
Tao Yue
Xiangyu Yue
Zihao Yue
Zongsheng Yue
Heeseung Yun
Jooyeol Yun
Juseung Yun
Kimin Yun
Se-Young Yun
Sukwon Yun
Tian Yun
Raza Yunus
Ekim Yurtsever
Eloi Zablocki
Riccardo Zaccone
Martin Zach
Muhammad Zaigham
    Zaheer
Ilya Zakharkin
Egor Zakharov
Abhaysinh S. Zala
Pierluigi Zama Ramirez
Eduard Sebastian Zamfir
Amir Zamir
Luca Zancato
Yuan Zang
Yuhang Zang
Zelin Zang
Pietro Zanuttigh
Giacomo Zara
Samira Zare
Olga Zatsarynna
Denis Zavadski
Vitjan Zavrtanik

Jan Zdenek
Yanjie Ze
Bernhard Zeisl
John Zelek
Oliver Zendel
Ailing Zeng
Chong Zeng
Dan Zeng
Fangao Zeng
Haijin Zeng
Huimin Zeng
Jia Zeng
Jiabei Zeng
Kuo-Hao Zeng
Libing Zeng
Ling-An Zeng
Ming Zeng
Pengpeng Zeng
Runhao Zeng
Tieyong Zeng
Wei Zeng
Yan Zeng
Yanhong Zeng
Yawen Zeng
Yuyuan Zeng
Zilai Zeng
Ziyao Zeng
Kaiwen Zha
Ruyi Zha
Yaohua Zha
Bohan Zhai
Qiang Zhai
Runtian Zhai
Wei Zhai
YiKui Zhai
Yuanhao Zhai
Yunpeng Zhai
De-Chuan Zhan
Fangneng Zhan
Guanqi Zhan
Huangying Zhan
Huijing Zhan
Kun Zhan
Xueying Zhan
Aidong Zhang

Baochang Zhang
Baoheng Zhang
Baoming Zhang
Biao Zhang
Bingfeng Zhang
Binjie Zhang
Bo Zhang
Bo Zhang
Borui Zhang
Bowen Zhang
Can Zhang
Ce Zhang
Chang-Bin Zhang
Chao Zhang
Chao Zhang
Chen-Lin Zhang
Cheng Zhang
Cheng Zhang
Chenghao Zhang
Chenyangguang Zhang
Chi Zhang
Chongyang Zhang
Chris Zhang
Chuhan Zhang
Chunhui Zhang
Chuyu Zhang
Congyi Zhang
Daichi Zhang
Dan Zhang
Daoan Zhang
Daoqiang Zhang
David Junhao Zhang
Dexuan Zhang
Dingwen Zhang
Dingyuan Zhang
Dongsu Zhang
Fan Zhang
Fan Zhang
Fang-Lue Zhang
Feilong Zhang
Frederic Z. Zhang
Fuyang Zhang
Gang Zhang
Gengwei Zhang
Gengyu Zhang

Gengyuan Zhang
Gongjie Zhang
GuiXuan Zhang
Guofeng Zhang
Guozhen Zhang
Hang Zhang
Hang Zhang
Hanwang Zhang
Hao Zhang
Hao Zhang
Hao Zhang
Haokui Zhang
Haonan Zhang
Haotian Zhang
Hengrui Zhang
Hongguang Zhang
Hongrun Zhang
Hongyuan Zhang
Howard Zhang
Huaidong Zhang
Huaiwen Zhang
Hui Zhang
Hui Zhang
Jason Y. Zhang
Ji Zhang
Jiahui Zhang
Jiakai Zhang
Jiaming Zhang
Jian Zhang
Jianfu Zhang
Jiangning Zhang
Jianhua Zhang
Jianming Zhang
Jianpeng Zhang
Jianping Zhang
Jianrong Zhang
Jichao Zhang
Jie Zhang
Jie Zhang
Jie Zhang
Jimuyang Zhang
Jing Zhang
Jing Zhang
Jinghao Zhang
Jingyi Zhang

Jinlu Zhang
Jiqing Zhang
Jiyuan Zhang
Junbo Zhang
Junge Zhang
Junyi Zhang
Juyong Zhang
Kai Zhang
Kai Zhang
Kaidong Zhang
Kaihao Zhang
Kaipeng Zhang
Kaiyi Zhang
Ke Zhang
Ke Zhang
Kui Zhang
Le Zhang
Le Zhang
Lefei Zhang
Lei Zhang
Leo Yu Zhang
Li Zhang
Lianbo Zhang
Liang Zhang
Liangpei Zhang
Lin Zhang
Linfeng Zhang
Liqing Zhang
Lu Zhang
Malu Zhang
Manyuan Zhang
Mengmi Zhang
Mengqi Zhang
Mi Zhang
Min Zhang
Min-Ling Zhang
Mingda Zhang
Mingfang Zhang
Minghui Zhang
Mingjin Zhang
Mingyuan Zhang
Minjia Zhang
Ni Zhang
Pan Zhang
Peiyan Zhang

Pengze Zhang
Pingping Zhang
Qi Zhang
Qi Zhang
Qian Zhang
Qiang Zhang
Qijian Zhang
Qiming Zhang
Qing Zhang
Qing Zhang
Renrui Zhang
Rongyu Zhang
Ruida Zhang
Ruimao Zhang
Ruixin Zhang
Runze Zhang
Sanyi Zhang
Shan Zhang
Shanghang Zhang
Shaofeng Zhang
Sheng Zhang
Shengping Zhang
Shengyu Zhang
Shimian Zhang
Shiwei Zhang
Shizhou Zhang
Shu Zhang
Shuo Zhang
Siwei Zhang
Song-Hai Zhang
Tao Zhang
Tianyun Zhang
Ting Zhang
Tong Zhang
Weixia Zhang
Wendong Zhang
Wenlong Zhang
Wenqiang Zhang
Wentao Zhang
Wentian Zhang
Wenxiao Zhang
Wenxuan Zhang
Xi Zhang
Xiang Zhang
Xiang Zhang

Xianling Zhang
Xiao Zhang
Xiaohan Zhang
Xiaoming Zhang
Xiaoran Zhang
Xiaowei Zhang
Xiaoyun Zhang
Xikun Zhang
Xin Zhang
Xinfeng Zhang
Xingchen Zhang
Xingguang Zhang
Xingxuan Zhang
Xiong Zhang
Xiuming Zhang
Xu Zhang
Xuanyang Zhang
Xucong Zhang
Xuying Zhang
Yabin Zhang
Yabo Zhang
Yachao Zhang
Yahui Zhang
Yan Zhang
Yan Zhang
Yanan Zhang
Yang Zhang
Yanghao Zhang
Yawen Zhang
Yechao Zhang
Yi Zhang
Yi Zhang
Yi Zhang
Yi-Fan Zhang
Yifan Zhang
Yifei Zhang
Yifeng Zhang
Yihao Zhang
Yihua Zhang
Yimeng Zhang
Yiming Zhang
Yin Zhang
Yinan Zhang
Yinda Zhang
Ying Zhang

Yingliang Zhang
Yitian Zhang
Yixin Zhang
Yiyuan Zhang
Yongfei Zhang
Yonggang Zhang
Yonghua Zhang
Youjian Zhang
Youmin Zhang
Youshan Zhang
Yu Zhang
Yu Zhang
Yuan Zhang
Yuechen Zhang
Yuexi Zhang
Yufei Zhang
Yuhan Zhang
Yuhang Zhang
Yunchao Zhang
Yunhe Zhang
Yunhua Zhang
Yunpeng Zhang
Yunzhi Zhang
Yuxin Zhang
Yuyao Zhang
Zaixi Zhang
Zeliang Zhang
Zewei Zhang
Zeyu Zhang
Zhang Zhang
Zhao Zhang
Zhaoxiang Zhang
Zhen Zhang
Zheng Zhang
Zheng Zhang
Zhenyu Zhang
Zhenyu Zhang
Zheyuan Zhang
Zhicheng Zhang
Zhilu Zhang
Zhishuai Zhang
Zhitian Zhang
Zhiwei Zhang
Zhixing Zhang
Zhiyuan Zhang

Zhong Zhang
Zhongping Zhang
Zhongqun Zhang
Zicheng Zhang
Zicheng Zhang
Zihao Zhang
Ziming Zhang
Ziqi Zhang
Qilong Zhangli
Bin Zhao
Bingchen Zhao
Bingyin Zhao
Cairong Zhao
Can Zhao
Chen Zhao
Dong Zhao
Dongxu Zhao
Fang Zhao
Feng Zhao
Fuqiang Zhao
Gangming Zhao
Ganlong Zhao
Guiyu Zhao
Haimei Zhao
Hanbin Zhao
Handong Zhao
Jian Zhao
Jiaqi Zhao
Jie Zhao
Kai Zhao
Kaifeng Zhao
Lei Zhao
Liang Zhao
Lirui Zhao
Long Zhao
Luxi Zhao
Mingyang Zhao
Minyi Zhao
Na Zhao
Nanxuan Zhao
Pu Zhao
Qi Zhao
Qian Zhao
Qibin Zhao
Qingsong Zhao

Qingyu Zhao
Qinyu Zhao
Rongchang Zhao
Rui Zhao
Rui Zhao
Rui Zhao
Ruiqi Zhao
Shanshan Zhao
Shihao Zhao
Shiyu Zhao
Shizhen Zhao
Shuai Zhao
Siheng Zhao
Tianchen Zhao
TianHao Zhao
Tiesong Zhao
Wang Zhao
Wangbo Zhao
Weichao Zhao
Weiyue Zhao
Wenda Zhao
Wenliang Zhao
Xiangyun Zhao
Xiaoming Zhao
Xiaonan Zhao
Xiaoqi Zhao
Xin Zhao
Xin Zhao
Xingyu Zhao
Xu Zhao
Yajie Zhao
Yang Zhao
Yifan Zhao
Ying Zhao
Yiqun Zhao
Yiqun Zhao
Yizhou Zhao
Yizhou Zhao
Yucheng Zhao
Yue Zhao
Yunhan Zhao
Yuyang Zhao
Yuzhi Zhao
Zelin Zhao
Zengqun Zhao

Zhen Zhao
Zhenghao Zhao
Zhengyu Zhao
Zhou Zhao
Zibo Zhao
Zimeng Zhao
Ziwei Zhao
Ziwei Zhao
Zixiang Zhao
Jin Zhe
Anling Zheng
Ce Zheng
Chaoda Zheng
Chengwei Zheng
Chuanxia Zheng
Duo Zheng
Ervine Zheng
Guangcong Zheng
Haitian Zheng
Haiyong Zheng
Hao Zheng
Haotian Zheng
Haoxin Zheng
Huan Zheng
Huan Zheng
Jia Zheng
Jian Zheng
Jian-Qing Zheng
Jianqiao Zheng
Jianwei Zheng
Jin Zheng
Jingxiao Zheng
Kaiwen Zheng
Kecheng Zheng
Léon Zheng
Meng Zheng
Naishan Zheng
Peng Zheng
Qi Zheng
Qian Zheng
Rongkun Zheng
Shen Zheng
Shuai Zheng
Shuhong Zheng
Shunyuan Zheng

Siming Zheng
Tianhang Zheng
Wenting Zheng
Wenzhao Zheng
Xiaozheng Zheng
Xiawu Zheng
Xu Zheng
Yajing Zheng
Yalin Zheng
Yang Zheng
Ye Zheng
Yinglin Zheng
Yinqiang Zheng
Yu Zheng
Zangwei Zheng
Zehan Zheng
Zerong Zheng
Zhaoheng Zheng
Zhedong Zheng
Zhuo Zheng
Zilong Zheng
Shuaifeng Zhi
Tiancheng Zhi
Bineng Zhong
Fangcheng Zhong
Fangwei Zhong
Guoqiang Zhong
Nan Zhong
Yaoyao Zhong
Yijie Zhong
Yiqi Zhong
Yiran Zhong
Yiwu Zhong
Yunshan Zhong
Zichun Zhong
Ziming Zhong
Brady Zhou
Chong Zhou
Chu Zhou
Chunluan Zhou
Da-Wei Zhou
Dawei Zhou
Dewei Zhou
Dingfu Zhou
Donghao Zhou

Dongzhan Zhou
Fan Zhou
Hang Zhou
Hang Zhou
Hao Zhou
Hao Zhou
Haoyi Zhou
Hong-Yu Zhou
Honglu Zhou
Huayi Zhou
Jiahuan Zhou
Jiaming Zhou
Jian Zhou
Jianan Zhou
Jiantao Zhou
Jianxiong Zhou
JIngkai Zhou
Junsheng Zhou
Kailai Zhou
Kaiyang Zhou
Keyang Zhou
Kun Zhou
Lei Zhou
Mingyang Zhou
Mingyi Zhou
Mingyuan Zhou
Mo Zhou
Pan Zhou
Peng Zhou
Peng Zhou
Qianyi Zhou
Qianyu Zhou
Qihua Zhou
Qin Zhou
Qinqin Zhou
Qunjie Zhou
Sheng Zhou
Shenglong Zhou
Shijie Zhou
Shuchang Zhou
Sihang Zhou
Tao Zhou
Tianfei Zhou
Xiangdong Zhou
Xiaoqiang Zhou

Xin Zhou

Xingyi Zhou

Yan-Jie Zhou

Yang Zhou

Yang Zhou

Yanqi Zhou

Yi Zhou

Yi Zhou

Yichao Zhou

Yichen Zhou

Yin Zhou

Yiyi Zhou

Yu Zhou

Yucheng Zhou

Yufan Zhou

Yunsong Zhou

Yuqian Zhou

Yuxiao Zhou

Yuxuan Zhou

Zhenyu Zhou

Zijian Zhou

Zikun Zhou

Ziqi Zhou

Zixiang Zhou

Zongwei Zhou

Alex Z. Zhu

Benjin Zhu

Bin Zhu

Bin Zhu

Chenming Zhu

Chenyang Zhu

Deyao Zhu

Dongxiao Zhu

Fangrui Zhu

Fei Zhu

Feida Zhu

Fengqing Maggie Zhu

Guibo Zhu

Haidong Zhu

Hanwei Zhu

Hao Zhu

Hao Zhu

Heming Zhu

Jiachen Zhu

Jianke Zhu

Jiawen Zhu

Jiayin Zhu

Jinjing Zhu

Junyi Zhu

Kai Zhu

Ke Zhu

Lanyun Zhu

Lin Zhu

Linchao Zhu

Liyuan Zhu

Meilu Zhu

Muzhi Zhu

Qingtian Zhu

Ronghang Zhu

Rui Zhu

Rui Zhu

Rui-Jie Zhu

Ruizhao Zhu

Shengjie Zhu

Sijie Zhu

Siyu Zhu

Tyler Zhu

Wang Zhu

Weicheng Zhu

Wenwu Zhu

Xiangyu Zhu

Xiaofeng Zhu

Xiaoguang Zhu

Xiaosu Zhu

Xiaoyu Zhu

Xingkui Zhu

Xinxin Zhu

Xiyue Zhu

Yangguang Zhu

Yanjun Zhu

Yao Zhu

Ye Zhu

Feida Zhu

Fengqing Maggie Zhu

Guibo Zhu

Haidong Zhu

Hanwei Zhu

Hao Zhu

Hao Zhu

Heming Zhu

Jiachen Zhu

Jianke Zhu

Jiawen Zhu

Jiayin Zhu

Jinjing Zhu

Junyi Zhu

Kai Zhu

Ke Zhu

Lanyun Zhu

Lin Zhu

Linchao Zhu

Liyuan Zhu

Meilu Zhu

Muzhi Zhu

Qingtian Zhu

Ronghang Zhu

Rui Zhu

Rui Zhu

Rui-Jie Zhu

Ruizhao Zhu

Shengjie Zhu

Sijie Zhu

Siyu Zhu

Tyler Zhu

Wang Zhu

Weicheng Zhu

Wenwu Zhu

Xiangyu Zhu

Xiaofeng Zhu

Xiaoguang Zhu

Xiaosu Zhu

Xiaoyu Zhu

Xingkui zhu

Xinxin Zhu

Xiyue Zhu

Yangguang Zhu

Yanjun Zhu

Yao Zhu

Ye Zhu

Ye Zhu

Yichen Zhu

Yingying Zhu

Yousong Zhu

Yuansheng Zhu

Yurui Zhu

Zhen Zhu
Zhenwei Zhu
Zhenyao Zhu
Zhifan Zhu
Zhigang Zhu
Zhihong Zhu
Zihan Zhu
Zixin Zhu
Zunjie Zhu
Bingbing Zhuang
Jia-Xin Zhuang
Jiafan Zhuang
Peiye Zhuang
Wanyi Zhuang
Weiming Zhuang
Yihong Zhuang
Yixin Zhuang
Mingchen Zhuge
Tao Zhuo
Wei Zhuo

Yaoxin Zhuo
Bartosz Zieliński
Wojciech Zielonka
Filippo Ziliotto
Karel Zimmermann
Primo Zingaretti
Nikolaos Zioulis
Liu Ziyin
Mohammad Zohaib
Yongshuo Zong
Zhuofan Zong
Maria Zontak
Gaspard Zoss
Changqing Zou
Chuhang Zou
Danping Zou
Dongqing Zou
Haoming Zou
Longkun Zou
Shihao Zou

Xingxing Zou
Xueyan Zou
Yang Zou
Yuexian Zou
Yuli Zou
Yuliang Zou
Yunhao Zou
Zhiming Zou
Zihang Zou
Silvia Zuffi
Idil Esen Zulfikar
Maria A. Zuluaga
Ronglai Zuo
Xingxing Zuo
Xinxin Zuo
Yifan Zuo
Yiming Zuo
Reyer Zwiggelaar
Vlas Zyrianov

Contents – Part XIII

COHO: Context-Sensitive City-Scale Hierarchical Urban Layout Generation

Liu He$^{(\boxtimes)}$ (iD) and Daniel Aliaga (iD)

Purdue University, West Lafayette, USA
{he425,aliaga}@purdue.edu

Abstract. The generation of large-scale urban layouts has garnered substantial interest across various disciplines. Prior methods have utilized procedural generation requiring manual rule coding or deep learning needing abundant data. However, prior approaches have not considered the context-sensitive nature of urban layout generation. Our approach addresses this gap by leveraging a canonical graph representation for the entire city, which facilitates scalability and captures the multi-layer semantics inherent in urban layouts. We introduce a novel graph-based masked autoencoder (GMAE) for city-scale urban layout generation. The method encodes attributed buildings, city blocks, communities and cities into a unified graph structure, enabling self-supervised masked training for graph autoencoder. Additionally, we employ scheduled iterative sampling for 2.5D layout generation, prioritizing the generation of important city blocks and buildings. Our approach achieves good realism, semantic consistency, and correctness across the heterogeneous urban styles in 330 US cities. Codes and datasets are released at https://github.com/Arking1995/COHO.

Keywords: Layout Generation · Urban Modeling · Masked Graph Autoencoder · Context Sensitivity

1 Introduction

Large-scale urban layout generation has received significant interest by computer vision, urban planning and related disciplines. In particular, cities easily have 1,000 to 50,000 city blocks spanning a mostly horizontal terrain. The configuration of each city block varies significantly across the city, yet there is some configuration similarity amongst subsets within and among cities. Reconstructing existing layouts and generating future planned layouts are crucial tasks for urban simulation, digital twins, and game/content design.

Prior works have generated urban layouts from a variety of sources. Procedural layout generation [39,57] used hand-coded rules to generate cities, and inverse procedural modeling [5] extracted the rules from data and then enabled

Supplementary Information The online version contains supplementary material available at https://doi.org/10.1007/978-3-031-72624-8_1.

A. Leonardis et al. (Eds.): ECCV 2024, LNCS 15071, pp. 1–18, 2025.
https://doi.org/10.1007/978-3-031-72624-8_1

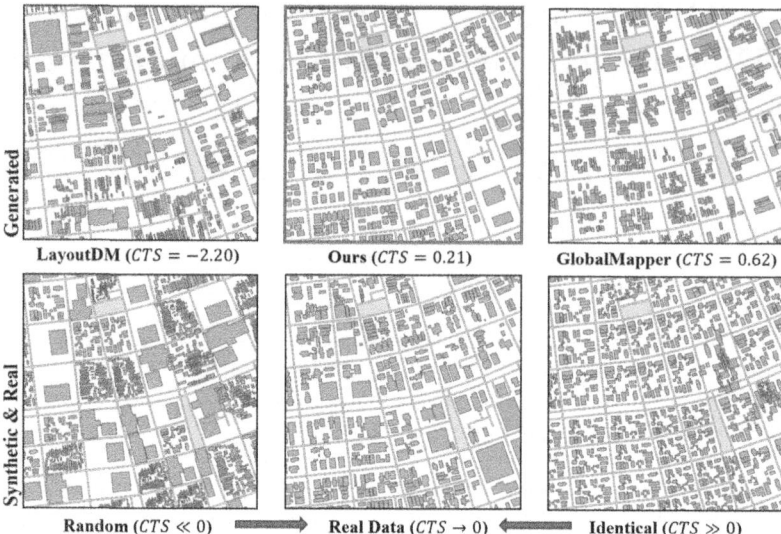

Fig. 1. Context-Sensitive Generation. Our method (Green) pursues realistic context harmonization among neighboring city blocks as real data (bottom-middle). Other methods (e.g. LayoutDM [29], GlobalMapper [20]) show over-diversity/-similarity in city-scale layout generation (evaluated by Context Score CTS as Eq. 4.). Fully random and identical layouts are synthetically generated to illustrate extreme cases. (Color figure online)

generation. More recently, deep learning has promoted data-driven approaches that produce either pixel-based layout generation (e.g., InfiniCity [38], City-Dreamer [61], CityGEN [13]) or graph-based approaches for general polygonal shape layouts (e.g., LayoutTransformer [17], VTN [2]) and for urban scenarios (e.g., BlockPlanner [62], GlobalMapper [20], BuildingGAN [11]). However, none of these techniques explicitly recover and consider the context-sensitive nature of urban layout generation (Fig. 1). In other words, the layout of buildings, blocks, and communities cannot be treated in isolation. The multi-layer semantics are critical to generating realistic and plausible urban layouts. While such multi-level structure (e.g., from single building to a community or subset of a city) has been specified via hand-coded rules, it has not been discovered, organized and used for generation in data-driven deep learning approaches.

Our approach builds on three key observations:

1. **Representation:** Cities, communities, blocks, buildings, and roads are arbitrarily shaped and with complex topology. Hence, a graph-based representation can better capture the regular and non-regular configurations, as opposed to the limiting regular structure of an image. Further, graphs can be more compact which in turn supports better scalability. Altogether, having buildings or city blocks as atomic units, instead of pixels, will contribute to improving correctness, realism, and consistency with priors.
2. **Context-Sensitivity:** Buildings, city blocks, and communities are built considering their neighboring structures and not in isolation. Hence, a genera-

tor should consider this multi-layer structure and inter-dependency. In order words, the style features at one level (e.g., size, shape, location) depend on the surrounding context.

3. **Prioritization:** The stylistic and semantic importance, or priority, of city blocks and buildings varies. This implies that a generation scheme should generate the most important city blocks and buildings first and then fill-in the rest, as opposed to a region growing approach for example. Overall, prioritization will improve correctness and realism as well as enable large-scale urban layout generation.

We propose a novel Graph-based Masked AutoEncoder (GMAE) for large-scale multi-layer 2.5D urban layout generation. *First*, we define a canonical graph-based representation capturing the multi-layer context sensitivity of any city's urban layouts (Fig. 2). The entire graph G corresponds to a city. Subsets of the graph map to communities and each node b amounts to a city block. Further, each node encodes its building layouts, shapes, and heights as a quantized feature by well-trained quantizor. *Second*, we utilize our GMAE to learn the multi-layer stylistic behavior of many urban layouts (i.e., 330 cities).

Using subsets of the graph (or communities), node features are randomly masked and self-supervised GMAE training is performed to learn node feature reconstruction (Fig. 3). *Third*, utilizing the well-trained GMAE as a generator, we perform an iterative priority-based scheduling to generate a large layout or to complete a city fragment; i.e., each iteration predicts node features for the entire graph but only the most confident ones are preserved for the next step (Fig. 4). This methodology is able to leverage any percentage of given prior (e.g., $[0, 100\%]$) for urban layout completion and generation.

To the best of our knowledge, our approach is the first data-driven deep-learning method to enable large-scale 2.5D layout generation with good realism, semantic consistency, and correctness. We show the capability to generalize to the heterogeneous urban layout styles of 330 cities in the US (e.g., all cities with a population >100K, totaling 833,473 city blocks and 17,663,607 buildings). In addition, we provide comparisons to several prior works including SDXL [48]), VTN [2], LayoutDM [29], and GlobalMapper [20] in Sect. 4 and exceed their performance in multiple well-known geometric and perceptual metrics.

Our contributions include the following:

- A canonical graph representation for large-scale 2.5D urban layouts. It encodes global and local node features and neighboring relations for buildings, city blocks, and communities within a city.
- A self-supervised Graph-based Masked AutoEncoder (GMAE) enabling arbitrarily shaped city layout generation with consistency, correctness, and realism.
- Priority-based scheduled iterative sampling for controllable urban layout synthesis including completion, refinement, and generation with any percentage of priors ($[0, 100\%]$).
- A open dataset of road network and urban layouts of 330 US cities for benchmarking current and future urban layout generation methods (https://huggingface.co/datasets/Arking95/COHO).

2   Related Work

Layout Generation. Numerous data-driven deep learning methods have been proposed for layout generation; for example, generalized layout synthesis [2,17, 31,34,64], document layouts [21,30,46,55,67], urban land lots [22,62,66], indoor scenes [42,44,60], and 3D buildings [3,11,45]. Recently, diffusion models have also been used for layout generation tasks (e.g., [9,21,28,29]). However, most of these works assume 2D layouts generated on a rectangular empty canvas. The position and size of each bounding box is often limited to a finite set of categories and to rectilinear alignments. This results in poor diversity and low realism for urban layout generation tasks. BlockPlanner [62] and GlobalMapper [20] adapt layout generation to either rectilinear or arbitrarily-shaped city block generation. However, all previous works assume each layout, or city block, is independent. This is not the case in urban environments where adjacent and nearby city blocks may share a common style or be positioned in a purposeful multi-block arrangement. Without a contextual guidance, city block generation may result in either unwanted harmonization or unexpected diversity between neighboring blocks (see Fig. 1). Thus, we pursue city-scale urban layout generation with the help of context-aware neighboring information.

Infinite Visual Synthesis. With bombing of visual deep learning works [27, 40,50,51,53,54], researchers have explored synthesizing large, potentially unbounded images. For example, omnidirectional image synthesis from multiple smaller field-of-view images [36,37]. Other works focus on unbounded 3D scene generation [8,12,49]. Particularly relevant are several recent infinite urban scene generators [13,38,61] that directly adapt image synthesis methods (e.g. InfinityGAN [37], MaskGIT [10]). The synthesis performs an autoregressive outpainting process using layout image patches. However, patch-based training and inference doesn't learn global features nor their interdependencies. Rather, it learns a very local behavior. Prior pixel-based methods acknowledge limited realism and semantics [13,38,61]. Often, post-processing [13,61], and human-in-the-loop editing [38] is required to refine the generated images to vector-based urban layouts. Even the latest stable diffusion model SDXL [48] seeks high-resolution image synthesis but may not produce realistic urban layouts (Fig. 5).

Learning Based on Quantized Representations. Vector quantization has been a prevailing stepping-stone technique for text (e.g., Word2Vec [41]), image (e.g., VQVAE/GAN [15,16,56]), and video synthesis (e.g., SORA [6,47,63]). Pretrained quantizers provide a scalable token representation benefiting high-level and large-scale synthesis. A well-trained quantizer combined with self-supervised masked training often performs well in reconstructing the original signal during generative tasks. BERT [14] provides plausible text-based representation learning. MAE [19] defines a masked autoencoder for image space which excels at multiple downstream tasks. MAE also provides representation learning for a broad range of node/graph feature synthesis [25,26]. Specifically, MaskGIT [10] and MAGE [35] utilize MAE frameworks for impressive image generation tasks. The idea of a generative masked autoencoder based on pretrained quantization was inspirational to our city-scale urban layout approach.

3   Method

First we describe our canonical graph representation of 2.5D urban layouts (Fig. 2). Second, we explain the training process for our GMAE based on the graphs for many cities (Fig. 3). Third, we show our priority-based scheduling for iteratively generating city-scale urban layouts given any amount of prior (Fig. 4).

3.1   Canonical Graph Representation

A city is represented by a graph $G = \{B, E\}$, where B is a set of nodes $B = \{b_i | i \in [1, N]\}$ such that b_i represents each city block (Fig. 2), and any connected subgraph of G corresponds to a community. $N = |B|$ is the total number of blocks for a given city. E is a set of edges $E = \{e_{ij} | i, j \in [1, N]\}$ representing mutually adjacent nodes (or city blocks). $K = |E|$ is the total number of edges.

Node (or City Block). Each node, or city block, b_i is a region defined by an enclosing road network and contains a set of block-level features and building-level features. The per-node features include:

- s_i is the shape and location feature vector of a city block contour. It includes four features: (1) aspect ratio, (2) total block area, (3) ratio of total block area over the block's convex-hull area, and (4) relative distance from current block to the centroid of the city.
- q_i is a 512-dimensional vector from a codebook C. It hierarchically captures the buildings layouts and their configuration within b_i.

We concatenate these node features to form a 516-dimensional vector ($[s_i, q_i]$), which is able to represent our large and heterogeneous set of blocks and buildings.

Building Layout Quantization. The 512-dimensional vector q_i describing the building layouts within a city block is obtained from a quantized codebook $C = \{1, 2, \ldots, L\}$ with quantization level $L = |C|$. To create the codebook, we train a block-level variational autoencoder (BVAE) for building layout within a city block, using a canonical graph structure for representing the multiple building shapes, positions, and heights inside a single city block. Using this pretrained encoder, the latent vector of all possible building layouts will form a numerical distribution for each latent dimension. We further quantize each of these distributions into L bins with equal percentiles. Then, each latent value is

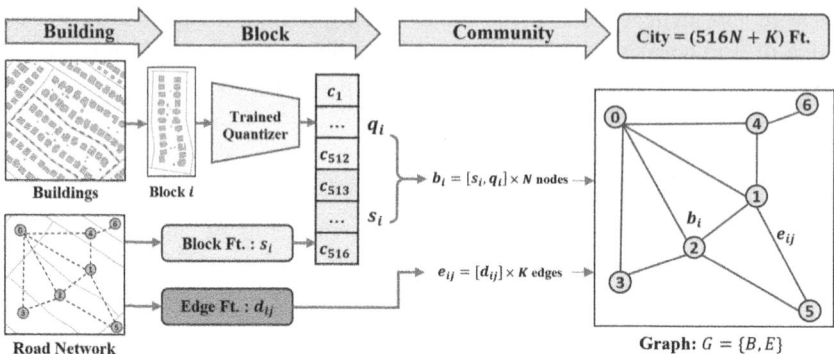

Fig. 2. Canonical Graph Representation. Our method represents a city as a canonical graph G. Each node b_i represent a single city block, and each edge e_{ij} connects spatially adjacent blocks. Each block/node corresponds to a set of node features s_i and a quantized vector q_i hierarchically capturing enclosed building layouts. Edge feature d_{ij} encodes distances between block centroids. Graph G is used for GMAE training.

replaced by the rank of the bin containing its value. Therefore, any building layout within a city block is quantized into a 512-dimensional categorical index vector $q_i = [c_1, \ldots, c_{512}]$, where each $c \in C$. Note that our codebook is defined after BVAE training and not dynamically trainable as in VQGAN/VQVAE [16,56]. After performing a set of detailed evaluations using different training schemes, encoding models, and quantization levels L (see Sect. 4.3), we selected the Graph Attention Network (GAT) [59] as the backbone for the BVAE. Further, we select $L = 20$ to balance compactness and representation ability after quantization. BVAE and quantization details are found in Supplementary Materials.

Edge. Each edge e_{ij} contains a feature d_{ij} storing the relative distance between the centroids of the adjacent city blocks b_i and b_j. This distance-weighted adjacency helps encode the relationship between neighboring blocks.

Altogether, the shape, style, and multi-layer urban layout of a city with N nodes and K edges is represented by a total of $(516N + K)$ variables. This canonical graph is used for the self-supervised GMAE training described below.

3.2   Graph-Based Masked AutoEncoder

We train our GMAE using a self-supervised process. A road-only city (or community) is a graph (or subgraph) G where building layout features q_i of every node (block) are missing, but s_i and e_{ij} are preserved. A context-sensitive node feature generation process corresponds to recreating the purposefully removed building layout features.

During training (Fig. 3), we use dynamic masking ratios of q_i in order to learn how to regenerate the original features. In each training iteration, the mask ratio m is sampled from a truncated Gaussian distribution as recommended by Li et al. [35]. The ratio is obtained from the range $m \in [0.5, 1.0]$ and is centered at

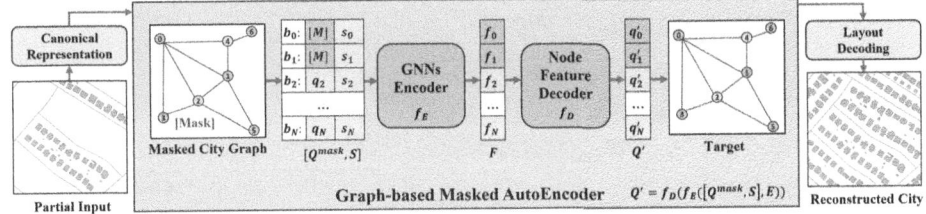

Fig. 3. Graph-based Masked Autoencoder. Given a canonical city graph, quantized building layout features Q are masked with dynamic masking ratios $m \in [0.5, 1.0]$, while block shape and location features S are kept. The GNN encoder uses message passing between neighboring nodes to obtain the context-aware node features F. The decoder uses F to reconstruct Q'. The predicted Q' are decoded to 2.5D urban layouts.

0.55. We denote building layout features as $Q = \{q_i | i \in [1, N]\}$ and those after masking as Q^{mask}. Block features are denoted as $S = \{s_i | i \in [1, N]\}$. The graph neural network (GNN) encoder is denoted as f_E, and node feature decoder as f_D. Thus the process of our GMAE can be written as:

$$F = f_E([Q^{mask}, S], E), \quad Q' = f_D(F) \tag{1}$$

where $F = \{f_i | i \in [1, N]\}$ indicates the context-sensitive node features obtained by message passing between neighboring nodes by f_E. Extracted F is processed by f_D to reconstruct building layout features Q'.

The backbones of f_E and f_D can be any GNN, including GAT [58], GCN [33], and GraphSAGE [18]. After experimentation, we select GAT as f_E because of its best context-sensitive feature extraction ability, and we choose a straightforward MLP as f_D for efficient processing. We observed that the re-masking trick and complex GNN decoder utilized by [25,26] requires more memory and calculation but has no net benefit in our application. Since Q contains quantized index vectors, we use cross entropy loss for self-supervised reconstruction training on masked nodes. Detailed ablations of masking strategy and model selections are provided in Sect. 4.3.

Our objective function is as follows, where L indicates the length of the quantized codebook C:

$$L_{recon} = -\sum_{i=1}^{L} [Q_i \log(Q'_i)]^{mask} \tag{2}$$

To capture community behavior (and also reduce GPU memory usage), we randomly sample a batch of subgraphs from G for each training iteration. The nodes in each subgraph are sampled using a chosen radius from a random center node. While community sizes vary significantly, we found that a practical compromise is to use 500 m as the radius value. This radius ensures to cover a reasonable number of the city blocks with a typical census tract from governmental CJEST dataset [1].

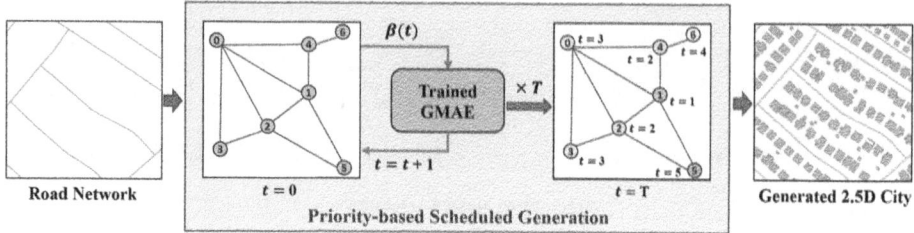

Fig. 4. Priority-based Scheduled Generation. We iteratively utilize pretrained GMAE to reconstruct masked node features. In each iteration, we accept a certain ratio of predicted nodes decided by the scheduling function $\beta(t) = 1 - cos(t/T)$. We obtain a full graph after T iterations.

3.3   Priority-Based Scheduled Generation

While in theory the trained GMAE can predict all nodes in a single pass, the generation quality is much better when performed iteratively. Recent published papers (e.g., [10, 35]) have found that gradually sampling with a reasonable number of iterations improves generation quality. In urban scenarios, small number of representative city blocks are important to indicate context behavior. We intend to generate those blocks at early iterations and then span neighboring blocks.

In Fig. 4, we propose a T-iteration sampling strategy similar to MaskGIT [10] and MAGE [35]. At each iteration, GMAE predicts per-node building layout features for all remaining nodes. Then we rank generated nodes by their prediction confidence. The generated nodes with the highest prediction confidence are accepted. The accepted nodes after the current iteration $t \in [1, T]$ are scheduled by a cosine function $\beta(t) = 1 - cos(t/T)$. When t is small, the ratio is near to 0 causing few nodes to be accepted in initial iterations. As iterations continue, t gets bigger and hence the number of accepted nodes during each iteration will also increase. This behavior follows the prioritization observation in Sect. 1 which causes distinguished, or unique, city blocks to be generated first and essentially guide the layout style for nearby city blocks. Typically we set $T = 12$ which results in dozens of nodes being generated as is the typical size of a community. Detailed ablations of alternative sampling schedules are provided in Sect. 4.3.

4   Experiments

4.1   Implementation Details

Open Dataset. We have created a vector-based urban layout dataset of all US cities with a population of 100K or more. *We believe this dataset will be of significant use to the community to further develop urban layout generation methods and thus we will release the dataset together with the paper.* For each city, we define a rectangular bounding box containing most of the metropolitan area. Then, we use this bounding box to extract data from OpenStreetMap

(OSM) [43], Microsoft Building Footprints (MSF) [23], and Topologically Integrated Geographic Encoding and Referencing (TIGER) dataset [7]. The data we collected includes: (1) road network vectors and attributes, (2) building footprints and heights, and (3) block-level social economic metrics. Specifically, we collect road networks and block contours from TIGER. Then we look for building footprints and heights from OSM and MSF within each city block contour. We choose the one with more and accurate details (e.g., containing more buildings, having features that match those in TIGER). Over all 330 cities, this dataset has a total of 833,473 city blocks, and 17,663,607 buildings. We set training, validation, testing split as 70%, 20%, and 10% respectively.

Training Details. The training has two stages. We first use all city blocks for self-supervised training of a graph-based Block-level Variational AutoEncoder (BVAE). Given this trained block-level encoder, we perform the latent quantization for all building layouts in each city block. Then our GMAE is trained based on canonical city graph representation. Training time for BVAE and GMAE typically takes 12 h and 15 h respectively by a single A5000 GPU. We will release our codes upon acceptance.

Inferencing using our graph-based approach is seemingly instantaneous for all the results shown in this paper. This is contrast to the VTN [2] and SDXL [48] approaches which take significantly more time to infer and to train.

4.2 Comparisons

We compare our method to several alternate city-scale layout generators (Table 1). Specifically, we randomly select 100 communities from among our test set, total 3700 city blocks. All methods generate the layout of the 100 communities in one pass, and without any post-processing or human-in-the-loop refinement. Implementations are downloaded from the author's repositories and trained with default parameter settings. For fairness, we convert our dataset to token sequences, graphs, or image patches (1024×1024) as needed. VTN [2], LayoutDM [29], GlobalMapper [20] are retrained for single block generation. SDXL [48] is fine-tuned for outpainting by using the upper half to predict the lower half.

Evaluation Metrics. We define a Context Score (CTS) to evaluate the context relationships among neighboring city blocks (i.e., a community). This metric is a numerical extension of LayoutSim [20] and DocSim index [46] which evaluate the similarity between two city blocks regarding location, size, and category. Specifically, for each generated city block b_i, we calculate the average LayoutSim with its neighbors $N(i)$. Then, the local average of LayoutSim represents the context similarity among the community (CT). High CT indicates identical blocks in a community and low CT corresponds to high diversity in a community. CTS is the subtraction of the real community's CT from the generated CT:

| SDXL | VTN | LayoutDM | GlobalMapper | Ours |

Fig. 5. Qualitative Comparisons. Given the same road network (except SDXL [48]), all above methods generate urban layouts in only one pass without post-processing or human-in-the-loop refinement. The "even rows" are a zoom-in of the highlighted areas in the "odd rows". Our method generates a realistic distribution of urban layouts with plausible context-dependent behaviors (as indicated by CTS score). VTN [2] and LayoutDM [29] show abnormal style shift among neighboring blocks, and no awareness of road networks. GlobalMapper [20] generates over-similar communities. SDXL [48] provides poor outpainting quality containing semantic errors and undesired overlaps.

Table 1. Quantitative Comparisons. All methods are compared to the same real urban layouts (except SDXL [48] which cannot take-in a road network). Best values are in bold, second best values are underlined. Our method outperforms other existing methods in all but the overlap metric. See text for an explanation of the metrics.

Method	$CTS_{x\to0}$	WD-5D↓	WD-CO↓	Overlap↓	O-Blk↓	FID↓	KID↓	LPIPS↓
SDXL [48]	–	–	–	–	–	120.24	0.079	0.48
VTN [2]	−1.14	3.18	5.81	**1.24**	7.35	69.14	0.047	<u>0.32</u>
LayoutDM [29]	−2.20	<u>2.92</u>	12.50	4.56	1.72	66.77	0.040	0.39
GlobalMP [20]	<u>0.62</u>	4.77	<u>4.14</u>	2.52	<u>0.68</u>	<u>49.55</u>	<u>0.024</u>	0.34
Ours	**0.21**	**2.28**	**1.91**	<u>1.27</u>	**0.42**	**23.63**	**0.005**	**0.20**

$$CT = \frac{1}{\|N(i)\|} \sum_{j}^{N(i)} LayoutSim(b_i, b_j), \quad N(i) = \{j \mid (i,j) \in E\} \qquad (3)$$

$$CTS = CT_{gen} - CT_{real} \qquad (4)$$

Given a real community, the metric indicates whether a generated community is over-diverse ($CTS < 0$), or over-similar ($CTS > 0$). A low value means more realism.

In our comparisons, we compute several metrics. One metric is the Wasserstein Distance between the distributions of generated communities and real communities in terms of 2D position (x, y) and 2.5D geometry (length, width, height), and building count (CO) – we name these WD-5D and WD-CO. We define two quality metrics as well: Overlap indicates the percentage by which two building overlaps, and O-Blk represents the percentage of building area that sticks outside of its city block. We also implement several perceptual metrics: FID [24], KID [4], and LPIPS [65] to evaluate the visual quality and realism of generated layouts. For all the above metrics, lower value indicates better quality.

Quantitative Results. As Table 1 shows, our approach performs better than existing approaches in all but one metric. Note that the poor quality of SDXL results makes it hard to decipher layouts. We only compare to it by pixel-based perceptual metrics.

Qualitative Results. We show in Fig. 5 qualitative results for several exemplary areas (more examples are in Supplemental Material). In general, our method shows good contextual harmonization and has awareness of arbitrary road networks. It generates a realistic distribution of urban layouts with plausible context dependent behaviors. Other methods show improper patterns (e.g., overly diverse or similar) and potentially containing semantic errors or undesired overlaps.

Table 2. Block Quantization Ablations. We report metrics (all in %) among all alternative ablations of our Block-level graph-based Variantional AutoEncoder (BVAE). Using GAT-backbone with post-trained dimensional quantization with $L = 20$ shows the best overall compromise. The best values are in boldface.

Encode Model	Overlap↓	Out-Blk↓	Pos-E.↓	Geom-E.↓	Ct-E.↓	Cov-E.↓
LayoutVAE [32]	13.84	11.15	15.09	37.35	-	24.34
VTN [2]	2.70	4.70	6.01	22.86	18.58	1.97
BlockPlanner [62]	2.93	1.30	4.78	12.06	3.63	5.66
GlobalMapper [20]	1.04	_1.20_	_3.25_	2.83	0.08	0.46
BVAE (SAGE [18])	1.33	1.70	5.01	3.25	0.10	0.35
BVAE (GCN [33])	_0.45_	1.21	4.25	_2.74_	_0.04_	_0.28_
BVAE* (GAT [58])	**0.17**	**0.28**	**1.00**	**0.71**	**0.006**	**0.04**

Quantization	Overlap↓	Out-Blk↓	Pos-E.↓	Geom-E.↓	Ct-E.↓	Cov-E.↓
Trainable-VQ	3.02	0.55	3.75	5.78	0.07	0.60
KMeans-Q	2.90	_0.02_	20.98	17.85	50.01	16.23
DIM-Q ($L = 5$)	0.28	0.62	7.71	16.20	25.0	14.68
DIM-Q ($L = 10$)	0.56	0.28	5.73	4.48	8.33	4.58
DIM-Q ($L = 30$)	_0.17_	**0.01**	**2.68**	**0.78**	**0.005**	_0.15_
DIM-Q* ($L = 20$)	**0.11**	**0.01**	_3.50_	_0.89_	_0.006_	**0.12**

4.3   Ablation Studies

Block Quantization. We ablate block quantization in order to understand and maximize the impact of changing its parameters and design strategy. First, we report in Table 2 (Encode Model Table) the reconstruction loss using different block-level variational autoencoders. We evaluate using various geometric metrics: Overlap (percent of building-to-building overlap), Out-Blk (percent of buildings that stick outside of its city block), Pos-E (positional error as a percentage of block diagonal), Geom-E (2.5D building geometry error as percentage of block scale), Ct-E (percentage of building count error), and Cov-E (total building coverage error). They clearly show our approach is the best overall. Thus we choose our BVAE based on Graph Attention Networks (GAT) [58].

Then, in Table 2 (Quantization Table), we ablate various codebook approaches and quantitization levels for block-level encoding. "Trainable-VQ" leverages trainable vector quantization codebooks introduced by VQVAE [56]. We also provide an alternative non-training based quantization method using KMeans ("KMeans-Q") to cluster BVAE latent space values to a reasonable number of categories in order to handle diverse urban layouts (e.g., 100 classes). The "DIM-Q" rows imply quantizing each of the 512 dimensions of the block-level code into a codebook $C = \{1, 2, ..., L\}$ as described in Sect. 3.1. This per-dimension quantization performs clearly the best. Specifically, setting L to infin-

Table 3. City-Scale GMAE Ablations. We ablate backbone message passing layers, maximum messaging passing hops (D: depth of encoder), masking strategy during training, and scheduling functions of iterative generation (T indicates total iterations). The overall best setting is marked as *.

Ablations	Context↓	WD-3D↓	WD-N↓	Overlap↓	Out-Blk↓	FID↓	KID↓	LPIPS↓
GMAE (GCN [33])	−0.30	2.74	2.91	**0.26**	0.36	52.95	0.025	0.32
GMAE (GTrans [52])	0.74	3.02	3.27	1.96	1.07	43.60	0.018	0.29
GMAE (SAGE [18])	0.97	2.98	2.36	1.75	2.60	43.15	0.017	0.29
GMAE (GAT, Dec)	0.94	2.45	2.96	2.13	1.10	48.24	0.019	0.30
GMAE (GAT, $D = 1$)	1.25	3.03	2.64	1.73	0.44	43.75	0.018	0.29
GMAE (GAT, $D = 2$)	0.33	2.66	<u>2.01</u>	1.31	0.57	34.82	0.014	0.27
GMAE (GAT, $D = 4$)	<u>0.24</u>	**1.92**	2.31	<u>0.63</u>	**0.28**	<u>24.59</u>	<u>0.009</u>	<u>0.22</u>
GMAE ($Mask \in [0.05, 1]$)	1.30	2.70	3.62	1.70	2.45	43.14	0.018	0.29
GMAE ($Mask \in [0.05, 0.5]$)	1.10	3.07	3.68	1.68	1.30	46.85	0.020	0.30
GMAE ($Mask = 0.15$)	1.59	4.91	3.41	1.83	1.20	49.50	0.022	0.31
GMAE (Linear, $T = 12$)	0.63	2.83	3.79	1.27	0.93	37.42	0.015	0.28
GMAE (Log, $T = 12$)	−1.31	2.77	7.27	7.41	3.14	63.25	0.038	0.33
GMAE (Cosine, $T = 1$)	−1.67	7.44	5.64	3.56	4.01	76.12	0.051	0.35
GMAE (Cosine, $T = 20$)	0.32	2.92	3.98	1.68	<u>0.31</u>	32.06	0.013	0.24
*(GAT-D3, [0.5,1], Cos-T12)	**0.21**	<u>2.28</u>	**1.91**	1.27	0.42	**23.63**	**0.005**	**0.20**

ity produces a lossless representation of BVAE latent vectors, but also causes the curse of dimensionality. We discover that $L = 20$ per dimension is the best overall compromise balancing compactness and representation ability.

City-Scale Graph Masked Autoencoder. In Table 3, We ablate various model parameters, together with significant training and inference strategies for our GMAE. Each alternatives is evaluated by the same set of metrics in Sect. 4.2.

By varying the message passing layers, we find GAT [58] provides the best performance. GCN [33] produces diverse and unrealistic urban layouts, while transformer-based [52], and GraphSAGE [18] backbones generate blocks that are too similar. The depth of the stacked layers in the encoder (D) dictates the maximum message passing hops from each node and thus the span of context behaviors captured for realistic generation. We find that 3 hops is promising and larger contexts cause over-similarity and unnecessary computing burden. We also observed that using stacked GAT layers for decoding with remasking tricks, as described by [25,26], didn't produce a net benefit towards city generation. So we choose MLP as the node feature decoder for both its performance and efficiency.

The masking strategy during training is crucial for GMAE performance. We find that dynamic and large masking ratios are beneficial, while a fixed small ratio (e.g., 0.15 as in [14]) leads to over similarity. This behavior is also observed by Li et al. [35]. We adopt a similar masking strategy by sampling a Gaussian distribution $N(0.55, 0.25)$, truncated by $[0.5, 1.0]$, for each training iteration.

Scheduling for iterative generation is also critical to capture representative city blocks (Sect. 3.3). We evaluate several functions $\beta(t)$ (e.g., linear, logarithmic, or cosine-based) that provide the scheduled acceptance ratio, and evaluate alternative total iteration counts T. Rapid iteration leads to over-diversity and long iteration results in over-similarity. Logarithmic functions with reverse pattern produces the worst realism. Metrics show that cosine-based function, which is conservative initially and generous in later iterations, outperforms others.

In summary, this study determines the best configuration for our GMAE as: GAT encoder with $D = 3$ and MLP decoder, training mask ratio $m \in [0.5, 1.0]$, and priority-based scheduling function $\beta(t) = 1 - cos(t/T)$ with $T = 12$.

4.4   Limitations

Our method faces limitations where shapes of city blocks or buildings are not simple polygons (e.g. the building with a garden inside). It also struggles to handle very irregular and concave shaped city blocks (e.g., cul-de-sac's). Also, road-network to contextual structure relations are not explicitly considered.

4.5   Additional Results

We place several other interesting results in Supplementary Materials.

- **Controllable Generation of Entire City.** Our method enables realistic and efficient city-scale controllable generation given any percentage of priors. We show several large examples spanning a subset of the many cities.
- **Socio-Economic Prediction.** Trained GMAE may act as a powerful encoder for downstream tasks. For example, the encoded GMAE latent vectors of urban layouts can indicate social-economic metrics for each city block.
- **Semantic Manipulations.** Our method enables semantic manipulations over various building layout styles across different cities.

5   Conclusions and Future Works

Large-scale urban layout generation has seen significant advancements, spurred by the intersection of computer vision, urban planning, and related disciplines. Previous methodologies often neglect the context-sensitive nature of urban layouts, failing to capture the multi-layer semantics crucial for realism and plausibility. Our proposed Graph-based Masked AutoEncoder (GMAE) addresses these limitations by leveraging a graph-based representation, acknowledging context-sensitivity, and prioritizing the generation process. By training on a diverse dataset of 330 cities, we demonstrate the efficacy of our approach in generating large-scale 2.5D layouts with realism and semantic consistency. Moreover, our method outperforms existing techniques in various metrics, marking a significant advancement in the realm of urban layout generation.

Moving forward, our approach paves the way for more accurate and efficient urban simulation, digital twin creation, and game/content design. As future work we intend to incorporate our method for large-scale photo-realistic multi-view scene synthesis, and city-scale 3D modeling. Those potentials will contribute to synthetic data generation in autonomous driving and world model training.

Acknowledgements. This project was funded in part by NSF Grant #2107096 and NSF Grant #1835739.

References

1. Climate and economic justice screening tool. https://screeningtool.geoplatform. gov/
2. Arroyo, D.M., Postels, J., Tombari, F.: Variational transformer networks for layout generation. In: Proceedings of the IEEE/CVF Conference on Computer Vision and Pattern Recognition, pp. 13642–13652 (2021)
3. Bhatt, M., et al.: Design and deployment of photo2building: a cloud-based procedural modeling tool as a service. In: Practice and Experience in Advanced Research Computing, pp. 132–138 (2020)
4. Bińkowski, M., Sutherland, D.J., Arbel, M., Gretton, A.: Demystifying mmd GANs. arXiv preprint arXiv:1801.01401 (2018)
5. Bokeloh, M., Wand, M., Seidel, H.P.: A connection between partial symmetry and inverse procedural modeling. In: ACM SIGGRAPH 2010 Papers, pp. 1–10 (2010)
6. Brooks, T., et al.: Video generation models as world simulators (2024). https:// openai.com/research/video-generation-models-as-world-simulators
7. Bureau, U.S.C.: Topologically integrated geographic encoding and referencing. https://www.census.gov/geographies/mapping-files/time-series/geo/tiger-line-file.html
8. Chai, L., Tucker, R., Li, Z., Isola, P., Snavely, N.: Persistent nature: a generative model of unbounded 3D worlds. In: Proceedings of the IEEE/CVF Conference on Computer Vision and Pattern Recognition, pp. 20863–20874 (2023)
9. Chai, S., Zhuang, L., Yan, F.: LayoutDM: transformer-based diffusion model for layout generation. In: Proceedings of the IEEE/CVF Conference on Computer Vision and Pattern Recognition, pp. 18349–18358 (2023)
10. Chang, H., Zhang, H., Jiang, L., Liu, C., Freeman, W.T.: MaskGIT: masked generative image transformer. In: Proceedings of the IEEE/CVF Conference on Computer Vision and Pattern Recognition, pp. 11315–11325 (2022)
11. Chang, K.H., Cheng, C.Y., Luo, J., Murata, S., Nourbakhsh, M., Tsuji, Y.: Building-GAN: graph-conditioned architectural volumetric design generation. In: Proceedings of the IEEE/CVF International Conference on Computer Vision, pp. 11956–11965 (2021)
12. Chen, Z., Wang, G., Liu, Z.: Scenedreamer: unbounded 3D scene generation from 2D image collections. arXiv preprint arXiv:2302.01330 (2023)
13. Deng, J., et al.: CityGen: infinite and controllable 3D city layout generation. arXiv preprint arXiv:2312.01508 (2023)
14. Devlin, J., Chang, M.W., Lee, K., Toutanova, K.: Bert: pre-training of deep bidirectional transformers for language understanding. arXiv preprint arXiv:1810.04805 (2018)
15. Dosovitskiy, A., et al.: An image is worth 16x16 words: transformers for image recognition at scale. arXiv preprint arXiv:2010.11929 (2020)

16. Esser, P., Rombach, R., Ommer, B.: Taming transformers for high-resolution image synthesis. In: Proceedings of the IEEE/CVF Conference on Computer Vision and Pattern Recognition, pp. 12873–12883 (2021)
17. Gupta, K., Lazarow, J., Achille, A., Davis, L.S., Mahadevan, V., Shrivastava, A.: Layouttransformer: Layout generation and completion with self-attention. In: Proceedings of the IEEE/CVF International Conference on Computer Vision, pp. 1004–1014 (2021)
18. Hamilton, W., Ying, Z., Leskovec, J.: Inductive representation learning on large graphs. In: Advances in Neural Information Processing Systems, vol. 30 (2017)
19. He, K., Chen, X., Xie, S., Li, Y., Dollár, P., Girshick, R.: Masked autoencoders are scalable vision learners. In: Proceedings of the IEEE/CVF Conference on Computer Vision and Pattern Recognition, pp. 16000–16009 (2022)
20. He, L., Aliaga, D.: Globalmapper: arbitrary-shaped urban layout generation. In: Proceedings of the IEEE/CVF International Conference on Computer Vision, pp. 454–464 (2023)
21. He, L., Lu, Y., Corring, J., Florencio, D., Zhang, C.: Diffusion-based document layout generation. In: Fink, G.A., Jain, R., Kise, K., Zanibbi, R. (eds.) Document Analysis and Recognition - ICDAR 2023. LNCS, pp. 361–378. Springer, Cham (2023). https://doi.org/10.1007/978-3-031-41676-7_21
22. He, L., Shan, J., Aliaga, D.: Generative building feature estimation from satellite images. IEEE Trans. Geosci. Remote Sens. **61**, 1–13 (2023)
23. Heris, M.P., Foks, N.L., Bagstad, K.J., Troy, A., Ancona, Z.H.: A rasterized building footprint dataset for the united states. Sci. Data **7**(1), 207 (2020)
24. Heusel, M., Ramsauer, H., Unterthiner, T., Nessler, B., Hochreiter, S.: GANs trained by a two time-scale update rule converge to a local NASH equilibrium. In: Advances in Neural Information Processing Systems, vol. 30 (2017)
25. Hou, Z., et al.: Graphmae2: a decoding-enhanced masked self-supervised graph learner. In: Proceedings of the ACM Web Conference 2023, pp. 737–746 (2023)
26. Hou, Z., et al.: GraphMAE: self-supervised masked graph autoencoders. In: Proceedings of the 28th ACM SIGKDD Conference on Knowledge Discovery and Data Mining, pp. 594–604 (2022)
27. Hua, H., et al.: Finematch: aspect-based fine-grained image and text mismatch detection and correction. arXiv preprint arXiv:2404.14715 (2024)
28. Hui, M., Zhang, Z., Zhang, X., Xie, W., Wang, Y., Lu, Y.: Unifying layout generation with a decoupled diffusion model. In: Proceedings of the IEEE/CVF Conference on Computer Vision and Pattern Recognition, pp. 1942–1951 (2023)
29. Inoue, N., Kikuchi, K., Simo-Serra, E., Otani, M., Yamaguchi, K.: LayoutDM: discrete diffusion model for controllable layout generation. In: Proceedings of the IEEE/CVF Conference on Computer Vision and Pattern Recognition, pp. 10167–10176 (2023)
30. Jiang, Z., et al.: Layoutformer++: conditional graphic layout generation via constraint serialization and decoding space restriction. In: Proceedings of the IEEE/CVF Conference on Computer Vision and Pattern Recognition, pp. 18403–18412 (2023)
31. Jyothi, A.A., Durand, T., He, J., Sigal, L., Mori, G.: LayoutVAE: stochastic scene layout generation from a label set. In: Proceedings of the IEEE/CVF International Conference on Computer Vision pp. 9895–9904 (2019)
32. Jyothi, A.A., Durand, T., He, J., Sigal, L., Mori, G.: LayoutVAE: stochastic scene layout generation from a label set. In: 2019 IEEE/CVF International Conference on Computer Vision (ICCV), pp. 9894–9903 (2019). https://doi.org/10.1109/ICCV.2019.00999

33. Kipf, T.N., Welling, M.: Semi-supervised classification with graph convolutional networks. arXiv preprint arXiv:1609.02907 (2016)
34. Li, J., Yang, J., Hertzmann, A., Zhang, J., Xu, T.: LayoutGAN: synthesizing graphic layouts with vector-wireframe adversarial networks. IEEE Trans. Pattern Anal. Mach. Intell. **43**(7), 2388–2399 (2020)
35. Li, T., Chang, H., Mishra, S., Zhang, H., Katabi, D., Krishnan, D.: Mage: masked generative encoder to unify representation learning and image synthesis. In: Proceedings of the IEEE/CVF Conference on Computer Vision and Pattern Recognition, pp. 2142–2152 (2023)
36. Li, Z., Wang, Q., Snavely, N., Kanazawa, A.: Infinitenature-zero: learning perpetual view generation of natural scenes from single images. In: Avidan, S., Brostow, G., Cissé, M., Farinella, G.M., Hassner, T. (eds.) ECCV 2022. LNCS, vol. 13661, pp. 515–534. Springer, Cham (2022)
37. Lin, C.H., Lee, H.Y., Cheng, Y.C., Tulyakov, S., Yang, M.H.: InfinityGAN: towards infinite-pixel image synthesis. arXiv preprint arXiv:2104.03963 (2021)
38. Lin, C.H., et al.: Infinicity: infinite-scale city synthesis. arXiv preprint arXiv:2301.09637 (2023)
39. Lipp, M., Scherzer, D., Wonka, P., Wimmer, M.: Interactive modeling of city layouts using layers of procedural content. In: Computer Graphics Forum, vol. 30, pp. 345–354. Wiley Online Library (2011)
40. Ma, H., Zeng, D., Liu, Y.: Learning individualized treatment rules with many treatments: a supervised clustering approach using adaptive fusion. Adv. Neural. Inf. Process. Syst. **35**, 15956–15969 (2022)
41. Mikolov, T., Chen, K., Corrado, G., Dean, J.: Efficient estimation of word representations in vector space. arXiv preprint arXiv:1301.3781 (2013)
42. Nauata, N., Chang, K.-H., Cheng, C.-Y., Mori, G., Furukawa, Y.: House-GAN: relational generative adversarial networks for graph-constrained house layout generation. In: Vedaldi, A., Bischof, H., Brox, T., Frahm, J.-M. (eds.) ECCV 2020, Part I. LNCS, vol. 12346, pp. 162–177. Springer, Cham (2020). https://doi.org/10.1007/978-3-030-58452-8_10
43. OpenStreetMap contributors (2017). Planet dump retrieved from https://planet.osm.org. https://www.openstreetmap.org
44. Para, W., Guerrero, P., Kelly, T., Guibas, L.J., Wonka, P.: Generative layout modeling using constraint graphs. In: Proceedings of the IEEE/CVF International Conference on Computer Vision, pp. 6690–6700 (2021)
45. Patel, P., Kalyanam, R., He, L., Aliaga, D., Niyogi, D.: Deep learning-based urban morphology for city-scale environmental modeling. PNAS Nexus **2**(3), pgad027 (2023)
46. Patil, A.G., Ben-Eliezer, O., Perel, O., Averbuch-Elor, H.: Read: recursive autoencoders for document layout generation. In: Proceedings of the IEEE/CVF Conference on Computer Vision and Pattern Recognition Workshops, pp. 544–545 (2020)
47. Peebles, W., Xie, S.: Scalable diffusion models with transformers. In: Proceedings of the IEEE/CVF International Conference on Computer Vision, pp. 4195–4205 (2023)
48. Podell, D., et al.: SDXL: improving latent diffusion models for high-resolution image synthesis. arXiv preprint arXiv:2307.01952 (2023)
49. Shen, Y., Ma, W.C., Wang, S.: SGAM: building a virtual 3d world through simultaneous generation and mapping. Adv. Neural. Inf. Process. Syst. **35**, 22090–22102 (2022)

50. Sheng, Y., et al.: Controllable shadow generation using pixel height maps. In: Avidan, S., Brostow, G., Cissé, M., Farinella, G.M., Hassner, T. (eds.) ECCV 2022. LNCS, vol. 13683, pp. 240–256. Springer, Cham (2022). https://doi.org/10.1007/978-3-031-20050-2_15
51. Sheng, Y., et al.: Dr. bokeh: differentiable occlusion-aware bokeh rendering. In: Proceedings of the IEEE/CVF Conference on Computer Vision and Pattern Recognition, pp. 4515–4525 (2024)
52. Shi, Y., Huang, Z., Feng, S., Zhong, H., Wang, W., Sun, Y.: Masked label prediction: Unified message passing model for semi-supervised classification. arXiv preprint arXiv:2009.03509 (2020)
53. Song, Y., et al.: Objectstitch: object compositing with diffusion model. In: Proceedings of the IEEE/CVF Conference on Computer Vision and Pattern Recognition, pp. 18310–18319 (2023)
54. Song, Y., et al.: Imprint: generative object compositing by learning identity-preserving representation. In: Proceedings of the IEEE/CVF Conference on Computer Vision and Pattern Recognition, pp. 8048–8058 (2024)
55. Tabata, S., Yoshihara, H., Maeda, H., Yokoyama, K.: Automatic layout generation for graphical design magazines. In: ACM SIGGRAPH 2019 Posters, pp. 1–2 (2019)
56. Van Den Oord, A., Vinyals, O., et al.: Neural discrete representation learning. In: Advances in Neural Information Processing Systems, vol. 30 (2017)
57. Vanegas, C.A., Kelly, T., Weber, B., Halatsch, J., Aliaga, D.G., Müller, P.: Procedural generation of parcels in urban modeling. In: Computer Graphics Forum, vol. 31, pp. 681–690. Wiley Online Library (2012)
58. Veličković, P., Cucurull, G., Casanova, A., Romero, A., Lio, P., Bengio, Y.: Graph attention networks. arXiv preprint arXiv:1710.10903 (2017)
59. Velickovic, P., Cucurull, G., Casanova, A., Romero, A., Lio, P., Bengio, Y., et al.: Graph attention networks. Stat **1050**(20), 10–48550 (2017)
60. Wu, W., Fu, X.M., Tang, R., Wang, Y., Qi, Y.H., Liu, L.: Data-driven interior plan generation for residential buildings. ACM Trans. Graph. (TOG) **38**(6), 1–12 (2019)
61. Xie, H., Chen, Z., Hong, F., Liu, Z.: Citydreamer: compositional generative model of unbounded 3d cities. arXiv preprint arXiv:2309.00610 (2023)
62. Xu, L., et al.: Blockplanner: city block generation with vectorized graph representation. In: Proceedings of the IEEE/CVF International Conference on Computer Vision, pp. 5077–5086 (2021)
63. Yan, W., Zhang, Y., Abbeel, P., Srinivas, A.: VideoGPT: video generation using VQ-VAE and transformers. arXiv preprint arXiv:2104.10157 (2021)
64. Yang, C.F., Fan, W.C., Yang, F.E., Wang, Y.C.F.: LayoutTransformer: scene layout generation with conceptual and spatial diversity. In: Proceedings of the IEEE/CVF Conference on Computer Vision and Pattern Recognition, pp. 3732–3741 (2021)
65. Zhang, R., Isola, P., Efros, A.A., Shechtman, E., Wang, O.: The unreasonable effectiveness of deep features as a perceptual metric. In: Proceedings of the IEEE Conference on Computer Vision and Pattern Recognition, pp. 586–595 (2018)
66. Zhang, X., Ma, W., Varinlioglu, G., Rauh, N., He, L., Aliaga, D.: Guided pluralistic building contour completion. Vis. Comput. **38**(9), 3205–3216 (2022)
67. Zheng, X., Qiao, X., Cao, Y., Lau, R.W.: Content-aware generative modeling of graphic design layouts. ACM Trans. Graph. (TOG) **38**(4), 1–15 (2019)

Joint RGB-Spectral Decomposition Model Guided Image Enhancement in Mobile Photography

Kailai Zhou[1], Lijing Cai[1], Yibo Wang[1], Mengya Zhang[1], Bihan Wen[3],
Qiu Shen[1,2(✉)], and Xun Cao[1,2]

[1] School of Electronic Sciene and Engineering, Nanjing University, Nanjing, China
{calayzhou,cailijing,ybwang,522023230161}@smail.nju.edu.cn,
{shenqiu,caoxun}@nju.edu.cn
[2] Key Laboratory of Optoelectronic Devices and Systems with Extreme
Performances of MOE, Nanjing University, Nanjing, China
[3] Nanyang Technological University, Singapore, Singapore
bihan.wen@ntu.edu.sg

Abstract. The integration of miniaturized spectrometers into mobile devices offers new avenues for image quality enhancement and facilitates novel downstream tasks. However, the broader application of spectral sensors in mobile photography is hindered by the inherent complexity of spectral images and the constraints of spectral imaging capabilities. To overcome these challenges, we propose a joint RGB-Spectral decomposition model guided enhancement framework, which consists of two steps: joint decomposition and prior-guided enhancement. Firstly, we leverage the complementarity between RGB and Low-resolution Multi-Spectral Images (Lr-MSI) to predict shading, reflectance, and material semantic priors. Subsequently, these priors are seamlessly integrated into the established HDRNet to promote dynamic range enhancement, color mapping, and grid expert learning, respectively. Additionally, we construct a high-quality Mobile-Spec dataset to support our research, and our experiments validate the effectiveness of Lr-MSI in the tone enhancement task. This work aims to establish a solid foundation for advancing spectral vision in mobile photography. The code is available at https://github.com/CalayZhou/JDM-HDRNet.

Keywords: Image enhancement · RGB-Spectral fusion · Spectral image intrinsic decomposition · Spectral dataset and application

1 Introduction

Recent advancements in miniaturized spectrometers [29,37] have facilitated their integration into mobile devices, thereby expanding the scope of possible applications such as medical diagnosis [15], food safety evaluation [10], anti-spoofing face

Supplementary Information The online version contains supplementary material available at https://doi.org/10.1007/978-3-031-72624-8_2.

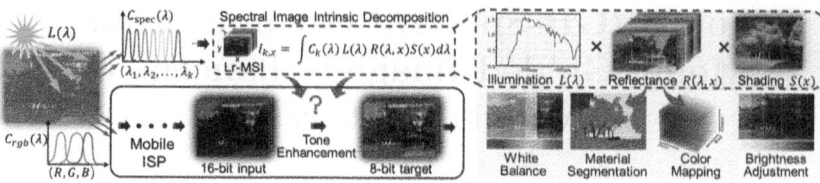

Fig. 1. The Lr-MSI is decomposed into the illumination $L(\lambda)$, reflectance $R(\lambda, x)$ and shading $S(x)$ terms. We then analyze their respective roles in the mobile ISP pipeline.

recognition [31] and color analysis [9]. Nevertheless, spectral sensors [2] in mobile photography are mainly employed for illuminant estimation in the Auto White Balance (AWB) operation at present. The spectral information holds potential to enhance the aesthetic appeal and perceptual quality of images. Exploring how to leverage additional spectral information to improve other image enhancement operations in the mobile Image Signal Processing (ISP) pipeline is an area requiring deeper investigation.

We attempt to make a deeper analysis of the role of Lr-MSI in mobile photography by drawing insights from the spectral image intrinsic decomposition. As illustrated in Fig. 1, the spectral image can be decomposed into its constituent terms [5]: illumination curve $L(\lambda)$, reflectance $R(\lambda, x)$, and shading $S(x)$. $L(\lambda)$ captures the spectral characteristics of the light source and has traditionally been applied for illuminant estimation [8]. $R(\lambda, x)$ embodies the albedo invariant color and texture of the material. The reflectance of spectral images which possesses fine-grained color channels shows promise for material segmentation and color mapping tasks. $S(x)$ represents interaction between the geometry of objects and illumination, which may contribute to the local brightness adjustment of high dynamic scenes. We contend that the potential of the reflectance $R(\lambda, x)$ and shading $S(x)$ terms has not been thoroughly explored.

In this paper, we explore the application of low-resolution multi-spectral images in the tone enhancement task within the mobile ISP pipeline, especially for outdoor High Dynamic Range (HDR) scenes. The objective of tone enhancement is to adjust the brightness, contrast, and color characteristics of an image to enhance details in both the highlights and shadows while preserving its natural appearance. The primary challenge is the heterogeneous information fusion involving the integration of extra Lr-MSI with the original RGB workflow. This challenge manifests in two aspects: Firstly, variations in imaging mechanisms lead to the inherent complexity of spectral images, encompassing factors like scene geometry, inter-reflections and intricate artificial illumination [5], making it difficult to directly integrate spectral information into the mobile ISP workflow. Secondly, despite commercial spectral sensors offering over ten spectral channels [2], their spatial resolution is often confined due to the constrained spectral imaging capabilities on mobile devices. This limited spatial resolution remains a problem for the application of Lr-MSI in tone enhancement.

To address the above challenges, we propose a joint RGB-Spectral decomposition model guided image enhancement framework which consists of two phases: joint decomposition and prior-guided enhancement. The effectiveness of this framework relies on three assumptions: (1) The near-infrared band in Lr-MSI can serve as an approximation of the shading term [7]. (2) Lr-MSI and RGB images exhibit complementary characteristics in both spatial and spectral resolutions. (3) The increased color channels in Lr-MSI contribute to material segmentation. Firstly, to mitigate limited spectral imaging capabilities, the joint decomposition phase leverages the complementarity between Lr-MSI and RGB images to predict shading, reflectance, and material semantic priors. Secondly, to overcome the inherent complexity of spectral images, the prior-guided enhancement phase seamlessly integrates these priors into the established HDRNet [11] to provide clear mechanistic guidance. Specifically, the shading component is separated from original RGB images to enhance the adaptability in dealing with localized brightness variations. Subsequently, the $R(\lambda, x)$ of Lr-MSI is employed to guide the prediction of bilateral grids via the spectral perception map, particularly within the reflectance domain. Harnessing the reflectance domain alongside higher-dimensional spectral channels facilitates enhanced learning of color relationships for tone mapping. Lastly, we introduce the mixture of semantic grid experts to specifically adapt to the distinct color characteristics of individual material categories. The Joint DecoMposition framework guided HDRNet (JDM-HDRNet) exhibits superior performance compared to previous enhancement methods in terms of PSNR, SSIM [35] and a color difference metric.

The contributions of this paper are as follows: (1) The pioneering Mobile-Spec dataset is meticulously constructed with high quality aligned RGB and hyperspectral images, which serves as a solid foundation for spectral applications. (2) We propose a joint RGB-Spectral decomposition model guided enhancement framework to address challenges of inherent complexity of spectral images and limited spectral imaging capabilities. The Lr-MSI and RGB images are decomposed into shading, reflectance and material semantic priors, and these priors are employed for promoting dynamic range enhancement, color mapping, and semantic grid expert learning, respectively. (3) Experiments demonstrate the effectiveness of introducing extra Lr-MSI in the tone enhancement task, which provides valuable exploration for the industrial application of spectral sensors.

2   Related Work

2.1   Potential of Mobile Spectral Sensors

As the cost of miniaturized spectrometers [29, 37] decreases and their availability becomes more widespread, the integration of these spectral sensors with mobile devices creates new opportunities. Studies have demonstrated the potential of hyperspectral imaging in medical applications [15, 16, 23, 30] such as hemodynamics monitoring [15] and skin analysis [16, 30]. Moreover, smartphones equipped with snapshot hyperspectral sensors can enhance anti-spoofing face recognition [31] by preventing high-quality 3D mask attacks. Other applications include food safety evaluation [10], color analysis [9], and water quality assessment [32]. Despite these advancements, the use of spectral sensors in mobile photography remains limited. As RGB sensors lack inherent illumination adaptation abilities to maintain color constancy, in recent years, most high-end smartphones are equipped with spectral sensors which aim to improve the accuracy of illuminant estimation [8,13,33] in an uncontrolled environment. The spectral sensor has evolved from its origins as an ambient light sensor to the refined capture of spectral signatures, primarily employed in auto white balance because its spatial resolution is often confined to a single pixel [2]. Our objective is to validate the applicability of Lr-MSI to other enhancement tasks within the mobile ISP pipeline, particularly under assumptions where the constraints of spectral imaging are mitigated by dual-camera acquisition in the Mobile-Spec dataset.

2.2   Enhancement in the Mobile ISP Pipeline

A standard mobile ISP pipeline [19] comprises both the Bayer processing routine and the subsequent photo-finishing routine. There are several enhancement procedures such as exposure correction [38], white-balance [12], contrast enhancement [4], color manipulation [20], local and global tone-mapping [28]. In this study, we aim to fully exploit the intrinsic decomposition components of spectral images by integrating Lr-MSI into the tone enhancement. This task involves adjusting the tonal range while preserving overall appearance and details, especially for HDR scenes. The commonly used method is the histogram equalization [21] which globally improves the contrast of entire images. However, global operations often fall short when dealing with scenes of high dynamic range. In the deep learning era, color transform-based methods are often employed for real-time processing of high-resolution images in a collaborative global-local manner. According to the color transform functions, these methods can be classified into affine transformation matrices [11,34], multilayer perceptrons (MLPs) [14], 3D LUTs [36,39,41], etc. The pioneering work is the HDRNet [11] which learns the bilateral grid of coefficients at low resolution and performs transformations from input to output at full resolution. 3D LUT [39] replaces the affine transformation matrices with image-adaptive 3-dimensional lookup tables, enabling rapid and robust photo enhancement. We choose the HDRNet as the baseline due to its simple and flexible network design. The key challenge lies in the heterogeneous information fusion involving the integration of extra Lr-MSI with the HDRNet.

2.3   Spectral Image Intrinsic Decomposition

To fuse the extra Lr-MSI into the original RGB workflow in HDRNet, we consider the role of spectral information in mobile photography from the perspective of Spectral Intrinsic Image Decomposition (SIID) [5]. The SIID problem aims to separate a spectral image into more elaborate intrinsic properties, such as the spectrum of the illumination, reflectance and shading components. This decomposition is typically performed in the spectral domain, which allows much more flexibility in the algorithm design and leads to a more accurate analysis. Zheng et al. [42] model illumination and reflectance spectra separation of a hyperspectral image into a low-rank matrix factorization problem. Chen et al. [5] resolve

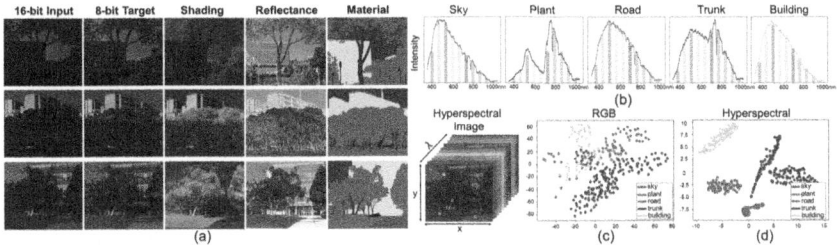

Fig. 2. (a) The Mobile-Spec dataset comprises RGB images (16-bit input and 8-bit target), hyperspectral images and their corresponding shading, reflectance and material segmentation images. Near-infrared images serve as the guide map to approximate the shading term. (b) Spectral responses of different material categories. (c-d) Visualizations of RGB and hyperspectral data for different categories with t-SNE [27].

a spectral image into independent intrinsic components: illumination, shading, and reflectance, and develop an effective algorithm which first reduces the spatial and spectral resolution with the super-pixel down-sampling, then the SIID problem is optimized under local constant and global sparse constraint. Huang et al. [17] extend the Retinex model to the multispectral domain based on the subspace constraint, which assumes the reflectance and shading vectors both live in a low dimensional subspace along the spectral domain. SIID is an ill-posed and under-constrained problem and previous algorithms are developed under the ideal lab environment, underscoring the challenges of tackling complex outdoor scenes. Fortunately, Cheng et al. [7] find that near-infrared images can function as reliable approximations of shading priors, and Zhang et al. [40] further perform intrinsic decomposition for outdoor scenes using aligned RGB and hyperspectral images. Based on this assumption, our joint decomposition model can be trained in an end-to-end manner, which will be discussed in Sect. 4.1.

3   The Mobile-Spec Dataset

To the best of our knowledge, no existing image enhancement dataset includes both aligned hyperspectral and RGB images. To bridge this data gap, we have constructed the high-quality Mobile-Spec dataset, consisting of 200 image groups. As shown in Fig. 2(a), each group represents an independent scene and includes a 16-bit input image, an 8-bit target image, a hyperspectral image, and the corresponding shading, reflectance, and material segmentation images.

Dual Camera System. To address the limited spectral imaging capabilities of sensors on smartphones, the Mobile-Spec is captured using a dual camera system: a high-end commercial smartphone and a Gaiasky-mini2 scanning hyperspectral camera [1]. The Gaiasky-mini2 covers a spectral range from 400 nm to 1000 nm with 176 spectral channels. The 16-bit input images are fused from multiple exposure frames, and the 8-bit target images are generated by a commercial privacy model that integrates the aesthetics of multiple experts.

Image Matching. The hyperspectral image size is $1057 \times 960 \times 176$, while the 16-bit input and 8-bit target RGB images are $4096 \times 3072 \times 3$. To align RGB images with a wide angle of view onto hyperspectral images, we employ an automatic alignment algorithm that computes SIFT descriptors [26] and estimates homography transformation matrices. Consequently, both RGB and hyperspectral images achieve a consistent spatial resolution of 1057×960.

Influence of Light Sources. The Mobile-Spec dataset is captured exclusively outdoors to avoid the impact of intricate indoor lighting. Solar radiation spans a broad range of wavelengths, ranging from ultraviolet to visible light and extending into the infrared region. It is reasonable to assume that near-infrared images can approximate shading priors in outdoor settings [7].

Dataset Collection and Filtering. Our dataset spans multiple seasons and includes five categories commonly observed in outdoor environments: sky, building, plant, trunk, and road. Their respective spectral curves are illustrated in Fig. 2(b). Mobile-Spec undergoes rigorous filtering to ensure high quality. Firstly, samples are selected based on dynamic range, determined by an automated algorithm analyzing the difference between maximum and minimum pixel intensities. Secondly, samples exhibiting significant alignment errors are filtered out. Thirdly, samples with unsatisfactory visual quality are excluded, considering factors such as chromatic aberration, sharpness, noise, and artifacts. Lastly, to augment dataset diversity, scenes with high similarity are eliminated.

Shading, Reflectance and Material Segmentation. To obtain targets of S, R and M for the joint decomposition model, shading images are approximated by averaging near-infrared bands from 850 nm to 1000 nm in hyperspectral images. Reflectance images can be derived based on the Retinex theory [3,22]. Ground truths of material segmentation are meticulously labeled by human annotators. More details about the Mobile-Spec are provided in the appendix.

4   Proposed Method

4.1   Joint RGB-Spectral Decomposition Model

To integrate low-resolution spectral images into the RGB tone enhancement workflow, we first analyze the differences between hyperspectral and RGB camera imaging mechanisms. The imaging model for spectral cameras can be represented as follows based on [5,18]:

$$I_{k,x} = \int_{400 \text{ nm}}^{1000 \text{ nm}} C_k(\lambda)L(\lambda)S(x)R(\lambda,x)d\lambda, \text{k} = 1,\, 2,\, 3\, \ldots \tag{1}$$

$I_{k,x}$ represents the pixel intensities at position x of the k-th band, $C_k(\lambda)$ denotes the camera sensitivity function, $L(\lambda)$ indicates the illumination curve of

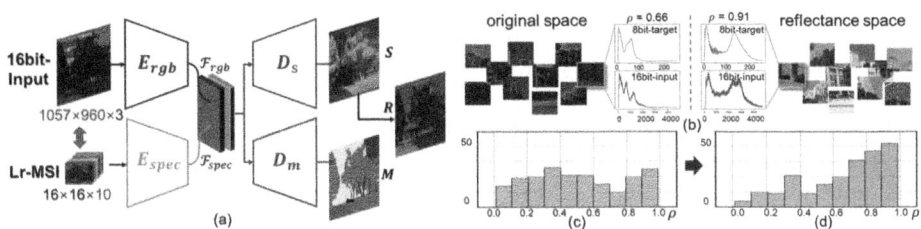

Fig. 3. (a) The joint RGB-Spectral decomposition model leverages the complementarity between RGB images and Lr-MSIs to predict shading (S), reflectance (R) and material category (M) priors. (b) The pixel intensity histogram. (c-d) The distribution of Pearson correlation coefficient ρ in the original RGB space and reflectance space.

the light source, $S(x)$ refers to shading, and $R(\lambda, x)$ is the reflectance of scenes. $L(\lambda)$ is usually utilized for illuminant estimation in the auto white balance procedure, while the role of the shading $S(x)$ and reflectance $R(\lambda, x)$ components has been rarely explored. The core difference between the RGB and hyperspectral cameras lies in $C_k(\lambda)$, where hyperspectral images have fine-grained color channels due to a higher spectral sampling rate. To simulate the limited spectral imaging capability on smartphones, we downsample the hyperspectral image ($1057 \times 960 \times 176$) to the Lr-MSI ($16 \times 16 \times 10$). This downsampling is conducted to match the typical 10-channel configuration of commercial smartphone spectral sensors. [2]. We maintain flexibility in setting the spatial resolution to a default of 16×16, which will be discussed in Sect. 5.2. We posit that the effect of Lr-MSI in the joint decomposition model can be observed in two aspects.

Approximate Estimation of Shading Priors with the Near-infrared Band. To obtain the shading term, traditional optimization-based methods design various hand-crafted priors to constrain the solution [5,17], which are often inapplicable in complex outdoor scenes. Recently, near-infrared images have offered a simple yet powerful approach for shading smoothness [7], as the spectral curves of different colors gradually flatten out and texture variation is considerably reduced in the near-infrared band (refer to the appendix). Given that the Mobile-Spec dataset is exclusively captured outdoors to avoid the impact of complex indoor light sources, and considering the wide spectrum of solar radiation covering the infrared region, the shading term can be predicted through the joint decomposition of Lr-MSI and RGB images, as they share the same $S(x)$.

Enhancing Material Segmentation with the Visible Band. We employ t-SNE [27] to analyze the complex spectral patterns and relationships between different spectral bands. In Fig. 2(c), clusters representing different categories exhibit denser aggregation in the t-SNE visualization of hyperspectral images compared to that of RGB images. This observation suggests that the increased granularity of spectral channels provides enhanced discriminative capacity among various materials, and Lr-MSI may potentially contribute to the material segmentation task.

In Fig. 3(a), we employ a joint RGB-Spectral decomposition model to leverage both near-infrared and visible bands in Lr-MSI for shading prediction and material segmentation. Lr-MSI provides high spectral resolution but limited spatial resolution, while the RGB image offers rich spatial information but lower spectral resolution. We transform the prediction of shading images from the regression problem to the classification problem by dividing shading images into eight brightness levels. Consequently, shading prediction and material segmentation can be decomposed in a collaborative manner. The joint RGB-Spectral decomposition model incorporates two independent encoders and decoders, adapted from the basic segmentation model FCN [25]. Specifically, Lr-MSI is resized to match the spatial resolution of the RGB image. The RGB encoder E_{rgb} and spectral encoder E_{spec} are trained to project the 16-bit input image I_{rgb} and Lr-MSI I_{msi} to the same latent space for representation alignment: $\mathcal{F}_{rgb} = E_{rgb}(I_{rgb})$ and $\mathcal{F}_{spec} = E_{spec}(I_{msi})$. $\mathcal{F}_{rgb}$ and $\mathcal{F}_{spec}$ are fused together with the concatenation to share the common and complementary feature representations. Material segmentation M and shading S are independently predicted with the decoder D_m and D_s based on the fused feature representation. The joint RGB-Spectral decomposition model can be formulated as:

$$M, S = D_{m,s}(concat(\mathcal{F}_{rgb}, \mathcal{F}_{spec})) \tag{2}$$

With the shading component S, the reflectance component R_{rgb}, R_{msi} can subsequently be obtained according to the Retinex theory [3, 22].

4.2   JDM-HDRNet

Based on the established HDRNet [11], the JDM-HDRNet comprehensively exploits the S, R and M priors to provide explicit guidance for tone enhancement.

S: **Localized Brightness Adaptation.** To adjust both highlight and shadow areas during tone enhancement, the HDRNet's bilateral grid depth is set to 8, accommodating eight different luminance levels. However, this design may not effectively handle various localized brightness variations in high dynamic range scenes. We propose that separating the shading component would yield benefits in such contexts. As shown in Fig. 3(b), the pixel histogram vectors between the 16-bit input and 8-bit target exhibits a Pearson correlation coefficient ρ of 0.66 in the original RGB space, which increases to 0.91 in the reflectance space. A larger ρ indicates greater similarity in pixel histogram characteristics. Analyzing the statistics of the ρ distribution in the Mobile-Spec dataset, we observe that the reflectance space (Fig. 3(d)) has a greater proportion of samples with high ρ than the RGB space (Fig. 3(c)). This finding indicates that separating the shading component from the RGB space to the reflectance space may reduce the difficulty of color mapping learning. In our implementation, the shading

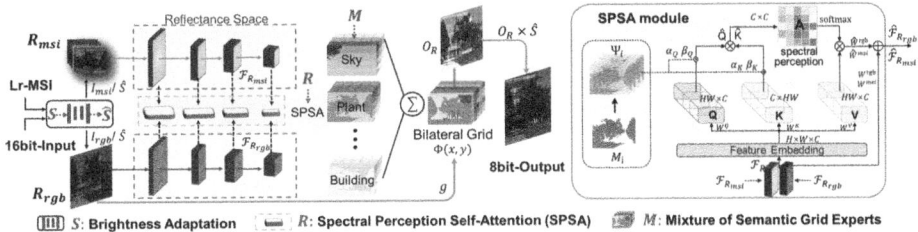

Fig. 4. The overall architecture of the proposed JDM-HDRNet. The S prior is employed to enhance localized dynamic ranges, while R and M priors contribute to the color mapping and grid expert learning, respectively.

component S is converted to the brightness representation $\hat{S}$ with a localized brightness adaptation module, comprising two layers of convolution and deconvolution. This lightweight module helps mitigate drastic pixel intensity changes in localized high dynamic range areas, thereby enhancing the adaptability. Subsequently, the reflectance images of 16-bit input $R_{rgb} = I_{rgb}/\hat{S}$ and Lr-MSI $R_{msi} = I_{msi}/\hat{S}$ are fed into the coefficient prediction part of the bilateral grid (Fig. 4).

R: **Spectral Perception Self-attention.** After transforming from RGB space to reflectance space, we aim to leverage the reflectance of Lr-MSI R_{msi} which has fine-grained spectral channels to enhance the color mapping learning of bilateral grid. It should be noted that R_{msi} has limited spatial resolution. We argue that downsampling in the spatial dimension will not cause significant degradation since bilateral grid coefficient prediction is carried out at a much lower spatial resolution. Hence, we design the Spectral Perception Self-Attention (SPSA) module which utilizes the R_{msi} to guide R_{rgb} for better bilateral coefficient prediction. R_{msi} is resized to match the spatial resolution of R_{rgb}, which is 256×256 by default in HDRNet. The R_{msi} and R_{rgb} are processed with a stack of strided convolutional layers to extract the low-level features and generate hierarchical feature maps. These feature maps $\mathcal{F}_{R_{msi}}$, $\mathcal{F}_{R_{rgb}}$ are concatenated along the channel dimension to form the cross-spectral feature embedding $\mathcal{F}_R$. Then 1×1 convolutions W_1 and 3×3 depth-wise convolutions W_3 are applied to the $\mathcal{F}_\mathcal{R}$ to generate query Q, key K and value V:

$$Q = W_1^Q W_3^Q \mathcal{F}_\mathcal{R}, K = W_1^K W_3^K \mathcal{F}_\mathcal{R}, V = W_1^V W_3^V \mathcal{F}_\mathcal{R} \qquad (3)$$

The query Q and key K are reshaped into $\hat{Q} \in \mathbb{R}^{HW \times C}$ and $\hat{K} \in \mathbb{R}^{C \times HW}$ so that their dot-product interaction generates a spectral perception map $A \in \mathbb{R}^{C \times C}$, which models the mutual information across the different spectral channels. The importance of different channels of the reshaped value $\hat{V} \in \mathbb{R}^{HW \times C}$ is reweighted by A, and the SPSA module can be formulated as follows:

$$A = softmax(\sigma \hat{K} \cdot \hat{Q}) \tag{4}$$

$$\hat{\mathcal{F}}_{R_{msi}} = \hat{W}_3^{msi}\hat{V} \cdot A + W_3^{msi}\mathcal{F}_{R_{msi}} \tag{5}$$

$$\hat{\mathcal{F}}_{R_{rgb}} = \hat{W}_3^{rgb}\hat{V} \cdot A + W_3^{rgb}\mathcal{F}_{R_{rgb}} \tag{6}$$

The σ is a learnable scaling parameter to control the magnitude of the dot product. The SPSA module serves as the residual learning which fuse the cross-spectral feature $\hat{V} \cdot A$ with the original feature $\mathcal{F}_{R_{msi}}$, $\mathcal{F}_{R_{rgb}}$ by the adaptive weighting of 3×3 convolution $\hat{W}_3^{msi}$, $\hat{W}_3^{rgb}$. The fused outputs $\hat{\mathcal{F}}_{R_{msi}}$, $\hat{\mathcal{F}}_{R_{rgb}}$ of the SPSA module are fed into next layer in a progressive way.

M: **Mixture of Semantic Grid Experts.** Common outdoor scenes can be roughly categorized into several primary types (e.g., sky, building, plant). As illustrated in Fig. 2(b), the plant spectral curves typically feature a reflection peak at around 550 nm, whereas the spectral curve of sky primarily resides within the blue wavelength range. Given these distinct characteristics, a multi-expert design can be introduced to effectively adapt to each material category's color preference. We believe that incorporating the material semantic prior M can offer robust constraints and yield visually pleasing results for tone enhancement. Consequently, we introduce a mixture of semantic grid experts, wherein each category is individually associated with a unique bilateral grid coefficient. For a specific semantic grid expert, the binarized material segmentation of the i-th category M_i is transformed into the probability map Ψ_i with a mapping function comprising two convolutional and deconvolutional layers. Subsequently, a pair of affine transformation parameters $(\alpha, \beta) \in \mathbb{R}^{HW \times C}$ is learned based on the probability map Ψ_i. The (α, β) is then introduced to the Q and K branch to modulate features conditioned on the probability map as follows:

$$\hat{Q} = (1 + \alpha_Q) \cdot Q + \beta_Q, \hat{K} = (1 + \alpha_K) \cdot K + \beta_K \tag{7}$$

In each semantic grid expert $\Phi_i(x, y)$, the material category M_i is introduced in the same way. The bilateral grids of different experts are dynamically fused based on the category-dependent weight coefficient denoted as w_i:

$$\Phi(x, y) = \sum_{i=1}^{N} w_i \Phi_i(x, y) \tag{8}$$

Here, N is the number of material categories, and $\Phi(x, y)$ denotes the final multi-expert bilateral grid. The full-resolution output of reflectance image O_R is obtained through interpolated affine transformations, which is operated based on the bilateral grid $\Phi(x, y)$ and the guidance map g. The g is learned from the R_{rgb} using the original guidance map auxiliary network. With the multiplication of O_R and $\hat{S}$, we get the final 8-bit output of JDM-HDRNet.

5   Experiments

5.1   Implementation Details

In the Mobile-Spec dataset, 80% of the samples are allocated for the training set, while the remaining 20% are designated for the test set. The JDM-HDRNet is trained with a batch size of 4 and a learning rate of 0.0001. Specifically, the joint decomposition model is first trained with cross-entropy loss for both shading and segmentation predictions, then the JDM-HDRNet is trained with MSE loss in a supervised manner. During the training, input images are cropped to the size of 512×512, and these images undergo further downsampling to 256×256 in the low-resolution stream.

Three metrics are employed to validate the effectiveness of our method, including the PSNR, SSIM [35], and $\triangle E^*$. $\triangle E^*$ measures the perceptual difference between two colors in the CIELAB space, a smaller $\triangle E^*$ value indicates better color accuracy or image quality. For the evaluation of the material segmentation, we adopt the standard metric as mean Intersection-over-Union (mIoU).

Table 1. Performance evaluation of HDRNet with shading S^* and reflectance prior R^*. Ablation studies are conducted on the spectral channel and spatial resolution.

(a) Performance evaluation of HDRNet and HDRNet with shading prior (HDRNet+S^*).

	PSNR↑	SSIM↑	$\triangle E^*$ ↓		PSNR↑	SSIM↑	$\triangle E^*$ ↓
HDRNet	27.75	0.939	5.12	HDRNet+S^*	28.68	0.957	4.29

(b) Ablation study of reflectance prior R^* on channels with different spectral ranges.

	PSNR↑	SSIM↑	$\triangle E^*$ ↓
baseline	28.68	0.957	4.29
400-520 nm	28.74	0.957	4.00
520-640 nm	29.09	0.964	3.82
640-760 nm	29.19	0.962	3.92
400-760 nm	29.24	0.967	3.72
760-1000 nm	29.07	0.965	3.88
400-1000 nm	29.68	0.968	3.55

(c) Ablation study of reflectance prior R^* on the spatial resolution, which is ranging from 1×1 to 256×256.

	PSNR↑	SSIM↑	$\triangle E^*$ ↓
baseline	28.68	0.957	4.29
1×1	29.05	0.963	4.06
4×4	29.21	0.963	3.80
16×16	29.68	0.968	3.55
64×64	29.81	0.969	3.47
256×256	29.78	0.967	3.49

Table 2. Performance evaluation of HDRNet with material semantic prior M^*.

(a) The effectiveness of incorporating extra Lr-MSI in the joint RGB-Spectral decomposition model for the material segmentation task.

	building	plant	sky	trunk	road	mIoU
RGB	81.99	87.63	94.84	10.96	83.87	71.86
+LrMSI	87.53	91.01	95.87	31.14	89.11	78.93

(b) Ablation study of M^* on the number of material categories.

	1	2	4	6
PSNR↑	27.75	27.98	28.47	29.11
SSIM↑	0.939	0.944	0.954	0.964
$\triangle E^*$ ↓	5.12	5.35	4.69	4.28

5.2 Effectiveness of S^*, R^*, M^* Priors

To avoid the impact of inaccurate decomposition, we begin by systematically integrating the ideal values of these priors (denoted as S^*, R^*, M^*) individually into the HDRNet baseline for validation.

Shading Prior S^*. Table 1(a) illustrates that the HDRNet baseline attains a PSNR of 27.75 dB, while the separation of the shading component (S^*) elevates the PSNR from 27.75 dB to 28.68 dB. Notably, the architecture of HDR-Net remains unchanged in this sub-experiment. This finding demonstrates that transforming from RGB into reflectance space of input images with S^* is a simple yet effective design for the tone enhancement task, which exhibits enhanced adaptability in diverse high dynamic range scenes.

Reflectance Prior R^*. Given the practical limitations of spectral imaging capability on mobile devices, we conduct ablation studies to examine the spectral and spatial resolution configurations of Lr-MSI. We first maintain the spatial resolution at the default setting of 16×16 and extract channels of different spectral ranges. The Lr-MSI has ten spectral channels, ranging from 400 nm to 1000 nm, with each channel featuring a 60 nm bandwidth. Notably, 400–760 nm encompasses six visible wavelength channels, while 760–1000 nm includes four near-infrared channels. As observed in Table 1(b), the closer to the wavelength of red channel, the higher PSNR it will achieve. The possible explanation lies in the composition of our Mobile-Spec dataset, where a larger proportion of the area is covered by the sky and plant, while the presence of red objects is comparatively limited. Consequently, the HDRNet baseline tends to prioritize the blue and green wavelengths over the red wavelength during the training. Furthermore, the combination of all ten channels (400–1000 nm) achieves the best results of 29.68 dB. In our view, fine-grained spectral information enhances bilateral grids with heightened color perception capabilities and compensates for the imbalanced learning of different color channels. Then we maintain the number of channels at ten, and the spatial resolution is resized from 1×1 to 256×256. As depicted in Table 1(c), there is a trend of improved performance in evaluation metrics with higher spatial resolution. Considering the inherent constraint of spatial resolution in Lr-MSI on mobile devices, we contend that the optimal parameter configuration for Lr-MSI may align to the spatial resolution of the bilateral grid (16×16), which is the 8x downsampling corresponding to the spatial resolution of low-resolution stream input (256×256). Our experimental results suggest that the 16×16 configuration yields satisfactory results, beyond which increasing the spatial resolution leads to diminishing marginal utility.

Material Semantic Prior M^*. Table 2(a) demonstrates that incorporating the Lr-MSI branch into the joint RGB-Spectral decomposition model enhances the mIoU of material segmentation from 71.86% to 78.93%. Despite Lr-MSI's notably lower spatial resolution compared to the 16-bit, it exhibits superior material identification capability owing to its fine-grained spectral channels. Subsequently, we conduct ablation experiments to evaluate the number of material categories. The category number of the original HDRNet baseline is set to one. In the two-category set, the sky, plant, and trunk are grouped into one category, while building, road, and others form another. For the four-category set, the plant and trunk, as well as building and road, are each combined into single categories, whereas the sky and others are treated as separate, independent categories. By refining the division of material categories, the mixture of semantic grid experts becomes more specialized in enhancing tones for specific materials. Table 2(b) demonstrates improved quantitative results across three evaluation metrics with the inclusion of six semantic grid experts.

5.3   Ablation Study of JDM-HDRNet

Table 3(a) presents the ablation study incorporating ideal values S^*, R^*, M^* into HDRNet (denoted as JDM-HDRNet*). The collaboration of shading S and reflectance R priors leads to a substantial decrease in $\triangle E^*$ from 5.12 to 3.55 in the CIELAB color space. Moreover, the design of the mixture of semantic grid experts enables specific experts to focus on individual material categories, resulting in a PSNR of 30.14 dB. The above experiments are conducted with ideal priors. Subsequently, the predictions for S, R, M from joint RGB-Spectral decomposition model are integrated with HDRNet (denoted as JDM-HDRNet). Table 3(b) shows that the JDM-HDRNet exhibits comparable performance across three metrics in the S set, indicating that the localized brightness adaptation module demonstrates adaptive adjustment capabilities in response to shading image degradation. In addition, we observe that the performance of the reflectance (R) set remains relatively unaffected, achieving an improvement of 0.98 dB. Because both the Lr-MSI and 16-bit RGB images have undergone consistent transformation into the reflectance space using the same S from the joint decomposition model. Regarding the material semantic prior (M), JDM-HDRNet dynamically fuses different grid experts at the bilateral grid level to avoid the inaccurate segmentation of M. Overall, JDM-HDRNet achieves a gain of 2.08 dB compared to the HDRNet baseline, which demonstrates that the low resolution spectral information can be effectively leveraged by the joint decomposition model.

Table 3. Ablation study of the proposed JDM-HDRNet on the S, R and M priors.

(a) Ablation study on the ideal values of S^*, R^* and M^* priors.

	S^*	R^*	M^*	PSNR↑	SSIM↑	$\triangle E^*$ ↓
HDRNet				27.75	0.939	5.12
JDM-HDRNet*	✓			28.68	0.957	4.29
	✓	✓		29.68	0.968	3.55
	✓	✓	✓	30.14	0.972	3.44

(b) Ablation study on S, R and M priors predicted from the joint decomposition model.

	S	R	M	PSNR↑	SSIM↑	$\triangle E^*$ ↓
HDRNet				27.75	0.939	5.12
JDM-HDRNet	✓			28.59	0.953	4.27
	✓	✓		29.57	0.966	3.62
	✓	✓	✓	29.83	0.967	3.60

Table 4. Comparisons of our JDM-HDRNet with previous enhancement methods.

	DPE	CSRNet	3D LUT	CLUT	SepLUT-S	SepLUT-L	4D LUT	HDRNet	UPE	Ours
PSNR↑	22.81	26.34	27.52	27.30	27.57	28.08	27.77	27.75	28.19	**29.83**
SSIM↑	0.806	0.923	0.926	0.938	0.933	0.944	0.935	0.939	0.946	**0.967**
$\triangle E^*$ ↓	11.06	6.44	5.39	4.63	5.37	4.26	4.32	5.12	4.79	**3.60**

5.4   Comparisons with Previous Methods

Quantitative Comparisons. We compare JDM-HDRNet with previous color transform-based methods for tone enhancement. In Table 4, DPE [6] exhibits inferior performance compared to other methods, attributed to its unpaired setting, leading to output images affected by undesired artifacts introduced by the generative adversarial network. Among the color transform functions, the MLPs based CSRNet [14] exhibits a slightly lower PSNR of 26.34 dB. In contrast, the 3D LUTs based methods, such as 3D LUT [39], CLUT [41], SepLUT-S [36], and 4D LUT [24], demonstrate the similar PSNR ranging from 27.30 dB to 27.77 dB. Notably, SepLUT-L [36] with larger parameters exhibits a superior performance because it simultaneously takes advantage of both 1D and 3D LUTs for enhancement, the 1D LUT adjusts image contrast to achieve a more uniform distribution in an image-adaptive manner. Additionally, UPE [34] estimates an image-to-illumination mapping with constraints of smoothness loss, which is exclusively designed to enhance images under diverse lighting conditions, and it achieves good performance in handling high dynamic range scenes in the Mobile-Spec dataset. Previous methods face challenges in learning accurate tone enhancement for HDR scenes, while our priors-guided JDM-HDRNet outperforms previous methods by effectively leveraging S, R, M priors predicted from the joint RGB-Spectral decomposition model to provide explicit guidance.

Qualitative Comparisons. Figure 5 presents qualitative comparisons of two representative examples. In the first row, JDM-HDRNet exhibits enhanced contrast and a wider dynamic range in the plant area, while other methods appear comparatively dim. In the second row, JDM-HDRNet shows reduced color deviation with the ground truth in the wall area, indicating the beneficial role of

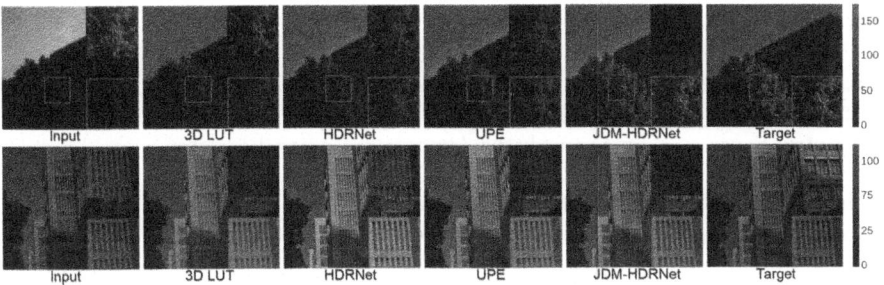

Fig. 5. Qualitative comparisons on the Mobile-Spec dataset. The error maps displayed at the top-right corner illustrate the differences with ground truth for each image.

spectral information integration in improving photographic realism. In summary, JDM-HDRNet, guided by S, R and M priors, achieves more accurate and pleasing colors for high dynamic range scenes.

Limitations. The spatial resolution of Lr-MSI in this study is fixed at 16×16, a setting not yet achievable on commercial smartphones. Increasing the spatial resolution would improve the accuracy of the joint decomposition model. Further exploration is required to address the constrained spectral imaging capabilities in the context of image enhancement and related applications on mobile devices.

6   Conclusion

We have investigated the effectiveness of Lr-MSI in the mobile ISP pipeline for enhancing tone enhancement. We first construct the high-quality Mobile-Spec, a pioneering dataset comprising both aligned spectral and RGB images. Then we analyze the role of Lr-MSI from the perspective of spectral image intrinsic decomposition. Shading, reflectance and material semantic priors are derived using the joint decomposition model to address the inherent complexity of spectral images and the constraints of spectral imaging capabilities. Subsequently, JDM-HDRNet is meticulously designed under the explicit guidance of above priors, which exhibits superior performance than the competitive methods. Our dataset and method are expected to attract further research into this area and explore the untapped potential of spectral information on mobile devices.

Acknowledgments. This research was supported by National Key Research and Development Program of China (2023YFF0713300), National Natural Science Foundation of China under Grant (62071216), Leading Technology of Jiangsu Basic Research Plan under Grant (BK20192003), and National Research Foundation Singapore Competitive Research Program (award number CRP29-2022-0003). We thank the device support from NJU-ZPTECH University-Enterprise Joint Laboratory for Computational Spectral Imaging.

References

1. Gaiasky-mini hyperspectral imaging camera. https://www.dualix.com.cn/en/Goods/desc/id/123/aid/954.html
2. Spectral Sensing — ams.com. https://ams.com/spectral-sensing
3. Barrow, H., Tenenbaum, J., Hanson, A., Riseman, E.: Recovering intrinsic scene characteristics. Comput. vis. Syst. **2**(3–26), 2 (1978)
4. Cai, J., Gu, S., Zhang, L.: Learning a deep single image contrast enhancer from multi-exposure images. IEEE Trans. Image Process. **27**(4), 2049–2062 (2018)
5. Chen, X., et al.: Intrinsic decomposition from a single spectral image. Appl. Opt. **56**(20), 5676–5684 (2017)
6. Chen, Y.S., Wang, Y.C., Kao, M.H., Chuang, Y.Y.: Deep photo enhancer: unpaired learning for image enhancement from photographs with GANs. In: Proceedings of the IEEE Conference on Computer Vision and Pattern Recognition, pp. 6306–6314 (2018)
7. Cheng, Z., Zheng, Y., You, S., Sato, I.: Non-local intrinsic decomposition with near-infrared priors. In: Proceedings of the IEEE/CVF International Conference on Computer Vision, pp. 2521–2530 (2019)
8. Erba, I., et al.: Computational color constancy beyond RGB images: multispectral and temporal extensions (2024)
9. Finlayson, G.D., Zhu, Y.: Designing color filters that make cameras more colorimetric. IEEE Trans. Image Process. **30**, 853–867 (2020)
10. Fu, X., Ying, Y.: Food safety evaluation based on near infrared spectroscopy and imaging: a review. Crit. Rev. Food Sci. Nutr. **56**(11), 1913–1924 (2016)
11. Gharbi, M., Chen, J., Barron, J.T., Hasinoff, S.W., Durand, F.: Deep bilateral learning for real-time image enhancement. ACM Trans. Graph. (TOG) **36**(4), 1–12 (2017)
12. Gijsenij, A., Gevers, T., Van De Weijer, J.: Computational color constancy: survey and experiments. IEEE Trans. Image Process. **20**(9), 2475–2489 (2011)
13. Glatt, O., et al.: Beyond RGB: a real world dataset for multispectral imaging in mobile devices. In: Proceedings of the IEEE/CVF Winter Conference on Applications of Computer Vision, pp. 4344–4354 (2024)
14. He, J., Liu, Y., Qiao, Yu., Dong, C.: Conditional sequential modulation for efficient global image retouching. In: Vedaldi, A., Bischof, H., Brox, T., Frahm, J.-M. (eds.) ECCV 2020. LNCS, vol. 12358, pp. 679–695. Springer, Cham (2020). https://doi.org/10.1007/978-3-030-58601-0_40
15. He, Q., Wang, R.: Hyperspectral imaging enabled by an unmodified smartphone for analyzing skin morphological features and monitoring hemodynamics. Biomed. Opt. Express **11**(2), 895–910 (2020)
16. He, Q., Wang, R.K.: Analysis of skin morphological features and real-time monitoring using snapshot hyperspectral imaging. Biomed. Opt. Express **10**(11), 5625–5638 (2019)
17. Huang, Q., et al.: Multispectral image intrinsic decomposition via subspace constraint. In: Proceedings of the IEEE Conference on Computer Vision and Pattern Recognition, pp. 6430–6439 (2018)
18. Jiang, J., Liu, D., Gu, J., Süsstrunk, S.: What is the space of spectral sensitivity functions for digital color cameras? In: 2013 IEEE Workshop on Applications of Computer Vision (WACV), pp. 168–179. IEEE (2013)
19. Karaimer, H.C., Brown, M.S.: A software platform for manipulating the camera imaging pipeline. In: Leibe, B., Matas, J., Sebe, N., Welling, M. (eds.) Computer

Vision – ECCV 2016: 14th European Conference, Amsterdam, The Netherlands, October 11–14, 2016, Proceedings, Part I, pp. 429–444. Springer International Publishing, Cham (2016). https://doi.org/10.1007/978-3-319-46448-0_26

20. Kim, S.J., Lin, H.T., Lu, Z., Süsstrunk, S., Lin, S., Brown, M.S.: A new in-camera imaging model for color computer vision and its application. IEEE Trans. Pattern Anal. Mach. Intell. **34**(12), 2289–2302 (2012)

21. Kim, Y.T.: Contrast enhancement using brightness preserving bi-histogram equalization. IEEE Trans. Consum. Electron. **43**(1), 1–8 (1997)

22. Land, E.H., McCann, J.J.: Lightness and retinex theory. Josa **61**(1), 1–11 (1971)

23. Lindholm, V., et al.: Differentiating malignant from benign pigmented or non-pigmented skin tumours-a pilot study on 3D hyperspectral imaging of complex skin surfaces and convolutional neural networks. J. Clin. Med. **11**(7), 1914 (2022)

24. Liu, C., Yang, H., Fu, J., Qian, X.: 4D LUT: learnable context-aware 4D lookup table for image enhancement. IEEE Trans. Image Process. **32**, 4742–4756 (2023)

25. Long, J., Shelhamer, E., Darrell, T.: Fully convolutional networks for semantic segmentation. In: Proceedings of the IEEE Conference on Computer Vision and Pattern Recognition, pp. 3431–3440 (2015)

26. Lowe, D.G.: Distinctive image features from scale-invariant keypoints. Int. J. Comput. Vision **60**(2), 91–110 (2004). https://doi.org/10.1023/B:VISI.0000029664.99615.94

27. Van der Maaten, L., Hinton, G.: Visualizing data using t-SNE. J. Mach. Learn. Res. **9**(11) (2008)

28. Mantiuk, R., Daly, S., Kerofsky, L.: Display adaptive tone mapping. In: ACM SIGGRAPH 2008 papers, pp. 1–10 (2008)

29. McGonigle, A.J., et al.: Smartphone spectrometers. Sensors **18**(1), 223 (2018)

30. Ng, P.C., Chi, Z., Verdie, Y., Lu, J., Plataniotis, K.N.: Hyper-skin: a hyperspectral dataset for reconstructing facial skin-spectra from RGB images. Adv. Neural Inf. Process. Syst. **36** (2024)

31. Rao, S., Huang, Y., Cui, K., Li, Y.: Anti-spoofing face recognition using a metasurface-based snapshot hyperspectral image sensor. Optica **9**(11), 1253–1259 (2022)

32. e Silva, G.M., et al.: Smartphone-based spectrometry system as a prescreening assessment of copper and iron for real time control of water pollution. J. Environ. Manag. **323**, 116214 (2022)

33. Thomas, J.B.: Illuminant estimation from uncalibrated multispectral images. In: 2015 Colour and Visual Computing Symposium (CVCS), pp. 1–6. IEEE (2015)

34. Wang, R., Zhang, Q., Fu, C.W., Shen, X., Zheng, W.S., Jia, J.: Underexposed photo enhancement using deep illumination estimation. In: Proceedings of the IEEE/CVF Conference on Computer Vision and Pattern Recognition, pp. 6849–6857 (2019)

35. Wang, Z., Bovik, A.C., Sheikh, H.R., Simoncelli, E.P.: Image quality assessment: from error visibility to structural similarity. IEEE Trans. Image Process. **13**(4), 600–612 (2004)

36. Yang, C., Jin, M., Xu, Y., Zhang, R., Chen, Y., Liu, H.: SepLUT: separable image-adaptive lookup tables for real-time image enhancement. In: Avidan, S., Brostow, G., Cissé, M., Farinella, G.M., Hassner, T. (eds.) Computer Vision – ECCV 2022, pp. 201–217. Springer Nature Switzerland, Cham (2022). https://doi.org/10.1007/978-3-031-19797-0_12

37. Yang, Z., Albrow-Owen, T., Cai, W., Hasan, T.: Miniaturization of optical spectrometers. Science **371**(6528), eabe0722 (2021)

38. Yuan, L., Sun, J.: Automatic exposure correction of consumer photographs. In: Fitzgibbon, A., Lazebnik, S., Perona, P., Sato, Y., Schmid, C. (eds.) ECCV 2012. LNCS, vol. 7575, pp. 771–785. Springer, Heidelberg (2012). https://doi.org/10.1007/978-3-642-33765-9_55
39. Zeng, H., Cai, J., Li, L., Cao, Z., Zhang, L.: Learning image-adaptive 3D lookup tables for high performance photo enhancement in real-time. IEEE Trans. Pattern Anal. Mach. Intell. **44**(4), 2058–2073 (2020)
40. Zhang, F., You, S., Li, Y., Fu, Y.: HSI-guided intrinsic image decomposition for outdoor scenes. In: Proceedings of the IEEE/CVF Conference on Computer Vision and Pattern Recognition, pp. 313–322 (2022)
41. Zhang, F., Zeng, H., Zhang, T., Zhang, L.: CLUT-Net: learning adaptively compressed representations of 3DLUTs for lightweight image enhancement. In: Proceedings of the 30th ACM International Conference on Multimedia, pp. 6493–6501 (2022)
42. Zheng, Y., Sato, I., Sato, Y.: Illumination and reflectance spectra separation of a hyperspectral image meets low-rank matrix factorization. In: Proceedings of the IEEE Conference on Computer Vision and Pattern Recognition, pp. 1779–1787 (2015)

SpatialFormer: Towards Generalizable Vision Transformers with Explicit Spatial Understanding

Han Xiao[1,2], Wenzhao Zheng[1,3], Sicheng Zuo[1], Peng Gao[2], Jie Zhou[1], and Jiwen Lu[1(✉)]

[1] Tsinghua University, Beijing, China
zsc23@mails.tsinghua.edu.cn, {jzhou,lujiwen}@tsinghua.edu.cn
[2] Shanghai AI Laboratory, Shanghai, China
{xiaohan,gaopeng}@pjlab.org.cn
[3] UC Berkeley, Berkeley, China

Abstract. Vision transformers have demonstrated promising results and become core components in many tasks. Most existing works focus on context feature extraction and incorporate spatial information through additional positional embedding. However, they only consider the local positional information within each image token and cannot effectively model the global spatial relations of the underlying scene. To address this challenge, we propose an efficient vision transformer architecture, SpatialFormer, with explicit spatial understanding for generalizable image representation learning. Specifically, we accompany the image tokens with adaptive spatial tokens to represent the context and spatial information respectively. We initialize the spatial tokens with positional encoding to introduce general spatial priors and augment them with learnable embeddings to model adaptive spatial information. For better generalization, we employ a decoder-only overall architecture and propose a bilateral cross-attention block for efficient interactions between context and spatial tokens. SpatialFormer learns transferable image representations with explicit scene understanding, where the output spatial tokens can further serve as enhanced initial queries for task-specific decoders for better adaptations to downstream tasks. Extensive experiments on image classification, semantic segmentation, and 2D/3D object detection tasks demonstrate the efficiency and transferability of the proposed SpatialFormer architecture. Code is available at https://github.com/Euphoria16/SpatialFormer.

1 Introduction

Feature extraction is at the core of computer vision and has been dominated by convolutional neural networks (CNNs) since the deep learning era. The

---

H. Xiao and W. Zheng—Equal contributions.

Supplementary Information The online version contains supplementary material available at https://doi.org/10.1007/978-3-031-72624-8_3.

A. Leonardis et al. (Eds.): ECCV 2024, LNCS 15071, pp. 37–54, 2025.
https://doi.org/10.1007/978-3-031-72624-8_3

monopoly has been broken by vision transformers (ViTs), which patchify each image into a sequence of tokens and then process them using alternating self-attention and multilayer perception (MLP) operations. ViTs are capable of modeling long-range global dependencies among tokens with great flexibility and thus empower the state-of-the-art performance for various tasks, such as image classification [29,41,42,50], object detection [4,65], and semantic segmentation [34,46,59].

Different from CNNs which implicitly encourage translation invariance by using local receptive fields, ViTs introduce less inductive bias by leveraging the self-attention mechanism to incorporate long-range relations between patch tokens. Recent methods improve the original ViT by designing various token interaction modules with better efficiency [8,12,29,52], yet most of them still focus on modeling the interactions between the image tokens, with little efforts on the spatial information incorporation. While existing methods employ the positional embeddings added directly to each token or incorporated into attention layers [14,29,42], these additional positional embeddings are designed primarily to inject the knowledge of local position to each token. This design constrains the modeling of the broader spatial scene, posing challenges in generalization to downstream tasks. For instance, on tasks demanding fine-tuning on high-resolution images, most ViTs require the upsampling of positional embeddings through bicubic interpolation, resulting in significant information loss. Furthermore, when adapting to 3D tasks, the learned features only capture 2D context information from the images, lacking the perceptual capacity for 3D scenes [61]. Although introducing spatial queries in task-specific decoders [4,24,45] can partially alleviate this issue, the image encoder still falls short in learning meaningful image features. This hampers task performance due to the restricted perception of the spatial scene. In contrast, we primarily focus on designing an efficient image backbone capable of directly learning spatial-aware image features that facilitate transferability to downstream tasks.

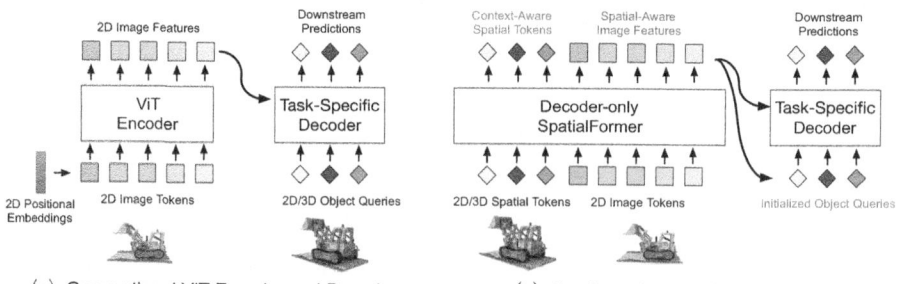

(a) Conventional ViT Encoder and Decoder   (b) Our Decoder-only SpatialFormer

Fig. 1. Motivation of our SpatialFormer. (a) Conventional ViT encoder has limitations in effectively modeling spatial scenes, relying on task-specific decoders to introduce spatial priors for downstream tasks. (b) Our SpatialFormer learns spatial-aware image representations through the incorporation of adaptive spatial tokens, which can serve as enhanced initial queries for task-specific decoders.

In this paper, we propose SpatialFormer, a spatial-aware vision transformer model as an effective solution, as shown in Fig. 1. In addition to the image tokens, we incorporate a set of spatial tokens to capture the spatial scene information. We initialize each spatial token with positional encoding to introduce general spatial prior and further incorporate learnable embeddings to encode adaptive spatial information. These spatial tokens provide essential spatial prior to guide the spatial-aware context feature learning from the images. Using the spatial tokens as queries, we perform cross-attention to transfer context information to obtain enhanced spatial representation. We employ a decoder-only architecture for more generalizable image backbones and propose bilateral cross-attention to enable information flow between context and spatial tokens. Notably, the adaptable representation of spatial tokens enables seamless transfer to images of varying sizes or those augmented with 3D information. The updated spatial tokens can also serve as enhanced initial queries for task-specific adaptations, facilitating the transfer of knowledge acquired from image pre-training to downstream tasks. Extensive experiments on ImageNet-1K [11] for image classification, MS-COCO [25] for 2D object detection, ADE20K [63] for 2D semantic segmentation, and nuScenes [2] for 3D object detection verify the effectiveness, efficiency, and generalization ability of the proposed SpatialFormer architecture.

2   Related Work

Vision Transformer. Since Vaswani et al. [40] first proposed the transformer architecture for machine translation, it has conquered almost all NLP tasks and become the new state-of-the-art model in the language domain. Inspired by its great success, Dosovitskiy et al. [14] introduced the transformer to computer vision by partitioning an image into a sequence of non-overlapping patches as input. Further works then designed more efficient self-attention modules and adopted a multi-stage pyramid architecture to gradually process tokens of lower resolutions [8,12,29,31,52]. Despite many efforts being expended on processing inputs of different formulations, there has been limited exploration into the integration of spatial information. Most existing ViTs resort to additional positional embedding to inject the knowledge of local position into each image token. Absolute positional embeddings are first introduced to be added into each token after the image patchfication step [14,40]. Relative positional embeddings are further proposed to directly incorporate relative positional information into the attention matrix [15,27,29]. However, the acquired positional bias proves insufficient in modeling the broader spatial scene, thereby posing challenges in generalizing to downstream tasks with varying image resolutions and 3D spatial information. Differently, the proposed SpatialFormer adopts a decoder-only architecture to efficiently perform interactions between image tokens and spatial tokens to obtain spatial-aware image representations.

Learning Spatial Information from Images. Learning spatial information from images is a long-standing problem in computer vision, particularly crucial for 3D perception tasks. Equipped with large pretrained image models, vision-based methods have achieved comparable performance to LiDAR-based methods in various 3D perception tasks such as 3D object detection [24,28,54,56] and 3D occupancy prediction [19,21,37,47,58,60,67]. With cameras being much cheaper than 3D sensors like LiDAR, vision-based scene perception has attracted intense attention and is increasingly important in real-world applications including autonomous driving [24,56,57,62]. Existing works follow a conventional paradigm, which first uses image backbones to extract 2D image features and then incorporate the spatial information to lift the 2D feature into the 3D space. DETR3D [45] employs 3D object queries and performs cross-attention to iteratively refine them to decode 3D object information. BEVFormer [24] and TPV-Former [20] further improved it using deformable cross-attention to obtain bird's eye view and tri-perspective view representations, respectively, to model the 3D scene. Different from vision-based scene perception, our objective is to design an efficient image backbone that directly learns spatial-aware image features encoding the 3D scene information when adapting to 3D perception tasks.

3   Proposed Approach

3.1   Learning Spatial-Aware Image Representation

Vision transformers (ViTs) process image patches as input, capturing their long-range relationships within the context feature space. While most ViTs incorporate additional positional embeddings to impart local positional awareness to each image token, this approach only captures limited positional information of the input 2D pixels during pre-training. As depicted in Fig. 1, it falls short of effectively modeling the spatial information inherent in the underlying scene, relying on task-specific decoders to introduce spatial priors for downstream tasks.

To address this, we propose SpatialFormer, an efficient architecture with adaptive spatial tokens to directly model the underlying 2D/3D scene from images. We leverage a set of adaptive spatial tokens to represent the spatial scene and decode the spatial information through interactions between spatial tokens and image tokens. To facilitate efficient interactions, we modify conventional vision transformer decoder blocks into bilateral cross-attention (BCA) blocks, as depicted in Fig. 2. By introducing this novel design of adaptive spatial tokens and a decoder-only architecture with BCA blocks, we enhance the ability to learn generalizable representations of the 2D/3D underlying scene from images.

Adaptive Spatial Tokens. To incorporate the 2D/3D spatial scene information, we introduce a set of adaptive spatial tokens capable of generalizing to various types of input images. Specifically, for a given input image, we initiate the position of spatial tokens by sampling a grid of $N \times N$ spatial points, denoted as $P = \{p_i \in \mathcal{R}^m, i = 1, 2, ..., N^2\}$.

For a common single-view image, we simply employ the pixel coordinate:

$$p_i = (u, v)^T \in \mathcal{R}^2. \tag{1}$$

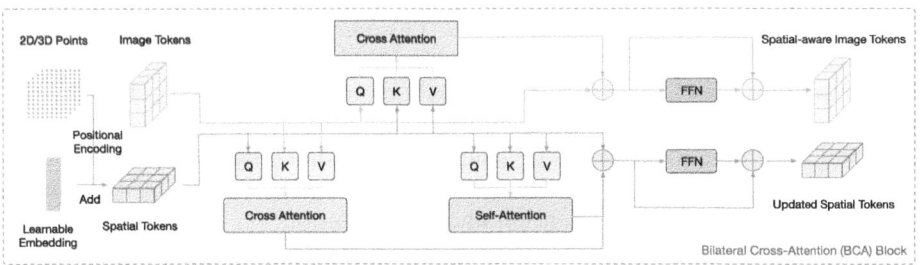

Fig. 2. Illustration of the bilateral cross-attention (BCA) block. We simultaneously perform image-to-spatial and spatial-to-image cross-attention, along with self-attention among spatial tokens in parallel to achieve low latency.

For multi-view images, we generate 3D coordinates $p_i = (x, y, z)^T \in \mathcal{R}^3$ within the real-world scene. The process involves lifting the 2D pixel to a 3D spatial point by introducing a depth dimension orthogonal to the image plane. By discretizing the camera frustum space and sampling candidate depth values along the axis, we obtain corresponding 3D points as $\{(u \times d_j, v \times d_j, d_j)^T\}_{j=1}^D$, where $\{d_j\}_{j=1}^D$ represents the set of candidate depth values. Subsequently, these pixel coordinates are transformed into 3D world coordinates using the corresponding camera intrinsic matrix K and extrinsic matrix E:

$$p_i = (x, y, z)^T = EK\{(u \times d_j, v \times d_j, d_j)^T\}_{j=1}^D. \tag{2}$$

We then employ the mapping function γ to transform the points coordinates into high-dimensional fourier features $\gamma(p) \in \mathcal{R}^{N^2 \times C}$ following [36]:

$$\gamma(P) = \{[\cos(2\pi \mathcal{S} p_i'), \sin(2\pi \mathcal{S} p_i')]^T, i = 1, 2, ..., N^2\}, \tag{3}$$

where $p_i' \in \mathcal{R}^m$ denotes the normalized coordinate in the $[-1, 1]$ of the spatial points (with $m = 2$ for 2D pixels and $m = 3$ for 3D points, respectively). Each entry in $\mathcal{S} \in \mathcal{R}^{C/2 \times m}$ is sampled from a Gaussian distribution $\mathcal{N}(0, \sigma^2)$.

Subsequently, we utilize the obtained positional encodings, summed with learnable embeddings, to initialize the corresponding spatial tokens:

$$\mathbf{Z}_S^0 = \gamma(P) + \mathbf{Q_p}, \tag{4}$$

where $\mathbf{Q_p}$ is the learnable embedding designed to enhance the representation capacity of the spatial tokens. By adopting the positional encoding format, our spatial tokens can effectively represent both 2D and 3D spatial scenes. This facilitates a versatile interaction with the image tokens in the subsequent blocks.

Bilateral Cross-Attention Block. To enable the efficient interaction between the spatial tokens and image tokens, we customize the conventional vision transformer decoder layers into Bilateral Cross-Attention (BCA) blocks. In the proposed BCA block, spatial tokens are first enhanced through cross-attention with the context feature from the image tokens. This mechanism enables the guidance of spatial tokens by context features, facilitating the acquisition of meaningful representations. We then perform self-attention among the spatial tokens to capture the spatial clues within the underlying spatial scene, followed by LayerNorm (LN) and feed-forward networks to generate the updated spatial tokens. For the l-th decoder block, we update the spatial scene representation by the spatial tokens $\mathbf{Z_S}$ using both cross-attention and self-attention:

$$\hat{\mathbf{Z}}_{\mathbf{S}}^{l} = \mathbf{CA}(\mathbf{Q}_{\mathbf{S}}^{l}, \mathbf{K}_{\mathbf{I}}^{l}, \mathbf{V}_{\mathbf{I}}^{l}) + \mathbf{Z}_{\mathbf{S}}^{l},$$
$$\hat{\mathbf{Z}}_{\mathbf{S}}^{l}{}' = \mathbf{SA}(\mathbf{Q}_{\mathbf{S}}^{l}, \mathbf{K}_{\mathbf{S}}^{l}, \mathbf{V}_{\mathbf{S}}^{l}) + \hat{\mathbf{Z}}_{\mathbf{S}}^{l}, \tag{5}$$
$$\mathbf{Z}_{\mathbf{S}}^{l+1} = \mathbf{FFN}(\mathbf{LN}(\hat{\mathbf{Z}}_{\mathbf{S}}^{l}{}')) + \hat{\mathbf{Z}}_{\mathbf{S}}^{l}{}',$$

where $\mathbf{K}_{\mathbf{I}}^{l}, \mathbf{V}_{\mathbf{I}}^{l}, \mathbf{K}_{\mathbf{S}}^{l}, \mathbf{V}_{\mathbf{S}}^{l}$ denote the keys and values obtained from the image tokens $\mathbf{Z}_{I}^{l}$ and input spatial tokens $\mathbf{Z}_{S}^{l}$.

The updated spatial tokens encode the spatial prior essential for the context feature extraction. Therefore, we introduce another cross-attention that utilizes the spatial prior provided by the spatial tokens to update the image tokens. Specifically, given the input image tokens $\mathbf{Z}_{I}^{l}$ in the l-th decoder block, we update them through cross-attention with the spatial tokens:

$$\hat{\mathbf{Z}}_{\mathbf{I}}^{l} = \mathbf{CA}(\mathbf{Q}_{\mathbf{I}}^{l}, \mathbf{K}_{\mathbf{S}}^{l}, \mathbf{V}_{\mathbf{S}}^{l}) + \mathbf{Z}_{\mathbf{I}}^{l}, \quad \mathbf{Z}_{\mathbf{I}}^{l+1} = \mathbf{FFN}(\mathbf{LN}(\hat{\mathbf{Z}}_{\mathbf{I}}^{l})) + \hat{\mathbf{Z}}_{\mathbf{I}}^{l}. \tag{6}$$

By introducing the learnable spatial tokens, our approach can produce spatial-aware features from images in an efficient manner. This approach enables us to obtain comprehensive scene representations directly from images, facilitating the downstream 2D and 3D visual perception task. The updated spatial tokens can function as initial queries for task-specific decoders, transferring the spatial perception capacity acquired from image pre-training to downstream tasks. Further details about task-specific adaptations are elaborated in Sect. 3.3.

3.2   SpatialFormer

We propose a decoder-only vision backbone, SpatialFormer, which leverages the Bilateral Cross-Attention (BCA) block as the fundamental building block for learning spatial-aware image representation efficiently, as shown in Fig. 3. Inspired by the recent state-of-the-art architectures [8,29], we design our models with a four-stage pyramid structure, ensuring that the tiny, small, and base models have comparable parameters and FLOPs to existing ones. We provide the detailed architectures of our SpatialFormer in the supplementary.

In the initial stage, we employ non-overlapping convolutional layers to extract patch embeddings from the input image. In the subsequent stages, we utilize the patch merging module to reduce spatial resolution while simultaneously increasing the dimensions of image features. As for the spatial tokens, the number can be adjusted flexibly according to the target tasks and datasets regardless of the input resolutions. Within the first two stages, we stack the proposed BCA

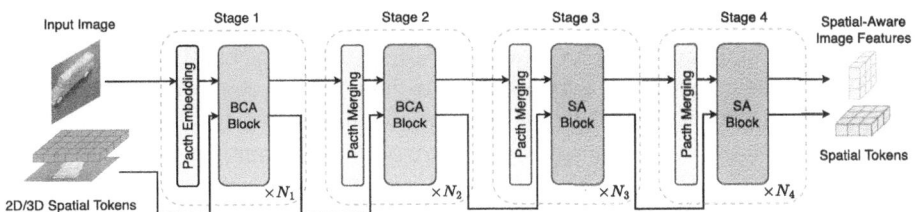

Fig. 3. Illustration of our SpatialFormer architecture. We adopt a four-stage pyramid structure and employ the bilateral attention block in the first two stages. With fewer image tokens in the latter stages, we add self-attention between image tokens to improve the performance without introducing too much computation load.

block, which facilitates both self-attention and cross-attention to update the spatial tokens, while only applying cross-attention to the image tokens. In the third and fourth stages, we introduce additional self-attention to the image tokens. We implement the Self-Attention (SA) block, where we concatenate the spatial tokens and image tokens and apply self-attention to them.

Our SpatialFormer achieves a favorable trade-off between accuracy and computational complexity with the BCA block and SA block. This allows for effective information exchange and integration between the 3D scene tokens and 2D image representations, enhancing the overall understanding of the 3D scene.

3.3   Adaptation to Downstream Tasks

In contrast to traditional vision transformer backbones, our SpatialFormer demonstrates remarkable adaptability for downstream tasks by incorporating spatial tokens to integrate spatial scene information. For instance, in the context of 3D perception tasks, we seamlessly integrate 3D scene knowledge into the backbone model by utilizing 3D positional encodings. Furthermore, the spatial tokens updated by our decoder-only architecture can be directly utilized by task-specific heads as initial queries to acquire dense predictions, enhancing spatial understanding capacities for tasks such as detection and segmentation.

Image Classification. We employ a classification head consisting of a fully connected layer to process the output tokens. The default token number is set to 8×8 to balance accuracy and computational complexity. Adjusting the number of spatial tokens allows for tailoring to specific needs, enabling improvements in accuracy for higher-resolution fine-tuning or reducing computational complexity.

2D Detection and Segmentation. Our SpatialFormer is a general vision backbone compatible with widely used frameworks like Mask-RCNN and also generalizes to multi-scale training by reconfiguring the initial positions of spatial tokens. Moreover, the generated spatial tokens can serve as initial object queries, feeding into multiple transformer decoder layers, thereby directly contributing to the acquisition of dense prediction results.

3D Perception. Traditional vision transformers are typically pretrained on 2D single-view images, resulting in feature extraction that lacks 3D knowledge. In contrast, our SpatialFormer processes multi-view images and encodes 3D scene representation directly by utilizing camera parameters to derive 3D positional encodings, initializing the spatial tokens for enhanced 3D perception capabilities.

4   Experiments

In this section, we conduct extensive experiments to evaluate our SpatialFormer. We provide more dataset and implementation details in the supplementary.

4.1   ImageNet Classification

Experimental Settings. We first evaluate our SpatialFormer on ImageNet [33], which is a widely used benchmark for image classification. ImageNet consists of around 1.2M training images and 50K validation images from 1K different categories. For a fair comparison, we follow the same training recipe as the DeiT model [38]. Specifically, we train the models from scratch for 300 epochs with an input size of 224×224. We use the default data augmentation and regularization strategy. Additionally, we finetune the SpatialFormer backbones at a resolution of 384 × 384 for 30 epochs following previous settings [30].

Results. Table 1 shows that our SpatialFormer outperforms state-of-the-art methods in terms of accuracy and computation efficiency across different model sizes. Specifically, our SpatialFormer-T achieves a top-1 accuracy of 81.5%, outperforming PVTv2-b1 [43] and Shunted-T [32] by 2.8% and 1.7%, respectively. Our small model SpatialFormer-S achieves a competitive result of 83.8% with only 4.8G FLOPs, demonstrating the best trade-offs between computational cost and accuracy among models of similar size. Our largest model, SpatialFormer-B, achieves 84.5% top-1 accuracy with only 50M parameters and 9.8G FLOPs, surpassing models with much higher FLOPs, such as Swin-B [29], ConvNeXt-B [30], and CSWin-B [12]. Furthermore, our SpatialFormer shows promising performance potential when fine-tuned at higher resolution. The performance of SpatialFormer-S and SpatialFormer-B can be further boosted to 84.7% and 85.3%, respectively, outperforming existing architectures with fewer FLOPs.

4.2   Object Detection and Instance Segmentation

Experimental Settings. To evaluate SpatialFormer on downstream dense prediction tasks, we conduct experiments on object detection and instance segmentation using the COCO 2017 dataset [25]. We use the SpatialFormer-S and SpatialFormer-B models pretrained on ImageNet as the backbones using Mask-RCNN with a 1x training schedule and Cascade Mask-RCNN [3] with a 3x training schedule and multi-scale training for further validation. Moreover, our

Table 1. Comparisons on ImageNet classification. We compare the parameters, FLOPs, and accuracy of our SpatialFormer with other state-of-the-art architectures.

Model	Image Size	Params	FLOPs	Top-1 Acc (%)	Top-5 Acc(%)
ResNet-18 [18]	224^2	11.7M	1.8G	69.8	89.1
DeiT-T [38]	224^2	5.7M	1.6G	72.2	91.3
PVT-T [42]	224^2	13.2M	1.9G	75.1	92.4
PVTv2-b1 [43]	224^2	13.1M	3.1G	78.7	-
Shunted-T [32]	224^2	11.5M	2.1G	79.8	-
SpatialFormer-T	224^2	11.6M	2.4G	**81.5**	**95.8**
ResNet-50 [18]	224^2	26M	4.1G	78.3	94.3
DeiT-S [38]	224^2	22M	4.6G	79.9	95.0
Swin-T [29]	224^2	29M	4.5G	81.2	95.5
PVT-S [42]	224^2	25M	3.8G	79.8	95.0
CrossFormer-S [44]	224^2	31M	4.9G	82.5	-
RegionViT-S [5]	224^2	31M	5.3G	82.6	96.1
CSWin-T [12]	224^2	23M	4.3G	82.7	-
UniFormer-S [22]	224^2	24M	4.2G	82.9	96.2
PaCa-Small [16]	224^2	21M	5.4G	83.2	-
CMT-S [17]	224^2	25M	4.0G	83.5	96.6
SpatialFormer-S	224^2	25M	4.8G	**83.8**	96.4
CvT-13 [48]	384^2	20M	16.3G	83.0	-
CaiT-XS24 [39]	384^2	27M	19.3G	84.1	96.9
CSWin-T [12]	384^2	23M	14.0G	84.3	-
SpatialFormer-S	384^2	25M	12.8G	**84.7**	**97.1**
ResNet-152 [18]	224^2	60M	11.6G	81.3	95.5
DeiT-B [38]	224^2	86 M	17.5G	81.8	95.6
Swin-B [29]	224^2	88M	15.4G	83.5	96.5
ConvNeXt-B [30]	224^2	89M	15.4G	83.8	-
CSWin-B [13]	224^2	78M	15.0G	84.2	-
DAT-B [49]	224^2	88M	15.8G	84.0	96.7
UniFormer-B [22]	224^2	50M	8.3G	83.9	96.7
RegionViT-B [5]	224^2	74M	13.6G	83.8	96.1
QFormer-B [55]	224^2	90M	15.7G	84.1	96.8
SpatialFormer-B	224^2	50M	9.8G	**84.5**	**96.8**
Swin-B [29]	384^2	88M	47.0G	84.5	97.0
DAT-B [49]	384^2	88M	49.8G	84.8	97.0
CaiT-S24 [39]	384^2	47M	32.2G	85.1	97.4
SpatialFormer-B	384^2	50M	23.5G	**85.3**	**97.5**

Table 2. Results on COCO for object detection and instance segmentation. The FLOPs are measured on input sizes of 1280×800.

(a) Mask-RCNN Object Detection and Instance Segmentation

Backbone	Params	FLOPs	Schedule	AP^b	AP^b_{50}	AP^b_{75}	AP^m	AP^m_{50}	AP^m_{75}
PVT-S [42]	44M	245G	1x	40.4	62.9	43.8	37.8	60.1	40.3
Swin-T [29]	48M	267G	1x	42.2	64.6	46.2	39.1	61.6	42.0
CrossFormer-S [44]	50M	301G	1x	45.4	68.0	49.7	41.4	64.8	44.6
UniFormer-S [22]	41M	269G	1x	45.6	68.1	49.7	41.6	64.8	45.0
SpatialFormer-S	42M	255G	1x	**47.8**	**70.1**	**52.5**	**42.5**	**66.6**	**45.8**
Swin-S [29]	69M	354G	1x	44.8	66.6	48.9	40.9	63.4	44.2
DAT-S [49]	69M	378G	1x	47.1	69.9	51.5	42.5	66.7	45.4
CrossFormer-B [44]	72M	408G	1x	47.2	69.9	51.8	42.7	66.6	46.2
UniFormer-B [22]	69M	399G	1x	47.4	69.7	52.1	43.1	66.0	46.5
SpatialFormer-B	59M	316G	1x	**49.2**	**71.1**	**54.4**	**43.7**	**67.7**	**47.1**

(b) Cascade Mask-RCNN Object Detection and Instance Segmentation

Backbone	Params	FLOPs	Schedule	AP^b	AP^b_{50}	AP^b_{75}	AP^m	AP^m_{50}	AP^m_{75}
Swin-T [29]	86 M	745G	3x	48.1	67.1	52.2	41.7	64.4	45.0
DAT-T [49]	86 M	750G	3x	49.1	68.2	52.9	42.5	65.4	45.8
Swin-S [29]	107 M	838G	3x	50.4	69.2	54.7	43.7	66.6	47.3
DAT-S [49]	107 M	857G	3x	51.3	70.1	55.8	44.5	67.5	48.1
SpatialFormer-S	82 M	734G	3x	**52.2**	**70.8**	**56.6**	**45.2**	**68.3**	**49.0**
Swin-S [29]	86 M	745G	3x	48.1	67.1	52.2	41.7	64.4	45.0
DAT-S [49]	86 M	750G	3x	49.1	68.2	52.9	42.5	65.4	45.8
Swin-S [29]	107 M	838G	3x	50.4	69.2	54.7	43.7	66.6	47.3
DAT-S [49]	107 M	857G	3x	51.3	70.1	55.8	44.5	67.5	48.1
SpatialFormer-B	98M	812G	3x	**52.9**	**71.3**	**57.4**	**45.9**	**68.9**	**49.9**

(c) Transformer Decoder-Based Object Detection

Framework	Backbone	Two-Stage	Epoch	AP^b	AP^b_{50}	AP^b_{75}	AP^b_S	AP^b_M	AP^b_L
DETR [4]	ResNet-50	×	12	15.5	29.4	14.5	4.3	15.1	26.7
DAB-DETR [26]	ResNet-50	×	12	38.0	60.3	39.8	19.2	40.9	55.4
Dynamic DETR [10]	ResNet-50	×	12	42.9	61.0	46.3	24.6	44.9	54.4
Deformable DETR [66]	ResNet-50	×	12	37.2	55.5	40.5	21.1	40.7	50.5
Deformable DETR [66]	ResNet-50	✓	12	43.7	62.9	47.2	26.7	46.9	57.2
Deformable DETR [66]	Swin-T	×	12	41.9	59.0	44.5	26.4	41.3	45.9
Deformable DETR [66]	Swin-T	✓	12	45.3	64.8	49.0	27.8	48.5	60.6
SpatialFormer	Ours-Small	×	12	43.5	61.8	47.0	27.3	46.3	58.2
SpatialFormer	Ours-Small	✓	12	**46.5**	**64.9**	**50.5**	**29.3**	**49.5**	**61.0**

SpatialFormer can utilize spatial tokens as initial object queries to build a transformer decoder-based detection framework. We duplicate the 8×8 spatial tokens, select the top 150 queries, and stack 6 deformable DETR decoder layers to obtain detection results. The model is trained using a 1x training schedule.

Table 3. Results on ADE20K for semantic segmentation. The FLOPs are measured at the input resolution of 512×2048.

(a) UpperNet Semantic Segmentation					
Backbone	Params	FLOPs	mIoU	mIoU (MS)	mAcc
Swin-T [29]	60M	945G	44.4	45.8	55.6
DAT-T [49]	60M	957G	45.5	46.4	58.0
CrossFormer-S [44]	62M	980G	47.6	48.4	-
SpatialFormer-S	52M	935G	**48.1**	**49.2**	**59.4**
Swin-B [29]	121M	1188G	48.1	49.7	59.1
DAT-B [49]	121M	1212G	49.4	50.5	61.8
CrossFormer-B [44]	84M	1079G	49.2	50.1	-
SpatialFormer-B	73M	1028G	**50.3**	**51.2**	61.8
(b) Transformer Decoder-Based Semantic Segmentation					
Framework	Backbone	Iterations	mIoU	mIoU (MS)	mAcc
MaskFormer [7]	Swin-T	160k	46.7	48.8	-
Mask2Former [6]	Swin-T	160k	47.7	49.6	-
kMaX-DeepLab [53]	ConvNeXt-T	100k	48.3	-	-
SpatialFormer	Ours-Small	160k	**50.0**	**52.8**	**63.1**

Results. We report the performance on COCO object detection and instance segmentation in Table 2. Our SpatialFormer outperforms all the other vision backbones under different model sizes. For Mask-RCNN, SpatialFormer-S outperforms UniFormer-S by 2.2 while SpatialFormer-B outperforms UniFormer-B by 1.8 in terms of AP^b, verifying the advantages of spatial-aware image representation learning. For Cascade Mask-RCNN, SpatialFormer surpasses models with much higher parameters and FLOPs such as Swin-S and DAT-S. Moreover, we observe a significant performance improvement (1.2 of AP^b) for SpatialFormer-S compared to the most competitive transformer decoder-based frameworks. Notably, SpatialFormer surpasses Deformable DETR with ResNet50 and Swin-T in the one-stage setting (43.5% vs 37.2% vs 41.9% in terms of mAP), even without spatial tokens as initialization. This results from the ability of SpatialFormer to capture spatial scene information, providing additional information about the object details. By encoding such spatial information, our model can effectively handle challenging scenarios in object detection.

4.3   Semantic Segmentation

Experimental Settings. We further evaluate the performance of our SpatialFormer on the ADE20K dataset [64] for semantic segmentation. We first adopt the UpperNet [51] framework with the integration of our SpatialFormer backbone. Additionally, we employ the generated spatial tokens as initial object queries to derive segmentation masks using the Mask2Former [6] decoder layers. For all the models, we train them using the AdamW optimizer for 160k iterations, with a batch size of 16 and input image cropped to 512×512. We follow the settings of Swin Transformer [29] and Mask2Former [6] for fair comparison.

Results. We present results using the UpperNet semantic segmentation framework on the ADE20K [64] dataset in Table 3. Our SpatialFormer-S outperforms CrossFormer-S by 0.5 mIoU and our SpatialFormer-B can attain 2.2 higher mIoU than Swin-B, which has much higher FLOPs. Furthermore, SpatialFormer surpasses other decoder-based segmentation frameworks by incorporating spatial tokens. Unlike conventional backbones relying on context feature extraction, SpatialFormer excels in capturing fine-grained spatial information and exhibits strong transferability to transformer-based segmentation architectures.

4.4   3D Object Detection on NuScenes

Experimental Settings. To evaluate the generalization of SpatialFormer for 3D perception tasks, we applied SpatialFormer with to BEVFormer [24] to perform 3D object detection tasks on nuScenes. To verify the effectiveness of 3D scene information incorporation, we experiment with both 2D and 3D types of positional encoding to initialize the spatial tokens. We train the models for 24 epochs, following the other hyperparameter settings of BEVFormer [24].

Results. Table 4 summarizes the detailed comparison results. Our SpatialFormer notably improves the NDS scores of baseline BEVFormer with Swin-T backbone from 0.369 to 0.389 on nuScenes val. This improvement underscores the effectiveness of the proposed spatial tokens in refining the localization accuracy of 3D object detection predictions. Furthermore, we observe our SpatialFormer can further boost the performance to 0.392 when adopting the 3D type of spatial tokens. These experimental results demonstrate the superior capabilities of our SpatialFormer for 3D scene understanding based on multi-view images.

Table 4. 3D object detection results on nuScenes.

Framework	Backbone	Epochs	NDS↑	mAP↑	mATE↓	mASE↓	mAOE↓	mAVE↓	mAAE↓
BEVDepth [23]	ResNet-50	24	0.367	0.315	0.702	0.271	0.621	1.042	0.315
BEVDepth [23]	ResNet-101	24	0.381	0.320	0.682	0.272	0.562	0.997	0.284
BEVFormer [24]	ResNet-50	24	0.354	0.252	0.900	0.294	0.655	0.657	0.216
BEVFormer [24]	Swin-T	24	0.369	0.265	0.898	0.284	0.631	0.594	0.224
SpatialFormer(2D)	Ours-Small	24	0.389	0.290	0.856	0.287	0.582	0.611	0.221
SpatialFormer(3D)	Ours-Small	24	**0.392**	0.297	0.848	0.285	0.610	0.608	0.217

Table 5. Ablation on the number of spatial tokens under various input resolutions of SpatialFormer-S.

#Token	Size	Params	FLOPs	Acc (%)
None	224^2	24M	4.0G	81.3
16	224^2	24M	4.3G	83.0
64	224^2	25M	4.8G	83.8
64	384^2	25M	12.8G	84.7
256	224^2	25M	7.0G	84.1
256	384^2	25M	15.0G	**85.0**

Table 6. Ablation on different architecture designs of patialFormer-S.

Stages	Params	FLOPs	Acc(%)
BCA-BCA-SA-SA	25M	4.8G	83.8
BCA-BCA-BCA-SA	31M	5.2G	84.0
BCA-BCA-BCA-BCA	37M	5.4G	**84.1**

4.5   Quantitive Analysis

Incorporation of Spatial Tokens. To investigate the impact of spatial tokens, we remove the positional encoding from the bilateral decoder layer and explore various numbers of spatial tokens across different input resolutions, as shown in Table 5. Without positional encoding, the spatial tokens devolve into random-initialized queries, lacking any spatial priors and resulting in performance drops. Decreasing the number of spatial tokens to 4×4 yields lower accuracy but reduces FLOPs. Conversely, increasing to 16×16 improves accuracy by incorporating more fine-grained spatial information, albeit at the cost of increased computational complexity. To achieve the best trade-off between performance and computation, we use 8×8 spatial tokens in our experiments as default.

Architecture Design . We further investigate the impact of different architectural designs of our SpatialFormer. We replace the Self-Attention block with our proposed Bilateral Cross-Attention (BCA) block in the third and fourth stages. Results in Table 6 demonstrate that increasing the number of BCA blocks improves performance. This highlights the effectiveness of incorporating more BCA blocks in learning the underlying 3D scene from images. However, it is less efficient compared to the SA block in the last two stages. Hence, we choose to use BCA blocks in the first two stages and SA blocks for the remaining stages.

Table 7. Comparisons with conventional positional embeddings/encodings.

Method	Params	FLOPs	Acc (%)
Baseline	24M	4.0G	81.3
Absolute PE [14]	24M	4.0G	81.7 (+0.4)
Relative PE [29]	25M	4.2G	82.0 (+0.7)
Conditional PE [9]	25M	4.7G	82.1 (+0.8)
Rotary PE [35]	24M	4.5G	81.8 (+0.5)
Dynamic Conv. [1]	25M	4.3G	81.7 (+0.4)
SpatialFormer-S	25M	4.8G	**83.8 (+2.5)**

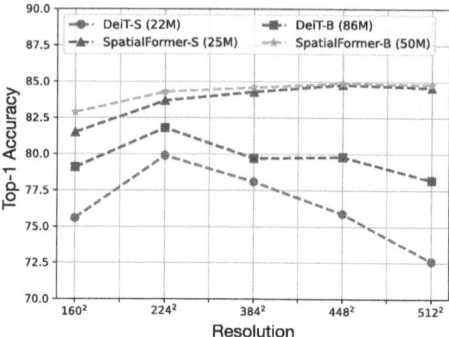

Fig. 4. Effect of different embedding sizes.

4.6 Qualitative Analysis

Attention Maps for a Certain Spatial Token. To better illustrate the interactions between image tokens and spatial tokens, we provide visualizations of attention maps generated from a spatial token in Fig. 5. The Cross-Attention maps showcase interactions between spatial and image tokens, emphasizing the focus on context features. Meanwhile, the Self-Attention maps illustrate dependencies among spatial tokens, revealing a concentration on location-related information. These visualizations demonstrate the ability of learned spatial tokens to discern context and location details through cross and self-attention computations.

Attention Maps for Multi-view Images. To better understand the spatial token, we further provide visualizations of attention maps for multi-view images from nuScenes [2]. As depicted in Fig. 6, the spatial token tends to pay attention to the same object, even in different views. This shows that spatial tokens encoding 3D information establish position correlations between different views and can enhance spatial reasoning capabilities.

Comparisons with Other Positional Embeddings. We provide comparisons with existing positional embedding/encoding (PE) methods in Table 7, including absolute positional embedding [14], relative positional embedding [29], conditional position encoding [9], rotary position embedding [35], and position encoding generated by dynamic convolution layers [1]. We establish a baseline model without spatial tokens, using only learnable embeddings to perform bilateral cross attentions with image tokens. In this configuration, the original spatial tokens degenerate into image feature queries, while maintaining the primary model architectures. We see that our method outperforms existing PEs by a large margin. While some methods explore input-relevant positional information, they only add positional embedding to individual image patches at early

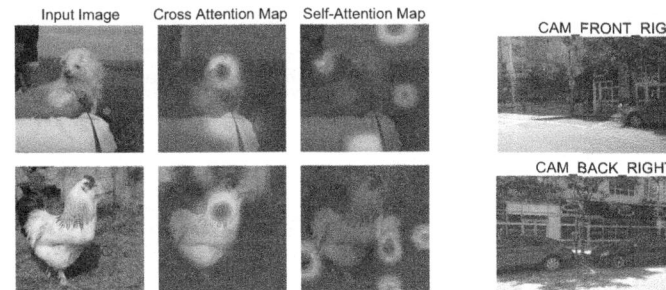

Fig. 5. Visualization of attention maps generated from a fixed spatial token.

Fig. 6. Visualization of attention maps on multi-view images.

stages. SpatialFormer enables a more spatial-aware representation by updating adaptive spatial tokens through bilateral cross-attention with image tokens.

Generalization Ability Without Fine-Tuning. To further verify the generalization ability of our model, we evaluate the transferring ability without fine-tuning. Specifically, we transfer models trained on 224^2 images directly to images of different resolutions without fine-tuning. We see that DeiT suffers significant performance drops on other resolutions while our method can directly generalize to larger image sizes without additional fine-tuning, as demonstrated in Fig. 4. This is because conventional positional embeddings defined on local image patches are inherently task-agnostic and may not effectively leverage spatial information during fine-tuning, resulting in inferior generalization. This motivates the use of adaptive spatial tokens for enhanced generalizability across various downstream tasks, as also demonstrated in Sects. 4.2, 4.3, and 4.4.

5 Conclusion

In this paper, we have introduced a SpatialFormer model to enhance spatial-aware image feature learning and improve generalization to downstream tasks. By incorporating adaptive spatial tokens alongside image tokens, SpatialFormer efficiently captures spatial scene information, effectively addressing limitations in conventional transformers that rely solely on local positional embeddings. Our approach, employing a decoder-only architecture, facilitates efficient interaction between image and spatial tokens, resulting in enhanced spatial representation. The proposed model not only advances standard image classification tasks but also exhibits promising performance across diverse downstream applications, including 2D dense prediction and 3D perception tasks. We hope our work can inspire future research to develop more spatial-aware image backbones.

Acknowledgement. This work was supported in part by the National Key Research and Development Program of China under Grant 2023YFB280690, and in part by the National Natural Science Foundation of China under Grant 62321005, Grant 62336004, and Grant 62125603.

References

1. Baevski, A., Zhou, Y., Mohamed, A., Auli, M.: wav2vec 2.0: a framework for self-supervised learning of speech representations. NeurIPS **33**, 12449–12460 (2020)
2. Caesar, H., et al.: nuScenes: a multimodal dataset for autonomous driving. In: CVPR (2020)
3. Cai, Z., Vasconcelos, N.: Cascade R-CNN: delving into high quality object detection. In: CVPR, pp. 6154–6162 (2018)
4. Carion, N., Massa, F., Synnaeve, G., Usunier, N., Kirillov, A., Zagoruyko, S.: End-to-end object detection with transformers. In: ECCV, pp. 213–229 (2020)
5. Chen, C.F., Panda, R., Fan, Q.: RegionViT: regional-to-local attention for vision transformers. arXiv preprint arXiv:2106.02689 (2021)
6. Cheng, B., Misra, I., Schwing, A.G., Kirillov, A., Girdhar, R.: Masked-attention mask transformer for universal image segmentation. In: CVPR, pp. 1290–1299 (2022)
7. Cheng, B., Schwing, A.G., Kirillov, A.: Per-pixel classification is not all you need for semantic segmentation. In: NeurIPS (2021)
8. Chu, X., et al.: Twins: revisiting the design of spatial attention in vision transformers. In: NeurIPS (2021)
9. Chu, X., Tian, Z., Zhang, B., Wang, X., Shen, C.: Conditional positional encodings for vision transformers. In: ICLR (2022)
10. Dai, X., Chen, Y., Yang, J., Zhang, P., Yuan, L., Zhang, L.: Dynamic DETR: end-to-end object detection with dynamic attention. In: ICCV, pp. 2988–2997 (2021)
11. Deng, J., Dong, W., Socher, R., Li, L.J., Li, K., Fei-Fei, L.: ImageNet: a large-scale hierarchical image database. In: CVPR, pp. 248–255. IEEE (2009)
12. Dong, X., et al.: CSWin transformer: a general vision transformer backbone with cross-shaped windows. arXiv preprint arXiv:2107.00652 (2021)
13. Dong, X., et al.: CSWin transformer: a general vision transformer backbone with cross-shaped windows. In: CVPR, pp. 12124–12134 (2022)
14. Dosovitskiy, A., et al.: An image is worth 16x16 words: transformers for image recognition at scale. In: ICLR (2020)
15. Graham, B., et al.: LeViT: a vision transformer in ConvNet's clothing for faster inference. In: ICCV, pp. 12259–12269 (2021)
16. Grainger, R., Paniagua, T., Song, X., Cuntoor, N., Lee, M.W., Wu, T.: PaCa-ViT: learning patch-to-cluster attention in vision transformers. In: CVPR, pp. 18568–18578 (2023)
17. Guo, J., et al.: CMT: convolutional neural networks meet vision transformers. In: CVPR, pp. 12175–12185 (2022)
18. He, K., Zhang, X., Ren, S., Sun, J.: Deep residual learning for image recognition. In: CVPR, pp. 770–778 (2016)
19. Huang, Y., Zheng, W., Zhang, B., Zhou, J., Lu, J.: SelfOcc: self-supervised vision-based 3D occupancy prediction. In: CVPR (2024)
20. Huang, Y., Zheng, W., Zhang, Y., Zhou, J., Lu, J.: Tri-perspective view for vision-based 3D semantic occupancy prediction. arXiv preprint arXiv:2302.07817 (2023)

21. Huang, Y., Zheng, W., Zhang, Y., Zhou, J., Lu, J.: Gaussianformer: scene as Gaussians for vision-based 3D semantic occupancy prediction. In: ECCV (2024)
22. Li, K., et al.: Uniformer: unifying convolution and self-attention for visual recognition. arXiv preprint arXiv:2201.09450 (2022)
23. Li, Y., et al.: BEVDepth: acquisition of reliable depth for multi-view 3D object detection. arXiv preprint arXiv:2206.10092 (2022)
24. Li, Z., et al.: BEVFormer: learning bird's-eye-view representation from multi-camera images via spatiotemporal transformers. In: ECCV (2022)
25. Lin, T.-Y., et al.: Microsoft COCO: common objects in context. In: Fleet, D., Pajdla, T., Schiele, B., Tuytelaars, T. (eds.) Computer Vision – ECCV 2014: 13th European Conference, Zurich, Switzerland, September 6-12, 2014, Proceedings, Part V, pp. 740–755. Springer International Publishing, Cham (2014). https://doi.org/10.1007/978-3-319-10602-1_48
26. Liu, S., et al.: DAB-DETR: dynamic anchor boxes are better queries for DETR. arXiv preprint arXiv:2201.12329 (2022)
27. Liu, Y., Sangineto, E., Bi, W., Sebe, N., Lepri, B., Nadai, M.: Efficient training of visual transformers with small datasets. NeurIPS **34**, 23818–23830 (2021)
28. Liu, Y., Wang, T., Zhang, X., Sun, J.: PETR: position embedding transformation for multi-view 3D object detection. arXiv preprint arXiv:2203.05625 (2022)
29. Liu, Z., et al.: Swin transformer: hierarchical vision transformer using shifted windows. In: ICCV (2021)
30. Liu, Z., Mao, H., Wu, C.Y., Feichtenhofer, C., Darrell, T., Xie, S.: A convNet for the 2020s. arXiv preprint arXiv:2201.03545 (2022)
31. Lu, J., et al.: SOFT: softmax-free transformer with linear complexity. In: NeurIPS (2021)
32. Ren, S., Zhou, D., He, S., Feng, J., Wang, X.: Shunted self-attention via multi-scale token aggregation. In: CVPR, pp. 10853–10862 (2022)
33. Russakovsky, O., et al.: Imagenet large scale visual recognition challenge. IJCV **115**(3), 211–252 (2015). https://doi.org/10.1007/s11263-015-0816-y
34. Strudel, R., Garcia, R., Laptev, I., Schmid, C.: Segmenter: transformer for semantic segmentation. In: ICCV (2021)
35. Su, J., Ahmed, M., Lu, Y., Pan, S., Bo, W., Liu, Y.: RoFormer: enhanced transformer with rotary position embedding. Neurocomputing **568**, 127063 (2024)
36. Tancik, M., et al.: Fourier features let networks learn high frequency functions in low dimensional domains. NeurIPS **33**, 7537–7547 (2020)
37. Tong, W., et al.: Scene as occupancy. In: ICCV, pp. 8406–8415 (2023)
38. Touvron, H., Cord, M., Douze, M., Massa, F., Sablayrolles, A., Jégou, H.: Training data-efficient image transformers and distillation through attention. In: ICML, pp. 10347–10357 (2021)
39. Touvron, H., Cord, M., Sablayrolles, A., Synnaeve, G., Jégou, H.: Going deeper with image transformers. In: ICCV, pp. 32–42 (2021)
40. Vaswani, A., et al.: Attention is all you need. In: NeurIPS, pp. 5998–6008 (2017)
41. Wang, C., Zheng, W., Zhu, Z., Zhou, J., Lu, J.: OPERA: omni-supervised representation learning with hierarchical supervisions. In: ICCV, pp. 5559–5570 (2023)
42. Wang, W., et al.: Pyramid vision transformer: a versatile backbone for dense prediction without convolutions. In: ICCV (2021)
43. Wang, W., et al.: PVT v2: improved baselines with pyramid vision transformer. Comput. Vis. Media **8**(3), 415–424 (2022). https://doi.org/10.1007/s41095-022-0274-8
44. Wang, W., et al.: CrossFormer: a versatile vision transformer hinging on cross-scale attention. In: ICLR (2023)

45. Wang, Y., Guizilini, V., Zhang, T., Wang, Y., Zhao, H., Solomon, J.M.: DETR3D: 3D object detection from multi-view images via 3D-to-2D queries. In: CoRL (2021)
46. Wang, Y., et al.: End-to-end video instance segmentation with transformers. In: CVPR, pp. 8741–8750 (2021)
47. Wei, Y., Zhao, L., Zheng, W., Zhu, Z., Zhou, J., Lu, J.: Surroundocc: multi-camera 3D occupancy prediction for autonomous driving. In: ICCV, pp. 21729–21740 (2023)
48. Wu, H., et al.: CvT: introducing convolutions to vision transformers. In: CVPR, pp. 22–31 (2021)
49. Xia, Z., Pan, X., Song, S., Li, L.E., Huang, G.: Vision transformer with deformable attention. In: CVPR, pp. 4794–4803 (2022)
50. Xiao, H., Zheng, W., Zhu, Z., Zhou, J., Lu, J.: Token-label alignment for vision transformers. In: ICCV, pp. 5495–5504 (2023)
51. Xiao, T., Liu, Y., Zhou, B., Jiang, Y., Sun, J.: Unified perceptual parsing for scene understanding. In: ECCV, pp. 418–434 (2018)
52. Yang, J., et al.: Focal self-attention for local-global interactions in vision transformers. arXiv preprint arXiv:2107.00641 (2021)
53. Yu, Q. et al.: K-means mask transformer. In: Avidan, S., Brostow, G., Cissé, M., Farinella, G.M., Hassner, T. (eds) Computer Vision – ECCV 2022. ECCV 2022. LNCS, vol 13689, pp. 288–307 (2022) Springer, Cham. https://doi.org/10.1007/978-3-031-19818-2_17
54. Zeng, S., Zheng, W., Lu, J., Yan, H.: Hardness-aware scene synthesis for semi-supervised 3D object detection. TMM (2024)
55. Zhang, Q., Zhang, J., Xu, Y., Tao, D.: Vision transformer with quadrangle attention. TPAMI (2024)
56. Zhang, Y., Zheng, W., Zhu, Z., Huang, G., Zhou, J., Lu, J.: A simple baseline for multi-camera 3D object detection. arXiv preprint arXiv:2208.10035 (2022)
57. Zhang, Y., et al.: BEVerse: unified perception and prediction in birds-eye-view for vision-centric autonomous driving. arXiv preprint arXiv:2205.09743 (2022)
58. Zhao, L., et al.: LowRankOcc: tensor decomposition and low-rank recovery for vision-based 3D semantic occupancy prediction. In: CVPR. pp. 9806–9815 (2024)
59. Zheng, S., et al.: Rethinking semantic segmentation from a sequence-to-sequence perspective with transformers. In: CVPR, pp. 6881–6890 (2021)
60. Zheng, W., Chen, W., Huang, Y., Zhang, B., Duan, Y., Lu, J.: OccWorld: learning a 3D occupancy world model for autonomous driving. In: ECCV (2024)
61. Zheng, W., Lu, J., Jie, Z.: Structural deep metric learning for room layout estimation. In: ECCV (2020)
62. Zheng, W., Song, R., Guo, X., Chen, L.: GenAD: Generative end-to-end autonomous driving. In: ECCV (2024)
63. Zhou, B., Zhao, H., Puig, X., Fidler, S., Barriuso, A., Torralba, A.: Scene parsing through ade20k dataset. In: Proceedings of the IEEE Conference on Computer Vision and Pattern Recognition, pp. 633–641 (2017)
64. Zhou, B., et al.: Semantic understanding of scenes through the ADE20K dataset. IJCV **127**, 302–321 (2019). https://doi.org/10.1007/s11263-018-1140-0
65. Zhu, X., Su, W., Lu, L., Li, B., Wang, X., Dai, J.: Deformable DETR: deformable transformers for end-to-end object detection. In: ICLR (2020)
66. Zhu, X., Su, W., Lu, L., Li, B., Wang, X., Dai, J.: Deformable DETR: deformable transformers for end-to-end object detection. In: ICLR (2021)
67. Zuo, S., Zheng, W., Huang, Y., Zhou, J., Lu, J.: PointOcc: cylindrical tri-perspective view for point-based 3D semantic occupancy prediction. arXiv preprint arXiv:2308.16896 (2023)

OccWorld: Learning a 3D Occupancy World Model for Autonomous Driving

Wenzhao Zheng[1,2], Weiliang Chen[1], Yuanhui Huang[1], Borui Zhang[1],
Yueqi Duan[1], and Jiwen Lu[1(✉)]

[1] Tsinghua University, Beijing, China
{chen-wl20,huangyh22,zhang-br21}@mails.tsinghua.edu.cn,
{duanyueqi,lujiwen}@tsinghua.edu.cn
[2] UC Berkeley, Berkeley, USA
https://wzzheng.net/OccWorld

Abstract. Understanding how the 3D scene evolves is vital for making decisions in autonomous driving. Most existing methods achieve this by predicting the movements of object boxes, which cannot capture more fine-grained scene information. In this paper, we explore a new framework of learning a world model, OccWorld, in the 3D occupancy space to simultaneously predict the movement of the ego car and the evolution of the surrounding scenes. We propose to learn a world model based on 3D occupancy rather than 3D bounding boxes and segmentation maps for three reasons: 1) **expressiveness**. 3D occupancy can describe the more fine-grained 3D structure of the scene; 2) **efficiency**. 3D occupancy is more economical to obtain (e.g., from sparse LiDAR points). 3) **versatility**. 3D occupancy can adapt to both vision and LiDAR. To facilitate the modeling of the world evolution, we learn a reconstruction-based scene tokenizer on the 3D occupancy to obtain discrete scene tokens to describe the surrounding scenes. We then adopt a GPT-like spatial-temporal generative transformer to generate subsequent scene and ego tokens to decode the future occupancy and ego trajectory. Extensive experiments on nuScenes demonstrate the ability of OccWorld to effectively model the driving scene evolutions. OccWorld also produces competitive planning results without using instance and map supervision. Code: https://github.com/wzzheng/OccWorld.

1 Introduction

Autonomous driving has been widely explored in recent years and demonstrated promising results in various scenarios [21,58,67,70]. While LiDAR-based models typically show strong performance and robustness in 3D perception due to its

---

W. Zheng and W. Chen—Equal contributions.

Supplementary Information The online version contains supplementary material available at https://doi.org/10.1007/978-3-031-72624-8_4.

A. Leonardis et al. (Eds.): ECCV 2024, LNCS 15071, pp. 55–72, 2025.
https://doi.org/10.1007/978-3-031-72624-8_4

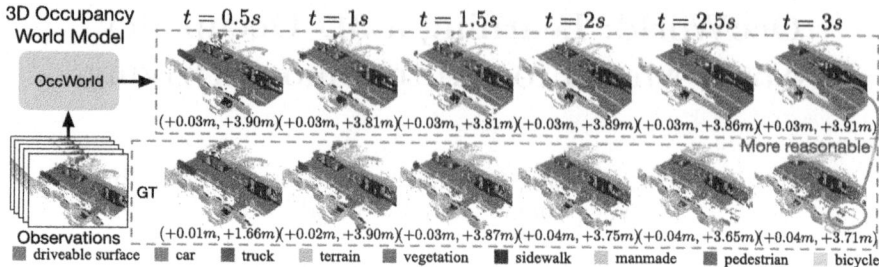

Fig. 1. Given past 3D occupancy observations, our self-supervised OccWorld can forecast future scene evolutions and ego movements jointly. This task requires a spatial understanding of the 3D scene and temporal modeling of how driving scenarios develop. We observe that OccWorld successfully forecasts the movements of surrounding agents and future map elements such as drivable areas. OccWorld even generates more reasonable drivable areas than the ground truth, demonstrating its ability to understand the scene rather than memorizing training data.

capture of structural information [7,36,52,62,63], the more hardware-economical vision-centric solutions have dramatically caught up with the increased perception ability of deep networks [19,33,34,43,45].

Forecasting future scene evolutions is important to the safety of autonomous driving vehicles. Most existing methods follow a conventional pipeline of perception, prediction, and planning [17,18,26]. Perception aims to obtain a semantic understanding of the surrounding scene such as 3D object detection [19,33,34,65] and semantic map construction [31,35,37,67]. The subsequent prediction module captures the motion of other traffic participants [11,14,25,67], and the planning module then makes decisions based on previous outputs [17,18,26,46]. However, this serial design usually requires ground-truth labels at each stage of training, yet the instance-level bounding boxes and high-definition maps are difficult to annotate. Furthermore, they usually only predict the motion of object bounding boxes, failing to capture more fine-grained information about the 3D scene.

In this paper, we explore a new paradigm to simultaneously predict the evolution of the surrounding scene and plan the future trajectory of the self-driving vehicle. We propose OccWorld, a world model in the 3D semantic occupancy space, to model the development of the driving scenes. We adopt 3D semantic occupancy as the scene representation over the conventional 3D bounding boxes and segmentation maps, which can describe the more fine-grained 3D structure of the scene. Moreover, 3D occupancy can be effectively learned from sparse LiDAR points [21], and thus is a potentially more economical way to describe the surrounding scenes. Given the 3D semantic occupancy representation of the current scene, OccWorld aims to predict how it evolves as the self-driving vehicle advances. To achieve this, we first employ a vector-quantized variational autoencoder (VQVAE) [42] to refine high-level concepts and obtain discrete scene tokens in a self-supervised manner. We then tailor the generative pre-training transformers (GPT) [2] architecture and propose a spatial-temporal generative transformer to predict the subsequent scene tokens and ego tokens to forecast the future occupancy and ego trajectory, respectively. We first perform spatial mixing to aggregate scene tokens and obtain multi-scale tokens to represent scenes at multiple levels. We then apply temporal attention to tokens at different levels to predict tokens for the next frame and use a U-net structure to integrate them. Finally, we use the trained VQVAE decoder to transform scene tokens to the occupancy space and learn a trajectory decoder to obtain ego planning results.

To demonstrate the effectiveness of OccWorld, we formulate a challenging task of 4D occupancy forecasting, which aims to predict the 3D occupancy of the following frames given a few past frames. Our OccWorld can effectively forecast future evolutions including moving agents and static elements as shown in Fig. 1, and achieves an average IoU of 26.63 and mIoU of 17.13 for 3 s future given 2 s history, OccWorld can also produce planning trajectories with an L2 error of 1.17 and a collision rate of 0.60% *without using any instance and map annotations.* Combined with self-supervised 3D occupancy learned from images [20], OccWorld achieves non-trivial 4D occupancy forecasting and planning results *without any human-annotated labels*, paving the way for scaling to large interpretable end-to-end autonomous driving models.

2 Related Work

3D Occupancy Prediction: 3D occupancy prediction aims to predict whether each voxel in the 3D space is occupied and its semantic label if occupied [21,22,53,54,57,58,68,71]. Early methods exploited LiDAR as inputs to complete the 3D occupancy of the entire 3D scene [6,30,47,60]. Recent methods began to explore the more challenging vision-based 3D occupancy prediction [4,21] or applying vision backbones to efficiently perform LiDAR-based 3D occupancy prediction [71]. 3D occupancy provides more comprehensive descriptions of the surrounding scene and includes both dynamic and static elements [21,58,71]. It can also be efficiently learned from sparse accumulated multiple LiDAR scans [58], LiDAR [21], or video sequences [5]. However, existing methods only focus on obtaining the 3D semantic occupancy and ignore its temporal evolution, which is vital to the safety of autonomous driving. In this paper, we explore the task of 4D occupancy forecasting and propose a 3D occupancy world model to for it.

World Models for Autonomous Driving: World models have a long history in control engineering and artificial intelligence [50], which are usually defined as producing the next scene observation given action and past observations [12]. The development of deep neural networks [13,49,51] promoted the use of deep generative models [10,29] as world models. Based on large pre-trained image generative models like StableDiffusion [48], recent methods [9,15,32,56,61] can generate realistic driving sequences of diverse scenarios. However, they produce future observations in the 2D image space, lacking understanding of the 3D surrounding scene. Some other methods explore forecasting point clouds using unannotated LiDAR scans [27,28,41,59], which ignore the semantic information and cannot be applied to vision-based or fusion-based autonomous driving.

Considering this, we explore a world model in the 3D occupancy space to more comprehensively model the 3D scene evolution.

End-to-End Autonomous Driving: The ultimate goal of autonomous driving is to obtain controlling signals based on observations of the surrounding scenes. Recent methods follow this concept to output planning results for the ego car given sensor inputs [17,18,26,54,64]. Most of them follow a conventional pipeline of perception [21,33,34,58,67], prediction [11,14,38,67], and planning [23,24,55,69]. They usually first perform BEV perception to extract relevant information (e.g., 3D agent boxes, semantic maps, tracklets) and then exploit them to infer future trajectories of agents and the ego vehicle. The following methods incorporated more data [64] or extracted more intermediate features [17,18,26] to provide more information for the planner, which achieved remarkable performance. Most methods only model object motions and cannot capture the fine-grained structural and semantic information of the surroundings [11,14,25,26,67]. Differently, we propose a world model to predict the evolution of both the surrounding dynamic and static elements.

3 Proposed Approach

3.1 World Model for Autonomous Driving

Autonomous driving aims to automatically steer a vehicle to fully prevent or partially reduce actions from human drivers [18]. Formally, the objective of autonomous driving is to obtain the control commands $\mathbf{c}^T$ (e.g., throttle, steer, break) for the present time stamp T given the sensor inputs $\{\mathbf{s}^T, \mathbf{s}^{T-1}, \cdots, \mathbf{s}^{T-t}\}$ from the current and past t frames.

As the mapping from trajectories to control signals is highly dependent on the vehicle specifications and status, the literature usually assumes a given satisfactory controller and focuses on trajectory planning. An autonomous driving model A then takes input as the sensor inputs and ego trajectory from the past T frames and predicts the ego trajectory of future f frames:

$$
\begin{aligned}
A(\{\mathbf{s}^T, \mathbf{s}^{T-1}, \cdots, \mathbf{s}^{T-t}\}, \{\mathbf{p}^T, \mathbf{p}^{T-1}, \cdots, \mathbf{p}^{T-t}\}) \\
= \{\mathbf{p}^{T+1}, \mathbf{p}^{T+2}, \cdots, \mathbf{p}^{T+f}\},
\end{aligned}
\tag{1}
$$

where $\mathbf{p}^t$ denotes the 3D ego position at the t-th time.

The conventional pipeline usually follows a design of perception, prediction, and planning [17,18,26]. The perception module p_{er} perceives the surrounding scenes and extracts high-level information $\mathbf{z}$ from the input sensor data $\mathbf{s}$. The prediction module p_{re} then integrates the high-level information $\mathbf{z}$ to predict the future trajectory $\mathbf{t}_i$ of each agent in the scene. The planning module p_{la} finally processes the perception and prediction results $\{\mathbf{z}, \{\mathbf{t}_i\}\}$ to plan the motion of the ego vehicle. The conventional pipeline can be formulated as:

$$
\begin{aligned}
p_{la}(p_{er}(\{\mathbf{s}^T, \cdots, \mathbf{s}^{T-t}\}), p_{re}(p_{er}(\{\mathbf{s}^T, \cdots, {}^{T-t}\}))) \\
= \{\mathbf{p}^{T+1}, \mathbf{p}^{T+2}, \cdots, \mathbf{p}^{T+f}\}.
\end{aligned}
\tag{2}
$$

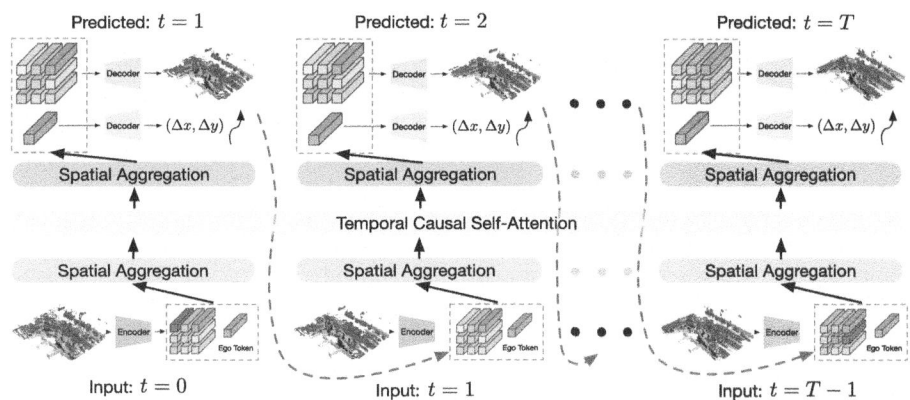

Fig. 2. Framework of our OccWorld for 4D semantic occupancy forecasting and motion planning. We adopt a GPT-like generative architecture to predict the next scene from previous scenes in an autoregressive manner. We adapt GPT [2] to the autonomous driving scenario with two key designs: 1) We train a 3D occupancy scene tokenizer to produce discrete high-level representations of the 3D scene; 2) We perform spatial mixing before and after spatial-wise temporal causal self-attention to efficiently produce globally consistent scene predictions. We use ground-truth and predicted scene tokens as inputs for future generations for training and inference, respectively.

Despite the promising performance of this framework [17,18,26], it usually requires labels at each stage, which can be laborious to annotate. It only considers object-level movement and fails to model more fine-grained evolutions.

Motivated by this, we explore a new world-model-based autonomous driving paradigm to comprehensively model the evolution of the surrounding scenes and the ego movements. Inspired by the success of generative pre-training transformers (GPT) [2] in natural language processing (NLP), we propose an auto-regressive generative modeling framework for autonomous driving scenarios. We define a world model w to act on scene representations $\mathbf{y}$ and be able to predict future scenes. Formally, we formulate the function of a world model w as follows:

$$w(\{\mathbf{y}^T, \cdots, \mathbf{y}^{T-t}\}, \{\mathbf{p}^T, \cdots, \mathbf{p}^{T-t}\}) = \mathbf{y}^{T+1}, \mathbf{p}^{T+1}. \tag{3}$$

Having obtained the predicted scene $\mathbf{y}^{T+1}$ and the ego position $\mathbf{p}^{T+1}$, we add them to the input and predict the next frame in an auto-regressive manner, as shown in Fig. 2. The world model w captures the joint distribution of the scene evolution and the ego movement, considering their high-order interactions.

3.2   3D Occupancy Scene Tokenizer

As the world model w operates on the scene representation $\mathbf{y}$, the choice of $\mathbf{y}$ is vital to w. We select the 3D scene representation $\mathbf{y}$ based on three principles: 1) **expressiveness**. It should be able to comprehensively contain the 3D structural and semantic information of the 3D scene; 2) **efficiency**. It should be economical to learn (e.g., from weak supervision or self-supervision); 3) **versatility**. It should be able to adapt to both vision and LiDAR modalities.

Considering these principles, we propose to adopt 3D occupancy as the 3D scene representation $\mathbf{y} \in \mathbb{R}^{H \times W \times D}$. 3D occupancy partitions the surrounding 3D space into $H \times W \times D$ voxels and assigns each voxel with a label l denoting whether it is occupied and which material it is occupied with. 3D occupancy provides a dense representation of the 3D scene and can describe both the 3D structural and semantic information of the scene. It can be effectively learned from sparse LiDAR annotations [21] or potentially from self-supervision of temporal frames [20]. 3D occupancy is also modality-agnostic and can be obtained from monocular camera [4], surrounding cameras [21,54,58], or LiDAR [71].

Despite its comprehensiveness, 3D occupancy only provides a low-level understanding of the scene, making it difficult to directly model its evolution. We therefore propose a self-supervised way to tokenize the scene into high-level tokens from 3D occupancy. We train a vector-quantized autoencoder (VQ-VAE) [42] on $\mathbf{y}$ to obtain discrete tokens $\mathbf{z}$ to better represent the scene, as shown in Fig. 3.

For efficiency, we first transform the 3D occupancy $\mathbf{y} \in \mathbb{R}^{H \times W \times D}$ to a BEV representation $\hat{\mathbf{y}} \in \mathbb{R}^{H \times W \times DC'}$ by assigning each category with a learnable class embedding $\in \mathbb{R}^{C'}$ and concatenating them in the height dimension. We then adopt a lightweight encoder composed of 2D convolutions to obtain downsampled features $\hat{\mathbf{z}} \in \mathbb{R}^{\frac{H}{d} \times \frac{W}{d} \times C}$, where d is the down-sampling factor.

To obtain a more compact representation, we simultaneously learn a codebook $\mathbf{C} \in \mathbb{R}^{N \times C}$ containing N codes. Each code $\mathbf{c} \in \mathbb{R}^{C}$ encodes a high-level concept of the scene, e.g., whether a position is occupied by a car. We quantized each spatial feature $\hat{\mathbf{z}}_{ij}$ in $\hat{\mathbf{z}}$ by classifying it to the nearest code $\mathcal{N}(\hat{\mathbf{z}}_{ij}, \mathbf{C})$:

$$\mathbf{z}_{ij} = \mathcal{N}(\hat{\mathbf{z}}_{ij}, \mathbf{C}) = \min_{\mathbf{c} \in \mathbf{C}} ||\hat{\mathbf{z}}_{ij} - \mathbf{c}||_2, \tag{4}$$

where $||\cdot||_2$ denotes the L2 norm. We then integrate the quantized features $\{\mathbf{z}_{ij}\}$ to obtain the final scene representation $\mathbb{R}^{\frac{H}{d} \times \frac{W}{d} \times C}$.

To reconstruct $\widetilde{\mathbf{y}}$ from the learned scene representation $\mathbf{z}$, we use a decoder of 2D deconvolution layers to progressively upsample $\mathbf{z}$ to its original BEV resolution $H \times W \times C''$. We then perform a split in the channel dimension to reconstruct the height dimension $H \times W \times D \times \frac{C''}{D}$ and apply a softmax layer on each spatial feature to classify them into semantic occupancy.

The scene tokenizer transforms 3D occupancy into a more compact discrete space to encode higher-level concepts. This refined compact space facilitates the modeling of scene evolution for the subsequent world model.

3.3 Spatial-Temporal Generative Transformer

The core of autonomous driving is the prediction of how the surrounding world evolves and planning the movement of the ego vehicle accordingly. While conventional methods consider them separately [17,18], we propose a world model w to jointly model the distributions of scene evolution and ego trajectory.

As defined in (3), a world model w takes as inputs the past scenes and ego positions and predicts their outcome after driving a certain time interval. Following the principles of expressiveness, efficiency, and versatility, we adopt 3D occupancy $\mathbf{y}$ as the scene representation and use a self-supervised tokenizer to obtain high-level scene tokens $\mathbf{T} = \{\mathbf{z}_i\}$. To integrate the ego movement, we further aggregate $\mathbf{T}$ with an ego token $\mathbf{z}_0 \in \mathbb{R}^C$ to encode the spatial ego position. The proposed OccWorld w then functions on the world tokens $\mathbf{T}$:

$$w(\mathbf{T}^T, \cdots, \mathbf{T}^{T-t}) = \mathbf{T}^{T+1}, \tag{5}$$

where T is the current time stamp, and t is the number of history frames.

Inspired by the remarkable sequential prediction performance of GPT [2], we adopt a GPT-like autoregressive transformer architecture to instantiate (5). However, the migration of GPT from natural language processing to autonomous driving is not trivial. GPTs predict a single token each time, while the world model w in autonomous driving is required to predict a set of tokens $\mathbf{T}$ as the next future. Due to the vast number of world tokens, directly leveraging the GPT architecture to predict each token $\in \mathbf{T}^{T+1}$ is both inefficient and ineffective.

Both the spatial relations of world tokens within each time stamp and the temporal relations of tokens across different time stamps should be considered to comprehensively model the world evolution. Therefore, we propose a spatial-temporal generative transformer architecture to effectively process past world tokens and make predictions of the next future, as shown in Fig. 3b.

We apply spatial aggregation (e.g., self-attention [8]) to world tokens $\mathbf{T}$ to enable interactions between scene tokens and ego tokens. We then merge the scene tokens in each 2×2 window with a stride of 2 to achieve a 1/4 downsampling. We repeat this procedure for K times to obtain world tokens of hierarchical scales $\{\mathbf{T}_0, \cdots, \mathbf{T}_K\}$ to describe the 3D scene at different levels.

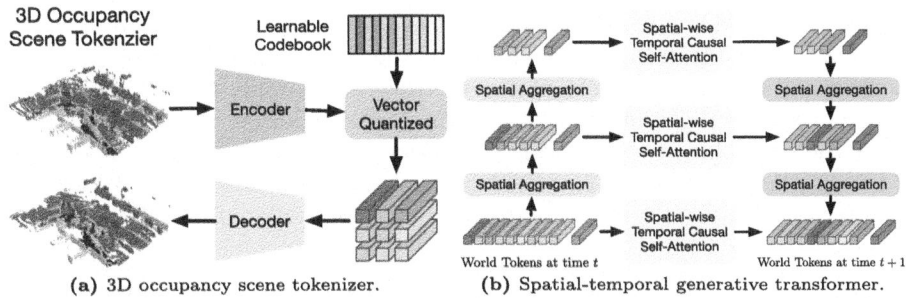

(a) 3D occupancy scene tokenizer.   (b) Spatial-temporal generative transformer.

Fig. 3. Illustration of the proposed modules. (a) We use CNNs to encode the 3D occupancy and perform vector quantization to obtain discrete tokens using a learnable codebook [42]. We then employ a decoder to reconstruct the input 3D occupancy using the quantized tokens and train the autoencoder and codebook simultaneously. (b) As each scene is represented by world tokens, we adopt spatial mixing to model their intrinsic dependencies and obtain multi-scale world tokens to capture multi-level information. We perform spatial-wise temporal causal self-attention at each level to forecast the next scene and employ a U-net structure to aggregate multi-scale predictions.

We use several sub-world models $w = \{w_0, \cdots, w_K\}$ to predict the future at different spatial scales. For each sub-world model w_i, we impose temporal attention on the tokens $\{\mathbf{z}_{j,i}^T, \cdots, \mathbf{z}_{j,i}^{T-t}\}$ at each position j to obtain the predicted corresponding token $\mathbf{z}_{j,i}^{T+1}$ of the next frame:

$$\hat{\mathbf{z}}_{j,i}^{T+1} = \text{TA}(\mathbf{z}_{j,i}^T, \cdots, \mathbf{z}_{j,i}^{T-t}), \tag{6}$$

where TA denotes masked temporal attention which blocks the effect of future tokens to previous tokens. $\mathbf{z}_{j,i}^t \in \mathbf{T}_i^t$ represents the j-th world token of the i-th scale at time stamp t. We finally employ a U-net structure to aggregate predicted tokens at different scales to ensure spatial consistency.

Our spatial-temporal generative transformer models the world evolution considering the joint distributions of world tokens within each time and across time. The temporal attention predicts the evolution of a fixed position in the surroundings, while the spatial aggregation makes each token aware of the global scene.

3.4   OccWorld: A 3D Occupancy World Model

In this subsection, we present the overall training framework of the proposed OccWorld model for autonomous driving. Having obtained the forecasted world tokens, we reuse the scene decoder d to decode the predicted 3D occupancy $\hat{\mathbf{y}}^{T+1} = d(\hat{\mathbf{z}}^{T+1})$ and additionally learn an ego decoder d_{ego} to produce the ego displacement $\hat{p}^{T+1} = d_{ego}(\hat{z}_0^{T+1})$ w.r.t the current frame.

We adopt a two-stage strategy to effectively train our OccWorld. For the first stage, we train the scene tokenizer e and decoder d using 3D occupancy loss [21]:

$$J_{e,d} = L_{soft}(d(e(\mathbf{y})), \mathbf{y}) + \lambda_1 \, L_{lovasz}(d(e(\mathbf{y})), \mathbf{y}), \tag{7}$$

where L_{soft} and L_{lovasz} is the softmax and lovasz-softmax loss [1], respectively.

For the second stage, we adopt the learned scene tokenizer e to obtain scene tokens $\mathbf{z}$ for all the frames and constrain the discrepancy between predicted tokens $\hat{\mathbf{z}}$ and $\mathbf{z}$. Due to the use of discrete tokens, we apply the softmax loss to enforce the correct classification of $\hat{\mathbf{z}}$ to the correct codes in the codebook $\mathbf{C}$ as $\mathbf{z}$. For the ego token, we simultaneously learn the ego decoder d_{ego} and apply L2 loss on the predicted displacement $\hat{p} = d_{ego}(\hat{\mathbf{z}}_0)$ and the ground-truth one $\mathbf{p}$. The overall objective for the second stage can be formulated as follows:

$$J_{w,d_{ego}} = \sum_{t=1}^{T} (\sum_{j=1}^{M_0} L_{soft}(\hat{\mathbf{z}}_{j,0}^t, \mathbf{C}(\mathbf{z}_{j,0}^t) + \lambda_2 \, L_{L2}(d_{ego}(\hat{\mathbf{z}}_0^t), \mathbf{p}^t)), \tag{8}$$

where T and M_0 are the number of frames and the number of spatial tokens of the original scale, respectively. $\mathbf{C}(\cdot)$ denotes the index of the corresponding code in the codebook $\mathbf{C}$. L_{L2} measures the L2 discrepancy between two trajectories.

For efficient training, we use tokens obtained by the scene tokenizer e as inputs but apply masked temporal attention [2] to block the effect of future tokens on previous ones. For inference, we progressively predict world tokens of the next frame using predicted past tokens instead of ground-truth ones.

The proposed OccWorld can be applied to different types of 3D occupancy to adapt to different settings (e.g., end-to-end autonomous driving). The scene representation model r can be an oracle that provides ground-truth occupancy, or a perception model that is trained using dense supervision (e.g., accumulated LiDAR [58]), sparse supervision (e.g., LiDAR [21]), or self-supervision (e.g., videos [5]). Different from the conventional perception, predicting, and planning pipeline for autonomous driving, OccWorld models the joint evolution of the surrounding scene and the ego movement to capture high-order interactions with the environment. Combined with machine-annotated [58], LiDAR-collected [21], or self-supervised [20] 3D occupancy, OccWorld has the potential to scale up to large-scale training data, paving the way for training large driving models.

4   Experiments

4.1   Task Descriptions

In this paper, we explore a world-model-based framework for autonomous driving and propose OccWorld to model the joint evolutions of ego trajectory and scene evolutions. We conduct two tasks to evaluate our OccWorld: 4D occupancy forecasting on Occ3D [53] and motion planning on nuScenes [3].

4D Occupancy Forecasting. 3D occupancy prediction aims to reconstruct the semantic occupancy for each voxel in the surrounding space and cannot capture the temporal evolution of the 3D occupancy. In this paper, we explore 4D occupancy forecasting, which aims to forecast future 3D occupancy given historical occupancy. We use the mean intersection of region (mIoU) of all the semantic categories between forecasted and ground-truth occupancy to measure the semantic forecast performance. We adopt the intersection of region (IoU) between occupied and unoccupied voxels to evaluate the structural forecast quality. We report the forecasting performance for future frames of 1 s, 2 s, and 3 s.

Motion Planning. Motion planning aims to produce safe future trajectories for the vehicle given ground-truth surrounding information or perception results. The planned trajectory is represented by a series of 2D waypoints in the BEV plane. We use L2 error and box collision rate to measure the quality of the planned trajectory. The L2 error computes the L2 distance between the planned waypoint and the ground-truth waypoint at a given time stamp. The box collision rate measures the frequency of intersection between the BEV box of the ego vehicle and other traffic agents for a certain period of time. We follow previous methods [17] and report planning performance for the future 1 s, 2 s, and 3 s.

4.2   Datasets Details

nuScenes [3] contains 1000 driving scenes, i.e., videos of 6 surrounding cameras with 360° horizontal FOV and 32-beam LiDAR point clouds for 20 s, and provides annotated at 2 Hz for keyframes. We follow the official split [3] and employ 700 and 150 scenes for training and validation, respectively.

Table 1. 4D occupancy forecasting performance. Aux. Sup. denotes auxiliary supervision apart from the ego trajectory. Avg. denotes the average performance of that in 1 s, 2 s, and 3 s. We use bold numbers to denote the best results.

Method	Input	Aux. Sup.	mIoU (%) ↑					IoU (%) ↑					FPS
			0 s	1 s	2 s	3 s	Avg.	0 s	1 s	2 s	3 s	Avg.	
Copy&Paste	3D-Occ	None	66.38	14.91	10.54	8.52	11.33	62.29	24.47	19.77	17.31	20.52	–
OccWorld-O	3D-Occ	None	**66.38**	**25.78**	**15.14**	**10.51**	**17.14**	**62.29**	**34.63**	**25.07**	**20.18**	**26.63**	18.0
OccWorld-D	Camera	3D-Occ	18.63	11.55	8.10	6.22	8.62	22.88	18.90	16.26	14.43	16.53	2.8
OccWorld-T	Camera	LiDAR	7.21	4.68	3.36	2.63	3.56	10.66	9.32	8.23	7.47	8.34	2.8
OccWorld-S	Camera	None	0.27	0.28	0.26	0.24	0.26	4.32	5.05	5.01	4.95	5.00	2.8

Occ3D [53] provides 3D semantic occupancy annotations for nuScenes. Each scene is split into $200 \times 200 \times 16$ voxels covering a -40 m$\sim$40 m area along the X and Y axis and -1 m$\sim$5.4 m along the Z axis. Each voxel is annotated as occupied or unoccupied and an additional semantic category if occupied.

4.3   Implementation Details

We followed existing works [18,26] and used a 2-s historical context to forecast the subsequent 3 s. We encode instructions and incorporate them via cross-attention to the ego token. The scene tokenizer employs a down-sampling factor of 4, featuring a codebook with a size of 512 and a dimension of 128. The spatial-temporal generative transformer comprises 3 scales, each incorporating 6 layers of spatial-wise temporal attention for scene tokens with 2 layers of spatial cross-attention and temporal cross-attention for ego planning tokens.

During training, we applied mask operations to all temporal attention mechanisms to prevent the influence of future information on forecasting. For inference, we employ autoregressive prediction to foresee 3 s into the future based on a 2-s historical context. We adopted AdamW [40] with a Cosine Annealing scheduler [39] for training. We set an initial learning rate of 10^{-3} and the weight decay at 0.01. We use a batch size of 1 per GPU on 8 NVIDIA 4090 GPUs.

4.4   Results and Analysis

4D Occupancy Forecasting. We evaluated our OccWorld in several settings: OccWorld-O (using ground-truth 3D occupancy), OccWorld-D (using predicted results of TPVFormer [21] trained with dense ground-truth 3D occupancy), OccWorld-T (using predicted results of TPVFormer [21] trained with sparse semantic LiDAR), and OccWorld-S (using predicted results of self-supervised SelfOcc [20]). Copy&Paste denotes copying the current ground-truth occupancy as future observations. The 0 s results represent the reconstruction accuracy.

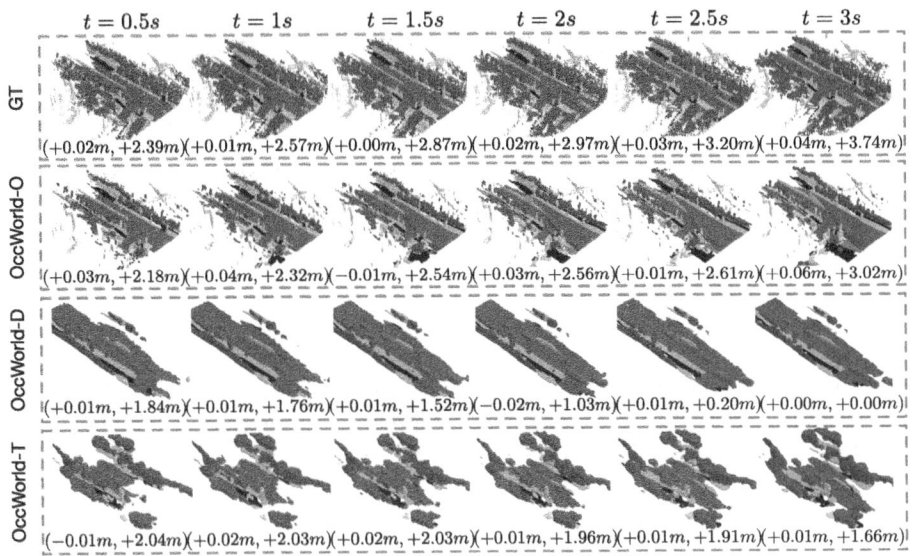

Fig. 4. Visualizations of the forecasting and planning results of OccWorld-O, OccWorld-D, and OccWorld-T.

Table 1 shows that OccWorld-O can learn the underlying scene evolution and generate non-trivial future 3D occupancy with much better results than Copy&Paste. OccWorld-D, OccWorld-T, and OccWorld-S can be seen as end-to-end vision-based 4D occupancy forecasting methods as they take surrounding images as input. This task is very challenging since it requires both 3D structure reconstruction and forecasting. It is especially difficult for OccWorld-S which exploits no 3D occupancy information during training. Still, OccWorld generates future 3D occupancy with non-trivial mIoU and IoU on the end-to-end setting.

The performance drop over time results from the generative nature of our method. Driving scenarios are essentially random distributions (i.e., drivers can make different yet reasonable decisions), while the ground-truth trajectory only provides one possible future. As shown in Fig. 1, the forecastings of OccWorld might not exactly match the ground truth but are still reasonable. Different from motion annotations, 3D occupancy only provides local perception, rendering the performance drop if the car advances a long distance. In real scenarios, the autonomous driving system continuously perceives the surroundings and only needs to make instant decisions, rendering near-future planning more important.

Visualizations. We visualize the output results of the proposed OccWorld in Fig. 4. We see that our models can successfully forecast the movements of cars and can complete unseen map elements in the inputs such as drivable areas. The planning trajectory is also more accurate with better 4D occupancy forecasting.

Motion Planning. We compare the motion planning performance of our Occ-World with state-of-the-art end-to-end methods, as shown in Table 2. We also

Table 2. Motion planning performance. We use bold and underlined numbers to denote the best and second-best results, respectively. B, Mo, Ma, D, T, O denote using the auxiliary supervision signal of box, motion, map, depth, tracklets, and occupancy, respectively. [†] denotes using the metric computation code adopted in VAD [26].

Method	Input	Aux. Sup.	L2 (m) ↓				Collision Rate (%) ↓				FPS
			1 s	2 s	3 s	Avg.	1 s	2 s	3 s	Avg.	
IL [44]	LiDAR	None	0.44	1.15	2.47	1.35	0.08	0.27	1.95	0.77	–
NMP [66]	LiDAR	B & Mo	0.53	1.25	2.67	1.48	**0.04**	0.12	0.87	0.34	–
FF [16]	LiDAR	Freespace	0.55	1.20	2.54	1.43	0.06	0.17	1.07	0.43	–
EO [27]	LiDAR	Freespace	0.67	1.36	2.78	1.60	**0.04**	**0.09**	0.88	0.33	–
ST-P3 [17]	Camera	Ma & B & D	1.33	2.11	2.90	2.11	0.23	0.62	1.27	0.71	1.6
UniAD [18]	Camera	Ma & B & Mo & T & O	0.48	**0.96**	**1.65**	**1.03**	0.05	0.17	**0.71**	**0.31**	1.8
VAD-Tiny [26]	Camera	Ma & B & Mo	0.60	1.23	2.06	1.30	0.31	0.53	1.33	0.72	16.8
VAD-Base [26]	Camera	Ma & B & Mo	0.54	1.15	1.98	1.22	**0.04**	0.39	1.17	0.53	4.5
OccNet [54]	Camera	3D-Occ & Ma & B	1.29	2.13	2.99	2.14	0.21	0.59	1.37	0.72	2.6
OccNet [54]	3D-Occ	Ma & B	1.29	2.31	2.98	2.25	0.20	0.56	1.30	0.69	–
OccWorld-O	3D-Occ	None	**0.43**	1.08	1.99	1.17	0.07	0.38	1.35	0.60	**18.0**
OccWorld-D	Camera	3D-Occ	0.52	1.27	2.41	1.40	0.12	0.40	2.08	0.87	2.8
OccWorld-T	Camera	LiDAR	0.54	1.36	2.66	1.52	0.12	0.40	1.59	0.70	2.8
OccWorld-S	Camera	None	0.67	1.69	3.13	1.83	0.19	1.28	4.59	2.02	2.8
VAD-Tiny[†] [26]	Camera	Ma & B & Mo	0.46	0.76	1.12	0.78	0.21	0.35	0.58	0.38	16.8
VAD-Base[†] [26]	Camera	Ma & B & Mo	0.41	0.70	1.05	0.72	0.07	0.17	0.41	0.22	4.5
OccWorld-O[†]	3D-Occ	None	0.32	0.61	0.98	0.64	0.06	0.21	0.47	0.24	18.0
OccWorld-D[†]	Camera	3D-Occ	0.39	0.73	1.18	0.77	0.11	0.19	0.67	0.32	2.8
OccWorld-T[†]	Camera	LiDAR	0.40	0.77	1.28	0.82	0.12	0.22	0.56	0.30	2.8
OccWorld-S[†]	Camera	None	0.49	0.95	1.55	0.99	0.19	0.56	1.54	0.76	2.8

Table 3. Effect of different hyperparameters for the scene tokenizer. We use bold numbers to denote the best results.

Setting	Reconstruction		Forecasting mIoU (%) ↑				Planning L2 (m) ↓				FPS
	mIoU ↑	IoU ↑	1 s	2 s	3 s	Avg.	1 s	2 s	3 s	Avg.	
$(50^2, 128, 512)$	66.38	62.29	**25.78**	**15.14**	10.51	**17.14**	0.43	**1.08**	1.99	1.17	18.0
$(50^2, 128, 256)$	63.40	60.33	24.25	14.34	10.13	16.24	**0.42**	**1.08**	**1.95**	**1.15**	17.8
$(50^2, 128, 1024)$	60.50	59.07	23.55	14.66	**10.68**	16.30	0.47	1.18	2.19	1.28	17.8
$(25^2, 256, 512)$	36.28	44.02	12.10	8.13	6.20	8.81	3.27	6.54	9.78	6.53	**28.1**
$(100^2, 128, 512)$	**78.12**	**71.63**	18.71	10.75	7.68	12.38	0.50	1.25	2.33	1.36	6.7
$(50^2, 64, 512)$	64.98	61.50	21.83	12.90	9.28	14.67	0.49	1.24	2.26	1.33	20.1

Table 4. Ablation study of the spatial-temporal generative transformer. We report average results over the 1 s, 2 s, and 3 s.

Method	Forecast		Planning		FPS
	mIoU (%) ↑	IoU (%) ↑	L2 (m)↓	Collision Rate (%)↓	
OccWorld-O	**17.14**	**26.63**	**1.17**	**0.60**	18.0
w/o spatial attn	10.07	21.44	1.42	1.21	**28.6**
w/o temporal attn	8.98	20.10	2.06	2.56	26.5
w/o ego	15.13	24.66	–	–	18.8
w/o ego temporal	12.07	23.09	5.89	6.23	18.5

evaluate our model under different settings (-O, -D, -T, -S). We see that UniAD achieves the best overall performance, which exploits various types of auxiliary supervision to improve its planning quality. Despite the strong performance, the additional annotations in the 3D space are very difficult to obtain, making it difficult to scale to large-scale driving data. Alternatively, OccWorld demonstrates competitive performance by employing 3D occupancy as the scene representation which can be efficiently obtained by accumulating LiDAR scans [58].

We observe that using ground-truth 3D occupancy as inputs, our OccWorld-O outperforms the previous perception-prediction-planning-based method Occ-Net [54] by a large margin without using maps and bounding boxes as supervision, demonstrating the superiority of the world-model paradigm for autonomous driving. Our end-to-end models OccWorld-D and OccWorld-T also demonstrate competitive performance using only 3D occupancy as supervision and OccWorld-S delivers non-trivial results with no supervision other than the future trajectory, showing the potential for interpretable end-to-end autonomous driving.

Though our model demonstrates competitive L2 error, it falls behind on the collision rate. This is because it is more difficult to learn safe trajectories without the guidance of freespace or bounding box. Still, OccWorld demonstrates comparable collision rates with OccNet which exploits map and box supervision, showing that OccWorld can learn the concept of freespace with 3D occupancy.

We also see that OccWorld shows excellent short-term planning performance (1 s), but worsens quickly when planning longer futures. For example, OccWorld-O achieves the best L2 error at 1 s among all the methods but reaches 1.99 at 3 s compared to 1.65 of UniAD. This might result from the diverse future generations of world models, which might deviate from the ground-truth trajectory.

Analysis of the Scene Tokenizer. We analyze the effect of different hyperparameters for the scene tokenizer in Table 3. The setting (S, C, N) denotes latent spatial resolution S, latent channel dimension C, and the codebook size N. We see that using a larger codebook than 512 leads to overfitting and using a smaller S, C, and N might not be enough to capture the scene distribution. The reconstruction accuracy improves with a larger spatial resolution, yet leading to poor

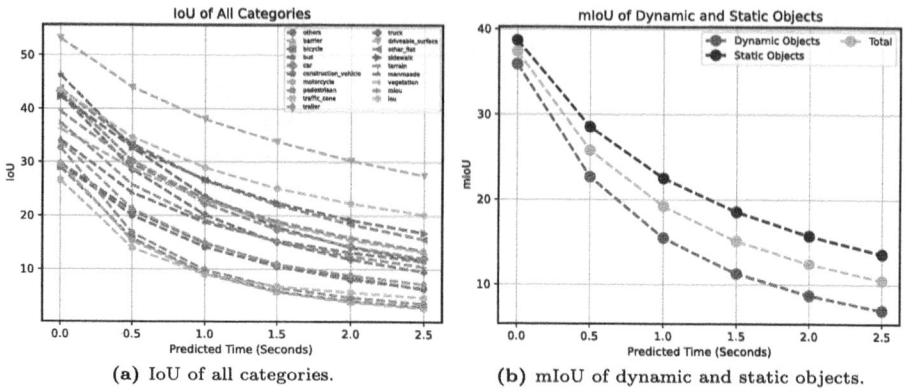

Fig. 5. Analysis of the performance for static and dynamic objects.

forecasting and planning performance. This is because the tokens cannot learn high-level concepts, resulting in more difficulty in forecasting the future.

Analysis of the Spatial-Temporal Generative Transformer. We conducted an ablation study on both 4D occupancy forecasting and motion planning to analyze the design of the proposed spatial-temporal generative transformer, as shown in Table 4. w/o spatial attn denotes discarding spatial aggregation and directly applying temporal attention to the input tokens. w/o temporal attn represents that we replace the temporal attention with a simple convolution to output the next scene using the current world tokens. w/o ego represents that we discard the ego token. w/o ego temporal represents that we replace the temporal attention of the ego token with a simple MLP. We observe that using spatial aggregation to model spatial dependencies and using temporal attention to integrate history information is vital to the performance of both 4D occupancy forecasting and motion planning tasks. Also, only performing the 4D occupancy forecasting task without predicting motion reduces the performance. This verifies the effectiveness of joint modeling of scene evolutions and ego trajectories. Finally, discarding the ego temporal attention leads to poor planning and surprisingly worse 3D forecast occupancy performance. We think this is because integrating a wrongly predicted ego trajectory will mislead the forecasting.

Analysis of Dynamic Object Prediction. We analyze the performance of our model on static and dynamic objects as shown in Fig. 5. We see that the performance for dynamic objects is lower, indicating modeling dynamic objects is more challenging. This is reasonable since the observations of dynamic objects results from both the movements of the ego vehicle and the objects themselves. Still, OccWorld can successfully predict the future trajectories of dynamic objects, which is more important for making decisions.

5   Conclusion and Discussions

In this paper, we have presented a 3D occupancy world model (OccWorld) to model the joint evolutions of ego movements and surrounding scenes. We have employed a 3D occupancy scene tokenizer to extract high-level concepts and used a spatial-temporal generative transformer for future prediction in an auto-regressive manner. Both quantitative and visualization results have shown that OccWorld can effectively predict future scene evolutions in the comprehensive 3D semantic occupancy space. We believe that OccWorld has paved the way for interpretable end-to-end autonomous driving without additional supervision signals and facilitated the scaling to large driving models.

Limitations. OccWorld simultaneously models the ego movements and scene evolutions, yet cannot predict the futures conditioned on certain driving commands. However, the ability to forecast multiple futures based on different conditions is important for a world model and is an interesting future direction.

Acknowledgements. This work was supported in part by the National Key Research and Development Program of China under Grant 2023YFB280690, and in part by the National Natural Science Foundation of China under Grant 62321005, Grant 62336004, and Grant 62125603.

References

1. Berman, M., Triki, A.R., Blaschko, M.B.: The lovász-softmax loss: a tractable surrogate for the optimization of the intersection-over-union measure in neural networks. In: CVPR, pp. 4413–4421 (2018)
2. Brown, T., et al.: Language models are few-shot learners. In: NeurIPS, vol. 33, pp. 1877–1901 (2020)
3. Caesar, H., et al.: nuScenes: a multimodal dataset for autonomous driving. In: CVPR (2020)
4. Cao, A.Q., de Charette, R.: MonoScene: monocular 3D semantic scene completion. In: CVPR, pp. 3991–4001 (2022)
5. Cao, A.Q., de Charette, R.: SceneRF: self-supervised monocular 3D scene reconstruction with radiance fields. In: ICCV, pp. 9387–9398 (2023)
6. Chen, X., Lin, K.Y., Qian, C., Zeng, G., Li, H.: 3D sketch-aware semantic scene completion via semi-supervised structure prior. In: CVPR, pp. 4193–4202 (2020)
7. Cheng, R., Razani, R., Taghavi, E., Li, E., Liu, B.: 2-S3Net: attentive feature fusion with adaptive feature selection for sparse semantic segmentation network. In: CVPR, pp. 12547–12556 (2021)
8. Dosovitskiy, A., et al.: An image is worth 16×16 words: transformers for image recognition at scale. In: ICLR (2020)
9. Gao, R., et al.: MagicDrive: street view generation with diverse 3D geometry control. arXiv preprint arXiv:2310.02601 (2023)
10. Goodfellow, I., et al.: Generative adversarial nets. In: NeurIPS, vol. 27 (2014)
11. Gu, J., et al.: ViP3D: end-to-end visual trajectory prediction via 3D agent queries. arXiv preprint arXiv:2208.01582 (2022)
12. Ha, D., Schmidhuber, J.: World models. arXiv preprint arXiv:1803.10122 (2018)

13. He, K., Zhang, X., Ren, S., Sun, J.: Deep residual learning for image recognition. In: CVPR, pp. 770–778 (2016)
14. Hu, A., et al.: FIERY: future instance prediction in bird's-eye view from surround monocular cameras. In: ICCV (2021)
15. Hu, A., et al.: GAIA-1: a generative world model for autonomous driving. arXiv preprint arXiv:2309.17080 (2023)
16. Hu, P., Huang, A., Dolan, J., Held, D., Ramanan, D.: Safe local motion planning with self-supervised freespace forecasting. In: CVPR (2021)
17. Hu, S., Chen, L., Wu, P., Li, H., Yan, J., Tao, D.: ST-P3: end-to-end vision-based autonomous driving via spatial-temporal feature learning. In: Avidan, S., Brostow, G., Cissé, M., Farinella, G.M., Hassner, T. (eds.) ECCV 2022. LNCS, vol. 13698, pp. 533–549. Springer, Cham (2022). https://doi.org/10.1007/978-3-031-19839-7_31
18. Hu, Y., et al.: Planning-oriented autonomous driving. In: CVPR, pp. 17853–17862 (2023)
19. Huang, J., Huang, G., Zhu, Z., Du, D.: BEVDet: high-performance multi-camera 3D object detection in bird-eye-view. arXiv preprint arXiv:2112.11790 (2021)
20. Huang, Y., Zheng, W., Zhang, B., Zhou, J., Lu, J.: SelfOcc: self-supervised vision-based 3D occupancy prediction. In: CVPR (2024)
21. Huang, Y., Zheng, W., Zhang, Y., Zhou, J., Lu, J.: Tri-perspective view for vision-based 3D semantic occupancy prediction. In: CVPR, pp. 9223–9232 (2023)
22. Huang, Y., Zheng, W., Zhang, Y., Zhou, J., Lu, J.: GaussianFormer: scene as gaussians for vision-based 3D semantic occupancy prediction. In: ECCV (2024)
23. Huang, Z., Liu, H., Lv, C.: GameFormer: game-theoretic modeling and learning of transformer-based interactive prediction and planning for autonomous driving. arXiv preprint arXiv:2303.05760 (2023)
24. Huang, Z., Liu, H., Wu, J., Lv, C.: Differentiable integrated motion prediction and planning with learnable cost function for autonomous driving. IEEE Trans. Neural Netw. Learn. Syst. (2023)
25. Jiang, B., et al.: Perceive, interact, predict: learning dynamic and static clues for end-to-end motion prediction. arXiv preprint arXiv:2212.02181 (2022)
26. Jiang, B., et al.: VAD: vectorized scene representation for efficient autonomous driving. arXiv preprint arXiv:2303.12077 (2023)
27. Khurana, T., Hu, P., Dave, A., Ziglar, J., Held, D., Ramanan, D.: Differentiable raycasting for self-supervised occupancy forecasting. In: Avidan, S., Brostow, G., Cissé, M., Farinella, G.M., Hassner, T. (eds.) ECCV 2022. LNCS, vol. 13698, pp. 353–369. Springer, Cham (2022). https://doi.org/10.1007/978-3-031-19839-7_21
28. Khurana, T., Hu, P., Held, D., Ramanan, D.: Point cloud forecasting as a proxy for 4D occupancy forecasting. In: CVPR, pp. 1116–1124 (2023)
29. Kingma, D.P., Welling, M.: Auto-encoding variational bayes. arXiv preprint arXiv:1312.6114 (2013)
30. Li, J., Han, K., Wang, P., Liu, Y., Yuan, X.: Anisotropic convolutional networks for 3D semantic scene completion. In: CVPR, pp. 3351–3359 (2020)
31. Li, Q., Wang, Y., Wang, Y., Zhao, H.: HDMapNet: an online HD map construction and evaluation framework. In: ICRA (2022)
32. Li, X., Zhang, Y., Ye, X.: DrivingDiffusion: layout-guided multi-view driving scene video generation with latent diffusion model. arXiv preprint arXiv:2310.07771 (2023)
33. Li, Y., et al.: BEVDepth: acquisition of reliable depth for multi-view 3D object detection. arXiv preprint arXiv:2206.10092 (2022)

34. Li, Z., et al.: BEVFormer: learning bird's-eye-view representation from multi-camera images via spatiotemporal transformers. In: Avidan, S., Brostow, G., Cissé, M., Farinella, G.M., Hassner, T. (eds.) ECCV 2022. LNCS, vol. 13669, pp. 1–18. Springer, Cham (2022). https://doi.org/10.1007/978-3-031-20077-9_1

35. Liao, B., et al.: MapTR: structured modeling and learning for online vectorized HD map construction. arXiv preprint arXiv:2208.14437 (2022)

36. Liong, V.E., Nguyen, T.N.T., Widjaja, S., Sharma, D., Chong, Z.J.: AMVNet: assertion-based multi-view fusion network for LiDAR semantic segmentation. arXiv preprint arXiv:2012.04934 (2020)

37. Liu, Y., Wang, Y., Wang, Y., Zhao, H.: VectorMapNet: end-to-end vectorized HD map learning. arXiv preprint arXiv:2206.08920 (2022)

38. Liu, Y., Zhang, J., Fang, L., Jiang, Q., Zhou, B.: Multimodal motion prediction with stacked transformers. In: CVPR (2021)

39. Loshchilov, I., Hutter, F.: SGDR: stochastic gradient descent with warm restarts. arXiv preprint arXiv:1608.03983 (2016)

40. Loshchilov, I., Hutter, F.: Decoupled weight decay regularization. arXiv preprint arXiv:1711.05101 (2017)

41. Mersch, B., Chen, X., Behley, J., Stachniss, C.: Self-supervised point cloud prediction using 3D spatio-temporal convolutional networks. In: CoRL, pp. 1444–1454 (2022)

42. Van Den Oord, A., Vinyals, O., Kavukcuoglu, K.: Neural discrete representation learning. arXiv preprint arXiv:1711.00937 (2017)

43. Philion, J., Fidler, S.: Lift, splat, shoot: encoding images from arbitrary camera rigs by implicitly unprojecting to 3D. In: Vedaldi, A., Bischof, H., Brox, T., Frahm, J.-M. (eds.) ECCV 2020. LNCS, vol. 12359, pp. 194–210. Springer, Cham (2020). https://doi.org/10.1007/978-3-030-58568-6_12

44. Ratliff, N.D., Bagnell, J.A., Zinkevich, M.A.: Maximum margin planning. In: Proceedings of the 23rd International Conference on Machine Learning, pp. 729–736 (2006)

45. Reading, C., Harakeh, A., Chae, J., Waslander, S.L.: Categorical depth distribution network for monocular 3D object detection. In: CVPR (2021)

46. Renz, K., Chitta, K., Mercea, O.B., Koepke, A., Akata, Z., Geiger, A.: PlanT: explainable planning transformers via object-level representations. arXiv preprint arXiv:2210.14222 (2022)

47. Roldao, L., de Charette, R., Verroust-Blondet, A.: LMSCNet: lightweight multiscale 3D semantic completion. In: 2020 International Conference on 3D Vision (3DV), pp. 111–119 (2020)

48. Rombach, R., Blattmann, A., Lorenz, D., Esser, P., Ommer, B.: High-resolution image synthesis with latent diffusion models. In: CVPR, pp. 10684–10695 (2022)

49. Simonyan, K., Zisserman, A.: Very deep convolutional networks for large-scale image recognition. arXiv abs/1409.1556 (2014)

50. Sutton, R.S.: Dyna, an integrated architecture for learning, planning, and reacting. ACM SIGART Bull. 2(4), 160–163 (1991)

51. Szegedy, C., et al.: Going deeper with convolutions. In: CVPR, pp. 1–9 (2015)

52. Tang, H., et al.: Searching efficient 3D architectures with sparse point-voxel convolution. In: Vedaldi, A., Bischof, H., Brox, T., Frahm, J.-M. (eds.) ECCV 2020. LNCS, vol. 12373, pp. 685–702. Springer, Cham (2020). https://doi.org/10.1007/978-3-030-58604-1_41

53. Tian, X., Jiang, T., Yun, L., Wang, Y., Wang, Y., Zhao, H.: Occ3D: a large-scale 3D occupancy prediction benchmark for autonomous driving. arXiv preprint arXiv:2304.14365 (2023)

54. Tong, W., et al.: Scene as occupancy. In: ICCV, pp. 8406–8415 (2023)
55. Vitelli, M., et al.: SafetyNet: safe planning for real-world self-driving vehicles using machine-learned policies. In: 2022 International Conference on Robotics and Automation (ICRA), pp. 897–904 (2022)
56. Wang, X., Zhu, Z., Huang, G., Chen, X., Lu, J.: DriveDreamer: towards real-world-driven world models for autonomous driving. arXiv preprint arXiv:2309.09777 (2023)
57. Wang, X., et al.: OpenOccupancy: a large scale benchmark for surrounding semantic occupancy perception. arXiv preprint arXiv:2303.03991 (2023)
58. Wei, Y., Zhao, L., Zheng, W., Zhu, Z., Zhou, J., Lu, J.: SurroundOcc: multi-camera 3D occupancy prediction for autonomous driving. In: ICCV, pp. 21729–21740 (2023)
59. Weng, X., Wang, J., Levine, S., Kitani, K., Rhinehart, N.: Inverting the pose forecasting pipeline with SPF2: sequential pointcloud forecasting for sequential pose forecasting. In: CoRL, pp. 11–20 (2021)
60. Yan, X., et al.: Sparse single sweep LiDAR point cloud segmentation via learning contextual shape priors from scene completion. In: AAAI, vol. 35, pp. 3101–3109 (2021)
61. Yang, K., Ma, E., Peng, J., Guo, Q., Lin, D., Yu, K.: BEVControl: accurately controlling street-view elements with multi-perspective consistency via BEV sketch layout. arXiv preprint arXiv:2308.01661 (2023)
62. Ye, D., et al.: LidarMultiNet: towards a unified multi-task network for LiDAR perception. arXiv preprint arXiv:2209.09385 (2022)
63. Ye, M., Wan, R., Xu, S., Cao, T., Chen, Q.: DRINet++: efficient voxel-as-point point cloud segmentation. arXiv preprint arXiv:2111.08318 (2021)
64. Ye, T., et al.: FusionAD: multi-modality fusion for prediction and planning tasks of autonomous driving. arXiv preprint arXiv:2308.01006 (2023)
65. Zeng, S., Zheng, W., Lu, J., Yan, H.: Hardness-aware scene synthesis for semi-supervised 3D object detection. TMM **26**, 9644–9656 (2024)
66. Zeng, W., et al.: End-to-end interpretable neural motion planner. In: CVPR (2019)
67. Zhang, Y., et al.: BEVerse: unified perception and prediction in birds-eye-view for vision-centric autonomous driving. arXiv preprint arXiv:2205.09743 (2022)
68. Zhao, L., et al.: LowRankOcc: tensor decomposition and low-rank recovery for vision-based 3D semantic occupancy prediction. In: CVPR, pp. 9806–9815 (2024)
69. Zhou, J., et al.: Exploring imitation learning for autonomous driving with feedback synthesizer and differentiable rasterization. In: IROS, pp. 1450–1457 (2021)
70. Zhu, X., et al.: Cylindrical and asymmetrical 3D convolution networks for LiDAR segmentation. In: CVPR, pp. 9939–9948 (2021)
71. Zuo, S., Zheng, W., Huang, Y., Zhou, J., Lu, J.: PointOcc: cylindrical tri-perspective view for point-based 3D semantic occupancy prediction. arXiv preprint arXiv:2308.16896 (2023)

MyVLM: Personalizing VLMs
for User-Specific Queries

Yuval Alaluf[1,2]([✉]), Elad Richardson[2], Sergey Tulyakov[1], Kfir Aberman[1],
and Daniel Cohen-Or[1,2]

[1] Snap Inc., Santa Monica, USA
yuvalalaluf@gmail.com
[2] Tel Aviv University, Tel Aviv-Yafo, Israel

Abstract. Recent large-scale vision-language models (VLMs) have demonstrated remarkable capabilities in understanding and generating textual descriptions for visual content. However, these models lack an understanding of *user-specific* concepts. In this work, we take a first step toward the *personalization* of VLMs, enabling them to learn and reason over user-provided concepts. For example, we explore whether these models can learn to *recognize* you in an image and *communicate* what you are doing, tailoring the model to reflect *your* personal experiences and relationships. To effectively recognize a variety of user-specific concepts, we augment the VLM with external concept heads that function as toggles for the model, enabling the VLM to identify the presence of specific target concepts in a given image. Having recognized the concept, we learn a new concept embedding in the intermediate feature space of the VLM. This embedding is tasked with guiding the language model to naturally integrate the target concept in its generated response. We apply our technique to BLIP-2 and LLaVA for personalized image captioning and further show its applicability for personalized visual question-answering. Our experiments demonstrate our ability to generalize to unseen images of learned concepts while preserving the model behavior on unrelated inputs. Code and data will be made available upon acceptance.

Keywords: Vision-Language Models · Personalization

1 Introduction

Large language models (LLMs) [85] have transformed human-computer interaction, offering users intuitive interfaces for interacting with textual information. The integration of vision into LLMs through vision-language models (VLMs) [81] has further enhanced this interaction, enabling these models to "see" and reason over visual content. However, current VLMs possess *generic* knowledge, lacking

Supplementary Information The online version contains supplementary material available at https://doi.org/10.1007/978-3-031-72624-8_5.

A. Leonardis et al. (Eds.): ECCV 2024, LNCS 15071, pp. 73–91, 2025.
https://doi.org/10.1007/978-3-031-72624-8_5

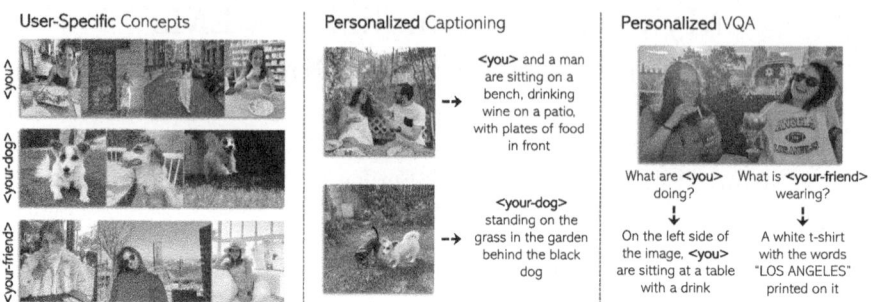

Fig. 1. Given a set of images depicting user-specific concepts such as ⟨you⟩, ⟨your-dog⟩ and ⟨your-friend⟩ (left), we teach a pretrained vision-language model (VLM) to understand and reason over these concepts. First, we enable the model to generate personalized captions incorporating the concept into its output text (middle). We further allow the user to ask subject-specific questions, querying the model with questions such as "What are ⟨you⟩ doing?" or "What is my ⟨your-friend⟩ wearing?" (right).

a personalized understanding of individual users. For example, the VLM can easily recognize an image of *a* dog but lacks the ability to understand that the depicted dog is *your* personal dog. This raises an intriguing question: can we equip these models with the ability to comprehend and utilize user-specific concepts, tailored specifically to *you*? That is, can we ask the model questions about you, such as what *you* are wearing or what *you* are doing in the image? By personalizing these models, we can offer more meaningful interactions, better reflecting individual experiences and relationships.

Introducing personalized concepts into existing models poses significant challenges. Attempting to fine-tune these models for each user is computationally expensive and prone to catastrophic forgetting [26,44]. In the context of LLMs, this has driven the development of model editing techniques designed to efficiently modify such large models [77]. Yet, these methods only focus on altering the model's response to *specific* user queries, for instance, editing the answer of "Where is ECCV this year?" from "Tel Aviv" to "Milan".

Successfully personalizing a VLM requires a deep understanding of how its visual and linguistic components interact. Intuitively, for a VLM to effectively respond to visual queries, it must not only *recognize* and extract the relevant visual elements but also meaningfully *communicate* them in its response. Introducing another layer of complexity to VLM personalization, we also find that the visual features extracted by pretrained VLMs are not expressive enough to effectively distinguish between semantically-similar objects.

To address these challenges, we propose augmenting the VLM with external heads that are trained to identify user-specific concepts within a scene. The signal from these heads is then used to add specific learnable vectors alongside the outputs of the vision encoder. In a sense, these learnable vectors are tasked with guiding the response generated by the language model to incorporate the matching personalized word in a way that is contextually accurate and aligned with the input image. To train this concept vector, we are given a small set of images (3–5) depicting the concept, each with a corresponding caption containing the personalized word. We then optimize the concept embedding such that when given an image from the training set, appending the concept's embedding to the output of the vision encoder results in the VLM generating the corresponding personalized target caption. To encourage the learnable embedding to remain in distribution with respect to the other image tokens, we incorporate an additional regularization over the attention assigned by the VLM to the concept embedding.

Our personalization technique, named MyVLM, enables users to personalize a pretrained VLM without altering the original weights, preserving the model's general capabilities. Focusing on personalized image captioning, we apply MyVLM to both BLIP-2 [48] and LLaVA [52], further demonstrating its applicability for visual-question answering, see Fig. 1. We show that MyVLM can effectively incorporate and contextualize personalized concepts, including specific objects and individuals, requiring only a few images of the concept. We introduce and assess alternative baselines, highlighting our ability to better generalize to new instances of previously learned concepts. To evaluate this new task, we introduce a new dataset containing various objects and individuals depicted in multiple contexts each with a corresponding personalized caption. The object dataset will be publicly available, aiming to facilitate further advancements in the personalization of VLMs.

2   Related Works

Vision-Language Models (VLMs). The recent remarkable progress of large language models (LLMs) [15,18,19,68,70], has spurred efforts to equip them with the ability to reason over visual content [1,3,8,37,46,49,59,67,76,79,82,86].

A key area of research on VLMs focuses on leveraging frozen LLMs to align images and text within unified models that support both visual and language inputs. For instance, Flamingo [3] fuses vision and language modalities using a cross-attention mechanism while keeping the vision encoder and language model fixed. BLIP-2 [48] introduces a Q-Former transformer to align visual features extracted from a fixed visual encoder with a large language model [19,84]. LLaVA [51,52] and MiniGPT-4 [86] employ instruction-tuned language models [22,57,75] and extract visual features from a pretrained visual encoder (e.g., CLIP [60]). Specifically, LLaVA [52] utilizes a simple linear layer to map the visual features to the input space of the language model.

Recently, VLMs have been adopted for guiding various downstream tasks such as reinforcement learning [16] and image generation [14,62]. In this work, our focus is on personalizing VLMs, enabling them to reason over user-specific concepts. Importantly, our approach does not modify the original weights of the VLM, preserving its strong visual and linguistic priors. We apply our method to both BLIP-2 [48] and LLaVA [52], demonstrating its effectiveness as a general framework applicable across various VLMs.

Personalization. In the task of personalization, we aim to adapt a given model to capture new user-specific concepts. Personalization has been explored for a range of tasks including recommendation systems [4, 13] and object retrieval [10, 21, 41, 64, 80]. PALAVRA [21] optimizes a new token embedding within the input space of a text encoder to represent a new concept while Yeh *et al.* [80] extend this for retrieving concepts in videos. Personalization has also been heavily studied in the context of image generation [2, 6, 29, 30, 43, 47, 61, 63, 71, 73, 78][?]. Most relevant to our work are inversion-based approaches where embeddings are optimized to capture the target concept.

Another line of work focuses on personalizing image captioning models [20, 58, 65, 74, 83]. Park *et al.* [58] employ a memory network to store a user's active vocabulary and utilizes it to generate captions reflecting the user's personal writing style. More recently, Wang *et al.* [74] employed a transformer to fuse visual features and text features encoding user-specific keywords. These features are then passed to a pretrained language model to generate personalized captions. Importantly, personalized captioning techniques focus on generating a specific *writing style*. In contrast, we aim to teach the model to incorporate a new user-specific concept into a personalized textual output aligned with a given image.

Model Editing. While modern machine learning systems excel in achieving state-of-the-art performance, their effectiveness can diminish post-deployment [9], leading to hallucinations [12, 39] and factual decay [38, 66]. Consequently, there is a growing need for model editing, which aims to make data-efficient modifications to a model's behavior while minimizing the impact on performance across other inputs. In the context of language models, several approaches incorporate hypernetworks [33] to predict edits for specific inputs [23, 55, 56] or perform parameter-efficient model tuning [36, 50, 53, 54]. One particular area of interest is enabling a large set of edits within a single model [34, 54]. Hartvigsen *et al.* [34] introduce a codebook within the language model's intermediate feature space, storing previously learned edits. For each new edit, a new key is added to the codebook, and its corresponding value is optimized such that the language model produces the desired output for the given query. Similar model editing techniques have been explored for generative image models [5, 11, 31, 42, 69] and multi-modal learning [17]. Recently, Retrieval-Augmented Generation (RAG) has also emerged as an alternative approach for injecting knowledge into LLMs [32, 45, 72]. We refer the reader to Yao *et al.* [77] for a comprehensive survey on model editing.

Our goal of personalizing VLMs necessitates a different approach from model editing. Model editing focuses on applying precise modifications to the model behavior (e.g., associating "What is the capital of France?" with "Paris"). In contrast, personalization requires the model to adapt to new images of the concept, which may vary significantly (e.g., recognizing an individual across diverse settings). Moreover, it is essential to disentangle the concept from its surroundings when teaching a model a new concept, such as separating an individual from the clothes they are wearing. Finally, the VLM must not only identify the concept but also contextualize it within the generated response. For example, instead of simply outputting the concept identifier "S_*", the model should produce a more descriptive response such as "S_* sitting on a bench, drinking wine on a patio".

3   Method

Our goal is to extend the capabilities of a vision-language model (VLM) by teaching it to generate personalized textual responses focusing on user-specific concepts. We begin by outlining the specific families of VLM models considered in this work, namely BLIP-2 [48] and LLaVA [52]. We then introduce our personalization technique, MyVLM, and demonstrate its application for both personalized captioning and visual question-answering.

3.1   Preliminaries

BLIP-2. The BLIP-2 model, introduced by Li *et al.* [48], is a VLM model that is built around three main components: (1) a pretrained ViT-L/14 [27] vision encoder, (2) a pretrained language model [19], and (3) a trainable Querying Transformer (Q-Former) model tasked with bridging the vision-language modality gap. The Q-Former receives as input 32 learnable query tokens, each of dimension $d = 768$, and is composed of three types of layers: self-attention, cross-attention, and feed-forward layers. Most relevant to our work are the cross-attention layers, placed at every other transformer block. These blocks are designed to capture the interaction between the extracted image features and the learnable query tokens (as well as our learned concept representations).

More specifically, at each cross-attention layer, the image features are first projected into a set of keys (K) and values (V) via learned linear projections. The intermediate representations of the 32 learned query tokens are similarly projected into a set of attention queries q_i. For each query q_i, a weighted average is then computed over these representations, as given by:

$$A_i = \text{softmax}\left(\frac{q_i \cdot K^T}{\sqrt{d}}\right) V.$$ (1)

Intuitively, the probability defined by the softmax indicates the amount of information that will be passed from each image feature to each query token.

LLaVA. Similar to BLIP, LLaVA [52] seeks to connect a fixed vision encoder with a fixed language model, in this case, CLIP ViT-L/14 [60] and Vicuna [18] models, respectively. To do this, LLaVA follows a simpler architecture where a single linear layer is used to map the image features into the token embedding space of the language model. This sequence of projected visual tokens is then fed directly to the language model, along with the encoded language instruction.

3.2   MyVLM

We now turn to describe our approach to personalizing vision-language models for user-specific concepts. For simplicity, we describe MyVLM applied over the BLIP-2 model [48], followed by a discussion of the adjustments necessary for integrating MyVLM with LLaVA [52]. Given only a few images ($\sim$3-5) of the

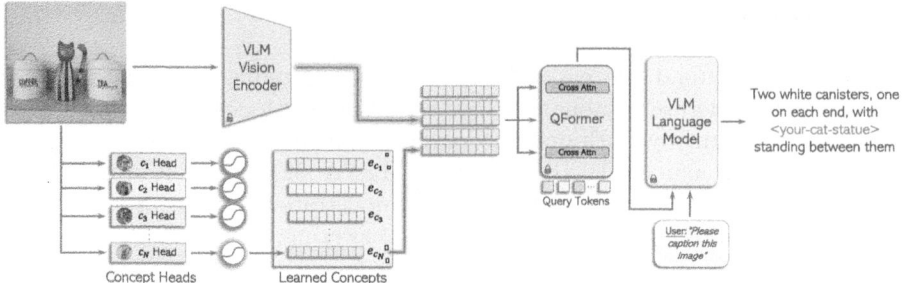

Fig. 2. MyVLM overview, applied over BLIP-2. Given an input image, we pass it through the frozen vision encoder of the VLM. In parallel, we pass the image through a set of learned *concept heads*, each tasked with recognizing a single user-specific concept. We append the *concept embedding* of the identified concept to the extracted vision features. These features are then passed to the Q-Former via a set of cross-attention layers to extract relevant information from the image features and concept embedding. Given the Q-Former outputs and language instruction, the frozen LLM outputs a response incorporating the concept identifier while remaining aligned with the input.

specific concept and corresponding captions that contain the concept identifier S_*, our objective is to augment the VLM with the ability to answer specific queries over new images depicting the concept. Our technique is comprised of two key stages: first *recognizing* the concept within the given scene, and then *communicating* information about the concept to the language model. To achieve this, we introduce a *concept head* designed to identify the presence of a personalized concept within an image. Then, a learned *concept embedding*, representing an object or individual, is used to guide the LLM in incorporating the concept into its personalized textual response (Fig. 2).

Recognizing. To enable the pretrained VLM to reason over personalized concepts, we must first identify their presence in a given scene. A direct approach for doing so is to consider the feature space of the VLM's vision encoder. However, we empirically observe that the feature space of the frozen vision encoder is not expressive enough to visually distinguish the target concept from similar concepts (see supplementary). While one can potentially fine-tune the vision encoder itself to better recognize our object of interest, this may naturally harm its strong general knowledge and impact its ability to extract information about the entire image, which is also crucial for generating accurate responses.

Instead, we augment the VLM with a set of external *concept heads*, with each head dedicated to recognizing a single personalized concept we wish to teach the model. These heads allow the model to identify the concepts of interest without hindering its ability to provide visual information about the entire scene depicted in the image. As the heads operate independently from the VLM model itself, we can support any specialized classification head to recognize our target concepts. Specifically, for identifying user-specific objects, we choose to employ a simple linear classifier trained over embeddings extracted from a pretrained CLIP model [28,60]. To generate personalized outputs tailored to specific individuals, we utilize a pretrained face recognition network [24,25] as an additional concept head. Importantly, defining a separate head for each concept provides additional flexibility, enabling one to naturally scale to additional concepts over time.

Communicating. Given the ability to *recognize* our concept of interest, we now turn to describe our approach for teaching the VLM to *communicate* responses about our target concepts. To do so, we learn a single concept embedding vector representing the concept within the intermediate feature space of the VLM. Intuitively, this embedding should guide the language model toward generating a text response incorporating the concept identifier that (1) is contextually correct and (2) aligns with both the provided image and language instruction.

To learn this embedding, we use a small set of images depicting the concept in various contexts, each with a corresponding target caption containing the concept identifier. For the identifier, we follow DreamBooth [63] and use an existing, uncommon word when personalizing outputs for objects and use a short name when personalizing individuals. We find the concept embedding e_* via direct optimization. The embedding e_* is appended to the image features extracted from the frozen vision encoder and fed to the Q-Former network via the cross-attention layers. The output of the Q-Former is then passed to the frozen language model that generates the predicted image caption. The optimization process aims to minimize the standard cross-entropy loss between the generated caption and the provided target caption. Our optimization can be defined as:

$$e_* = \arg\min_e \sum_{i=1}^{N} \mathcal{L}_{CE}\left(t_i, o(I_i, e)\right), \qquad (2)$$

where N is the number of training samples, t_i represents our target caption of the i-th sample, and $o(I_i, e)$ is the generated output caption of the i-th image I_i, given the concept embedding e. At inference, the embedding of a concept recognized by our concept heads is appended to the output of the vision encoder.

Improving Generalization. While the approach described above allows for generating personalized captions, we observe that directly appending the concept embedding to the image features may lead to unnatural captions being generated by the language model. This issue arises from two primary observations.

First, within the cross-attention layers of the Q-Former, we observed that the vector norms of the key (k_*) and value (v_*) corresponding to the concept embedding were significantly larger compared to the norms of the frozen image features. This behavior was also previously observed in text-to-image personalized techniques [2,69]. Therefore, before computing the cross-attention with the Q-Former query tokens, we normalize k_* and v_* to match the average norm of the original keys and values, denoted as n_k and n_v, respectively. The modified key and value of our embedding are then given by:

$$\hat{k}_* = (k_*/\|k_*\|) \cdot n_k \qquad\qquad \hat{v}_* = (v_*/\|v_*\|) \cdot n_v \qquad (3)$$

"S_*, dressed in a blue jacket and a green sweater..."   "S_* and a black dog running in a yard"   "S_* and a Chinese doll standing next to a gold gong..."   "S_* is sitting next to a coffee mug with a cartoon character..."

Fig. 3. Self-attention visualization. We examine the self-attention of LLaVA's language model to visualize the attention weights assigned from the concept embedding to each image feature. As can be seen, the concept embedding attends to relevant regions within the images, assigning higher weights to areas where the concept is located.

Second, in the attention weights computed in the Q-Former cross-attention layers (Eq. (1)), we observe that the concept token tended to dominate the attention distribution, causing the query tokens to no longer attend meaningfully to the image tokens. By failing to adequately attend to the original image tokens, the relevant visual information may no longer be passed to the language model, leading to a possible misalignment between the generated caption and the image. To encourage a more balanced distribution of attention across all tokens, we introduce an L2 regularization over the attention probabilities assigned to the concept embedding by all 32 Q-Former query tokens. That is, we compute:

$$\mathcal{L}_{reg} = \left\| \operatorname{softmax}\left(Q \cdot \hat{k}_* \right) \right\|_2^2. \tag{4}$$

By encouraging the tokens to attend to the original image features, we found the outputs to be more coherent and aligned with the image (see supplementary).

3.3   MyVLM over LLaVA

To apply MyVLM over LLaVA [52] we make the following adjustments to the scheme presented above. First, we append the concept embeddings to the output of the linear projection rather than directly after the vision encoder. We find that this resulted in faster, more stable convergence. Second, since LLaVA does not utilize a cross-attention mechanism, we omit the normalization of keys and values as presented in Eq. (3). Instead, we rescale the concept embedding such that its vector norm is equal to that of the [CLS] token outputted by the vision encoder. Finally, we modify the attention-based regularization defined in Eq. (4). Here, we apply an L2 regularization that encourages low attention to be assigned from the other input tokens to the concept embedding, including from both the language tokens and from the other projected image tokens.

Interestingly, since our concept embedding is passed as input to the language model along with the other projected image features, we have a natural way to investigate whether our learned concept embeddings attend to meaningful regions within the input images. Specifically, we examine the self-attention layers of LLaVA's language model and visualize the attention weights assigned by the concept embedding to each of the image patches, as illustrated in Fig. 3.

3.4   MyVLM for VQA

For applying MyVLM for personalized visual question-answering, we follow a similar approach as introduced above, but modify the language instructions and target outputs used for defining our objective function.

Observe that in personalized captioning, the language instruction passed to the language model when optimizing the concept embedding remains fixed. However, for visual question-answering, we are interested in generalizing to any question the user may ask over a given image. Therefore, we expand the set of instructions and targets used during the optimization process described above. Specifically, we define a set of 10 pairs of questions and answers related to the target concept. Then, at each optimization step, we randomly sample one question-answer pair to use for the current step. Intuitively, by optimizing the embedding vector through questions aimed specifically at the target concept, the embedding should better generalize to new questions the user may ask about the concept.

4   Experiments

Dataset. As there are no existing datasets for VLM personalization, we introduce a new dataset for evaluating this task. The dataset is split into two categories: objects and people. For objects, we curate a set of 29 objects including various toys, statues, mugs, and pets. For each concept, we collected at least 10 images containing the subject in diverse scenes alongside other objects and set against interesting backgrounds. For people, we collect images of 16 individuals ranging from ages 25 to 80. Each individual is represented by a minimum of 15 images, showcasing them in a range of scenarios, attire, and sometimes alongside other people in the same image. For each image, we wrote a corresponding personalized caption incorporating the concept identifier. In total, the dataset comprises over 680 pairs of images and captions. The subset of the 29 objects will be publicly available to facilitate further research into VLM personalization.

Evaluation Metrics. In this work, we focus on quantitatively evaluating personalized image captioning, as data for this task is more readily available. We evaluate the personalized captions along two fronts. First, we measure recall and validate whether the concept identifier appears at least once in the generated caption. This evaluates both our ability to recognize the concept in new images and our ability to incorporate the concept in the output via its embedding.

Second, we assess the alignment of the generated caption with the input image and target caption, considering two metrics. We first compute the CLIP-Score [35] between the generated captions and input images. We additionally compute a sentence similarity measure, computing the average cosine similarity

LLaVA	**LLaVA**	**LLaVA**	**LLaVA**	**LLaVA**
"Friends enjoying a day out in the city, posing for a photo on a cobblestone street"	"Friends sharing a moment by the water, enjoying a coffee break and a laugh"	"Sipping on sunshine: A moment of joy under the blue sky"	"A cat's curious paw reaches out to a laptop keyboard. The laptop displays a question..."	"A well-stocked refrigerator, ready for a weekend of culinary adventures!"
MyVLM	**MyVLM**	**MyVLM**	**MyVLM**	**MyVLM**
"S_*, dressed in a blue jacket and a green sweater, takes a selfie with his friends, who are also bundled up against the chilly weather... "	"S_*, a man and a woman are posing for a photograph with a table between them. S_* is wearing a denim jacket and a necklace, ..."	"Sitting at a table on a patio, S_* wearing a yellow dress, smiling at the camera, with the city skyline in the background"	"S_* sitting in front of a laptop on a wooden table with a question about how to write papers fast and efficiently?"	"S_* sits comfortably on the second shelf of an open refrigerator, ready to be stocked with a variety of food and drink items"

Fig. 4. Personalized captioning results obtained by MyVLM, applied over LLaVA [52]. Text in green highlights the description of the target concept in the image. (Color figure online)

between sentence embeddings extracted from the target caption and the generated caption. For both, we replace the concept identifier with the concept's category. In the supplementary, we present standard captioning metrics, showing that MyVLM preserves the general captioning capabilities of the underlying vision-language model.

Baselines. Since there are currently no existing baselines focusing on generating personalized captions for a target concept, we introduce several alternative approaches for doing so. First, we generate captions using the frozen VLM model. Then, for each concept, we define a set of three keywords describing the concept, obtained using GPT-4V [1] by providing it a cropped image of the concept. For people, we designate a single keyword per concept, either "man" or "woman". Given the caption generated by the VLM, we then search the caption for the keyword, and if found, we replace the keyword with the concept identifier.

Additionally, we introduce an LLM-guided baseline. Here, given the captions generated by the frozen VLM, we pass the caption into a language model [40] and ask it to integrate the concept identifier into the caption if one of the keywords is present. This approach offers a more flexible constraint, allowing the language model to more freely incorporate the concept into the caption.

Finally, we compare MyVLM with GPT-4V [1] by showing GPT-4V an image of the concept and its identifier and then asking it questions over new images.

LLM-Guided	**LLM-Guided**	**LLM-Guided**	**LLM-Guided**	**LLM-Guided**
"A cute cavalier king charles spaniel relaxing in a blue polka dot S_* bed"	"a cozy scene with a soft, pink S_* and a white lamb, ready for a nap on a gray couch"	"friendly fidos: two S_*s, one white and one black, pose for a photo on a grassy lawn..."	"Friends celebrating with funny hats and mustaches, S_* ready to party"	"Two S_* sitting at an outdoor table with food and drinks"
MyVLM	**MyVLM**	**MyVLM**	**MyVLM**	**MyVLM**
"A happy S_* laying in his blue dog bed on a white office floor"	"S_* sitting on the couch with a pink and white stuffed animal next to it"	"S_* is standing on the grass with a big smile and a wagging his tongue"	"In her living room, S_* and two friends are dressed in party hats and mustaches."	"S_* and a friend enjoying coffee and a sandwich at a cafe"

Fig. 5. Comparison to the LLM-guided captioning baseline. Results are obtained over LLaVA [52]. Sample images of the target concept are shown in the top row.

GPT-4V	"S_* placed next to a bottle of "Supreme Cabernet Sauvignon" wine..."	"A whimsically designed mug with a face which could be referred to as S_*"	"S_* is the small cup with a blue eye design on it, located on the right side of the image"	"S_* is in this image, identifiable as the cup with the blue eye design..."	"S_* is the figurine in the foreground , the background shows a scenic landscape..."	"S_* is the figurine in the image, depicted standing on a green base..."
MyVLM	"A shelf with S_* and wine glasses and a bottle of supreme wine"	"Whimsical Woodland Creature Sipping Tea"	"S_* is sitting next to a cup of coffee with a "bottomless cup" sign..."	"A whimsical tea party setup with a trio of coffee cups..."	"S_* in front of a picture of the grand canyon"	"Ready to score!"

Fig. 6. Comparison to GPT-4V [1]. We provide GPT-4V an image of the target concept (shown at the bottom left of each image) and ask whether the concept is present in new images. Results shown in red indicate incorrect false positives while results in green are correctly captioned negative images that do not contain the concept. (Color figure online)

Similarly, in the supplementary, we quantitatively compare MyVLM to Open-Flamingo [3, 7], which also supports interleaved image-text inputs.

Table 1. Quantitative Comparison: Recall. We compute the percent of generated captions that contain the concept identifier. Results are averaged over all concepts and 5 validation sets.

		Objects	People	All
BLIP	Simple Replace	29.30	**84.33**	<u>59.33</u>
	LLM-Guided	<u>51.55</u>	56.91	54.37
	MyVLM	**95.10**	<u>79.76</u>	**87.11**
LLaVA	Simple Replace	25.86	18.13	21.68
	LLM-Guided	<u>65.38</u>	<u>29.11</u>	<u>46.23</u>
	MyVLM	**94.76**	**97.08**	**95.97**

Table 2. Ablation Study: Number of Training Samples. We compute the average recall, text-to-image similarity, and text-to-text similarity, comparing results obtained using 1, 2, and 4 images for training the concept embedding. Results are averaged across all concepts and all five validation sets.

		Recall ↑	Image ↑	Text ↑
BLIP	MyVLM (1)	75.42	24.20	57.37
	MyVLM (2)	<u>84.27</u>	<u>24.91</u>	<u>61.01</u>
	MyVLM (4)	**87.11**	**25.42**	**62.61**
LLaVA	MyVLM (1)	88.93	23.44	50.39
	MyVLM (2)	<u>92.88</u>	<u>24.43</u>	<u>53.32</u>
	MyVLM (4)	**95.97**	**25.24**	**56.98**

4.1 Personalized Captioning

Qualitative Evaluation. In Fig. 4, we present personalized captioning for various user-provided concepts generated by our method applied to LLaVA [52]. Captions generated by MyVLM emphasize the target subject rather than offering a generic or abstract description of the entire scene, as generated by the original VLM. Moreover, MyVLM naturally integrates the concept identifier into the generated output while remaining aligned with the input image. In particular, even in scenes where multiple individuals are present in the image, MyVLM successfully focuses on the target identity when generating its caption. For instance, notice the man in the green sweater in the first column or the woman in the yellow dress in the third column. This is also evident when creating personalized captions for a user-provided object placed around numerous other objects in a scene. For instance, in the rightmost column, the original caption generated by LLaVA ignores the target ceramic mug entirely, whereas our personalized caption accurately communicates its location in the image.

Qualitative Comparison. In Fig. 5, we provide a visual comparison with our LLM-guided baseline. As can be seen, this baseline heavily relies on the original captions generated by the VLM. The baseline struggles when the target concept appears in the same image with another subject sharing the same keyword, resulting in an unnatural caption. In contrast, MyVLM successfully identifies the target subject and generates captions that accurately contextualize the concept within its surroundings. Importantly, we do so when multiple subjects are present and when the concept comprises a small region of the image.

Next, we compare our method to GPT-4V in Fig. 6. We provide it with an image of the target concept along with its identifier. We then ask it to caption images that may contain the concept. As can be seen, GPT-4V can generalize to new images of the concept. However, when presented with images of negative examples that have a similar textual description, GPT-4V misidentifies them as the target concept. For example, in the middle example, it incorrectly associates "a cup with a blue eye design" with the concept. In contrast, MyVLM can distinguish between these hard negative examples and the target concepts.

Interestingly, the fact that GPT-4V misidentifies visually distinct objects that share a similar textual description may hint that it heavily relies on the textual description of the object, even when prompted with an image of it. This emphasizes the advantage of *learning* a dedicated embedding to represent our concept instead of relying solely on natural language, where describing our *exact* target concept may be challenging.

Quantitative Comparison. We now turn to quantitatively compare MyVLM with the alternative baselines. To provide a larger validation sample size, we perform bootstrapping without replacement over our constructed dataset. For each concept, we randomly sample five different training sets, each containing four images, and set the remaining images as the corresponding validation set. We then train MyVLM on each training set and generate captions for all validation images. This results in a total of $2,430$ validation images, out of which $1,265$ contain user-specific objects, while the remaining images depict individuals.

We begin by measuring each baseline's ability to incorporate the concept identifier within the generated caption. Results are summarized in Table 1. As can be seen, for user-specific objects, trying to simply insert the concept identifier into the caption via a closed set of keywords is ineffective, with a notable gap in recall compared to MyVLM. While incorporating an external language model greatly improves recall, MyVLM still outperforms the LLM-guided baseline by 44% when using BLIP-2 and 30% for LLaVA. When considering individuals, although the keyword-replacement baseline and MyVLM achieve comparable results when applied over BLIP, MyVLM significantly outperforms both baselines when applied to LLaVA. The large gap to LLaVA appears to stem from the abstract-like captions generated by LLaVA, whereas BLIP-2 tends to generate simpler captions more likely to incorporate the predefined keywords. This highlights the robustness of MyVLM to different VLM models, whereas the hand-crafted baselines heavily rely on the captioning styles of the underlying VLM.

Next, we investigate MyVLM's performance when training the concept embedding using 4, 2, and only 1 image. Results, averaged across all 45 concepts, are presented in Table 2. In terms of recall, results over both BLIP-2 and LLaVA consistently improve when adding more training samples. Observe that even when trained using a single sample, MyVLM still outperforms all baselines by significant margins. We additionally compute the average similarities between our personalized captions and (1) the input images and (2) the target captions. As can be seen, adding additional training samples improves both the image similarity and text similarity, indicating improved generalization.

In the supplementary, we provide additional ablation studies on the contribution of our augmentations and regularization techniques. We additionally explore the output space of the VLM vision encoder and validate the use of our concept heads, showing that they attain both high recall over new images of the target concept and high precision over negative samples, demonstrating our ability to support multiple concepts in a single VLM.

4.2 Personalized Visual Question-Answering

Finally, we demonstrate that MyVLM can also be used for personalized visual question-answering. In Fig. 7, we demonstrate results across several user-specific concepts. MyVLM correctly answers questions related to the target concept, even within complex scenes containing multiple individuals (columns one and two), and in scenes where the subject occupies a small area of the image (columns three and four). For instance, MyVLM not only correctly identifies that the dangling child toy is located in the refrigerator but also its precise location on the top shelf. This highlights that MyVLM can faithfully capture distinctive features associated with the target concept, allowing it to correctly identify and localize the concept in a new scene.

"Can you describe what S_* is wearing?"	"Can you describe what S_* is wearing?"	"Where is S_* positioned in image?"	"From left to right, where is S_* located?"	"What is standing next to S_*?"
"In the image, S_* is wearing a white sleeveless top"	"S_* is wearing a brown sweater and has curly hair"	"S_* is positioned at the top of the refrigerator, sitting on a shelf with various food items and containers"	"S_* is located on the right side of the image, sitting in a green bowl next to a wooden bowl containing oranges and pears"	"S_* is a small figurine of a character wearing a pink hat with a blue flower on it. Next to S_*, there is a blue mug with pens and pencils"

Fig. 7. Personalized VQA results obtained by MyVLM over LLaVA [52]. Text in green highlights the description of the target concept in the image. (Color figure online)

5 Conclusions

In this paper, we introduce the idea of vision-language personalization, enabling VLMs to understand and reason over user-specific concepts, such as unique objects and individuals. As a first step in this endeavor, we present MyVLM, focusing on personalized captioning and VQA. Given only a few images of the concept, we augment the frozen VLM with a set of modular concept heads, enabling it to *recognize* user-specific concepts. We then train an embedding vector within the VLM's intermediate feature space, tasked with guiding the language model in incorporating the concept into the generated response in a natural and contextually accurate manner. We believe that the personalization of vision-language models opens up new opportunities for more meaningful human-computer interactions, and hope MyVLM will inspire additional advancements in this field.

Acknowledgements. We would like to thank Assaf Ben-Kish, Or Patashnik, Moran Yanuka, Morris Alper, Yonatan Biton, and Yuwei Fang for their fruitful discussions and valuable input which helped improve this work. This work was supported by the Israel Science Foundation under Grant No. 2366/16 and Grant No. 2492/20.

References

1. GPT-4 technical report (2023)
2. Alaluf, Y., Richardson, E., Metzer, G., Cohen-Or, D.: A neural space-time representation for text-to-image personalization (2023)
3. Alayrac, J.B., et al.: Flamingo: a visual language model for few-shot learning. In: Advances in Neural Information Processing Systems, vol. 35, pp. 23716–23736 (2022)
4. Amat, F., Chandrashekar, A., Jebara, T., Basilico, J.: Artwork personalization at Netflix. In: Proceedings of the 12th ACM Conference on Recommender Systems, pp. 487–488 (2018)
5. Arad, D., Orgad, H., Belinkov, Y.: ReFACT: updating text-to-image models by editing the text encoder (2023)
6. Arar, M., et al.: Domain-agnostic tuning-encoder for fast personalization of text-to-image models (2023)
7. Awadalla, A., et al.: OpenFlamingo: an open-source framework for training large autoregressive vision-language models. arXiv preprint arXiv:2308.01390 (2023)
8. Bai, J., et al.: Qwen-VL: a frontier large vision-language model with versatile abilities. arXiv preprint arXiv:2308.12966 (2023)
9. Balachandran, V., Hajishirzi, H., Cohen, W.W., Tsvetkov, Y.: Correcting diverse factual errors in abstractive summarization via post-editing and language model infilling. arXiv preprint arXiv:2210.12378 (2022)
10. Baldrati, A., Agnolucci, L., Bertini, M., Del Bimbo, A.: Zero-shot composed image retrieval with textual inversion. arXiv preprint arXiv:2303.15247 (2023)
11. Bau, D., Liu, S., Wang, T., Zhu, J.-Y., Torralba, A.: Rewriting a deep generative model. In: Vedaldi, A., Bischof, H., Brox, T., Frahm, J.-M. (eds.) ECCV 2020. LNCS, vol. 12346, pp. 351–369. Springer, Cham (2020). https://doi.org/10.1007/978-3-030-58452-8_21
12. Ben-Kish, A., Yanuka, M., Alper, M., Giryes, R., Averbuch-Elor, H.: MOCHa: multi-objective reinforcement mitigating caption hallucinations (2023)
13. Benhamdi, S., Babouri, A., Chiky, R.: Personalized recommender system for e-Learning environment. Educ. Inf. Technol. **22**, 1455–1477 (2017)
14. Black, K., Janner, M., Du, Y., Kostrikov, I., Levine, S.: Training diffusion models with reinforcement learning. arXiv preprint arXiv:2305.13301 (2023)

15. Brown, T., et al.: Language models are few-shot learners. In: Advances in Neural Information Processing Systems, vol. 33, pp. 1877–1901 (2020)
16. Chen, W., Mees, O., Kumar, A., Levine, S.: Vision-language models provide promptable representations for reinforcement learning (2024)
17. Cheng, S., et al.: Can we edit multimodal large language models? arXiv preprint arXiv:2310.08475 (2023)
18. Chiang, W.L., et al.: Vicuna: an open-source chatbot impressing GPT-4 with 90%* ChatGPT quality (2023). https://vicuna.lmsys.org. Accessed 14 Apr 2023
19. Chung, H.W., et al.: Scaling instruction-finetuned language models. arXiv preprint arXiv:2210.11416 (2022)
20. Chunseong Park, C., Kim, B., Kim, G.: Attend to you: personalized image captioning with context sequence memory networks. In: Proceedings of the IEEE Conference on Computer Vision and Pattern Recognition, pp. 895–903 (2017)
21. Cohen, N., Gal, R., Meirom, E.A., Chechik, G., Atzmon, Y.: "This is my unicorn, fluffy": personalizing frozen vision-language representations. In: Avidan, S., Brostow, G., Cissé, M., Farinella, G.M., Hassner, T. (eds.) ECCV 2022, Part XX. LNCS, vol. 13680, pp. 558–577. Springer, Cham (2022). https://doi.org/10.1007/978-3-031-20044-1_32
22. Dai, W., et al.: InstructBLIP: towards general-purpose vision-language models with instruction tuning (2023)
23. De Cao, N., Aziz, W., Titov, I.: Editing factual knowledge in language models. In: Moens, M.F., Huang, X., Specia, L., Yih, S.W.T. (eds.) Proceedings of the 2021 Conference on Empirical Methods in Natural Language Processing, Punta Cana, Dominican Republic, pp. 6491–6506. Association for Computational Linguistics (2021). https://doi.org/10.18653/v1/2021.emnlp-main.522. https://aclanthology.org/2021.emnlp-main.522
24. Deng, J., Guo, J., Ververas, E., Kotsia, I., Zafeiriou, S.: RetinaFace: single-shot multi-level face localisation in the wild. In: Proceedings of the IEEE/CVF Conference on Computer Vision and Pattern Recognition, pp. 5203–5212 (2020)
25. Deng, J., Guo, J., Yang, J., Xue, N., Kotsia, I., Zafeiriou, S.: ArcFace: additive angular margin loss for deep face recognition. IEEE Trans. Pattern Anal. Mach. Intell. 44(10), 5962–5979 (2022). https://doi.org/10.1109/TPAMI.2021.3087709
26. Ding, Y., Liu, L., Tian, C., Yang, J., Ding, H.: Don't stop learning: towards continual learning for the CLIP model (2022)
27. Dosovitskiy, A., et al.: An image is worth 16×16 words: transformers for image recognition at scale. arXiv preprint arXiv:2010.11929 (2020)
28. Fang, A., Jose, A.M., Jain, A., Schmidt, L., Toshev, A., Shankar, V.: Data filtering networks. arXiv preprint arXiv:2309.17425 (2023)
29. Gal, R., et al.: An image is worth one word: personalizing text-to-image generation using textual inversion. In: The Eleventh International Conference on Learning Representations (2023). https://openreview.net/forum?id=NAQvF08TcyG
30. Gal, R., Arar, M., Atzmon, Y., Bermano, A.H., Chechik, G., Cohen-Or, D.: Encoder-based domain tuning for fast personalization of text-to-image models. ACM Trans. Graph. (2023). https://doi.org/10.1145/3592133
31. Gandikota, R., Materzynska, J., Fiotto-Kaufman, J., Bau, D.: Erasing concepts from diffusion models. arXiv preprint arXiv:2303.07345 (2023)
32. Gao, Y., et al.: Retrieval-augmented generation for large language models: a survey. arXiv preprint arXiv:2312.10997 (2023)
33. Ha, D., Dai, A.M., Le, Q.V.: Hypernetworks. In: International Conference on Learning Representations (2017). https://openreview.net/forum?id=rkpACe1lx

34. Hartvigsen, T., Sankaranarayanan, S., Palangi, H., Kim, Y., Ghassemi, M.: Aging with grace: lifelong model editing with discrete key-value adaptors. In: Advances in Neural Information Processing Systems (2023)

35. Hessel, J., Holtzman, A., Forbes, M., Le Bras, R., Choi, Y.: CLIPScore: a reference-free evaluation metric for image captioning. In: Moens, M.F., Huang, X., Specia, L., Yih, S.W.T. (eds.) Proceedings of the 2021 Conference on Empirical Methods in Natural Language Processing, Punta Cana, Dominican Republic, pp. 7514–7528. Association for Computational Linguistics (2021). https://doi.org/10.18653/v1/2021.emnlp-main.595. https://aclanthology.org/2021.emnlp-main.595

36. Hu, E.J., et al.: LoRA: low-rank adaptation of large language models. arXiv preprint arXiv:2106.09685 (2021)

37. Huang, S., et al.: Language is not all you need: aligning perception with language models. arXiv preprint arXiv:2302.14045 (2023)

38. Muneeswaran, I., et al.: Minimizing factual inconsistency and hallucination in large language models (2023)

39. Ji, Z., et al.: Survey of hallucination in natural language generation. ACM Comput. Surv. **55**(12), 1–38 (2023)

40. Jiang, A.Q., et al.: Mistral 7B (2023)

41. Karthik, S., Roth, K., Mancini, M., Akata, Z.: Vision-by-language for training-free compositional image retrieval. arXiv preprint arXiv:2310.09291 (2023)

42. Kumari, N., Zhang, B., Wang, S.Y., Shechtman, E., Zhang, R., Zhu, J.Y.: Ablating concepts in text-to-image diffusion models. In: Proceedings of the IEEE/CVF International Conference on Computer Vision, pp. 22691–22702 (2023)

43. Kumari, N., Zhang, B., Zhang, R., Shechtman, E., Zhu, J.Y.: Multi-concept customization of text-to-image diffusion (2023)

44. Lee, C., Cho, K., Kang, W.: Mixout: effective regularization to finetune large-scale pretrained language models. arXiv preprint arXiv:1909.11299 (2019)

45. Lewis, P., et al.: Retrieval-augmented generation for knowledge-intensive NLP tasks. In: Advances in Neural Information Processing Systems, vol. 33, pp. 9459–9474 (2020)

46. Li, B., Zhang, Y., Chen, L., Wang, J., Yang, J., Liu, Z.: Otter: a multi-modal model with in-context instruction tuning (2023)

47. Li, D., Li, J., Hoi, S.C.H.: BLIP-Diffusion: pre-trained subject representation for controllable text-to-image generation and editing (2023)

48. Li, J., Li, D., Savarese, S., Hoi, S.: BLIP-2: bootstrapping language-image pre-training with frozen image encoders and large language models. arXiv preprint arXiv:2301.12597 (2023)

49. Li, W., et al.: UNIMO: towards unified-modal understanding and generation via cross-modal contrastive learning. arXiv preprint arXiv:2012.15409 (2020)

50. Li, X., Li, S., Song, S., Yang, J., Ma, J., Yu, J.: PMET: precise model editing in a transformer. arXiv preprint arXiv:2308.08742 (2023)

51. Liu, H., Li, C., Li, Y., Lee, Y.J.: Improved baselines with visual instruction tuning (2023)

52. Liu, H., Li, C., Wu, Q., Lee, Y.J.: Visual instruction tuning. In: NeurIPS (2023)

53. Meng, K., Bau, D., Andonian, A., Belinkov, Y.: Locating and editing factual associations in GPT. In: Advances in Neural Information Processing Systems, vol. 36 (2022)

54. Meng, K., Sen Sharma, A., Andonian, A., Belinkov, Y., Bau, D.: Mass editing memory in a transformer. In: The Eleventh International Conference on Learning Representations (ICLR) (2023)

55. Mitchell, E., Lin, C., Bosselut, A., Finn, C., Manning, C.D.: Fast model editing at scale. In: International Conference on Learning Representations (2022). https://openreview.net/pdf?id=0DcZxeWfOPt

56. Mitchell, E., Lin, C., Bosselut, A., Manning, C.D., Finn, C.: Memory-based model editing at scale. In: Chaudhuri, K., Jegelka, S., Song, L., Szepesvari, C., Niu, G., Sabato, S. (eds.) Proceedings of the 39th International Conference on Machine Learning. Proceedings of Machine Learning Research, vol. 162, pp. 15817–15831. PMLR (2022). https://proceedings.mlr.press/v162/mitchell22a.html

57. Ouyang, L., et al.: Training language models to follow instructions with human feedback. In: Advances in Neural Information Processing Systems, vol. 35, pp. 27730–27744 (2022)

58. Park, C.C., Kim, B., Kim, G.: Towards personalized image captioning via multi-modal memory networks. IEEE Trans. Pattern Anal. Mach. Intell. **41**(4), 999–1012 (2018)

59. Peng, Z., et al.: Kosmos-2: grounding multimodal large language models to the world. arXiv preprint arXiv:2306.14824 (2023)

60. Radford, A., et al.: Learning transferable visual models from natural language supervision. In: International Conference on Machine Learning, pp. 8748–8763. PMLR (2021)

61. Rahman, T., Lee, H.Y., Ren, J., Tulyakov, S., Mahajan, S., Sigal, L.: Make-a-story: visual memory conditioned consistent story generation (2023)

62. Richardson, E., Goldberg, K., Alaluf, Y., Cohen-Or, D.: ConceptLab: creative concept generation using VLM-guided diffusion prior constraints (2023)

63. Ruiz, N., Li, Y., Jampani, V., Pritch, Y., Rubinstein, M., Aberman, K.: Dream-Booth: fine tuning text-to-image diffusion models for subject-driven generation (2022)

64. Saito, K., et al.: Pic2Word: mapping pictures to words for zero-shot composed image retrieval. In: Proceedings of the IEEE/CVF Conference on Computer Vision and Pattern Recognition, pp. 19305–19314 (2023)

65. Shuster, K., Humeau, S., Hu, H., Bordes, A., Weston, J.: Engaging image captioning via personality. In: Proceedings of the IEEE/CVF Conference on Computer Vision and Pattern Recognition, pp. 12516–12526 (2019)

66. Sinitsin, A., Plokhotnyuk, V., Pyrkin, D., Popov, S., Babenko, A.: Editable neural networks. arXiv preprint arXiv:2004.00345 (2020)

67. Sun, Q., et al.: Generative multimodal models are in-context learners. arXiv preprint arXiv:2312.13286 (2023)

68. Taori, R., et al.: Stanford alpaca: an instruction-following llama model (2023). https://github.com/tatsu-lab/stanford_alpaca

69. Tewel, Y., Gal, R., Chechik, G., Atzmon, Y.: Key-locked rank one editing for text-to-image personalization. In: ACM SIGGRAPH 2023 Conference Proceedings, pp. 1–11 (2023)

70. Touvron, H., et al.: LLaMA: open and efficient foundation language models. arXiv preprint arXiv:2302.13971 (2023)

71. Voynov, A., Chu, Q., Cohen-Or, D., Aberman, K.: $p+$: extended textual conditioning in text-to-image generation. arXiv preprint arXiv:2303.09522 (2023)

72. Vu, T., et al.: FreshLLMs: refreshing large language models with search engine augmentation (2023)

73. Wang, Q., Bai, X., Wang, H., Qin, Z., Chen, A.: InstantID: zero-shot identity-preserving generation in seconds. arXiv preprint arXiv:2401.07519 (2024)

74. Wang, X., Wang, G., Chai, W., Zhou, J., Wang, G.: User-aware prefix-tuning is a good learner for personalized image captioning. In: Liu, Q., et al. (eds.) PRCV 2023. LNCS, vol. 14431, pp. 384–395. Springer, Cham (2023). https://doi.org/10. 1007/978-981-99-8540-1_31
75. Wei, J., et al.: Finetuned language models are zero-shot learners. In: International Conference on Learning Representations (2022). https://openreview.net/forum? id=gEZrGCozdqR
76. Wu, C., Yin, S., Qi, W., Wang, X., Tang, Z., Duan, N.: Visual ChatGPT: talking, drawing and editing with visual foundation models. arXiv preprint arXiv:2303.04671 (2023)
77. Yao, Y., et al.: Editing large language models: problems, methods, and opportunities (2023)
78. Ye, H., Zhang, J., Liu, S., Han, X., Yang, W.: IP-Adapter: text compatible image prompt adapter for text-to-image diffusion models (2023)
79. Ye, Q., et al.: mPLUG-Owl: modularization empowers large language models with multimodality. arXiv preprint arXiv:2304.14178 (2023)
80. Yeh, C.H., Russell, B., Sivic, J., Heilbron, F.C., Jenni, S.: Meta-personalizing vision-language models to find named instances in video. In: Proceedings of the IEEE/CVF Conference on Computer Vision and Pattern Recognition, pp. 19123–19132 (2023)
81. Yin, S., et al.: A survey on multimodal large language models (2023)
82. Yu, J., Wang, Z., Vasudevan, V., Yeung, L., Seyedhosseini, M., Wu, Y.: CoCa: contrastive captioners are image-text foundation models. arXiv preprint arXiv:2205.01917 (2022)
83. Zeng, W., Abuduweili, A., Li, L., Yang, P.: Automatic generation of personalized comment based on user profile. arXiv preprint arXiv:1907.10371 (2019)
84. Zhang, S., et al.: OPT: open pre-trained transformer language models. arXiv preprint arXiv:2205.01068 (2022)
85. Zhao, W.X., et al.: A survey of large language models. arXiv preprint arXiv:2303.18223 (2023)
86. Zhu, D., Chen, J., Shen, X., Li, X., Elhoseiny, M.: MiniGPT-4: enhancing vision-language understanding with advanced large language models. arXiv preprint arXiv:2304.10592 (2023)

AMEGO: Active Memory from Long EGOcentric Videos

Gabriele Goletto[1]([✉])(ID), Tushar Nagarajan[2](ID), Giuseppe Averta[1](ID), and Dima Damen[3](ID)

[1] Politecnico di Torino, Turin, Italy
gabriele.goletto@polito.it
[2] FAIR, Meta, Austin, USA
[3] University of Bristol, Bristol, UK
https://gabrielegoletto.github.io/AMEGO/

Abstract. Egocentric videos provide a unique perspective into individuals' daily experiences, yet their unstructured nature presents challenges for perception. In this paper, we introduce AMEGO, a novel approach aimed at enhancing the comprehension of very-long egocentric videos. Inspired by the human's ability to maintain information from a single watching, AMEGO focuses on constructing a self-contained representations from one egocentric video, capturing key locations and object interactions. This representation is semantic-free and facilitates multiple queries without the need to reprocess the entire visual content. Additionally, to evaluate our understanding of very-long egocentric videos, we introduce the new Active Memories Benchmark (AMB), composed of more than 20K of highly challenging visual queries from EPIC-KITCHENS. These queries cover different levels of video reasoning (sequencing, concurrency and temporal grounding) to assess detailed video understanding capabilities. We showcase improved performance of AMEGO on AMB, surpassing other video QA baselines by a substantial margin.

Keywords: Long video understanding · Egocentric vision

1 Introduction

Episodic memory is a fundamental aspect of human cognition, which allows us to remember and recall our unique personal experiences [52]. Recently, there has been growing interest in leveraging first-person or *egocentric* videos to develop artificial episodic memory systems [11] that identify temporal segments from the video that contain answers to questions [38] or occurrences of objects [28,61] and activities [30,69].

Critically, these approaches build representations of long videos from uniformly sampled frame or clip features, and then train a model to retrieve salient

---

Supplementary Information The online version contains supplementary material available at https://doi.org/10.1007/978-3-031-72624-8_6.

A. Leonardis et al. (Eds.): ECCV 2024, LNCS 15071, pp. 92–110, 2025.
https://doi.org/10.1007/978-3-031-72624-8_6

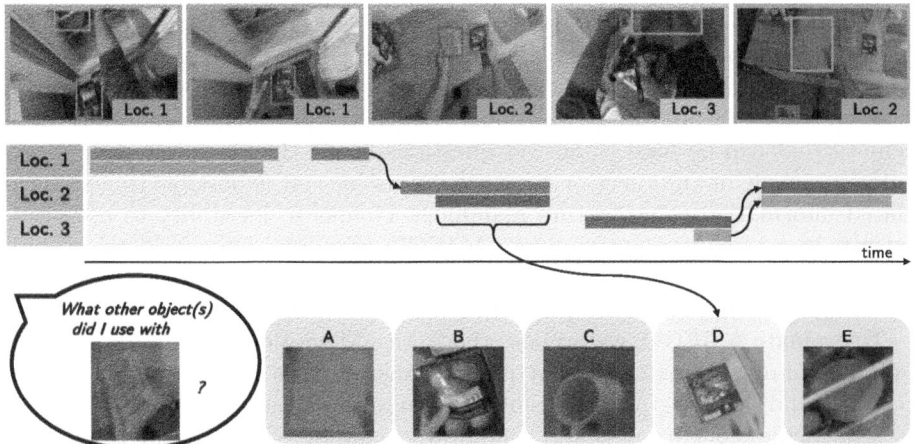

Fig. 1. AMEGO captures key locations and object interactions in a structured representation. In the each frame on top, the external border colour refers to a specific location in AMEGO while colours of objects define specific instances. AMEGO unlocks fine-grained long video understanding allowing multiple queries, such as the one depicted at the bottom of the figure, without reprocessing the long input video. (Color figure online)

moments from the video using them. This has three drawbacks: (1) they are human activity agnostic—the simplistic uniform sampling of frames is done without an understanding of where the camera-wearer is, when object interactions occur, or what hand the camera-wearer uses, which are key parameters of human activity, (2) they rely on semantically labelled training data—explicitly training encoders to relate the query to the input video representations, and (3) they are difficult to interpret—the implicit representations do not directly reveal human activity leaving such approaches largely as *black boxes*.

To address these issues, we present AMEGO, an Active Memory of the EGOcentric video, which serves as an explicit, structured representation that captures both objects interacted with, locations visited, and the interplay between the two (see Fig. 1). Specifically, AMEGO is composed of (i) a collection of hand-object interaction (HOI) tracklets, which contain consistent interactions between the camera wearer and objects, and (ii) location segments, representing temporal intervals during which the camera wearer engages in activities within specific locations. Importantly, the tracklets and segments are built using visual perception models of human activity and motion rather than a naive sampling of frames. Such models capture information like "has an object-interaction begun or ended?", "is an interaction ongoing, despite hands not visible?", and so on, leading to representations that are directly tied to activities.

We populate our AMEGO memory following a three-step process: first, we identify the onset of object interactions; second, we detect the conclusion of ongoing interactions; and finally, we match concluded interactions to previously observed object or location instances. This online pipeline is preferred for efficiently storing and preserving only the relevant information, mirroring the way humans build episodic memories in everyday activities [51]. Each step uses off-the-shelf visual perception models, resulting in a training-free approach.

Noticeably, the resulting tracklets and segments are not associated with any fixed taxonomy of objects or locations, resulting in a semantic-free, queryable memory that can be used to answer questions about the video. By simply providing an image of an object or a location of interest, it is possible to access the related information using feature matching. Thanks to its ability to use visual features in a semantic-free manner, AMEGO is more adept at distinguishing objects with subtle differences. This enhances its robustness and flexibility in capturing a wide range of interactions.

To evaluate AMEGO, we propose the Active Memories Benchmark (AMB) – composed of more than 20K visual question-answer pairs covering active objects, locations, and their interplay. We center the questions around 3 levels of reasoning, i.e. sequencing, concurrency, and temporal grounding. Notably, our benchmark is the first to tackle simultaneously interacting objects and locations. AMEGO achieves state of the art results on AMB, surpassing other common Video QA baselines by 12.7%. This performance underlines its capabilities across the three reasoning levels.

2   Related Works

Long Video Understanding. Long videos have gained significant interest recently, largely due to the emergence of large-scale egocentric datasets [5,11,12]. The task involves understanding videos lasting for minutes or even hours, leading to the creation of specialised models [14,58]. Several question-answering benchmarks have been introduced to evaluate models' proficiency in understanding long videos [8,10,19,22,39,48,53,63,66]. Among these, the widely adopted benchmark EgoSchema [29] focuses on videos of up to 3 min in duration. Another benchmark, ReST [65], shares similarities with ours as it focuses on visual queries over long egocentric videos. However, it does not target locations and only emphasises object interactions. Different approaches have tackled long video understanding: some treat it as a natural language question answering task by first captioning the video and then using LLMs to answer queries [26,34,54,56,57,68]. Others integrate LLMs with a video encoder, leveraging the powerful comprehension and generation capabilities of LLMs [21,36,40,45]. Our approach is similar to [9], proposing a structured representation of the video, but we specifically focus on interactions rather than indiscriminately memorising all the objects in the video.

Structured Video Representations. Various studies have explored methods for enhancing video representations by incorporating structured information. Contextual relationships have been a key focus, with numerous works investigating relationships between objects and actors [1,2,4,15,17,25,46,55], as well as among actions [3,14] using graph-based models. In the realm of egocentric vision, efforts have been made to construct structured representations of videos. For instance, [35] proposes grouping clips by activity threads, while [42] introduces egocentric scene graphs to capture interactions of the camera wearer. [32] focuses on constructing a human-centric representation of scenes by capturing the spatial locations of interactions, while [7] builds an allocentric top-down semantic scene representations, grounding the position of objects, from a video capturing a tour of the environment. Despite addressing various aspects of activities, these approaches do not capture the multiple dimensions inherent in egocentric videos—namely, object interactions, key locations, and their interplay.

Video Summarisation. Another related task is video summarisation [27,31, 41,73] whose aim is to generate a shorter version of the video in the form of key frames or key shots. Egocentric summarisation approaches consider important people and objects [18], essential events [24] or aesthetic characteristics of key frames [59]. Some works have also proposed generating the summaries in an online fashion [23,70] but do not target a structured representation of the video. [62] proposes a generic object finder, which automatically detects and clusters manipulated objects generating a timeline of the interactions. However, they do not exploit the temporal dimension proposing a system which is affected by noise coming from the detector. Another work which is related to ours is [60]. The method introduces a storyline representation for egocentric videos, summarising them based on actors, events, locations, and objects. It allows querying across dimensions using boolean operators. However, it mainly detects predefined attractions and supporting objects, which are visually distinct. Our work focuses on finer activities in cluttered scenes.

3   Method - AMEGO

Given a long and untrimmed egocentric video, we aim to capture the knowledge of active objects, key locations and their interplay using a unified structured representation. Such a representation must be *self-contained*—providing a full description of the camera-wearer's interactions with objects and locations—and *queryable*—as it should help retrieve temporal segments in the video indicating when an object was used, when a location was visited and their intersection (i.e. when an object was used in a specific location). In short, it is an Active Memory of the EGOcentric video, named hereinafter AMEGO.

We decompose the long egocentric video $\mathcal{V}$ into a set of hand-object interaction (HOI) tracklets ($\mathcal{O}$) and location segments ($\mathcal{L}$). Each **HOI tracklet** is a spatio-temporal representation of an object consistently interacting with at least one hand of the subject. It is characterised by spatio-temporal bounding boxes and their appearance features. Each **location** segment corresponds to the window of time where the camera-wearer *visits* a location to *perform* interactions, i.e. we are interested in activity-centric zones or hot-spots for interactions.

Put together, the HOI tracklets and location segments form a memory of what objects the camera-wearer interacts with over time, in which locations, and how those objects are moved around the scene. This memory $\mathcal{E} = \{\mathcal{O}, \mathcal{L}\}$ is built online, eliminating the need for reprocessing past visual information, and then queried to answer a variety of questions about objects, locations and their interplay, as our experiments will show. Critically, our representations are

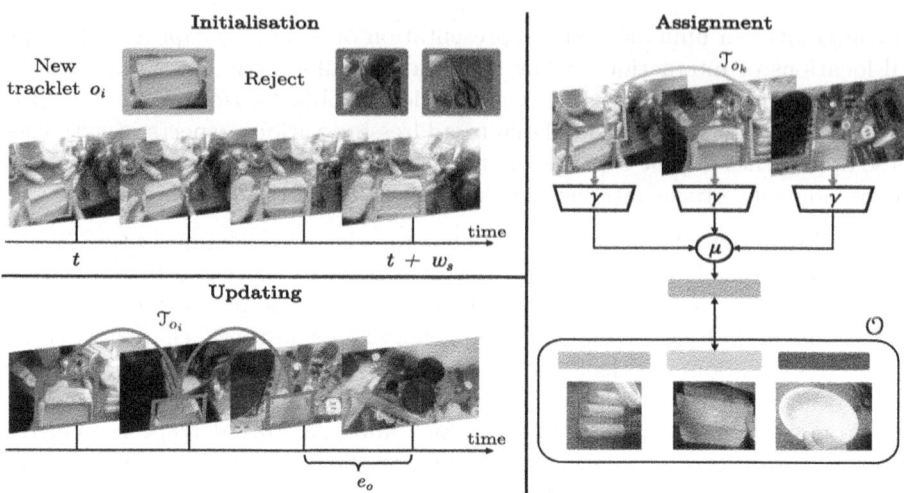

Fig. 2. We build $\mathcal{O}$ in an online manner, performing the 3 steps depicted at each frame of our video. (i) **Initialisation** We use consistent active object detections to generate new HOI tracklets. We thus discard noise resulting in sparse detections. (ii) **Updating** Once a new tracklet is initialised, we use a SOT tracker ($\mathcal{T}_{o_i}$) to update its detections even when hands go out of the field of view. We end the tracklet when there are e_o consecutive frames with a free hand or a distinctive new object interaction. (iii) **Assignment** Once a tracklet terminates, we assign it an object instance based on the similarity between its visual features wrt those in memory $\mathcal{O}$.

semantic-free—they represent instances of objects and locations but are not tied to a fixed taxonomy of labels or a known vocabulary. They are tied to the visuals of objects and not to discrete categories, allowing a more fine-grained distinction.

In the following sections, we describe our pipeline to characterise and store object interactions (Sect. 3.1), to identify location segments (Sect. 3.2), and then to put them together to form our AMEGO representation. Finally, we describe how to query it to answer various questions (Sect. 3.3).

3.1   Object Interactions

We begin by characterising object interactions as HOI tracklets $\mathcal{O}$. Each tracklet $o_i \in \mathcal{O}$ is a tuple (t_s, t_e, b_t, h, id) where (t_s, t_e) are the start and end frames of the interaction, b_t is the sequence of bounding boxes representing the object in each frame, h is the hand side that performs the interaction (i.e., left or right), and id is the object instance associated to the tracklet.

We iteratively build $\mathcal{O}$. At each frame $\mathcal{V}_t$ we perform 3 main steps: (1) initialise possible new candidate HOI tracklets, (2) update the HOI tracklets that are active (i.e. corresponding to ongoing interactions), and (3) store the ones that terminate in the memory $\mathcal{E}$, and assign their corresponding object instance.

Initialisation. We use a class-agnostic hand-object interaction detector [43], which provides a set of active object and hand bounding boxes denoted as $\mathcal{B}_t^o$ and $\mathcal{B}_t^h$ respectively. We initiate a new HOI tracklet o_i for each new hand-object interaction, defined as a tubelet comprising at least s_o bounding boxes exhibiting strong spatial overlap within a temporal window of w_s frames (Fig. 2, top-left). This spatio-temporal filtering allows us to account for noise as the result of hand-object detectors applied independently over frames. By leveraging the duration of natural hand-object interactions, we can reliably identify new active HOI tracklets, ensuring spatio-temporal consistency in the detections. The HOI tracklet o_i is now considered *active* and is added to $\mathcal{O}$.

For all subsequent frames, we calculate the intersection over union (IoU) for each object interacting with the same hand side. Matching bounding boxes over a threshold, θ, are assigned to the tracklet o_i. When bounding boxes cannot be assigned to the tracklet, it is considered complete.

Updating. Next, we need to capture the entire duration of the interaction, and concurrently, capture all spatial occurrences of the object, performing interaction-aware tracking. While the frame-level HOI detectors are sufficient to identify new interactions, they are unable to reliably extend tracks over time where hands or objects exit the egocentric field of view. Instead, we use an off-the-shelf single-object tracker (SOT) [47] which can reliably track the object across the whole interaction. (Fig. 2, bottom-left).

Specifically, for each active object track o_i we initialise a SOT. We consider the track o_i completed if there are no associated detections $\mathcal{B}^o$ for e_o consecutive frames, while the hand h remains visible. This is because when the hand is out of view, it is likely to still be holding the object. This results in a spatio-temporal track $\mathcal{T}_{o_i}$ which tracks the object's position, but lacks information about the interaction itself.

At this point, o_i contains information about its temporal duration (start and end time) and spatial bounding boxes corresponding to the active object, by combining the strengths of frame-based HOI detection and the SOT.

Assignment and Storing. Finally, we match o_i to already seen object instances in our memory. Specifically, given the set of stored HOI tracklets $\mathcal{O}_t$ observed so far, and the set of running SOT tracks $\mathcal{T}_t$, we check whether o_i can be matched to an existing object instance or if we need to start a new one. To do this, we first compute the visual features of o_i:

$$f(o_i) = \frac{1}{|\mathcal{V}_{o_i}|} \sum_{k \in \mathcal{V}_{o_i}} \gamma(k, b_k^o) \tag{1}$$

where $\mathcal{V}_{o_i}$ is the set of frames associated with o_i, b_k^o is the detection for frame k and γ is a visual feature extractor (in our experiments, DINOv2 [33]). To match o_i with instances in $\mathcal{O}_t$, we use an online clustering approach based on $f(o_i)$. The similarity between o_i and a specific object instance id_j is computed as follows:

$$s(o_i, id_j) = \frac{1}{|\mathcal{O}_t \in id_j|} \sum_{\mathcal{O}_t \in id_j} <f_{\mathcal{O}_t}, f_{o_i}> \tag{2}$$

where $\mathcal{O}_t \in id_j$ are the HOI tracklets belonging to instance id_j and $<.>$ measures the cosine similarity. We assign o_i to the object instance id_j^* that maximizes Eq. 2, and is above a specified threshold, θ. Note that if any tracker in $\mathcal{T}_t$ overlaps significantly with o_i, and the tracker confidence is higher than the maximum similarity above, then it is assigned to the tracker's instance. Otherwise, o_i is assigned to the corresponding instance id_j^*. If the maximum similarity is below the threshold, a new instance is created for o_i (Fig. 2, bottom).

At the end of this stage we associate $f(o_i)$ and the assigned instance to o_i, and store it into $\mathcal{E}$. We will refer to this *confirmed* tracklet as $\mathcal{O}_i$. It becomes part of AMEGO and can be consequently used in the querying process.

3.2   Location Segments

We define the set of location segments $\mathcal{L}$ as the temporal segments when the subject is carrying out interactions at different activity-centric zones. As a subject may interact with multiple objects simultaneously but can only be present in one hot-spot at a time, each location segment $l_i \in \mathcal{L}$ is modelled as a temporal interval corresponding to the start and end of an interaction. Like object interactions, $\mathcal{L}$ is populated online, and in two steps as follows.

Temporal Segmentation. Given the egocentric frame $\mathcal{V}_t$ and the hand detections $\mathcal{B}_t^h$, to understand whether the hand is interacting with an object while being in a location, we compute the optical flow between $\mathcal{V}_{t-1}$ and $\mathcal{V}_t$ and check hand presence via $|\mathcal{B}_t^h| > 0$. We consider the subject carrying out a task if both optical flow has low norm and there is at least one detected hand. We used the criteria discussed above as proxies to determine whether the subject has paused (through low optical flow) and is actively interacting with the scene (through a detected hand). Similar to the process for determining HOI tracklets, we adopt temporal filtering and consider a location segment, l_j, to be active only if these two conditions are verified for a consecutive number of frames, s_l. Similarly, we terminate l_j when we observe a consecutive number of frames, e_l, with either optical flow norm above the threshold or absent of hand detections.

At the end of this stage, we have temporally defined l_j but we still need to match it to previous location segments at the same hot-spot.

Assignment and Storing. We utilise a visual feature extractor σ for locations, to compute average features for the location segment denoted as $g(l_j)$. Next, we calculate similarity scores between the stored location instances and l_j by computing the average cosine similarity. We assign l_j to the instance that maximizes the similarity beyond a specified threshold, τ. If the threshold is not met, a new instance is created. Finally, we pair $g(l_j)$ and the assigned instance to l_j, and store it into $\mathcal{E}$. We will refer to this *confirmed* location segment as $\mathcal{L}_j$.

3.3   Querying AMEGO Representations

After processing the whole video we obtain our *AMEGO*: a complete set of HOI tracklets $\mathcal{O}$ and Location segments $\mathcal{L}$ (see Fig. 3). Utilising AMEGO, we can

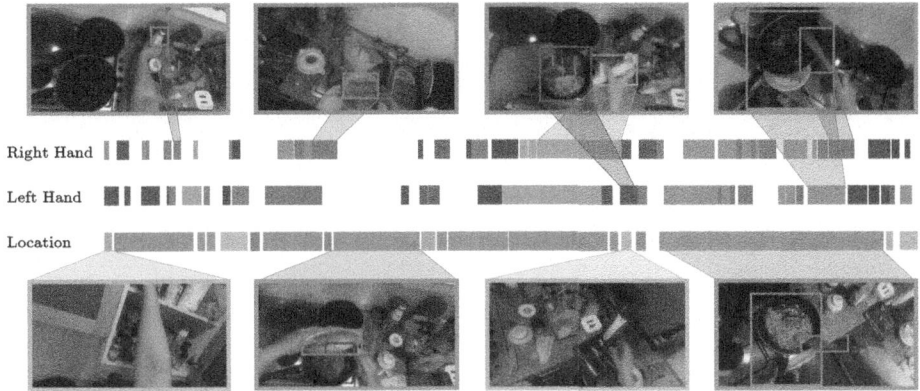

Right Hand

Left Hand

Location

Fig. 3. An example of AMEGO on a long egocentric video depicting objects interacting with the left and right hand of the subject and the visited locations.

determine whether any object has been in use and if the person has interacted at any locations. We achieve this in a semantic-free retrieval manner. Given an object image, q_o, we first extract its visual features, $f(q_o)$, using γ. Subsequently, we assign it an object instance, q_{id}, based on the similarity between its visual features and those in $\mathcal{O}$. With the obtained q_{id}, we can query $\mathcal{E}$ to retrieve information about its interactions. For instance, by searching for all tracklets in $\mathcal{O} \in q_{id}$, and their interaction intervals (t_s, t_e), we can identify all the temporal segments when q_{id} has been used.

Similarly, using σ, we can match any input location image to $\mathcal{L}$ in an identical manner. Consequently, leveraging the common temporal dimension, we can understand where query objects been used or what objects have been used at the query location. This process allows us to answer any set of queries involving objects and locations without reprocessing the entire video. Inside $\mathcal{E}$, we encapsulate all the information about what occurred in the video. This transforms AMEGO into an active memory of the video that, regardless of queries, is aware of what interactions took place at any point in the video.

4   Active Memories Benchmark

We propose the Active Memories Benchmark (AMB)—a comprehensive framework to study the interaction between active objects, locations, and their interplay in long egocentric videos, which form key components of daily human activity. The benchmark consists of 20.5k queries covering various levels of reasoning. The queries take the form of multiple-choice questions ranging from simple questions about object use (e.g., What did I use with [VQ]? where [VQ] is a visual crop of an object). Given a set of [VA] visual answers, the task is to select the correct representation of an object that has been used at the same time as the object represented in [VQ]. Similarly, questions can be answered on the interplay

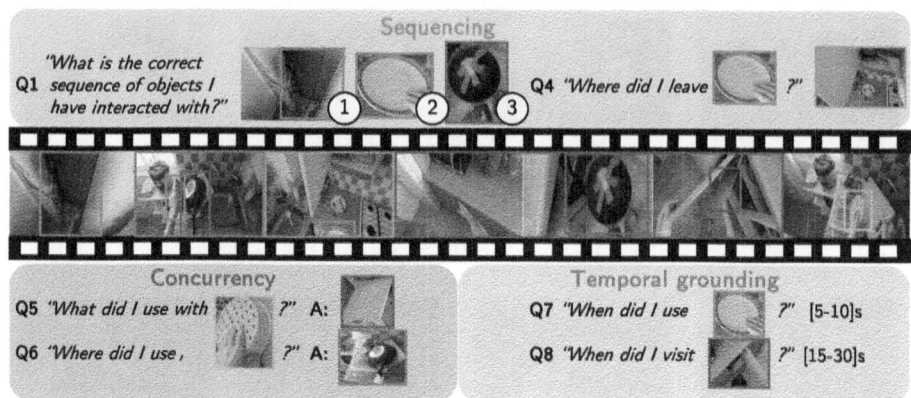

Fig. 4. Examples queries of Active Memories Benchmark on an egocentric video (in the middle). We build our benchmark around 3 different levels of reasoning, i.e. Sequencing, **Concurrency** and Temporal grounding.

between locations and objects, e.g. What locations did I use [VQ] in? The answer here would be a set of correct location representations [{LA, ...}]. Critically, each visual query of an object [VQ], visual object answer [VA], location query [LQ] or location answer [LA] are parameterised as *visual crops* [11,65,67] to mitigate the need for a fixed vocabulary or taxonomy resulting in biases associated with language. Forming a language-free benchmark avoids models that neglect visual data when answering the questions [16,29,64]. See Fig. 4 for a visual example.

4.1 Query Criteria

To construct our benchmark, we build a set of visual query templates that involve objects, locations, and their natural interplay (See Table 1). We structure our benchmark evaluation around three main reasoning levels, which serve as essential building blocks to enable higher-level activity understanding.

Sequencing (SQ) questions assess the ability to discriminate the temporal order of events. For example, can the model order interactions in time and identify which object the subject used before or after using another object? These are captured by templates Q1–4.

Concurrency (CO) questions assess the ability to capture multiple interactions happening at the same time. For example, can the model reason about whether different objects have been used together (i.e. object-object concurrency), as well as whether an object interaction took place in a specific location (i.e. object-location concurrency)? These are captured by Q5–6.

Temporal grounding (TG) questions assess the model's ability to retrieve all intervals of interactions with an active object or a location within the long video. For example, can the model identify when a given object was used or a location was visited (i.e. the start and end time). These are captured by Q7–8.

Table 1. The question templates proposed in our benchmark, along with the corresponding required reasoning, dimensions, types of answers, and number of questions in Active Memories Benchmark. SQ, **CO**, and TG represent sequencing, concurrency, and temporal grounding respectively. [VQ] and [LQ] represent object and location crops, while O and L stand for object and location.

Reasoning	Query	Template	Dim.	Answer	Qs
SQ	Q1	What is the correct sequence of objects I have interacted with?	O	Obj. seqs	4464
	Q2	What did I use with the left/right hand *after* [VQ]?	O	Obj.	3466
	Q3	What did I use with the left/right hand *before* [VQ]?	O	Obj.	3466
	Q4	Where did I take/leave [VQ]?	O, L	Loc.	1266
CO	Q5	What did I use with [VQ]?	O	Obj. sets	2105
	Q6	Where did I use [VQ]?	O, L	Loc. sets	2320
TG	Q7	When did I use [VQ]?	O	Intervals	2614
	Q8	When did I visit [LQ]?	L	Intervals	809

These three aspects provide a holistic view of information stored in the active memory when observing the long video, serving as the foundational elements for task understanding and causal inference within procedural egocentric videos.

4.2 Benchmark Construction

We build our queries adopting the templates listed in Table 1 and express them as multiple-choice questions.

Egocentric Videos. We construct our benchmark using 100 videos sourced from the EPIC-KITCHENS dataset [5]. This dataset is composed of long, unscripted egocentric recordings of participants performing daily living activities in a kitchen environment. On average, the selected videos are 14 min long. More specifically, 18 videos are shorter than 3 min, 35 videos are of medium length (between 3 and 10 min), 35 videos are long (10–30 minutes) and 12 are very long (>30 min). To define our ground truth, we leverage the publicly available dense camera poses from EPIC Fields [50] and active object masks from VISOR [6]. We segmented videos into activity-centric zones by leveraging camera positions from EPIC Fields to track the subject's attention on the scene. We merged EPIC-KITCHENS actions involving prolonged object usage and aligned VISOR masks with class semantics to obtain object bounding boxes. It is important to note that while we utilised these annotations for benchmark construction, our focus was solely on their application in evaluation. Additional details can be found in Supp.

Query Generation. We generate queries in a semi-automatic manner from our templates. AMB consists of 20.5K multiple choice question answer pairs. Each question consists of five possible options which are extracted semi-automatically from ground truths, to increase the challenge of these questions. In particular, we select candidate answers differently according to the type of question. For instance, for question Q6, options include locations visited immediately after the subject interacted with a specific object. Similarly, for questions Q3–4, options might include objects used with the opposite hand. This design makes AMB particularly challenging, demanding a detailed understanding of the events in the long video. Similarly for Q2–3, the query time t is set such that [VQ] is yet to be used – requiring the search for the interaction with [VQ] first before finding interactions before/after [VQ]. This systematic approach enabled us to create over 20,500 questions, of which 61.7% are sequencing questions, 21.6% concurrency questions, and 16.7% temporal grounding questions, see Table 1. Our dataset comprises 2614 object instances across videos and 809 activity-centric locations. On average, short videos (<3 mins) contain 62 questions, medium-length videos (3–10 mins) contain 134 questions, and long videos (>10 mins) have 313 questions.

5   Experiments

5.1   Experimental Setup

Implementation Details We use the hand-object interaction detector from [43] for identifying object-hand interactions at the frame level. Visual features of objects are extracted using the DINO-v2 pre-trained model [33] (γ), with resizing to 224×224 and evaluation on ViT-S and ViT-L versions. Object tracking during interactions employ the EgoSTARK tracker [47] and we set $\theta = 0.6$, $w_s = 30$, $s_o = 20$, and $e_o = 20$.

For locations we use SWAG [44], σ, as the visual feature extractor, trained for image classification using weak supervision of hashtags. This model is currently state of the art in scene classification on Places-365 [72]. We evaluate on ViT-B and ViT-L versions with frames resized to 384 × 384 and 512 × 512. We estimate optical flow with the Flowformer model [13], and use a threshold of 2000 for the optical flow L2 norm. We set $s_l = e_l = 5$ and $\tau = 0.5$.

Baselines. Our approach is the first able to create a complete representation of the long video which captures multifaced interacting elements. Prior works in this direction focused just on one specific dimension (e.g. locations [32] or activities [35]). SOT trackers would be able to track the object in the video but this would happen regardless of whether the object was interacted with. Consequently, we compare AMEGO against common baselines adopted for video QA on the proposed Active Memories Benchmark:

- **Semantic-free QA (SF-QA)** uses vision-language models, i.e. CLIP [37], to map the query, the video, and the answers into the same embedding space. This process involves extracting visual features from frames of the long video, query patches, and answers, while textual features are obtained from the question. The query embedding is generated by averaging the features from the video, patches, and question. Then, the similarity between this embedding and all answer embeddings is computed. The answer with the highest similarity score is selected.

- **SF-QA (obj)** is a variant of SF-QA, with visual features extracted also from active objects detected by [43].
- **Semantic QA (S-QA)** uses off-the-shelf captioners to generate a semantic summary of our video. We use the egocentric video captioner, i.e. LaViLa [71] at 1 fps, as in [68], and an image captioner, BLIP-2 [20], for generating captions of both the video and of the query patches. Increasing the captioning rate for video would only introduce redundancy in the resulting textual summary. Then, we input captions into an LLM and prompt it with the question. We use LLaMA-2-7B [49] because of its public availability, ensuring the reproducibility of our results. Due to the limited context window, we uniformly subsample textual summaries when they contain more than 4096 tokens.
- **Multi-round semantic QA (LLoVi)** [68] is similar to the previous one but it queries the LLM twice. First to summarise the video captions given the question, and then to answer the actual query based on the previously generated summary.

We evaluate AMEGO alongside the baselines in a zero-shot setting, measuring the accuracy over the queries provided in AMB.

5.2 Standalone Performance

We first evaluate the effectiveness of the different AMEGO components against the ground truth. To do so, we manually annotate temporal interactions of objects and locations for two long videos: a 20-min video from EPIC-KITCHENS [5] (which is not included in AMB) and a 10-min video from Ego4D [11]. Our annotations identify 22 distinct location instances and 67 different objects, allowing for a temporal comparison with the capabilities of AMEGO in defining interaction intervals.

Table 2. Standalone evaluation for HOI tracklets (left) and location (right) segments

	AIoU P ↑	AIoU GT ↑	$\Delta N \to 0$
$s_o = 1$	0.08	0.49	3249
$e_o = 1$	0.13	0.34	537
Track w/o hand detections	0.19	0.38	222
No tracker	0.19	0.39	218
AMEGO	0.20	0.41	210

	AIoU P ↑	AIoU GT ↑	$\Delta N \to 0$
$s_l = e_l = 1$	0.14	0.48	163
No flow filter	0.35	0.35	−1
No hand filter	0.34	0.49	27
AMEGO	0.36	0.50	44

We evaluate AMEGO using three metrics: (i) AIoU P, Average Intersection over Union between each predicted segment and its best-matching ground truth segment, indicating the precision of the predicted tracklets; (ii) AIoU GT, Average Intersection over Union between each temporal ground truth segment and its best-matching predicted segment, evaluating the recall of AMEGO; (iii) ΔN, the difference between the number of predicted and ground truth segments. Performance is improved when this number is closer to 0 $\Delta N \rightarrow 0$; Table 2 show the results for the both object and location interactions. In particular, it is possible to see the negative effect of noisy detections either at the beginning ($s_o = 1$) or at the end ($e_o = 1$) of the HOI tracklet. While AIoU GT is high for $s_o = 1$, this is due to the large number of segments predicted ($\Delta N > 3K$). Without using the hand detections, the tracking is stopped after e_o consecutive missing matches, regardless of hand presence. Accordingly, leveraging hand detections as a proxy to terminate the tracklet helps in detecting long interactions. Without using the tracker, the method performs worse as it is unable to track the object when the hands exits the field of view. Similar results for location segments show the importance of the various design decisions. It can be noticed how both flow and hand detection help to detect visited locations and, they are complementary.

5.3   Results on Active Memories Benchmark

To query AMEGO on AMB, we follow simple processes. As an example, to answer temporal grounding queries (Q7–8), we compare the query patch with instances in $\mathcal{E}$, as explained in Sect. 3.3, then extract the intervals in $\mathcal{E}$ corresponding to the matched instance. Additional details can be found in the Supplementary material. We report results on AMEGO - S, and AMEGO - L, depending on the size of the visual feature extractors adopted (ViT-S/B vs ViT-L).

Table 3 shows the main results on Active Memories Benchmark. All the baselines struggle to perform slightly better than random among the five answers. Particularly, it is noticeable that despite reaching high results on high-level understanding datasets [29], Semantic-QA approaches show limited understanding of fine-grained details on long videos. All the baselines show better results

Table 3. Accuracy results (%) over the different queries of AMB. Best in **bold**.

Method	SQ				CO		TG		Total
	Q1	Q2	Q3	Q4	Q5	Q6	Q7	Q8	
Random	20.0	20.0	20.0	20.0	20.0	20.0	20.0	20.0	20.0
SF-QA	13.7	21.6	22.5	26.8	22.1	31.9	23.7	26.2	22.0
SF-QA (obj)	13.1	23.4	22.6	23.2	21.7	26.1	23.8	25.2	21.2
S-QA (LaViLa)	20.9	20.6	21.2	24.6	24.9	27.1	21.4	22.6	22.4
S-QA (BLIP-2)	23.9	22.0	22.5	23.3	27.5	27.0	20.2	24.1	23.6
S-QA (LaViLa+BLIP-2)	22.8	22.2	21.4	22.6	25.1	26.1	21.4	24.5	22.9
LLoVi (LaViLa)	21.1	20.2	20.8	21.0	21.2	20.3	20.5	21.6	20.8
LLoVi (BLIP-2)	22.3	21.4	21.8	22.2	25.6	26.7	18.1	22.2	22.4
LLoVi (LaViLa+BLIP-2)	22.8	21.9	21.5	24.6	25.3	26.5	18.5	19.8	22.6
AMEGO - S	32.0	35.1	34.8	35.8	24.7	37.8	33.6	44.3	33.8
AMEGO - L	**33.7**	**36.3**	**37.2**	**38.3**	**27.6**	**44.3**	**34.7**	**48.9**	**36.3**

on concurrency-related questions (wrt the other reasoning proposed), which may hint at the fact that they might leverage training patterns, e.g. a pan often used at the cooktop. The semantic-free QA baseline performs the worst, demonstrating that features by themselves, without a proper representation, are not enough. On average, BLIP-2 performs better on object-related queries. Indeed, differently from LaViLa, it has been trained on object-centric datasets and therefore shows superior capability to recognise them. Finally, we observe that multi-stage LLM pipelines, such as [68], perform worse than standard-QA. This likely depends on the fact that directly processing the textual summary reduces the amount of information at subsequent stages for correctly answering the query.

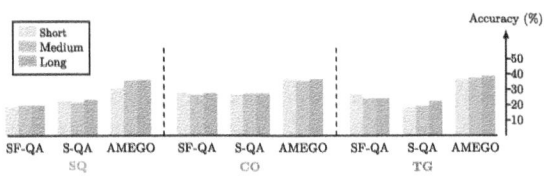

Fig. 5. Quantitative results depending on the temporal duration of the queried video.

AMEGO achieves good results on the whole set of queries, outperforming baselines by a large margin (+12.7%). It can be noticed that Q5 is the question where AMEGO struggles the most. This difficulty arises from current hand-object interaction detectors facing obstacles in predicting concurrent objects interacting with the same hand of the subject. Hence, despite our initialisation and tracking process allowing multiple objects to interact with the same hand, further improvements are needed in this regard.

Does Video Duration Impact Performance? We separately evaluate short (<3 min), medium (3–10 min) and long (>10 min) videos. Figure 5 compares the best performing semantic-free QA, semantic QA and AMEGO - L. In general, concurrency and temporal grounding questions are the ones that create more difficulty in long videos for SF-QA and AMEGO. This is reasonable as temporally locating objects and locations in longer videos is intuitively harder.

Qualitative Results. Figure 6 shows two examples of concurrency and sequencing queries with the answers obtained querying AMEGO (in green) against the ones obtained via Semantic-QA (in red). AMEGO can understand the correct order of usage of items. Indeed it is possible to observe the steps performed by the camera wearer for preparing a coffee (upper part). The S-QA approach is not able to capture all fine-grained details in the video and is limited only to part of the sequence. In the bottom query, instead, it is possible to observe the training biases of LLMs preferring a cupboard (typically used to store a pan) rather than a washing machine.

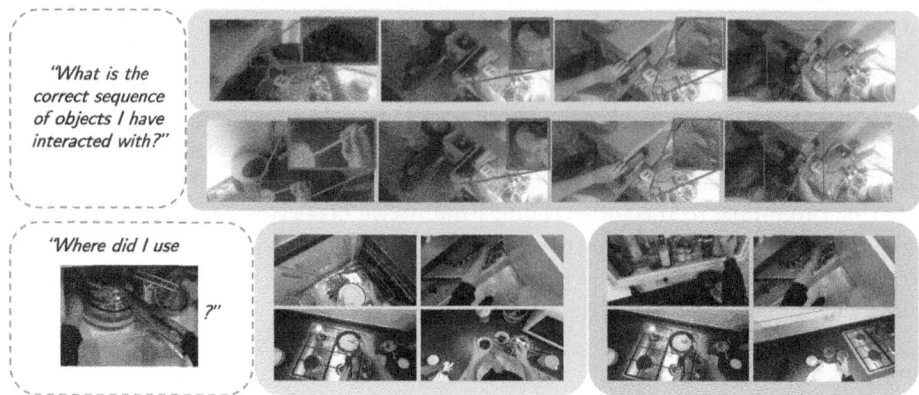

Fig. 6. Qualitative results are presented with sequencing and concurrency queries. Correct answers obtained from querying AMEGO have green background, while incorrect answers from Semantic-QA have red background. (Color figure online)

6 Conclusion

In this work, we introduce AMEGO, an innovative Active Memory approach tailored for egocentric videos. By dynamically organising interactions and activities into a structured representation, in an online manner, AMEGO mimics the episodic memory cognition. Through semantic-free querying, AMEGO offers a powerful solution for efficient video comprehension without the need for exhaustive reprocessing.

We evaluate AMEGO on a newly proposed Active Memories Benchmark which underscores the effectiveness of AMEGO, showcasing its superior performance over common Video QA baselines. This highlights its ability to capture and represent intricate interactions within egocentric videos, paving the way for enhanced video understanding and analysis.

Acknowledgements. G. Goletto is supported by PON "Ricerca e Innovazione" 2014–2020 - DM 1061/2021 funds and acknowledges travel support from ELISE (GA no 951847). G. Averta is supported by the project FAIR and Next-GenerationEU (PIANO NAZIONALE DI RIPRESA E RESILIENZA (PNRR) - MISSIONE 4 COMPONENTE 2, INVESTIMENTO 1.3 - D.D. 1555 11/10/2022, PE00000013). This manuscript reflects only the authors' views and opinions, neither the European Union nor the European Commission can be considered responsible for them. D. Damen is supported by EPSRC Visual AI EP/T028572/1 and UMPIRE EP/T004991/1.

We thank Chiara Plizzari for help in extracting the ground truths for AMB, Jian Ma for assistance in computing hand-object detections, and the members of the MaVi group for helpful discussions.

References

1. Arnab, A., Sun, C., Schmid, C.: Unified graph structured models for video understanding. In: Proceedings of the International Conference on Computer Vision, pp. 8117–8126 (2021)
2. Baradel, F., Neverova, N., Wolf, C., Mille, J., Mori, G.: Object level visual reasoning in videos. In: Proceedings of the European Conference on Computer, pp. 105–121 (2018)
3. Brendel, W., Todorovic, S.: Learning spatiotemporal graphs of human activities. In: Proceedings of the International Conference on Computer Vision, pp. 778–785 (2011)
4. Cong, Y., Liao, W., Ackermann, H., Rosenhahn, B., Yang, M.Y.: Spatial-temporal transformer for dynamic scene graph generation. In: Proceedings of the International Conference on Computer Vision, pp. 16372–16382 (2021)
5. Damen, D., et al.: Rescaling egocentric vision: collection, pipeline and challenges for EPIC-KITCHENS-100. IJCV **130**, 33–55 (2022)
6. Darkhalil, A., et al.: EPIC-KITCHENS VISOR benchmark: video segmentations and object relations. In: NeurIPS (2022)
7. Datta, S., et al.: Episodic memory question answering. In: CVPR (2022)
8. Du, Y., et al.: Towards event-oriented long video understanding. arXiv preprint arXiv:2406.14129 (2024)
9. Fan, Y., et al.: VideoAgent: a memory-augmented multimodal agent for video understanding. In: ECCV (2024)
10. Fang, X., et al.: MMBench-Video: a long-form multi-shot benchmark for holistic video understanding. arXiv preprint arXiv:2406.14515 (2024)
11. Grauman, K., et al.: Ego4D: around the world in 3,000 hours of egocentric video. In: CVPR (2022)
12. Grauman, K., et al.: Ego-Exo4D: understanding skilled human activity from first- and third-person perspectives. In: CVPR (2024)
13. Huang, Z., et al.: FlowFormer: a transformer architecture for optical flow. In: Avidan, S., Brostow, G., Cissé, M., Farinella, G.M., Hassner, T. (eds.) ECCV 2022. LNCS, vol. 13677, pp. 668–685. Springer, Cham (2022). https://doi.org/10.1007/978-3-031-19790-1_40
14. Hussein, N., Gavves, E., Smeulders, A.W.: VideoGraph: recognizing minutes-long human activities in videos. arXiv preprint arXiv:1905.05143 (2019)
15. Jain, A., Zamir, A.R., Savarese, S., Saxena, A.: Structural-RNN: deep learning on spatio-temporal graphs. In: CVPR (2016)
16. Jasani, B., Girdhar, R., Ramanan, D.: Are we asking the right questions in MovieQA? In: ICCV Workshop (2019)
17. Ji, J., Krishna, R., Fei-Fei, L., Niebles, J.C.: Action genome: actions as compositions of spatio-temporal scene graphs. In: CVPR (2020)
18. Lee, Y.J., Ghosh, J., Grauman, K.: Discovering important people and objects for egocentric video summarization. In: CVPR (2012)
19. Lei, J., Yu, L., Bansal, M., Berg, T.L.: TVQA: localized, compositional video question answering. In: EMNLP (2018)
20. Li, J., Li, D., Savarese, S., Hoi, S.: BLIP-2: bootstrapping language-image pretraining with frozen image encoders and large language models. In: ICML (2023)
21. Li, Y., Chen, X., Hu, B., Zhang, M.: LLMs meet long video: advancing long video comprehension with an interactive visual adapter in LLMs. arXiv preprint arXiv:2402.13546 (2024)

22. Li, Y., Chen, X., Hu, B., Wang, L., Shi, H., Zhang, M.: VideoVista: a versatile benchmark for video understanding and reasoning. arXiv preprint arXiv:2406.11303 (2024)
23. Lin, Y.L., Morariu, V.I., Hsu, W.: Summarizing while recording: context-based highlight detection for egocentric videos. In: ICCV Workshop (2015)
24. Lu, Z., Grauman, K.: Story-driven summarization for egocentric video. In: CVPR (2013)
25. Ma, C.Y., Kadav, A., Melvin, I., Kira, Z., AlRegib, G., Graf, H.P.: Attend and interact: higher-order object interactions for video understanding. In: CVPR (2018)
26. Ma, Z., et al.: DrVideo: document retrieval based long video understanding. arXiv preprint arXiv:2406.12846 (2024)
27. Mahasseni, B., Lam, M., Todorovic, S.: Unsupervised video summarization with adversarial LSTM networks. In: CVPR (2017)
28. Mai, J., Hamdi, A., Giancola, S., Zhao, C., Ghanem, B.: EgoLoc: revisiting 3D object localization from egocentric videos with visual queries. In: ICCV (2023)
29. Mangalam, K., Akshulakov, R., Malik, J.: EgoSchema: a diagnostic benchmark for very long-form video language understanding. In: NeurIPS (2024)
30. Mavroudi, E., Afouras, T., Torresani, L.: Learning to ground instructional articles in videos through narrations. In: ICCV (2023)
31. Meena, P., Kumar, H., Yadav, S.K.: A review on video summarization techniques. Eng. Appl. Artif. Intell. **118**, 105667 (2023)
32. Nagarajan, T., Li, Y., Feichtenhofer, C., Grauman, K.: EGO-TOPO: environment affordances from egocentric video. In: CVPR (2020)
33. Oquab, M., et al.: DINOv2: learning robust visual features without supervision. arXiv preprint arXiv:2304.07193 (2023)
34. Park, J., Ranasinghe, K., Kahatapitiya, K., Ryoo, W., Kim, D., Ryoo, M.S.: Too many frames, not all useful: efficient strategies for long-form video QA. arXiv preprint arXiv:2406.09396 (2024)
35. Price, W., Vondrick, C., Damen, D.: UnweaveNet: unweaving activity stories. In: CVPR (2022)
36. Qian, R., et al.: Streaming long video understanding with large language models. arXiv preprint arXiv:2405.16009 (2024)
37. Radford, A., et al.: Learning transferable visual models from natural language supervision. In: ICML (2021)
38. Ramakrishnan, S.K., Al-Halah, Z., Grauman, K.: NaQ: leveraging narrations as queries to supervise episodic memory. In: CVPR (2023)
39. Rawal, R., Saifullah, K., Basri, R., Jacobs, D., Somepalli, G., Goldstein, T.: CinePile: a long video question answering dataset and benchmark. arXiv preprint arXiv:2405.08813 (2024)
40. Ren, S., Yao, L., Li, S., Sun, X., Hou, L.: TimeChat: a time-sensitive multimodal large language model for long video understanding. In: CVPR (2024)
41. Rochan, M., Ye, L., Wang, Y.: Video summarization using fully convolutional sequence networks. In: Ferrari, V., Hebert, M., Sminchisescu, C., Weiss, Y. (eds.) ECCV 2018. LNCS, vol. 11216, pp. 358–374. Springer, Cham (2018). https://doi.org/10.1007/978-3-030-01258-8_22
42. Rodin, I., Furnari, A., Min, K., Tripathi, S., Farinella, G.M.: Action scene graphs for long-form understanding of egocentric videos. In: CVPR (2024)
43. Shan, D., Geng, J., Shu, M., Fouhey, D.F.: Understanding human hands in contact at internet scale. In: CVPR (2020)
44. Singh, M., et al.: Revisiting weakly supervised pre-training of visual perception models. In: CVPR (2022)

45. Song, E., et al.: MovieChat: from dense token to sparse memory for long video understanding. In: CVPR (2024)
46. Sun, C., Shrivastava, A., Vondrick, C., Murphy, K., Sukthankar, R., Schmid, C.: Actor-centric relation network. In: Ferrari, V., Hebert, M., Sminchisescu, C., Weiss, Y. (eds.) ECCV 2018. LNCS, vol. 11215, pp. 335–351. Springer, Cham (2018). https://doi.org/10.1007/978-3-030-01252-6_20
47. Tang, H., Liang, K.J., Grauman, K., Feiszli, M., Wang, W.: EgoTracks: a long-term egocentric visual object tracking dataset. In: NeurIPS (2024)
48. Tapaswi, M., Zhu, Y., Stiefelhagen, R., Torralba, A., Urtasun, R., Fidler, S.: MovieQA: understanding stories in movies through question-answering. In: CVPR (2016)
49. Touvron, H., et al.: Llama 2: open foundation and fine-tuned chat models. arXiv preprint arXiv:2307.09288 (2023)
50. Tschernezki, V., et al.: EPIC fields: marrying 3D geometry and video understanding. In: NeurIPS (2024)
51. Tulving, E.: Episodic memory: from mind to brain. Annu. Rev. Psychol. **53**(1), 1–25 (2002)
52. Tulving, E., et al.: Episodic and semantic memory. In: Organization of Memory, pp. 381–403 (1972)
53. Wang, W., et al.: LVBench: an extreme long video understanding benchmark. arXiv preprint arXiv:2406.08035 (2024)
54. Wang, X., Zhang, Y., Zohar, O., Yeung-Levy, S.: VideoAgent: long-form video understanding with large language model as agent. arXiv preprint arXiv:2403.10517 (2024)
55. Wang, X., Gupta, A.: Videos as space-time region graphs. In: Ferrari, V., Hebert, M., Sminchisescu, C., Weiss, Y. (eds.) ECCV 2018. LNCS, vol. 11209, pp. 413–431. Springer, Cham (2018). https://doi.org/10.1007/978-3-030-01228-1_25
56. Wang, Y., Yang, Y., Ren, M.: LifelongMemory: leveraging LLMs for answering queries in egocentric videos. arXiv preprint arXiv:2312.05269 (2023)
57. Wang, Z., et al.: VideoTree: adaptive tree-based video representation for LLM reasoning on long videos. arXiv preprint arXiv:2405.19209 (2024)
58. Wu, C.Y., et al.: MeMViT: memory-augmented multiscale vision transformer for efficient long-term video recognition. In: CVPR (2022)
59. Xiong, B., Grauman, K.: Detecting snap points in egocentric video with a web photo prior. In: Fleet, D., Pajdla, T., Schiele, B., Tuytelaars, T. (eds.) ECCV 2014. LNCS, vol. 8693, pp. 282–298. Springer, Cham (2014). https://doi.org/10.1007/978-3-319-10602-1_19
60. Xiong, B., Kim, G., Sigal, L.: Storyline representation of egocentric videos with an applications to story-based search. In: ICCV (2015)
61. Xu, M., Li, Y., Fu, C.Y., Ghanem, B., Xiang, T., Pérez-Rúa, J.M.: Where is my wallet? Modeling object proposal sets for egocentric visual query localization. In: CVPR (2023)
62. Yagi, T., Nishiyasu, T., Kawasaki, K., Matsuki, M., Sato, Y.: GO-finder: a registration-free wearable system for assisting users in finding lost objects via hand-held object discovery. In: International Conference on Intelligent User Interfaces (2021)
63. Yang, A., Miech, A., Sivic, J., Laptev, I., Schmid, C.: Just ask: learning to answer questions from millions of narrated videos. In: CVPR (2021)
64. Yang, J., Zhu, Y., Wang, Y., Yi, R., Zadeh, A., Morency, L.P.: What gives the answer away? Question answering bias analysis on video QA datasets. arXiv preprint arXiv:2007.03626 (2020)

65. Yang, X., Chu, F.J., Feiszli, M., Goyal, R., Torresani, L., Tran, D.: Relational space-time query in long-form videos. In: CVPR (2023)
66. Yu, Z., et al.: ActivityNet-QA: a dataset for understanding complex web videos via question answering. In: Conference on Artificial Intelligence (2019)
67. Zellers, R., Bisk, Y., Farhadi, A., Choi, Y.: From recognition to cognition: visual commonsense reasoning. In: CVPR (2019)
68. Zhang, C., et al.: A simple LLM framework for long-range video question-answering. arXiv preprint arXiv:2312.17235 (2023)
69. Zhang, C.L., Wu, J., Li, Y.: ActionFormer: localizing moments of actions with transformers. In: Avidan, S., Brostow, G., Cissé, M., Farinella, G.M., Hassner, T. (eds.) ECCV 2022. LNCS, vol. 13664, pp. 492–510. Springer, Cham (2022). https://doi.org/10.1007/978-3-031-19772-7_29
70. Zhao, B., Xing, E.P.: Quasi real-time summarization for consumer videos. In: CVPR (2014)
71. Zhao, Y., Misra, I., Krähenbühl, P., Girdhar, R.: Learning video representations from large language models. In: Proceedings of the IEEE/CVF Conference on Computer Vision and Pattern Recognition, pp. 6586–6597 (2023)
72. Zhou, B., Lapedriza, A., Xiao, J., Torralba, A., Oliva, A.: Learning deep features for scene recognition using places database. In: NeurIPS (2014)
73. Zhou, K., Qiao, Y., Xiang, T.: Deep reinforcement learning for unsupervised video summarization with diversity-representativeness reward. In: Conference on Artificial Intelligence (2018)

Power Variable Projection for Initialization-Free Large-Scale Bundle Adjustment

Simon Weber[1,2]([✉]), Je Hyeong Hong[3], and Daniel Cremers[1,2]

[1] Technical University of Munich, Munich, Germany
sim.weber@tum.de
[2] Munich Center for Machine Learning, Munich, Germany
[3] Department of Electronic Engineering, Hanyang University, Seoul, South Korea

Abstract. Most Bundle Adjustment (BA) solvers like the Levenberg-Marquardt algorithm require a good initialization. Instead, initialization-free BA remains a largely uncharted territory. The under-explored Variable Projection algorithm (VarPro) exhibits a wide convergence basin even without initialization. Coupled with object space error formulation, recent works have shown its ability to solve small-scale initialization-free bundle adjustment problem. To make such initialization-free BA approaches scalable, we introduce Power Variable Projection (PoVar), extending a recent inverse expansion method based on power series. Importantly, we link the power series expansion to Riemannian manifold optimization. This projective framework is crucial to solve large-scale bundle adjustment problems without initialization. Using the real-world BAL dataset, we experimentally demonstrate that our solver achieves state-of-the-art results in terms of speed and accuracy. To our knowledge, this work is the first to address the scalability of BA without initialization opening new venues for initialization-free structure-from-motion.

Keywords: Bundle Adjustment · Initialization-Free · Schur Complement · Riemannian Manifold Optimization

1 Introduction

Bundle adjustment (BA) is the key component of many structure-from-motion and 3D reconstruction algorithms. With the recent emergence of large-scale internet photo collections [3] and new applications (mixed reality, autonomous driving, digital twins), the need to solve large-scale BA has become an important challenge. Traditional BA addresses the following question: *Given image measurements and approximate landmark positions and camera parameters, can we derive the exact positions and parameters?* The gold standard is to use the Levenberg-Marquardt algorithm [25] coupled with the Schur complement trick

Supplementary Information The online version contains supplementary material available at https://doi.org/10.1007/978-3-031-72624-8_7.

and a scalable solver for the reduced camera system, which is often the preconditioned conjugate gradient algorithm. Recent work [23] achieves outstanding
speed for large-scale BA by using a power series expansion of the inverse Schur
complement.

Recently, a new line of works [11,13,14] has attempted to solve the BA problem *without* careful initialization: *Given only image measurements, how do we
derive pose parameters and 3D landmark positions?* This challenge is largely
uncharted, and its scalability a blind spot. In particular, most existing works
aim to formalize the problem into a stratified BA formulation, and none of them
try to design effective solvers. It is noteworthy that even the most recent works
only use direct factorization which becomes impractical for large-scale problems
with several hundreds of cameras. In contrast to the traditional BA problem,
the deficiency of competitive solvers for initialization-free BA can be broadly
explained by the difference of convergence behaviour between a well-initialized
problem and an initialization-free problem.

Following up on the recent findings concerning inverse expansion methods,
we address the scalability of initialization-free BA. Our new solver based on the
Variable Projection algorithm overcomes the issues of convergence of the scalable
preconditioned conjugate gradients algorithm, while being efficient for thousand
of camera viewpoints. In summary, we make the following contributions:

- We introduce Power Variable Projection (*PoVar*) for efficient large-scale bundle adjustment *without* good initialization of camera poses and 3D landmarks.
  To the best of our knowledge, we are the first to address the scalability of
  initialization-free bundle adjustment formulation.
- We provide theoretical proofs that justify the extension of recent *inverse
  expansion* method to the variable projection algorithm. While sharing a close
  algorithmic structure, the proposed extension and the existing *power-series-
  based* method largely differ in the theory, in the applications and in the convergence behaviour.
- We theoretically extend the power series expansion for bundle adjustment to
  Riemannian manifold optimization. We take advantage of the matrix-specific
  structure to propose an efficient storage and memory-efficient computation
  for such optimization.
- We perform extensive evaluation of the proposed approach on the real-world
  BAL dataset. We emphasize the benefits of *PoVar* in terms of scalability,
  speed and accuracy. In contrast to state-of-the-art solvers, our work is the
  first that solves large-scale bundle adjustment without initialization.
- We release our solver as open source to facilitate further research: https://
  github.com/tum-vision/povar.

2   Related Work

As we address the scalability of the variable projection (VarPro) algorithm for
initialization-free bundle adjustment (BA), we review works on VarPro and on

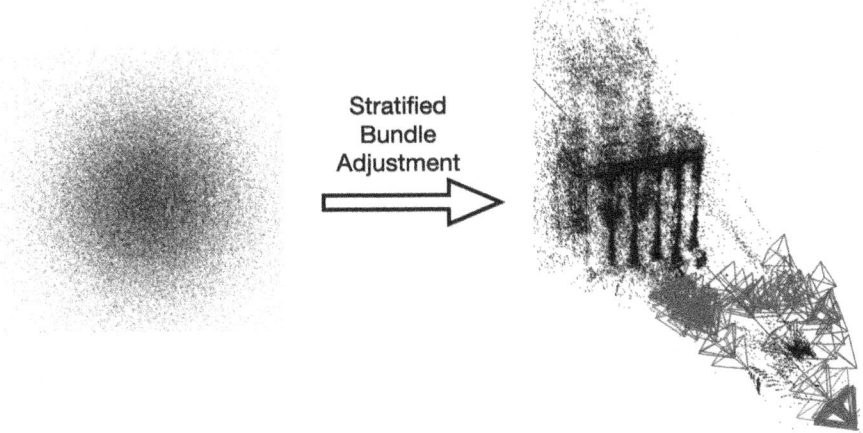

Fig. 1. In contrast to traditional bundle adjustment problem, initialization-free BA is a largely under-explored problem. It does not assume any approximation of pose and landmark parameters, making the problem much harder to solve. From a random initialization (*left figure*), and given only image measurements, we aim to recover pose and landmark parameters. Our approach, that extends inverse expansion method, is motivated by the lack of scalability of existing solvers. On the real-world BAL problems (e.g. *Venice-89, right figure*), we demonstrate the efficiency of the proposed combination of our novel solver Power Variable Projection (PoVar) and Riemannian manifold optimization framework for expansion method to solve the stratified BA problem.

BA from arbitrary initialization. We also provide some background on the inverse expansion methods. A more general description of BA can be found in [22].

Variable Projection (VarPro) Algorithm. VarPro is an optimization approach for solving bivariate problems that can be formulated as minimizing a cost function $f(u, v)$ with $f : \mathbb{R}^m \times \mathbb{R}^n \to \mathbb{R}$ over two sets of variables $u \in \mathbb{R}^m$ and $v \in \mathbb{R}^n$. Unlike alternation, which fixes u and optimizes over v and vice versa, or joint optimization, which optimizes the stack of u and v simultaneously, variable projection replaces v with $v^*(u) := \arg\min_v f(u, v)$ (v always optimal over u) such that the optimized cost function $f(u, v^*(u)) =: f^*(u)$ becomes a function of u only. The original VarPro algorithm by Golub and Pereyra [9] and its approximations by Ruhe and Wedin [19] assume solving a separable nonlinear least squares (SNLS) problem where $v^*(u)$ can be obtained in closed form. VarPro was consistently ignored in the computer vision community and even misidentified as a form of alternating optimization [5]. It first receives proper attention when Okatani et al. [17] demonstrates VarPro equipped with a trust-region approach such as Levenberg-Marquardt can yield a wide basin of convergence for several toy problems that can be formulated as a SNLS problem such as affine structure-from-motion and factorization-based non-rigid structure-from-motion. Shortly after, Strelow [20,21] extends VarPro to the nonlinear case where $v^*(u)$

is not in closed form such as bundle adjustment. Nevertheless, these works do not address the issue of increased algorithmic complexity. Later, it has been shown by Hong et al. [12] that VarPro can be efficiently implemented by performing inner iterations [2] (also known as embedded point iterations [15]) over the set of eliminated variables u followed by performing a joint optimization step with no damping on u. While this enables faster runtime compared to previous studies, it has only been tested up to small-medium sized problems with around 300 camera views. To this date, no work to the best of our knowledge has improved the scalability of the VarPro algorithm beyond [12].

Initialization-Free Bundle Adjustment. While traditional (large-scale) bundle adjustment is regularly studied [3,4,6,7,18,24,28], initialization-free BA is a recent research topic. In a seminal work, Hong et al. [13] propose to solve projective bundle adjustment from arbitrary initialization with the Variable Projection algorithm. Notably, they propose a stratified bundle adjustment formulation with increasing difficulty. In a follow-up work, pOSE [11] incorporates the equivalence between nonlinear VarPro and the Schur complement recently identified in [12]. Additionally, they define an objective in-between affine and projective models that leads to a wide convergence basin. Iglesias et al. [14] complements this *pseudo object space error* formulation with an exponential regularization term, and show interesting results on very-small-scale problems. Nevertheless, to our knowledge, none of these studies look closely into the scalability of the proposed frameworks for initialization-free bundle adjustment.

Inverse Expansion Method. Despite its short recent appearance in the literature, inverse expansion method is a highly efficient and competitive approach for solving the linearized system of equations for bundle adjustment that is already challenging established factorization and iterative methods. It links the Schur complement [26] to the power series expansion of its inverse. Using the *power Schur complement* as a preconditioner for normal equations leads to good results for physics problems such as convection-diffusion [27]. PoBA [23] proposes to directly apply the power Schur complement to the right-hand side of the reduced camera system. Their solver results in improved speed and accuracy for traditional bundle adjustment formulation with respect to the existing methods.

3   Problem Statement and Motivation

We consider a typical form of bundle adjustment. Given observations m_{ij} for pose i and landmark j, K_i, $R_i \in SO(3)$, $t_i \in \mathbb{R}^3$ the intrinsics, rotation and translation of pose i, and $x_j \in \mathbb{R}^3$, $\tilde{x}_j \in S^4$ the inhomogeneous and homogeneous landmark 3D positions (where S^n denotes the set of all vectors on the unit n-sphere), we aim to solve:

$$\min_{\{R_i\}\in SO(3),\{t_i\},\{\tilde{x}_j\}} \sum_{(i,j)\in\Omega} \|\pi(K_i[R_i|t_i]\tilde{x}_j) - m_{ij}\|_2^2, \tag{1}$$

where π is the perspective projection $\pi([x, y, z]^\top) := [x/z, y/z]^\top$.

Recently, Eq. 1 with good initialization has been solved efficiently with a novel inverse expansion method.

3.1   Inverse Expansion Method

Inverse expansion method [23,27] relates on the expansion of the inverse of a matrix into a power series, as stated in the following proposition:

Proposition 1. *Let M be an $n \times n$ matrix. If the spectral radius of M satisfies $\|M\| < 1$, then*

$$(I - M)^{-1} = \sum_{i=0}^{m} M^i + R, \tag{2}$$

where the error matrix

$$R = \sum_{i=m+1}^{\infty} M^i, \tag{3}$$

satisfies

$$\|R\| \leq \frac{\|M\|^{m+1}}{1 - \|M\|}. \tag{4}$$

Given a linear approximation of Eq. 1 followed by the Schur complement trick, Weber et al. [23] relates the inverse Schur complement of the Levenberg-Marquardt algorithm to its power series. They show significant improvement in terms of speed and accuracy to solve BA problem with good initialization. In addition to this empirical insights, some convergence behaviour concerning the approximated results for the BA problem are theoretically proved.

However, it is well-known in the literature (see e.g. [13]) that solving Eq. 1 from arbitrary initialization is non-feasible. Hong et al. [11] override this challenge by proposing a stratified bundle adjustment problem. Let us revisit the formulation of this approach.

3.2   Initialization-Free Bundle Adjustment

The stratified BA problem is decomposed in two minimization problems, followed by a metric upgrade. Although we consider a pinhole camera model, the first stage assumes a projective model.

First Stage: Separable Nonlinear Optimization with Projective Camera. Given n_p poses and n_l landmarks, $x = (x_p, x_l)$ contains all the optimization variables. We model the camera as a projective one. For pose i, we consider the camera parameters $x_p^i \in \mathbb{R}^{3 \times 4}$ and solve the following generic nonlinear separable problem:

$$\min_{x_p, \tilde{x}_l} F(x_p, \tilde{x}_l) = \left\| r(x_p, \tilde{x}_l) \right\|_2^2 = \left\| G(x_p) \tilde{x}_l - z(x_p) \right\|_2^2, \tag{5}$$

with $G(.)$ and $z(.)$ some linear operators, and the last coefficient of $\tilde{x}_l^j$ fixed to 1. For instance, pOSE [11] – extensively used in our analysis (see Supplemental), proposes the following minimization problem:

$$F_{pOSE}(x_p, x_l) = \sum_{(i,j)\in\Omega} \left\| \begin{array}{c} \sqrt{1-\eta}(x_p^{i,1:2}\tilde{x}_l^j - (x_p^{i,3}\tilde{x}_l^j)m_{ij}) \\ \sqrt{\eta}(x_p^{i,1:2}\tilde{x}_l^j - m_{ij}) \end{array} \right\|_2^2, \tag{6}$$

with $x_p^{i,1:2}$ and $x_p^{i,3}$ respectively the first two rows and the third row of x_p^i, and $\eta \in [0, 1]$.

Hong et al. [12] argue the superiority of VarPro over joint optimization to solve the previous equation, due to the random initialization of this stage.

Second Stage: Projective Refinement. The cameras and landmarks parameters obtained by solving Eq. 5 are refined by minimizing the projective standard objective [13] over the projective camera models in homogeneous form ($\{\tilde{x}_p^i \mid \text{vec}(\tilde{x}_p^i) \in S^{12}\}$) and the 3D landmarks in homogeneous coordinates ($\{\tilde{x}_l^j \mid \tilde{x}_l^j \in S^4\}$):

$$\sum_{(i,j)\in\Omega} \left\| \pi(\tilde{x}_p^i \tilde{x}_l^j) - m_{ij} \right\|_2^2. \tag{7}$$

Importantly, the optimization is performed in homogeneous coordinates. Especially, Riemannian manifold optimization [1] has to be incorporated.

Metric Upgrade. The last stage is a minimization problem to enforce the projective camera matrices to satisfy $SE(3)$ properties. Note that our work mostly focuses on the first two stages. The proposed implementation for this third stage has illustrative purpose, see Fig. 1. We refer the reader to Supplemental for further details.

3.3   Limitations and Proposed Method

In contrast to the Levenberg-Marquardt algorithm, few solvers have been designed to efficiently solve VarPro. In practice, even the most recent works [11,14] use a direct factorization (e.g. Cholesky decomposition, QR factorization), that is well-known to be poorly scalable (see e.g. [3]), to solve Eq. 5. In particular, we note that these works consider bundle adjustment problems with only few tens of cameras – sometimes less than ten. On the other hand, Hong and Fitzgibbon [10] show that coupling VarPro with the popular preconditioned conjugate gradients algorithm may not converge efficiently.

We propose to build on recent inverse expansion method. We first adapt the power series expansion to the VarPro algorithm (Sect. 4.2). We show that the so-called *PoVar* efficiently solves Eq. 5. Moving forward, we extend the power series expansion to Riemannian manifold optimization (Sect. 4.3). This new Riemannian framework, that we call *RiPoBA*, is necessary to use expansion method for solving Eq. 7. We demonstrate that the combination of this two solvers is highly competitive (Sect. 5).

4   Power Variable Projection

We start by revisiting the VarPro algorithm. We refer to [9] for further details.

4.1   Revisited Variable Projection

In contrast to joint optimization, VarPro optimizes over landmark parameters and camera parameters, but in a way different from the standard alternating least squares such that landmarks are not assumed to be fixed when updating the camera parameters. It first considers Eq. 5 as a nonlinear minimization problem over $\tilde{x}_l$ only. Due to the separability of the equation, a closed-form solution for optimal $\tilde{x}_l^*(x_p)$ is straightforward:

$$\tilde{x}_l^*(x_p) = \arg\min_{\tilde{x}_l} \|G(x_p)\tilde{x}_l - z(x_p)\|_2^2 = G(x_p)^\dagger z(x_p), \tag{8}$$

with $G(x_p)^\dagger$ the pseudo-inverse of $G(x_p)$. As we set the last coordinate of $\tilde{x}_l$ to 1, we normalize $\tilde{x}_l^*(x_p)$ by its last coordinate and its substitution in Eq. 5 leads to the following reduced problem:

$$\min_{x_p} r^*(x_p) = \min_{x_p} \|(G(x_p)G^\dagger(x_p) - I)z(x_p)\|_2^2. \tag{9}$$

Following [12], Eq. 9 can be solved with LM algorithm over x_p. The Jacobian of r^* is approximated with the so-called RW2 approximation [16], that leads to the normal equation:

$$\begin{pmatrix} U_\lambda & W \\ W^\top & V_0 \end{pmatrix} \begin{pmatrix} \Delta x_p \\ \Delta x_l \end{pmatrix} = - \begin{pmatrix} b_p \\ b_l \end{pmatrix}, \tag{10}$$

where

$$U_\lambda = J_p^\top J_p + \lambda D_p^\top D_p, \tag{11}$$

$$V_0 = J_l^\top J_l, \quad W = J_p^\top J_l, \tag{12}$$

$$b_p = J_p^\top r^0, \quad b_l = J_l^\top r^0, \tag{13}$$

with J_l and J_p respectively the landmark and pose Jacobians of the original residual (Eq. 5) around an equilibrium r^0, and D_p diagonal damping matrix for pose variables. In contrast to joint optimization, only the pose Jacobian is damped in the associated Hessian. It follows that U_λ is symmetric positive-definite [22], whereas V_0 is only guaranteed to be symmetric positive-semidefinite. Nevertheless, we observe V_0 is usually of full rank unless a 3D landmark is observed by few cameras with narrow baselines. By using the Schur complement trick, the update equation for VarPro becomes:

$$(U_\lambda - WV_0^{-1}W^\top)\Delta x_p = b_p - WV_0^{-1}b_l. \tag{14}$$

Note that the Schur complement associated to VarPro:

$$S^V = U_\lambda - WV_0^{-1}W^\top, \tag{15}$$

while sharing a close structure to the Schur complement of the traditional BA problem, has a different convergence behaviour, due to the undamped landmark Jacobian V_0.

4.2   Power Series for VarPro

Inspired by Weber et al. [23], the following lemma holds, even if V_0 is only symmetric positive-semidefinite:

Lemma 1. *Let μ be an eigenvalue of $U_\lambda^{-1} W V_0^\dagger W^\top$. Then*

$$0 \le \mu < 1. \tag{16}$$

Proof. We refer the reader to the Supplemental.

It follows that the pose update Δx_p in Eq. 14 can be directly approximated[1] by $x(m)$ with Proposition 1 applied to the Schur complement S^V:

$$x(m) = -\sum_{i=0}^{m} (U_\lambda^{-1} W V_0^{-1} W^\top)^i U_\lambda^{-1} (b_p - W V_0^{-1} b_l). \tag{17}$$

Once the pose update is estimated, the landmark update is derived following closed-form Eq. 8. That extends the inverse expansion method to Variable Projection algorithm. The difference between both solvers is the role of the damped parameters. While it seems slight, this is enough to offer a dissimilar convergence behaviour, as we will see in the experiments.

Before that, let us investigate Riemannian manifold optimization, that is necessary to solve the second stage of the stratified BA problem. In particular, we show that we can link this framework to expansion method.

4.3   Power Riemannian Manifold Optimization

As the projective refinement step in Eq. 7 involves both camera matrices and 3D landmarks in homogeneous forms, we exhibit local scale freedom for both camera and landmark parameters. This necessitates incorporation of the Riemannian manifold optimization framework without which the linearized system of equations is always rank-deficient and unsolvable. While a complete theoretical overview of such framework can be found in [1], we formalize Eq. 7 as:

$$\underset{\Delta \tilde{x}_p, \Delta \tilde{x}_l}{\arg \min} \| f(\tilde{x}_p + \Delta \tilde{x}_p, \tilde{x}_l + \Delta \tilde{x}_l) \|_2^2, \tag{18}$$

where $\tilde{x}_p \in \mathbb{R}^{12 n_p}$ denotes the stack of vectorized homogeneous camera parameters, $\tilde{x}_l \in \mathbb{R}^{4 n_l}$ denotes the stack of homogeneous 3D landmarks and $\Delta \tilde{x}_p \in \mathbb{R}^{12 n_p}$ and $\Delta \tilde{x}_l \in \mathbb{R}^{4 n_l}$ are the updates in homogeneous camera parameters and 3D landmarks respectively. The unknowns are searched in the tangent space of current $\tilde{x} = [\tilde{x}_p^\top, \tilde{x}_l^\top]^\top$, that we note $\tilde{x}^\perp \in \mathbb{R}^{(12 n_p + 4 n_l) \times (11 n_p + 3 n_l)}$ such that $(\tilde{x}^\perp)^\top x = 0$. To simplify notations, $\tilde{x}_p$ is considered as a vector in this section,

---

[1] As in practice V_0 is full-rank, and to avoid encumbering notations, we write V_0^{-1} instead of the pseudo-inverse in the rest of the paper.

and $\tilde{x}_p^\perp \in \mathbb{R}^{12n_p \times 11n_p}$, $\tilde{x}_l^\perp \in \mathbb{R}^{4n_l \times 3n_l}$ are block-diagonal, each block corresponding to the associated pose i and landmark j, respectively. The projective refinement becomes:

$$\underset{\Delta\tilde{x} \perp \tilde{x}}{\arg\min} \|f(\tilde{x}_p + \Delta\tilde{x}_p, \tilde{x}_l + \Delta\tilde{x}_l)\|_2^2. \tag{19}$$

By coupling Riemannian manifold optimization and LM algorithm, and according to [1], we get the following normal equation, projected onto the tangent space of $\tilde{x}$:

$$\begin{pmatrix} (\tilde{x}_p^\perp)^\top U_\lambda \tilde{x}_p^\perp & (\tilde{x}_p^\perp)^\top W \tilde{x}_l^\perp \\ (\tilde{x}_l^\perp)^\top W^\top \tilde{x}_p^\perp & (\tilde{x}_l^\perp)^\top V_\lambda \tilde{x}_l^\perp \end{pmatrix} \begin{pmatrix} \Delta x_p \\ \Delta x_l \end{pmatrix} = - \begin{pmatrix} (\tilde{x}_p^\perp)^\top b_p \\ (\tilde{x}_l^\perp)^\top b_l \end{pmatrix}. \tag{20}$$

where $\Delta x_p \in \mathbb{R}^{11n_p}$ and $\Delta x_l \in \mathbb{R}^{3n_l}$ are the camera update and the landmark update respectively made on the tangent space of x. By keeping coherent notations we can note the projected Jacobians and the projected damping parameters onto the tangent space of x as:

$$\tilde{J}_p = J_p \tilde{x}_p^\perp, \quad \tilde{J}_l = J_l \tilde{x}_l^\perp, \quad \tilde{\lambda}_p = (\tilde{x}_p^\perp)^\top \lambda \tilde{x}_p^\perp, \quad \tilde{\lambda}_l = (\tilde{x}_l^\perp)^\top \lambda \tilde{x}_l^\perp, \tag{21}$$

and then Eq. 20 becomes:

$$\begin{pmatrix} \tilde{U}_{\tilde{\lambda}} & \tilde{W} \\ \tilde{W}^\top & \tilde{V}_{\tilde{\lambda}} \end{pmatrix} \begin{pmatrix} \Delta x_p \\ \Delta x_l \end{pmatrix} = - \begin{pmatrix} \tilde{b}_p \\ \tilde{b}_l \end{pmatrix}, \tag{22}$$

where

$$\tilde{U}_{\tilde{\lambda}} = \tilde{J}_p^\top \tilde{J}_p + D_p^\top \tilde{\lambda}_p D_p, \tag{23}$$

$$\tilde{V}_{\tilde{\lambda}} = \tilde{J}_l^\top \tilde{J}_l + D_l^\top \tilde{\lambda}_l D_l, \tag{24}$$

$$\tilde{W} = \tilde{J}_p^\top \tilde{J}_l, \tag{25}$$

$$\tilde{b}_p = \tilde{J}_p^\top r^0, \quad \tilde{b}_l = \tilde{J}_l^\top r^0, \tag{26}$$

We have unified the notations of bundle adjustment with Riemannian manifold optimization. As the projection $x^\perp$ is full-rank, it follows that $\tilde{U}_{\tilde{\lambda}}$ and $\tilde{V}_{\tilde{\lambda}}$ are symmetric positive-definite (see Supplemental), and then the associated Riemannian Schur complement:

$$\tilde{S} = \tilde{U}_{\tilde{\lambda}} - \tilde{W}\tilde{V}_{\tilde{\lambda}}^{-1}\tilde{W}^\top \tag{27}$$

satisfies the assumption of Proposition 1:

Lemma 2. *Let $\tilde{\mu}$ be an eigenvalue of $\tilde{U}_{\tilde{\lambda}}^{-1}\tilde{W}\tilde{V}_{\tilde{\lambda}}^{-1}\tilde{W}^\top$. Then*

$$0 \leq \tilde{\mu} < 1. \tag{28}$$

The power series expansion can be applied to the inverse Riemannian Schur complement:

$$\tilde{S}^{-1} \approx \sum_{i=0}^{m} (\tilde{U}_{\tilde{\lambda}}^{-1}\tilde{W}\tilde{V}_{\tilde{\lambda}}^{-1}\tilde{W}^\top)^i \tilde{U}_{\tilde{\lambda}}^{-1}, \tag{29}$$

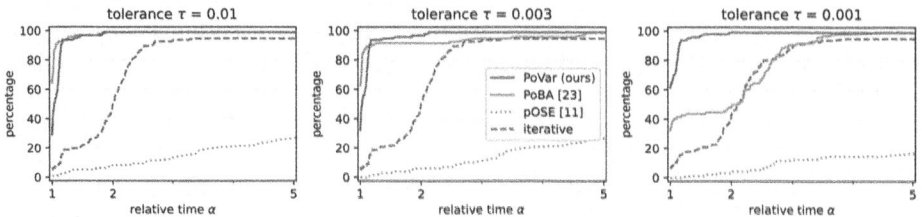

Fig. 2. Average performance profiles across all BAL problems for solving the first stage Eq. 5. Given a tolerance $\tau \in \{0.01, 0.003, 0.001\}$, it represents the percentage of solved problems (y-axis) with relative runtime α (x-axis). Expansion methods *PoVar* and *PoBA* show outstanding speed-accuracy results. Our solver *PoVar* is competitive, and most notably for the highest accuracy $\tau = 0.001$. (Color figure online)

to get pose updates, and then landmark updates by back-substitution.

Finally, the homogeneous pose and landmark updates $\Delta \tilde{x}_p$ and $\Delta \tilde{x}_l$ are retrieved by back-projecting pose and landmark updates in the tangent space to the original vector space dimension as follows:

$$\Delta \tilde{x}_p = \tilde{x}_p^\perp \Delta x_p, \quad \Delta \tilde{x}_l = \tilde{x}_l^\perp \Delta x_l.$$

After above updates are added to the pose and landmark parameters x, we carry out manifold retraction by normalizing individual vectors of camera parameters and 3D landmarks to maintain their normalized homogeneous forms.

We call this projective framework *Riemannian PoBA* (RiPoBA), that extends PoBA [23] to the Riemannian manifold optimization framework.

5   Experiments

5.1   Implementation

We implement $pOSE^2$ [11], *PoVar* and *RiPoBA* framework in C++, directly on the publicly available implementation of PoBA[3] [23]. That leads to fair comparisons with this recent and challenging solver. *pOSE* differs from *PoVar* by the use of a direct sparse Cholesky factorization. We also compare to VarPro with the conjugate gradients algorithm, preconditioned with Schur-Jacobi preconditioner, called *iterative* in our experiments. For the second stage, we compare *RiPoBA* to the conjugate gradients algorithm preconditioned by Schur-Jacobi preconditioners with Riemannian manifold optimization framework, called *RiPCG*. Except the solver itself, all implementations share much of the code with [23]. We run experiments on MacOS 14.2.1 with an Intel Core i5 (4 cores at 2 GHz).

---

[2] We use our custom implementation as the official code of [11] is not publicly available.
[3] https://github.com/simonwebertum/poba.

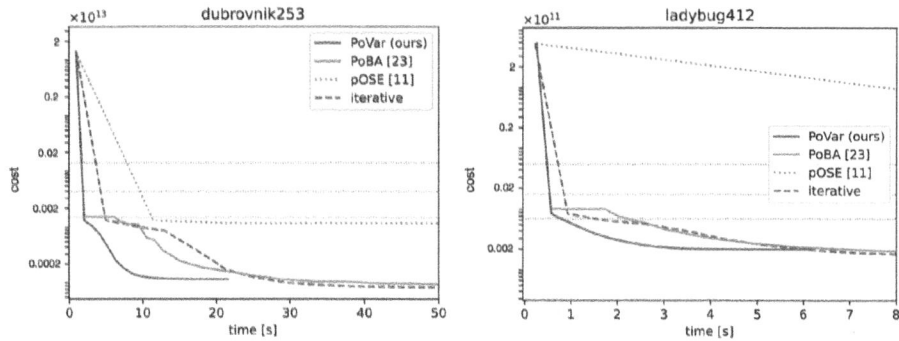

Fig. 3. Convergence plots of *Dubrovnik-253* (left) from BAL datasets with 253 poses and *Ladybug-412* with 412 poses, for solving the first stage (Eq. 5). The dotted lines correspond to cost thresholds for tolerance $\tau \in \{0.01, 0.003, 0.001\}$. (Color figure online)

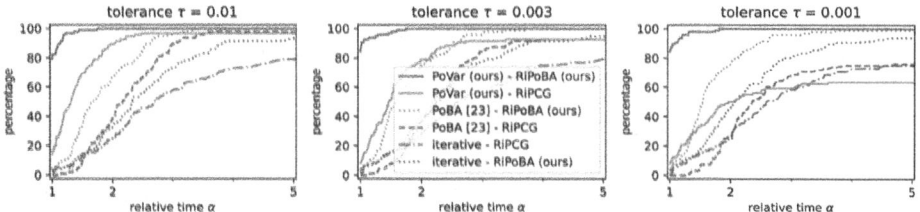

Fig. 4. Performance profiles for all BAL problems for solving the first two stages (Eq. 5 and Eq. 7). In each combination, the first solver is used to solve the first stage, and the second solver is used for the second stage. Our proposed combination *PoVar* followed by a Riemannian expansion method outperforms by a large margin compared to other competitors. Also, our proposed Riemannian-expansion method *RiPoBA* outperforms the iterative baseline in all cases, given a same solver for the first stage. (Color figure online)

5.2 Experimental Settings

Setup. For each stage, we set the maximum number of iterations to 50, stopping earlier if a relative function tolerance of 10^{-6} is achieved. The damping factor λ starts for each stage at 10^{-4} and is updated accordingly to the success or failure of the iteration. For expansion methods, we set the maximal order of power series to 20 and a threshold to 0.01. For iterative methods, we set the maximum number of inner iterations to 500. We set η for pOSE (Eq. 6) to 0.1.

Efficient Storage for Riemannian Manifold Optimization Framework. As in [23], we leverage the special structure of BA problem and propose a memory-efficient storage. We organize the landmarks into dense blocks. In particular, we apply on each row associated to a landmark the block-matrices of the projection $\tilde{x}_{l_j}^{\perp}, \tilde{x}_{p_i}^{\perp}$ corresponding to the landmark and to the cameras stored in the considered dense landmark block (see Fig. 6).

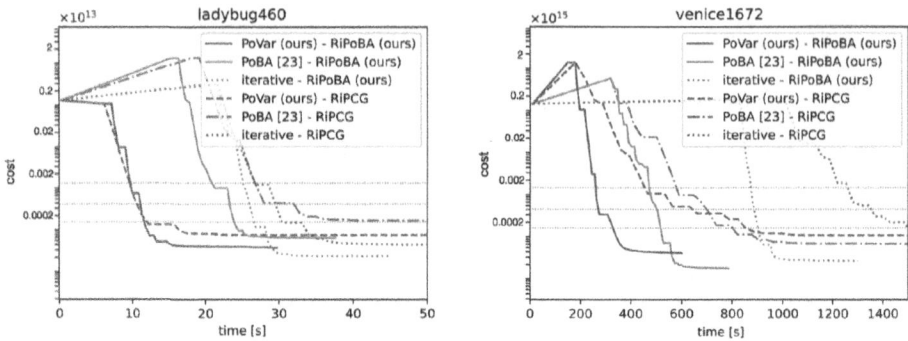

Fig. 5. Convergence plots of *Ladybug-460* (left) from BAL datasets with 460 poses and *Venice-1672* with 1672 poses, for solving the second stage (Eq. 7). The dotted lines correspond to cost thresholds for tolerance $\tau \in \{0.01, 0.003, 0.001\}$. Note that for fair comparison, the initial cost is derived before the first stage. The second cost – that may be higher than the initial one, is the initial cost of the second stage, after the first stage has been run. The runtime includes the time spent to solve the first stage. (Color figure online)

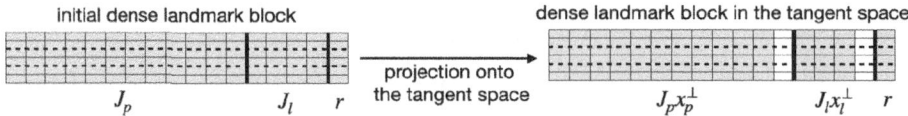

Fig. 6. Example of dense landmark block in the tangent space with 3 observations from a single landmark. We project the initial dense landmark block (*left*) onto the tangent space by applying to each i-th pose Jacobian $J_{p_i} \in \mathbb{R}^{2 \times 12}$ the projection $\tilde{x}_{p_i}^\perp$, and to each j-th landmark Jacobian $J_{l_j} \in \mathbb{R}^{2 \times 4}$ the projection $\tilde{x}_{l_j}^\perp$. The resulting pose and landmark Jacobians in the tangent space (*right*) belong to $\mathbb{R}^{2 \times 11}$ and $\mathbb{R}^{2 \times 3}$ respectively.

Dataset. We extensively evaluate our solver and the baselines on the 97 real-world bundle adjustment problems from the BAL project page [3]. The number of poses goes from 16 to 13682. We refer the reader to Supplemental for further details about these problems. For each problem, we only keep the observation measurements. Pose parameters are randomly drawn from an isotropic Gaussian distribution with mean 0 and variance 1, and landmark parameters are deduced with Eq. 8. Notably and contrary to previous works on initialization-free BA, each solver is ran on the same randomized problem, for fair comparisons.

5.3   Performance Profile

We jointly evaluate both runtime and accuracy with performance profiles [8]. Given a solver, the performance profile maps the relative runtime α to the percentage of problems solved with accuracy τ. Graphically, the performance profile of a given solver is the percentage of problems solved faster than the relative runtime α on the x-axis. Let be S and P the sets of solvers and problems, respectively. In practice, we can define the objective threshold for a problem p by:

$$f_\tau(p) = f^*(p) + \tau(f^0(p) - f^*(p)), \tag{30}$$

with $f^0(p)$ the initial objective and $f^*(p)$ the smallest error reached by the family of solvers. The runtime a solver s needs to reach this threshold is noted $T_\tau(p, s)$. The performance profile of a solver for a relative runtime α is defined as:

$$\rho(s, \alpha) = \frac{100}{|P|} |\{p \in P | T_\tau(p, s) \le \alpha \min_{s \in S} T_\tau(p, s)\}|. \tag{31}$$

Graphically, a curve on the left of the performance profile is linked to better runtime, whereas a curve on the right is related to better accuracy. Note that for meaningful comparison, all solvers should have the same initial objective.

5.4   Analysis

First Stage. Figure 2 shows the performance profiles for all BAL datasets with tolerances $\tau \in \{0.01, 0.003, 0.001\}$ to solve Eq. 5. As expected, the direct factorization solver (dashed green) used in Hong et al. [11] shows poor performance. Our solver *PoVar* (blue) challenges *PoBA* for the largest tolerance 0.01, and is by far the most competitive solver for the smallest tolerance $\tau = 0.001$, that is for the highest accuracy. For $\tau = 0.003$, *PoVar* slightly outperforms *PoBA*. We also highlight that our solver outperforms the two competitors associated to the VarPro algorithm, iterative (dashed red) and direct factorization.

Figure 3 shows two examples of the cost decrease during the first stage. Notably, *PoVar* (blue) converges much more smoothly than its main challenger *PoBA* which gets stuck in early iterations. By considering the intersections of the solvers with the cost thresholds (dashed grey lines), it also demonstrates the slowness of the iterative method compared to expansion methods in terms of runtime.

First and Second Stage. Figure 4 shows the performance profiles for all BAL datasets with tolerances $\tau \in \{0.01, 0.003, 0.001\}$ to solve the first two stages, that are Eq. 5 followed by Eq. 7. Note that we compare in this experiment the cost of the second stage only, as the first stage is only used to get an approximated initialization for the projective formulation. As direct factorization shows poor performance during the first stage, we only take into account the most promising combinations of solvers. *PoVar* followed by *RiPoBA* (blue) outperforms all the competitors both in terms of runtime and accuracy. The combinations with *RiPoBA* outperform *RiPCG* for the highest accuracy $\tau = 0.001$ for all relative time greater than $\alpha = 2$, that reflects the better convergence of Riemannian expansion method compared to *RiPCG*. We also note, given a same solver for first stage, *RiPoBA* outperforms *RiPCG* during the second stage across all tolerances.

Figure 5 illustrates on two examples the cost decrease during the second stage. On the left figure (*Ladybug-460*), the best two solvers in terms of final convergence are built on our framework *RiPoBA*. On the right figure (*Venice-1672*), all the solvers using *RiPoBA* converge to a smaller error than their iterative competitor *RiPCG*. By considering the intersection between the solvers and the cost thresholds (dashed grey lines), the combination *PoVar-RiPoBA* outperforms all other combinations, and most notably for the largest problem with 1672 poses.

Conclusive Remark. The experiments emphasize the high efficacy of our solvers *PoVar* and *RiPoBA*, during both first and second stages of the stratified BA problem. Concerning the first stage, the convergence of *PoVar* is much smoother than its competitors, that explains its larger speed with respect to *PoBA* to reach the cost thresholds, albeit both are built with power series. Regarding the second stage, for a same given solver in the first stage, our *RiPoBA* outperforms the preconditioned conjugate gradients with Riemannian manifold optimization framework in terms of speed and accuracy, especially when coupled with *PoVar*.

6   Conclusion

We have introduced a novel approach to address the scalability challenge for initialization-free bundle adjustment. Our proposed Power Variable Projection (PoVar) algorithm, theoretically justified, offers new insights to this uncharted problem. By extending recent inverse expansion techniques to the VarPro algorithm on one hand, and to Riemannian manifold optimization on the other hand, we have demonstrated the capability to efficiently solve large-scale stratified BA problem with thousands of cameras. Notably, we achieve state-of-the-art results in terms of speed and accuracy on the real-world BAL dataset. While initialization-free BA is still in its nascent stage, we hope that our method will pave the way for further exploration of this difficult optimization problem, and will generate further steps towards initialization-free structure-from-motion.

Limitations and Future Work. First, our analysis applies to the BA problem in which excess outlier point tracks are assumed to have been filtered out as in previous formulations [11–14]. Second, the 3D reconstruction in Fig. 1 assumes that the intrinsics are known during the metric upgrade stage. In practice, we use the approximated focal lengths given in the BAL dataset, that results in imperfectly accurate illustration. However, some formulations based on pOSE handle unknown intrinsics and could be easily adapted to *PoVar*.

Acknowledgements. This work was supported in part by the NRF grants funded by the Korea government (MSIT) (No. 2022R1C1C1004907) and in part by the Institute of Information & communications Technology Planning & Evaluation (IITP) under the artificial intelligence semiconductor support program to nurture the best talents (IITP-(2024)-RS-2023-00253914) grant funded by the Korea government (MSIT).

References

1. Absil, P.A., Mahony, R., Sepulchre, R.: Optimization Algorithms on Matrix Manifolds. Princeton University Press (2008)
2. Agarwal, S., Mierle, K., et al.: Ceres solver (2024). http://ceres-solver.org
3. Agarwal, S., Snavely, N., Seitz, S.M., Szeliski, R.: Bundle adjustment in the large. In: Daniilidis, K., Maragos, P., Paragios, N. (eds.) ECCV 2010, Part II. LNCS, vol. 6312, pp. 29–42. Springer, Heidelberg (2010). https://doi.org/10.1007/978-3-642-15552-9_3
4. Belder, A., Vivanti, R., Tal, A.: A game of bundle adjustment-learning efficient convergence. In: Proceedings of the IEEE/CVF International Conference on Computer Vision, pp. 8428–8437 (2023)
5. Buchanan, A.M., Fitzgibbon, A.W.: Damped Newton algorithms for matrix factorization with missing data. In: 2005 IEEE Conference on Computer Vision and Pattern Recognition (CVPR), vol. 2, pp. 316–322 (2005). https://doi.org/10.1109/CVPR.2005.118
6. Demmel, N., Gao, M., Laude, E., Wu, T., Cremers, D.: Distributed photometric bundle adjustment. In: 2020 International Conference on 3D Vision (3DV), pp. 140–149. IEEE (2020)
7. Demmel, N., Sommer, C., Cremers, D., Usenko, V.: Square root bundle adjustment for large-scale reconstruction. In: Proceedings of the IEEE/CVF Conference on Computer Vision and Pattern Recognition, pp. 11723–11732 (2021)
8. Dolan, E.D., Moré, J.J.: Benchmarking optimization software with performance profiles. Math. Program. **91**, 201–213 (2002)
9. Golub, G.H., Pereyra, V.: The differentiation of pseudo-inverses and nonlinear least squares problems whose variables separate. SIAM J. Numer. Anal. **10**(2), 413–432 (1973)
10. Hong, J.H., Fitzgibbon, A.: Secrets of matrix factorization: approximations, numerics, manifold optimization and random restarts. In: Proceedings of the IEEE International Conference on Computer Vision, pp. 4130–4138 (2015)
11. Hong, J.H., Zach, C.: pOSE: pseudo object space error for initialization-free bundle adjustment. In: Proceedings of the IEEE Conference on Computer Vision and Pattern Recognition, pp. 1876–1885 (2018)
12. Hong, J.H., Zach, C., Fitzgibbon, A.: Revisiting the variable projection method for separable nonlinear least squares problems. In: 2017 IEEE Conference on Computer Vision and Pattern Recognition (CVPR), pp. 5939–5947. IEEE (2017)
13. Hong, J.H., Zach, C., Fitzgibbon, A., Cipolla, R.: Projective bundle adjustment from arbitrary initialization using the variable projection method. In: Leibe, B., Matas, J., Sebe, N., Welling, M. (eds.) ECCV 2016, Part I. LNCS, vol. 9905, pp. 477–493. Springer, Cham (2016). https://doi.org/10.1007/978-3-319-46448-0_29
14. Iglesias, J.P., Nilsson, A., Olsson, C.: expOSE: accurate initialization-free projective factorization using exponential regularization. In: Proceedings of the IEEE/CVF Conference on Computer Vision and Pattern Recognition, pp. 8959–8968 (2023)
15. Jeong, Y., Nister, D., Steedly, D., Szeliski, R., Kweon, I.S.: Pushing the envelope of modern methods for bundle adjustment. In: 2010 IEEE Conference on Computer Vision and Pattern Recognition (CVPR), pp. 1474–1481 (2010). https://doi.org/10.1109/CVPR.2010.5539795
16. Kaufman, L.: A variable projection method for solving separable nonlinear least squares problems. BIT Numer. Math. **15**, 49–57 (1975)

17. Okatani, T., Yoshida, T., Deguchi, K.: Efficient algorithm for low-rank matrix factorization with missing components and performance comparison of latest algorithms. In: 2011 IEEE International Conference on Computer Vision (ICCV), pp. 842–849 (2011). https://doi.org/10.1109/ICCV.2011.6126324
18. Ren, J., Liang, W., Yan, R., Mai, L., Liu, S., Liu, X.: MegBA: a GPU-based distributed library for large-scale bundle adjustment. In: Avidan, S., Brostow, G., Cissé, M., Farinella, G.M., Hassner, T. (eds.) ECCV 2022. LNCS, vol. 13697, pp. 715–731. Springer, Cham (2022). https://doi.org/10.1007/978-3-031-19836-6_40
19. Ruhe, A., Wedin, P.Å.: Algorithms for separable nonlinear least squares problems. SIAM Rev. (SIREV) **22**(3), 318–337 (1980). https://doi.org/10.1137/1022057
20. Strelow, D.: General and nested Wiberg minimization. In: 2012 IEEE Conference on Computer Vision and Pattern Recognition (CVPR), pp. 1584–1591 (2012). https://doi.org/10.1109/CVPR.2012.6247850
21. Strelow, D.: General and nested Wiberg minimization: L_2 and maximum likelihood. In: Fitzgibbon, A., Lazebnik, S., Perona, P., Sato, Y., Schmid, C. (eds.) ECCV 2012. LNCS, vol. 7578, pp. 195–207. Springer, Heidelberg (2012). https://doi.org/10.1007/978-3-642-33786-4_15
22. Triggs, B., McLauchlan, P.F., Hartley, R.I., Fitzgibbon, A.W.: Bundle adjustment—a modern synthesis. In: Triggs, B., Zisserman, A., Szeliski, R. (eds.) IWVA 1999. LNCS, vol. 1883, pp. 298–372. Springer, Heidelberg (2000). https://doi.org/10.1007/3-540-44480-7_21
23. Weber, S., Demmel, N., Chan, T.C., Cremers, D.: Power bundle adjustment for large-scale 3D reconstruction. In: Proceedings of the IEEE/CVF Conference on Computer Vision and Pattern Recognition, pp. 281–289 (2023)
24. Weber, S., Demmel, N., Cremers, D.: Multidirectional conjugate gradients for scalable bundle adjustment. In: Bauckhage, C., Gall, J., Schwing, A. (eds.) DAGM GCPR 2021. LNCS, vol. 13024, pp. 712–724. Springer, Cham (2021). https://doi.org/10.1007/978-3-030-92659-5_46
25. Nocedal, J., Wright, S.J.: Numerical Optimization. Springer, New York (2006). https://doi.org/10.1007/978-0-387-40065-5
26. Zhang, F.: The Schur Complement and Its Applications, vol. 4. Springer, New York (2006). https://doi.org/10.1007/b105056
27. Zheng, Q., Xi, Y., Saad, Y.: A power Schur complement low-rank correction preconditioner for general sparse linear systems. SIAM J. Matrix Anal. Appl. **42**(2), 659–682 (2021)
28. Zhou, L., et al.: Stochastic bundle adjustment for efficient and scalable 3D reconstruction. In: Vedaldi, A., Bischof, H., Brox, T., Frahm, J.-M. (eds.) ECCV 2020. LNCS, vol. 12360, pp. 364–379. Springer, Cham (2020). https://doi.org/10.1007/978-3-030-58555-6_22

Collaborative Control for Geometry-Conditioned PBR Image Generation

Shimon Vainer[1]([✉]), Mark Boss[2][ID], Mathias Parger[1], Konstantin Kutsy[1], Dante De Nigris[1], Ciara Rowles[1], Nicolas Perony[1], and Simon Donné[1][ID]

[1] Unity Technologies, San Francisco, USA
shimon.vainer@unity3d.com
[2] San Francisco, USA

Abstract. Graphics pipelines require physically-based rendering (PBR) materials, yet current 3D content generation approaches are built on RGB models. We propose to model the PBR image distribution directly, avoiding photometric inaccuracies in RGB generation and the inherent ambiguity in extracting PBR from RGB. As existing paradigms for cross-modal fine-tuning are not suited for PBR generation due to both a lack of data and the high dimensionality of the output modalities, we propose to train a new PBR model that is tightly linked to a frozen RGB model using a novel cross-network communication paradigm. As the base RGB model is fully frozen, the proposed method retains its general performance and remains compatible with *e.g.* IPAdapters for that base model.

Keywords: Image Generation · Material Properties · Multi-Modal Generation · Physically-Based Rendering

1 Introduction

The recent meteoric rise of diffusion models has made at-scale generation of high-quality RGB image content more accessible than ever and Text-to-Texture and Text-to-3D approaches successfully lift this to 3D [34]. But to maximize the usefulness of the generated textures in downstream 3D workflows, generated content must be compatible with physically-based rendering (PBR) pipelines for proper shading and relighting. Current approaches rely on generated RGB images and subsequent PBR extraction through inverse rendering, suffering from the physically **in**accurate lighting in the generated RGB images as well as from significant ambiguities in the inverse rendering. We propose a solution

---

S. Vainer and S. Donné—Equal Contributions.
M. Boss—Stability AI, work done while at Unity Technologies.
Core Technical Contributions.

Supplementary Information The online version contains supplementary material available at https://doi.org/10.1007/978-3-031-72624-8_8.

A. Leonardis et al. (Eds.): ECCV 2024, LNCS 15071, pp. 127–145, 2025.
https://doi.org/10.1007/978-3-031-72624-8_8

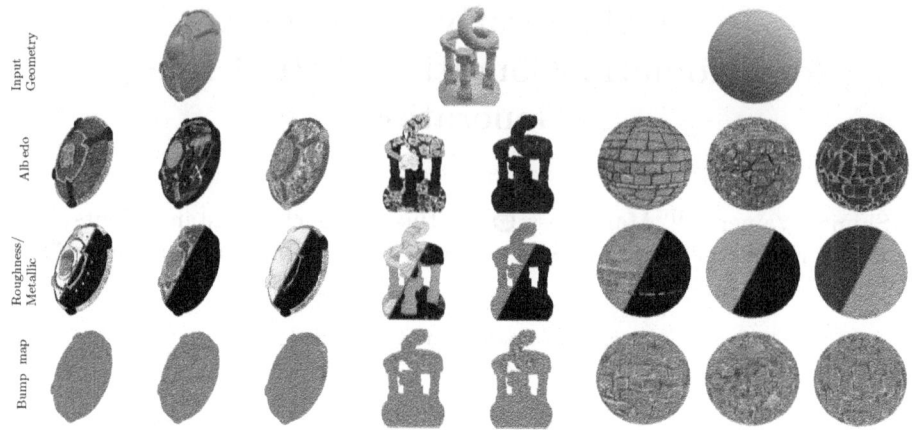

Fig. 1. Generated PBR materials. By tightly linking the PBR diffusion model with a frozen RGB model, we produce high-quality PBR images conditioned on geometry and prompts. Visit the project page at https://unity-research.github.io/holo-gen.

for geometry-conditioned generation of PBR images by modeling the joint distribution directly, avoiding the issues around photometric consistency and inverse rendering (Fig. 1).

To model the distribution of non-RGB modalities, existing approaches typically fine-tune the weights of a base RGB model [11,29,33,40,41,58]. Applied to PBR images, this means either directly predicting the entire PBR image stack or sequentially predicting them conditioned on one another. Neither is sufficient for our use-case: jointly predicting the entire PBR image stack is problematic as the higher-dimensional modality does not compress well into the established latent spaces (as we show in Sect. 5), and sequentially predicting the elements of the PBR image stack is significantly more expensive and risks compounding errors in the sequential generation. Furthermore, while state-of-the-art RGB diffusion models are trained on billions of images [52], there is unfortunately no dataset of such size at our disposal for PBR content generation. Instead, the largest available dataset of PBR content is Objaverse [9], containing around 800,000 objects with associated PBR textures, limited to "everyday" appearances of the objects. In light of the restricted training data available, fine-tuning the base model results in catastrophic forgetting, forfeiting generalizability, as we illustrate in Sect. 5.

Instead, we keep a pre-trained RGB image model frozen and train a parallel model to generate PBR images, as shown in Fig. 2. We tightly link the PBR model to the frozen RGB model using our proposed cross-network control paradigm, in order to leverage its expressivity and rich internal state. As a result we are able to generate qualitative and diverse PBR content, even for unlikely appearances of objects (far out-of-distribution for the Objaverse dataset). Crucially, the frozen RGB model safeguards against catastrophic forgetting *and* remains compatible with techniques such as IPAdapter [73]. In summary, we:

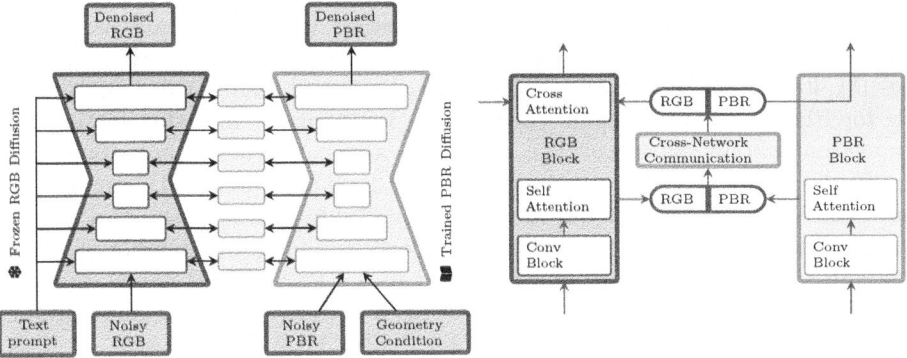

Fig. 2. Collaborative Control. Two parallel models collaborate to generate pixel-aligned outputs of different modalities. We freeze the left pre-trained RGB model and train the right PBR model with its cross-network communication layers. The cross-communication concatenates the states of both models, processes them with a small MLP, and residually distributes the result back to the respective models. As discussed in Sect. 5, prompt cross-attention in the PBR model is counter-productive.

1. Propose the novel *Collaborative Control* paradigm to tightly link the PBR generator to a fully frozen pre-trained RGB model, modeling the joint distribution of RGB and PBR images directly (see Sect. 4.1),
2. Illustrate that the proposed control mechanism is data-efficient, and generates high-quality images even from a very restricted training set,
3. Demonstrate the compatibility with IPAdapter [73] specifically, and
4. Ablate our design choices to show the improvement over existing paradigms in literature and the issues with existing paradigms.

2 Related Work

Generating Natural Images from Text Prompts. Natural image generation has a long history: from GANs [14,25–28] and VAEs [30], to autoregressive models [45,64]. More recently, the introduction of diffusion models [18,56,57] was a breakthrough in the generative field—far more stable than typical GAN training, albeit slow and computationally expensive, and easier to control and condition. Unfortunately, these approaches require billions of images to train from scratch [52], and for PBR image generation the largest commonly available dataset is Objaverse [9]. With 800,000+ objects it is still several orders of magnitude smaller than LAION-5B [52] and proves insufficient to train generative models that can generalize to unlikely semantics (as illustrated in Sect. 5). While pre-trained RGB models encode rich prior knowledge around structure, semantics, and materials [11,53,59], Sarkar *et al.* warn that the models are often still geometrically inaccurate [50]: we have found that this extends to material properties, as diffusion models prefer idealized and artistic appearances over photometric accuracy.

Generating Non-RGB Modalities. Existing works fine-tune pre-trained RGB models to predict *e.g.* Depth [29,58], semantics [33] or intrinsic properties [11,40,41], either directly or through LoRA's [19]. Sadly, this is not plausible for PBR image generation: compressing PBR images into the existing low-dimensional latent space overloads it, and the alternative of sequentially predicting channel triplets is too costly and slow. Wonder3D [41] and UniDream [40] perform joint RGB and normal diffusion using a cross-domain self-attention aligning the two parallel branches, yet this scales poorly an increasing number of output modalities. Our proposed approach instead uses a frozen RGB model, negating the risk of catastrophic forgetting, and trains a parallel branch for all additional modalities jointly (in the latent space of a PBR VAE), reducing cost.

Image-Based Conditioning. Existing pixel-accurate control techniques come in two flavors: re-training of the base model with modified input spaces [12,29], and training of a parallel model that affects the base model's state [10,20,79]. We find in Sect. 5 that the former risks losing the base model's expressiveness and quality. In ControlNet [79] and ControlNet-XS [10], the controlling model only influences the base RGB model's output while in AnimateAnyone [20] the parallel model is only tasked with generating its own output. Instead, we leave the RGB base model's weights fully frozen and residually edit its internal states from a parallel model that is itself tasked with generating PBR images: our PBR model both *controls* the base model (to guide it towards the domain of rendered images), and *generates* its own PBR output (based on the RGB model's internal state); therefore, our proposed approach requires full bidirectional connections between both branches as shown in Fig. 5. To condition on input geometry we concatenate it to the input of the PBR branch, as Ke *et al.* [29].

Text-to-3D describes the task of generating 3D objects from text prompts, often with the aim to support downstream graphics pipelines such as game engines. Earlier methods leverage Score Distillation Sampling [46] (SDS) to iteratively optimize a 3D representation by backpropagating the diffusion model's noise predictions [15,22,35,37,42,43,55,61,62,66–68,71,80,81] through the RGB model, or building on viewpoint-aware image models [21,38,39,41,47,54,76] for direct fusion. Such RGB methods ignore that object appearance varies with viewing angle, often resulting in artifacts around highlights, and their RGB output is not useful in graphics pipelines. More recent work generates PBR properties so using inverse rendering with a differentiable renderer [7,40,70,72,74]: a major concern is lighting being baked into the material channels (*e.g.* HyperDreamer [70] uses an ad-hoc regularization to reduce these artifacts). **Text-to-Texture** methods restrict the Text-to-3D problem to objects with known structure by conditioning the diffusion model on the object geometry [4,6,31,32,75,77,78], but face similar issues by operating in the RGB domain. Paint3D [77] also discusses the lighting artifacts typical with inverse rendering and introduces a custom post-processing diffusion model to alleviate these. By directly generating PBR content, our proposed technique promises to resolve issues related to inverse rendering in the latter methods, all the while retaining the simplicity of the former methods.

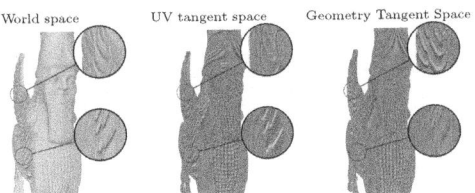
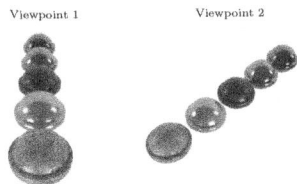

Fig. 3. Bump map. Similar surface bumps in world space (left) are dissimilar in the UV tangent space (middle) because of the arbitrary UV mapping. Representing the bump map in a tangent space solely dependent on the geometry (right) resolves this issue.

Fig. 4.   Rendering   function. The dataset is constructed so that the lighting remains constant with respect to the camera, simplifying the rendering function f_{RGB}: notice the similar highlight location.

Evaluation metrics for generative methods compare the output distributions with known ground-truth distributions, typically with the Inception Score [49] (IS) or the Fréchet Inception Distance [17] (FID). CMMD [23] argues that neither is well suited to modern generative models, and compare the distributions of CLIP embeddings rather than Inceptionv3 [60] internal states.

Aside from comparing modelled distributions with the ground truth, we also wish to evaluate the alignment of the generated images to their text prompts. CLIPScore [16] compares the image's CLIP embedding with that of the prompt: whether all the relevant elements are represented and whether any extraneous elements were introduced. We also report the OneAlign *aesthetics* and *quality* metrics of the generated images [69], which have been shown to align well with human perception, to provide a more quantitative indication of quality.

3   Preliminaries

PBR materials are a compact representation of the bidirectional reflectance distribution function (BRDF), which describes how light is reflected from the surface of an object. We use the popular Cook-Torrance analytical BRDF model [8], using specifically the Disney BRDF Basecolor-Metallic parametrization [3] as it inherently promotes physical correctness. In this parametrization, the BRDF comprises *Albedo* ($b_a \in \mathbb{R}^3$), *Metallic* ($b_m \in \mathbb{R}$), and *Roughness* ($b_r \in \mathbb{R}$) components. To increase realism during rendering beyond the resolution of the underlying geometry (often a mesh), graphics pipelines add small details such as wood grain or grout between tiles by encoding them as offsets to the surface normals in an additional *bump map* ($b_n \in \mathbb{R}^3$). As this bump map is typically defined in a tangent space based on an arbitrary UV-unwrapping, it entangles the surface property with this arbitrary UV mapping. Instead, we propose to predict the bump map defined in a tangent space based solely on the object geometry, disentangling the texture from the UV mapping as shown in Fig. 3. To construct this geometry tangent space for a point $p = [p_x, p_y, p_z]^T$ with geometry normal

n, we construct the local tangent vector as $t = n \times ([-p_y, p_x, 0]^T \times n)$, corresponding to Blender's *Radial Z* geometry tangent. The geometry tangent space is then constructed as $(t/\|t\|, n \times t/\|t\|, n)^T$.

Diffusion models [18,56] iteratively invert a forward degradation process to generate high-quality images from pure noise (typically white Gaussian noise). Formally, the forward process iteratively degrades images from the data distribution $z_0 \sim p(z)$ to standard-normal samples $z_T \sim \mathcal{N}(0, I)$ over the course of T degradation steps as $z_t \sim \mathcal{N}(\alpha_t z_{t-1}, (1 - \alpha_t)I)$, where α_t denotes the noise schedule for timestep t. Practically, the forward process can be condensed into the direct distribution $z_t \sim \mathcal{N}(\sqrt{\bar{\alpha}_t}z_0, (1 - \bar{\alpha}_t)I)$ with the appropriate choice of $\bar{\alpha}_t$. The diffusion model $\mathcal{D}$ is trained to sample the stochastic reverse process $\mathcal{D}_t(z_t) \sim p(z_{t-1}|z_t)$ to iteratively generate z_0 from z_T.

4   Method

We wish to train a PBR diffusion model $\mathcal{D}_{pbr}$ that models the reverse denoising process for PBR images as represented in the latent space of a VAE [48], representing the data distribution $p(z_{pbr})$. We find that we lack the data required to train this model directly, and instead propose to model $p(z'_{rgb} := f_{rgb}(z_{pbr}), z_{pbr})$ based on an RGB diffusion model $\mathcal{D}_{rgb}$ for the RGB data distribution $p(z_{rgb})$; f_{rgb} is a rendering function that projects the PBR images onto the RGB domain. To motivate this, we split the joint reverse process into two separate processes using Bayes' rule:

$$p(z'_{rgb,t-1}, z_{pbr,t-1}|z'_{rgb,t}, z_{pbr,t})$$
$$\sim p(z'_{rgb,t-1}|z'_{rgb,t}, z_{pbr,t})p(z_{pbr,t-1}|z'_{rgb,t-1}, z'_{rgb,t}, z_{pbr,t}) \quad (1)$$

The RGB model is implemented based on $\mathcal{D}_{rgb}(z_{rgb,t-1}) \sim p(z_{rgb,t-1}|z_{rgb,t})$: we align the current RGB sample with the PBR sample and restrict it to $\mathrm{Im}(f)$ (the domain of rendered images with the fixed environment map) so that its internal states are more easily interpreted by the PBR branch[1]. To simplify this alignment problem, the rendering function f_{rgb} uses fixed camera settings and a fixed environment map as shown in Fig. 4. The PBR model now no longer models $p(z_{pbr,t-1}|z_{pbr,t})$: it additionally has access to the RGB context $(z'_{rgb,t-1}, z'_{rgb,t})$ which simplifies the problem. The RGB and PBR models are in practice much more intertwined than Eq. (1) implies: this derivation serves mostly as an intuitive indication for why the joint problem is more tractable. Note that $z'_{rgb,t}$ is a degraded version of $z'_{rgb,0}$, and *not* a rendered version of $z_{pbr,t}$: the PBR model

---

[1] Our intuition as to why a fixed environment map is beneficial is that it makes the RGB model's internal states more consistent to interpret, and makes the control problem of projecting to $\mathrm{Im}(f)$ simpler. Early in training, generated sample quality can be boosted significantly by applying the foreground mask to the RGB estimate for the first few timesteps; a rough projection to bring the estimate much closer to $\mathrm{Im}(f)$. After longer training, this is no longer necessary, as the PBR branch is capable enough to restrict the RGB branch to $\mathrm{Im}(f)$.

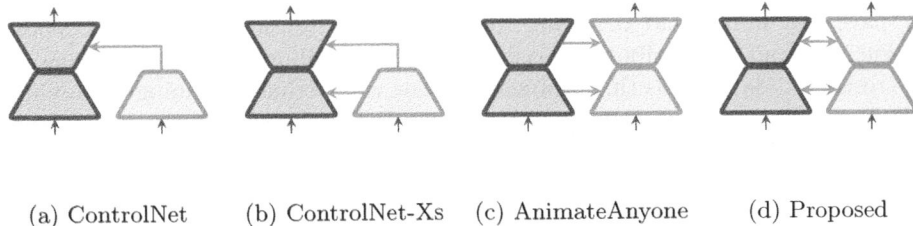

(a) ControlNet   (b) ControlNet-Xs   (c) AnimateAnyone   (d) Proposed

Fig. 5. High-level overview of communication in (a) ControlNet [79], (b) ControlNet-XS [10], (c) AnimateAnyone [20] and (d) our proposed Collaborative Control approach. Blue represents frozen blocks, while orange elements are optimized during training. (Color figure online)

does not learn to do inverse rendering in degraded image space but rather learns to denoise PBR images given additional RGB context.

4.1   Collaborative Control

In summary, our proposed approach comprises two models working in tandem: a pre-trained RGB image model and a new PBR model (see Fig. 5 for a high-level overview of our proposed control scheme). The previous section identifies two tasks for this cross-network communication: aligning the RGB model's output with both the PBR model's output and the map of the rendering function $Im(f)$, and communicating knowledge in the RGB model to the PBR model. ControlNet [79] and ControlNet-XS [10] discuss solutions to the former control problem—the authors conclude that communication from the base model's encoder to the controlling model's encoder, and from the controlling model's decoder to the base model's decoder, is sufficient. AnimateAnyone [20] addresses the latter problem and concludes that, there, uni-directional communication from the left model to the right model is sufficient. We have found that full bidirectional communication is crucial for our approach: the PBR branch needs to extract relevant information from the RGB model's hidden state, while simultaneously guiding the RGB output towards render-like images (*i.e.* with a black background and compatible lighting) to ensure those hidden states are consistent with its own expectations. We dub this *Collaborative Control*.

We implement the cross-network communication as a connecting layer between the two models after every self-attention module; its inputs are the concatenation of the model states and its outputs are residually distributed to both models again. During training, we only optimize the weights of the PBR model and the cross-network communication links against both models' outputs, while the RGB model remains fully frozen. By adopting this approach, we safeguard the base RGB model's weights, and do not risk catastrophic forgetting for that base model. As we discuss in Sect. 5, we have found that a single per-pixel linear layer is sufficient, although we also evaluate the other control schemes from Fig. 5 as well as an attention-based communication layer. Notably, we have also found that disabling the text cross-attention in the PBR model is crucial to out-of-distribution performance; we attribute this to overfitting on the restricted dataset, as this problem worsens with reduced training data. Only allowing prompt attention through the frozen RGB model prevents such overfitting.

4.2   Implementation

Compressing PBR Images into Latent Space. RGB diffusion models benefit immensely from a dedicated VAE to encode the images into a lower-dimensional latent space [48]. Existing solutions that generate an alternate modality typically encode that modality with the RGB VAE, but PBR images cannot be compressed into the same latent space due to the higher dimensionality. Instead, we could select channel triplets b_a, $[b_m, b_r, 0]$, and b_n and process those with the RGB VAE, but we instead choose to train a dedicated PBR VAE—our ablation studies indicate that the distribution mismatch between the PBR channels and the RGB space is too large, and performance would otherwise suffer. We adopt the VAE architecture and training code from StableDiffusion v1.5 [48], although following Vecchio et al. [65] we set the latent space channel count to 14 for the optimal balance between quality and compression when processing PBR images.

Conditioning on Existing Geometry. We concatenate the screen-space geometry normals to the PBR model's inputs to condition the joint output. Referring to Fig. 5, Collaborative Control encapsulates the ControlNet scheme that would typically be used for this conditioning [10]: as we jointly train from scratch, this does not introduce additional cost.

Generating Training Data. Our dataset for training both the PBR VAE and the Collaborative Control scheme is based on Objaverse [9]: a dataset containing 800,000+ 3D models with annotations for what the models represent (describing both shape and texture). After sanitizing and filtering the dataset we retain roughly 300,000 objects. Each of the objects is rendered with Blender from 16 viewpoints encircling the object using a fixed pinhole camera model and a fixed (camera colocated) environment map[2] as in Fig. 4. For the evaluations in Sect. 5, we leave out 2% randomly selected elements.

Training Collaborative Control. For most of the experiments in Sect. 5, ZeroDiffusion [36,63] is the base RGB model, a zero-terminal-SNR version fine-tuned from StableDiffusion v1.5 [48]. As Collaborative Control is agnostic to the base model, we also illustrate StableDiffusion v1.5 and v2.1 as base models in Sect. 5. We optimize the PBR model's weights as well as the cross-network communication layers to minimize the training loss for the RGB and PBR denoising jointly, while keeping the RGB model fully frozen. Unless otherwise stated, we directly train on 512×512 resolution for a total of 200,000 update steps with a

---

[2] https://polyhaven.com/a/studio_small_08.

batch size of 12 and a learning rate of 3×10^{-5} (on one 80 GB VRAM A100, taking roughly two days). We evaluate the effect of a larger training budget by training on 8 A100's for the same number of steps, increasing the batch size by a factor of 8 without affecting training time—for environmental and cost purposes, the training budget is kept low for the main ablation study.

5   Results

Distribution Match Metrics. As an evaluation of how well the data distribution is modeled, distribution match is considered a proxy to both quality and diversity. The Inception Score (IS [49]), which checks the distribution match against ImageNet, is not relevant in a PBR context as it applies only to RGB images. The Fréchet Inception Distance (FID [17]), which compares the distributions of the last hidden state of the Inceptionv3 [60] network on both real and generated images, has been found to better align to perceptual quality. Finally, the recently introduced CLIP Maximum-Mean Discrepancy (CMMD [23]) compares the distribution of the CLIP embeddings of generated images to that of a reference dataset. It offers significantly improved sample efficiency, and was shown by the authors to be a better indicator of low-level image quality than FID. However, as these metrics are intended for three-channel color images, we evaluate them on PBR images following Chambon et al. [5], by averaging the relevant scores of multiple triplets. We report as PBR distribution match the average of the scores over each of the PBR channels independently, as well as over three additional triplets, as the full set of triplets is prohibitively expensive to compute (the supplementary contains all the constituting scores). The additional triplets are (grayscale albedo, roughness, metallic), (roughness, metallic, normal XY norm) and (grayscale albedo, normal X, normal Y) for a balance between the full cartesian product (which is costly) and mixing channels that are normally relatively independent.

Out-of-distribution (OOD) performance metrics indicate the level to which our generator can align to conditioning that it was not trained on. Recent work has introduced the CLIP alignment score [13,16], which estimates the average distance between the text prompt CLIP embedding and the generated image's CLIP embedding, indicating how faithfully the prompt was followed. Additionally, OneAlign [69] is a neural model that estimates aesthetics and quality scores for images, shown to align well with human opinions, summarized in a QAlign score for both aesthetics and general quality. In order to evaluate the OOD performance, we randomly select 50 objects from Objaverse, and generate unlikely appearance prompts for them using ChatGPT4 [1]. These results, as well as a t-SNE plot of the embeddings of the original and OOD prompts, are integrally shown in the supplementary material.

5.1   Comparisons and Ablations

To the best of our knowledge, there are no published PBR generation models that generate PBR images for entire objects or scenes (only for generation of single materials [51,65]). Therefore, we perform an extensive ablation study on our design choices, taking care to include typical approaches from techniques that generate other modalities than PBR. Please refer to Fig. 6 for the qualitative comparisons, while Table 1 contains quantitative results.

Comparison Between Control Paradigms. We compare the performance of the proposed bidirectional cross-network communication layer against two other paradigms: one inspired by ControlNet-XS [10], and one inspired by AnimateAnyone [20]. In the former, dubbed *one-way* communication, the communication layers receive as input only the RGB model's internal state, and they only affect the PBR model's internal state. The latter, dubbed *clockwise* communication, functions in the same way for the encoder part of the architecture, but reverses the information flow to go from the PBR model to the RGB model for the decoder half of the architecture. We see that the *one-way* attention does not perform well, with lower distribution match scores as well as OOD performance scores; the frozen RGB model cannot realign to the conditional distribution required from it in Eq. (1), and we see that the positions it generates for the objects does not align at all with the mask from the normal image. The *clockwise* attention performs significantly better, but is likely still hampered by $z'_{rgb,t-1}$ not being easily available to the PBR model—a similar reasoning as to why the authors of ControlNet-XS included the direct communication link between the base and controlling models' encoders.

Comparison Between Communication Types. In terms of the type of communication, we compare the proposed single-layer per-pixel communication against a per-pixel MLP-based communication layer, and a global attention layer. The latter performs surprisingly well considering that it lacks pixel correspondences; it is hard to enforce pixel-wise alignment through a global attention layer, which we hypothesize to be the reason for the lower quantitative performance. As Jin et al. [24] discuss, an attention-based architecture is also less robust to resolution changes. The per-pixel MLP, containing four hidden per-pixel linear layers with normalization layers [2] in-between, does not qualitatively perform notably better than the single-layer communication layer, so that we settle for the simpler and more computationally efficient choice.

Comparison Against Fine-Tuning. We also compare Collaborative Control against the alternative where we edit the first and last layers of the pre-trained network to match the dimensionality of the PBR images (optionally with the rendered image), and then fine-tune the entire network end-to-end. Although the distribution match scores for these fine-tuning variants are similar to Collaborative Control, the fine-tuning methods strongly overfit to the training data and perform very poorly in a qualitative OOD comparison.

PBR-Specific VAE vs RGB VAE. We compare the performance of Collaborative Control with a PBR-specific VAE against a version that uses the triplets-based RGB VAE mentioned in Sect. 4 to encode the PBR channels (encoding albedo, roughness+metallic, and bump maps in separate triplets and concatenating their latent representations). The mismatch within the PBR domain is

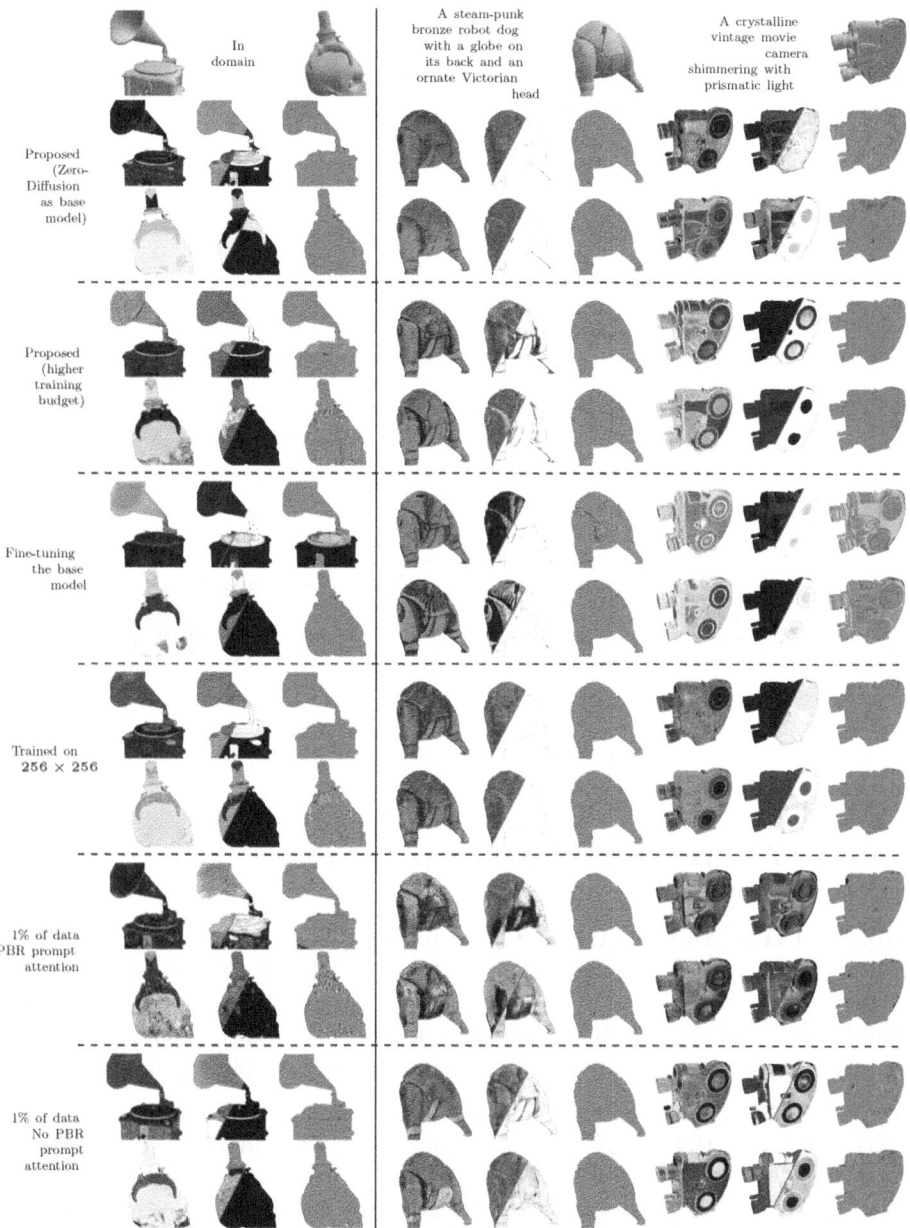

Fig. 6. Generated albedo, roughness/metallic and bump map images from the ablation studies. While significant quality differences are visible, only the fine-tuning approach and the data-sparse regime *with* PBR prompt cross-attention fail completely. The version that was trained on a smaller resolution does not break but does not result in maximum detail either. Best viewed digitally.

Table 1. Quantative results for all evaluated variants. The ablation baseline is high-lighted in bold, duplicated for easier comparisons within the individual ablations.

	CMMD ↓		FID ↓		2% held-out QAlign ↑ Ae		2% held-out QAlign ↑ Q		OOD QAlign ↑ Ae		OOD QAlign ↑ Q		CLIPScore ↑	
	PBR	Relit	PBR	Relit	Albedo	Relit	Albedo	Relit	Albedo	Relit	Albedo	Relit	Albedo	Relit
Communication														
one-way	16.44	13.38	20.90	16.39	1.95	1.97	2.37	2.48	1.91	1.59	2.35	1.69	23.08	23.40
clockwise	6.78	2.76	12.21	11.53	2.04	2.02	2.63	2.64	**2.14**	**1.70**	2.77	**1.74**	26.45	24.53
bi-directional	**6.30**	**1.79**	**11.65**	**10.64**	**2.11**	2.03	**2.75**	**2.66**	2.12	1.76	**2.78**	1.73	**26.76**	**25.41**
Pixel-wise zero-conv	6.30	1.79	11.65	**10.64**	**2.11**	**2.03**	**2.75**	**2.66**	2.12	1.76	2.78	1.73	26.76	25.41
Pixel-wise MLP	**5.43**	1.87	**11.43**	10.67	2.10	2.02	2.74	**2.66**	**2.26**	1.75	**2.96**	**1.81**	**27.15**	**25.95**
Global Attention	7.60	5.22	13.61	11.93	1.94	1.98	2.51	2.60	1.99	1.72	2.71	1.80	24.50	24.01
Collaborative Control	6.30	**1.79**	11.65	**10.64**	**2.11**	2.03	**2.75**	**2.66**	2.12	1.76	2.78	1.73	**26.76**	**25.41**
Fine-tuning (with RGB output)	13.40	2.78	14.42	10.79	2.05	2.02	2.60	2.61	2.10	1.76	2.62	**1.86**	25.04	22.69
Fine-tuning (without RGB output)	**5.25**	2.88	**11.41**	11.37	2.03	1.99	2.58	2.58	**2.26**	1.71	**2.97**	1.81	25.66	23.31
PBR VAE	**6.30**	**1.79**	**11.65**	**10.64**	2.11	**2.03**	**2.75**	**2.66**	2.12	1.76	2.78	1.73	**26.76**	**25.41**
RGB VAE on triplets	84.66	5.99	25.81	11.63	**2.16**	1.99	2.67	2.55	**2.30**	**1.71**	**2.95**	**1.80**	25.27	23.98
Training budget														
1 A100, two days	6.30	1.79	11.65	10.64	**2.11**	2.03	**2.75**	2.66	**2.12**	**1.76**	2.78	1.73	26.76	**25.41**
8 A100s, two days	**2.96**	**1.12**	**9.55**	**9.76**	2.08	2.04	2.68	2.67	2.01	1.73	**2.82**	**1.82**	**26.78**	25.22
Training resolution														
256×256	**2.23**	**1.44**	**9.82**	10.20	2.10	2.04	2.73	2.68	2.20	1.78	**3.13**	**1.80**	26.71	25.21
512×512	6.30	1.79	11.65	10.64	**2.11**	2.03	**2.75**	2.66	**2.12**	1.76	2.78	1.73	**26.76**	**25.41**
Training data														
No PBR prompt attention — 1%	6.25	**1.43**	11.87	10.79	2.18	2.04	**2.86**	2.69	**2.35**	1.76	**3.35**	1.89	26.58	25.11
No PBR prompt attention — 5%	**5.77**	1.45	**11.49**	**10.54**	2.13	2.04	2.78	2.69	2.09	1.73	3.01	1.84	**27.28**	25.04
No PBR prompt attention — 20%	5.97	1.68	11.50	10.61	2.12	2.03	2.78	2.67	2.23	1.76	3.23	1.88	25.72	24.99
No PBR prompt attention — 98%	6.30	1.79	11.65	10.64	2.11	2.03	2.75	2.66	2.12	1.76	2.78	1.73	26.76	25.41
PBR prompt attention — 1%	20.61	4.25	18.35	12.16	**2.19**	2.03	2.76	2.62	2.25	1.61	3.83	1.80	24.75	23.25
PBR prompt attention — 5%	12.17	2.58	14.95	10.97	2.13	2.04	2.71	2.65	2.27	1.80	2.97	1.81	26.89	**25.53**
PBR prompt attention — 20%	11.35	2.33	14.78	10.78	2.13	2.03	2.74	2.65	2.29	1.76	3.07	1.77	26.79	25.39
PBR prompt attention — 98%	9.18	2.57	13.25	11.02	2.10	2.03	2.68	2.64	2.17	1.80	2.84	1.87	25.80	24.83

clear, both quantitatively through the worse distribution matching scores, and qualitatively in the produced images.

Impact of the Training Budget. Comparing the version training on a single A100 with the version trained with 8 A100s (for eight times the batch size), we see that the latter performs significantly better quantitatively in terms of distribution match, but not quality. Visually, the differences are less clear, although the higher-budget model appears to follow complex prompts slightly better.

Impact of the Training Resolution. We compare the performance of Collaborative Control with two training resolutions: 256×256 and 512×512, both evaluated on 512×512 (ZeroDiffusion's native resolution). While the low-resolution model quantitatively performs better, visually it is clear that it does not capture the same level of detail as the high-resolution model—we explain this through the metrics focusing on high-level encoding of the images, and the lower resolution enables smoother training through a larger batch size (42).

Impact of Training Dataset Size. Now, we evaluate the performance of Collaborative Control under data sparsity by evaluating models trained on 98%, 20%, 5% and 1% of the full 6M training images. The proposed approach proves

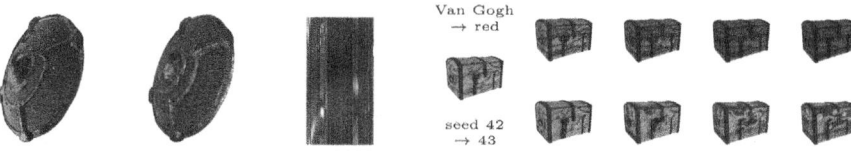

Fig. 7. A re-lit generated texture in Peppermint Powerplant and Pine Attic and lightfield slice under environment rotation.

Fig. 8. Interpolation on text embeddings and initial noise shows the stability of the proposed approach in both of these spaces.

Fig. 9. Our PBR diffusion remains compatible with control techniques trained for the base frozen RGB model. We illustrate this using StableDiffusion 1.5 as the base model using a public IP-Adapter [73].

Fig. 10. Our most common failure case is the constant roughness, metallic or bump maps. Prompting a *porcelain barrel with intricate designs* for two different random seeds illustrates this behaviour.

very data-efficient and performs well even when trained on only a few thousand images (1%). We observe that it is necessary to disable prompt cross-attention in the PBR model, and that this effect gets more pronounced with fewer data: we hypothesize that the model overfits to the training data and that forcing prompt attention to occur through the frozen RGB base model prevents this overfitting.

Compatibility with Other Control Techniques. As a closing experiment, we illustrate that Collaborative Control is compatible with other control techniques [10,44,73], which drastically expands the practical applications of our proposed method. We demonstrate this specifically with IP-Adapter [73], which allows us to condition the final output on a style image by introducing additional style cross-attention layers within the base model. We can apply an IP-Adapter to the base model and still generate PBR content, as illustrated in Fig. 9.

Relighting. For a small qualitative indication that the PBR materials we generate also look natural under different environment lighting, Fig. 7 shows a generated texture when relit in two novel environment maps. Furthermore, a slice of the lightfield under rotation of the second environment map shows that the highlights and shadows move smoothly (hinting that the bump map is meaningful).

Interpolation. To illustrate the stability of our proposed approach in terms of both initial noise and the text prompt, we (separately) interpolate between two prompts and between two initial noises in Fig. 8. We perform prompt interpolation in CLIP space, and for noise interpolation we rescale the final image to have unit standard deviation and zero mean after blending. The resulting images appear natural yet meaningfully interpolate between both extremes.

5.2   Limitations and Failure Cases

We identify two major failure cases: lack of detail in the roughness, metallic, and bump maps, and a failure to follow OOD prompts. In the former, we see (*e.g.* Fig. 10) that the model outputs a constant (though varying per instance) roughness and metallic value, and a flat bump map. We attribute this to the training data: Objaverse contains many objects with constant roughness and metallic properties and a flat bump map—likely biasing the model towards such outputs. Anecdotally, we have found that selecting a different random seed will often succeed where the first generation disappointed—practically, the model produces very diverse results even for the same prompt and the same conditioning geometry, so that we argue that this is either not a significant issue or can be resolved with better training data.

A failure to follow out-of-distribution prompts happens mostly when structural features in the prompt are incompatible with the conditioning geometry, such as for example *a gilded lion* for a table mesh. We hypothesize that the control signal from the PBR model conflicts with the text cross-attention in the frozen RGB model, resulting in lackluster outputs. Different random seeds occasionally resolve this issue, albeit more rarely.

Finally, we note that the model does come at the cost of executing two parallel diffusion models. We note that the motivation for this was mainly to retain a frozen copy of the base RGB model: in applications where these requirements are prohibitively expensive, distilling our approach into a direct PBR model is likely to bring relief.

6   Conclusion

In this work, we have proposed Collaborative Control, a new paradigm for leveraging a pre-trained image-based RGB diffusion model for generating high-quality PBR image content conditioned on object geometry. We have shown that this bi-directional control paradigm is extremely data-efficient while retaining the high quality and expressiveness of the base RGB model, even when faced with text queries completely out of distribution for the PBR training data. The plug-and-play nature of our proposed approach is compatible with existing adaptations of the base RGB model, which we have illustrated with IP-Adapter for style guidance of the PBR content. The availability of high-quality PBR content generation as offered by our proposed approach opens up new avenues for graphics applications, specifically in Text-to-Texture.

Acknowledgements. This work was supported fully by Unity Technologies, without external funding. We would like to thank the reviewers for their valuable feedback and suggestions.

References

1. Achiam, J., et al.: GPT-4 technical report. arXiv preprint arXiv:2303.08774 (2023)
2. Ba, J.L., Kiros, J.R., Hinton, G.E.: Layer normalization. stat **1050**, 21 (2016)
3. Burley, B.: Physically based shading at disney. ACM Trans. Graph. (SIGGRAPH) (2012)
4. Cao, T., Kreis, K., Fidler, S., Sharp, N., Yin, K.: TexFusion: synthesizing 3D textures with text-guided image diffusion models. In: Proceedings of the IEEE/CVF International Conference on Computer Vision, pp. 4169–4181 (2023)
5. Chambon, T., Heitz, E., Belcour, L.: Passing multi-channel material textures to a 3-channel loss. In: ACM SIGGRAPH 2021 Talks, pp. 1–2 (2021)
6. Chen, D.Z., Siddiqui, Y., Lee, H.Y., Tulyakov, S., Nießner, M.: Text2Tex: text-driven texture synthesis via diffusion models. In: Proceedings of the IEEE/CVF International Conference on Computer Vision, pp. 18558–18568 (2023)
7. Chen, R., Chen, Y., Jiao, N., Jia, K.: Fantasia3D: disentangling geometry and appearance for high-quality text-to-3D content creation. In: Proceedings of the IEEE/CVF International Conference on Computer Vision, pp. 22246–22256 (2023)
8. Cook, R.L., Torrance, K.E.: A reflectance model for computer graphics. ACM Trans. Graph. (ToG) **1**(1), 7–24 (1982)
9. Deitke, M., et al.: Objaverse: a universe of annotated 3D objects. In: Proceedings of the IEEE/CVF Conference on Computer Vision and Pattern Recognition, pp. 13142–13153 (2023)
10. Zavadski, D., Feiden, J.F., Rother, C.: ControlNet-XS: designing an efficient and effective architecture for controlling text-to-image diffusion models (2023)
11. Du, X., Kolkin, N., Shakhnarovich, G., Bhattad, A.: Intrinsic LoRA: a generalist approach for discovering knowledge in generative models. In: Synthetic Data for Computer Vision Workshop, CVPR 2024 (2024)
12. Duan, Y., Guo, X., Zhu, Z.: DiffusionDepth: diffusion denoising approach for monocular depth estimation. arXiv preprint arXiv:2303.05021 (2023)
13. Foong, T.Y., Kotyan, S., Mao, P.Y., Vargas, D.V.: The challenges of image generation models in generating multi-component images. arXiv preprint arXiv:2311.13620 (2023)
14. Goodfellow, I., et al.: Generative adversarial networks. Commun. ACM **63**(11), 139–144 (2020)
15. Guo, P., et al.: StableDreamer: taming noisy score distillation sampling for text-to-3D. arXiv preprint arXiv:2312.02189 (2023)
16. Hessel, J., Holtzman, A., Forbes, M., Bras, R.L., Choi, Y.: CLIPScore: a reference-free evaluation metric for image captioning. arXiv preprint arXiv:2104.08718 (2021)
17. Heusel, M., Ramsauer, H., Unterthiner, T., Nessler, B., Hochreiter, S.: GANs trained by a two time-scale update rule converge to a local Nash equilibrium. In: Advances in Neural Information Processing Systems, vol. 30 (2017)
18. Ho, J., Jain, A., Abbeel, P.: Denoising diffusion probabilistic models. In: Advances in Neural Information Processing Systems, vol. 33, pp. 6840–6851 (2020)
19. Hu, E.J., et al.: LoRA: low-rank adaptation of large language models. arXiv preprint arXiv:2106.09685 (2021)

20. Hu, L.: Animate anyone: consistent and controllable image-to-video synthesis for character animation. In: Proceedings of the IEEE/CVF Conference on Computer Vision and Pattern Recognition, pp. 8153–8163 (2024)
21. Huang, T., et al.: DreamControl: control-based text-to-3D generation with 3D self-prior. In: Proceedings of the IEEE/CVF Conference on Computer Vision and Pattern Recognition, pp. 5364–5373 (2024)
22. Huang, Y., Wang, J., Shi, Y., Qi, X., Zha, Z.J., Zhang, L.: DreamTime: an improved optimization strategy for text-to-3D content creation. arXiv preprint arXiv:2306.12422 (2023)
23. Jayasumana, S., Ramalingam, S., Veit, A., Glasner, D., Chakrabarti, A., Kumar, S.: Rethinking FID: towards a better evaluation metric for image generation. In: Proceedings of the IEEE/CVF Conference on Computer Vision and Pattern Recognition, pp. 9307–9315 (2024)
24. Jin, Z., Shen, X., Li, B., Xue, X.: Training-free diffusion model adaptation for variable-sized text-to-image synthesis. In: Advances in Neural Information Processing Systems, vol. 36 (2024)
25. Karras, T., Aila, T., Laine, S., Lehtinen, J.: Progressive growing of GANs for improved quality, stability, and variation. arXiv preprint arXiv:1710.10196 (2017)
26. Karras, T., et al.: Alias-free generative adversarial networks. In: Advances in Neural Information Processing Systems, vol. 34, pp. 852–863 (2021)
27. Karras, T., Laine, S., Aila, T.: A style-based generator architecture for generative adversarial networks. In: Proceedings of the IEEE/CVF Conference on Computer Vision and Pattern Recognition, pp. 4401–4410 (2019)
28. Karras, T., Laine, S., Aittala, M., Hellsten, J., Lehtinen, J., Aila, T.: Analyzing and improving the image quality of StyleGAN. In: Proceedings of the IEEE/CVF Conference on Computer Vision and Pattern Recognition, pp. 8110–8119 (2020)
29. Ke, B., Obukhov, A., Huang, S., Metzger, N., Daudt, R.C., Schindler, K.: Repurposing diffusion-based image generators for monocular depth estimation. In: Proceedings of the IEEE/CVF Conference on Computer Vision and Pattern Recognition, pp. 9492–9502 (2024)
30. Kingma, D.P., Welling, M.: Auto-encoding variational bayes. arXiv preprint arXiv:1312.6114 (2013)
31. Knodt, J., Gao, X.: Consistent mesh diffusion. arXiv preprint arXiv:2312.00971 (2023)
32. Le, C., Hetang, C., Cao, A., He, Y.: EucliDreamer: fast and high-quality texturing for 3D models with stable diffusion depth. arXiv preprint arXiv:2311.15573 (2023)
33. Lee, H.Y., Tseng, H.Y., Yang, M.H.: Exploiting diffusion prior for generalizable dense prediction. In: Proceedings of the IEEE/CVF Conference on Computer Vision and Pattern Recognition, pp. 7861–7871 (2024)
34. Li, X., et al.: Advances in 3D generation: a survey. arXiv preprint arXiv:2401.17807 (2024)
35. Liang, Y., Yang, X., Lin, J., Li, H., Xu, X., Chen, Y.: LucidDreamer: towards high-fidelity text-to-3D generation via interval score matching. In: Proceedings of the IEEE/CVF Conference on Computer Vision and Pattern Recognition, pp. 6517–6526 (2024)
36. Lin, S., Liu, B., Li, J., Yang, X.: Common diffusion noise schedules and sample steps are flawed. In: Proceedings of the IEEE/CVF Winter Conference on Applications of Computer Vision, pp. 5404–5411 (2024)
37. Liu, F., Wu, D., Wei, Y., Rao, Y., Duan, Y.: Sherpa3D: boosting high-fidelity text-to-3D generation via coarse 3D prior. In: Proceedings of the IEEE/CVF Conference on Computer Vision and Pattern Recognition, pp. 20763–20774 (2024)

38. Liu, R., Wu, R., Van Hoorick, B., Tokmakov, P., Zakharov, S., Vondrick, C.: Zero-1-to-3: zero-shot one image to 3D object. In: Proceedings of the IEEE/CVF International Conference on Computer Vision, pp. 9298–9309 (2023)

39. Liu, Y.T., Guo, Y.C., Luo, G., Sun, H., Yin, W., Zhang, S.H.: PI3D: efficient text-to-3D generation with pseudo-image diffusion. In: Proceedings of the IEEE/CVF Conference on Computer Vision and Pattern Recognition, pp. 19915–19924 (2024)

40. Liu, Z., et al.: UniDream: unifying diffusion priors for relightable text-to-3D generation. arXiv preprint arXiv:2312.08754 (2023)

41. Long, X., et al.: Wonder3D: single image to 3D using cross-domain diffusion. In: Proceedings of the IEEE/CVF Conference on Computer Vision and Pattern Recognition, pp. 9970–9980 (2024)

42. Ma, B., Deng, H., Zhou, J., Liu, Y.S., Huang, T., Wang, X.: GeoDream: disentangling 2D and geometric priors for high-fidelity and consistent 3D generation. arXiv preprint arXiv:2311.17971 (2023)

43. Ma, Y., et al.: X-Dreamer: creating high-quality 3D content by bridging the domain gap between text-to-2D and text-to-3D generation. arXiv preprint arXiv:2312.00085 (2023)

44. Mou, C., et al.: T2I-Adapter: learning adapters to dig out more controllable ability for text-to-image diffusion models. In: Proceedings of the AAAI Conference on Artificial Intelligence, vol. 38, pp. 4296–4304 (2024)

45. Van den Oord, A., Kalchbrenner, N., Espeholt, L., Vinyals, O., Graves, A., et al.: Conditional image generation with PixelCNN decoders. In: Advances in Neural Information Processing Systems, vol. 29 (2016)

46. Poole, B., Jain, A., Barron, J.T., Mildenhall, B.: DreamFusion: text-to-3D using 2D diffusion. In: The Eleventh International Conference on Learning Representations (2023)

47. Raj, A., et al.: DreamBooth3D: subject-driven text-to-3D generation. In: Proceedings of the IEEE/CVF International Conference on Computer Vision, pp. 2349–2359 (2023)

48. Rombach, R., Blattmann, A., Lorenz, D., Esser, P., Ommer, B.: High-resolution image synthesis with latent diffusion models. In: Proceedings of the IEEE/CVF Conference on Computer Vision and Pattern Recognition, pp. 10684–10695 (2022)

49. Salimans, T., Goodfellow, I., Zaremba, W., Cheung, V., Radford, A., Chen, X.: Improved techniques for training GANs. In: Advances in Neural Information Processing Systems, vol. 29 (2016)

50. Sarkar, A., Mai, H., Mahapatra, A., Lazebnik, S., Forsyth, D.A., Bhattad, A.: Shadows don't lie and lines can't bend! Generative models don't know projective geometry... for now. In: Proceedings of the IEEE/CVF Conference on Computer Vision and Pattern Recognition, pp. 28140–28149 (2024)

51. Sartor, S., Peers, P.: MatFusion: a generative diffusion model for SVBRDF capture. In: SIGGRAPH Asia 2023 Conference Papers, SA 2023. ACM (2023). https://doi.org/10.1145/3610548.3618194

52. Schuhmann, C., et al.: LAION-5B: an open large-scale dataset for training next generation image-text models. In: Advances in Neural Information Processing Systems, vol. 35, pp. 25278–25294 (2022)

53. Sharma, P., et al.: Alchemist: parametric control of material properties with diffusion models. In: Proceedings of the IEEE/CVF Conference on Computer Vision and Pattern Recognition, pp. 24130–24141 (2024)

54. Shi, R., et al.: Zero123++: a single image to consistent multi-view diffusion base model. arXiv preprint arXiv:2310.15110 (2023)

55. Shi, Y., Wang, P., Ye, J., Mai, L., Li, K., Yang, X.: MVDream: multi-view diffusion for 3D generation. In: The Twelfth International Conference on Learning Representations (2024)
56. Sohl-Dickstein, J., Weiss, E., Maheswaranathan, N., Ganguli, S.: Deep unsupervised learning using nonequilibrium thermodynamics. In: International Conference on Machine Learning, pp. 2256–2265. PMLR (2015)
57. Song, J., Meng, C., Ermon, S.: Denoising diffusion implicit models. In: International Conference on Learning Representations (2020)
58. Stan, G.B.M., et al.: LDM3D: latent diffusion model for 3D. arXiv preprint arXiv:2305.10853 (2023)
59. Subias, J.D., Lagunas, M.: In-the-wild material appearance editing using perceptual attributes. In: Computer Graphics Forum, vol. 42, pp. 333–345. Wiley Online Library (2023)
60. Szegedy, C., Vanhoucke, V., Ioffe, S., Shlens, J., Wojna, Z.: Rethinking the inception architecture for computer vision. In: Proceedings of the IEEE Conference on Computer Vision and Pattern Recognition, pp. 2818–2826 (2016)
61. Tang, B., Wang, J., Wu, Z., Zhang, L.: Stable score distillation for high-quality 3D generation. arXiv preprint arXiv:2312.09305 (2023)
62. Tang, J., Ren, J., Zhou, H., Liu, Z., Zeng, G.: DreamGaussian: generative gaussian splatting for efficient 3D content creation. In: The Twelfth International Conference on Learning Representations (2024)
63. Huggingface zerodiffusion model weights v0.9. https://huggingface.co/drhead. https://huggingface.co/drhead/ZeroDiffusion. Accessed 08 Feb 2024
64. Van Den Oord, A., Kalchbrenner, N., Kavukcuoglu, K.: Pixel recurrent neural networks. In: International Conference on Machine Learning, pp. 1747–1756. PMLR (2016)
65. Vecchio, G., et al.: ControlMat: a controlled generative approach to material capture. arXiv preprint arXiv:2309.01700 (2023)
66. Wang, P., et al.: SteinDreamer: variance reduction for text-to-3D score distillation via stein identity. arXiv preprint arXiv:2401.00604 (2023)
67. Wang, Z., Li, M., Chen, C.: LucidDreaming: controllable object-centric 3D generation. arXiv preprint arXiv:2312.00588 (2023)
68. Wang, Z., et al.: ProlificDreamer: high-fidelity and diverse text-to-3D generation with variational score distillation. In: Advances in Neural Information Processing Systems, vol. 36 (2024)
69. Wu, H., et al.: Q-Align: teaching LMMs for visual scoring via discrete text-defined levels. In: Forty-First International Conference on Machine Learning (2024)
70. Wu, T., et al.: HyperDreamer: hyper-realistic 3D content generation and editing from a single image. In: SIGGRAPH Asia 2023 Conference Papers, pp. 1–10 (2023)
71. Wu, Z., Zhou, P., Yi, X., Yuan, X., Zhang, H.: Consistent3D: towards consistent high-fidelity text-to-3D generation with deterministic sampling prior. In: Proceedings of the IEEE/CVF Conference on Computer Vision and Pattern Recognition, pp. 9892–9902 (2024)
72. Xu, X., Lyu, Z., Pan, X., Dai, B.: MATLABER: material-aware text-to-3D via latent BRDF auto-encoder. arXiv preprint arXiv:2308.09278 (2023)
73. Ye, H., Zhang, J., Liu, S., Han, X., Yang, W.: IP-Adapter: text compatible image prompt adapter for text-to-image diffusion models. arXiv preprint arXiv:2308.06721 (2023)
74. Yeh, Y.Y., et al.: TextureDreamer: image-guided texture synthesis through geometry-aware diffusion. In: Proceedings of the IEEE/CVF Conference on Computer Vision and Pattern Recognition, pp. 4304–4314 (2024)

75. Youwang, K., Oh, T.H., Pons-Moll, G.: Paint-it: text-to-texture synthesis via deep convolutional texture map optimization and physically-based rendering. In: Proceedings of the IEEE/CVF Conference on Computer Vision and Pattern Recognition, pp. 4347–4356 (2024)
76. Yu, K., Liu, J., Feng, M., Cui, M., Xie, X.: Boosting3D: high-fidelity image-to-3D by boosting 2D diffusion prior to 3D prior with progressive learning. arXiv preprint arXiv:2311.13617 (2023)
77. Zeng, X., et al.: Paint3D: paint anything 3D with lighting-less texture diffusion models. In: Proceedings of the IEEE/CVF Conference on Computer Vision and Pattern Recognition, pp. 4252–4262 (2024)
78. Zhang, J., et al.: Repaint123: fast and high-quality one image to 3D generation with progressive controllable 2D repainting. arXiv preprint arXiv:2312.13271 (2023)
79. Zhang, L., Rao, A., Agrawala, M.: Adding conditional control to text-to-image diffusion models. In: Proceedings of the IEEE/CVF International Conference on Computer Vision, pp. 3836–3847 (2023)
80. Zhou, L., Shih, A., Meng, C., Ermon, S.: DreamPropeller: supercharge text-to-3D generation with parallel sampling. In: Proceedings of the IEEE/CVF Conference on Computer Vision and Pattern Recognition, pp. 4610–4619 (2024)
81. Zhuang, J., Wang, C., Lin, L., Liu, L., Li, G.: DreamEditor: text-driven 3D scene editing with neural fields. In: SIGGRAPH Asia 2023 Conference Papers, pp. 1–10 (2023)

Co-synthesis of Histopathology Nuclei Image-Label Pairs Using a Context-Conditioned Joint Diffusion Model

Seonghui Min, Hyun-Jic Oh, and Won-Ki Jeong[(⊠)]

Department of Computer Science and Engineering, College of Informatics, Korea University, Seoul, South Korea
wkjeong@korea.ac.kr

Abstract. In multi-class histopathology nuclei analysis tasks, the lack of training data becomes a main bottleneck for the performance of learning-based methods. To tackle this challenge, previous methods have utilized generative models to increase data by generating synthetic samples. However, existing methods often overlook the importance of considering the context of biological tissues (e.g., shape, spatial layout, and tissue type) Moreover, while generative models have shown superior performance in synthesizing realistic histopathology images, none of the existing methods are capable of producing image-label pairs at the same time. In this paper, we introduce a novel framework for co-synthesizing histopathology nuclei images and paired semantic labels using a context-conditioned joint diffusion model. We propose conditioning of a diffusion model using nucleus centroid layouts with structure-related text prompts to incorporate spatial and structural context information into the generation targets. Moreover, we enhance the granularity of our synthesized semantic labels by generating instance-wise nuclei labels using distance maps synthesized concurrently in conjunction with the images and semantic labels. We demonstrate the effectiveness of our framework in generating high-quality samples on multi-institutional, multi-organ, and multi-modality datasets. Our synthetic data consistently outperforms existing augmentation methods in the downstream tasks of nuclei segmentation and classification.

Keywords: Joint diffusion model · Data augmentation · Histopathology nuclei segmentation

1 Introduction

Cell nuclei segmentation and classification are crucial tasks in digital pathology for examining nuclear characteristics, such as size, shape, density, etc., which

---

S. Min and H.-J. Oh—Equal contribution.

Supplementary Information The online version contains supplementary material available at https://doi.org/10.1007/978-3-031-72624-8_9.

A. Leonardis et al. (Eds.): ECCV 2024, LNCS 15071, pp. 146–162, 2025.
https://doi.org/10.1007/978-3-031-72624-8_9

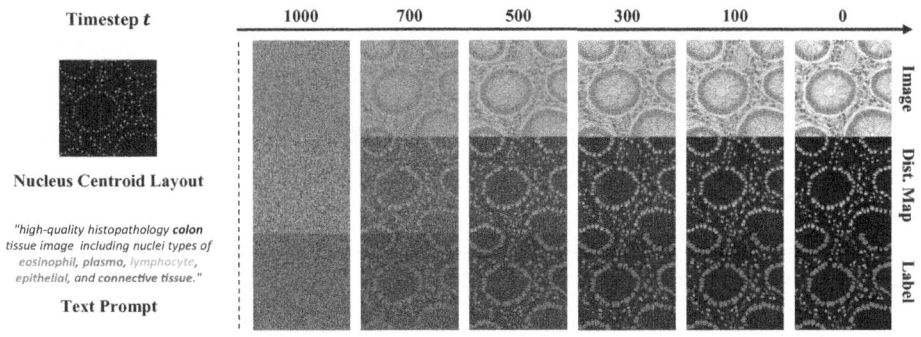

Fig. 1. Given a nucleus centroid layout and structure-related text prompt, our proposed approach generates a pair of histopathology nuclei image, distance map, and semantic label that aligns with the specified conditions. This process is concurrently performed by a single joint diffusion model.

provide important evidence for disease diagnosis [6]. Given the complicated nature of histopathology images, the adoption of deep learning-based computer-aided diagnosis has now become the de facto standard in computational pathology [30]. Indeed, numerous fully-supervised learning approaches have been introduced and proven to be effective in nuclei analysis [8,13,14]. However, the potential of learning algorithms is limited by the lack of appropriate training data. This is primarily attributed to the labor-intensive nature of manually generating annotations, necessitating the domain expertise of pathologists for accurate labeling (Fig. 1).

One notable research direction to address data insufficiency is synthetically generating the training data (i.e., data augmentation). Recently, there has been a growing interest in using generative models based on GANs [11] and diffusion models [16] for data synthesis. Specifically, diffusion models have demonstrated superior performance compared to GANs in both natural and histopathology image synthesis [7,21]. Furthermore, the incorporation of semantic label conditional diffusion models [33] has gained prominence for generating images that seamlessly align with provided labels. In histopathology nuclei image synthesis, the majority of current research relies on pixel-level segmentation labels to generate synthetic histopathology images that are accurately aligned with the provided labels. Some studies [28,32] employed the original reference labels as is, while others introduced random perturbations to the original labels [10,25] or generated randomly distributed labels prior to image generation [5,18,38]. Although these methods are effective in generating histopathology image-label pairs, they may not be as versatile in producing diverse labels and do not faithfully reconstruct spatial and structural contexts in real histopathology specimens.

The primary motivation driving our work arises from the observation of a critical gap in existing methods - none are capable of producing image-label pairs with user-controllable image content (e.g., tissue type) in a highly reliable spatial context. For example, Semantic-Palette [19] allows for the manipulation

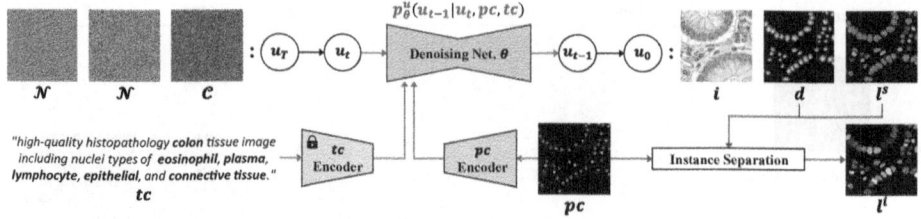

Fig. 2. Overview of the proposed method. We formulate joint diffusion process to synthesize multiple targets: image i, distance map d, and semantic label l^s. We utilize conditioning of text tc and point map pc to improve the sample quality and provide controllable capacity. We generate highly accurate instance label l^i using pc, d, and l^s.

of class proportions during layout generation, but it cannot precisely control the spatial placement of labels. On the other hand, Abousamra *et al.* [1] introduced a technique for generating cell layouts alongside corresponding histopathology images, adhering to specified spatial contexts. However, its applicability is limited to detection and classification tasks due to the inherent nature of point labels. Therefore, our idea is to bridge the gap between these two methodologies by incorporating a point map condition representing nucleus centroids and a text condition representing tissue and nuclei types in the generation of histopathology images and their corresponding multi-class nuclei segmentation labels, empowering users with complete control over the spatial layout and content of cell images. Additionally, we also observed a common issue in conventional nuclei label synthesis - nuclei instances in the resulting semantic labels tend to be closely located and clustered into a larger one. To address this issue, we propose the generation of distance maps alongside images and labels, which can be directly used to separate individual instances. The overview of the proposed method is shown in Fig. 2. The main contributions of our work are several-fold as follows:

- We propose a co-synthesis framework for multi-class nuclei datasets that generates images with semantic and instance labels using a single diffusion model. The proposed method models the joint distribution of histopathology images, semantic labels, and distance maps, enabling the simultaneous generation of the whole targets.
- We incorporate two conditions, a nucleus centroid layout and a text prompt, to enhance the model's capacity to capture the critical and intricate visual context of real histopathology specimens. This not only gives the user full control over the tissue and nuclei types but also provides flexibility in the design of the spatial arrangement.
- We generate high-quality instance labels by separating individual nuclei in the semantic labels using distance maps and the conditioned nucleus centroid layouts. The generated instance labels are necessary for high-level nuclei analysis such as state-of-the-art nuclei segmentation algorithms.
- We demonstrate the efficacy of our approach using multi-institutional, multi-organ, and multi-modality datasets through quantitative and quali-

tative assessment of downstream tasks, including nuclei segmentation and classification.

2   Related Work

2.1   Generative Models for Image-Label Synthesis

The task of generating image-label pairs aims to model the joint distribution between images and their associated labels. This research area has seen significant developments, greatly influenced by generative adversarial neural networks (GANs) [11]. For instance, Dataset-GAN [39] initially generates images and then uses GAN latent codes to generate corresponding semantic labels. Alternatively, SB-GAN [2] reverses this process: starting with label generation and then synthesizing images based on these generated labels. Semantic-Palette [19] introduces controllable class proportions in the process of generating semantic layouts.

A recent milestone in generative models has been the emergence of the Denoising Diffusion Probabilistic Model (DDPM) [16]. DDPM stochastically simulates the denoising process and has showcased superior performance over state-of-the-art generative models [7]. Notably, the diffusion model has consistently outperformed traditional GAN-based approaches in various studies [3,23] on generating image-label pairs. Recently, Park et al. [26] have introduced an approach for co-synthesizing image-label pairs using a single diffusion model for the text-to-image synthesis task. Their method efficiently captures the joint distribution of image-label pairs by applying a Gaussian diffusion process to the images and a categorical diffusion process to the labels.

2.2   Histopathology Image Synthesis

Extensive research has been dedicated to this field, with a focus on leveraging generative models for data synthesis. Notably, recent diffusion model-based approaches have exhibited remarkable superiority over GAN-based methods [21,28,36].

For nuclei image synthesis, previous research has often used original labels as a reference due to the requirement of domain-specific knowledge to annotate histopathology images. For instance, SIAN [32] employed the original reference labels as conditioning, yielding a diverse set of stylized images using encoded style vectors for a multi-organ, single-class nuclei dataset. NASDM [28], on the other hand, synthesized images by referencing the original labels, leveraging a semantic label conditional diffusion model (SDM) [33]. Alternatively, InsMix [20] and DiffMix [25] applied random label perturbations, such as copying and pasting or random adjustments to nuclei positions, before generating images based on these modified labels. In addition, some methods [5,18,38] generated randomly distributed labels and used them as the basis for image generation. For example, Yu et al. [38] introduced a two-stage diffusion model framework, comprising an unconditional training approach utilizing a diffusion model for label generation followed by SDM. However, this approach was designed for single-class nuclei data, lacked fine control over the label synthesis step, and involved inference from two diffusion models, resulting in prolonged data synthesis times. Moreover, these approaches do not deeply consider the spatial details in the original labels when altering or generating new ones. In contrast, Abousamra et al. [1] introduced a method focused on generating structure-aware point layouts, emphasizing the significance of capturing spatial and structural correlations related to nuclei positions. Nevertheless, this approach generated point labels rather than complete pixel-level labels.

In this paper, we introduce a unified framework to generate histopathology images and their corresponding labels simultaneously using a single joint diffusion model. Furthermore, we introduce nucleus centroid layout and text conditioning for better control over nuclei positioning and enhance the structural realism of the synthesized pairs. Lastly, we incorporate highly accurate nuclei instance labels through post-processing, expanding the applicability of the generated dataset to high-level nuclei analysis.

3   Methods

3.1   Background: Diffusion Models

Denoising diffusion probabilistic models convert noise with a specific simple distribution into data sampled from a more intricate distribution [16]. Diffusion models employ a noise schedule denoted as β during the forward process to add noise to the actual data. Sequential denoising then occurs in the reverse process across time steps $t \in [1, 2, ..., T]$, resulting in the generation of synthetic data x_0 from the noise x_T. Various noise distributions have been systematically investigated to align with the characteristics of the target data.

The Gaussian diffusion model is commonly used for synthesizing continuous distribution data such as images. The forward process with Gaussian noise that follows a normal distribution can be described as:

$$q(x_t \mid x_{t-1}) = \mathcal{N}(x_t; \sqrt{1 - \beta_t}x_{t-1}, \beta_t I). \tag{1}$$

The categorical diffusion model proposed by Hoogeboom et al. [17] is designed to synthesize discrete distribution data such as texts and segmentation labels. For K categories, using the categorical distribution $\mathcal{C}$, the forward process is defined as:

$$q(x_t \mid x_{t-1}) = \mathcal{C}(x_t; (1 - \beta_t)x_{t-1} + \beta_t/K). \tag{2}$$

The reverse process $p_\theta(x_{t-1} \mid x_t)$ unfolds with a deep neural network ϵ_θ. Different types of diffusion models primarily focus on learning the denoising step transitioning from t to $t - 1$. Consequently, the definition of the training loss is formulated as follows:

$$\mathcal{L} = \mathbb{E}_{t, x_0, \epsilon} \left[\| \epsilon - \epsilon_\theta(x_t, t) \|_2^2 \right], \tag{3}$$

where the objective is to minimize the discrepancy between the predicted noise $\epsilon_\theta(x_t, t)$ and the real noise ϵ in x_t.

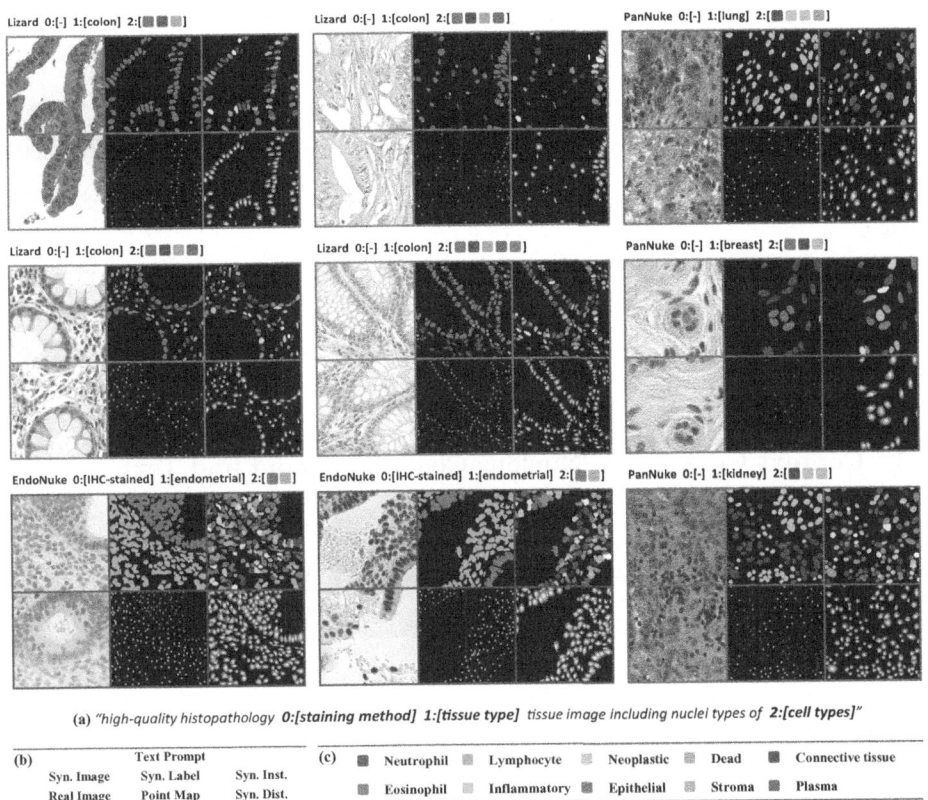

(a) "high-quality histopathology **0:[staining method]** **1:[tissue type]** tissue image including nuclei types of **2:[cell types]**"

Fig. 3. Synthetic samples on the Lizard [12], PanNuke [9], and EndoNuke [22] datasets generated by our method. Each set of paired synthetic images, semantic labels, instance masks, and distance maps is shown with conditional point maps and text prompts. Real images are also included for color comparison with the corresponding synthetic images. (a) is a frame for the text prompting. (b) indicates the arrangement of each component. (c) provides a color legend for the classes of nuclei.

3.2 Joint Diffusion Process

The integrative deployment of appropriate distributions facilitates the concurrent generation of multiple targets in different modalities. We simultaneously generate images, multi-class semantic labels, and distance maps to provide a dataset with high utility for histopathology nuclei image analysis. Distance maps are used to separate indistinguishable nuclei in semantic labels, providing specific labels for each instance. Further details on leveraging distance maps for instance separation are described in Sect. 3.4.

Let us denote the image, distance map, and semantic label by i, d, and l^s, respectively. These multi-modal elements collectively form a tripartite data unit $u := (i, d, l^s)$. Considering the properties of each modality, we model the continuous variables i and d with Gaussian distributions, and the discrete variable l^s

with a categorical distribution. Subsequently, we define the reverse process for u, where each component undergoes an independent forward process (Eq. (1) and Eq. (2)), as follows:

$$p_\theta^u(u_{t-1} \mid u_t) = p_\theta^i(i_{t-1} \mid u_t) \cdot p_\theta^d(d_{t-1} \mid u_t) \cdot p_\theta^{l^s}(l^s_{t-1} \mid u_t). \tag{4}$$

To train the joint diffusion model, we utilize a composite objective function, defined as:

$$\mathcal{L}_{total} = \lambda_i \cdot \mathcal{L}_i + \lambda_d \cdot \mathcal{L}_d + \lambda_{l^s} \cdot \mathcal{L}_{l^s}, \tag{5}$$

where λ_i, λ_d, and λ_{l^s} are weighting factors that balance the contribution of each generation target to the overall training objective.

3.3   Context Conditions: Nucleus Centroid Layout and Text Prompt

To generate highly realistic images and precisely control the generation process, we incorporate two types of nuclei image context conditions: nucleus centroid layout (in the form of a point map) and structure-related text prompts.

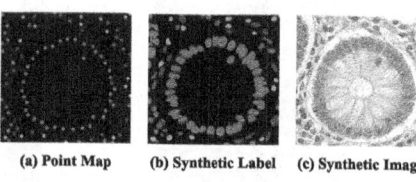

(a) Point Map   (b) Synthetic Label   (c) Synthetic Image

Point map condition, indicated by pc, defines the centroids of the nuclei instances, providing information on their spatial positioning and class distribution, as illustrated

Fig. 4. Example of a point map conditioned synthetic image-label pair. Glands and lumens similar to those observed in real histopathology images were generated for epithelial cell points arranged in circular patterns.

in Fig. 4. This contextual overview is vital for the precise delineation and comprehension of the complex nuclear patterns found in histopathology specimens. Moreover, pc, in contrast to pixel-wise constraints, provides flexibility of label generation. While full-label conditioned image synthesis can only diversify the generated images, our approach can generate a variety of images and labels, as shown in Fig. 5. In terms of steerability, pc enables customizable data generation, allowing control over the type, quantity, and spatial configuration of nuclei. The encoding of pc is achieved through RRDBs proposed by [34].

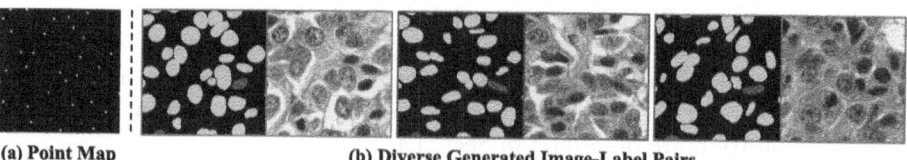

(a) Point Map   (b) Diverse Generated Image-Label Pairs

Fig. 5. Examples of diverse image-label pairs generated from a same point layout condition.

Text condition, represented as tc, includes information on the tissue type of the synthetic sample and the categories of nuclei it encompasses. Further details on generating text prompts are described in Sect. 4. We utilize PLIP [37] to encode tc, which is a vision-language foundation model specialized for pathology images.

With these conditioning schemes, the output of the model is redefined as $\epsilon_\theta(u_t, t, pc, tc)$. During the sampling, we employ classifier-free guidance [7] to adjust the predictive noise as follows:

$$\tilde{\epsilon}_\theta(u_t, t, pc, tc) = \omega\epsilon_\theta(u_t, t, pc, tc) + (1 - \omega)\epsilon_\theta(u_t, t, pc), \tag{6}$$

where ω represents the guidance scale for tc, balancing the trade-off between overfitting and semantic alignment.

3.4 Nuclei Instance Separation

In this section, we delineate the methodology employed to derive instance labels, l^i, from l^s, utilizing d and pc. As depicted in Fig. 2, d quantifies the Euclidean distance from the centroid of each nucleus, normalized to a scale from 0 to 1. To separate l^s at the instance level, we apply the marker-controlled watershed algorithm, as referenced in [35], to d and l^s with pc as a marker map. This method allows for identifying adjacent yet separate structures within the histological samples. The integration of the synthesized i, l^s, and l^i results in a data structure that encompasses not only semantic information but also instance-specific details. This comprehensive dataset is hence suited for a wide array of downstream applications, including segmentation and classification tasks, thus enhancing the utility of the generated synthetic histopathology images.

4 Experiments

Datasets. We tested our method on three multi-class histopathology nuclei segmentation datasets: Lizard [12], PanNuke [9], and EndoNuke [22]. Each dataset comprises image regions derived from histopathology slides.

Lizard is a large multi-institutional collection of six different datasets, containing colon tissue samples with 495,179 nuclei categorized into six classes. We applied Vahadane stain normalization [29] to standardize the color distribution to that of a reference slide. We cropped patches from image regions to a size of 256 × 256 pixels, using a stride of 128 pixels as done by NASDM [28]. We prepared 13,064 patches for the experiment with a distribution of 95% for the training set and 5% for the test set.

PanNuke is a multi-organ H&E-stained dataset consisting of 19 different tissue types with 189,744 nuclei categorized into five classes. We have omitted color normalization to preserve the color variations between the different tissue types. The dataset consists of 7,901 patches and we divided it into 80% of the patches for training and the remaining 20% for testing (Table 1).

Table 1. Comparative overview of generative models by conditional inputs and output targets.

Method	Condition	Generation Target		
		Image	Dist.Map	Label
Yu *et al.* [38]	None	✓	✓	✓
SemanticPalette [19]	Class Proportion	✓	–	✓
Park *et al.* [26]	Text	✓	–	✓
SDM [33]	Label, Instance Edge	✓	–	–
Ours	Text, Point Map	✓	✓	✓

Table 2. Comparative results of generative models, evaluated by FID, IS, and FSD. The best results are in **bold** and the second best are in underlined.

Method	Lizard			PanNuke			EndoNuke		
	FID↓	IS↑	FSD↓	FID↓	IS↑	FSD↓	FID↓	IS↑	FSD↓
Yu *et al.* [38]	–	–	963.36	–	–	1292.05	–	–	931.21
SemanticPalette [19]	86.17	2.11	0.55	109.23	3.36	**1.23**	90.00	1.40	**1.88**
Park *et al.* [26]	52.65	2.22	65.06	61.16	3.48	34.43	**52.99**	1.88	110.00
SDM [33]	45.99	2.35	–	107.80	**3.82**	–	105.17	**2.27**	–
Ours w/o *pc*	69.10	2.02	109.18	–	–	–	–	–	–
Ours	**38.78**	**2.40**	**0.13**	**37.35**	3.77	1.44	69.94	2.17	29.57

EndoNuke is an IHC-stained dataset comprising 245,120 nuclei in endometrial tissue samples, categorized into three classes. This dataset is designed primarily for nuclei detection tasks, but it also provides coarse semantic labels, automatically generated by watershed [31] algorithm. We selected this dataset to evaluate the applicability of our method across multiple image modalities, as it uses a different staining technique than other datasets. We divided a total of 1,780 patches, with a distribution of 85% for the training set and 15% for the test set. For the PanNuke and EndoNuke, we used provided patches of 256 × 256 pixels.

Text Prompt Generation. Since none of the datasets provide text descriptions for each sample, we generated our own prompts for text conditioning. The text prompts include information on the tissue type and the cell types it contains, as follows: "high-quality histopathology [tissue type] tissue image including nuclei types of [list of cell types]." Specifically for EndoNuke, we added information on the staining method, which differs from the most commonly used H&E staining as: "high-quality histopathology IHC-stained [tissue type] tissue image including nuclei types of [list of cell types]."

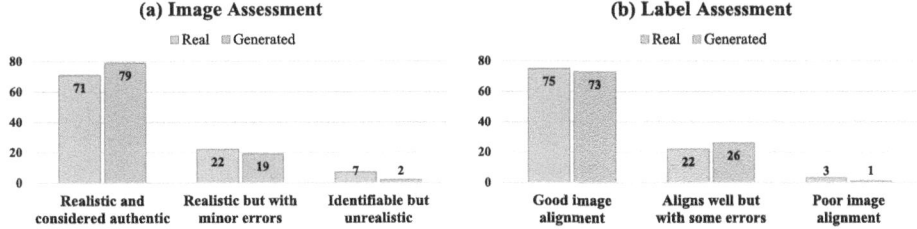

Fig. 6. Pathologists' evaluation of authenticity and alignment for real image-label pairs and those generated by our method.

Implementation Details. We used the Adam optimizer with $\beta_1 = 0.9$ and $\beta_2 = 0.99$ for model training. The learning rates were set at 10^{-4} for Lizard and PanNuke, and at 10^{-5} for EndoNuke. The weighting factors λ_i, λ_d, and λ_{l^s} were set to 9, 1, and 3, respectively, across all datasets. The training batch size was 16, and we employed three separate cosine schedules, one for each output type: image, distance map, and semantic label. The sampling step T was set to 1000. The guidance scale for the text condition was set at 3 for Lizard, 2 for PanNuke, and 0.5 for EndoNuke. All experiments were conducted using NVIDIA RTX A6000 GPUs.

4.1 Quantitative Evaluation

We compare the quality of generated image-label pairs quantitatively with other methods. Since our approach is not directly comparable to existing work, i.e., no other work uses point conditions and generates image-label pairs, we compare the results with the methods using various conditioning techniques. We employ three metrics for the quality assessment, Fréchet Inception Distance (FID) [15] and Inception Score (IS) [27] for the image, and Fréchet Segmentation Distance (FSD) [4] for the label.

The comparative experiment was conducted for Yu *et al.* [38], Semantic-Palette [19], Park *et al.* [26], SDM [33], and ours, as shown in Table 2. The SDM composite data is guided pixel-wise from full semantic labels, resulting in high correspondence with real data in the Lizard dataset. The instance edges generated based on instance maps are also utilized as conditions, contributing to good alignment between images and labels and enabling the generation of high-resolution images. Furthermore, since only images are the sole generation target, this approach focuses entirely on generating compliant quality images. However, SDM produced unrealistic colored images with high IS scores and poor FID scores on PanNuke, a dataset with a wide color distribution, and EndoNuke, a dataset with relatively little training data. Yu *et al.* generates images, labels, and distance maps without any conditions. Despite the similarity to the SDM in the image generation process, since Yu *et al.* was originally proposed for a single

Table 3. Comparative results of instance separation algorithms, evaluated by mDice. The best results are in **bold** and the second best are underlined.

Method	mDice		
	Lizard	PanNuke	EndoNuke
Connectivity-based	0.9383	0.9146	0.5524
Yu *et al.* [38]	0.9374	0.9462	0.9268
Ours	**0.9754**	**0.9980**	**0.9634**

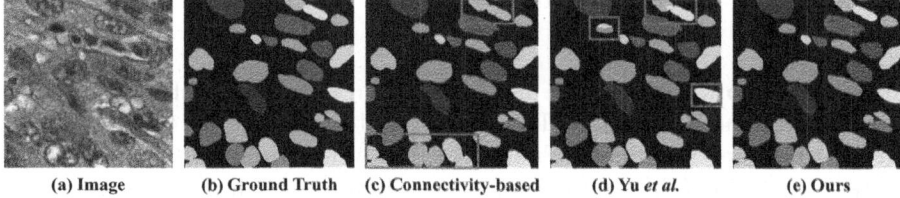

(a) Image   (b) Ground Truth   (c) Connectivity-based   (d) Yu *et al.*   (e) Ours

Fig. 7. Comparison of instance separation algorithms. Under-, over-, or mis-separated instances are shown in red boxes. (Color figure online)

class, the quality of both images and labels deteriorated when extended to a multi-class task. We did not proceed with the image generation step because Yu *et al.* produces noisy and unrealistic labels. SemanticPalette is a method that conditions on the pixel proportion of different classes in the label. It has achieved commendable FSD scores because its conditioning approach and the measurement principle of FSD are similar.

Our approach involves generating image-label pairs simultaneously with a single model, incorporating appropriate noise design. We have achieved superior performance compared to other methods for generating image-label pairs. Particularly, through point conditioning, we demonstrated remarkable FSD with guidance of only 1 pixel per instance. Our approach demonstrated superior performance even compared to SDM, which uses full pixel labels. Even though ours achieved second place in some cases, ours scored better overall compared to other methods.

4.2 Qualitative Assessment by Pathologists

As illustrated in Fig. 3, our method effectively generates samples that satisfy the given conditions. The synthetic images exhibit colors that closely resemble those of real images, and the synthetic labels are well-aligned with the paired images. To further validate our method, we conducted an expert analysis involving the evaluation of both synthetic and real image-label pairs by five pathologists. Our study included 20 synthetic images and 20 authentic pairs, focusing on evaluating their realism and alignment with their corresponding labels. The pathologists evaluated the images based on their expertise in pathology, including determining correspondence to actual histological structures, diagnostic quality of the tissue, color accuracy, and presence of artifacts, etc. When evaluating the labels, they considered various criteria, such as the accuracy of cell location and type, as well as the distinction between difficult-to-identify cell types. As shown in Fig. 6a, the synthetic images received higher realism ratings than the real images. Meanwhile, Fig. 6b shows that our synthetic data achieved image-label alignment ratings comparable to the real data. In addition, the pathologists rated the quality of the synthetic data pairs as good overall. This finding highlights the potential of generative models for histopathology data augmentation.

4.3   Comparison of Instance Separation Methods

We evaluate the efficiency of the point condition for instance separation using the average of the Dice coefficient (mDice) [24], which calculates Dice score per nucleus. Table 3 shows the results compared to traditional connectivity-based algorithms and the distance map-based watershed approach of Yu *et al.*. Our point condition-based instance separation performed best on all datasets. The other methods struggled to split the nuclei clusters, resulting in under-, over-, or incorrect separation as shown in Fig. 7. Connectivity-based separation failed to separate nuclei clusters as shown in Fig. 7c. Distance map-based watershed algorithm worked well, but it tends to over-separate instances as shown in Fig. 7d.

4.4   Downstream Tasks

We evaluated the effectiveness of synthetic data for data augmentation by assessing performance on downstream tasks such as nuclei segmentation and classification. For training the downstream models, we used the image patches not utilized in training the diffusion model. We synthesized image-label pairs based on the point layout extracted from these patches and balanced the number of real and synthetic patches at a 50:50 ratio. Additionally, we set aside 25% of the patches used in training the diffusion model for inference in downstream tasks. This approach ensures the inference set includes a sufficient proportion of each nucleus class, enabling a comprehensive comparison of classification results. We used Hover-Net [13] as a baseline network, a neural network designed to segment clustered nuclei by predicting the horizontal and vertical distances between individual nucleus pixels and their respective centers of mass. Since Hover-Net aims to predict horizontal and vertical maps to improve nuclei instance segmentation performance, it requires instance labels to generate ground truth distance masks. Therefore, we excluded methods that do not generate distance maps, such as SemanticPalette [19] and Park *et al.* [26]. In addition, even though Yu *et al.* [38] generates distance maps during unconditional label generation, the label quality was poor, so we excluded it from the comparison.

Table 4. Downstream segmentation and classification performance comparison for various augmentation methods using Hover-Net [13] as a baseline network. The best results are in **bold** and the second best are underlined.

Dataset	Method	Segmentation		Classification							
		Dice	AJI	F_d	Acc	F^{c1}	F^{c2}	F^{c3}	F^{c4}	F^{c5}	F^{c6}
Lizard	Baseline	0.620	0.383	0.619	0.763	0.012	0.548	0.318	0.146	0.050	0.252
	w/ Aug.	0.676	0.425	0.646	0.818	0.062	0.599	0.351	0.268	0.264	0.367
	w/ SDM	**0.718**	**0.488**	**0.699**	<u>0.862</u>	**0.185**	**0.679**	<u>0.413</u>	**0.350**	<u>0.333</u>	**0.455**
	w/ **Ours**	<u>0.716</u>	<u>0.484</u>	0.694	**0.866**	0.161	<u>0.676</u>	**0.434**	<u>0.346</u>	**0.341**	<u>0.447</u>
PanNuke	Baseline	0.782	0.598	0.763	0.668	0.420	0.356	0.102	0.301	0.475	–
	w/ Aug.	0.816	0.641	0.791	<u>0.708</u>	<u>0.492</u>	0.394	0.107	<u>0.380</u>	0.524	–
	w/ SDM	<u>0.821</u>	<u>0.654</u>	<u>0.800</u>	0.702	0.481	<u>0.398</u>	**0.130**	0.336	<u>0.528</u>	–
	w/ **Ours**	**0.824**	**0.662**	**0.806**	**0.736**	**0.516**	**0.434**	<u>0.127</u>	**0.420**	**0.561**	–
EndoNuke	Baseline	0.878	0.594	0.815	0.891	0.734	0.504	0.013	–	–	–
	w/ Aug.	0.889	0.602	0.820	0.905	0.747	0.598	0.008	–	–	–
	w/ SDM	**0.900**	<u>0.642</u>	**0.848**	<u>0.909</u>	<u>0.768</u>	<u>0.654</u>	0.013	–	–	–
	w/**Ours**	<u>0.899</u>	**0.645**	<u>0.844</u>	**0.926**	**0.787**	**0.665**	0.008	–	–	–

Therefore, we chose to compare our method with SDM [33] (see Table 4). We conducted experiments with the following configurations: Baseline (with conventional augmentations), SDM, and our method. For the dataset used in this task, we generated an equal number of patches using both the SDM method and our approach. Table 4 shows the nuclei segmentation and classification performance using the baseline method. For segmentation, the Dice coefficient and Aggregated Jaccard Index (AJI) metrics are employed to measure performance for semantic and instance segmentation performance, respectively. For classification, F^{ci} represents the F_1 score for the i-th nulceus class (type) and F_d indicates the detection quality to measure the quality of instance detection. We analyzed the performance improvements resulting from the application of conventional augmentation (denoted as w/ Aug.), as well as the addition of SDM and our synthetic data. In particular, our synthetic data led to significant improvements across all datasets compared to using conventional augmentation alone. In the Lizard dataset, our method secured second place for most metrics by small margins, typically less than 0.5%. Especially, in the case of AJI metrics which implies instance segmentation performance, there was a gap of around 0.4%. Given that SDM relies on complete labeling for conditions and demonstrates high efficiency in generating images matching these label conditions, this suggests that our masks, employed in this downstream task, are effectively created, contributing to robust performance. In the PanNuke and EndoNuke datasets, our approach predominantly achieved first place, demonstrating robustness across different tissue types and staining modalities in histopathology datasets.

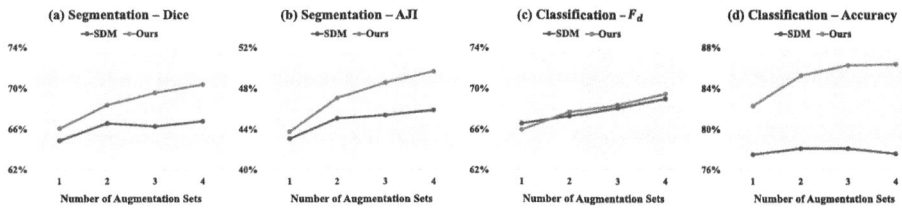

Fig. 8. Benefits of point condition-based diverse label generation on downstream tasks

Furthermore, we demonstrate the effectiveness of generating diverse labels by our scheme (see Fig. 5) for data augmentation, as shown in Fig. 8. In this experiment, we trained the models exclusively with synthetic data by increasing the number of augmentation sets. Our approach is compared to SDM, which generates images based on full-pixel labels. To perform this comparison, we prepared point layouts for our method and full-pixel labels for SDM. As we increased the number of synthetic sets, we evaluated the downstream segmentation and classification performance on the Lizard dataset. As shown in Fig. 8 (a) and (b), SDM exhibits performance saturation in both the Dice and AJI metrics at a lower point compared to our method, demonstrating continuous improvement. Moreover, as shown in Fig. 8 (c) and (d), our method achieves higher accuracy (an increase by over 10%) compared to SDM when the number of synthetic sets reaches 4, which also reflects a higher F_d value. These results indicate that our synthetic data, with its diverse labels, leads to a more diverse set of synthetic image samples, effectively improving the data distribution.

5   Conclusion

In this work, we introduced a novel approach to concurrently generate image-label pairs for histopathology nuclei images. We model the joint distribution of image, semantic label, and distance map using a single joint diffusion model. In addition, we introduced two context conditioning methods, a point map and text prompts, to enable precise control over the label synthesis process and faithful synthesis of histopathology images. Lastly, we use the synthesized distance mask to obtain instance label maps which are useful for downstream tasks such as nuclei instance segmentation.

For future work, we plan to reduce the time cost for data synthesis while maintaining the quality of the sampling to address the challenges of time-efficient synthesized data collection. Although we can employ existing methods (e.g., Abousamra *et al.* [1]) to generate the input point layout, developing a generative method for synthesizing more realistic point layouts is another research direction to explore.

Acknowledgements. This work was partially supported by the National Research Foundation of Korea (RS-2024-00349697, NRF-2021R1A6A1A13044830), the Institute for Information & Communications Technology Planning & Evaluation (RS-2020-II201819), the Technology development Program(RS-2024-00437796) funded by the Ministry of SMEs and Startups(MSS, Korea), and a Korea University Grant.

References

1. Abousamra, S., Gupta, R., Kurc, T., Samaras, D., Saltz, J., Chen, C.: Topology-guided multi-class cell context generation for digital pathology. In: Proceedings of the IEEE/CVF Conference on Computer Vision and Pattern Recognition, pp. 3323–3333 (2023)
2. Azadi, S., Tschannen, M., Tzeng, E., Gelly, S., Darrell, T., Lucic, M.: Semantic bottleneck scene generation. arXiv preprint arXiv:1911.11357 (2019)
3. Baranchuk, D., Rubachev, I., Voynov, A., Khrulkov, V., Babenko, A.: Label-efficient semantic segmentation with diffusion models. arXiv preprint arXiv:2112.03126 (2021)
4. Bau, D., et al.: Seeing what a GAN cannot generate. In: Proceedings of the IEEE/CVF International Conference on Computer Vision, pp. 4502–4511 (2019)
5. Butte, S., Wang, H., Xian, M., Vakanski, A.: Sharp-GAN: sharpness loss regularized GAN for histopathology image synthesis. In: 2022 IEEE 19th International Symposium on Biomedical Imaging (ISBI), pp. 1–5. IEEE (2022)
6. Cheng, J., et al.: Computational analysis of pathological images enables a better diagnosis of tfe3 xp11. 2 translocation renal cell carcinoma. Nat. Commun. **11**(1), 1778 (2020)
7. Dhariwal, P., Nichol, A.: Diffusion models beat GANs on image synthesis. Adv. Neural Inf. Process. Syst. **34**, 8780–8794 (2021)
8. Doan, T.N., Song, B., Vuong, T.T., Kim, K., Kwak, J.T.: Sonnet: a self-guided ordinal regression neural network for segmentation and classification of nuclei in large-scale multi-tissue histology images. IEEE J. Biomed. Health Inform. **26**(7), 3218–3228 (2022)
9. Gamper, J., et al.: Pannuke dataset extension, insights and baselines. arXiv preprint arXiv:2003.10778 (2020)
10. Gong, X., Chen, S., Zhang, B., Doermann, D.: Style consistent image generation for nuclei instance segmentation. In: Proceedings of the IEEE/CVF Winter Conference on Applications of Computer Vision, pp. 3994–4003 (2021)
11. Goodfellow, I., et al.: Generative adversarial nets. Adv. Neural Inf. Process. Syst. **27** (2014)
12. Graham, S., et al.: Lizard: a large-scale dataset for colonic nuclear instance segmentation and classification. In: Proceedings of the IEEE/CVF International Conference on Computer Vision, pp. 684–693(2021)
13. Graham, S.: Hover-net: simultaneous segmentation and classification of nuclei in multi-tissue histology images. Med. Image Anal. **58**, 101563 (2019)
14. He, Z., Unberath, M., Ke, J., Shen, Y.: TransNuSeg: a lightweight multi-task transformer for nuclei segmentation. In: Greenspan, H., et al. (eds.) Medical Image Computing and Computer Assisted Intervention – MICCAI 2023. MICCAI 2023. LNCS, vol. 14223, pp. 206–215. Springer, Cham (2023). https://doi.org/10.1007/978-3-031-43901-8_20

15. Heusel, M., Ramsauer, H., Unterthiner, T., Nessler, B., Hochreiter, S.: GANs trained by a two time-scale update rule converge to a local nash equilibrium. Adv. Neural Inf. Process. Syst. **30** (2017)

16. Ho, J., Jain, A., Abbeel, P.: Denoising diffusion probabilistic models. Adv. Neural Inf. Process. Syst. **33**, 6840–6851 (2020)

17. Hoogeboom, E., Nielsen, D., Jaini, P., Forré, P., Welling, M.: Argmax flows and multinomial diffusion: learning categorical distributions. Adv. Neural Inf. Process. Syst. **34**, 12454–12465 (2021)

18. Hou, L., Agarwal, A., Samaras, D., Kurc, T.M., Gupta, R.R., Saltz, J.H.: Robust histopathology image analysis: to label or to synthesize? In: Proceedings of the IEEE/CVF Conference on Computer Vision and Pattern Recognition, pp. 8533–8542 (2019)

19. Le Moing, G., Vu, T.H., Jain, H., Pérez, P., Cord, M.: Semantic palette: guiding scene generation with class proportions. In: Proceedings of the IEEE/CVF Conference on Computer Vision and Pattern Recognition, pp. 9342–9350 (2021)

20. Lin, Y., Wang, Z., Cheng, KT., Chen, H.: InsMix: towards realistic generative data augmentation for nuclei instance segmentation. In: Wang, L., Dou, Q., Fletcher, P.T., Speidel, S., Li, S. (eds.) Medical Image Computing and Computer Assisted Intervention – MICCAI 2022. MICCAI 2022. LNCS, vol. 13432, pp. 140–149. Springer, Cham (2022). https://doi.org/10.1007/978-3-031-16434-7_14

21. Moghadam, P.A., et al.: A morphology focused diffusion probabilistic model for synthesis of histopathology images. In: Proceedings of the IEEE/CVF Winter Conference on Applications of Computer Vision, pp. 2000–2009 (2023)

22. Naumov, A., et al.: Endonuke: nuclei detection dataset for estrogen and progesterone stained IHC endometrium scans. Data **7**(6), 75 (2022)

23. Nguyen, Q., Vu, T., Tran, A., Nguyen, K.: Dataset diffusion: diffusion-based synthetic dataset generation for pixel-level semantic segmentation. arXiv preprint arXiv:2309.14303 (2023)

24. Nishimura, K., Ker, D.F.E., Bise, R.: Weakly supervised cell instance segmentation by propagating from detection response. In: Shen, D., et al. (eds.) MICCAI 2019. LNCS, vol. 11764, pp. 649–657. Springer, Cham (2019). https://doi.org/10.1007/978-3-030-32239-7_72

25. Oh, H.J., Jeong, W.K.: DiffMix: diffusion model-based data synthesis for nuclei segmentation and classification in imbalanced pathology image datasets. In: Greenspan, H., et al. (eds.) Medical Image Computing and Computer Assisted Intervention – MICCAI 2023. MICCAI 2023. LNCS, vol. 14222, pp. 337–345. Springer, Cham (2023). https://doi.org/10.1007/978-3-031-43898-1_33

26. Park, M., Yun, J., Choi, S., Choo, J.: Learning to generate semantic layouts for higher text-image correspondence in text-to-image synthesis. In: Proceedings of the IEEE/CVF International Conference on Computer Vision, pp. 7591–7600 (2023)

27. Salimans, T., Goodfellow, I., Zaremba, W., Cheung, V., Radford, A., Chen, X.: Improved techniques for training GANs. Adv. Neural Inf. Process. Syst. **29** (2016)

28. Shrivastava, A., Fletcher, P.T.: NASDM: nuclei-aware semantic histopathology image generation using diffusion models. In: Greenspan, H., et al. (eds.) Medical Image Computing and Computer Assisted Intervention – MICCAI 2023. MICCAI 2023. LNCS, vol. 14225, pp. 786–796. Springer, Cham (2023). https://doi.org/10.1007/978-3-031-43987-2_76

29. Vahadane, A., et al.: Structure-preserving color normalization and sparse stain separation for histological images. IEEE Trans. Med. Imaging **35**(8), 1962–1971 (2016)

30. Verghese, G., et al.: Computational pathology in cancer diagnosis, prognosis, and prediction-present day and prospects. J. Pathol. **260**(5), 551–563 (2023)
31. Van der Walt, S., et al.: Scikit-image: image processing in python. PeerJ **2**, e453 (2014)
32. Wang, H., Xian, M., Vakanski, A., Shareef, B.: Sian: style-guided instance-adaptive normalization for multi-organ histopathology image synthesis. In: 2023 IEEE 20th International Symposium on Biomedical Imaging (ISBI). pp. 1–5. IEEE (2023)
33. Wang, W., et al.: Semantic image synthesis via diffusion models. arXiv preprint arXiv:2207.00050 (2022)
34. Wang, X., et al.: Esrgan: enhanced super-resolution generative adversarial networks. In: Proceedings of the European Conference on Computer Vision (ECCV) Workshops (2018)
35. Yang, X., Li, H., Zhou, X.: Nuclei segmentation using marker-controlled watershed, tracking using mean-shift, and kalman filter in time-lapse microscopy. IEEE Trans. Circuits Syst. I Regul. Pap. **53**(11), 2405–2414 (2006)
36. Ye, J., Xue, Y., Liu, P., Zaino, R., Cheng, K.C., Huang, X.: A multi-attribute controllable generative model for histopathology image synthesis. In: de Bruijne, M., et al. (eds.) MICCAI 2021. LNCS, vol. 12908, pp. 613–623. Springer, Cham (2021). https://doi.org/10.1007/978-3-030-87237-3_59
37. Yellapragada, S., Graikos, A., Prasanna, P., Kurc, T., Saltz, J., Samaras, D.: Pathldm: text conditioned latent diffusion model for histopathology. In: Proceedings of the IEEE/CVF Winter Conference on Applications of Computer Vision, pp. 5182–5191 (2024)
38. Yu, X., et al.: Diffusion-Based Data Augmentation for Nuclei Image Segmentation. In: Greenspan, H., et al. (eds.) Medical Image Computing and Computer Assisted Intervention – MICCAI 2023. MICCAI 2023. LNCS, vol. 14227, pp. 592–602. Springer, Cham (2023). https://doi.org/10.1007/978-3-031-43993-3_57
39. Zhang, Y., et al.: Datasetgan: efficient labeled data factory with minimal human effort. In: Proceedings of the IEEE/CVF Conference on Computer Vision and Pattern Recognition, pp. 10145–10155 (2021)

One-Stage Prompt-Based Continual Learning

Youngeun Kim$^{(\boxtimes)}$ ⓘ, Yuhang Li ⓘ, and Priyadarshini Panda ⓘ

Yale University, New Haven, CT, USA
{youngeun.kim,yuhang.li,priya.panda}@yale.edu

Abstract. Prompt-based Continual Learning (PCL) has gained considerable attention as a promising continual learning solution because it achieves state-of-the-art performance while preventing privacy violations and memory overhead problems. Nonetheless, existing PCL approaches face significant computational burdens because of two Vision Transformer (ViT) feed-forward stages; one is for the query ViT that generates a prompt query to select prompts inside a prompt pool; the other one is a backbone ViT that mixes information between selected prompts and image tokens. To address this, we introduce a one-stage PCL framework by directly using the intermediate layer's token embedding as a prompt query. This design removes the need for an additional feed-forward stage for query ViT, resulting in ∼50% computational cost reduction for both training and inference with marginal accuracy drop ($\leq$1%). We further introduce a Query-Pool Regularization (QR) loss that regulates the relationship between the prompt query and the prompt pool to improve representation power. The QR loss is only applied during training time, so there is no computational overhead at inference from the QR loss. With the QR loss, our approach maintains ∼50% computational cost reduction during inference as well as outperforms the prior two-stage PCL methods by ∼1.4% on public class-incremental continual learning benchmarks including CIFAR-100, ImageNet-R, and DomainNet.

Keywords: Efficient learning · Continual learning · Transfer learning

1 Introduction

Training models effectively and efficiently on a continuous stream of data presents a significant practical hurdle. A straightforward approach would entail accumulating both prior and new data and then updating the model using this comprehensive dataset. However, as data volume grows, fully retraining a model on such extensive data becomes increasingly impractical [11,25]. Additionally, storing past data can raise privacy issues, such as those highlighted by the EU

Supplementary Information The online version contains supplementary material available at https://doi.org/10.1007/978-3-031-72624-8_10.

A. Leonardis et al. (Eds.): ECCV 2024, LNCS 15071, pp. 163–179, 2025.
https://doi.org/10.1007/978-3-031-72624-8_10

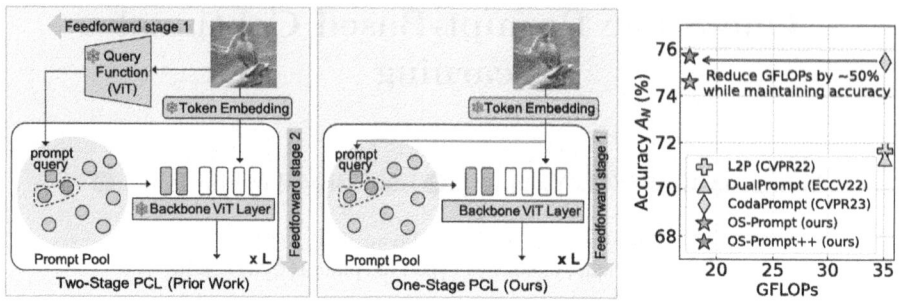

Fig. 1. Difference between prior PCL work and ours. Prior PCL work (**Left**) has two feed-forward stages for (1) a query function (ViT) to select input-specific prompts and (2) a backbone ViT layer to perform prompt learning with the selected prompts and image tokens. On the other hand, our one-stage PCL framework (**Middle**) uses an intermediate layer's token embedding as a prompt query so that it requires only one backbone ViT feed-forward stage. As a result, our method reduces GFLOPs by ∼50% compared to prior work while maintaining accuracy (**Right**).

General Data Protection Regulation (GDPR) [36]. An alternative solution is to adapt the model based solely on currently incoming data, eschewing any access to past data. This paradigm is termed as *rehearsal-free continual learning* [7,9,32,33,39–41], and the primary goal is to diminish the effects of catastrophic forgetting on previously acquired data.

Among the rehearsal-free continual learning methods, Prompt-based Continual Learning (PCL) stands out as it has demonstrated state-of-the-art performance in image classification tasks, even surpassing rehearsal-based methods [33,37,39,40]. PCL utilizes a pre-trained Vision Transformer (ViT) [8] and refines the model by training learnable tokens on the given data. PCL adopts a prompt pool-based training scheme where different prompts are selected and trained for each continual learning stage. This strategy enables a model to learn the information of the training data in a sequential manner with less memory overhead, as the prompt pool requires minimal resources.

Although PCL methods show state-of-the-art performance, huge computational costs from the two ViT feed-forward stages make the model difficult to deploy into resource-constrained devices [12,26,38]. Specifically, the PCL method requires two-stage ViT feedforward steps. One is for the query function that generates a prompt query. The other one is a backbone ViT that mixes information between selected prompts and input image tokens. We refer to this approach as a *two-stage PCL* method, illustrated in Fig. 1 Left.

To address this limitation, we propose a *one-stage PCL* framework with only one ViT feed-forward stage (Fig. 1 Middle), called OS-Prompt. Rather than deploying a separate ViT feed-forward phase to generate a prompt query, we directly use the intermediate layer's token embedding as a query. This is based on our observation that early layers show marginal shifts in the feature space during continual prompt learning (details are in Sect. 4.1). This observation enables us to use the intermediate token embedding as a prompt query which requires consistent representation across continual learning to minimize catastrophic forgetting. Surprisingly, OS-Prompt shows a marginal performance drop (less than 1%) while saving ∼50% of computational cost (Fig. 1 Right).

As our OS-Prompt uses the intermediate layer's token embedding as a query instead of the last layer's token embedding (which is used in prior PCL methods), there is a slight performance drop due to the lack of representation power. To address this, we introduce a Query-Pool Regularization (QR) loss, which is designed to enhance the representation power of a prompt pool by utilizing the token embedding from the last layer. Importantly, the QR loss is applied only during training, ensuring no added computational burden during inference. We refer to our enhanced one-stage framework with QR loss as OS-Prompt++. Our OS-Prompt++ bridges the accuracy gap, performing better than OS-Prompt.

Overall, our contribution can be summarized as follows: (1) We raise a drawback of the current PCL methods - high computational cost, particularly due to the two distinct ViT feed-forward stages. As a solution to the computational inefficiency in existing PCL techniques, we propose OS-Prompt that reduces the computational cost by nearly 50% without any significant performance drop. (2) To counter the slight performance degradation observed in our one-stage PCL framework, we introduce a QR loss. This ensures that the prompt pool maintains similarity with token embedding from both intermediate and final layers. Notably, our QR loss avoids any additional computational overhead during inference. (3) We conduct experiments with rehearsal-free continual learning setting on CIFAR-100 [21], ImageNet-R [16], and DomainNet [27] benchmarks, outperforming the previous SOTA method CodaPrompt [33] by ∼1.4% with ∼50% computational cost saving.

2   Related Work

2.1   Continual Learning

For a decade, continual learning has been explored as an important research topic, concentrating on the progressive adaptation of models over successive tasks or datasets. One of the representative methods to address continual learning is the regularization-based method [1,23,44]. By adding a regularization loss between the current model and the previous model, these methods aim to minimize catastrophic forgetting. However, their performance is relatively low on challenging datasets compared to the other continual learning methods. An alternative approach proposes to expand the network architecture as it progresses through different continual learning stages [22,30,31,42]. Though these typically surpass the results of regularization-based methods, they come at the expense of significant memory consumption due to the increased parameters. Rehearsal-based continual learning has introduced a component for archiving previous data [2,5,6,13,29]. By leveraging accumulated data during the subsequent stages, they often outperform other continual learning methods. However, saving images causes memory overhead and privacy problems. Given the inherent limitations, a growing interest is observed in rehearsal-free continual learning. These are crafted to address catastrophic forgetting without access to past data. Most of them propose methods to generate rehearsal images [7,9,32,41], but generating images is resource-heavy and time-consuming. Recently, within the rehearsal-free domain, Prompt-based Continual Learning (PCL) methods have gained considerable attention because of their performance and memory efficiency. By utilizing a pre-trained ViT model, they train only a small set of parameters called prompts to maintain the information against catastrophic forgetting. For example, L2P [40] utilizes a prompt pool, selecting the appropriate prompts for the given image. Extending this principle, DualPrompt [39] proposes general prompts for encompassing knowledge across multiple continual learning stages. Advancing this domain further, [33] facilitates end-to-end training of the prompt pool, achieving state-of-the-art performance. However, huge computational costs from the two ViT feed-forward stages make the model difficult to deploy into resource-constrained devices. Our work aims to resolve such computational complexity overhead in PCL. Furthermore, Tang *et al.* [35] have proposed a prompt-based continual learning that generates prompts based on the intermediate features from the transformer. However, this prior work focuses on maximizing continual learning performance without a pre-trained model, while ours aims to leverage the power of pre-trained models. Also, the prior work focuses on the prompt generation method, while we propose a framework compatible with different prompt learning techniques.

2.2   Prompt-Based Learning

The efficient fine-tuning method for large pre-trained models has shown their practical benefits across various machine learning tasks [14,18,28,45–47]. Instead of adjusting all parameters within neural architectures, the emphasis has shifted to leveraging a minimal set of weights for optimal transfer outcomes. In alignment with this, multiple methodologies [3,30] have integrated a streamlined bottleneck component within the transformer framework, thus constraining gradient evaluations to select parameters. Strategies like TinyTL [3] and BitFit [43] advocate for bias modifications in the fine-tuning process. In a recent shift, prompt-based learning [19,20] captures task-specific knowledge with much smaller additional parameters than the other fine-tuning methods [40]. Also, prompt-based learning only requires storing several prompt tokens, which are easy to plug-and-play, and hence, they are used to construct a prompt pool for recent continual learning methods.

3   Preliminary

3.1   Problem Setting

In a continual learning setting, a model is trained on T continual learning stages with dataset $D = \{D_1, D_2, ..., D_T\}$, where D_t is the data provided in t-th stage. Our problem is based on an image recognition task, so each data D_t consists of pairs of images and labels. Also, data from the previous tasks is not accessible for future tasks. Following the previous rehearsal-free continual learning settings [33,39], we focus on the class-incremental continual learning setting where task identity is unknown at inference. This is a challenging scenario compared to others such as task-incremental continual learning where the task labels are provided for both training and test phases [17]. For the experiments, we split classes into T chunks with continual learning benchmarks including CIFAR-100 [21] and ImageNet-R [16].

3.2   Two-Stage Prompt-Based Continual Learning

In our framework, we focus on improving efficiency through a new query selection rather than changing the way prompts are formed from the prompt pool with the given query, a common focus in earlier PCL studies [33,39,40]. For a fair comparison with earlier work, we follow the overall PCL framework established in previous studies. Given that our contribution complements the prompt formation technique, our method can seamlessly work with future works that propose a stronger prompt generation method.

The PCL framework selects the input-aware prompts from the layer-wise prompt pool and adds them to the backbone ViT to consider task knowledge. The l-th layer has the prompt pool $P_l = \{k_l^1 : p_l^1, ..., k_l^M : p_l^M\}$ which contains M prompt components $p \in \mathbb{R}^{L_p \times D}$ and the corresponding key $k \in \mathbb{R}^D$. Here, L_p and D stand for the prompt length and the feature dimension, respectively.

The prior PCL method consists of two feed-forward stages. In the first stage, a prompt query $q \in \mathbb{R}^D$ is extracted from pre-trained query ViT $Q(\cdot)$, utilizing the $[CLS]$ token from the final layer.

$$q = Q(x)_{[CLS]}. \tag{1}$$

Here x is the given RGB image input. The extracted prompt query is used to form a prompt $\phi_l \in \mathbb{R}^{L_p \times D}$ from a prompt pool P_l, which can be formulated as:

$$\phi_l = g(q, P_l). \tag{2}$$

The prompt formation function $g(\cdot)$ has been a major contribution in the previous literature. L2P [40] and DualPrompt [39] select prompts having top-N similarity (e.g. cosine similarity) between query q and prompt key k_l. The recent state-of-the-art CodaPrompt [33] conducts a weighted summation of prompt components based on their similarity to enable end-to-end training.

The obtained prompt ϕ_l for layer l is given to backbone ViT layer f_l with input token embedding x_l. This is the second feed-forward stage of the prior PCL methods.

$$x_{l+1} = f_l(x_l, \phi_l). \tag{3}$$

Following DualPrompt [39] and CodaPrompt [33], we use prefix-tuning to add the information of a prompt ϕ_l inside f_l. The prefix-tuning splits prompt ϕ_l into $[\phi_k, \phi_v] \in \mathbb{R}^{\frac{L_p}{2} \times D}$, and then prepends them to the key and value inside the self-attention block of ViT. We utilize Multi-Head Self-Attention (MHSA) [8]

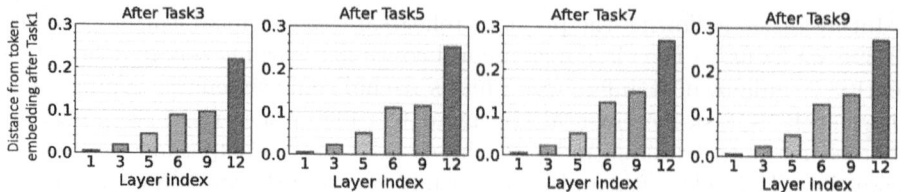

Fig. 2. We measure layer-wise feature distances between token embeddings of the model after training on task 1 and when a new task is learned. Each column represents the embeddings after training a model on Tasks 3, 5, 7, and 9. For instance, the left figure represents the distance between the token embeddings of Task 1 and those learned when Task 3 is completed. We train prompts using CodaPrompt [33] and use $1 - CosSim(x, y)$ to measure the layer-wise distance on the training dataset. We use a CIFAR100 10-task setting. We provide more examples in the Supplementary Materials.

like the prior PCL work, which computes the outputs from multiple single-head attention blocks.

$$MHSA(x, \phi_k, \phi_v) = Concat[head_1, ..., head_H]W_o. \tag{4}$$

$$head_i = Attention(xW_q^i, [\phi_k; x]W_k^i, [\phi_v; x]W_v^i). \tag{5}$$

Here, W_o, W_q, W_k, W_v are the projection matrices. This prefix-tuning method is applied to the first 5 layers of backbone ViT. The leftover layers conduct a standard MHSA without prompts.

However, the prior PCL approach requires two-stage ViT feedforward (Eq. 1 and Eq. 3), which doubles computational cost. In our work, we aim to improve the computational efficiency of the PCL framework without degrading the performance.

4   One-Stage Prompt-Based Continual Learning

We propose to reduce the computational cost of PCL by restructuring two-step feedforward stages. To this end, we propose a new one-stage PCL framework called *OS-Prompt*. Instead of using a separate feedforward step to compute the prompt query q from Eq. 1, we take a token embedding from the intermediate layer of a backbone ViT as the query (illustrated in Fig. 1).

4.1   How Stable are Token Embeddings Across Continual Learning?

We first ensure the validity of using intermediate token embedding as a prompt query. The original two-stage design employs a frozen pre-trained ViT, ensuring consistent prompt query representation throughout continual learning. It is essential to maintain a consistent (or similar) prompt query representation because changes in prompt query would bring catastrophic forgetting. In our approach, token embeddings in the backbone ViT continually change as prompt tokens are updated during learning. For instance, after training on tasks 1 and 2, prompts have different values, resulting in varying token embeddings for an identical image. Consequently, it is crucial to assess if token embeddings sustain a consistent representation throughout the continual learning.

To address the concern, we validate the deviation of the token embedding as prompt continual learning progresses. In Fig. 2, we show how this change happens at each layer during continuous learning tasks. From our study, two main observations can be highlighted: (1) When we add prompts, there is a larger difference in the deeper layers. For example, layers $1 \sim 5$ have a small difference (≤ 0.1). However, the last layer shows a bigger change (≥ 0.1). (2) As learning continues, the earlier layers remain relatively stable, but the deeper layers change more. This observation concludes that although prompt tokens are included, the token embedding of early layers shows minor changes during training. Therefore, using token embeddings from these earlier layers would give a stable and consistent representation throughout the continual learning stages.

4.2   One-Stage PCL Framework

Building on these observations, we employ the token embedding from the early layers as a prompt query to generate layer-wise prompt tokens. For a fair comparison, we implement prompts across layers 1 to 5, in line with prior work. The proposed OS-Prompt framework is illustrated in Fig. 3. For each layer, given the input token embedding, we directly use the $[CLS]$ token as the prompt query. The original query selection equation (Eq. 1) can be reformulated as:

$$q_l = x_{l_{[CLS]}}. \tag{6}$$

Using the provided query q_l, we generate a prompt from the prompt pool following the state-of-the-art CodaPrompt [33]. It is worth highlighting that our primary contribution lies not in introducing a new prompt generation technique but in presenting a more efficient framework for query selection. We first measure the cosine similarity $\gamma(\cdot)$ between q_l and keys $\{k_l^1, ..., k_l^M\}$, and then we perform a weighted summation of the corresponding value p_l^m (*i.e.* prompt) based on the similarity.

$$\phi_l = \sum_m \gamma(q_l, k_l^m) p_l^m. \tag{7}$$

The generated prompt is then prepended to the image tokens. Notably, since we produce a prompt query without the need for an extra ViT feedforward, the computational overhead is reduced by approximately 50% for both training and inference.

4.3   Query-Pool Regularization Loss

Our OS-Prompt relies on intermediate token embeddings. As a result, the prompt query exhibits diminished representational capacity compared to previous PCL approaches that utilize the $[CLS]$ token from the final layer. This

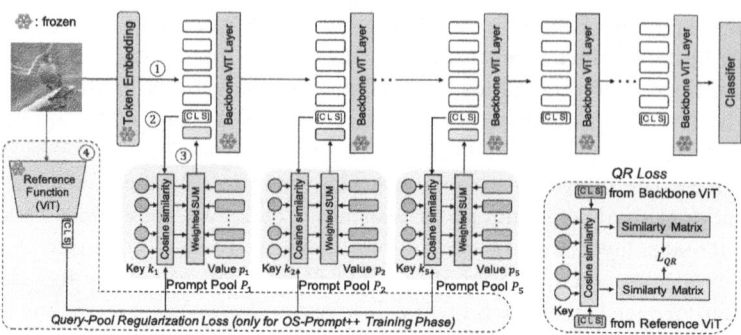

Fig. 3. Our OS-Prompt (OS-Prompt++) framework. An image passes through the backbone ViT layers to get the final prediction. From layer 1 to layer 5, we prepend prompt tokens to the image tokens, which can be obtained in the following progress: ① We first compute the image token embedding from the previous layer. ② We use $[CLS]$ token as a prompt query used for measuring cosine similarity between prompt keys inside the prompt pool. ③ Based on the similarity, we do weighted sum values to obtain prompt tokens. ④ To further improve the accuracy, we present OS-Prompt++. We integrate the query-pool regularization loss (dotted line), enabling the prompt pool to capture a stronger representation from the reference ViT.

reduced capacity for representation brings performance degradation. To mitigate this, we introduce a Query-Pool Regularization (QR) loss. The QR loss ensures that the query-pool relationship becomes similar to that of the final layer's $[CLS]$ token, thereby improving representation power. We extract the $[CLS]$ token from the last layer of a reference ViT architecture $R(\cdot)$ and employ it as our reference prompt query $r \in \mathbb{R}^{1 \times D}$ where D is the feature dimension:

$$r = R(x). \tag{8}$$

Let $K_l \in \mathbb{R}^{M \times D}$ be the matrix representation of prompt keys at layer l. We then define a similarity matrix $A_{query}^l \in \mathbb{R}^{M \times 1}$ to capture the relationship between query $q_l \in \mathbb{R}^{1 \times D}$ (from Eq. 6) and prompt keys. Similarly, we compute the similarity matrix A_{ref}^l to represent the relationship between the reference prompt query r and the prompt keys.

$$A_{query}^l = Softmax(\frac{K_l q_l^T}{||K_l||_2 ||q_l||_2}) \qquad A_{ref}^l = Softmax(\frac{K_l r^T}{||K_l||_2 ||r||_2}). \tag{9}$$

We apply $Softmax()$ to measure the relative similarity among query-key pairs, considering that q and r show distinct feature distributions originating from different layers. The QR loss penalizes the distance between two similarity matrices.

$$\mathcal{L}_{QR} = \sum_l ||A_{query}^l - A_{ref}^l||_2^2. \tag{10}$$

Table 1. Results on ImageNet-R for different task configurations: 5 tasks (40 classes/task), 10 tasks (20 classes/task), and 20 tasks (10 classes/task). A_N represents the average accuracy across tasks, and F_N indicates the mean forgetting rate. ↑ and ↓ indicate whether a metric is better with a higher or lower value, respectively. We average values over five runs with different seeds.

Setting	Task-5		Task-10		Task-20	
Method	$A_N(\uparrow)$	$F_N(\downarrow)$	$A_N(\uparrow)$	$F_N(\downarrow)$	$A_N(\uparrow)$	$F_N(\downarrow)$
UB	77.13		77.13	–	77.13	
FT	18.74 ± 0.44	41.49 ± 0.52	10.12 ± 0.51	25.69 ± 0.23	4.75 ± 0.40	16.34 ± 0.19
ER	71.72 ± 0.71	13.70 ± 0.26	64.43 ± 1.16	10.30 ± 0.05	52.43 ± 0.87	7.70 ± 0.13
LwF	74.56 ± 0.59	4.98 ± 0.37	66.73 ± 1.25	3.52 ± 0.39	54.05 ± 2.66	2.86 ± 0.26
L2P	70.83 ± 0.58	3.36 ± 0.18	69.29 ± 0.73	2.03 ± 0.19	65.89 ± 1.30	1.24 ± 0.14
Deep L2P	73.93 ± 0.37	2.69 ± 0.10	71.66 ± 0.64	1.78 ± 0.16	68.42 ± 1.20	1.12 ± 0.13
DualPrompt	73.05 ± 0.50	2.64 ± 0.17	71.32 ± 0.62	1.71 ± 0.24	67.87 ± 1.39	1.07 ± 0.14
CodaPrompt	76.51 ± 0.38	2.99 ± 0.19	75.45 ± 0.56	1.64 ± 0.10	72.37 ± 1.19	0.96 ± 0.15
OS-Prompt	75.74 ± 0.58	3.32 ± 0.31	74.58 ± 0.56	1.92 ± 0.15	72.00 ± 0.60	1.09 ± 0.11
OS-Prompt++	**77.07 ± 0.15**	**2.23 ± 0.18**	**75.67 ± 0.40**	**1.27 ± 0.10**	**73.77 ± 0.19**	**0.79 ± 0.07**

Table 2. Accuracy comparison on CIFAR-100 10-task setting.

Method	$A_N(\uparrow)$	$F_N(\downarrow)$
UB	89.30	–
ER	76.20 ± 1.04	8.50 ± 0.37
Deep L2P	84.30 ± 1.03	1.53 ± 0.40
DualPrompt	83.05 ± 1.16	1.72 ± 0.40
CodaPrompt	86.25 ± 0.74	1.67 ± 0.26
OS-Prompt	86.42 ± 0.61	1.64 ± 0.14
OS-Prompt++	**86.68 ± 0.67**	**1.18 ± 0.21**

Table 3. GFLOPs comparison of PCL works. We provide relative cost (%) with respect to L2P.

Method	Training GFLOPs	Inference GFLOPs
L2P	52.8 (100%)	35.1 (100%)
Deep L2P	52.8 (100%)	35.1 (100%)
DualPrompt	52.8 (100%)	35.1 (100%)
CodaPrompt	52.8 (100%)	35.1 (100%)
OS-Prompt	**35.4 (66.7%)**	**17.6 (50.1%)**
OS-Prompt++	52.8 (100%)	**17.6 (50.1%)**

The QR loss is added to the cross-entropy loss $\mathcal{L}_{CE}$ for classification. The total loss function can be written as:

$$\mathcal{L}_{total} = \mathcal{L}_{CE} + \lambda\mathcal{L}_{QR}, \qquad (11)$$

where λ is a hyperparameter to balance between two loss terms. Importantly, the QR loss is applied during the training phase, ensuring that there is no computational overhead during inference. Note that we only train prompts within a prompt pool while freezing the other weight parameters.

5 Experiments

5.1 Experiment Setting

Dataset. We utilize the Split CIFAR-100 [21] and Split ImageNet-R [15] benchmarks for class-incremental continual learning. The Split CIFAR-100 partitions the original CIFAR-100 dataset into 10 distinct tasks, each comprising 10 classes. The Split ImageNet-R benchmark is an adaptation of ImageNet-R, encompassing diverse styles, including cartoon, graffiti, and origami. For our experiments, we segment ImageNet-R into 5, 10, or 20 distinct class groupings. Given its substantial intra-class diversity, the ImageNet-R benchmark is viewed as a particularly challenging benchmark. We also provide experiments on DomainNet [27], a large-scale domain incremental learning dataset. The dataset consists of around 0.6 million images distributed across 345 classes. We use 5 different domains (Clipart $\rightarrow$ Real $\rightarrow$ Infograph $\rightarrow$ Sketch $\rightarrow$ Painting) as a continual learning task where each task consists of 69 classes. The DomainNet results are provided in the Supplementary Materials.

Experimental Details. We performed our experiments using the ViT-B/16 [8] pre-trained on ImageNet-1k, a standard backbone in earlier PCL research. To ensure a fair comparison with prior work, we maintain the same prompt length (8) and number of prompt components (100) as used in CodaPrompt. Like CodaPrompt, we divide the total prompt components into the number of tasks, and provide the partial component for each task. In task N, the prompt components from task $1 \sim N$ are frozen, and we only train the key and prompt components from task N. We split 20% of the training set for hyperparameter λ tuning in Eq. 11. Our experimental setup is based on the PyTorch. The experiments utilize four Nvidia RTX2080ti GPUs. For robustness, we conducted our benchmarks with five different permutations of task class order, presenting both the average and standard deviation of the results. More detailed information can be found in the Supplementary Materials.

Evaluation Metrics. We use two metrics for evaluation: (1) Average final accuracy A_N, which measures the overall accuracy across N tasks. [33,39,40] (2) Average forgetting F_N, which tracks local performance drop over N tasks [24,33,34]. Note, A_N is mainly used for performance comparison in literature.

5.2 Comparison with Prior PCL Works

We compare our OS-Prompt with prior continual learning methods. This includes non-PCL methods such as ER [6] and LWF [23], as well as PCL methods like L2P [40], DualPrompt [39], and CodaPrompt [33]. We present both the upper bound (UB) performance and results from fine-tuning (FT). UB refers to training a model with standard supervised training using all data (no continual learning). FT indicates that a model undergoes sequential training with continual learning data without incorporating prompt learning. Moreover, we report L2P/Deep L2P implementation from CodaPrompt [33], both in its original form and improved version by applying prompt pool through layers 1 to 5. In our comparisons, *OS-Prompt* represents our one-shot prompt framework without the QR loss, while *OS-Prompt++* is with the QR loss.

Table 1 shows the results on ImageNet-R. Our OS-Prompt indicates only a slight performance drop ($\leq 1\%$) across the 5-task, 10-task, and 20-task settings. The reason for the performance drop could be the reduced representational

Table 4. Relative training computational cost with repect to ER.

CL strategy	Method	Training Complexity
Replay-based	ER	1
Regularization-based	LWF	4/3
Prompt-based	L2P	1
Prompt-based	DualPrompt	1
Prompt-based	CodaPrompt	1
Prompt-based	OS-Prompt (Ours)	2/3
Prompt-based	OS-Prompt++ (Ours)	1

Table 5. ImageNet-R 10-task results with unsupervised pre-trained model [4].

Method	$A_N(\uparrow)$	$F_N(\downarrow)$
ER	60.43 ± 1.16	13.30 ± 0.12
LwF	62.73 ± 1.13	4.32 ± 0.63
L2P	60.32 ± 0.56	2.30 ± 0.11
DualPrompt	61.77 ± 0.61	2.31 ± 0.23
CodaPrompt	67.61 ± 0.19	2.23 ± 0.29
OS-Prompt	67.52 ± 0.34	2.32 ± 0.13
OS-Prompt++	67.92 ± 0.42	2.19 ± 0.26

capacity of a prompt query from the intermediate token embedding. However, OS-Prompt++ effectively counters this limitation by incorporating the QR loss. Such an observation underscores the significance of the relationship between the query and the prompt pool in PCL. OS-prompt++ shows slight performance improvement across all scenarios. This implies a prompt query that contains task information helps to enhance representation in the prompt pool, suggesting that exploring methods to integrate task information inside the prompt selection process could be an interesting research direction in PCL. Table 2 provides results on 10-task CIFAR-100, which shows a similar trend with ImageNet-R.

In Table 3, we provide GFLOPs for various PCL methods, rounding GFLOPs to one decimal place. It is worth noting that while prior PCL methods might have marginally different GFLOP values, they are close enough to appear identical in the table. This implies that different prompt formation schemes proposed in prior works do not substantially impact GFLOPs. The table shows OS-Prompt operates at 66.7% and 50.1% of the GFLOPs, during training and infer-

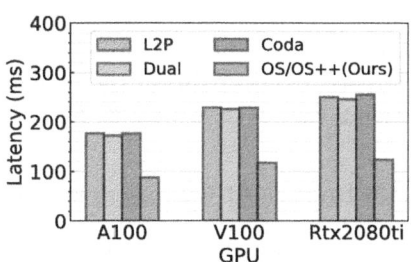

Fig. 4. Comparison of latency across PCL methods on different GPUs.

ence, respectively, relative to prior methods. While OS-Prompt++ maintains a training GFLOPs count comparable to earlier work due to the added feedforward process in the reference ViT, it does not employ the reference ViT during inference, bringing it down to 50.1% GFLOPs. In Fig. 4, we compare the inference GPU latency of previous PCL methods with our OS-Prompt across three distinct GPU setups. All measurements are taken using a batch size of 32. Similar to the trend observed in FLOPs, our approach reduces the latency by ~50%.

5.3   Discussion on Training Computational Cost

In this section, we provide an in-depth discussion of the training computational cost of PCL methods. Following [10], Table 4 presents the relative training complexity of each method with respect to ER [6], which is the simple baseline with standard gradient-based training.

Here, we would like to clarify the training computational cost of prompt learning. In general neural network training, the forward-backward computational cost ratio is approximately 1:2. This is due to gradient backpropagation ($\frac{dL}{da_l} = W_{l+1}\frac{dL}{da_{l+1}}$) and weight-updating ($\frac{dL}{dW_l} = \frac{dL}{da_{l+1}}a_l$), where L represents the loss, W denotes the weight parameter, a is the activation, and l is the layer index. In prompt learning, the forward-backward computational cost ratio is approximately 1:1. This is in contrast to general neural network training, as only a small fraction of the weight parameters (less than 1%) are updated.

Bearing this in mind, we present the observations derived from the table, assuming that all methods employ the same architecture. (1) Previous PCL (L2P, DualPrompt, CodaPrompt) consists of two steps: First, the query ViT requires only feedforward without backpropagation, Second, the backbone ViT requires feedforward-backward training with prompt tuning. This results in the relative training computational cost 1 ($= \frac{\text{QueryViT}_{fw}(1) + \text{BackboneViT}_{fw}(1) + \text{BackboneViT}_{bw}(1)}{\text{ER}_{fw}(1) + \text{ER}_{bw}(2)}$). Here, fw and bw denote forward and backward propagation respectively. (2) Our OS-Prompt requires one feedforward-backward step like standard training. Therefore, relative training cost becomes $\frac{2}{3}$ ($= \frac{\text{BackboneViT}_{fw}(1) + \text{BackboneViT}_{bw}(1)}{\text{ER}_{fw}(1) + \text{ER}_{bw}(2)}$). This shows the potential of our OS-Prompt on online continual learning. (3) Our OS-Prompt++ has a reference ViT to perform the regularization, bringing 1 relative training computational cost with respect to ER.

5.4   Impact of Unsupervised Pre-trained Weights

One underlying assumption of existing PCL methods is their reliance on supervised pretraining on ImageNet-1k. While this pre-training has a substantial impact on model performance, it may not always be feasible for a general continual learning task. One possible solution is using unsupervised pre-trained models. To explore this, we use DINO pre-trained weights [4] instead of ImageNet-1k pre-trained weights, and compare the performance across ER, LwF, and PCL works. We set the other experimental settings as identical. We report the results with ImageNet-R 10-task in Table 5. Compared to the ImageNet-1k pre-trained model (with supervised training), performances across all methods degrade. This observation suggests that the backbone model plays a crucial role in PCL. Moreover, similar to the ImageNet-1k pre-trained model, CodaPrompt and our OS-Prompt outperform the other methods by a significant margin. This indicates that our method continues to perform well with various backbones.

5.5   Analysis of Design Components

QR Loss Design. To understand the effect of different components in our proposed QR loss (Eq. 10), we conducted an ablation study with different settings, summarized in Table 6. There are two main components in the QR loss: *Cosine Similarity* and *Softmax*, and we provide the accuracy of different four combinations. Without Cosine Similarity and Softmax, our framework yields a performance of 75.00% for A_N. Adding Cosine Similarity or Softmax improves the performance, suggesting their collaborative role in enhancing the model's performance in our QR loss design. We also provide the sensitivity analysis of our method with respect to the hyperparameter λ in Table 7. Across the three distinct task configurations on ImageNet-R, we note that the performance fluctuations are minimal, despite the variation in λ values. These results underline the robustness of our method with respect to λ value, suggesting that our approach is not sensitive to hyperparameter.

Prompt Design. We further examine the impact of both the number of prompt components and the prompt length on accuracy. In Fig. 5 (left), we measure accuracy across configurations with $\{10, 20, 50, 100, 200, 500\}$ prompts. The OS-prompt demonstrates a consistent accuracy enhancement as the number of prompts increases. Notably, OS-prompt++ reaches a performance plateau after just 50 prompts. Our ablation study of prompt length, presented in Fig. 5 (right), reveals that ours maintains stable accuracy across various prompt lengths.

Table 6. Analysis of QR loss design on 10-task ImageNet-R setting.

CosSim	Softmax	$A_N(\uparrow)$	$F_N(\downarrow)$
		75.00 ± 0.53	1.68 ± 0.12
	✓	75.47 ± 0.42	1.38 ± 0.16
✓		75.51 ± 0.33	1.28 ± 0.06
✓	✓	75.67 ± 0.40	1.27 ± 0.10

Table 7. Hyperparameter sensitivity study of λ on ImageNet-R 5/10/20-task.

λ	Task-5	Task-10	Task-20
1e-5	77.03 ± 0.10	75.63 ± 0.39	73.63 ± 0.21
5e-5	77.02 ± 0.13	75.62 ± 0.41	73.62 ± 0.19
1e-4	77.07 ± 0.15	75.67 ± 0.40	73.77 ± 0.19
5e-4	77.13 ± 0.24	75.68 ± 0.38	73.68 ± 0.17

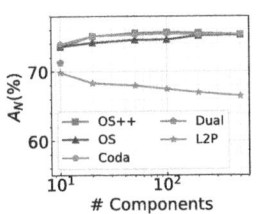
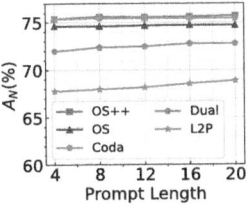
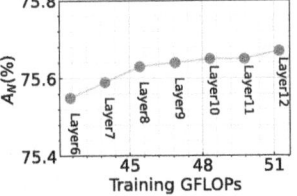

Fig. 5. Analysis of accuracy A_N with respect to prompt components (**Left**) and prompt length (**Right**). We use 10-taks ImageNet-R setting.

Fig. 6. Trade-off between GFLOPs and accuracy within a reference ViT.

5.6   Accuracy-Efficiency Trade-Off Within a Reference ViT

During the training of OS-prompt++, our method does not improve the energy efficiency (Note, we achieve ~50% computational saving during inference). This arises from our approach of extracting the reference prompt query r from the final layer of the reference ViT. To further enhance energy efficiency during the OS-prompt++ training, we delve into the trade-off between accuracy and efficiency within the Reference ViT. Instead of relying on the last layer's $[CLS]$ token embedding for the reference prompt query, we opt for intermediate token embeddings. Since our prompt pool is applied in layers 1 to 5 of the backbone ViT, we utilize the intermediate token embeddings from layers deeper than 5 within the reference ViT. Figure 6 illustrates this trade-off with respect to the layer index where we get the reference prompt. Our findings show that utill layer 8, there is only a slight increase in accuracy, which suggests a potential to reduce GFLOPs without performance drop.

Table 8. Analysis on prompt formation methods. Results on ImageNet-R for different task configurations: 5 tasks (40 classes/task), 10 tasks (20 classes/task), and 20 tasks (10 classes/task). A_N represents the average accuracy across tasks, and F_N indicates the mean forgetting rate. We average values over five runs with different seeds.

Setting	Task-5		Task-10		Task-20	
Method	$A_N(\uparrow)$	$F_N(\downarrow)$	$A_N(\uparrow)$	$F_N(\downarrow)$	$A_N(\uparrow)$	$F_N(\downarrow)$
UB	77.13	–	77.13	–	77.13	–
Deep L2P	73.93 ± 0.37	2.69 ± 0.10	71.66 ± 0.64	1.78 ± 0.16	68.42 ± 1.20	1.12 ± 0.13
OS-Prompt (L2P)	74.61 ± 0.29	2.29 ± 0.21	73.43 ± 0.60	1.42 ± 0.10	71.42 ± 0.24	0.95 ± 0.06
DualPrompt	73.05 ± 0.50	2.64 ± 0.17	71.32 ± 0.62	1.71 ± 0.24	67.87 ± 1.39	1.07 ± 0.14
OS-Prompt (Dual)	74.65 ± 0.15	2.06 ± 0.20	72.57 ± 0.23	1.13 ± 0.03	70.55 ± 0.48	0.81 ± 0.10
CodaPrompt	76.51 ± 0.38	2.99 ± 0.19	75.45 ± 0.56	1.64 ± 0.10	72.37 ± 1.19	0.96 ± 0.15
OS-Prompt (Coda)	75.74 ± 0.58	3.32 ± 0.31	74.58 ± 0.56	1.92 ± 0.15	72.00 ± 0.60	1.09 ± 0.11

5.7   Analysis on Prompt Formation Strategy

Here, we delve into the impact of varying prompt formation strategies on performance. For comparative insight, we present results from two previous PCL methodologies: L2P and DualPrompt. While L2P selects 5 prompts from the internal prompt pool, DualPrompt distinguishes between general and task-specific prompts. Aside from the distinct prompt formation strategies, the configurations remain consistent. In Table 8, we adopt the prompt formation strategies proposed in L2P and Dual, resulting in configurations denoted as OS-Prompt (L2P) and OS-Prompt (Dual). Our findings reveal the following: (1) The effectiveness of OS-prompt varies based on the prompt formation technique employed. Notably, OS-Prompt (L2P) and OS-Prompt (Dual) yield lower accuracy than our original approach, which relies on CodaPrompt. This suggests that our performance could benefit from refined prompt formation techniques in future iterations. (2) The OS-prompt framework, when integrated with L2P and Dual prompt formation strategies, outperforms the original Deep L2P and

DualPrompt, a trend not observed with CodaPrompt. This may indicate that the prompt formation strategies of L2P and DualPrompt, which rely on hard matching (i.e., top-k), are more resilient than CodaPrompt's. Conversely, CodaPrompt employs soft matching, utilizing a weighted summation of all prompts based on proximity.

6 Conclusion

We introduce the OS-Prompt framework where we improve the efficiency of the conventional two-stage PCL in a simple yet effective manner. One of the potential limitations of our OS-Prompt++ is, in comparison to OS-Prompt (our light version), it introduces an increase in training computational cost from the reference ViT feedforward. This computational efficiency during training becomes important in the context of addressing the online continual learning problem [10] where the model needs rapid training on the streaming data.

Acknowledgment. This work was supported in part by CoCoSys, a JUMP2.0 center sponsored by DARPA and SRC, the National Science Foundation (CAREER Award, Grant #2312366, Grant #2318152), and the DoE MMICC center SEA-CROGS (Award #DE-SC0023198).

References

1. Aljundi, R., Babiloni, F., Elhoseiny, M., Rohrbach, M., Tuytelaars, T.: Memory aware synapses: learning what (not) to forget. In: Ferrari, V., Hebert, M., Sminchisescu, C., Weiss, Y. (eds.) ECCV 2018. LNCS, vol. 11207, pp. 144–161. Springer, Cham (2018). https://doi.org/10.1007/978-3-030-01219-9_9
2. Buzzega, P., Boschini, M., Porrello, A., Abati, D., Calderara, S.: Dark experience for general continual learning: a strong, simple baseline. In: NeurIPS (2020)
3. Cai, H., Gan, C., Zhu, L., Han, S.: Tinytl: reduce memory, not parameters for efficient on-device learning. Adv. Neural Inf. Process. Syst. **33**, 11285–11297 (2020)
4. Caron, M., et al.: Emerging properties in self-supervised vision transformers. In: Proceedings of the IEEE/CVF International Conference on Computer Vision, pp. 9650–9660 (2021)
5. Chaudhry, A., Ranzato, M., Rohrbach, M., Elhoseiny, M.: Efficient lifelong learning with a-gem. arXiv preprint arXiv:1812.00420 (2018)
6. Chaudhry, A., et al.: On tiny episodic memories in continual learning. arXiv preprint arXiv:1902.10486 (2019)
7. Choi, Y., El-Khamy, M., Lee, J.: Dual-teacher class-incremental learning with data-free generative replay. In: Proceedings of the IEEE/CVF Conference on Computer Vision and Pattern Recognition, pp. 3543–3552 (2021)
8. Dosovitskiy, A., et al.: An image is worth 16 × 16 words: transformers for image recognition at scale. ICLR (2021)
9. Gao, Q., Zhao, C., Ghanem, B., Zhang, J.: R-DFCIL: relation-guided representation learning for data-free class incremental learning. In: Avidan, S., Brostow, G., Cissé, M., Farinella, G.M., Hassner, T. (eds.) Computer Vision – ECCV 2022. ECCV 2022. LNCS, vol. 13683, pp. 423–439. Springer, Cham (2022). https://doi.org/10.1007/978-3-031-20050-2_25

10. Ghunaim, Y., et al.: Real-time evaluation in online continual learning: a new hope. In: Proceedings of the IEEE/CVF Conference on Computer Vision and Pattern Recognition, pp. 11888–11897 (2023)
11. Hadsell, R., Rao, D., Rusu, A.A., Pascanu, R.: Embracing change: continual learning in deep neural networks. Trends Cogn. Sci. (2020)
12. Harun, M.Y., Gallardo, J., Hayes, T.L., Kanan, C.: How efficient are today's continual learning algorithms? In: Proceedings of the IEEE/CVF Conference on Computer Vision and Pattern Recognition, pp. 2430–2435 (2023)
13. Hayes, T.L., Cahill, N.D., Kanan, C.: Memory efficient experience replay for streaming learning. In: ICRA (2019)
14. He, X., Li, C., Zhang, P., Yang, J., Wang, X.E.: Parameter-efficient fine-tuning for vision transformers. arXiv preprint arXiv:2203.16329 (2022)
15. Hendrycks, D., et al.: The many faces of robustness: a critical analysis of out-of-distribution generalization. arXiv preprint arXiv:2006.16241 (2020)
16. Hendrycks, D., et al.: The many faces of robustness: a critical analysis of out-of-distribution generalization. In: ICCV, pp. 8340–8349 (2021)
17. Hsu, Y.C., Liu, Y.C., Ramasamy, A., Kira, Z.: Re-evaluating continual learning scenarios: a categorization and case for strong baselines. arXiv preprint arXiv:1810.12488 (2018)
18. Hu, E.J., et al.: Lora: low-rank adaptation of large language models. arXiv preprint arXiv:2106.09685 (2021)
19. Jia, M., et al.: Visual prompt tuning. In: Avidan, S., Brostow, G., Cissé, M., Farinella, G.M., Hassner, T. (eds.) Computer Vision – ECCV 2022. ECCV 2022. LNCS, vol. 13693, pp. 709–727. Springer, Cham (2022). https://doi.org/10.1007/978-3-031-19827-4_41
20. Khattak, M.U., Rasheed, H., Maaz, M., Khan, S., Khan, F.S.: Maple: multi-modal prompt learning. In: Proceedings of the IEEE/CVF Conference on Computer Vision and Pattern Recognition, pp. 19113–19122 (2023)
21. Krizhevsky, A., Hinton, G., et al.: Learning multiple layers of features from tiny images (2009)
22. Li, X., Zhou, Y., Wu, T., Socher, R., Xiong, C.: Learn to grow: a continual structure learning framework for overcoming catastrophic forgetting. In: ICML, pp. 3925–3934. PMLR (2019)
23. Li, Z., Hoiem, D.: Learning without forgetting. TPAMI 40(12), 2935–2947 (2017)
24. Lopez-Paz, D., Ranzato, M.: Gradient episodic memory for continual learning. NeurIPS (2017)
25. Mai, Z., Li, R., Jeong, J., Quispe, D., Kim, H., Sanner, S.: Online continual learning in image classification: an empirical survey. Neurocomputing 469, 28–51 (2022)
26. Pellegrini, L., Lomonaco, V., Graffieti, G., Maltoni, D.: Continual learning at the edge: real-time training on smartphone devices. arXiv preprint arXiv:2105.13127 (2021)
27. Peng, X., Bai, Q., Xia, X., Huang, Z., Saenko, K., Wang, B.: Moment matching for multi-source domain adaptation. In: Proceedings of the IEEE/CVF International Conference on Computer Vision, pp. 1406–1415 (2019)
28. Rebuffi, S.A., Bilen, H., Vedaldi, A.: Efficient parametrization of multi-domain deep neural networks. In: Proceedings of the IEEE Conference on Computer Vision and Pattern Recognition, pp. 8119–8127 (2018)
29. Rebuffi, S.A., Kolesnikov, A., Sperl, G., Lampert, C.H.: iCaRL: incremental classifier and representation learning. In: CVPR, pp. 2001–2010 (2017)
30. Rusu, A.A., et al.: Progressive neural networks. arXiv preprint arXiv:1606.04671 (2016)

31. Serra, J., Suris, D., Miron, M., Karatzoglou, A.: Overcoming catastrophic forgetting with hard attention to the task. In: ICML, pp. 4548–4557 (2018)

32. Smith, J., Hsu, Y.C., Balloch, J., Shen, Y., Jin, H., Kira, Z.: Always be dreaming: a new approach for data-free class-incremental learning. In: Proceedings of the IEEE/CVF International Conference on Computer Vision, pp. 9374–9384 (2021)

33. Smith, J.S., et al.: Coda-prompt: continual decomposed attention-based prompting for rehearsal-free continual learning. In: Proceedings of the IEEE/CVF Conference on Computer Vision and Pattern Recognition, pp. 11909–11919 (2023)

34. Smith, J.S., Tian, J., Halbe, S., Hsu, Y.C., Kira, Z.: A closer look at rehearsal-free continual learning. In: Proceedings of the IEEE/CVF Conference on Computer Vision and Pattern Recognition, pp. 2409–2419 (2023)

35. Tang, Y.M., Peng, Y.X., Zheng, W.S.: When prompt-based incremental learning does not meet strong pretraining. In: Proceedings of the IEEE/CVF International Conference on Computer Vision, pp. 1706–1716 (2023)

36. Voigt, P., Von dem Bussche, A.: The Eu general data protection regulation (GDPR). A Practical Guide, 1st Ed., Springer International Publishing, Cham (2017). **10**(3152676), 10–5555

37. Wang, Y., Huang, Z., Hong, X.: S-Prompts learning with pre-trained transformers: an Occam's razor for domain incremental learning. Adv. Neural Inf. Process. Syst. **35**, 5682–5695 (2022)

38. Wang, Z., et al.: SparCL: sparse continual learning on the edge. Adv. Neural Inf. Process. Syst. **35**, 20366–20380 (2022)

39. Wang, Z., et al.: DualPrompt: complementary prompting for rehearsal-free continual learning. In: Avidan, S., Brostow, G., Cissé, M., Farinella, G.M., Hassner, T. (eds.) Computer Vision – ECCV 2022. ECCV 2022. LNCS, vol. 13686, pp. 631–648. Springer, Cham (2022). https://doi.org/10.1007/978-3-031-19809-0_36

40. Wang, Z., et al.: Learning to prompt for continual learning. CVPR (2022)

41. Yin, H., et al.: Dreaming to distill: data-free knowledge transfer via deepinversion. In: Proceedings of the IEEE/CVF Conference on Computer Vision and Pattern Recognition, pp. 8715–8724 (2020)

42. Yoon, J., Yang, E., Lee, J., Hwang, S.J.: Lifelong learning with dynamically expandable networks. arXiv preprint arXiv:1708.01547 (2017)

43. Zaken, E.B., Ravfogel, S., Goldberg, Y.: Bitfit: simple parameter-efficient fine-tuning for transformer-based masked language-models. arXiv preprint arXiv:2106.10199 (2021)

44. Zenke, F., Poole, B., Ganguli, S.: Continual learning through synaptic intelligence. In: ICML (2017)

45. Zhang, J.O., Sax, A., Zamir, A., Guibas, L., Malik, J.: Side-Tuning: a baseline for network adaptation via additive side networks. In: Vedaldi, A., Bischof, H., Brox, T., Frahm, J.-M. (eds.) ECCV 2020. LNCS, vol. 12348, pp. 698–714. Springer, Cham (2020). https://doi.org/10.1007/978-3-030-58580-8_41

46. Zhang, R., et al.: Tip-adapter: training-free clip-adapter for better vision-language modeling. arXiv preprint arXiv:2111.03930 (2021)

47. Zhou, K., Yang, J., Loy, C.C., Liu, Z.: Learning to prompt for vision-language models. Int. J. Comput. Vis. **130**(9), 2337–2348 (2022)

SpaceJAM: a Lightweight and Regularization-Free Method for Fast Joint Alignment of Images

Nir Barel⬤, Ron Shapira Weber$^{(\boxtimes)}$⬤, Nir Mualem⬤, Shahaf E. Finder⬤, and Oren Freifeld⬤

The Department of Computer Science, Ben-Gurion University of the Negev, Beersheba, Israel
{banir,ronsha,nirmu,finders}@post.bgu.ac.il, orenfr@cs.bgu.ac.il

Abstract. The unsupervised task of Joint Alignment (JA) of images is beset by challenges such as high complexity, geometric distortions, and convergence to poor local or even global optima. Although Vision Transformers (ViT) have recently provided valuable features for JA, they fall short of fully addressing these issues. Consequently, researchers frequently depend on expensive models and numerous regularization terms, resulting in long training times and challenging hyperparameter tuning. We introduce the Spatial Joint Alignment Model (SpaceJAM), a novel approach that addresses the JA task with efficiency and simplicity. Space-JAM leverages a compact architecture with only ∼16K trainable parameters and uniquely operates without the need for regularization or atlas maintenance. Evaluations on SPair-71K and CUB datasets demonstrate that SpaceJAM matches the alignment capabilities of existing methods while significantly reducing computational demands and achieving at least a 10x speedup. SpaceJAM sets a new standard for rapid and effective image alignment, making the process more accessible and efficient. Our code is available at: https://bgu-cs-vil.github.io/SpaceJAM/.

Keywords: Joint Alignment · Congealing · Regularization-free · STN

1 Introduction

Joint alignment (JA) of images is the task of geometrically transforming an image collection in a way that optimizes some criterion of mutual similarity. JA is useful for reducing uninformative intra-dataset (or intra-class) variability, thereby facilitating more accurate analysis and interpretation across various applications,

---

N. Barel and R. S. Weber—Contributed equally.

Supplementary Information The online version contains supplementary material available at https://doi.org/10.1007/978-3-031-72624-8_11.

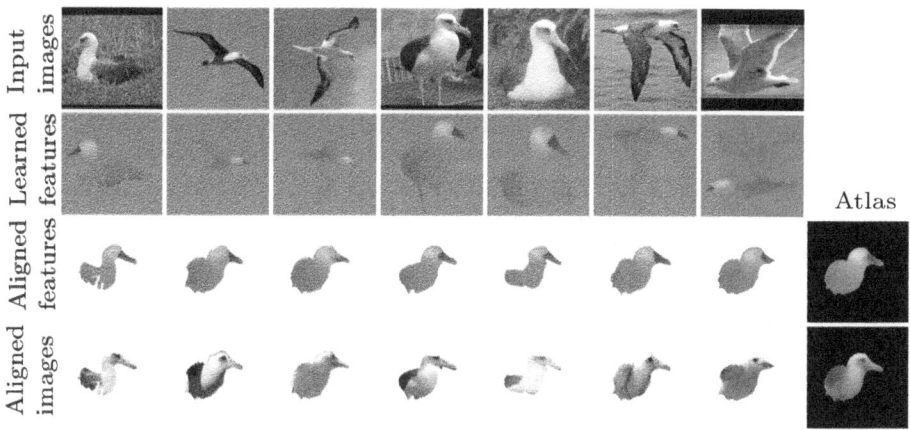

Fig. 1. SpaceJAM joint alignment Our framework jointly aligns a set of images of an object category in only a few minutes. Top-to-bottom: 1) input images; 2) learned low-dimensional representations; 3) aligned features; 4) aligned images. The last column depicts the average representation (atlas) obtained after training.

ranging from medical imaging to automated recognition systems. Unfortunately, achieving a successful JA is fraught with challenges, including poor local optima or even poor *global* optima (*i.e.*, trivial non-useful solutions; *e.g.*, shrinking all images to zero size), intra-class variations in pose/orientation/illumination, complex geometric deformations, visual clutter, and more. A JA method typically consists of four components: 1) features to be aligned; 2) a criterion of joint similarity; 3) an optimization process; 4) warping the images via either a parametric transformation family (*e.g.*, homographies) or dense pixel-to-pixel correspondences.

Traditionally, JA has relied on classical methods such as congealing [8,9,31, 44], which gradually aligns one image towards the rest, or alignment to a centroid which uses a reference image or a (possibly-latent) template. Other methods have utilized feature-based approaches, such as SIFT [40], to establish correspondences between images or key-points. While effective in certain domains (*e.g.*, medical scans), those classical approaches often struggle with more diverse image collections.

With the integration of Deep Learning (DL) into JA (*e.g.*, Chelly *et al.* [7]), especially via the use of Vision Transformers (ViT) [11] features, the field has witnessed a leap forward. ViTs, DINO specifically [6], offer rich semantic features that partially mitigate some of the challenges that hindered traditional JA methods. However, many issues persist even when using ViT features. This is arguably what led recent works [19,48] to over-rely on high-capacity-but-expensive models, as well as extensive regularization strategies. The latter, besides adding to the computational burden, require regularization hyperparameter tuning. Together, that approach led to methods that are slow and often too brittle.

We take a different approach. While we are happy to use DINO features, *we argue that feeding such features to high-capacity models is not, in fact, needed for the JA task and that rather than using such models here, it is better to explore what the best ways are to formulate, and then solve, this task.* With this in mind, this paper shows how to effectively, and efficiently, solve the JA task. To that end, we introduce the Spatial Joint Alignment Model (SpaceJAM), a regularization-free JA approach that does not require explicitly maintaining an atlas. See also Fig. 1. SpaceJAM utilizes a lightweight recurrent Spatial Transformer Network (STN) [25], a Lie-algebraic framework, and a novel loss function. The proposed approach not only ensures fast JA (*e.g.*, at least a 10x speedup in comparison to contemporary methods [19,48]) but also maintains the geometric integrity of the aligned images. We evaluate our method on the SPair-71K and CUB datasets, and show performance either better than, or on par with, state-of-the-art models but with significantly reduced computational demands, marking a significant advancement in the field of JA.

To summarize, our key contributions are as follows: 1) we introduce SpaceJAM, a lightweight method for JA of in-the-wild images which is much faster than contemporary approaches; 2) a novel inverse-compositional loss function that obviates the need of using regularization terms and does not require maintaining an atlas; 3) an efficient solution for handling reflections (also known as flips).

2   Related Work

Classical JA Methods focused on techniques that leverage geometric transformations and hand-crafted features. Congealing, a seminal concept introduced in [31,44], refers to gradually aligning an image set by iteratively aligning one image towards the others, to minimize a global cost function, typically the entropy or least-squares of the raw pixel values [8,9,31] or SIFT descriptors [23,37,60]. Other approaches use clustering to simultaneously partition the data and align all class members to their respective mean [38,43] or via template matching [13,17,26]. The success of those classical JA methods, however, was hindered by the quality of the to-be-aligned features. Another classical approach seeks to model an image set as a low-rank linear subspace[20,29,68], the most noticeable work being RASL [52], whose main problem lies in its reliance on a good initial alignment.

Deep Learning. The advent of deep learning has significantly expanded the capabilities of image alignment techniques. Huang *et al.* applied the classical congealing algorithm to the features of a Convolutional Neural Net (CNN) [22]. The Spatial Transformer Net (STN) [25] introduced a differentiable module that can be integrated into larger neural nets to predict and apply spatial transformations to input images or feature maps, enabling end-to-end learning of the alignment process. Since their inception, STNs have proven an invaluable tool for the JA task and were used in both the congealing framework [3], explicit Atlas building [10,34,61,65], joint clustering and alignment tasks [39,46], and for a JA step before building moving-camera background models [7,12]. Coupled with Generative Adversarial Networks (GAN) [18], STNs were also shown to produce high-fidelity class-category canonical spaces (or atlases) for natural images [47,51]. However, GAN-based methods requires large amount of training data.

Table 1. A comparison with recent JA methods (SPair's 'Cat' category [45]).

Method	# Params	# Losses	#HP	Atlas-free learning	#epochs	Time	PCK@0.10
NeuCongealing [48]	28.7 M	8	8	✗	8K	1:17:02	53.3
ASIC [19]	7.9 M	4	5	✗	20K	1:06:48	54.8
SpaceJAM (Ours)	**0.016M**	**1**	**0**	✓	**0.7K**	**0:05:58**	**60.8**

Semantic Correspondence Through Self-supervision. Self-supervised learning (SSL) approaches have recently gained traction for the task of correspondence discovery between images. By leveraging the representations learned by ViTs (such as DINO [6]) or text-to-image diffusion models [57], several works were able to use these features for the task of **pairwise** image correspondence and alignment [27,28,42,63,67]. While useful in the pairwise case, finding *joint dense correspondence* between N images is intractable for a large N, as the complexity is $O(N^2)$. Additionally, the Nearest Neighbor (NN) algorithm requires a comparison between every pixel in the source image and all of the pixels in all of the other images. Finally, methods such as [63,67] require multiple t steps for the diffusion model, which further contributes to the computation burden. In contrast, our proposed JA method produces such a mapping without performing NN searches.

JA Using DINO Features. DINO features offer robust and semantically meaningful representations for many computer-vision tasks, JA included. The Neural Congealing algorithm [48] utilizes a test-time training framework to build an atlas for a given class (*e.g.*, birds) by matching DINO features via rigid and non-rigid warps, using a ResNet-based STN for each [21]. We remark that despite its name, Neural Congealing [48] is more related to atlas-based methods than to congealing methods. Such ambivalent terminology is prevalent, unfortunately, as sometimes people mistakenly refer to the JA *task* itself as congealing, regardless whether the JA *method* is congealing-based or atlas-based.

ASIC [19] uses DINO features to perform dense warping to map every pixel in the input to a canonical space using a U-net architecture [56]. However, both [48] and [19] grapple with computational overhead, stability issues, and the necessity for extensive regularization to prevent model collapse or trivial solutions. Additionally, ASIC's dense warping approach is prone to incoherent global alignment, often resulting in fragmented or an inconsistent alignment. In contrast, SpaceJAM reaches competitive results much faster and with orders of magnitude fewer trainable parameters. See, *e.g.*, Table 1, as well as the supplementary material (**SupMat**) for additional running times.

In Summary, while classical methods have provided a strong foundation for image alignment, DL has significantly enhanced the ability to handle complex and varied alignment tasks. The move towards SSL and semantic correspondence methods further illustrates the evolving landscape of the image alignment field, highlighting ongoing challenges and the need for innovative solutions to

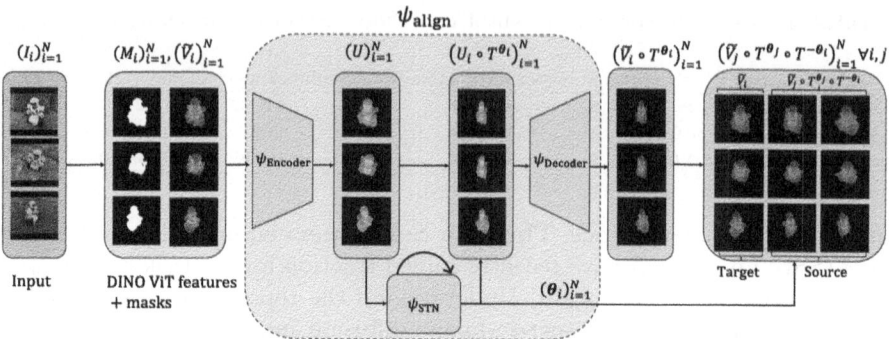

Fig. 2. Framework overview. Given a set of images,$(I_i)_{i=1}^{N}$, their DINO-ViT representations and coarse masks , $(V_i, M_i)_{i=1}^{N}$, SpaceJAM learns an inverse-compositional pairwise alignment between each image pair, and consequently, features, $(U_i)_{i=1}^{N}$, in a shared semantic space, where they are warped (according to learned warping parameters, $(\theta_i)_{i=1}^{N}$) to produce their aligned versions, $(U_i \circ T^{\theta_i})_{i=1}^{N}$. Pairwise alignment of I_j to a target image I_i, is achieved by warping to the shared space and then wapring the result by the inverse transformation of the target image, yielding $I_j \circ T^{\theta_j} \circ T^{-\theta_i}$.

address computational efficiency, coherence, and optimization stability. These needs motivated our proposed framework, whose overview is depicted in Fig. 2.

3   Background: Typical Challenges in Joint Alignment

Let $\mathcal{T}$ be a family of spatial transformations, from $\mathbb{R}^2$ to $\mathbb{R}^2$. For concreteness, our notation below assumes $\mathcal{T}$ is parametric (as is the case, *e.g.*, for affine transformations or homographies) but most of the discussion holds for the nonparametric case as well. Let $T^\theta \in \mathcal{T}$ denote the generic element of $\mathcal{T}$ where θ is the corresponding parameter vector. Let (I_i, I_j) be an image pair. The **pairwise alignment** problem is

$$\underset{T^{\theta_{ji}} \in \mathcal{T}}{\arg\min} \, D(I_i, I_j \circ T^{\theta_{ji}}) \tag{1}$$

where $\circ$ is function composition, $T^{\theta_{ji}}$ is the transformation warping I_j towards I_i, and D is a discrepancy measure, based on raw pixels or image features.

Joint alignment. Given N images, $(I_i)_{i=1}^{N}$, the underlying notion behind the JA problem (see, *e.g.*, [16]) is that for each observed image I_i there is a latent transformation, $T^{\theta_i} \in \mathcal{T}$, such that $(I_i \circ T^{\theta_i})_{i=1}^{N}$ are essentially different realizations of some shared canonical representation. In other words, the JA solution is the set of warping parameters, $(\theta_i)_{i=1}^{N}$, that warps the entire image ensemble to some shared representation, usually referred to as an *atlas* (or *prototypical image*). Thus, the JA problem is often formulated as finding a latent atlas, I_μ, together with the aforementioned parameters:

$$\arg\min_{I_\mu,(T^{\theta_i})_{i=1}^N \in \mathcal{T}} \sum\nolimits_{i=1}^{N} D(I_\mu, I_i \circ T^{\theta_i}). \tag{2}$$

JA Approach #1: Alignment to a Shared Atlas. One JA approach relies on building a shared atlas, I_μ, that minimizes Eq. 2. The atlas could be a learnable parameter [46] or the average of the warped images [7,13,26,52], $I_\mu = \frac{1}{N}\sum_{i=1}^{N} I_i \circ T^{\theta_i}$. A key challenge in that approach is that since the $(\theta_i)_{i=1}^N$ are unknown, so is I_μ and thus the latter must be found during the optimization process together with $(\theta_i)_{i=1}^N$. Note that the optimization may also be amortized via the training of a neural net [12].

Unfortunately, Eq. 2 can be undesirably minimized by, *e.g.*, (i) removing all images from the domain of I_μ or (ii) severely shrinking or stretching the images. Both (i) and (ii) represent poor *global* minima: the nonnegative loss becomes zero, yet the solution is useless. Additionally, the process heavily relies on a good initial alignment of the set and is prone to converge to poor local minima, which could be visually seen as a blurred representation of the set. See Erez *et al.* [12] for a detailed discussion on problems associated with that approach. One possible remedy to the issues above is to use a regularization over the predicted transformations (not to be confused with regularization over the network's weights). That is, the JA problem becomes:

$$\arg\min_{I_\mu,(T^{\theta_i})_{i=1}^N \in \mathcal{T}} \sum\nolimits_{i=1}^{N} D(I_\mu, I_i \circ T^{\theta_i}) + \mathcal{R}(T^{\theta_i}; \lambda), \tag{3}$$

where the regularization term, $T^{\theta_i} \mapsto \mathcal{R}(T^{\theta_i}; \lambda)$ is parameterized by *hyperparameters* (HP), λ. For instance, one may hope, against hope, that penalizing large deviations from the identity transformation would guide the optimization process to a more plausible atlas. However, *there is a major problem here whose severity is often swept under the rug*: the need to determine good values for λ.

JA Approach #2: Congealing. The discussion above leads to another JA approach: congealing [31,44]. The congealing algorithm seeks to minimize some JA criterion (*e.g.*, in its original formulation it was entropy) by iteratively warping each single image towards the other images but only after those images were already warped using the current estimates of their respective transformations. *One appealing property of congealing is that it defines a notion of central tendency of the data without having to explicitly build an atlas or to rely on a reference image* [31]. Congealing avoids some of the pitfalls of aligning to a shared atlas, though sometimes regularization is still used. It was later shown that least squares congealing (LSC) has several advantages over the entropy-based algorithm in terms of optimization and convergence [8]. Finally, the clever Inverse-Compositional LSC (IC-LSC) method was introduced to handle outlier images and objects being warped outside of the image domain when applying a single warp to the entire set [9]. Although not DL-based, IC-LSC is, in some sense, the closest to our approach.

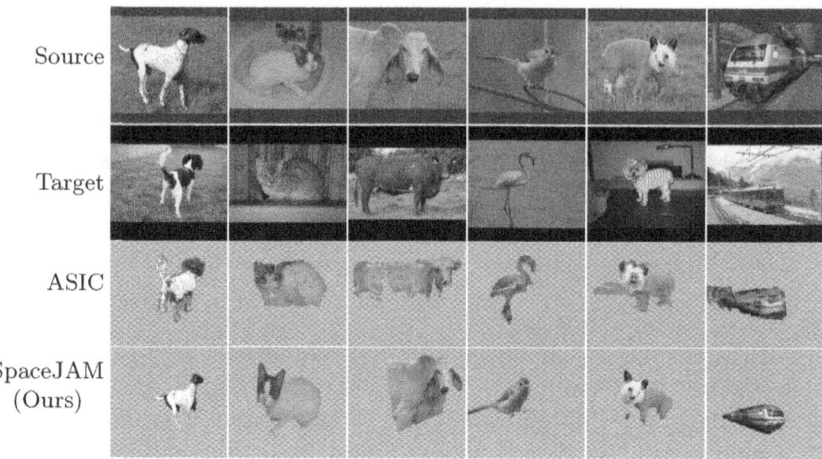

Fig. 3. Global alignment vs. dense warping evaluated on several classes of the SPair-71 K dataset [45]. A source image (1st row) is mapped to the target image (2nd row) via either dense warping (3rd row) achieved by ASIC [19] (as presented in their paper) or a parametric alignment by SpaceJAM (ours, 4th row). Dense mapping is prone to produce incoherent results (please zoom in to better see that effect) and heavily relies on regularization. In comparison, our regularization-free method produces geometrically-coherent results and is also much faster.

Regularization-based DL Approaches for Joint Alignment. The aforementioned challenges in solving the JA problem still exist even when considering rich semantically-meaningful representations such as DINO [6]. Contemporary methods handle these issues by employing extensive regularization on the predicted warps and/or learned atlas. Consider ASIC [19], for instance, which seeks to densely map every input pixel to a canonical shared space. Dense mapping strategies are inherently susceptible to breaking the object geometry (as shown in Fig. 3). To mitigate this problem, the authors of [19] enforce the following regularization and auxiliary losses: (1) an equivariance loss to handle geometric variations, (2) total-variation regularization to encourage smooth mappings, (3) consistent part alignment to map semantically similar parts, combined with additional two losses that are not directly related to the object geometry, for a total of 5 regularization terms (while [48] has 6 reg. terms and 8 HPs). Each loss term requires its own HP, λ. *The reason that the crucial dependency on regularization terms, let alone so many of them, is much worse than one might expect is threefold.* First, the optimal choice of λ is usually dataset-specific and this renders such methods either too brittle or limited. Second, even on a single dataset, determining a good value typically requires some supervision – at odds with the unsupervised nature of the JA task. Third, the need to have several (sometimes many) loss terms, can complicate and prolong the optimization process. Thus, it is unsurprising that current DL-based JA methods such as [19,48] must use expensive architectures and are slow.

4   Method

Given a collection of images, $(I_i)_{i=1}^N$, along with their DINO representations, $(V_i)_{i=1}^N$, of an object (*e.g.*, a particular bird) or an object class (*e.g.*, multiple birds), our goal is to jointly align the images in a semantically-meaningful manner. Before describing the proposed method we emphasize that all of our design choices below were made with the goal of making the JA process easier, faster, and more stable. Together, those choices have a surprisingly strong synergistic effect, yielding a lightweight JA method that is *much* faster and its number of trainable parameters is $< 1\%$ of the numbers in contemporary methods.

4.1   Preprocessing

The goal of our preprocessing procedure is to extract object masks and reduce the dimensionality of the DINO features. Similarly to [19,48], and to keep the comparison with those methods fair, we use [2] to produce the saliency-based **masks** for the image collection. Let M_i denote the binary mask for I_i. Next, we apply **dimensionality reduction** to the DINO features. Let $V_i \in \mathbb{R}^{d \times H \times W}$ where d is typically high (*e.g.*, $d = 384$) and (H, W) are the (height,width) of the DINO representation. Our dimensionality reduction has two steps.

To reduce the computational burden, we apply Principal Component Analysis (PCA) using the first K principal components, where $K < d$ (in our experiments, $K = 25$). Let $\hat{V}_i \in \mathbb{R}^{K \times H \times W}$ denote the PCA representation of V_i, and let $\tilde{V}_i = \hat{V}_i \cdot M_i$ denote its masked version.

In the second step, we train a fully convolutional, low-capacity autoencoder (only $\sim$3K parameters) to reduce the number of channels to 3. We chose 3 since it is low enough to serve as an information bottleneck that removes redundant details and since it allows for interpretability when visualizing the encoded features using RGB colors. The AE is trained to reconstruct $\tilde{V}_i$ using the following loss,

$$\mathcal{L}_{\text{AE}} = \sum_{i=1}^N \left\| \tilde{V}_i - \psi_{\text{dec}}(\psi_{\text{enc}}(\tilde{V}_i)) \right\|_{\ell_2}^2 . \tag{4}$$

As we explain later, the obtained latent representation of the AE, $U_i \triangleq \psi_{\text{enc}}(\tilde{V}_i) \in \mathbb{R}^{3 \times H \times W}$, will serve a purpose in our alignment network. We find that the combination of PCA and AE to reduce the input's number of channels is highly effective from both computational and representational points of view.

4.2   Alignment Network

Our alignment network, ψ_{align}, is based on an STN [25]. Below we provide a brief overview of that STN and how we handle common pitfalls in STN training. Let ψ_{STN} denote the STN and, for now, let X denote its generic input. The core of the STN is the so-called **localization network**, denoted by ψ_{loc}, that predicts transformation parameters from X; *i.e.*, $\theta = \psi_{\text{loc}}(X)$. The output of the entire

STN is $(\theta, X \circ T^\theta) = \psi_{\text{STN}}(X)$. That is, the output is both θ and the warped image obtained by applying T^θ to X. STNs have been shown to be an effective tool for various tasks, albeit difficult to optimize [34, 62]. The issue stems from a few reasons, which we address in a theoretically-grounded manner.

The first difficulty is in predicting large transformations. To address this issue, the Inverse-Compositional STN (IC-STN) [34] was proposed, which, based on the classical Inverse-Compositional Lucas & Kanade algorithm [41], predicts a cascade of smaller warps and composes them. In our case, we do so by recursively feeding the STN several times with its own output (see **SupMat** for details). Empirically, 5 recurrences suffice (a higher number yielded a very small gain) so this is what we used in our experiments. Since the recurrences share the same STN, the number of trainable parameters does not increase with the number of recurrences. As we apply several transformations in cascade, each predicted transformation can be relatively small. A fine nuance is that while usually each STN call involves interpolation (due to the resulting irregular grid), here whenever we compose transformations we perform only the last interpolation; this improves the quality of the resulting image as fewer interpolation artifacts occur. Our entire model, ψ_{align}, consists of the AE and 5 recurrences of ψ_{STN}.

4.3   Lie Algebras, Lie Groups, and the Matrix Exponential

Another difficulty is that, without special care, an STN might predict non-invertibile transformation matrices, which could hinder the optimization process; *e.g.*, [62] showed that restricting an affine STN to invertible affine transformations allows for more robust and stable training. Thus, our alignment network, ψ_{align}, uses a cascade of transformations that are members of a Lie group. This formulation is more stable and robust compared to the vanilla STN. We represent spatial transformations as elements of Lie groups via a Lie-algebraic parameterization, as we now explain. Consider these 6 spaces of 3-by-3 real matrices:

$$\text{se}(2) = \left\{ A : A = \begin{bmatrix} 0 & \theta_2 & \theta_3 \\ -\theta_2 & 0 & \theta_6 \\ 0 & 0 & 0 \end{bmatrix} \right\}, \ \text{SE}(2) = \left\{ T : T = \begin{bmatrix} T_1 & T_2 & T_3 \\ T_4 & T_5 & T_6 \\ 0 & 0 & 1 \end{bmatrix} \& \begin{bmatrix} T_1 & T_2 \\ T_4 & T_5 \end{bmatrix} \in \text{SO}(2) \right\},$$

$$\text{aff}(2) = \left\{ A : A = \begin{bmatrix} \theta_1 & \theta_2 & \theta_3 \\ \theta_4 & \theta_5 & \theta_6 \\ 0 & 0 & 0 \end{bmatrix} \right\}, \ \text{Aff}(2) = \left\{ T : T = \begin{bmatrix} T_1 & T_2 & T_3 \\ T_4 & T_5 & T_6 \\ 0 & 0 & 1 \end{bmatrix} \& \det T > 0 \right\},$$

$$\text{sl}(3) = \left\{ A : A = \begin{bmatrix} \theta_1 & \theta_2 & \theta_3 \\ \theta_4 & \theta_5 & \theta_6 \\ \theta_7 & \theta_8 & -(\theta_1+\theta_5) \end{bmatrix} \right\}, \ \text{SL}(3) = \left\{ H : H = \begin{bmatrix} H_1 & H_2 & H_3 \\ H_4 & H_5 & H_6 \\ H_7 & H_8 & H_9 \end{bmatrix} \& \det H = 1 \right\}.$$

The shortcuts SE (or se) Aff (or aff), SL (or sl), and SO stand for "Special Euclidean", "Affine", "Special Linear", and "Special Orthogonal", respectively. The following are well-known mathematical facts, widely used in computer vision (see, *e.g.*, [7, 12, 35, 36]): 1) se(2), aff(2), and sl(3) are *linear* spaces, called (matrix) **Lie algebras**. 2) SE(2), Aff(2), and SL(3) are *nonlinear* spaces. Moreover, they are (matrix) **Lie groups**; namely, each of these spaces is both a group (the binary operator being matrix multiplication) and a smooth manifold, and the group operations (multiplication; inversion) are smooth. 3) Each element of each of those groups is an **invertible matrix**. 4) **Nesting property:**

se(2) is a linear subspace of aff(2) and the latter is a linear subspace of sl(3). Likewise, SE(2) is a matrix subgroup of Aff(2) and the latter is a matrix subgroup of SL(3). 5) SE(2) is the group of rigid-body transformations in $\mathbb{R}^2$. Aff(3) is the group of (orientation-preserving) **invertible affine transformations** in $\mathbb{R}^2$. SL(3), the group of volume-preserving linear transformations in $\mathbb{R}^3$, can also be identified with the group of **homographies** in $\mathbb{R}^2$ (this is why we denoted its generic element by H). 6) The **matrix exponential** maps se(2) → SE(2), aff(2) → Aff(2), and sl(3) → SL(3).

The matrix exponential provides a differentiable map from a linear space (the algebra) into its corresponding nonlinear space (the group). This implies that, using the Lie-algebraic parametrization, an STN [25] can be easily restricted to yield only elements of the Lie group of interest [62]. This is useful since while optimization over nonlinear manifolds can be hard (*e.g.*, even a simple gradient-descent step might bring us outside the manifold), the Lie-algebraic parametrization, together with the matrix exponential, lets us perform gradient-based optimization in a linear space (for other methods for optimization on manifolds, see Boumal's textbook [4]) In fact, [62] showed that an STN restricted to *invertible* affine transformations was more stable and yielded better results than the commonly-used unrestricted STN. In our case, there is also an additional reason why we use Lie groups and this is the fact that our approach relies directly on Lie groups' closure under inversion and composition (as detailed below).

4.4   Loss Function

Recall that, for image I_i, we let V_i, U_i, and M_i denote its DINO representation, an encoded 3-channel representation, and mask map, respectively. We train ψ_{align} to find the JA of $(I_i)_{i=1}^{N}$ in a forward-inverse compositional manner. Part of the motivation behind our loss is to avoid the drawbacks of explicitly warping the images to a shared space (see Sect. 3). In particular, we avoid using regularization terms over the transformations. Our loss consists of two conceptual steps: (i) for each image, I_i, in batch k – denoted by $B_k = (I_i)_{i=1}^{N_b}$ (where N_b is the # of images in the batch) – predict the forward warping parameters, θ_i, using U_i as the recurrent STN's input, and then (ii) compose the forward warp with the inverse warp of every other image in the batch, to build the following loss:

$$\mathcal{L}_{\text{IC}} = \sum_{i=1}^{N_b} \sum_{j:j\neq i} \left\| \tilde{V}_j - (\tilde{V}_i \circ T^{\theta_i} \circ T^{-\theta_j}) \right\|_{\ell_2}^2 . \tag{5}$$

Minimizing this loss implicitly maps each image to the shared space, since

$$\tilde{V}_j \approx \tilde{V}_i \circ T^{\theta_i} \circ T^{-\theta_j} \Leftrightarrow \tilde{V}_j \circ T^{\theta_j} \approx \tilde{V}_i \circ T^{\theta_i} . \tag{6}$$

The loss in Eq. 5 relies on Lie groups' closure under inversion (*i.e.*, the matrix associated with T^θ is invertible) and composition. Notably, the proposed loss effectively frees us from having to use regularization. This is because, by construction, a "bad" predicted transformation would directly increase the loss as the corresponding warped image would fail to "explain away" the other images, unless all the predicted transformations would be "bad" in *exactly* the same way (which is unlikely and indeed never happens in practice).

Training is done via standard DL gradient-based optimization, except we also use a Lie-algebraic curriculum learning to ease the optimization further. Meaning, exploiting the nested structure of the Lie algebras, we start the training with SE(2) and end with homographies. See **SupMat** for details. After training, SpaceJAM jointly aligns the entire image collection using its forward pass.

Reflections. The outline of our proposed solution for handling reflections (*i.e.*, flips), whose full details are in our **SupMat**, is as follows. During training, we dynamically change the selected flip configuration and train on all configurations simultaneously (this yields smooth and robust training) but save computations and increase stability by *calculating gradients only for the best flip configuration*.

Atlas Building. Upon learning the JA, the method can also quickly generate an atlas as follows. Given the images $(I_i)_{i=1}^N$, we predict the forward warping parameters via SpaceJAM's forward pass and obtain the aligned images via $(I_i \circ T^{\theta_i})_{i=1}^N$. The atlas can be taken as the sample mean of either the warped features, $\frac{1}{N}\sum_{i=1}^N V_i \circ T^{\theta_i}$, or the warped PCA-related versions, $\frac{1}{N}\sum_{i=1}^N \tilde{V}_i \circ T^{\theta_i}$, Either way results in a semantically-meaningful atlas; see, Fig. 4.

4.5 Implementation

We use simple architectures. ψ_{loc} is a 2-layer CNN with $\sim$13 K trainable parameters and ψ_{AE} is a fully-convolutional AE with $\sim$3 K trainable parameters, so the total is only 16 K. The final layer predicts 8 parameters (the number of degrees of freedom in a homography) in total. We predict a cascade of 5 warps in total, where ψ_{align} is shared throughout the process (akin to standard IC-STN [34]). SpaceJAM is implemented in PyTorch [50] and optimized for 300 epochs for ψ_{AE} and then an additional 400 for the entire framework ($\psi_{\text{AE}}+\psi_{\text{align}}$), using the Adam optimizer [30] and a StepLR scheduler ($\gamma = 0.9$, every 50 epochs).

5 Results

We evaluate our method on several real-world benchmarks, containing object categories with different illuminations, visual clutter, and occlusions, and compare it with several JA methods. The **Datasets** include **SPair-71 k** [45], which consists of 1,800 images from 18 categories (*i.e.*, bird). As is common in the weakly-supervised test-time optimization scenario, we train (in a weakly-supervised manner; *i.e.* only the categories are known, but the latent alignments are not) SpaceJAM on each of the SPair-71 k test sets independently and report results for each category. For **CUB-200** [64], we 1) report results on its first 3 categories, consisting of 25–30 images each when comparing with [19] and 2) the average of 14 subsets, following the protocol in [48,51]. For a fair comparison, in the quantitative comparison our method, like [19,48], used DINOv1 (ViT-S/8) [6]. For completeness and future comparisons, we also report our results with DINOv2 [49].

5.1 Qualitative Results

Recall that SpaceJAM is trained in an inverse-compositional framework which first aligns an image to a shared space and then warps it to all other images in the batch using the inverse transform of the target. This allows us to compute (after the training is done) a shared atlas without explicitly maintaining one during training. Examples of our method's JA and resulting atlas can be seen in Fig. 1 and Fig. 4. Figure 3 provides a visual comparison between SpaceJAM and ASIC in terms of pairwise alignment (for AISC, we use the visual results appearing in [19]). Evidently, SpaceJAM achieves high fidelity and geometrically-coherent alignment under challenging conditions. Figure 4 shows how the learned low-dimensional features ($U_i \in \mathbb{R}^{3 \times H \times W}$; second row) allow us to perform both pairwise alignment (3rd row) and JA (4th row) in an interpretable manner, where semantically-related parts (e.g., faces) share the same color across images. The visual results indicate that SpaceJAM can align diverse collections of images efficiently and accurately (see the **SupMat** for more visual results).

5.2 Quantitative Evaluation

We assess our method on the semantic point correspondence task using the SPair-71 k [45] and CUB-200 [64] datasets, which are widely recognized benchmarks

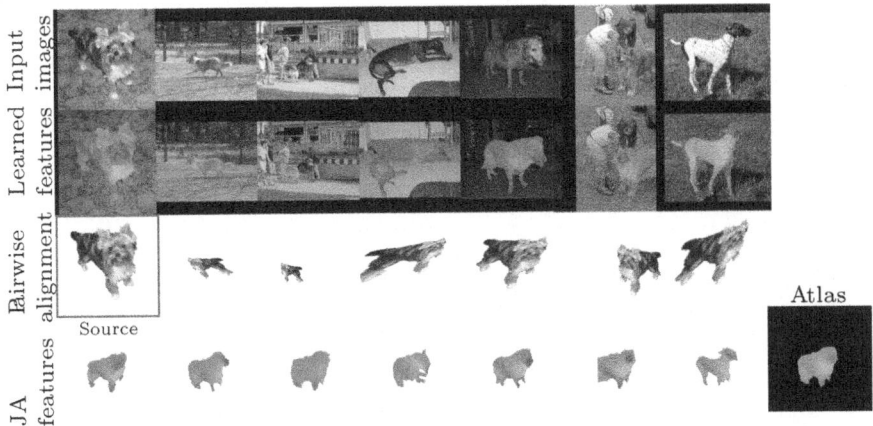

Fig. 4. Pairwise and joint alignment using SpaceJAM. The input images (1th row) overlayed by the learned features (2nd row) which are used to predict the warping parameters. The 3rd row shows source-to-target alignment under severe conditions and the 4th row shows the jointly aligned features and the category atlas (right column).

in both pairwise and JA domains. The performance is quantified by the PCK-Transfer metric, which calculates the proportion of keypoints accurately aligned within a threshold of $\alpha \cdot \max(h, w)$ relative to the true position. Following established protocols [48], $\alpha = 0.1$ (denoted as PCK@0.10) for both datasets, with (h, w) representing the dimensions of the object's bounding box. We perform pairwise alignment using Eq. 6.

In line with the comparative framework established by [19], we categorize existing methods into 3 groups: (1) *Strong supervision*, which relies on manually annotated keypoints; (2) *GAN supervision*, using category-specific GANs and; (3) *Weak supervision*, using category-level supervision without explicit keypoints. A subset of these methods adopts a train/test paradigm, using a large amount of data. Unlike those works, SpaceJAM involves training a compact model and leverages a test-time optimization scheme. The baseline scores, obtained from [19], include: 1) DINO nearest neighbor (DINO+NN; [2]); 2) Neural Congealing [48], which builds a shared atlas; and 3) ASIC [19] which densely maps pixels to a canonical grid (both are weakly-supervised JA methods).

Table 2 shows PCK@0.1 for all classes of the SPair-71k dataset [45]. We report the average of 3 runs for SpaceJAM. Our method performs best on average for all object classes (**All**) in the *weak supervision* category, using a much lighter model that trains 10x faster. It also performs best and second-best in 9 out of 18 classes. The results are consistent for both rigid objects (*i.e.*, 'Train') and ones with extreme variations ('Cat') with the exception of the 'Chair' category, where it was unable to overcome the initial coarse masks provided by [2]. Table 3 shows the results on the CUB-200 dataset [64] as reported in [19], where our method

Table 2. SPair-71 k Dataset results: Per-class and overall mean PCK@0.10 evaluated on the test set. The best result among *weakly-supervised* methods is boldfaced (the runner-up is underlined). (⋆) indicates that the method used a reference image. Missing values (−) indicate unreported results (we, like [19], found that running the code from [48] usually did not converge successfully).

Supervision	Method	Aero	Bike	Bird	Boat	Bottle	Bus	Car	Cat	Chair	Cow	Dog	Horse	Motor	Person	Plant	Sheep	Train	TV	All
Strong Supervision	SCorrSAN [24]	57.1	40.3	78.3	38.1	51.8	57.8	47.1	67.9	25.2	71.3	63.9	49.3	45.3	49.8	48.8	40.3	77.7	69.7	55.3
GAN supervision	GANgealing [51]	–	37.5	–	–	–	–	–	67.0	–	–	23.1	–	–	–	–	–	–	57.9	–
Weak supervision (train/test)	CNNGeo [53]	23.4	16.7	40.2	14.3	36.4	27.7	26.0	32.7	12.7	27.4	22.8	13.7	20.9	21.0	17.5	10.2	30.8	34.1	20.6
	A2Net [59]	22.6	18.5	42.0	16.4	37.9	<u>30.8</u>	26.5	35.6	13.3	29.6	24.3	16.0	21.6	22.8	<u>20.5</u>	13.5	31.4	36.5	22.3
	WeakAlign [54]	22.2	17.6	41.9	15.1	38.1	27.4	27.2	31.8	12.8	26.8	22.6	14.2	20.0	22.2	17.9	10.4	32.2	35.1	20.9
	NCNet [55]	17.9	12.2	32.1	11.7	29.0	19.9	16.1	39.2	9.9	23.9	18.8	15.7	17.4	15.9	14.8	9.6	24.2	31.1	20.1
	SFNet [32]	26.9	17.2	45.5	14.7	**38.0**	22.2	16.4	<u>55.3</u>	13.5	33.4	27.5	17.7	20.8	21.1	16.6	15.6	32.2	35.9	26.3
	PMD [33]	26.2	18.5	48.6	15.3	**38.0**	21.7	17.3	51.6	13.7	34.3	25.4	18.0	20.0	24.9	15.7	16.3	31.4	<u>38.1</u>	26.5
	PSCNet-SE [28]	28.3	17.7	45.1	15.1	37.5	30.1	<u>27.5</u>	47.4	14.6	32.5	26.4	17.7	24.9	24.5	19.9	16.9	34.2	37.9	27.0
Weak supervision (test-time optimization)	VGG+MLS [1]	29.5	22.7	61.9	**26.5**	20.6	25.4	14.1	23.7	14.2	27.6	30.0	29.1	24.7	27.4	19.1	19.3	24.4	22.6	27.4
	DINO+MLS [1,5]	49.7	20.9	63.9	19.1	32.5	27.6	22.4	48.9	14.0	36.9	39.0	30.1	21.7	41.1	17.1	18.1	35.9	21.4	31.1
	DINO+NN [2]	<u>57.2</u>	24.1	<u>67.4</u>	24.5	26.8	29.0	27.1	52.1	<u>15.7</u>	<u>42.4</u>	<u>43.3</u>	30.1	23.2	40.7	16.6	24.1	31.0	24.9	33.3
	NeuCongeal [48]	–	29.1*	–	–	–	–	–	53.3	–	–	35.2	–	–	–	–	–	–	–	–
	ASIC [19]	**57.9**	<u>25.2</u>	**68.1**	<u>24.7</u>	35.4	28.4	**30.9**	54.8	**21.6**	**45.0**	**47.2**	<u>39.9</u>	26.2	**48.8**	14.5	<u>24.5</u>	<u>49.0</u>	24.6	<u>36.9</u>
	SpaceJAM (DINOv1, ViT-S)	43.4	22.0	34.5	18.8	32.4	**36.3**	24.1	**60.8**	06.5	37.1	35.9	**46.7**	**53.6**	46.9	**36.0**	**24.6**	**75.0**	**46.0**	**37.8**
	SpaceJAM (DINOv2, ViT-B)	52.9	48.9	48.1	25.1	45.8	53.7	47.9	61.2	18.7	40.5	51.5	49.8	58.3	56.3	31.8	42.7	59.9	59.7	42.0
	SpaceJAM (DINOv2, ViT-L)	53.6	53.4	45.4	47.5	71.0	66.8	59.5	62.7	19.8	59.2	56.2	63.8	62.1	59.6	33.1	48.1	68.0	61.6	45.7

is the runner-up after [19]. That dataset consists of only birds and offers less variety than SPair-71 K. When comparing to [48,51], on 14 subsets, our method outperforms both. Please see our **SupMat** for further discussion of the results.

5.3   Complexity

SpaceJAM is a lightweight model, consisting of only 16K (*i.e.*, 0.016 M) trainable parameters. We compare our method to [19,48] using the official implementations. As Table 1 shows, SpaceJAM's # of trainable parameters is *much* lower than those of [48] and [19] which are 28.7 M and 7.9 M, respectively. Table 1 also shows that SpaceJAM achieves convergence within only 300 AE pretraining epochs and 400 AE+STN epochs, translating to a training time of $\approx$ 6 minutes on a single RTX4090, which starkly contrasts with the *hours* required by the competing methods. Despite its compact design, SpaceJAM attains a PCK@0.1 score of 60.8 on the 'Cats' dataset, outperforming both competitors. Evidently, SpaceJAM's efficiency does not hurt performance, as it not only accelerates the training process but also enhances alignment accuracy.

5.4   Ablation Study & Limitations

Ablation. We conducted an ablation study for SpaceJAM (using DINOv2 [49]) on subsets of the CUB-200 and SPair-71k datasets and evaluated them using PCK@0.1, as shown in Table 4. Due to the low-capacity of the models used in SpaceJAM, the AE pre-training proves essential before the JA. Predicting a cascade of 5 small transformations shows significant improvement over a single large one (STN) or three (ICSTN-3), when the # of trainable parameters is the same. An alignment to an atlas, even when using the robust Inverse Consistency Averaging Error (ICAE [66]) performs worse than our inverse-compositional framework. Table 4 also shows that using the Lie groups is crucial and that the curriculum learning's effect, while positive, is less significant.

Table 3. A comparison on CUB-200.

Method	CUB-200 (3 cate.)	Method	CUB-200 (Subsets)
VGG+MLS [1]	25.8	–	–
DINO+MLS [1,5]	67.0	–	–
DINO+NN [2]	68.3	GANgealing [51]	56.8
ASIC [19]	**75.9**	NeuCongeal [48]	63.6
SpaceJAM	69.6	SpaceJAM	**69.9**

Table 4. Ablation Study.

Ablation	CUB-200 (3 cate.)	SPair-71k (3 cate.)
Alignment to atlas	58.5	54.6
No AE pre-training	54.9	56.6
No Curiculum	70.8	57.2
No Lie Groups	65.6	57.1
STN	61.6	43.5
ICSTN-3	72.6	57.6
Complete model	**73.3**	**58.1**

Limitations. SpaceJAM, while being an JA effective method, was not designed for pairwise dense correspondences. Thus, it was not optimized for maximizing the per-pixel agreement between an image pair. That said, Sect. 5 shows that the speed gains in speed are drastic and that the method is much faster than contemporary JA methods [19,48] while its dense mapping score (a standard evaluation metric for both pairwise and JA tasks) is still comparable (and in fact, in some cases better). This trade-off also comes into play in the choice of the transformation family. As with other STNs [25,58,62], it is possible to incorporate more expressive transformations such as diffeomorphisms (as in, *e.g.*, [14,15,62]) which are richer than homographies. Lastly, our performance also partially relies on the quality of the initial masks, a limitation we share with [19,48].

6   Conclusion

SpaceJAM is a lightweight method that effectively performs JA with neither regularization terms nor building an explicit atlas during the process. It demonstrates excellent performance on benchmarks like SPair-71 K and CUB, on par with, or outperforming, existing methods despite having far less trainable parameters and a much shorter running time. While here we mentioned only 3 matrix groups, both our method and code support additional ones (*e.g.*, the similarity group).

Acknowledgements. This work was supported by the Lynn and William Frankel Center at BGU CS, by the Israeli Council for Higher Education via the BGU Data Science Research Center, and by Israel Science Foundation Personal Grant #360/21. R.S.W. and S.E.F. were also funded in part by the BGU Kreitman School Negev Scholarship.

References

1. Aberman, K., Liao, J., Shi, M., Lischinski, D., Chen, B., Cohen-Or, D.: Neural best-buddies: sparse cross-domain correspondence. In: ACM TOG (2018)
2. Amir, S., Gandelsman, Y., Bagon, S., Dekel, T.: Deep VIT features as dense visual descriptors. In: ECCV Workshops (2022)
3. Annunziata, R., Sagonas, C., Cali, J.: Jointly aligning millions of images with deep penalised reconstruction congealing. In: ICCV (2019)
4. Boumal, N.: An introduction to Optimization on Smooth Manifolds. Cambridge University Press, Cambridge (2023)
5. Caron, M., Misra, I., Mairal, J., Goyal, P., Bojanowski, P., Joulin, A.: Unsupervised learning of visual features by contrasting cluster assignments. In: NeurIPS (2020)
6. Caron, M., Touvron, H., Misra, I., Jégou, H., Mairal, J., Bojanowski, P., Joulin, A.: Emerging properties in self-supervised vision transformers. In: ICCV (2021)
7. Chelly, I., Winter, V., Litvak, D., Rosen, D., Freifeld, O.: JA-POLS: a moving-camera background model via joint alignment and partially-overlapping local subspaces. In: CVPR (2020)
8. Cox, M., Sridharan, S., Lucey, S., Cohn, J.: Least squares congealing for unsupervised alignment of images. In: CVPR (2008)

9. Cox, M., Sridharan, S., Lucey, S., Cohn, J.: Least-squares congealing for large numbers of images. In: ICCV, IEEE (2009)
10. Dalca, A., Rakic, M., Guttag, J., Sabuncu, M.: Learning conditional deformable templates with convolutional networks. In: NeurIPS (2019)
11. Dosovitskiy, A., et al.: An image is worth 16×16 words: transformers for image recognition at scale. arXiv preprint arXiv:2010.11929 (2020)
12. Erez, G., Weber, R.S., Freifeld, O.: A deep moving-camera background model. In: Avidan, S., Brostow, G., Cissé, M., Farinella, G.M., Hassner, T. (eds.) Computer Vision - ECCV 2022, ECCV 2022, LNCS, vol. 13695, pp. 177–194. Springer, Cham (2022). https://doi.org/10.1007/978-3-031-19833-5_11
13. Felzenszwalb, P.F., Schwartz, J.D.: Hierarchical matching of deformable shapes. In: CVPR, pp. 1–8. IEEE (2007)
14. Freifeld, O., Hauberg, S., Batmanghelich, K., Fisher III, J.W.: Highly-expressive spaces of well-behaved transformations: Keeping it simple. In: ICCV (2015)
15. Freifeld, O., Hauberg, S., Batmanghelich, K., Fisher III, J.W.: Transformations based on continuous piecewise-affine velocity fields. In: IEEE TPAMI (2017)
16. Frey, B.J., Jojic, N.: Estimating mixture models of images and inferring spatial transformations using the EM algorithm. In: CVPR, IEEE (1999)
17. Gavrila, D.M.: Multi-feature hierarchical template matching using distance transforms. In: ICPR, IEEE (1998)
18. Goodfellow, I., et al.: Generative adversarial nets. In: NeurIPS (2014)
19. Gupta, K., et al.: ASIC: aligning sparse in-the-wild image collections. In: ICCV (2023)
20. He, J., Zhang, D., Balzano, L., Tao, T.: Iterative grassmannian optimization for robust image alignment. Image Vis. Comput. **32**(10), 800–813 (2014)
21. He, K., Zhang, X., Ren, S., Sun, J.: Identity mappings in deep residual networks. In: Leibe, B., Matas, J., Sebe, N., Welling, M. (eds.) ECCV 2016. LNCS, vol. 9908, pp. 630–645. Springer, Cham (2016). https://doi.org/10.1007/978-3-319-46493-0_38
22. Huang, G., Mattar, M., Lee, H., Learned-Miller, E.G.: Learning to align from scratch. In: NeurIPS (2012)
23. Huang, G.B., Jain, V., Learned-Miller, E.: Unsupervised joint alignment of complex images. In: ICCV, IEEE (2007)
24. Huang, S., Yang, L., He, B., Zhang, S., He, X., Shrivastava, A.: Learning semantic correspondence with sparse annotations. In: Avidan, S., Brostow, G., Cissé, M., Farinella, G.M., Hassner, T. (eds.) Computer Vision - ECCV 2022, ECCV 2022, LNCS, vol. 13674, pp. 267–284. Springer, Cham (2022). https://doi.org/10.1007/978-3-031-19781-9_16
25. Jaderberg, M., Simonyan, K., Zisserman, A., et al.: Spatial transformer networks. In: NeurIPS (2015)
26. Jain, A.K., Zhong, Y., Lakshmanan, S.: Object matching using deformable templates. In: IEEE TPAMI (1996)
27. Jeon, S., Kim, S., Min, D., Sohn, K.: Parn: Pyramidal affine regression networks for dense semantic correspondence. In: ECCV (2018)
28. Jeon, S., Kim, S., Min, D., Sohn, K.: Pyramidal semantic correspondence networks. In: IEEE TPAMI (2021)
29. Kemelmacher-Shlizerman, I., Seitz, S.M.: Collection flow. In: CVPR, IEEE (2012)
30. Kingma, D.P., Ba, J.: Adam: A method for stochastic optimization. CoRR (2014). http://arxiv.org/abs/1412.6980
31. Learned-Miller, E.G.: Data driven image models through continuous joint alignment. In: IEEE TPAMI (2006)

32. Lee, J., Kim, D., Ponce, J., Ham, B.: SFNet: learning object-aware semantic correspondence. In: CVPR (2019)
33. Li, X., Fan, D.P., Yang, F., Luo, A., Cheng, H., Liu, Z.: Probabilistic model distillation for semantic correspondence. In: CVPR (2021)
34. Lin, C.H., Lucey, S.: Inverse compositional spatial transformer networks. In: CVPR (2017)
35. Lin, D., Grimson, E., Fisher III, J.: Learning visual flows: a lie algebraic approach. In: CVPR (2009)
36. Lin, D., Grimson, E., Fisher III, J.: Modeling and estimating persistent motion with geometric flows. In: CVPR (2010)
37. Lin, W.Y., Liu, L., Matsushita, Y., Low, K.L., Liu, S.: Aligning images in the wild. In: CVPR, IEEE (2012)
38. Liu, X., Tong, Y., Wheeler, F.W.: Simultaneous alignment and clustering for an image ensemble. In: ICCV, IEEE (2009)
39. Loiseau, R., Monnier, T., Aubry, M., Landrieu, L.: Representing shape collections with alignment-aware linear models. In: 3DV, IEEE (2021)
40. Lowe, D.G.: Object recognition from local scale-invariant features. In: ICCV, IEEE (1999)
41. Lucas, B.D., Kanade, T.: An iterative image registration technique with an application to stereo vision. In: IJCAI (1981)
42. Mariotti, O., Mac Aodha, O., Bilen, H.: Improving semantic correspondence with viewpoint-guided spherical maps. arXiv preprint arXiv:2312.13216 (2023)
43. Mattar, M.A., Hanson, A.R., Learned-Miller, E.G.: Unsupervised joint alignment and clustering using bayesian nonparametrics. arXiv preprint arXiv:1210.4892 (2012)
44. Miller, E.G., Matsakis, N.E., Viola, P.A.: Learning from one example through shared densities on transforms. In: CVPR, IEEE (2000)
45. Min, J., Lee, J., Ponce, J., Cho, M.: Spair-71k: a large-scale benchmark for semantic correspondence. arXiv preprint arXiv:1908.10543 (2019)
46. Monnier, T., Groueix, T., Aubry, M.: Deep transformation-invariant clustering. In: NeurIPS (2020)
47. Mu, J., De Mello, S., Yu, Z., Vasconcelos, N., Wang, X., Kautz, J., Liu, S.: CoordGAN: self-supervised dense correspondences emerge from GANs. In: CVPR (2022)
48. Ofri-Amar, D., Geyer, M., Kasten, Y., Dekel, T.: Neural congealing: aligning images to a joint semantic atlas. In: CVPR (2023)
49. Oquab, M., et al.: Dinov2: learning robust visual features without supervision. arXiv preprint arXiv:2304.07193 (2023)
50. Paszke, A., et al.: Pytorch: an imperative style, high-performance deep learning library. In: NeurIPS (2019)
51. Peebles, W., Zhu, J.Y., Zhang, R., Torralba, A., Efros, A.A., Shechtman, E.: Gan-supervised dense visual alignment. In: CVPR (2022)
52. Peng, Y., Ganesh, A., Wright, J., Xu, W., Ma, Y.: RASL: robust alignment by sparse and low-rank decomposition for linearly correlated images. In: IEEE TPAMI (2012)
53. Rocco, I., Arandjelovic, R., Sivic, J.: Convolutional neural network architecture for geometric matching. In: CVPR (2017)
54. Rocco, I., Arandjelović, R., Sivic, J.: End-to-end weakly-supervised semantic alignment. In: CVPR (2018)

55. Rocco, I., Cimpoi, M., Arandjelović, R., Torii, A., Pajdla, T., Sivic, J.: NCNet: Neighbourhood consensus networks for estimating image correspondences. In: IEEE TPAMI, pp. 1020–1034 (2020)
56. Ronneberger, O., Fischer, P., Brox, T.: U-net: convolutional networks for biomedical image segmentation. In: Navab, N., Hornegger, J., Wells, W., Frangi, A. (eds.) Medical Image Computing and Computer-Assisted Intervention - MICCAI 2015, MICCAI 2015, LNCS, vol. 9351, pp. 234–241. Springer, Cham (2015). https://doi.org/10.1007/978-3-319-24574-4_28
57. Saharia, C., et al.: Photorealistic text-to-image diffusion models with deep language understanding. In: NeurIPS (2022)
58. Schwöbel, P., Warburg, F.R., Jørgensen, M., Madsen, K.H., Hauberg, S.: Probabilistic spatial transformer networks. In: UAI (2022)
59. Seo, P.H., Lee, J., Jung, D., Han, B., Cho, M.: Attentive semantic alignment with offset-aware correlation kernels. In: ECCV (2018)
60. Shokrollahi Yancheshmeh, F., Chen, K., Kamarainen, J.K.: Unsupervised visual alignment with similarity graphs. In: CVPR (2015)
61. Sinclair, M., et al.: Atlas-ISTN: joint segmentation, registration and atlas construction with image-and-spatial transformer networks. Med. Image Anal. **78**, 102383 (2022)
62. Skafte Detlefsen, N., Freifeld, O., Hauberg, S.: Deep diffeomorphic transformer networks. In: CVPR (2018)
63. Tang, L., Jia, M., Wang, Q., Phoo, C.P., Hariharan, B.: Emergent correspondence from image diffusion. In: NeurIPS (2024)
64. Wah, C., Branson, S., Welinder, P., Perona, P., Belongie, S.: The caltech-ucsd birds-200-2011 dataset (2011)
65. Weber, R.S., Eyal, M., Skafte Detlefsen, N., Shriki, O., Freifeld, O.: Diffeomorphic temporal alignment nets. In: NeurIPS (2019)
66. Weber, R.S., Freifeld, O.: Regularization-free diffeomorphic temporal alignment nets. In: ICML, PMLR (2023)
67. Zhang, J., et al.: A tale of two features: Stable diffusion complements dino for zero-shot semantic correspondence. In: NeurIPS (2024)
68. Zhang, X., Wang, D., Zhou, Z., Ma, Y.: Robust low-rank tensor recovery with rectification and alignment. In: IEEE TPAMI (2019)

APL: Anchor-Based Prompt Learning for One-Stage Weakly Supervised Referring Expression Comprehension

Yaxin Luo[1], Jiayi Ji[2], Xiaofu Chen[1], Yuxin Zhang[2], Tianhe Ren[3], and Gen Luo[4(✉)]

[1] Technical University of Denmark, Lyngby, Denmark
[2] Xiamen University, Xiamen, China
[3] International Digital Economy Academy, Shanghai, China
[4] Shanghai Artificial Intelligence Laboratory, Shanghai, China
luogen@stu.xmu.edu.cn

Abstract. Referring Expression Comprehension (REC) aims to ground the target object based on a given referring expression, which requires expensive instance-level annotations for training. To address this issue, recent advances explore an efficient one-stage weakly supervised REC model called RefCLIP. Particularly, RefCLIP utilizes anchor features of pre-trained one-stage detection networks to represent candidate objects and conducts anchor-text ranking to locate the referent. Despite the effectiveness, we identify that visual semantics of RefCLIP are ambiguous and insufficient for weakly supervised REC modeling. To address this issue, we propose a novel method that enriches visual semantics with various prompt information, called *anchor-based prompt learning* (APL). Specifically, APL contains an innovative *anchor-based prompt encoder* (APE) to produce discriminative prompts covering three aspects of REC modeling, *e.g.,* position, color and category. These prompts are dynamically fused into anchor features to improve the visual description power. In addition, we propose two novel auxiliary objectives to achieve accurate vision-language alignment in APL, namely text reconstruction loss and visual alignment loss. To validate APL, we conduct extensive experiments on four REC benchmarks, namely RefCOCO, RefCOCO+, RefCOCOg and ReferIt. Experimental results not only show the state-of-the-art performance of APL against existing methods on four benchmarks, *e.g.,* +6.44% over RefCLIP on RefCOCO, but also confirm its strong generalization ability on weakly supervised referring expression segmentation. Source codes released at: https://github.com/Yaxin9Luo/APL.

Keywords: Weakly Supervised Referring Expression Comprehension · Anchor-based Prompt Learning

1 Introduction

Referring Expression Comprehension (REC) aims to locate the target object based on a free-form language description [7,29,42]. As a fundamental vision-

A. Leonardis et al. (Eds.): ECCV 2024, LNCS 15071, pp. 198–215, 2025.
https://doi.org/10.1007/978-3-031-72624-8_12

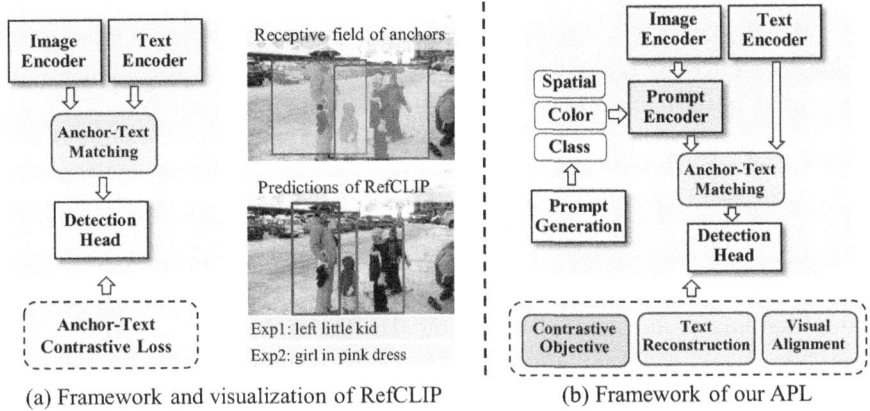

(a) Framework and visualization of RefCLIP   (b) Framework of our APL

Fig. 1. Comparison of RefCLIP and APL. (a) RefCLIP adopts the anchor-text matching framework to conduct weakly supervised REC. However, anchor features often lead to the visual misleading in anchor-text matching. (b) APL overcomes the visual shortcoming of RefCLIP by fusing rich prompts into anchor features, and conduct anchor-based prompt learning via two auxiliary losses.

language task, REC has gained increasing attention and achieved significant progress recently [11,29,31,38,56,64]. Driven by the great success, numerous efforts have been devoted to weakly supervised REC to reduce the expensive annotation costs [9,33–35,60,62,66]. Among them, most methods aim to directly apply weakly supervised objectives to a two-stage REC model, *e.g.*, MAttNet [64]. Despite the effectiveness, their two-stage modeling relies on region proposals and incurs expensive computational overhead.

Recently, Jin *et al.* [15] provided an efficient one-stage framework for weakly supervised REC, termed RefCLIP. As shown in Fig. 1, RefCLIP employs one-stage detectors to formulate weakly supervised REC as an anchor-text matching problem. Specifically, anchor (grid) features are first extracted from a pre-trained one-stage detector and then ranked based on the text features. To accomplish the REC task, RefCLIP will select the best matched anchor and decodes it to bounding box via the pre-trained detection head. During weakly supervised learning, RefCLIP adopts the anchor-text contrastive loss to achieve vision-language alignments using massive image-text pairs. Compared to two-stage methods, RefCLIP removes the expensive region proposal stage and achieves real-time inference speed, *e.g.*, +31.6 fps.

However, we identify that the visual semantics of RefCLIP are ambiguous and insufficient for weakly supervised REC modeling. Compared to region features, anchor features fall short in determining the actual visual object they represent. Despite the large receptive field of an anchor, it often encompasses noisy visual content and incomplete object information, which hinders the accurate anchor-text matching. As shown in Fig. 1, anchor features of the *"woman"* also capture the misleading information of the *"kid"*, thereby matching with the incorrect expression *"left little kid"*. Besides, we also notice that anchor features struggle to describe fine-grained visual semantics, *e.g.*, color, which are crucial for understanding diverse expressions. To explain, anchor features are pre-trained in common detection tasks, and the learned knowledge is fixed and limited, *e.g.*, 80 classes in COCO [27]. As shown in Fig. 1 (Exp2), RefCLIP fails to distinguish two kids that have fine-grained differences based on the expression.

To overcome these limitations, we propose a novel *anchor-based prompt learning* (APL) for one-stage weakly supervised REC. In particular, APL aims to improve the visual description ability of RefCLIP with an innovative *anchor-based prompt encoder* (APE). As shown in Fig. 1, APE can generate discriminative prompts that cover various knowledge of REC, *i.e.*, position, color and category. By injecting these prompts into anchor features, APE can greatly alleviate visual misleading in anchor-text matching. To effectively optimize APE, we further propose two novel auxiliary objectives in APL, namely text reconstruction loss and visual alignment loss. Specifically, text reconstruction loss minimizes the distance between text features and anchor features. And the visual alignment loss encourages anchor features to directly regress the pseudo-box[1] of the referent. With these training objectives, APL can achieve fine-grained vision-language alignments through weakly supervised training on massive image-text pairs.

To validate APL, we conduct extensive experiments on four common REC benchmarks, *i.e.*, RefCOCO [45], RefCOCO+ [45], RefCOCOg [44] and Refer-ItGame [17]. Experimental results show that APL achieves state-of-the-art performance on four datasets, *e.g.*, +6.44% over RefCLIP [15] on RefCOCO. Besides, we also conduct a bunch of ablation studies to validate our designs in APL. To validate the generalization ability of APL, we extend it to weakly supervised referring expression segmentation (RES) and also observe promising performance of APL against existing methods. In summary, our contributions are three folds:

- We identify the visual shortcoming in existing one-stage weakly supervised REC. To address this issue, we propose a novel *anchor-based prompt learning* (APL) to improve the visual description ability for the popular one-stage weakly supervised REC model termed RefCLIP.
- We propose an innovative *anchor-based prompt encoder* (APE) in APL, which generates and fuses rich multimodal prompts into anchor features. To achieve effective anchor-based prompt learning, we further equip APL with two auxiliary objectives, namely text reconstruction loss and visual alignment loss.
- APL achieves state-of-the-art results on four weakly supervised REC benchmark datasets. In addition, APL can be directly applied to weakly supervised RES and outperforms existing methods on three benchmark datasets.

2   Related Work

2.1   Referring Expression Comprehension

Referring Expression Comprehension (REC) aims to locate the target instance based on the given expression. Early REC methods [11,29,31,56,64] mainly fol-

---

[1] The pseudo-box is produced via the anchor-text matching.

low the two-stage pipeline, which first generates candidate regions using detection networks such as Faster-RCNN [54] and then selects the target one that best matches the referring expression. In spite of their success, two-stage methods are often criticized for their slow inference speed, which greatly limits their applications. To overcome this limitation, researchers have shifted their attention to one-stage REC [25,39,42,70,71]. In particular, these methods often embed text features into a one-stage detection network like YOLOv3 [53]. By directly predicting the target box via the detection head, one-stage REC models can achieve real-time inference speed. Based on this paradigm, existing methods further improve the reasoning ability via deep fusions [70] or attentions [39,42,71]. Subsequently, driven by the progress of Transformers [18,40,61], recent endeavors [7,36,39,43] also explore their applications to one-stage REC, which often stacks multiple Transformer layers for better cross-modal interactions. In this paper, we explore weakly supervised learning for one-stage REC from a novel perspective of anchor-based prompt learning.

2.2   Weakly Supervised Referring Expression Comprehension

Compared to fully supervised REC, weakly supervised REC limits the access to ground-truth annotations. Most existing methods [9,33–35,60,62,66] often explore weakly training objectives to optimize traditional two-stage REC models with image-text pairs. Among them, sentence reconstruction [34,35,62] selects the best matched region to reconstruct the given input expression. Contrastive learning [9,66] constructs positive and negative pairs from a set of regions and expressions and computes the InfoNCE loss [47]. Despite their effectiveness, these two-stage methods also suffer from expensive computational overhead. Therefore, researchers attempt to explore weakly supervised learning for one-stage REC [15,67]. Among them, the advanced method called RefCLIP [15] adopts anchor-text matching based on the one-stage detector. To achieve weakly supervised learning, RefCLIP conducts anchor-based contrastive learning with massive image-text pairs.

In this paper, we identify that anchor representations often contain ambiguous object information and greatly hinder weakly supervised learning. To address this issue, we propose a novel *anchor-based prompt learning* (APL) for one-stage weakly supervised REC. APL aims to fuse rich prompt information into anchors and prompts vision-language alignments via two novel auxiliary objectives, *i.e.*, text reconstruction loss and visual alignment loss.

2.3   Prompt Learning

Prompt learning is an emerging research hot topic in natural language processing (NLP), which inserts text instructions into the input of a pre-trained language model for a better understanding of the task [14,48,58]. Early works [4,6,14,48,51] focus on manually selected prompts to improve the zero-shot and few-shot performance of language models. Recently, most works regard learnable vectors as prompts and optimize them via task-specific fine-tuning.

These methods can greatly improve the adaptation ability of language models to various downstream tasks [10,12,24,37,68,69]. Inspired by these progresses, prompt learning has been a popular transfer learning scheme for pre-trained vision models. For example, VPT [68] adopts the deep prompt tuning strategy to transfer ViTs [8] to downstream tasks efficiently. CoOp [68] significantly improves the generalization ability of CLIP [50] on various out-domain tasks.

Different from previous work, APL dynamically constructs rich multimodal prompts, e.g., position and color, to improve the anchor representations. We also introduce two novel objectives to achieve accurate anchor-based prompt learning.

3 Preliminary

We first recap the framework of RefCLIP [15], which defines weakly supervised REC as an anchor-text matching problem. In particular, given an input image $I \in \mathbb{R}^{H \times W \times 3}$, anchor features $F_a \in \mathbb{R}^{(h \times w) \times d}$ are extracted from the last convolution feature map in YOLOv3 [53]. Based on anchor features, YOLOv3 employs the detection head to predict their corresponding bounding boxes. To accomplish REC, RefCLIP selects the target anchor that best matches with the given expression $T \in \mathbb{R}^L$, and predicts the bounding box of the referent via the detection head. This process can be formulated by

$$b = \mathcal{F}_{\text{det}}(\arg\max_{f_a \in F_a} \phi(f_a, f_t)), \tag{1}$$

where $f_a \in \mathbb{R}^d$ and $f_t \in \mathbb{R}^d$ denote anchor features and expression features, respectively. ϕ denotes the dot product similarity. And $\mathcal{F}_{\text{det}}(\cdot)$ is the detection head of YOLOv3 [53]. As defined in Eq. 1, once the target anchor is correctly selected, RefCLIP can directly predict the bounding box of the referent. Compared to two-stage methods, RefCLIP is much more efficient due to the elimination of the region proposal stage.

To achieve weakly supervised training, RefCLIP adopts anchor-text contrastive learning, defined by

$$\mathcal{L}_{\text{atc}} = -\log \frac{\exp\left(\phi(\hat{f}_{a_i}, f_{t_i})/\tau\right)}{\sum_{j=0}^{N} \mathbb{I}_{(i \neq j)} \exp\left(\phi(f_{a_j}, f_{t_i})/\tau\right)}, \tag{2}$$

where $\hat{f}_{a_i}$ denotes the best matched anchor features in i-th image, and N is the batch size. τ is the temperature for contrastive learning. With Eq. 2, RefCLIP can be directly optimized with massive image-text pairs.

As shown in Eq. 1, the effectiveness of RefCLIP lies in the accurate anchor-text matching. Nevertheless, anchor feature f_a suffers from the shortcoming of object representation. Compared to instance-level region features, anchor features are more fragmented and noisy, where an anchor often contains incomplete object information. Besides, anchor features also lack sufficient visual semantics for REC modeling, e.g., color. Therefore, the visual shortcoming of anchor features inevitably hinders the vision-language alignment of RefCLIP.

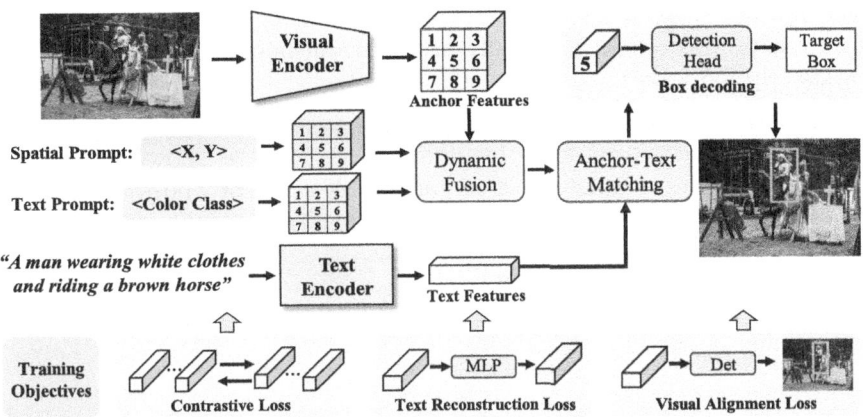

Fig. 2. Illustration of the proposed anchor-based prompt learning (APL). APL contains a novel anchor-based prompt encoder (APE), which fuses rich prompts into anchor features for accurate anchor-text matching. To promote anchor-based prompt learning, APL is equipped with two auxiliary losses, namely text reconstruction loss and visual alignment loss.

4   The APL Framework

4.1   Overview

To address the above issues, we propose a novel weakly supervised REC framework, namely *anchor-based prompt learning* (APL). The core idea of APL is to improve object representations of RefCLIP [15] with various prompts. To achieve this target, APL is equipped with a novel *anchor-based prompt encoder* (APE) to generate discriminative prompts covering position, color and category. Then, these prompts are dynamically fused into anchor features. Therefore, the anchor-text matching process can be re-written by

$$b = \mathcal{F}_{\text{dec}}(\underset{f_a \in F_a, f_p \in F_p}{\arg\max} \ \phi(\mathcal{F}_{\text{prompt}}(f_a, f_p), f_t)). \tag{3}$$

Here, $\mathcal{F}_{\text{prompt}}(\cdot)$ denotes the anchor-based prompt encoder. $f_p \in \mathbb{R}^{h \times w \times d}$ is prompt features that describe position, attribute and category for anchor features $f_a \in \mathbb{R}^{h \times w \times d}$. Similar to RefCLIP, APL conducts the anchor-text ranking and the box decoding to locate the referent. For weakly supervised training, in addition to the contrastive objective of RefCLIP, we further propose two novel objectives to promote vision-language alignment in APL, *i.e.*, text reconstruction loss and visual alignment loss.

4.2   Anchor-Based Prompt Encoder

As discussed above, anchor features often fall short of encoding discriminative object information. To tackle this challenge, the anchor-based prompt encoder

(APE) first generates a set of prompt information for each candidate anchor. Then, these prompts are dynamically fused into anchor features to improve the visual description power.

Anchor-based prompt generation. The key to APE is how to generate valuable and discriminative prompts for different anchors. As shown in Fig. 2, we first define a prompt template that contains three slots, *i.e.*, position, color and category. Then, we extract these pieces of information from the image to obtain the anchor-based prompt.

In particular, given an anchor feature $f_a \in \mathbb{R}^d$, we obtain its corresponding detected box $b \in \mathbb{R}^4$ and category $c \in \mathbb{R}^1$ via the detection head. Then, the spatial prompt $f_{ps} \in \mathbb{R}^d$ is defined by

$$f_{ps} = \rho_{\text{pos}}(\frac{b_0 + b_2}{2}, \frac{b_1 + b_3}{2}), \tag{4}$$

where $\rho_{\text{pos}}(\cdot)$ is the positional embedding function [61]. f_{ps} indicates the spatial information for anchor a. As defined in Eq. 4, we first obtain the center coordinates of the detected bxo b and then transform it to spatial prompt via the positional embedding. In practice, the spatial prompt can also be represented as a word, *e.g.*, "left" and "right", but it can only reflect the rough location.

Afterward, we define the color and category prompts as natural language descriptions, *e.g.*, "black sofa" and "white cup". Specifically, the category can be directly obtained through the class name of c. The color information is calculated based on the pixel values of the image region b^2. Then, we transform the obtained RGB color to natural words via a pre-defined color table. Finally, the color and the category are combined and processed to obtain textual features $f_{pt} \in \mathbb{R}^d$, which is defined by

$$f_{pt} = \mathcal{F}_{\text{text}}(t_p W_p), \tag{5}$$

where $t_p \in \mathbb{R}^{l_p}$ and $W_p \in \mathbb{R}^{l_p \times d}$ are tokenized prompt words and weights of the word embedding, respectively. As defined in Eq. 4 and 5, the generated prompts contain detailed object information, which can be combined with anchor features for better visual understanding.

Anchor-prompt Fusion. To achieve the above target, we propose a dynamic weighting strategy to fuse the prompt features, *i.e.*, f_{ps} and f_{pt}, into the anchor features f_a, which is formulated by

$$f_p = w_0 f_a + w_1 f_{ps} + w_2 f_{pt}, \tag{6}$$

where $f_p \in \mathbb{R}^d$ is the anchor-based prompt features that are used to conduct anchor-text matching. $w_0, w_1, w_2 \in \mathbb{R}^1$ are three attention weights, defined by

$$w_0, w_1, w_2 = \text{softmax}(\sigma((f_a + f_{ps} + f_{pt})W_1)W_2). \tag{7}$$

Here, $W_1 \in \mathbb{R}^{d \times d}$ and $W_2 \in \mathbb{R}^{d \times 2}$ are two projection weights. $\sigma(\cdot)$ denotes the activation function of ReLU [1]. According to Eq. 7, weights of different features are dynamically adjusted for different samples.

---

[2] We adopt average pooling and K-means to capture colors of objects accurately.

4.3 Anchor-Based Prompt Learning

Similar to RefCLIP, we adopt the contrastive loss to achieve weakly supervised learning. Besides, we propose two novel objectives to further facilitate the anchor-based prompt learning, namely text reconstruction loss and visual alignment loss. Therefore, the weakly supervised learning of APL can be written by

$$\min_{\theta} \mathcal{L}_{\text{ptc}}(F_p, f_t; \theta) + \mathcal{L}_{\text{tr}}(F_p, f_t; \theta) + \mathcal{L}_{\text{va}}(F_p; \theta), \tag{8}$$

where $\mathcal{L}_{\text{ptc}}$ is the anchor-text contrastive loss defined in Eq. 2. $\mathcal{L}_{\text{tr}}$ and $\mathcal{L}_{\text{va}}$ denote the text reconstruction loss and the visual alignment loss, respectively. θ denotes the model parameters.

In Eq. 8, $\mathcal{L}_{\text{tr}}$ aims to reconstruct expression features f_t with the best matched anchor-based prompt features $\hat{f}_p$, which is defined by

$$\mathcal{L}_{\text{tr}} = (\mathcal{F}_{\text{mlp}}(\hat{f}_p) - f_t)^2. \tag{9}$$

Here, $\mathcal{F}_{\text{mlp}}$ is a multi-layer MLP to project $\hat{f}_p$ into a latent space. In practice, $\hat{f}_p$ is selected from F_p based on the similarity of $\phi(f_p, f_t)$. With Eq. 9, anchor-based prompt features can learn fine-grained visual knowledge from diverse expressions and achieve better vision-language alignments.

In addition, we adopt the visual alignment loss to directly reconstruct the pseudo bounding box of the referent b based on F_p and f_t, which is defined by

$$\mathcal{L}_{\text{va}} = \mathcal{L}_{\text{det}}(\mathcal{F}_{\text{rec}}(F_p, f_t), b). \tag{10}$$

Here, $\mathcal{F}_{\text{rec}}$ denotes the REC decoder of SimREC [41], which contains a multimodal fusion layer, a GARAN layer [70] and a detection head [53]. $\mathcal{L}_{\text{det}}$ is the detection losses [26], which consists of the IoU loss [55] and the confidence loss [52]. Note that the bounding box b is generated via Eq. 3, which is still in line with the definition of weakly supervised learning. Since the bounding box b is often noisy at the beginning of training, we apply $\mathcal{L}_{\text{va}}$ after a short training phase. Compared to the text reconstruction loss, the visual alignment loss encourages anchor-based prompt features to encode discriminative semantics so that the target box can be directly predicted in a regression manner.

4.4 Network Settings

Feature Extraction. We deploy APL based on RefCLIP [15]. In particular, we use a bi-directional GRU layer [3] and a self-attention layer [61] to extract text features $f_t \in \mathbb{R}^{l \times d}$ from the expression T. For the visual backbone, we employ DarkNet-53 [53] to process the input image I and obtain the anchor features $F_a \in \mathbb{R}^{h \times w \times d}$. Following RefCLIP, we also adopt multi-scale fusion to fuse anchor features of different layers. Then, we filter most anchor features with low detection confidences. And the remaining anchor features are fused with prompt features to obtain the anchor-based prompt features F_p.

Table 1. Ablation study of APL on *val* set of RefCOCO and RefCOCO+.

APE	$\mathcal{L}_{ptc}$	$\mathcal{L}_{tr}$	$\mathcal{L}_{va}$	RefCOCO	RefCOCO+
–	✓	–	–	60.36	40.39
✓	✓	–	–	62.89	41.54
✓	✓	✓	-	63.01	42.31
✓	✓	✓	✓	**64.51**	**42.71**

Training and Inference. For weakly supervised training, we directly sum three losses and optimize the model via backpropagation. During inference, APL has two different ways to predict the bounding box of the referent. The first way is based on the anchor-text ranking, as defined in Eq. 3. The second way adopts the predictive branch using Eq. 10, which calculates the visual alignment loss. We adopt the second one for inference, which performs slightly better.

5   Experiments

5.1   Datasets and Metrics

RefCOCO [45] contains 142,210 referring expressions and 50,000 objects in 19,994 images from MSCOCO [27]. Expressions of RefCOCO mainly describe about absolute position. **RefCOCO+** [45] consists of 141,564 referring expressions for 49,856 bounding boxes in 19,992 MSCOCO images. Different from Ref-COCO, RefCOCO+ contains more descriptions of relationships and attributes. **RefCOCOg** [44] has 104,560 referring expressions and 54,822 bounding boxes for 26,711 images, where its expressions are longer and more complex than that of RefCOCO and RefCOCO+. RefCOCOg includes two different splits, *i.e.,* umd split [45] and google split [44]. We use the Google split in our experiments. Refer-ItGame [17] includes 120,072 referring expressions for 99,220 bounding boxes of 19,997 images. ReferItGame includes background descriptions, making it more challenging than RefCOCO and RefCOCO+.

For REC task, we use IoU@0.5 as the metric. In particular, a prediction is considered correct when the intersection over union (IoU) between the prediction and the ground truth is larger than 0.5. For the RES task, we follow previous works [2,21,28,30,49,57,63] to use mIoU as the metric, which averages the IoU scores of all testing samples.

5.2   Implementation Details

Following RefCLIP [15], the image resolution is set to 416 × 416, and the text length is 15, 15 and 20 for RefCOCO, RefCOCO+ and RefCOCOg, respectively. We use YOLOv3 [53] pre-trained on MSCOCO[3] [27] as the detection network.

---

[3] Validation and testing images in REC task are removed.

Table 2. Ablation study of the anchor-based prompt encoder (APE) on Ref-COCO and RefCOCO+. "S-prompt" and "T-prompt" denote the spatial prompt and the text prompt, respectively. Our final choice is colored in grey.

Settings	Choices	RefCOCO			RefCOCO+		
		val	testA	testB	val	testA	testB
S-prompt	Text	57.88	56.34	57.98	41.17	42.07	39.46
	Pos. Func.	**64.51**	**61.91**	**63.57**	**42.71**	**42.84**	**39.80**
T-prompt	BLIP [22]	63.47	60.72	63.16	42.11	41.84	39.44
	Template	**64.51**	**61.91**	**63.57**	**42.71**	**42.84**	**39.80**
Fusion	Add	**64.69**	61.75	63.67	42.10	42.00	38.89
	Concat	64.23	61.53	**64.08**	39.59	39.49	38.54
	Dynamic Sum	64.51	**61.91**	63.57	**42.71**	**42.84**	**39.80**

Table 3. Comparison of different prompt learning methods on three REC datasets. For fair comparisons, RefCLIP is used as the structure for VPT and CoOP.

Method	RefCOCO			RefCOCO+			RefCOCOg
	val	testA	testB	val	testA	testB	val-g
VPT [13]	55.39	53.30	53.94	33.56	33.74	31.72	40.79
CoOP [68]	60.33	59.11	59.63	37.63	36.87	37.84	49.38
APL (ours)	**64.51**	**61.91**	**63.57**	**42.70**	**42.84**	**39.80**	**50.22**

In APL, dimensions of prompt features, anchor features and text features are set to 512. For the text reconstruction objective, we adopt a three-layer MLP with a hidden size 512 for projecting anchor features. During weakly supervised training, we use Adam [19] as the optimizer. And the learning rate and the batch size are set to 1e-4 and 64, respectively. Training consists of 25 epochs, and the visual alignment loss is applied after 9,000 steps. The remaining settings are kept the same with RefCLIP.

5.3   Quantitative Results

Ablation Studies. We conduct extensive experiments to validate designs of APL in Table 1, 2 and 3. In particular, Table 1 shows the cumulative ablations of APL. From this table, the first observation is that all designs obviously contribute to the final performance. Specifically, APE provides the most apparent gains of all designs, *e.g.*, +2.53% on RefCOCO, suggesting the significance of discriminative anchor semantics. With the help of auxiliary losses, the performance of APL can be further boosted, *e.g.*, +1.62% on RefCOCO. Besides, We also notice that the text reconstruction loss yields more significant improvements on the dataset containing more diverse expressions, *i.e.*, RefCOCO+. In

Table 4. Comparison with state-of-the-art methods on four REC benchmark datasets. *GT proposals* means that official annotations of MSCOCO are used as candidates. *Pseudo Label* denotes that the student REC model is trained using pseudo-labels generated by a teacher model. For example, RefCLIP_SimREC means that RefCLIP and SimREC are the teacher and the student, respectively.

Method	RefCOCO			RefCOCO+			RefCOCOg	ReferIt	Inference
	val	testA	testB	val	testA	testB	val-g	test	speed
GT Proposals:									
VC [46]$_{CVPR'18}$	-	33.29	30.13	-	34.60	31.58	30.26	-	-
ARN [33]$_{ICCV'19}$	38.05	36.43	36.47	34.53	36.40	36.12	39.62	-	-
KPRN [34]$_{MM'19}$	36.34	35.28	37.72	37.16	36.06	39.29	38.37	33.87	-
DTWREG [60]$_{TPAMI'21}$	39.21	41.14	37.72	39.18	40.01	38.08	43.24	-	-
EARN [32]$_{TPAMI'22}$	38.08	38.25	38.59	37.54	37.58	37.92	45.33	36.86	-
RefCLIP$_{-MAttNet}$	69.31	67.23	71.27	43.01	44.80	41.09	51.31	-	-
APL$_{-MAttNet}$(ours)	**74.24**	**73.29**	**76.39**	**48.59**	**53.02**	**44.04**	**57.08**	**-**	**-**
Det Proposals:									
VC[46]$_{CVPR'18}$	-	32.68	27.22	-	34.68	28.10	29.65	14.50	-
KAC Net [5]$_{CVPR'18}$	-	-	-	-	-	-	-	15.83	-
MATN [67]$_{CVPR'18}$	-	-	-	-	-	-	-	13.61	-
ARN [33]$_{ICCV'19}$	32.17	35.25	30.28	32.78	34.35	32.13	33.09	26.19	5.7fps
IGN [66]$_{NeurIPS'20}$	34.78	37.64	32.59	34.29	36.91	33.56	34.92	-	-
DTWREG [60]$_{TAPAMI'21}$	38.35	39.51	37.01	38.91	39.91	37.09	42.54	-	5.9fps
RelR [35]$_{CVPR'21}$	-	-	-	-	-	-	-	37.68	-
NCE+Dist [62]$_{CVPR'21}$	-	-	-	-	-	-	-	38.39	-
RefCLIP [15]$_{CVPR'23}$	60.36	58.58	57.13	40.39	40.45	38.86	47.87	39.58	31.3fps
APL (ours)	**64.51**	**61.91**	**63.57**	**42.70**	**42.84**	**39.80**	**50.22**	**41.80**	26.7fps
Pesudo Labels:									
RefCLIP$_{-SimREC}$ [15]	62.57	62.70	61.22	39.13	40.81	36.59	45.68	42.33	54.8fps
RefCLIP$_{-Transvg}$ [15]	64.08	63.67	63.93	39.32	39.54	36.29	45.70	42.64	19.3fps
APL$_{-SimREC}$ (ours)	63.94	**64.72**	61.21	**42.11**	**44.85**	**38.31**	**48.35**	**45.22**	54.8fps
APL$_{-Transvg}$ (ours)	**64.86**	**64.89**	63.87	39.28	41.08	36.45	46.11	43.25	19.3fps

contrast, the visual alignment loss offers more benefits on RefCOCO, where visual understanding is the main challenge. These results extensively validate the effectiveness of the proposed auxiliary losses for vision-language alignments.

In Table 2, we further compare different designs for anchor-based prompt encoder (APE). For the choice of the spatial prompt, we observe that the text description, *e.g.*, "left", performs much worse than the positional function, *e.g.*, −6.63% on RefCOCO. In practice, the positional function can provide more accurate and continuous spatial information than the text description. Besides, we attempt to use BLIP [22] to generate region captions as the text prompt, but the performance declines. To explain, the generated captions often include noisy and unrelated descriptions, which potentially causes the visual misleading. In Table 2, we compare different strategies for anchor-prompt fusions, and our dynamic fusion still outperforms other methods.

Table 5. Comparison of APL and existing methods on weakly supervised RES. PKS [21] uses click annotations for supervision, so we mark it in gray.

Method	RefCOCO			RefCOCO+			RefCOCOg
	val	testA	testB	val	testA	testB	val-g
AMR [49]ₐₐₐᵢ'₂₂	14.12	11.69	17.47	14.13	11.47	18.13	15.83
GroupViT [63]ᴄᴠᴘʀ'₂₂	18.03	18.13	19.33	18.15	17.65	19.53	19.97
CLIP-ES [28]ᴄᴠᴘʀ'₂₃	13.79	15.23	12.87	14.57	16.01	13.53	14.16
GbS [2]ɪᴄᴄᴠ'₂₁	14.59	14.60	14.97	14.49	14.49	15.77	14.21
WWbL [57]ɴₑᵤᵣɪᴘs'₂₂	18.26	17.37	19.90	19.85	18.70	21.64	21.84
TSEG [59]ₐᵣₓᵢᵥ'₂₀	30.12	-	-	25.95	-	-	22.62
ALBEF [23]ɴₑᵤᵣɪᴘs'₂₁	23.11	22.79	23.42	22.44	22.07	22.51	24.18
I-Chunk [20]ɪᴄᴄᴠ'₂₃	31.06	32.30	30.11	31.28	32.11	30.13	32.88
TRIS [30]ɪᴄᴄᴠ'₂₃	31.17	32.43	29.56	30.90	30.42	30.80	36.00
PKS [21]	19.27	22.23	15.04	22.73	22.80	22.37	26.13
APL (ours)	**55.92**	**54.84**	**55.64**	**34.92**	**34.87**	**35.61**	**40.13**

In Table 3, we compare APL with existing prompt learning methods, *i.e.*, VPT [13] and CoOP [68]. The first observation is that APL greatly outperforms the other two methods on three datasets, *e.g.*, up to +5.97% on RefCOCO+ *testA*. From Table 3, we also find that VPT performs worse on three datasets, which achieves 55.39% on RefCOCO. To explain, prompts of VPT are not conditioned on anchors, which may produce useless and harmful prompt information. In contrast, CoOP can generate anchor-based prompts and demonstrate promising performance on RefCOCOg. Nevertheless, CoOP is limited in weak spatial and attribute information, so it still lags behind APL by large margins on Ref-COCO and RefCOCO+. These results greatly validate the design of APL.

Comparison with Existing Methods on REC. In Table 4, we compare APL with a set of methods on four REC benchmark datasets. From this table, we find that most methods adopt the two-stage modeling and their inference speed is vastly inferior to the one-stage one, *e.g.*, 5.9 fps of DTWREG [60] *vs.* 26.7 fps of APL. Regarding performance, one-stage models have obvious advantages in RefCOCO, suggesting its better spatial understanding ability. Nevertheless, on more challenging datasets like RefCOCO+ and ReferIt, one-stage models perform similarly to the two-stage ones. As discussed in Sect. 1, existing one-stage models often adopt a simple anchor-text matching, which lacks sufficient visual semantics for fine-grained REC modeling. Compared to these methods, APL achieves the best performances on four datasets, *e.g.*, +6.44% over RefCLIP on RefCOCO, and also maintains remarkable inference efficiency, *i.e.*, 26.7 fps. Compared to RefCLIP, APL has obvious performance gains on some complex splits, *e.g.*, RefCOCO *testB* and RefCOCOg *val*. Notably, APL can even achieve comparable performance with early supervised REC models on RefCOCO, *e.g.*, 63.57 of APL *vs.* 64.85 of Spe+Lis+Rl [65] on *testB*. These results further confirm that our APL can greatly promote fine-grained vision-language alignments.

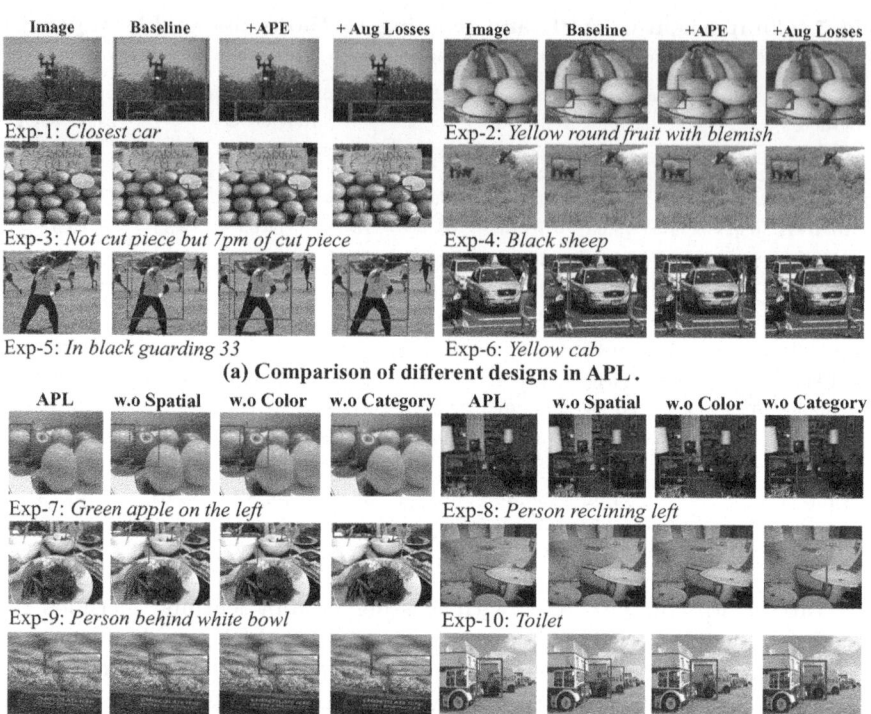

Exp-1: *Closest car*

Exp-2: *Yellow round fruit with blemish*

Exp-3: *Not cut piece but 7pm of cut piece*

Exp-4: *Black sheep*

Exp-5: *In black guarding 33*

Exp-6: *Yellow cab*

(a) Comparison of different designs in APL .

Exp-7: *Green apple on the left*

Exp-8: *Person reclining left*

Exp-9: *Person behind white bowl*

Exp-10: *Toilet*

Exp-11: *From bottom right row second up*

Exp-12: *Yellow and blue truck second from left*

(b) Predictions of APL with different anchor-based prompts.

Fig. 3. Visualizations of APL. Subfig-(a) compares different designs of APL and visualizes their predictions. Subfig-(b) compares predictions of APL with different prompts. Predictions and ground-truths are colored in red and blue, respectively. (Color figure online)

In Table 4, we further validate APL under the pseudo-label learning setups. Following RefCLIP [15], we use the well-trained APL to generate pseudo-labels for training common REC models. As shown in Table 4, the REC model taught by APL greatly outperforms the one taught by RefCLIP, *e.g.,* up to +4.04% for SimREC [41]. Besides, we also notice that APL brings more benefits for the student REC model on challenging datasets like ReferIt and RefCOCOg. In particular, SimREC supervised by APL can even outperform APL by +3.42% in ReferIt, suggesting that APL can produce high-quality pseudo-labels in this dataset. Nevertheless, we also find that performance gains of TransVG are not obvious as SimREC. We conjecture that the Transformer structure of TransVG has higher requirements for data quality.

Generalization Results on RES. In Table 5, we extend APL to weakly super-vised RES and compare it with a set of existing methods. To accomplish the RES task, we use the YOLOv5-Seg [16] as the detection network in APL, which can produce segmentation masks based on anchor features. From Table 5, we observe that APL demonstrates obvious performance gains against existing methods. For example, APL outperforms the state-of-the-art method, *i.e.*, TRIS [30], by up to 26.08%, which is a remarkable improvement. In other RES datasets, the obvious benefits of APL can also be witnessed, *e.g.*, +4.81% and +4.13% in RefCOCO+ and RefCOCOg, respectively. Compared to PKS [21], which uses additional click annotations for supervision, APL still performs better in Ref-COCO and RefCOCOg.

5.4   Qualitative Analysis

To gain in-depth insights into APL, we visualize and compare predictions of APL in Fig. 3. In Fig. 3 (a), we ablate the proposed APE and auxiliary losses and visualize the predictions. This figure shows that the default baseline often falls short in fine-grained recognition, *e.g.*, "*black sheep*" of Exp-4. With the help of APE, APL can better capture visual appearances such as color. However, APL still fails in some examples that involve complex descriptions, *e.g.*, Exp-2. These examples are correctly predicted after adopting two auxiliary losses. As shown in Fig. 3 (a) Exp-2, the "*orange*" can be located from a bunch of fruits.

In Fig. 3 (b), we further compare the effects of different prompt information in APE. As shown in Fig. 3 (b), removing the spatial information leads to the failure of APL on spatial-related expressions, *e.g.*, "*person reclining left*" of Exp-8. Besides, we observe that the color prompt is significant for locating the referent in a diverse scenario. Without the color prompt, APL can not correctly ground the "*green apple*" from various fruits in Exp-7. Moreover, the category prompt helps APL distinguish objects of similar appearances, *e.g.*, the "*toilet*".

To better understand the limitation of APL, we visualize its failure cases in Fig. 4. From this figure, we observe that APL still struggles to address long expressions, *e.g.*, Exp-3 and Exp-10, which requires strong reasoning ability. Besides, we can also see that complex visual scenes cause some failure cases, *e.g.*, the occluded objects in Exp-2. From these examples, we believe that APL has much room for improvement in reasoning ability and visual understanding.

Exp-1: *Bride throwing bouquet*   Exp-2: *Far-right Carrot just the orange part*   Exp-3: *Gray top jeans on left face partially hidden*   Exp-4: *Frontmost folded down chair*   Exp-5: *Bottom section third from left*

Exp-6: *Left top donut with chunk of sugar*   Exp-7: *Yucky broccoli front most piece*   Exp-8: *Elephant just to right of baby*   Exp-9: *Standing chappie far left*   Exp-10: *Chair of guy looking at you in front*

Fig. 4. Failure cases of APL on RefCOCO, RefCOCO+ and RefCOCOg.

6 Conclusion

In this paper, we focus on one-stage weakly supervised REC and identify that existing methods suffer from the ambiguous visual semantics in their REC modeling. To address this issue, we propose a novel approach, namely anchor-based prompt learning (APL). APL formulates weakly supervised REC as an anchor-text matching problem, equipped with an innovative anchor-based prompt encoder (APE) to enrich anchor semantics with a set of prompts. Moreover, we propose two novel auxiliary losses to achieve the effective anchor-based prompt learning, namely text reconstruction loss and visual alignment loss. Experimental results not only validate the state-of-the-art performance of APL in REC but also confirm the strong generalization ability of APL in RES.

Acknowledgements. This work was supported by the National Natural Science Foundation of China (No. 623B2088).

References

1. Agarap, A.F.M.: Deep learning using rectified linear units (relu) (2018)
2. Arbelle, A., et al.: Detector-free weakly supervised grounding by separation (2021)
3. Bahdanau, D., Cho, K., Bengio, Y.: Neural machine translation by jointly learning to align and translate (2014)
4. Brown, T.B., et al.: Language models are few-shot learners, pp. 1877–1901 (2020)
5. Chen, K., Gao, J., Nevatia, R.: Knowledge aided consistency for weakly supervised phrase grounding (2018)
6. Chen, Y., Liu, Y., Dong, L., Wang, S., Zhu, C., Zeng, M., Zhang, Y.: AdaPrompt: adaptive model training for prompt-based NLP. In: Findings of the Association for Computational Linguistics: EMNLP 2022, pp. 6057–6068. Association for Computational Linguistics (2022)
7. Deng, J., Yang, Z., Chen, T., Zhou, W., Li, H.: TransVG: end-to-end visual grounding with transformers, pp. 1769–1779 (2021)
8. Dosovitskiy, A., et al.: An image is worth 16×16 words: transformers for image recognition at scale (2020)
9. Gupta, T., Vahdat, A., Chechik, G., Yang, X., Kautz, J., Hoiem, D.: Contrastive learning for weakly supervised phrase grounding, pp. 752–768 (2020)
10. Houlsby, N., et al.: Parameter-efficient transfer learning for NLP (2019)
11. Hu, R., Rohrbach, M., Darrell, T.: Segmentation from natural language expressions, pp. 108–124 (2016)
12. Huang, T., Chu, J., Wei, F.: Unsupervised prompt learning for vision-language models (2022)
13. Jia, M., et al.: Visual prompt tuning (2022)
14. Jiang, Z., Xu, F.F., Araki, J., Neubig, G.: How can we know what language models know? Trans. Assoc. Comput. Linguist. (TACL) **8**, 423–438 (2020)
15. Jin, L., Luo, G., Zhou, Y., Sun, X., Jiang, G., Shu, A., Ji, R.: RefCLIP: a universal teacher for weakly supervised referring expression comprehension, pp. 2681–2690 (2023)
16. Jocher, G., et al.: Ultralytics/yolov5: V7.0 - yolov5 sota realtime instance segmentation. Zenodo (2022)

17. Kazemzadeh, S., Ordonez, V., Matten, M., Berg, T.: ReferItGame: referring to objects in photographs of natural scenes. In: Proceedings of the 2014 Conference on Empirical Methods in Natural Language Processing (EMNLP), pp. 787–798 (2014)

18. Kim, W., Son, B., Kim, I.: ViLT: vision-and-language transformer without convolution or region supervision, pp. 5583–5594 (2021)

19. Kingma, D.P., Ba, J.: Adam: a method for stochastic optimization (2014)

20. Lee, J., Lee, S., Nam, J., Yu, S., Do, J., Taghavi, T.: Weakly supervised referring image segmentation with intra-chunk and inter-chunk consistency. In: Proceedings of the IEEE/CVF International Conference on Computer Vision (ICCV), pp. 21870–21881 (2023)

21. Li, H., Sun, M., Xiao, J., Lim, E.G., Zhao, Y.: Fully and weakly supervised referring expression segmentation with end-to-end learning (2022)

22. Li, J., Li, D., Xiong, C., Hoi, S.: Blip: bootstrapping language-image pre-training for unified vision-language understanding and generation (2022)

23. Li, J., Selvaraju, R., Gotmare, A., Joty, S., Xiong, C., Hoi, S.C.H.: Align before fuse: vision and language representation learning with momentum distillation. In: Ranzato, M., Beygelzimer, A., Dauphin, Y., Liang, P., Vaughan, J.W. (eds.) Advances in Neural Information Processing Systems, vol. 34, pp. 9694–9705. Curran Associates, Inc. (2021)

24. Li, S.M., Nie, W., Huang, D.A., Yu, Z., Goldstein, T., Anandkumar, A., Xiao, C.: Test time prompt tuning for zero-shot generalization in vision language models (2022)

25. Liao, Y., Liu, S., Li, G., Wang, F., Chen, Y., Qian, C., Li, B.: A real-time cross-modality correlation filtering method for referring expression comprehension, pp. 10880–10889 (2020)

26. Lin, T.Y., Goyal, P., Girshick, R., He, K., Dollár, P.: Focal loss for dense object detection. In: Proceedings of the IEEE International Conference on Computer Vision, pp. 2980–2988 (2017)

27. Lin, T.-Y.: Microsoft COCO: common objects in context. In: Fleet, D., Pajdla, T., Schiele, B., Tuytelaars, T. (eds.) ECCV 2014. LNCS, vol. 8693, pp. 740–755. Springer, Cham (2014). https://doi.org/10.1007/978-3-319-10602-1_48

28. Lin, Y., et al.: Clip is also an efficient segmenter: a text-driven approach for weakly supervised semantic segmentation (2022)

29. Liu, D., Zhang, H., Wu, F., Zha, Z.J.: Learning to assemble neural module tree networks for visual grounding, pp. 4670–4679 (2019)

30. Liu, F., et al.: Referring image segmentation using text supervision (2023)

31. Liu, X., Wang, Z., Shao, J., Wang, X., Li, H.: Improving referring expression grounding with cross-modal attention-guided erasing, pp. 1950–1959 (2019)

32. Liu, X., Li, L., Wang, S., Zha, Z.J., Li, Z., Tian, Q., Huang, Q.: Entity-enhanced adaptive reconstruction network for weakly supervised referring expression grounding (2022)

33. Liu, X., Li, L., Wang, S., Zha, Z.J., Meng, D., Huang, Q.: Adaptive reconstruction network for weakly supervised referring expression grounding, pp. 2611–2620 (2019)

34. Liu, X., Li, L., Wang, S., Zha, Z.J., Su, L., Huang, Q.: Knowledge-guided pairwise reconstruction network for weakly supervised referring expression grounding, pp. 539–547 (2019)

35. Liu, Y., Wan, B., Ma, L., He, X.: Relation-aware instance refinement for weakly supervised visual grounding, pp. 5612–5621 (2021)

36. Lu, J., Batra, D., Parikh, D., Lee, S.: Vilbert: pretraining task-agnostic visiolin-guistic representations for vision-and-language tasks, pp. 13–23 (2019)
37. Lu, Y., Liu, J., Zhang, Y., Liu, Y., Tian, X.: Prompt distribution learning. In: Proceedings of the IEEE/CVF Conference on Computer Vision and Pattern Recognition, pp. 5206–5215 (2022)
38. Luo, G., Zhou, Y., Ji, R.: Towards language-guided visual recognition via dynamic convolutions. Int. J. Comput. Vision **132**(1), 1–19 (2024)
39. Luo, G., Zhou, Y., Ji, R., Sun, X., Su, J., Lin, C.W., Tian, Q.: Cascade grouped attention network for referring expression segmentation, pp. 1274–1282 (2020)
40. Luo, G., Zhou, Y., Ren, T., Chen, S., Sun, X., Ji, R.: Cheap and quick: Efficient vision-language instruction tuning for large language models. In: Advances in Neural Information Processing Systems 36 (NeurIPS 2023) (2023)
41. Luo, G., et al.: What goes beyond multi-modal fusion in one-stage referring expression comprehension: an empirical study (2022)
42. Luo, G., et al.: Multi-task collaborative network for joint referring expression comprehension and segmentation, pp. 10034–10043 (2020)
43. Luo, G., et al.: Towards lightweight transformer via group-wise transformation for vision-and-language tasks. IEEE Trans. Image Process. **31**, 3386–3398 (2022)
44. Mao, J., Huang, J., Toshev, A., Camburu, O., Yuille, A.L., Murphy, K.: Generation and comprehension of unambiguous object descriptions. In: Proceedings of the IEEE Conference on Computer Vision and Pattern Recognition, pp. 11–20 (2016)
45. Nagaraja, V.K., Morariu, V.I., Davis, L.S.: Modeling context between objects for referring expression understanding. In: Leibe, B., Matas, J., Sebe, N., Welling, M. (eds.) ECCV 2016. LNCS, vol. 9908, pp. 792–807. Springer, Cham (2016). https://doi.org/10.1007/978-3-319-46493-0_48
46. Niu, Y., Zhang, H., Lu, Z., Chang, S.F.: Variational context: Exploiting visual and textual context for grounding referring expressions. IEEE Trans. Pattern Anal. Mach. Intell. **43**(1), 347–359 (2021)
47. van den Oord, A., Li, Y., Vinyals, O.: Representation learning with contrastive predictive coding. arXiv preprint arXiv:1807.03748 (2018)
48. Petroni, F., et al.: Language models as knowledge bases? (2019)
49. Qin, J., Wu, J., Xiao, X., Li, L., Wang, X.: Activation modulation and recalibration scheme for weakly supervised semantic segmentation (2021)
50. Radford, A., et al.: Learning transferable visual models from natural language supervision (2021)
51. Raffel, C., et al.: Exploring the limits of transfer learning with a unified text-to-text transformer. J. Mach. Learn. Res. (JMLR) **21**(140), 1–67 (2019)
52. Redmon, J., Divvala, S., Girshick, R., Farhadi, A.: You only look once: Unified, real-time object detection. In: Proceedings of the IEEE Conference on Computer Vision and Pattern Recognition, pp. 779–788 (2016)
53. Redmon, J., Farhadi, A.: Yolov3: an incremental improvement. arXiv preprint arXiv:1804.02767 (2018)
54. Ren, S., He, K., Girshick, R., Sun, J.: Faster r-cnn: towards real-time object detection with region proposal networks. In: Advances in Neural Information Processing Systems, vol. 28 (2015)
55. Rezatofighi, H., Tsoi, N., Gwak, J., Sadeghian, A., Reid, I., Savarese, S.: Generalized intersection over union: a metric and a loss for bounding box regression. In: Proceedings of the IEEE Conference on Computer Vision and Pattern Recognition (CVPR) (2019)
56. Rohrbach, A., Rohrbach, M., Hu, R., Darrell, T., Schiele, B.: Grounding of textual phrases in images by reconstruction, pp. 817–834 (2016)

57. Shaharabany, T., Tewel, Y., Wolf, L.: What is where by looking: weakly-supervised open-world phrase-grounding without text inputs (2022)
58. Shin, T., Razeghi, Y., IV, R.L.L., Wallace, E., Singh, S.: AutoPrompt: eliciting knowledge from language models with automatically generated prompts, pp. 4222–4235 (2020)
59. Strudel, R., Laptev, I., Schmid, C.: Weakly-supervised segmentation of referring expressions (2022)
60. Sun, M., Xiao, J., Lim, E.G., Liu, S., Goulermas, J.Y.: Discriminative triad matching and reconstruction for weakly referring expression grounding. IEEE Trans. Pattern Anal. Mach. Intell. **43**(1), 4189–4195 (2021)
61. Vaswani, A., et al.: Attention is all you need. In: Advances in Neural Information Processing Systems, vol. 30 (2017)
62. Wang, L., Huang, J., Li, Y., Xu, K., Yang, Z., Yu, D.: Improving weakly supervised visual grounding by contrastive knowledge distillation, pp. 14090–14100 (2021)
63. Xu, J., De Mello, S., Liu, S., Byeon, W., Breuel, T., Kautz, J., Wang, X.: GroupViT: semantic segmentation emerges from text supervision (2022)
64. Yu, L., Lin, Z., Shen, X., Yang, J., Lu, X., Bansal, M., Berg, T.L.: MAttNet: modular attention network for referring expression comprehension, pp. 1307–1315 (2018)
65. Yu, L., Tan, H., Bansal, M., Berg, T.L.: A joint speaker-listener-reinforcer model for referring expressions. In: Proceedings of the IEEE Conference on Computer Vision and Pattern Recognition (CVPR) (2017)
66. Zhang, Z., Zhao, Z., Lin, Z., He, X., et al.: Counterfactual contrastive learning for weakly-supervised vision-language grounding. Adv. Neural. Inf. Process. Syst. **33**, 18123–18134 (2020)
67. Zhao, F., Li, J., Zhao, J., Feng, J.: Weakly supervised phrase localization with multi-scale anchored transformer network, pp. 5696–5705 (2018)
68. Zhou, K., Yang, J., Loy, C.C., Liu, Z.: Learning to prompt for vision-language models (2021)
69. Zhou, K., Yang, J., Loy, C.C., Liu, Z.: Conditional prompt learning for vision-language models. In: Proceedings of the IEEE/CVF Conference on Computer Vision and Pattern Recognition, pp. 16816–16825 (2022)
70. Zhou, Y., et al.: A real-time global inference network for one-stage referring expression comprehension. IEEE Trans. Neural Netw. Learn. Syst. **34**(1), 134–143 (2021)
71. Zhu, C., et al.: SeqTR: a simple yet universal network for visual grounding. In: Avidan, S., Brostow, G., Cissé, M., Farinella, G.M., Hassner, T. (eds.) Computer Vision - ECCV 2022, ECCV 2022, LNCS, vol. 13695, pp. 598–615. Springer, Cham (2022). https://doi.org/10.1007/978-3-031-19833-5_35

$GenQ$: Quantization in Low Data Regimes with Generative Synthetic Data

Yuhang Li[1]([✉])[ID], Youngeun Kim[1][ID], Donghyun Lee[1], Souvik Kundu[2], and Priyadarshini Panda[1][ID]

[1] Department of Electrical Engineering, Yale University, New Haven, USA
{yuhang.li,youngeun.kim,donghyun.lee,priya.panda}@yale.edu
[2] Intel Labs, San Diego, USA
souvikk.kundu@intel.com

Abstract. In the realm of deep neural network deployment, low-bit quantization presents a promising avenue for enhancing computational efficiency. However, it often hinges on the availability of training data to mitigate quantization errors, a significant challenge when data availability is scarce or restricted due to privacy or copyright concerns. Addressing this, we introduce $GenQ$, a novel approach employing an advanced Generative AI model to generate photorealistic, high-resolution synthetic data, overcoming the limitations of traditional methods that struggle to accurately mimic complex objects in extensive datasets like ImageNet. Our methodology is underscored by two robust filtering mechanisms designed to ensure the synthetic data closely aligns with the intrinsic characteristics of the actual training data. In case of limited data availability, the actual data is used to guide the synthetic data generation process, enhancing fidelity through the inversion of learnable token embeddings. Through rigorous experimentation, $GenQ$ establishes new benchmarks in data-free and data-scarce quantization, significantly outperforming existing methods in accuracy and efficiency, thereby setting a new standard for quantization in low data regimes. Code is released at https://github.com/Intelligent-Computing-Lab-Yale/GenQ.

1 Introduction

A variety of model compression techniques have been developed to deploy large deep learning models on embedded/mobile devices without significant accuracy drops. One prominent method is neural network quantization [18], which involves converting 32-bit floating-point models into a compact low-bit fixed-point format. This transformation leverages the efficiency of fixed-point computation and the benefits of reduced memory usage. Other notable techniques include pruning [5,44] and knowledge distillation [24], both of which aim to reduce network size while preserving as much of the original model's performance as possible.

Supplementary Information The online version contains supplementary material available at https://doi.org/10.1007/978-3-031-72624-8_13.

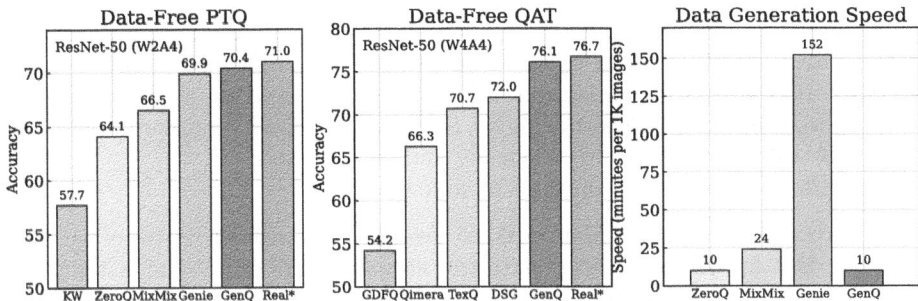

Fig. 1. Comparison of *GenQ* with existing methods on ImageNet. (1) Data-Free PTQ, (2) Data-Free QAT, (3) Data Generation Speed. Real* denotes using real ImageNet data in zero-shot quantization.

While these methods effectively accelerate neural networks, they typically require finetuning on the original training dataset to mitigate any accuracy loss in the compressed model. However, accessing the original training data can be challenging due to privacy concerns or Intellectual Property (IP) protection issues [43]. This challenge has led to the emergence of *data-free quantization* [7,21,33]. This approach involves generating synthetic data from a pretrained network and then finetuning the compressed or quantized model using this synthetic data, thereby circumventing the need for original training data.

Several prior works generate synthetic data by inverting the knowledge from pre-trained full-precision models. For instance, [7,21,69] propose to optimize the images by aligning them with the training data in terms of activation statistics. [39,62] apply generative adversarial networks [20] to synthesize the images. However, these approaches often struggle to match the compression performance achievable with real training data. The primary challenge lies in the inherent complexity of reverse mapping from a lower to a higher dimension [35]. For example, in the context of the ImageNet dataset, this involves mapping from a 1000-dimensional space to an image space of $224 \times 224 \times 3$. Such transformation makes accurate recreation of complex objects a hard problem. Additionally, these methods cost a significant amount of time to generate the images.

In this paper, we introduce *GenQ*, a new pipeline utilizing the advanced Generative AI model such as Stable Diffusion [56] to generate high-quality data for quantization purposes. This approach aligns with the growing trend of applying AI-generated content (AIGC) in deep learning and vision applications [6,58,61,67]. While direct application of generative AI data risks distribution shifts from original training data, we address this through two filtering mechanisms, ensuring the selection of in-distribution synthetic data. Our method excels in data-free quantization scenarios. More remarkably, in a majority of the model deployment scenarios with limited training data (*e.g.* one image per class), *GenQ* is significantly effective via prompt tuning to align generated images more closely with the real data. We conduct comprehensive experiments to validate the effectiveness of *GenQ*, including both Post-Training Quantization (PTQ) and Quantization-Aware Training (QAT), on various neural network architectures (*e.g.* CNNs [22,54] and ViTs [14]). Figure 1 showcases effectiveness of *GenQ* on data-free quantization. Meanwhile, compared to other data generation approaches, our method achieves faster generation of synthetic data (up to 15× faster), increasing both efficiency and effectiveness.

We summarize our contributions as follows:

1. We propose *GenQ*, the first work to leverage advanced text-to-image synthetic data for quantization in data-free scenarios. To reduce the distribution gap between synthetic data and real data, we introduce a suite of filtering mechanisms to select in-distribution synthetic data, including energy score filtering and BatchNorm Distribution filtering.
2. In case of limited data availability, we also propose to learn the token embedding to guide the synthesis using real data.
3. We conduct extensive experiments to show the usefulness of *GenQ* in generating synthetic data that helps achieve SoTA quantization performance in both PTQ and QAT. For example, our 4-bit QAT-based ResNet-50 achieves 76.10% accuracy on the ImageNet dataset, outperforming the latest existing method [8] by 5.4%.

2   Related Work

Quantization Methods. Quantization approaches to compress pretrained Deep Neural Networks (DNNs) can be broadly divided into PTQ and QAT. PTQ performs *calibration* on a pre-trained DNN after quantizing its weights and activations to low bits to maintain the original accuracy [16,41,46]. For example, [3,17] propose bias correction in the convolutional layers after quantization. [71] splits the outliers into additional channels to reduce the quantization error. Recently, a line of works [28,34,45,60] leverage the weight rounding optimization to reconstruct the activation of the original model. While PTQ mostly adopts ad hoc strategies to prevent accuracy loss from quantization, QAT can significantly improve the accuracy of the quantized model through end-to-end finetuning. The success of QAT is mainly based on the usage of Straight-Through Estimator (STE) [65], enabling gradient-based optimization [9,15,27,63,68,73]. QAT also explores optimizing the step size [15], clipping threshold [9], non-linear interval [33], non-uniform quantization levels [68] together with the weights.

Data in Quantization. To quantize a model to below 8 bits, data is essential, especially in QAT where end-to-end finetuning occurs. When access to real images is restricted due to privacy and copyright issues, ZeroQ [7] proposes to synthesize images in replacement of real data. There are two categories of creating synthetic images: (1) Inverting the images directly via gradient descent [7,21,32,35,69], and (2) Learning a Generative Adversarial Network (GAN) to continuously generate images [39,62].

The image inversion [64] relies on Batch Normalization (BN) [7,21] as an optimization metric to distill the data and obtain high-fidelity images. Other inversion methods based on loss function [72], model ensembling [35], and advanced convolutional layers [32] have been proposed to obtain better synthetic data. However, the inversion approach is limited in certain ways: (1) Inverting images adds a significant computation burden, which makes QAT with synthetic data prohibitive; (2) Models that do not have BN layers, such as ViTs, cannot invert images [14]. The second category proposes to use GAN as an image synthesis engine [10, 39, 49, 62, 72]. These works finetune the generator and the quantized model simultaneously.

Our method opens a third category for quantization in low data regime, *i.e.*, using text-to-image diffusion models. Additionally, we investigate how to use limited real data to guide the generation of synthetic data. To the best of our knowledge, this scenario is being studied for the first time.

Synthetic Data in Deep Learning. Initial studies, such as those by [4, 31, 70], explored the use of GANs for assisting DNNs in classifier tuning, object segmentation, and contrastive learning. Recently, the emergence of text-to-image models [13, 42, 52, 53] has revolutionized the synthesis of high-quality data in deep learning, owing to their effectiveness and efficiency. For example, [23] utilized GLIDE [47] for image synthesis in classifier tuning on the CLIP model [51]. In a more advanced application, StableRep [58] leverages Stable Diffusion to generate datasets for contrastive learning, using synthetic data from different seeds as positive pairs. Meanwhile, [2] investigates the synthesis of data within the ImageNet label space observing accuracy improvements. Furthermore, Fill-up [55] employs synthetic data from text-to-image models to balance the long-tail distribution in training datasets.

3   Preliminaries

3.1   Quantization

Uniform quantization maps full-precision weights into fixed-point numbers. For a step size $s \in \mathbb{R}$, the integer value of weights $\boldsymbol{w}$ are given by

$$\boldsymbol{w}_{\text{int}} = \text{clip}\left(\left\lfloor\frac{\boldsymbol{w}}{s}\right\rceil + \boldsymbol{z}, n, p\right), \tag{1}$$

where $\lfloor \cdot \rceil$ denotes the rounding-to-nearest operation, and n and p represent the lower and upper bound of the integers, respectively. For example, under the b-bit uniform quantization, n and p are set to 0 and 2^{b-1}. The zero-point vector $\boldsymbol{z}$ is an all-z vector, where $z = -\lfloor\frac{\min(\boldsymbol{w})}{s}\rceil$. Thus, the quantized weights $\boldsymbol{w}^q$ are given by

$$\boldsymbol{w}^q = s(\boldsymbol{w}_{\text{int}} - \boldsymbol{z}). \tag{2}$$

Quantization-Aware Training (QAT). We utilize the Learned Step Size Quantization (LSQ) [15] to update the step size s and $\boldsymbol{w}$ with gradient descent. The gradient to step size can be computed with Straight-Through Estimator (STE) [65], *i.e.*, $\frac{\partial\lfloor x\rceil}{\partial x} = 1$. QAT requires a significant amount of data and GPU resources to perform end-to-end fine-tuning.

Post-Training Quantization (PTQ). Under this PTQ regime, the training data is much less than QAT, hindering the learning of step size. To obtain the desired step size s under PTQ, we perform layerwise quantization-error minimization [3]. In addition, we also conduct layer-wise tuning of the weight rounding during PTQ as did in [32,34,45,60], which can significantly improve the model performance with only 1K images.

3.2 Stable Diffusion Data Generation

Stable Diffusion utilizes the Denoising Diffusion Probabilistic Model (DDPM) [25] to implement the training and inference. It contains a text encoder $\tau_\theta(\cdot)$, image encoder $\mathcal{E}(\cdot)$ and image decoder $\mathcal{D}(\cdot)$, and finally a denoising U-Net $\epsilon_\theta(\cdot)$.

Inference Process. During inference, a random noise image latent is sampled from Gaussian distribution $z \sim \mathcal{N}(0,1)$. Suppose the text prompt is $vs.$, the denoising U-Net fuses the text embedding $\tau_\theta(vs.)$ and the image and visual embedding through cross-attention layers and denoises the image latent gradually. After T timesteps of denoising, the image latent is decoded by $\mathcal{D}(\cdot)$ to generate the high-resolution image.

Training Process. Training the denoising U-Net essentially involves encoding the images first ($\hat{x}_0 = \mathcal{E}(x)$), and then diffusing the image through a Markov chain process $q(\hat{x}_t|\hat{x}_{t-1}) = \mathcal{N}(\hat{x}_t; \sqrt{1-\beta}\hat{x}_{t-1}, \beta I)$. Note that $t \in [1,T]$ is the number of diffusion timesteps and β is a hyper-parameter to control the perturbation. When $t = T$, then $\hat{x}_t = z$. Now the denoising U-Net can be trained by a reverse process by matching the output and the noises added before:

$$\min_\theta \mathbb{E}_{t \sim \mathcal{U}(1,T), \epsilon \sim \mathcal{N}(0,\mathbf{I})} \lambda(t) ||\epsilon - \epsilon_\theta(\hat{x}_t, t)||_F^2, \tag{3}$$

where $\lambda(t)$ is a positive weighting function [25], ϵ is a noise vector predicted from x_t. We use Stable Diffusion v1-5 [56] in our pipeline. We refer to [52] for more details about its principle.

4 Methodology

In this section, we introduce our methodology to synthesize the training data with Stable Diffusion. We discuss two quantization scenarios in low data regimes: (1) Data-Free Quantization (DFQ) where only the label space and a pre-trained full-precision model are provided, and (2) Data-Scarce Quantization (DSQ) where limited training data is provided in addition to the pre-trained model and label space. We first discuss *GenQ*'s data generation method under DFQ and DSQ and then introduce the quantization techniques we used.

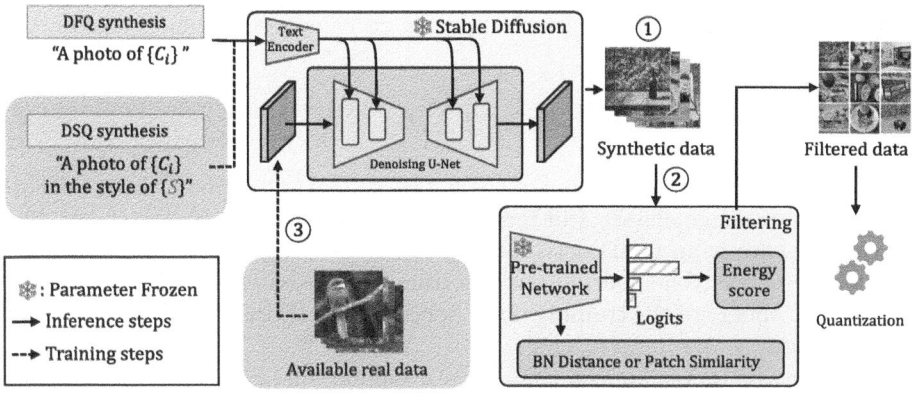

Fig. 2. The overall image synthesis and filtering procedure of *GenQ*. ① For DFQ synthesis, we directly use the label as the text prompt, and ② we use several metrics to filter our out-of-distribution synthetic images. ③ For data-scarce synthesis, real images are used to optimize prompt {S} (Eq. (11)). We then generate the synthetic images with the optimized prompt.

4.1 Data-Free *GenQ*

We propose a *label prompt* method that only uses the label as the prompt to synthesize the images without accessing the pre-trained model. For example, for an ImageNet pre-trained model [12], we can synthesize the images based on 1000 class names.

$$\text{Prompt} = \texttt{"A photo of a \{D\} \{C\}"}, \tag{4}$$

where {D} is a template adjective (*e.g.,* "nice, dark, small") derived from the CLIP ImageNet template [51][1]. {C} is a randomly chosen class label name. As an example, suppose we choose the class `hamster`, then a random prompt can be generated as `"A photo of a small hamster"`. Subsequently, we directly use the text-to-image model like Stable Diffusion [52] to synthesize the data.

In our *label prompt* method, we generate the images relying solely on the label name of each class and the ability of Stable Diffusion (Fig. 2①). However, this operation does not leverage any prior knowledge in the original training data, which might cause a distribution shift from original training data to synthetic training data. To generate better-quality data for quantization, we propose *model-dependent selection* to leverage the prior knowledge embedded in the pre-trained full-precision model.

We propose a set of filtering mechanisms for model-dependent selection. We argue that the synthetic data selection can be treated as an Out-of-Distribution (OOD) detection problem [48]. However, contrary to OOD detection, we intend to select in-distribution data from the generated synthetic data for quantization purposes (Fig. 2②).

---

[1] We list the full template in Appendix A.

Energy Filtering. We use energy filtering to calculate the energy score [38] as,

$$E(\boldsymbol{x}, f) = -\alpha \sum_{i=1}^{C} e^{-f_i(\boldsymbol{x})/\alpha}, \tag{5}$$

where $f_i(\boldsymbol{x})$ denotes the ith value of the network's output logits. C and α denote the number of image classes and temperature, respectively. [38] has shown that the energy score is an OOD detector due to its theoretical connection with the likelihood function. The neural networks trained by cross-entropy loss inherently decrease the energy of the training data, hence, the OOD data will have relatively higher energy.

To select synthetic images that have a similar quantization effect to that of real image data, we pass the synthetic data through the full-precision pre-trained model. Then, we calculate the energy scores of all synthetic images and only select those images for quantization that yield energy scores lower than a certain user-defined threshold. We will experiment with the choice of this threshold in Sect. 5.4.

BatchNorm Distribution Filtering. Originally proposed in [7,64], the channel mean and channel variance distance computed in the BN layers [30] are regarded as a standard metric for optimizing the synthetic data in Convolutional Neural Networks (CNNs). We use this metric to evaluate the distance between synthetic images and original training images on the full-precision pre-trained model. We calculate the BN distance by

$$D_{BN} = \sum_{\ell=1}^{n} \left(||\mu_\ell^s - \mu_\ell||_F + ||\sigma_\ell^s - \sigma_\ell||_F \right), \tag{6}$$

where μ_l, σ_l denote the running mean and variance of the layer l activation from original training images, and μ_l^s, σ_l^s are the current mean and variance of the layer l activation from synthetic images. Naively, for each generated synthetic data, we can evaluate the BN distance D_{BN} and filter out the ones with large D_{BN}. However, we find this approach does not bring improvement in the final data quality due to the difference between single data and batched data. Evaluating D_{BN} on a single image will lead to biased estimation as it ignores its interaction with other data. Ioffe [29] also demonstrates that with small batch sizes, the estimation of the batch mean and variance used during training become a less accurate approximation of the mean and variance used for testing.

To deal with this problem, we define a *BN sensitivity* metric, which measures the independence of one image from other images in a batch. Formally, given a batch of input images $\{\boldsymbol{x}_i\}_{i=1}^{B}$ where B indicates the batch size, we define the BN sensitivity of the i-th image $S(\boldsymbol{x}_i)$ as

$$S(\boldsymbol{x}_i) = D_{BN}(\{\boldsymbol{x}_j\}_{j=1}^{B}) - D_{BN}(\{\boldsymbol{x}_j\}_{j=1, j\neq i}^{B}). \tag{7}$$

Here, the BN sensitivity indicates the change in BN distance after removing the selected image. As such, a large BN sensitivity means the current image could potentially damage the internal distribution when it is batched with other images.

Patch Similarity Filtering for ViTs. As we discussed in Sect. 2, since ViTs do not have BN layers, we adopt the patch similarity metric in [36] to do additional filtering of the synthetic images after the energy filtering. Formally, given the output features tensor o, we first calculate the cosine similarity matrices between any two sub-tensors in the patch dimension, given by

$$\Gamma(o_i, o_j) = \frac{o_i \cdot o_j}{||o_i|| \, ||o_j||}, \tag{8}$$

where o_i denotes the i-th patch of the output featuremap. After calculation, we get the $N \times N$ (N is the #patches) matrix Γ. To measure the diversity of the patch similarities, we calculate differential entropy as follows

$$H = -\int \hat{f}_h(x) \cdot \log[\hat{f}_h(x)] dx. \tag{9}$$

where $\hat{f}_h(x)$ is the continuous probability density function of Γ, which can be obtained by the kernel density estimation [36]. A low entropy value indicates that the patch similarity distribution is less diverse. Hence, we can select the synthetic data that has the highest diversity in patch similarity.

Filtering Pipeline. The overall filtering pipeline is a two-stage procedure. First, regardless of the type of pre-trained model, we apply energy score filtering. Second, based on the category of the model (CNNs or ViTs), we either apply BatchNorm Distribution Filtering or Patch Similarity Filtering. The selected synthetic data is then used for quantization.

4.2   Data-Scarce *GenQ*

In this section, we describe how to synthesize and select synthetic data given a limited amount of training data (*i.e.*, data-scarce quantization). This case could be more common than the DFQ scenario as in practice it is not hard to obtain some permitted training data. We hypothesize that synthesizing a number of images based on this limited dataset can enhance quantization performance.

Considering the ImageNet-1k dataset [12] as the original training dataset, we assume that one image per class (*i.e.*, 1-shot) can be accessed by the cloud server. In this case, we propose to synthesize the images utilizing the existing information from the training data. Specifically, we optimize a text token embedding {S} with the prompt shown below

$$\text{Prompt}_{[i]} = \text{"A photo of a } \{C_{[i]}\} \text{ in the style of } \{S\}\text{"}. \tag{10}$$

Here, the learnable token embedding {S} indicates the dataset characteristics of ImageNet. To optimize this learnable token embedding, we associate each class name and the corresponding image into pairs ($\{C_{[i]}, x_{[i]}\}$, where $[i]$ is the class index), and let the Stable Diffusion generate $\boldsymbol{x}_{[i]}$ given $\text{Prompt}_{[i]}$. The optimization objective is given by:

$$\mathbb{E}_{t\sim\mathcal{U}(1,T),i\sim\mathcal{U}(1,M),\epsilon\sim\mathcal{N}(\mathbf{0},\mathbf{I})}\lambda(t)||\epsilon - \epsilon_\theta(\boldsymbol{x}_{[i]t},t,\text{Prompt}_{[i]})||_F^2, \tag{11}$$

where M is the number of all object classes. This method learns the token across multiple objects, aiming to characterize the whole dataset[2]. Figure 2③ illustrates the data-scarce prompt optimization. After optimization, we directly used the learned token embedding and the class label to generate images and filter the output images using the same technique in Fig. 2②.

4.3   Quantization with Synthetic Data from *GenQ*

The selected synthetic data from *GenQ* is then used to quantize a full-precision pre-trained model using QAT or PTQ. For PTQ, we adopt the state-of-the-art reconstruction-based rounding optimization methods [32,34]. As for QAT, we propose to finetune the quantized network on top of a PTQ model. Specifically, after the rounding optimization in PTQ, the quantization becomes

$$\boldsymbol{w}_{\text{int}} = \text{clip}\left(\left\lfloor\frac{\boldsymbol{w}}{s}\right\rfloor + \text{sgn}(\boldsymbol{v}) + \boldsymbol{z}, n, p\right), \tag{12}$$

where $\text{sgn}(\boldsymbol{v})$ is the learned rounding indicating up or down, which is already well-optimized in the PTQ reconstruction stage. To initialize from the PTQ model and stabilize the training process, we freeze all previous learnable variables including $\boldsymbol{w}$ and $\boldsymbol{v}$, and reinitialize an all-zero vector $\boldsymbol{u}$. The new quantization function is thus given by

$$\boldsymbol{w}_{\text{int}} = \text{clip}\left(\left\lfloor\frac{\boldsymbol{w}}{s}\right\rfloor + \text{sgn}(\boldsymbol{v}) + \left\lfloor\frac{\boldsymbol{u}}{s}\right\rfloor + \boldsymbol{z}, n, p\right). \tag{13}$$

The de-quantization step remains the same with Eq. (2). By introducing an additional variable $\boldsymbol{u}$ and freezing the previous variable, we start from the original PTQ model and avoid passing the gradient through floor operation. As a result, the $\boldsymbol{u}$ can be safely updated through STE in finetuning. Note, during PTQ or QAT, we only use the synthetic data from *GenQ* to perform quantization.

5   Experiments

In this section, we empirically demonstrate the effectiveness and the efficiency of our *GenQ* synthetic data in both a qualitative and a quantitative way. For the PTQ scenario, we follow the conventional setup [32,34,35] to synthesize 1k images. As for the QAT scenarios, we synthesize 1.2M images with 1200 images in each object class to match the original ImageNet dataset volume. Unless specified, we use Stable Diffusion v1-5 [1] and set the guidance scale to 3.5. We will first provide the visualization of our synthetic data, the latency

---

[2] We provide more technical details in Appendix A.

comparison of generating the data, and then compare them against existing state-of-the-art methods across multiple setups. Finally, we analyze *GenQ* by conducting various ablation studies. All experiments and accuracy noted are for the ImageNet dataset.

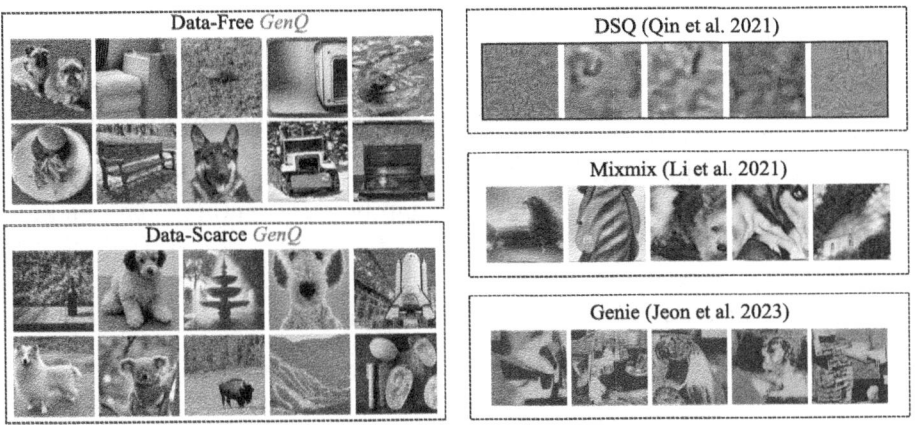

Fig. 3. Visualization of DF/DS *GenQ* and existing data synthesis method for quantization in low data regimes.

Table 1. Comparison of synthetic data generation approaches w.r.t. (a) generation speed and (b) data quality.

(a) Latency

Method	Type	Latency	Reusable
GENIE [32]	Image inversion	152 min	✓
TexQ [8]	GAN-based	10 min	✗
GenQ	Text-to-image	24 min	✓

(b) BatchNorm Distance

Data Source	Res50	MBV2
GENIE [32]	0.014±0.001	1.03±0.296
TexQ [8]	0.045±0.013	N/A
GenQ	0.024±0.002	0.78±0.106

5.1 Analysis on Synthetic Data

Visualization. As shown in Fig. 3, the *GenQ* images after filtering are high-quality and visually similar to real images. To compare the synthetic data with existing data-free quantization, we also provide the images from [32,35,50]. It can be observed that our method can generate much higher quality synthetic data than previous inversion-based methods. For example, [50] only demonstrates certain textures rather than objects, [35] shows some objects but the scene shows less resolution and clarity.

Data Synthesis Cost. We further study the cost of each synthetic data generation method. Specifically, we measure the latency of Genie [32], TexQ [8], and our method corresponding to image inversion, GAN-based, and text-to-image synthesis approaches, respectively, for 1000 images on the ImageNet dataset. The pre-trained model used for filtering or input optimization is ResNet-50. We report the latency in Table 1a, from which we can find using text-to-image synthesis provides faster data generation than image inversion-based techniques. Although GAN-based data generation is faster than *GenQ* , their data is not reusable across model types, which increases the overall latency if many models have to be quantized.

Data Quality Assessment. We further measure the quality of the synthetic data through some quantitative evaluation metrics. Specifically, we measure the average BN distances of the generated data as well as the original training data on ResNet-50 and MobileNetV2. The results are shown in Table 1b, from which we can find that *GenQ* largely closes the gap between synthetic data and real training data.

5.2   Quantization Performance Evaluation

In this section, we test the quantization accuracy with our synthetic data and compare it with various existing methods. We will introduce the experiment setup in each scenario individually.

Comparisons on Data-Free PTQ. We start with evaluating our proposed method by testing it on data-free PTQ including (1) CNNs such as ResNet [19], MobileNet [26,54], and MnasNet [57] and (2) Vision Transformers like original ViT [14]. For PTQ on CNNs, we employ the state-of-the-art PTQ algorithms, BRECQ [34] and GENIE-M [32] to perform the W4A4 and W2A4 quantization, which reconstruct the activation output in a block-wise manner. In this case, we use 1024 synthetic images for quantization evaluation. For PTQ on ViT, we perform PTQ4ViT [66] and RepQ-ViT [37] across 5 different models, including ViT-M [14], DeiT-M [59], and Swin-B [40]. We follow the open-source code implementation and generate 32 images as the calibration dataset to obtain the quantized ViTs. For all cases, we generate 2× synthetic data and filter 50% of them to match the volume.

For CNNs, we select SOTA existing methods including ZeroQ [7], the Knowledge Within [21], IntraQ [72], Qimera [10], MixMix [35], GENIE-D [32]. For reference, we also include the performance of PTQ using 1024 real ImageNet-1k images. We summarize all the accuracy results in Table 2. It can be observed that our *GenQ* achieves the highest accuracy in most cases. Interestingly, we find that the network architecture affects the performance of synthetic data. On ResNet architectures, *GenQ* and GENIE-D have similar accuracy (0.1–0.5%) levels, however, *GenQ* largely outperforms other data-free algorithms on lightweight network architectures due to their lower quantization resilience. As an example, *GenQ* increases 5% accuracy compared to GENIE-D on MobileNet-b W2A4 quantization and 4.4% accuracy on MnasNet W2A4 quantization.

Given that there is only one data-free algorithm for ViTs, PSAQ-ViT [36], we compare *GenQ* with PSAQ-ViT in Table 4. *GenQ* consistently outperforms PSAQ-ViT [36]. For example, on ViT-B and Swin-B, our method has 7.3% and

Table 2. Evaluation of data-free PTQ on CNN models (top-1 accuracy (%)).

Quant. Method	Syn. Method	#Bits W/A	Res18	Res50	MBV2	MB-b	MNas-1
Full Prec.	N/A	32/32	71.08	77.00	72.49	74.53	73.52
BRECQ [34]	Real Data	4/4	69.62	75.45	68.84	-	-
	ZeroQ[‡] [7]		69.32	73.73	49.83	55.93	52.04
	KW[‡] [21]		69.08	74.05	59.81	61.94	55.48
	IntraQ [72]		68.77	68.16	63.78	-	-
	Qimera [10]		67.86	72.90	58.33	-	-
	MixMix[‡] [35]		69.46	74.58	64.01	65.38	57.87
	GENIE-D		69.40	75.35	67.81	64.24	65.02
	GenQ		**69.52**	**75.47**	**68.34**	**67.08**	**66.33**
GENIE-M* [32]	Real Data	4/4	69.82	75.51	69.11	69.26	68.40
	GENIE-D [32]		69.72	**75.61**	68.62	67.31	67.03
	GenQ		**69.77**	75.50	**68.96**	**68.74**	**68.06**
BRECQ [34]	Real Data	2/4	65.25	70.65	54.22	-	-
	ZeroQ [7]		61.63	64.16[‡]	34.39	23.53	13.83
	KW[‡] [21]		-	57.74	-	-	-
	IntraQ [72]		55.39	44.78	35.38	-	-
	Qimera [10]		47.80	49.13	3.73	-	-
	MixMix[‡] [35]		-	66.49	-	-	-
	GENIE-D [32]		63.93	69.72	49.75	38.01	45.53
	GenQ		**65.04**	**69.90**	**53.08**	**47.31**	**50.84**
GENIE-M* [32]	Real Data	2/4	66.05	70.96	56.42	55.00	54.66
	GENIE-D [32]		64.86	69.89	51.47	47.69	48.38
	GenQ		**65.72**	**70.35**	**54.82**	**52.77**	**52.76**

[‡] The figures are taken from [35].
[*] Denotes our implementation based on the open-source code.

15% accuracy improvement using the RepQ-ViT method. This result proves that *GenQ* can achieve state-of-the-art performance on both CNNs and ViTs.

Comparisons on Data-Free QAT. We then compare *GenQ* with existing data-free QAT methods, such as GDFQ [62], ZAQ [39], Qimera [10], IntraQ [72], ARC [11], AdaDFQ [49], TexQ [8]. Note that these methods jointly optimize GAN and quantized model. Thus, they can generate *unlimited* synthetic data during finetuning. We originally have 1.6M synthetic images (1600 images/class), and then filter 0.4M (400 images/class) images using energy score and BN distance filtering. We use the SGD optimizer with a learning rate of 0.001 followed by a cosine annealing decay schedule for 50 epochs of QAT. Additionally, we report the LSQ results using 1.2M real ImageNet data. The results are summarized in Table 3. Our method largely outperforms other data-free QAT methods and nearly approaches the accuracy of LSQ baseline. Remarkably, *GenQ*

Table 3. Evaluation of data-free QAT on CNN models (top-1 accuracy (%)).

Method	#Real Data	#Syn Data	#Bits W/A	Res18	MBV2	Res50
LSQ [15]	1.2M	0	4/4	71.10	69.50	76.70
GDFQ [62]	0	1.2M	4/4	60.60	59.43	54.16
ZAQ [39]	0	4.6M		52.64	0.10	53.02
Qimera [10]	0	1.2M		63.84	61.62	66.25
IntraQ [72]	0	5k		66.47	65.10	-
ARC+AIT [11]	0	1.2M		65.73	66.47	68.27
DSG [50]	0	1.2M		62.18	60.46	71.96
AdaDFQ [49]	0	1.2M		66.53	65.41	68.38
TexQ [8]	0	1.2M		67.73	67.07	70.72
GenQ	0	1.2M		**70.03**	**69.65**	**76.10**
LSQ [15]	1.2M	0	3/3	70.20	65.30	75.80
GDFQ [62]	0	1.2M	3/3	20.23	1.46	0.31
AdaDFQ [49]	0	1.2M		38.10	28.99	17.63
TexQ [8]	0	1.2M		50.28	32.80	25.27
GenQ	0	1.2M		**68.18**	**59.15**	**73.99**

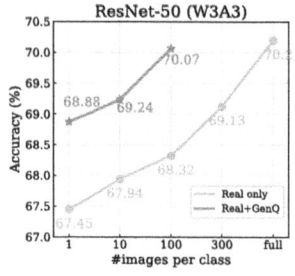

Fig. 4. Evaluation of data-scarce *GenQ* .

Table 4. Evaluation of data-free PTQ on ViTs (top-1 accuracy (%))

Quant. Method	Syn. Method	#Bits W/A	ViT-B	DeiT-B	Swin-B
Full Prec.	N/A	32/32	84.54	81.80	85.27
PTQ4ViT [66]	Real Data	4/4	58.07	63.57	75.20
	PSAQ-ViT [36]		60.26	65.18	72.12
	GenQ		**63.17**	**67.08**	**74.27**
RepQ-ViT [37]	Real Data	4/4	67.93	75.99	72.80
	PSAQ-ViT [36]		60.18	74.69	54.21
	GenQ		**67.50**	**76.10**	70.08

improves the accuracy of W3A3 quantization by a large margin, for instance, *GenQ* exceeds the accuracy of TexQ, the best-performing method, by 48% on ResNet-50.

Comparisons on Data-Scarce QAT. In this section, we evaluate the data-scarce QAT scenarios. Our main comparison is LSQ [15] using different amounts of real data. We show how much of the real data can our *GenQ* generated synthetic data match in practice. We experiment this on a ResNet-50 with W3A3 quantization. We initialize the QAT model by performing GENIE-M on 1k real data and then finetune the model using {1k, 10k, 100k, 300k, 1.2M} real data, corresponding to 1-shot, 10-shot, 100-shot, 300-shot, and full-shot data regimes, respectively. To generate *GenQ* data, we initialize the token embedding as

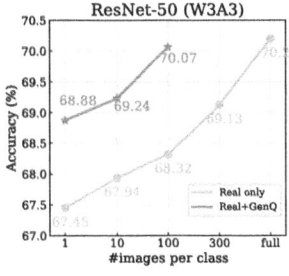

Fig. 5. Accuracy vs # syn. data with data-free PTQ using Genie-M.

Table 5. Cross-model evaluation on synthetic data. We compare the data transferability between (a) existing methods and (b) *GenQ* .

(a) GENIE-D & PSAQ-ViT			(b) *GenQ*		
Source model	Res18	MBV2	Source model	Res18	MBV2
Res18	64.86	48.37	Res18	65.72	54.32
MBV2	64.21	51.47	MBV2	65.52	54.82
ViT-B-16	47.27	17.96	ViT-B-16	65.44	54.24

ImageNet, and train the embedding for 50k iterations, (see Appendix A for more details).

As demonstrated in Fig. 4, the performance of QAT is highly correlated with the number of real data. Our method can boost the performance of QAT under the low data regime. For example, when only 1k real images are provided (*i.e.*, 1-shot), *GenQ* can achieve 68.88% ImageNet accuracy, similar to the QAT performance that needs 300× more training data.

5.3 Data Transferability Evaluation

It is demonstrated that the synthetic data extracted from one model has low transferability on other models [35]. To demonstrate that *GenQ* has relatively high transferability, we conduct a cross-model evaluation, *e.g.*, synthesizing images based on model A, and evaluate the quantization on model B. In Table 5a we provide the results of GENIE-D and PSAQ-ViT, tested on data-free PTQ W2A4 scenarios. We then summarize the results of *GenQ* in Table 5b. We use Genie-M as our PTQ method for CNN architectures and PASQ-ViT for ViT. We observe that the *GenQ* data, although filtered by a different model, only drops 0.2–0.5% final accuracy, while the existing methods drop 0.6–33% accuracy, especially when using ViT synthetic data for MobileNetV2 (MBV2) quantization.

5.4 Ablation Study

In this section, we conduct ablation studies on two variables, (1) the number of synthetic data, and (2) the image filtering strength.

Number of Synthetic Data in PTQ. In practice, synthesizing images using *GenQ* requires very low latency (1 image/second). Hence, we can safely increase the number of synthetic data with minimal overhead in PTQ cases. Yet retrieving more real data seems much more difficult if the training data is private or under IP protection. In Fig. 5, we show that increasing the synthetic data can consistently improve the PTQ performances of W2A4 quantized ResNet-50. *GenQ* even outperforms the quantized model baseline (optimized using PTQ on the 1k real training data) when using only 4k synthetic input images.

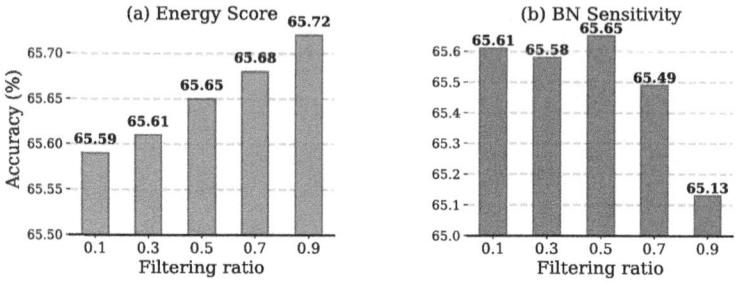

Fig. 6. Accuracy vs filtering strength with data-free PTQ.

Energy Score Filtering Strength. Given a fixed number of synthetic data (N) used for quantization and a percentage r representing how many images are filtered out, the total amount of generated images is $\frac{N}{1-r}$. We test different options of r from $\{0.1, 0.3, 0.5, 0.7, 0.9\}$ to test the effect of energy score filtering. We choose ResNet-18 and use W2A4 data-free PTQ (with Genie-M) for evaluation. The test performance is shown in Fig. 6(a). We generally find that a higher filtering ratio leads to better test accuracy. However, it will also increase the number of generated images. Nevertheless, the accuracy variation is rather small.

BN Sensitivity Filtering Strength. We run the same test for the BN sensitivity filtering mechanism. The results are shown in Fig. 6(b). Unlike the energy score, setting the ratio to 50% has the best performance. We hypothesize that the networks need more diverse synthetic data to effectively quantize a model.

6   Conclusion

In this paper, we have introduced *GenQ*, a novel attempt to synthesize images with text-to-image models for data-scarce quantization. Our *GenQ* generates images in both data free (no real data) and data scarce (few real data) regimes and refines the images with several filtering mechanisms and a token embedding learning algorithm. Extensive experiments show that *GenQ* establishes a new state of the art in both PTQ and QAT.

Acknowledgment. This work was supported in part by CoCoSys, a JUMP2.0 center sponsored by DARPA and SRC, the National Science Foundation (CAREER Award, Grant #2312366, Grant #2318152), and the DoE MMICC center SEA-CROGS (Award #DE-SC0023198).

References

1. runwayml/stable-diffusion-v1-5   Hugging Face — huggingface.co. https://huggingface.co/runwayml/stable-diffusion-v1-5. Accessed 13 Nov 2023

2. Azizi, S., Kornblith, S., Saharia, C., Norouzi, M., Fleet, D.J.: Synthetic data from diffusion models improves imagenet classification. arXiv preprint arXiv:2304.08466 (2023)

3. Banner, R., Nahshan, Y., Soudry, D.: Post training 4-bit quantization of convolutional networks for rapid-deployment. In: Advances in Neural Information Processing Systems, vol. 32 (2019)

4. Besnier, V., Jain, H., Bursuc, A., Cord, M., Pérez, P.: This dataset does not exist: training models from generated images. In: ICASSP 2020-2020 IEEE International Conference on Acoustics, Speech and Signal Processing (ICASSP), pp. 1–5. IEEE (2020)

5. Blalock, D., Gonzalez Ortiz, J.J., Frankle, J., Guttag, J.: What is the state of neural network pruning? Proc. Mach. Learn. Syst. **2**, 129–146 (2020)

6. Bommasani, R., et al.: On the opportunities and risks of foundation models. arXiv preprint arXiv:2108.07258 (2021)

7. Cai, Y., Yao, Z., Dong, Z., Gholami, A., Mahoney, M.W., Keutzer, K.: Zeroq: a novel zero shot quantization framework. In: Proceedings of the IEEE/CVF Conference on Computer Vision and Pattern Recognition, pp. 13169–13178 (2020)

8. Chen, X., Wang, Y., Yan, R., Liu, Y., Guan, T., He, Y.: Texq: zero-shot network quantization with texture feature distribution calibration. In: Thirty-seventh Conference on Neural Information Processing Systems (2023)

9. Choi, J., Wang, Z., Venkataramani, S., Chuang, P.I.J., Srinivasan, V., Gopalakrishnan, K.: Pact: parameterized clipping activation for quantized neural networks. arXiv preprint arXiv:1805.06085 (2018)

10. Choi, K., Hong, D., Park, N., Kim, Y., Lee, J.: Qimera: data-free quantization with synthetic boundary supporting samples. Adv. Neural. Inf. Process. Syst. **34**, 14835–14847 (2021)

11. Choi, K., et al.: It's all in the teacher: Zero-shot quantization brought closer to the teacher. In: Proceedings of the IEEE/CVF Conference on Computer Vision and Pattern Recognition, pp. 8311–8321 (2022)

12. Deng, J., Dong, W., Socher, R., Li, L.J., Li, K., Fei-Fei, L.: Imagenet: a large-scale hierarchical image database. In: 2009 IEEE Conference on Computer Vision and Pattern Recognition, pp. 248–255. IEEE (2009)

13. Dhariwal, P., Nichol, A.: Diffusion models beat GANs on image synthesis. Adv. Neural. Inf. Process. Syst. **34**, 8780–8794 (2021)

14. Dosovitskiy, A., et al.: An image is worth 16x16 words: Transformers for image recognition at scale. arXiv preprint arXiv:2010.11929 (2020)

15. Esser, S.K., McKinstry, J.L., Bablani, D., Appuswamy, R., Modha, D.S.: Learned step size quantization. arXiv preprint arXiv:1902.08153 (2019)

16. Fang, J., Shafiee, A., Abdel-Aziz, H., Thorsley, D., Georgiadis, G., Hassoun, J.H.: Post-training piecewise linear quantization for deep neural networks. In: Computer Vision–ECCV 2020: 16th European Conference, Glasgow, UK, August 23–28, 2020, Proceedings, Part II 16. pp. 69–86. Springer (2020). https://doi.org/10.1007/978-3-030-58536-5_5

17. Finkelstein, A., Almog, U., Grobman, M.: Fighting quantization bias with bias. arXiv preprint arXiv:1906.03193 (2019)

18. Gholami, A., Kim, S., Dong, Z., Yao, Z., Mahoney, M.W., Keutzer, K.: A survey of quantization methods for efficient neural network inference. arXiv preprint arXiv:2103.13630 (2021)
19. Goodfellow, I., Bengio, Y., Courville, A., Bengio, Y.: Deep learning, vol. 1. MIT Press (2016)
20. Goodfellow, I., et al.: Generative adversarial networks. Commun. ACM **63**(11), 139–144 (2020)
21. Haroush, M., Hubara, I., Hoffer, E., Soudry, D.: The knowledge within: Methods for data-free model compression. In: Proceedings of the IEEE/CVF Conference on Computer Vision and Pattern Recognition, pp. 8494–8502 (2020)
22. He, K., Zhang, X., Ren, S., Sun, J.: Deep residual learning for image recognition. In: Proceedings of the IEEE Conference on Computer Vision and Pattern Recognition, pp. 770–778 (2016)
23. He, R., et al.: Is synthetic data from generative models ready for image recognition? arXiv preprint arXiv:2210.07574 (2022)
24. Hinton, G., Vinyals, O., Dean, J.: Distilling the knowledge in a neural network. arXiv preprint arXiv:1503.02531 (2015)
25. Ho, J., Jain, A., Abbeel, P.: Denoising diffusion probabilistic models. Adv. Neural. Inf. Process. Syst. **33**, 6840–6851 (2020)
26. Howard, A.G., et al.: Mobilenets: efficient convolutional neural networks for mobile vision applications. arXiv preprint arXiv:1704.04861 (2017)
27. Hubara, I., Courbariaux, M., Soudry, D., El-Yaniv, R., Bengio, Y.: Quantized neural networks: training neural networks with low precision weights and activations. J. Mach. Learn. Res. **18**(1), 6869–6898 (2017)
28. 1 Hubara, I., Nahshan, Y., Hanani, Y., Banner, R., Soudry, D.: Improving post training neural quantization: layer-wise calibration and integer programming. arXiv preprint arXiv:2006.10518 (2020)
29. Ioffe, S.: Batch renormalization: towards reducing minibatch dependence in batch-normalized models. In: Advances in Neural Information Processing Systems, vol. 30 (2017)
30. Ioffe, S., Szegedy, C.: Batch normalization: accelerating deep network training by reducing internal covariate shift. In: International Conference on Machine Learning, pp. 448–456. pmlr (2015)
31. Jahanian, A., Puig, X., Tian, Y., Isola, P.: Generative models as a data source for multiview representation learning. arXiv preprint arXiv:2106.05258 (2021)
32. Jeon, Y., Lee, C., Kim, H.y.: Genie: show me the data for quantization. In: Proceedings of the IEEE/CVF Conference on Computer Vision and Pattern Recognition, pp. 12064–12073 (2023)
33. Jung, S., et al.: Learning to quantize deep networks by optimizing quantization intervals with task loss. In: Proceedings of the IEEE/CVF Conference on Computer Vision and Pattern Recognition, pp. 4350–4359 (2019)
34. Li, Y., et al.: Brecq: pushing the limit of post-training quantization by block reconstruction. arXiv preprint arXiv:2102.05426 (2021)
35. Li, Y., et al.: Mixmix: all you need for data-free compression are feature and data mixing. In: Proceedings of the IEEE/CVF International Conference on Computer Vision, pp. 4410–4419 (2021)
36. Li, Z., Ma, L., Chen, M., Xiao, J., Gu, Q.: Patch similarity aware data-free quantization for vision transformers. In: European Conference on Computer Vision. pp. 154–170. Springer (2022). https://doi.org/10.1007/978-3-031-20083-0_10

37. Li, Z., Xiao, J., Yang, L., Gu, Q.: Repq-vit: scale reparameterization for post-training quantization of vision transformers. In: Proceedings of the IEEE/CVF International Conference on Computer Vision, pp. 17227–17236 (2023)

38. Liu, W., Wang, X., Owens, J., Li, Y.: Energy-based out-of-distribution detection. Adv. Neural. Inf. Process. Syst. **33**, 21464–21475 (2020)

39. Liu, Y., Zhang, W., Wang, J.: Zero-shot adversarial quantization. In: Proceedings of the IEEE/CVF Conference on Computer Vision and Pattern Recognition, pp. 1512–1521 (2021)

40. Liu, Z., et al.: Swin transformer: hierarchical vision transformer using shifted windows. In: Proceedings of the IEEE/CVF International Conference on Computer Vision, pp. 10012–10022 (2021)

41. Liu, Z., Wang, Y., Han, K., Zhang, W., Ma, S., Gao, W.: Post-training quantization for vision transformer. Adv. Neural. Inf. Process. Syst. **34**, 28092–28103 (2021)

42. Lugmayr, A., Danelljan, M., Romero, A., Yu, F., Timofte, R., Van Gool, L.: Repaint: inpainting using denoising diffusion probabilistic models. In: Proceedings of the IEEE/CVF Conference on Computer Vision and Pattern Recognition, pp. 11461–11471 (2022)

43. Mehmood, A., Natgunanathan, I., Xiang, Y., Hua, G., Guo, S.: Protection of big data privacy. IEEE access **4**, 1821–1834 (2016)

44. Molchanov, P., Tyree, S., Karras, T., Aila, T., Kautz, J.: Pruning convolutional neural networks for resource efficient inference. arXiv preprint arXiv:1611.06440 (2016)

45. Nagel, M., Amjad, R.A., Van Baalen, M., Louizos, C., Blankevoort, T.: Up or down? adaptive rounding for post-training quantization. In: International Conference on Machine Learning, pp. 7197–7206. PMLR (2020)

46. Nahshan, Y., Chmiel, B., Baskin, C., Zheltonozhskii, E., Banner, R., Bronstein, A.M., Mendelson, A.: Loss aware post-training quantization. Mach. Learn. **110**(11–12), 3245–3262 (2021)

47. Nichol, A., et al.: Glide: towards photorealistic image generation and editing with text-guided diffusion models. arXiv preprint arXiv:2112.10741 (2021)t

48. Pimentel, M.A., Clifton, D.A., Clifton, L., Tarassenko, L.: A review of novelty detection. Signal Process. **99**, 215–249 (2014)

49. Qian, B., Wang, Y., Hong, R., Wang, M.: Adaptive data-free quantization. In: Proceedings of the IEEE/CVF Conference on Computer Vision and Pattern Recognition, pp. 7960–7968 (2023)

50. Qin, H., Ding, Y., Zhang, X., Wang, J., Liu, X., Lu, J.: Diverse sample generation: pushing the limit of generative data-free quantization. IEEE Transactions on Pattern Analysis and Machine Intelligence (2023)

51. Radford, A., et al.: Learning transferable visual models from natural language supervision. In: International Conference on Machine Learning, pp. 8748–8763. PMLR (2021)

52. Rombach, R., Blattmann, A., Lorenz, D., Esser, P., Ommer, B.: High-resolution image synthesis with latent diffusion models. In: Proceedings of the IEEE/CVF Conference on Computer Vision and Pattern Recognition, pp. 10684–10695 (2022)

53. Saharia, C., et al.: Palette: Image-to-image diffusion models. In: ACM SIGGRAPH 2022 Conference Proceedings, pp. 1–10 (2022)

54. Sandler, M., Howard, A., Zhu, M., Zhmoginov, A., Chen, L.C.: Mobilenetv2: inverted residuals and linear bottlenecks. In: Proceedings of the IEEE Conference on Computer Vision and Pattern Recognition, pp. 4510–4520 (2018)

55. Shin, J., Kang, M., Park, J.: Fill-up: balancing long-tailed data with generative models. arXiv preprint arXiv:2306.07200 (2023)

56. StabilityAI: Stable diffusion public release (Aug 2023). https://stability.ai/blog/stable-diffusion-public-release

57. Tan, M., Chen, B., Pang, R., Vasudevan, V., Le, Q.V.: Mnasnet: platform-aware neural architecture search for mobile. corr abs/1807.11626 (2018). arXiv preprint arXiv:1807.11626 (2018)

58. Tian, Y., Fan, L., Isola, P., Chang, H., Krishnan, D.: Stablerep: synthetic images from text-to-image models make strong visual representation learners. arXiv preprint arXiv:2306.00984 (2023)

59. Touvron, H., Cord, M., Douze, M., Massa, F., Sablayrolles, A., Jégou, H.: Training data-efficient image transformers and distillation through attention. In: International Conference on Machine Learning, pp. 10347–10357. PMLR (2021)

60. Wei, X., Gong, R., Li, Y., Liu, X., Yu, F.: Qdrop: randomly dropping quantization for extremely low-bit post-training quantization. arXiv preprint arXiv:2203.05740 (2022)

61. Wu, W., et al.: Datasetdm: synthesizing data with perception annotations using diffusion models. In: Advances in Neural Information Processing Systems, vol. 36 (2024)

62. Xu, S., et al.: Generative low-bitwidth data free quantization. In: Vedaldi, A., Bischof, H., Brox, T., Frahm, J.-M. (eds.) ECCV 2020. LNCS, vol. 12357, pp. 1–17. Springer, Cham (2020). https://doi.org/10.1007/978-3-030-58610-2_1

63. Yamamoto, K.: Learnable companding quantization for accurate low-bit neural networks. In: Proceedings of the IEEE/CVF Conference on Computer Vision and Pattern Recognition, pp. 5029–5038 (2021)

64. Yin, H., et al.: Dreaming to distill: data-free knowledge transfer via deepinversion. In: Proceedings of the IEEE/CVF Conference on Computer Vision and Pattern Recognition, pp. 8715–8724 (2020)

65. Yin, P., Lyu, J., Zhang, S., Osher, S., Qi, Y., Xin, J.: Understanding straight-through estimator in training activation quantized neural nets. arXiv preprint arXiv:1903.05662 (2019)

66. Yuan, Z., Xue, C., Chen, Y., Wu, Q., Sun, G.: Ptq4vit: post-training quantization for vision transformers with twin uniform quantization. In: European Conference on Computer Vision, pp. 191–207. Springer (2022). https://doi.org/10.1007/978-3-031-19775-8_12

67. Zhang, C., et al.: A complete survey on generative ai (aigc): Is chatgpt from gpt-4 to gpt-5 all you need? arXiv preprint arXiv:2303.11717 (2023)

68. Zhang, D., Yang, J., Ye, D., Hua, G.: Lq-nets: learned quantization for highly accurate and compact deep neural networks. In: Proceedings of the European conference on computer vision (ECCV), pp. 365–382 (2018)

69. Zhang, X., et al.: Diversifying sample generation for accurate data-free quantization. In: Proceedings of the IEEE/CVF Conference on Computer Vision and Pattern Recognition, pp. 15658–15667 (2021)

70. Zhang, Y., et al.: Datasetgan: Efficient labeled data factory with minimal human effort. In: Proceedings of the IEEE/CVF Conference on Computer Vision and Pattern Recognition. pp. 10145–10155 (2021)

71. Zhao, R., Hu, Y., Dotzel, J., De Sa, C., Zhang, Z.: Improving neural network quantization without retraining using outlier channel splitting. In: International Conference on Machine Learning, pp. 7543–7552. PMLR (2019)

72. Zhong, Y., et al.: Intraq: learning synthetic images with intra-class heterogeneity for zero-shot network quantization. In: Proceedings of the IEEE/CVF Conference on Computer Vision and Pattern Recognition, pp. 12339–12348 (2022)
73. Zhou, S., Wu, Y., Ni, Z., Zhou, X., Wen, H., Zou, Y.: Dorefa-net: training low bitwidth convolutional neural networks with low bitwidth gradients. arXiv preprint arXiv:1606.06160 (2016)

MVDD: Multi-view Depth Diffusion Models

Zhen Wang[1,2]([✉]), Qiangeng Xu[1], Feitong Tan[1], Menglei Chai[1], Shichen Liu[1], Rohit Pandey[1], Sean Fanello[1], Achuta Kadambi[1,2], and Yinda Zhang[1]

[1] Google, San Jose, USA
[2] University of California, Los Angeles, USA
zhenwang@ucla.edu
https://mvdepth.github.io/

Abstract. Denoising diffusion models have demonstrated outstanding results in 2D image generation, yet it remains a challenge to replicate its success in 3D shape generation. In this paper, we propose leveraging multi-view depth, which represents complex 3D shapes in a 2D data format that is easy to denoise. We pair this representation with a diffusion model, MVDD, that is capable of generating high-quality dense point clouds with 20K+ points with fine-grained details. To enforce 3D consistency in multi-view depth, we introduce an epipolar line segment attention that conditions the denoising step for a view on its neighboring views. Additionally, a depth fusion module is incorporated into diffusion steps to further ensure the alignment of depth maps. When augmented with surface reconstruction, MVDD can also produce high-quality 3D meshes. Furthermore, MVDD stands out in other tasks such as depth completion, and can serve as a 3D prior, significantly boosting many downstream tasks, such as GAN inversion. State-of-the-art results from extensive experiments demonstrate MVDD's excellent ability in 3D shape generation, depth completion, and its potential as a 3D prior for downstream tasks.

Keywords: 3D Shape Generation · Diffusion Model

1 Introduction

3D shape generative models have made remarkable progress in the wave of AI-Generated Content. A powerful 3D generative model is expected to possess the following attributes: (i) *Scalability*. The model should be able to create objects with fine-grained details; (ii) *Faithfulness*. The generated 3D shapes should

---

Z. Wang—Work done while the author was an intern at Google.

Supplementary Information The online version contains supplementary material available at https://doi.org/10.1007/978-3-031-72624-8_14.

A. Leonardis et al. (Eds.): ECCV 2024, LNCS 15071, pp. 236–253, 2025.
https://doi.org/10.1007/978-3-031-72624-8_14

exhibit high fidelity and resemble the objects in the dataset; and (iii) *Versatility*. The model can be plugged in as a 3D prior in various downstream 3D tasks through easy adaptation. Selecting suitable probabilistic models becomes the key factor in achieving these criteria. Among popular generative methods such as GANs [10,20], VAEs [17], and normalizing flows [33], denoising diffusion models [11,45] explicitly model the data distribution; therefore, they are able to faithfully generate samples that reflect content diversity.

It is also important to choose suitable 3D representations for shape generation. While delivering high geometric quality and infinite resolution, implicit function-based models [5,6,19,29,31,51,52] tend to be computationally expensive. This is due to the fact that the number of inferences increases cubically with the resolution and the time-consuming post-process, e.g., marching cubes. On the other hand, studies [26,56,60] learn diffusion models on a point cloud by adding noise and denoising either directly on point positions or their latent embeddings. Due to the irregular data format of the point set, it requires over 10,000 epochs for these diffusion models to converge on a single ShapeNet [4] category, while the number of points that can be generated by these models typically hovers around 2048.

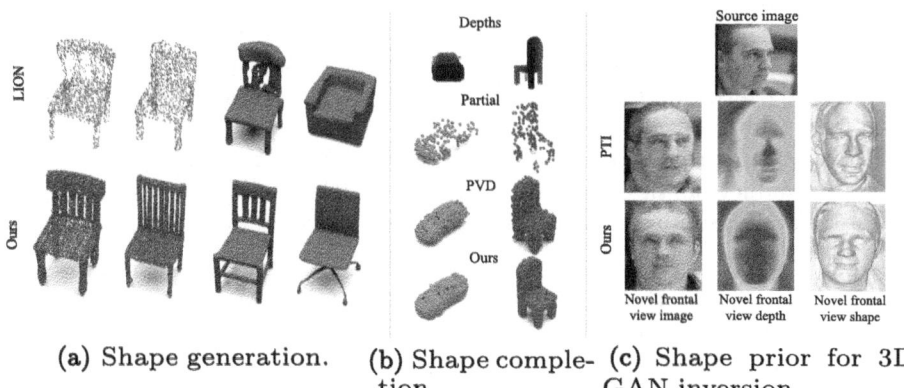

(a) Shape generation.   (b) Shape completion.   (c) Shape prior for 3D GAN inversion.

Fig. 1. Our proposed MVDD is versatile and can be utilized in various applications: (a) 3D shape generation: our model generates high-quality 3D shape with approximately 10X more points than diffusion-based point cloud generative models e.g., LION [56] and PVD [60] and contains diverse and fine-grained details. (b) Shape completion: we showcase shape completion results from partial inputs, highlighting the higher fidelity compared to PVD [60]. (c) Our model can serve as a powerful shape prior for downstream tasks such as 3D GAN inversion [3,39].

In this work, we investigate a multi-view depth representation and propose a novel diffusion model, namely MVDD, which generates 3D consistent multi-view depth maps for 3D shape generation. The benefits of using the multi-view depth representation with diffusion models come in three folds: 1) The representation is naturally supported by diffusion models. The 2D data format conveniently allows the direct adoption of powerful 2D diffusion architectures [21,37,42]; 2) Multi-view depth registers complex 3D surfaces onto 2D grids, essentially reducing the dimensionality of the 3D generation space to 2D. As a result, the generated 2D depth map can have higher resolution than volumetric implicit representations [29] and produce dense point clouds with a much higher point count; 3) Depth is a widely used representation; therefore, it is easy to use it as a 3D prior to support downstream applications.

While bearing this many advantages, one key challenge of using multi-view depths for 3D shape generation is cross-view consistency. Even with a well-trained diffusion model that learns the depth distribution from 3D consistent depth maps, the generated multi-view depth maps are not guaranteed to be consistent after ancestral sampling [26]. To tackle this challenge, our proposed MVDD conditions diffusion steps for each view on neighboring views, allowing different views to exchange information. This is realized by a novel epipolar "line segment" attention, which benefits from epipolar geometry. Differing from full attention [41] and epipolar attention [23], our epipolar "line segment" attention leverages the depth estimation in our diffusion process. Therefore, it only attends to features at the most relevant locations, making it both effective and efficient. However, even with relatively consistent multi-view maps, back-projected 3D points from each depth map are still not guaranteed to be perfectly aligned, resulting in "double layers" in the 3D shapes (see Fig. 7(c)). To address this issue, MVDD incorporates depth fusion in denoising steps to explicitly align depth from multiple views.

Empowered by these modules, MVDD can generate high-quality 3D shapes, faithfully conduct depth completion, and distill 3D prior knowledge for downstream tasks. We summarize our contributions as follows:

- To the best of our knowledge, we propose the first multi-view depth representation in the generative setting with a diffusion model MVDD. The representation reduces the dimension of the generation space and avoid unstructured data formats such as point set. Therefore, it is more scalable and suitable for diffusion frameworks and is easier to converge.
- We also propose an epipolar "line segment" attention and denoising depth fusion that could effectively enforce 3D consistency for multi-view depth maps.
- Through extensive experiments, we show the flexibility and versatility of MVDD in various tasks such as 3D shape generation and shape completion. Our method outperforms compared methods in both shape generation and shape completion by substantial margins.

2   Related Work

2.1   3D Shape Generative Models

Representations such as implicit functions, voxels, point clouds, and tetrahedron meshes have been used for 3D shape generation in previous studies.

Implicit-based models, such as AutoSDF [29], infer SDF from feature volumes. Since the computation for volumes grows cubically with resolution, the volume resolution is often limited. Voxel-based models, such as [30,60], face the same challenge. Other implicit-based models, such as 3D-LDM [31], IM-GAN [5], and Diffusion-sdf [6], generate latent codes and use auto-encoders to infer SDFs. The latent solution helps avoid the limitation of resolution but is prone to generate overly smoothed shapes. When combined with tetrahedron mesh, implicit methods [9,24] are able to generate compact implicit fields and achieve high-quality shape generation. However, unlike multi-view depth, it is non-trivial for them to serve as a 3D prior in downstream tasks that do not use tetrahedron grids.

Point cloud-based methods avoid modeling empty space inherently. Previous explorations include SetVAE [15] and VG-VAE [1], which adopt VAEs for point latent sampling. GAN-based models [43,50] employ adversarial loss to generate point clouds. Flow-based models [18,54] use affine coupling layers to model point distributions. To enhance generation diversity, some studies leverage diffusion [11,45] to generate 3D point cloud distributions. ShapeGF [2] applies the score-matching gradient flow to move the point set. DPM [26] and PVD [60] denoise Gaussian noise on point locations. LION [56] encodes the point set into latents and then conducts latent diffusion. Although these models excel in producing diverse shapes, the denoising scheme on unstructured point cloud data limits the number of points that can be generated. Our proposed model leverages multi-view depth representation, which can generate high-resolution point clouds, leading to 3D shapes with fine details.

2.2   Multi-view Diffusion Models

The infamous Janus problem [27,36] and 3D inconsistency have plagued SDS-based [36] 3D content generation. MVDream [41] connects rendered images from different views via a 3D self-attention fashion to constrain multi-view consistency in the generated multi-view images. SyncDreamer [23] builds a cost volume that correlates the corresponding features across different views to synchronize the intermediate states of all the generated images at each step of the reverse process. EfficientDreamer [59] and TextMesh [48] concatenate canonical views either channel-wise or spatially into the diffusion models to enhance 3D consistency. Wonder3D [25] generates both RGB and normal maps for the task of single image to 3d. SweetDreamer [22] proposes aligned geometry priors by fine-tuning the 2D diffusion models to be viewpoint-aware and to produce view-specific coordinate maps. Our method differs from them in that we generate multi-view depth maps, instead of RGB images, and thus propose an efficient epipolar line segment attention tailored for depth maps to enforce 3D consistency.

3   Method

In this section, we introduce our Multi-View Depth Diffusion Models (MVDD). We first provide an overview of MVDD in Sect. 3.1, a model that aims to produce multi-view depth. After that, we illustrate how multi-view consistency is enforced among different views of depth maps in our model by using epipolar "line segment" attention (Sect. 3.1) and denoising depth fusion (Sect. 3.1). Finally, we introduce the training objectives in Sect. 3.2 and implementation details in Sect. 3.3.

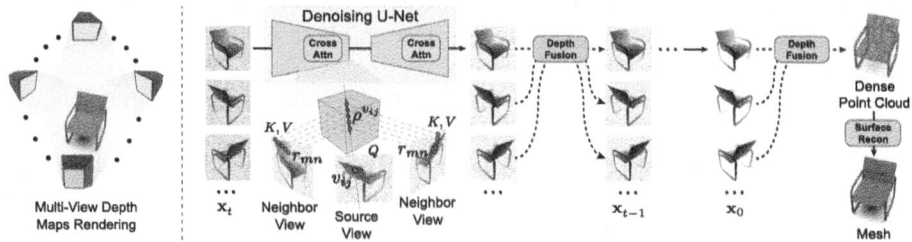

Fig. 2. Our method collects ground truth from multi-view rendered depth maps (left). Starting with multiple 2D maps with randomly sampled noise, MVDD generates diverse multi-view depth through an iterative denoising diffusion process (right). To enforce multi-view 3D consistency, MVDD denoises each depth map with an efficient epipolar "line segment" attention (Sect. 3.1). Specifically, by leveraging the denoised value from the current step, MVDD only needs to attend to features on a line segment centered around the back-projected depth (the red dot), rather than the entire epipolar line. To further align the denoised multi-view depths, depth fusion (Sect. 3.1) is incorporated after the U-Net in a denoising step. The final multi-view depth can be fused together to obtain a high-quality dense point cloud, which can then be reconstructed into high quality 3D meshes with fine-grained details.

3.1   Multi-View Depth Diffusion

Our method represents a 3D shape $\mathcal{X}$ by using its multi-view depth maps $\mathbf{x} \in \mathbb{R}^{N \times H \times W} = \{\mathbf{x}^v | v = 1, 2, ..., N\}$, where v is the index of the view, N is the total number of views, and H and W are the depth map resolution. To generate a 3D shape that is both realistic and faithful to the diversity distribution, we adopt the diffusion process [11, 44] that gradually denoise N depth maps. These depth maps can be fused to obtain a dense point cloud, which can optionally be used to reconstruct [13, 35] a high-quality mesh model. We illustrate the entire pipeline in Fig. 2.

In the diffusion process, we first create the ground truth multi-view depth diffusion distribution $q(\mathbf{x}_{0:T})$ in a *forward process*. In this process, we gradually add Gaussian noise to each ground truth depth map $\mathbf{x}_0^v$ for T steps, obtaining N depth maps of pure Gaussian noise $\mathbf{x}_T = \{\mathbf{x}_T^v | v = 1, 2, ..., N\}$. The joint distributions can be factored into a product of per-view Markov chains:

$$q(\mathbf{x}_{0:T}) = q(\mathbf{x}_0) \prod_{t=1}^{T} q(\mathbf{x}_t|\mathbf{x}_{t-1})$$

$$= q(\mathbf{x}_0^{1:N}) \prod_{v=1}^{N} \prod_{t=1}^{T} q(\mathbf{x}_t^v|\mathbf{x}_{t-1}^v), \tag{1}$$

$$q(\mathbf{x}_t^v|\mathbf{x}_{t-1}^v) := \mathcal{N}(\mathbf{x}_t^v; \sqrt{1-\beta_t}\mathbf{x}_{t-1}^v, \beta_t\mathbf{I}), \tag{2}$$

where β_t is the step t variance schedule at step t shared across views.

We then learn a diffusion denoising model to predict the distribution of a *reverse process* $p_\theta(\mathbf{x}_{0:T})$ to iteratively denoise the $\mathbf{x}_T$ back to the ground truth $\mathbf{x}_0$. The joint distribution can be formulated as:

$$p_\theta(\mathbf{x}_{0:T}) = p(\mathbf{x}_T) \prod_{t=1}^{T} p_\theta(\mathbf{x}_{t-1}|\mathbf{x}_t)$$

$$= p(\mathbf{x}_T^{1:N}) \prod_{v=1}^{N} \prod_{t=1}^{T} p_\theta(\mathbf{x}_{t-1}^v|\mathbf{x}_t^v), \tag{3}$$

$$p_\theta(\mathbf{x}_{t-1}^v|\mathbf{x}_t^v) := \mathcal{N}(\mathbf{x}_{t-1}^v; \mu_\theta(\mathbf{x}_t^v, t), \beta_t\mathbf{I}), \tag{4}$$

where $\mu_\theta(\mathbf{x}_t^v, t)$ estimates the mode of depth map distribution for view v at step $t-1$.

However, following Eq. (3) and Eq. (4), diffusion process denoises each view independently. Starting from N maps of pure random noise, a well-trained model of this kind would generate realistic depth maps $\mathbf{x}_0^{1:N}$, which however could not be fused into an intact shape due to no 3D consistency across views. Therefore, we propose to condition denoising steps for each view on its R neighboring views $\mathbf{x}_t^{r_1:r_R}$ and replace Eq. (3) and Eq. (4) with:

$$p_\theta(\mathbf{x}_{0:T}) = p(\mathbf{x}_T^{1:N}) \prod_{t=1}^{T} \prod_{v=1}^{N} p_\theta(\mathbf{x}_{t-1}^v|\mathbf{x}_t^v, \mathbf{x}_t^{r_1:r_R}), \tag{5}$$

$$p_\theta(\mathbf{x}_{t-1}^v|\mathbf{x}_t^v, \mathbf{x}_t^{r_1:r_R}) := \mathcal{N}(\mathbf{x}_{t-1}^v; \mu_\theta(\mathbf{x}_t^v, \mathbf{x}_t^{r_1:r_R}, t), \beta_t\mathbf{I}). \tag{6}$$

MVDD achieves this through an efficient epipolar 'line segment' attention (Sect. 3.1). Additionally, even though the denoising process is multi-view conditioned, back-projected depth maps are still not guaranteed to be perfectly aligned in 3D space. Inspired by multi-view stereo methods [8,40,53], MVDD conducts denoising depth fusion (Sect. 3.1) in each diffusion step (Eq. (6)).

Epipolar Line Segment Attention. To promote consistency across all depth maps, we introduce an attention module named epipolar "line segment" attention. With the depth value of current step, MVDD leverages this information and attends only to features from visible locations on other views. To be specific, we sample on the line segment centered by the back-projected depth, rather than on the entire epipolar line [41,49]. This design allows the proposed attention to obtain more relevant cross-view features, making it excel in both efficiency and effectiveness. The attention is defined as:

$$Q \in \mathbb{R}^{(B \times N \times H \times W) \times 1 \times F},$$

$$K, V \in \mathbb{R}^{(B \times N \times H \times W) \times (R \times k) \times F},$$

$$\text{Cross-Attn}(Q, K, V) = \text{softmax}\left(\frac{QK^T}{\sqrt{d_k}}\right)V, \tag{7}$$

where B is the batch size, N is the total number of views, k is the number of samples along the epipolar "line segment", R is the number of neighboring views and F is the number of feature channels. At denoising step t, for any pixel v_{ij} at a source depth map $\mathbf{x}_t^v$, we first back project its depth value $\mathbf{x}_t^{v_{ij}}$ into the 3D space to obtain a 3D point $\rho^{v_{ij}}$, and project it to a coordinate r_{mn} on neighboring view r:

$$\rho^{v_{ij}} = \mathbf{x}_t^{v_{ij}} A^{-1} v_{ij}, \text{where } v_{ij} := [i, j, 1]^T, \tag{8}$$

$$r_{mn} = A \, \pi_{v \to r} \rho^{v_{ij}}, \tag{9}$$

where $\rho^{v_{ij}}$ is in the camera coordinate of view v, $\pi_{v \to r}$ is the relative pose transformation matrix and A is the intrinsic matrix. Since $\mathbf{x}_t^{v_{ij}}$ is noisy, we select another $k-1$ evenly spaced points around $\rho^{v_{ij}}$ along the ray and project these points, $\{\rho_1^{v_{ij}}, ..., \rho_k^{v_{ij}}\}$, into each neighboring view, as shown in Fig. 2 (right). The k projected pixels lay on a epipolar "line segment" on view r and provides features for K, V in Eq. (7).

Cross Attention Thresholding. To ensure that depth features from a neighboring view r are relevant to $\mathbf{x}_t^{v_{ij}}$, we need to cull only the r_{mn} that are also visible from source view v. Let $z(\cdot)$ denote the operator to extract the z value from a vector $[x, y, z]$, we create the visibility mask by thresholding the Euclidean distance between the depth value of the 3D point in r's camera coordinate, $\rho^{r_{mn}} = \pi_{v \to r} \rho^{v_{ij}}$, and the predicted depth value on the pixel r_{mn} that ρ^r projects onto:

$$M(r_{mn}) = \|z(\pi_{v \to r} \rho^{v_{ij}}) - \mathbf{x}_t^{r_{mn}}\| < \tau. \tag{10}$$

For projected pixels that do not satisfy the above requirement, in Eq. (7), we manually overwrite their attention weights as a very small value to minimize its effect.

Depth Concatenation. For pixel v_{ij}, since the sampled points $\{\rho_1^{v_{ij}}, ..., \rho_k^{v_{ij}}\}$ query geometric features K, V from neighboring views, the attention mechanism conditions the denoising step of $\mathbf{x}_t^{v_{ij}}$ with the features V weighted by the similarity between Q and K. To enhance awareness of the locations of these points, we propose concatenating the depth values $\{z(\rho_1^{v_{ij}}), ..., z(\rho_k^{v_{ij}})\}$ to the feature dimension of V, resulting in the last dimension of V being $F + 1$.

The intuition behind this is if the geometric features of v_{ij} are very similar to features queried by $\rho_1^{v_{ij}}$, the depth value $\mathbf{x}_{t-1}^{v_{ij}}$ should move toward $z(\rho_1^{v_{ij}})$. We empirically verify the effectiveness of the depth concatenation in Table 3.

Denoising Depth Fusion. To further enforce alignment across multi-view depth maps, MVDD incorporates depth fusion in diffusion steps during ancestral sampling.

Assuming we have multi-view depth maps $\{\mathbf{x}_1, ..., \mathbf{x}_N\}$, following multi-view stereo methods [28,55], a pixel v_{ij} will be projected to another view r at r_{mn} as described in Eq. (8). Subsequently, we reproject r_{mn} with its depth value $\mathbf{x}^{r_{mn}}$ towards view v:

$$\rho^{v_{\tilde{i}\tilde{j}}} = \pi_{r \to v} \mathbf{x}^{r_{mn}} A^{-1} r_{mn}, \tag{11}$$

$$v_{\tilde{i}\tilde{j}} = A\rho^{v_{\tilde{i}\tilde{j}}}, \tag{12}$$

where $\rho^{v_{\tilde{i}\tilde{j}}}$ is the reprojected 3D point in view v's camera coordinate. To determine the visibility of pixel v_{ij} from view r, we set two thresholds:

$$\left\| v_{ij} - v_{\tilde{i}\tilde{j}} \right\| < \psi_{\max}, \quad \frac{|\mathbf{x}^{v_{ij}} - z(\rho^{v_{\tilde{i}\tilde{j}}})|}{\mathbf{x}^{v_{ij}}} < \epsilon_\theta, \tag{13}$$

where $z(\rho^{v_{\tilde{i}\tilde{j}}})$ represents the reprojected depth, $\psi_{\max}$ and ϵ_θ are the thresholds for discrepancies between reprojected pixel coordinates and depth compared to the original ones.

Integration with Denosing Steps. For a diffusion step t described in Eq. (6), after obtaining $\mu_\theta(\mathbf{x}_t, t)$, we apply *depth averaging*. For each pixel, we average the reprojected depths from other visible views to refine this predicted value. Subsequently, we add $\mathcal{N}(\mathbf{0}, \beta_t \mathbf{I})$ on top to obtain $\{\mathbf{x}_{t-1}^v | v = 1, 2, ..., N\}$. Only at the last step, we also apply *depth filtering* to X_0 to filter out the back-projected 3D points that are not visible from neighboring views.

3.2   Training Objectives

Aiming to maximize $p_\theta(\mathbf{x}_{0:T})$, we can minimize the objective, following DDPM [11]:

$$\begin{aligned} L_t &= \mathbb{E}_{t \sim [1,T], \mathbf{x}_0, \epsilon_t} \left[\left\| \epsilon_t - \epsilon_\theta(\mathbf{x}_t, t) \right\|^2 \right] \\ &= \mathbb{E}_{t \sim [1,T], \mathbf{x}_0, \epsilon_t} \left[\left\| \epsilon_t - \epsilon_\theta(\sqrt{\bar{\alpha}_t} \mathbf{x}_0 + \sqrt{1 - \bar{\alpha}_t} \epsilon_t, t) \right\|^2 \right], \end{aligned} \tag{14}$$

where $\mathbf{x}_0$ is the ground-truth multiview depth maps, β_t and $\bar{\alpha}_t := \prod_{s=1}^{t}(1 - \beta_s)$ are predefined coefficients of noise scheduling at step t.

3.3   Implementation Details

Our model is implemented in PyTorch [34] and employs the Adam optimizer [16] with the first and the second momentum set to 0.9 and 0.999, respectively, and a learning rate of $2e^{-4}$ to train all our models. Unless otherwise noted, we set the height H and width W of depth map to both be 128 and number of views of

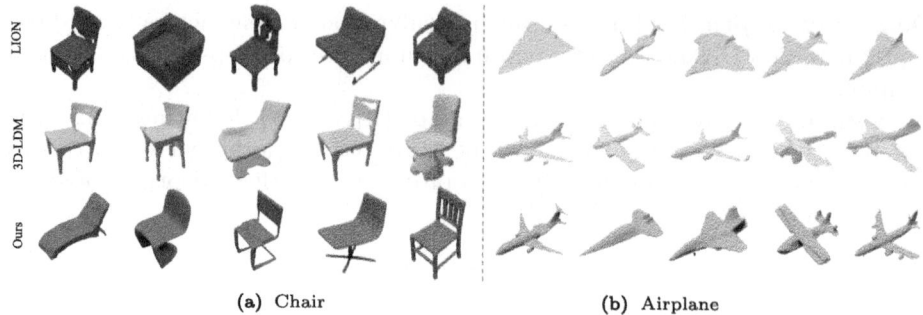

(a) Chair   (b) Airplane

Fig. 3. Our generated meshes exhibit superior quality compared to point cloud diffusion model [56] and implicit diffusion model [31].

Table 1. Unconditional generation on ShapeNet categories. MMD (EMD) is multiplied by 10^2. ⊛ represents the best result.

		Vox-diff [60]	DPM [26]	3D-LDM [31]	IM-GAN [5]	PVD [60]	LION [56]	MVDD (Ours)
Airplane	MMD (EMD)	1.561	0.990	3.520	0.980	1.000	0.920 ⊛	0.920 ⊛
	COV (EMD)	25.43	40.40	42.60	52.07	49.33	48.27	53.00 ⊛
	1-NNA (EMD)	98.13	73.47	80.10	64.04	64.89	63.49	62.50 ⊛
Car	MMD (EMD)	1.551	0.710	–	0.640	0.820	0.900	0.620 ⊛
	COV (EMD)	22.15	36.05	–	47.27	39.51	42.59	49.53 ⊛
	1-NNA (EMD)	96.83	80.33	–	57.04	71.29	65.70	56.80 ⊛
Chair	MMD (EMD)	2.930	2.140	8.200	2.200	2.330	1.720 ⊛	2.110
	COV (EMD)	21.75	46.17	42.20	49.51	46.47	50.52	51.55 ⊛
	1-NNA (EMD)	96.74	65.73	65.30	55.54	56.14	57.31	54.51 ⊛

depth map N to be 8. The first camera is free to be placed anywhere on a sphere, facing the object center, and form a cube with the other 7 cameras. The number of sampled points along the epipolar line segment k is 10. The threshold τ for cross attention thresholding is 0.15. We apply denoising depth fusion only in the last 20 steps. For training, we uniformly sample time steps $t = 1, ..., T = 1000$ for all experiments with cosine scheduling [32]. We train our model on 8 Nvidia A100-80GB and the model usually converges within 3000 epochs. Please refer to supplemental material for more details on network architecture, camera setting, and other aspects.

4   Application

4.1   3D Shape Generation

Inference strategy. Initialized as 2D maps of pure Gaussian noise, the multi-view depth maps can be generated by MVDD following ancestral sampling [11]:

$$\mathbf{x}_{t-1} = \frac{1}{\sqrt{\alpha_t}} \left(\mathbf{x}_t - \frac{1 - \alpha_t}{\sqrt{1 - \bar{\alpha}_t}} \epsilon_\theta \left(\mathbf{x}_t, t \right) \right) + \sqrt{\beta_t}\epsilon, \tag{15}$$

where ϵ follows a isotropic multivariate normal distribution. We iterate the above process for $T = 1000$ steps, utilizing the effective epipolar "line segment" attention and denoising depth fusion. Finally, we back-project the multi-view depth maps to form a dense ($>20K$) 3D point cloud with fine-grained details. Optionally, high-quality meshes can be created with SAP [35] from the dense point cloud.

Datasets and Comparison Methods. To assess the performance of our method compared to established approaches, we employ the ShapeNet dataset [4], which is the commonly adopted benchmark for evaluating 3D shape generative models. In line with previous studies of 3D shape generation [5,54,56,60], we evaluate our model on standard shape generation bench mark categories: airplanes, chairs, and cars, with the same train/test split. We compare MVDD with state-of-the-art point cloud generation methods such as DPM [26], PVD [60] and LION [56], implicit functions-based methods such as IM-GAN [5] and 3D-LDM [31], as well as a voxel diffusion model Vox-diff [60]. As our method generates varying number of points and point cloud backprojected from depth maps is not uniform, we sample 2048 points from meshes using SAP [35] and measure against ground-truth points with inner surface removed. For those implicit methods that are not impacted by inner surface, we directly use the number reported for comparison.

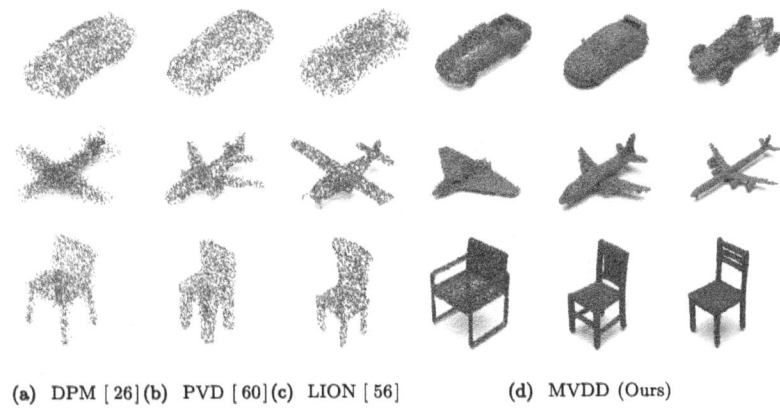

(a) DPM [26] (b) PVD [60] (c) LION [56]   (d) MVDD (Ours)

Fig. 4. Unconditional generation on ShapeNet car, airplane and chair category.

Metrics. We follow [5,56,60] and primarily employ: 1) Minimum Matching Distance (MMD), which calculate the average distance between the point clouds in the reference set and their closest neighbors in the generated set; 2) Coverage (COV), which measures the number of reference point clouds that are matched to at least one generated shape; 3) 1-Nearest Neighbor Alignment (1-NNA), which measures the distributional similarity between the generated shapes and the validation set. MMD focus on the shape fidelity and quality, while COV

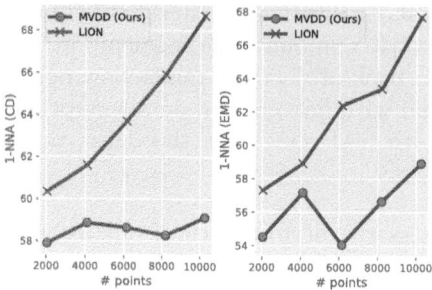

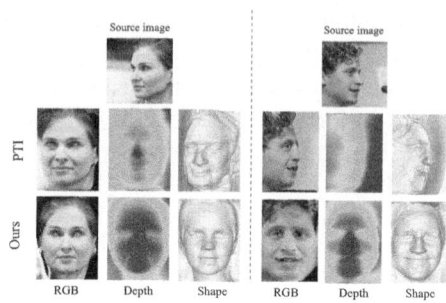

Fig. 5. We report the performance of our method and LION with varying number of point clouds measured by 1-NNA with CD and EMD, respectively, in the ShapeNet [4] chair category.

Fig. 6. Without proper shape regularization, 3D GAN inversion [3] using PTI [39] fails to reconstruct input image under extreme pose. Our model can serve as a shape prior for 3D GAN inversion and yield better reconstruction performance in novel frontal view.

focus on the shape diversity. 1-NNA can assess both quality and diversity of the generation results. For methods generate mesh or voxel, we transform it to point cloud and apply these metrics. Please refer to supplemental materials for more details.

Evaluation. We report the quantitative results of all methods in Table 1. Due to space constraints, we defer the performance in metric CD to the supplemental material. Our method MVDD exhibits strong competitiveness across all categories and surpassed comparison methods, particularly excelling in the 1-NNA (EMD) metric. This metric holds significant importance as it addresses the limitations of MMD and COV [56]. We augmented our generated point cloud and visualize the mesh quality in Fig. 3 together with LION [56] and 3D-LDM [31]. Our method generates more diverse and plausible 3D shapes compared with all baselines. The visualization of our meshes shows that our method excels in synthesizing intricate details, e.g. slats of the chair and thin structure in chair base. We also visualize point clouds in Fig. 1(a) and Fig. 4. The clean point cloud backprojected from our generated depth maps demonstrates 3D consistency and also validates the effectiveness of the proposed epipolar "line segment" attention and denoising depth fusion. In contrast, the number of points (2048) that can be generated by point cloud-based diffusion models [26,56,60] limits their capabilities to capture fine-grained details of the 3D shapes. To show the generalizability of our method, we also train our model on ShapeNet 13 classes simultaneously and compare the performance with the model trained separately in the supplemental material.

Generated Dense Point Cloud vs Up-sampled Sparse Point Cloud. Since our method can directly generate 20K points, while LION [56] is limited to producing sparse point cloud with 2048 points, we up-sample varying number of points from LION's meshes. We then compare the performance of our method with LION. As shown in Fig. 5, the performance of LION deteriorates significantly as the number of points increases. It is because LION struggles to faithfully capture necessary 3D shape details with its sparse point cloud. In contrast, the performance of our method is robust with the increased number of 3D points and outperforms LION by larger margins as the point cloud density increases. We also utilize a SOTA point cloud upsampler and please see supplemental materials for the comparison.

4.2 Depth Completion

Inference Strategy. We reuse an unconditional generative model to perform shape completion task, where depth maps from other views $\mathbf{x}^{\text{other}}$ can be generated conditioned on the single input view of depth map $\mathbf{x}^{\text{in}}$. In each reverse step of the diffusion model, we have:

$$\mathbf{x}^{\text{in}}_{t-1} \sim \mathcal{N}\left(\sqrt{\bar{\alpha}_t}\mathbf{x}^{\text{in}}_0, (1-\bar{\alpha}_t)\mathbf{I}\right),$$

$$\text{1st pass: } \hat{\mathbf{x}}^{\text{other}}_{t-1} \sim \mathcal{N}(\sqrt{1-\beta_t}\,\mu_\theta(\mathbf{x}^{r_1:r_R}_t, t), \beta_t\mathbf{I}), \tag{16}$$

$$\text{2nd pass: } \mathbf{x}^{\text{other}}_{t-1} \sim \mathcal{N}(\mu_\theta(\hat{\mathbf{x}}^{\text{other}}_{t-1}, \mathbf{x}^{\text{in}}_0, t), \beta_t\mathbf{I}),$$

where $\mathbf{x}^{\text{in}}_{t-1}$ is sampled using the given depth map $\mathbf{x}^{\text{in}}$, while $\mathbf{x}^{\text{other}}_{t-1}$ is sampled from the model, given the previous iteration $\mathbf{x}_t$. Different from unconditional generation, to enhance the consistency with the input view, we do two passes to denoise the other views. In the first pass each view attends to every other views and in the second pass each view only attends to the input view $\mathbf{x}^{\text{in}}$. We scale back noise at first pass, following the Langevin dynamics steps [46,47].

Table 2. Depth completion comparison against baselines. EMD is multiplied by 10^2. ⊛ represents the best result.

	SoftFlow [14]	PointFlow [54]	DPF-Net [18]	PVD [60]	MVDD (Ours)
Airplane	1.198	1.180	1.105	1.030	0.900 ⊛
Chair	3.295	3.649	3.320	2.939	2.400 ⊛
Car	2.789	2.851	2.318	2.146	1.460 ⊛

Datasets and Comparison Methods. Following the experimental setup of PVD [60], we use the benchmark provided by GenRe [58], which contains renderings of shapes in ShapeNet from 20 random views. For shape completion [7,38], as the ground-truth data are involved, Chamfer Distance and Earth Mover's Distance suffice to evaluate the reconstruction results. We select models Point-Flow [54], DPF-Net [18], SoftFlow [14], and PVD [60] for comparison.

Evaluation. We show the quantitative results of our method and baselines in Table 2. Our method consistently outperforms all the baselines with EMD metric on all categories. The qualitative results in Fig. 1(b) also showcases that our inference strategy for depth completion can effectively "pull" the learned depth map of other views to be geometrically consistent with the input depth map.

4.3   3D Prior for GAN Inversion

We illustrate how our trained multi-view depth diffusion model can be plugged into downstream tasks, such as 3D GAN inversion [3]. As in the case of 2D GAN inversion, the goal of 3D GAN inversion is to map an input image I onto the space represented by a pre-trained unconditional 3D GAN model, denoted as $G_{3D}(\cdot; \theta)$, which is defined by a set of parameters θ. Upon successful inversion, G_{3D} has the capability to accurately recreate the input image when provided with the corresponding camera pose. One specific formulation of the 3D GAN inversion problem [39] can be defined as follows:

$$w^*, \theta^* = \arg\max_{w, \theta} = \mathcal{L}\left(G_{3D}(w, \pi; \theta), I\right), \tag{17}$$

where w is the latent representation in $\mathcal{W}^+$ space and π is the corresponding camera matrix of input image. w and θ are optimized alternatively, i.e., w is optimized first and then θ is also optimized together with the photometric loss:

$$\begin{aligned}
\mathcal{L}_{\text{photo}} &= \mathcal{L}_2\left(G_{3D}\left(w, \pi_s; \theta\right), I_s\right) \\
&+ \mathcal{L}_{\text{LPIPS}}\left(G_{3D}\left(w, \pi_s; \theta\right), I_s\right),
\end{aligned} \tag{18}$$

where $\mathcal{L}_{\text{LPIPS}}$ is the perceptual similarity loss [57]. However, with merely supervision from single or sparse views, this 3D inversion problem is ill-posed without proper regularization, so that the geometry could collapse (shown in Fig. 1(c) and Fig. 6 2nd row). To make the inversion look plausible from other views, a 3D geometric prior is needed, as well as a pairing regularization method which can preserve diversity. Score distillation sampling has been proposed in DreamFusion [36] to utilize a 2D diffusion model as a 2D prior to optimize the parameters of a radiance field. In our case, we use our well-trained MVDD model as a 3D prior to regularize on the multi-view depth maps extracted from the tri-plane radiance fields. As a result, the following gradient direction would not lead to collapsed geometry after inversion:

$$\nabla\mathcal{L} = \nabla\mathcal{L}_{\text{photo}} + \nabla\lambda_{\text{SDS}}\mathcal{L}_{\text{SDS}}, \tag{19}$$

where λ_{SDS} is the weighting factor of $\mathcal{L}_{\text{SDS}}$ [36].

To learn the shape prior for this 3D GAN inversion task, we render multi-view depth maps from the randomly generated radiance fields of EG3D [3] trained with FFHQ [12] dataset. We then use them as training data and train our multi-view depth diffusion model. Using Eq. (19), we perform test-time optimization for each input image to obtain the optimized radiance field. In Fig. 1(c) and Fig. 6, we show the rendering and geometry results of 3D GAN inversion with and without regularization by MVDD. With the regularization of our model, the "wall" artifact is effectively removed and it results in better visual quality in the rendered image from novel frontal view.

4.4 Ablation Study

We perform ablation study to further examine the effectiveness of each module described in the method section. Specifically, in Table 3 we report the ablated results of epipolar "line segment" attention, depth concatenation, and cross attention thresholding (Sect. 3.1) and depth fusion (Sect. 3.1) in ShapeNet chair category for the unconditional generation task as we describe in Sect. 4.1. Without the designed cross attention, the model could barely generate plausible 3D shapes as measured by 1NN-A metric. With designs such as depth concatenation and cross attention thresholding being added, the 3D consistency along with the performance of our model is progressively improving. Last but not least, denoising depth fusion align the depth maps and further boost the performance. Qualitatively, Sect. 7 illustrates how the denoising depth fusion help eliminate double layers in depth completion task.

We report the shape generation results with full self-attention (as in MVDream [41]) and epipolar attention in the supplemental material. This abla-

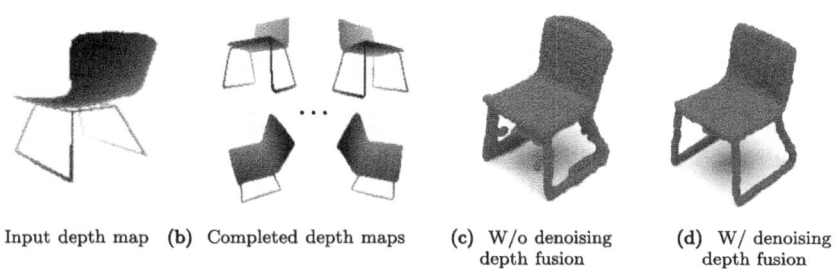

(a) Input depth map (b) Completed depth maps (c) W/o denoising depth fusion (d) W/ denoising depth fusion

Fig. 7. Depth completion results prove the effectiveness of the proposed denoising depth fusion strategy (Sect. 3.1).

Table 3. Ablation study on the chair category. ● is the top result.

Cross attn. (Sect. 3.1)	Depth concat. (Sect. 3.1)	Cross attn. thresholding (Sect. 3.1)	Depth fusion (Sect. 3.1)	1NN-A	
				CD	EMD
✗	✗	✗	✗	92.00	90.00
✓	✗	✗	✗	61.78	59.65
✓	✓	✗	✗	60.72	59.00
✓	✓	✓	✗	59.82	57.75
✓	✓	✓	✓	57.90 ●	54.51 ●

tion study shows that our model excels in both efficiency and effectiveness. The proposed attention obtains more useful cross-view features by avoiding the irrelevant pixels in calculating the keys and values.

5   Conclusion

We leveraged multi-view depth representation in 3D shape generation and proposed a novel denoising diffusion model MVDD. To enforce 3D consistency among different view of depth maps, we proposed an epipolar "line segment" attention and denoising depth fusion technique. Through extensive experiments in various tasks such as shape generation, shape completion and shape regularization, we demonstrated the scalability, faithfulness and versatility of our multi-view depth diffusion model.

Acknowledgement. A.K. was supported by a National Science Foundation (NSF) CAREER award IIS-2046737, Army Young Investigator Program Award, and Defense Advanced Research Projects Agency (DARPA) Young Faculty Award.

References

1. Anvekar, T., Tabib, R.A., Hegde, D., Mudengudi, U.: VG-VAE: a venatus geometry point-cloud variational auto-encoder. In: Proceedings of the IEEE/CVF Conference on Computer Vision and Pattern Recognition, pp. 2978–2985 (2022)
2. Cai, R., et al.: Learning gradient fields for shape generation. In: Vedaldi, A., Bischof, H., Brox, T., Frahm, J.-M. (eds.) ECCV 2020. LNCS, vol. 12348, pp. 364–381. Springer, Cham (2020). https://doi.org/10.1007/978-3-030-58580-8_22
3. Chan, E.R., et al.: Efficient geometry-aware 3D generative adversarial networks. In: Proceedings of the IEEE/CVF Conference on Computer Vision and Pattern Recognition, pp. 16123–16133 (2022)
4. Chang, A.X., et al.: ShapeNet: an information-rich 3D model repository. arXiv preprint arXiv:1512.03012 (2015)
5. Chen, Z., Zhang, H.: Learning implicit fields for generative shape modeling. In: Proceedings of the IEEE/CVF Conference on Computer Vision and Pattern Recognition, pp. 5939–5948 (2019)
6. Chou, G., Bahat, Y., Heide, F.: Diffusion-SDF: conditional generative modeling of signed distance functions. In: Proceedings of the IEEE/CVF International Conference on Computer Vision, pp. 2262–2272 (2023)
7. Chu, R., et al.: Diffcomplete: diffusion-based generative 3D shape completion. In: Advances in Neural Information Processing Systems, vol. 36 (2024)
8. Galliani, S., Lasinger, K., Schindler, K.: Massively parallel multiview stereopsis by surface normal diffusion. In: Proceedings of the IEEE International Conference on Computer Vision, pp. 873–881 (2015)
9. Gao, J., et al.: Get3d: a generative model of high quality 3D textured shapes learned from images. Adv. Neural. Inf. Process. Syst. **35**, 31841–31854 (2022)
10. Goodfellow, I., et al.: Generative adversarial networks. Commun. ACM **63**(11), 139–144 (2020)
11. Ho, J., Jain, A., Abbeel, P.: Denoising diffusion probabilistic models. Adv. Neural. Inf. Process. Syst. **33**, 6840–6851 (2020)

12. Karras, T., Laine, S., Aila, T.: A style-based generator architecture for generative adversarial networks. In: Proceedings of the IEEE/CVF Conference on Computer Vision and Pattern Recognition, pp. 4401–4410 (2019)
13. Kazhdan, M., Bolitho, M., Hoppe, H.: Poisson surface reconstruction. In: Proceedings of the fourth Eurographics Symposium on Geometry Processing, vol. 7, p. 0 (2006)
14. Kim, H., Lee, H., Kang, W.H., Lee, J.Y., Kim, N.S.: Softflow: probabilistic framework for normalizing flow on manifolds. Adv. Neural. Inf. Process. Syst. **33**, 16388–16397 (2020)
15. Kim, J., Yoo, J., Lee, J., Hong, S.: SETVAE: learning hierarchical composition for generative modeling of set-structured data. In: Proceedings of the IEEE/CVF Conference on Computer Vision and Pattern Recognition, p. 15059–15068 (2021)
16. Kingma, D.P., Ba, J.: Adam: a method for stochastic optimization. arXiv preprint arXiv:1412.6980 (2014)
17. Kingma, D.P., Welling, M.: Auto-encoding variational Bayes. arXiv preprint arXiv:1312.6114 (2013)
18. Klokov, R., Boyer, E., Verbeek, J.: Discrete point flow networks for efficient point cloud generation. In: Vedaldi, A., Bischof, H., Brox, T., Frahm, J.-M. (eds.) ECCV 2020. LNCS, vol. 12368, pp. 694–710. Springer, Cham (2020). https://doi.org/10.1007/978-3-030-58592-1_41
19. Lan, Y., et al.: Ln3diff: scalable latent neural fields diffusion for speedy 3D generation. arXiv preprint arXiv:2403.12019 (2024)
20. Lan, Y., Meng, X., Yang, S., Loy, C.C., Dai, B.: E3dge: self-supervised geometry-aware encoder for style-based 3D GAN inversion (2023)
21. Lan, Y., et al.: Gaussian3diff: 3D gaussian diffusion for 3D full head synthesis and editing. arXiv preprint arXiv:2312.03763 (2023)
22. Li, W., Chen, R., Chen, X., Tan, P.: Sweetdreamer: aligning geometric priors in 2D diffusion for consistent text-to-3D. arXiv preprint arXiv:2310.02596 (2023)
23. Liu, Y., et al.: Syncdreamer: generating multiview-consistent images from a single-view image. arXiv preprint arXiv:2309.03453 (2023)
24. Liu, Z., Feng, Y., Black, M.J., Nowrouzezahrai, D., Paull, L., Liu, W.: Meshdiffusion: score-based generative 3D mesh modeling. arXiv preprint arXiv:2303.08133 (2023)
25. Long, X., et al.: Wonder3D: single image to 3D using cross-domain diffusion. arXiv preprint arXiv:2310.15008 (2023)
26. Luo, S., Hu, W.: Diffusion probabilistic models for 3D point cloud generation. In: Proceedings of the IEEE/CVF Conference on Computer Vision and Pattern Recognition, pp. 2837–2845 (2021)
27. Melas-Kyriazi, L., Laina, I., Rupprecht, C., Vedaldi, A.: Realfusion: 360deg reconstruction of any object from a single image. In: Proceedings of the IEEE/CVF Conference on Computer Vision and Pattern Recognition, pp. 8446–8455 (2023)
28. Merrell, P., et al.: Real-time visibility-based fusion of depth maps. In: 2007 IEEE 11th International Conference on Computer Vision, pp. 1–8. IEEE (2007)
29. Mittal, P., Cheng, Y.C., Singh, M., Tulsiani, S.: AutoSDF: shape priors for 3D completion, reconstruction and generation. In: Proceedings of the IEEE/CVF Conference on Computer Vision and Pattern Recognition, pp. 306–315 (2022)
30. Mo, S., Xie, E., Chu, R., Hong, L., Niessner, M., Li, Z.: Dit-3D: exploring plain diffusion transformers for 3D shape generation. In: Advances in Neural Information Processing Systems, vol. 36 (2024)

31. Nam, G., Khlifi, M., Rodriguez, A., Tono, A., Zhou, L., Guerrero, P.: 3D-LDM: neural implicit 3D shape generation with latent diffusion models. arXiv preprint arXiv:2212.00842 (2022)
32. Nichol, A.Q., Dhariwal, P.: Improved denoising diffusion probabilistic models. In: International Conference on Machine Learning, pp. 8162–8171. PMLR (2021)
33. Papamakarios, G., Nalisnick, E., Rezende, D.J., Mohamed, S., Lakshminarayanan, B.: Normalizing flows for probabilistic modeling and inference. J. Mach. Learn. Res. **22**(1), 2617–2680 (2021)
34. Paszke, A., et al.: Pytorch: an imperative style, high-performance deep learning library. In: Advances in Neural Information Processing Systems, vol. 32 (2019)
35. Peng, S., Jiang, C., Liao, Y., Niemeyer, M., Pollefeys, M., Geiger, A.: Shape as points: a differentiable Poisson solver. Adv. Neural. Inf. Process. Syst. **34**, 13032–13044 (2021)
36. Poole, B., Jain, A., Barron, J.T., Mildenhall, B.: Dreamfusion: Text-to-3D using 2D diffusion. arXiv preprint arXiv:2209.14988 (2022)
37. Radford, A., et al.: Learning transferable visual models from natural language supervision. In: International Conference on Machine Learning, pp. 8748–8763. PMLR (2021)
38. Rao, Y., Nie, Y., Dai, A.: Patchcomplete: learning multi-resolution patch priors for 3D shape completion on unseen categories. Adv. Neural. Inf. Process. Syst. **35**, 34436–34450 (2022)
39. Roich, D., Mokady, R., Bermano, A.H., Cohen-Or, D.: Pivotal tuning for latent-based editing of real images. ACM Trans. Graphics (TOG) **42**(1), 1–13 (2022)
40. Schönberger, J.L., Zheng, E., Frahm, J.-M., Pollefeys, M.: Pixelwise view selection for unstructured multi-view stereo. In: Leibe, B., Matas, J., Sebe, N., Welling, M. (eds.) ECCV 2016. LNCS, vol. 9907, pp. 501–518. Springer, Cham (2016). https://doi.org/10.1007/978-3-319-46487-9_31
41. Shi, Y., Wang, P., Ye, J., Long, M., Li, K., Yang, X.: MvDream: multi-view diffusion for 3D generation. arXiv preprint arXiv:2308.16512 (2023)
42. Shonenkov, A., Konstantinov, M., Bakshandaeva, D., Schuhmann, C., Ivanova, K., Klokova, N. (2023). https://github.com/deep-floyd/IF/tree/develop
43. Shu, D.W., Park, S.W., Kwon, J.: 3D point cloud generative adversarial network based on tree structured graph convolutions. In: Proceedings of the IEEE/CVF International Conference on Computer Vision, pp. 3859–3868 (2019)
44. Sohl-Dickstein, J., Weiss, E., Maheswaranathan, N., Ganguli, S.: Deep unsupervised learning using nonequilibrium thermodynamics. In: International Conference on Machine Learning, pp. 2256–2265. PMLR (2015)
45. Song, J., Meng, C., Ermon, S.: Denoising diffusion implicit models. arXiv preprint arXiv:2010.02502 (2020)
46. Song, Y., Ermon, S.: Generative modeling by estimating gradients of the data distribution. In: Advances in Neural Information Processing Systems, vol. 32 (2019)
47. Song, Y., Sohl-Dickstein, J., Kingma, D.P., Kumar, A., Ermon, S., Poole, B.: Score-based generative modeling through stochastic differential equations. arXiv preprint arXiv:2011.13456 (2020)
48. Tsalicoglou, C., Manhardt, F., Tonioni, A., Niemeyer, M., Tombari, F.: Textmesh: generation of realistic 3D meshes from text prompts. arXiv preprint arXiv:2304.12439 (2023)
49. Tseng, H.Y., Li, Q., Kim, C., Alisan, S., Huang, J.B., Kopf, J.: Consistent view synthesis with pose-guided diffusion models. In: Proceedings of the IEEE/CVF Conference on Computer Vision and Pattern Recognition, pp. 16773–16783 (2023)

50. Valsesia, D., Fracastoro, G., Magli, E.: Learning localized generative models for 3D point clouds via graph convolution. In: International Conference on Learning Representations (2018)
51. Wang, Z., et al.: Alto: alternating latent topologies for implicit 3D reconstruction. In: Proceedings of the IEEE/CVF Conference on Computer Vision and Pattern Recognition, pp. 259–270 (2023)
52. Xu, Q., Wang, W., Ceylan, D., Mech, R., Neumann, U.: DISN: deep implicit surface network for high-quality single-view 3D reconstruction. In: Advances in Neural Information Processing Systems, vol. 32 (2019)
53. Xu, Q., Xu, Z., Philip, J., Bi, S., Shu, Z., Sunkavalli, K., Neumann, U.: Point-nerf: point-based neural radiance fields. In: Proceedings of the IEEE/CVF Conference on Computer Vision and Pattern Recognition, pp. 5438–5448 (2022)
54. Yang, G., Huang, X., Hao, Z., Liu, M.Y., Belongie, S., Hariharan, B.: Pointflow: 3D point cloud generation with continuous normalizing flows. In: Proceedings of the IEEE/CVF International Conference on Computer Vision, pp. 4541–4550 (2019)
55. Yao, Y., Luo, Z., Li, S., Fang, T., Quan, L.: MVSNet: depth inference for unstructured multi-view stereo. In: Ferrari, V., Hebert, M., Sminchisescu, C., Weiss, Y. (eds.) ECCV 2018. LNCS, vol. 11212, pp. 785–801. Springer, Cham (2018). https://doi.org/10.1007/978-3-030-01237-3_47
56. Zeng, X., et al.: Lion: latent point diffusion models for 3D shape generation. arXiv preprint arXiv:2210.06978 (2022)
57. Zhang, R., Isola, P., Efros, A.A., Shechtman, E., Wang, O.: The unreasonable effectiveness of deep features as a perceptual metric. In: Proceedings of the IEEE Conference on Computer Vision and Pattern Recognition, pp. 586–595 (2018)
58. Zhang, X., Zhang, Z., Zhang, C., Tenenbaum, J., Freeman, B., Wu, J.: Learning to reconstruct shapes from unseen classes. In: Advances in Neural Information Processing Systems, vol. 31 (2018)
59. Zhao, M., et al.: Efficientdreamer: high-fidelity and robust 3D creation via orthogonal-view diffusion prior. arXiv preprint arXiv:2308.13223 (2023)
60. Zhou, L., Du, Y., Wu, J.: 3D shape generation and completion through point-voxel diffusion. In: Proceedings of the IEEE/CVF International Conference on Computer Vision, pp. 5826–5835 (2021)

Rethinking Video-Text Understanding: Retrieval from Counterfactually Augmented Data

Wufei Ma[1,2]([✉]), Kai Li[1], Zhongshi Jiang[1], Moustafa Meshry[1], Qihao Liu[2], Huiyu Wang[3], Christian Häne[1], and Alan Yuille[2]

[1] Meta Reality Labs, Burlingame, USA
wufeim@gmail.com
[2] Johns Hopkins University, Burlingame, USA
[3] Meta AI, Burlingame, USA

Abstract. Recent video-text foundation models have demonstrated strong performance on a wide variety of downstream video understanding tasks. Can these video-text models genuinely understand the contents of natural videos? Standard video-text evaluations could be misleading as many questions can be inferred merely from the objects and contexts in a single frame or biases inherent in the datasets. In this paper, we aim to better assess the capabilities of current video-text models and understand their limitations. We propose a novel evaluation task for video-text understanding, namely *retrieval from counterfactually augmented data* (RCAD), and a new *Feint6K* dataset. To succeed on our new evaluation task, models must derive a comprehensive understanding of the video from cross-frame reasoning. Analyses show that previous video-text foundation models can be easily fooled by counterfactually augmented data and are far behind human-level performance. In order to narrow the gap between video-text models and human performance on RCAD, we identify a key limitation of current contrastive approaches on video-text data and introduce *LLM-teacher*, a more effective approach to learn action semantics by leveraging knowledge obtained from a pretrained large language model. Experiments and analyses show that our approach successfully learn more discriminative action embeddings and improves results on Feint6K when applied to multiple video-text models. Our Feint6K dataset and project page is available here.

Keywords: Video-Text Understanding · Retrieval

W. Ma—All data collection and experiments were conducted at JHU.

Supplementary Information The online version contains supplementary material available at https://doi.org/10.1007/978-3-031-72624-8_15.

1   Introduction

Video-text foundation models have gained increasing attention due to their simple formulation and strong transferability [12,16,23]. By pretraining on web scale video-text datasets, these models demonstrate strong performance across a wide range of downstream tasks, such as video-text retrieval [22,28] and video question answering [25].

As video-text foundation models evolve and achieve increasingly better performance on various benchmarks, we raise the following questions: Can these video-text model truly grasp the semantics of natural videos? Are these models genuinely reaching a level of understanding comparable to humans? Despite the remarkable achievements made in previous works, our study suggests that current video-text models still fall far behind human-level understanding.

We argue that existing prominent results on standard video-text tasks can be misleading as models largely exploit the shortcuts and biases inherent in the dataset. Many of the questions can be answered by objects or context extracted from a single frame without capturing cross-frame relations. For the examples in Fig. 2a, the video-text alignment can be easily inferred from shortcuts such as "cymbals" or "football". Moreover, the models may utilize biases in the datasets, such as the spurious correlation between "outdoor" and "football". Current evaluations of video-text understanding are compromised by shortcuts and biases, which obscure us from analyzing the limitations of current models. As we proceed from image-text understanding to video-text understanding, we should focus on more challenging semantics in the video domain that require cross-frame reasoning to solve, such as the interaction between person and object or the change of appearances over a sequence of frames (see Fig. 1a).

In this paper we propose a novel evaluation paradigm of video-text understanding, i.e., *retrieval from counterfactually augmented data* (RCAD). As demonstrated in Fig. 2b, the goal is to retrieve the only *positive caption* with matched semantics among "hard" negative captions. *Negative captions* are counterfactually augmented so that video-text models must derive a holistic understanding of the video sequence for both objects and actions, in order to retrieve the correct caption. As a comparison, negative examples from standard video-to-text retrieval [22,24,28] are captions of different videos in the same dataset, often containing different object entities that are easy to distinguish.

We follow the previous human-in-the-loop system [9] and develop a benchmark dataset for our new RCAD task, i.e., the *Feint6K* dataset. We test a wide range of public video-text methods with various pretraining strategies on this new benchmark. Two failure examples are visualized in Fig. 1a and quantitative results are summarized in Fig. 1b. Note how state-of-the-art video-text model, InternVideo [23], features a 87.9% rank-1 accuracy on standard video-to-text retrieval but drops by 32.8% (pretraining only) and 29.7% (finetuned) when tested on our benchmark with counterfactually augmented data. By establishing a human-level baseline on our benchmark, the InternVideo model surprisingly falls far behind human performance by 38.6%. Our findings sharply contrast to the prominent performance obtained on previous video-text benchmarks and

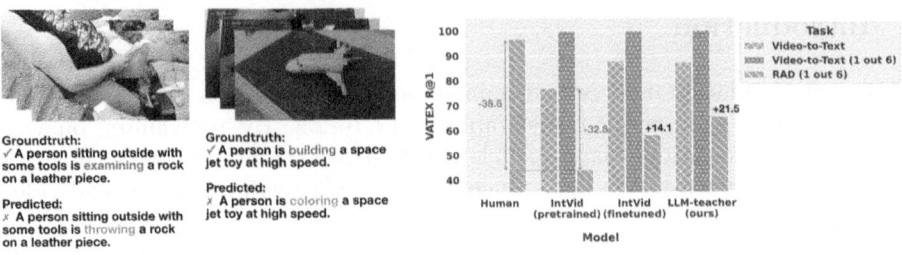

(a) Failures of InternVideo [23] on Feint6K   (b) Quantitative results on Feint6K dataset

Fig. 1. **(a):** Although with large-scale pretraining on web-scale data, current video-text model like [23] can be easily fooled by counterfactually augmented data. **(b):** The performance of InternVideo on *retrieval from counterfactually augmented data* (RCAD) drops by over 30% when compared to the standard video-to-text retrieval and by 38.6% when compared to human-level performance. We also evaluate models on standard video-text retrieval from only 6 sampled candidates and show that our RCAD task is indeed much more challenging. Our LLM-teacher successfully improves the performance on RCAD by enforcing a more effective learning of action semantics.

question the common belief that latest video-text models can develop a fairly effective representation for texts and videos in existing datasets.

In light of our evaluation results on Feint6K, we identify a significant limitation of the widely adopted contrastive representation learning on video-text data, which is the issue of shortcut learning (see Sect. 4.1). To address this, we propose *LLM-teacher* that enables a more effective learning of action semantics by introducing extra knowledge from pretrained large language models (LLMs). Specifically we generate synthetic captions by modifying the contents of an existing caption and LLM serves as the teacher to determine if a synthetic caption matches the semantics in the original caption using binary pseudo-labels or continuous logits. This approach enables a more effective learning of action embeddings from the available video-text data. Experimental results show that LLM-teacher learns more discriminative action embeddings from our advanced contrastive objectives and improves the performance of retrieval from counterfactually augmented data when applied to multiple video-text models.

In summary, our key contributions are as follows. (1) We propose to evaluate video-text models on questions that require cross-frame reasoning and develop a new task, retrieval from counterfactually augmented data, and a new dataset, Feint6K (Sect. 3). (2) Extensive experimental results on our new evaluation paradigm suggest that existing video-text models demonstrate very limited understanding of the action semantics in a video, which is contrary to the prominent performance achieved on standard video-text retrieval benchmarks. (3) From our results on Feint6K dataset, we identify a key shortcoming of contrastive learning approaches on video-text data. We present LLM-teacher to enforce a more effective learning of action embeddings by injecting knowledge from pretrained LLMs (Sect. 4). Our approach effectively improves the results on Feint6K when applied to multiple text-video models.

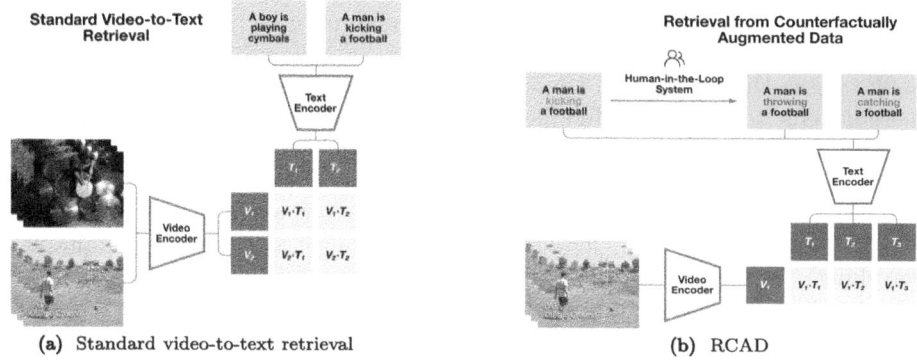

(a) Standard video-to-text retrieval   (b) RCAD

Fig. 2. Different evaluations of video-text understanding. (a): In standard video-to-text retrieval, negative captions are sampled from different videos in the same dataset. Image-text models can achieve good performance by exploiting shortcuts (*e.g.*, "football" and "cymbals") and biases (*e.g.*, spurious correlation between "outdoor" and "football"). **(b):** In our proposed RCAD paradigm, we adopt a human-in-the-loop system (see Sect. 3.2) to obtain "hard" negative captions with unchanged object entities but modified actions. Models must develop a holistic understanding of the semantics from the sequence of frames to retrieve the matched caption.

2   Related Works

Video-Text Pretraining. With the availability of web-scale video-text paired datasets, such as WebVid2M [1] and HowTo100M [17], recent developments of video-text pretraining models achieved improved results on a wide range of video understanding tasks. Visual and textual embeddings jointly learned from the large-scale data demonstrate strong transferability and largely reduce the efforts in downstream tasks. Mainstream pretraining objectives can be categorized into discriminative and generative. Discriminative approaches extend the objective of CLIP [18] to the video-text domain and learn multi-modal representations by contrasting between matched and unmatched pairs [15,16,23,26] or simply predicting if a video-text pair is matched [16]. Generative methods follow the masked modeling idea in BERT [5] and adopted masked language modeling (MLM) and masked video modeling (MVM) for video-text pretraining [16,21,23]. These generative approaches are often considered superior for action recognition, while discriminative approaches learn better semantics from texts [21].

Evaluation of Video-Text Models. In order to analyze the effectiveness of video-text pretraining, previous works [15,16,21,23,26,29] focused on the following tasks: (1) *video-text retrieval (zero-shot or finetuned)* [22,28]: retrieving the matched video (or text) given a text (or video) as query where the negative candidates are unmatched pairs from the same dataset, (2) *video question answering (finetuned)* [25]: predicting the answer with a classification head, (3) *video classification (zero-shot or finetuned)* [6,10,20]: classifying videos using label names as text prompts, and (4) *video captioning (zero-shot or finetuned)* [11,28]: summarizing the contents of a video.

We argue that results on these video-text tasks could be deceitful as most tasks heavily rely on the alignment of object entities in the video and text (which image-text models are also capable of), and "action understanding" can often be developed through shortcuts and biases. For instance, a commonly-used evaluation for "action recognition" is video classification on ActivityNet [6], where a model classifies a given video into classes such as "playing badminton", "kayaking", or "volleyball". However, video-text models would classify the videos by exploiting objects and contexts as shortcuts (*e.g.*, "badminton" and "kayak"), as well as other biases (*e.g.*, spurious correlation between "kayak" and "on the water") without genuinely understanding the semantics of the action represented by interactions between the person and the objects over time. For instance, classifying the action as "kayaking" is essentially a "kayak" detection problem in ActivityNet. Therefore we propose *retrieval from counterfactually augmented data*, a new evaluation paradigm where we aim to eliminate the shortcuts from the questions so models must establish a comprehensive understanding of the semantics from cross-frame reasoning in order to predict the correct answer.

3   Retrieval from Counterfactually Augmented Data

Previous evaluation tasks of video-text models focused on the alignment of feature embeddings of the video-text pairs (*e.g.*, video-text retrieval [22,28] and video classification [10,20]) or reconstruction of the semantics with text (*e.g.*, video question answering [25] and video captioning [11,28]). However, these evaluation tasks are largely limited by the paired data available in existing video-text datasets. As demonstrated by in Fig. 2a, we can extract most of the semantics of a video, such as the objects, people, and contexts in the video, by looking at only one frame from the video. The action in the video can be further inferred by exploiting biases in the datasets. Therefore, models pretrained on image-text data often perform surprisingly well on existing video-text tasks.

To effectively evaluate how video-text models can understand the contents and semantics of videos beyond images, we must develop video understanding tasks that are free from shortcuts and require cross-frame reasoning to solve. To this end, we propose a new evaluation task for video-text models, *i.e.*, retrieval from counterfactually augmented data (Sect. 3.1), and a new dataset Feint6K (Sect. 3.2). In contrast to previous datasets with matched video-text pairs scrapped from the web [24,28] or annotated by human [22], we adopt a human-in-the-loop system and annotate counterfactually manipulated texts as negative pairs with the original video. This allows us to evaluate video-text models with novel tasks and better understand their limitations. Lastly we measure human performance on our dataset as a direct comparison with the state-of-the-art video-text models (Sect. 3.3).

3.1   Task Formulation

Retrieval from counterfactually augmented data is a variant of the standard video-to-text retrieval. As shown in Fig. 2b, given a video sequence and a list of

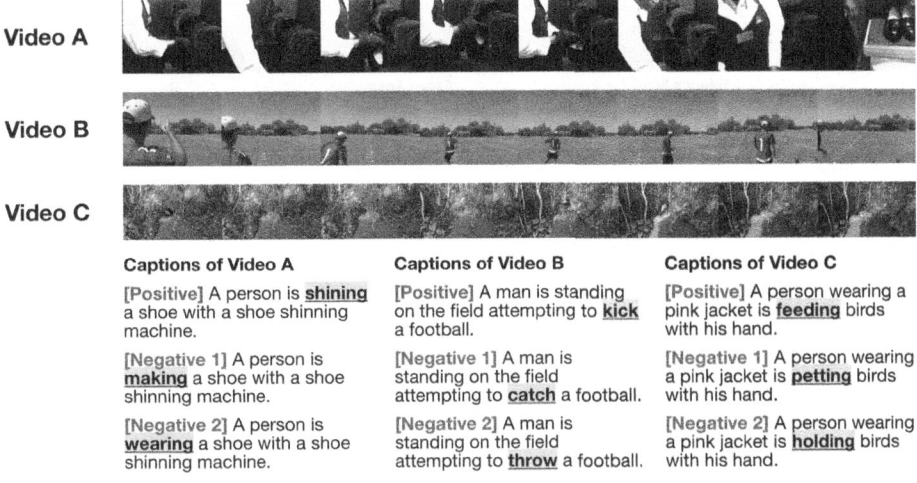

Video A

Video B

Video C

Captions of Video A	Captions of Video B	Captions of Video C
[Positive] A person is **shining** a shoe with a shoe shinning machine.	[Positive] A man is standing on the field attempting to **kick** a football.	[Positive] A person wearing a pink jacket is **feeding** birds with his hand.
[Negative 1] A person is **making** a shoe with a shoe shinning machine.	[Negative 1] A man is standing on the field attempting to **catch** a football.	[Negative 1] A person wearing a pink jacket is **petting** birds with his hand.
[Negative 2] A person is **wearing** a shoe with a shoe shinning machine.	[Negative 2] A man is standing on the field attempting to **throw** a football.	[Negative 2] A person wearing a pink jacket is **holding** birds with his hand.

Fig. 3. Examples of *RCAD* on our Feint6K dataset.

candidate captions, the model is asked to retrieve the caption that best matches the semantics in the video. However, unlike standard video-to-text retrieval where negative captions come from other video-text pairs in the same dataset, negative captions in our task are modified from the positive captions, with the same text structure and object entities but different actions (see Fig. 3).

Similar to video-text retrieval, our proposed task provides the benefit of zero-shot evaluation. This allows us to quantitatively analyze the effectiveness of video-text pretraining without downstream finetuning.

What distinguishes our task from all previous video-text tasks is its focus on questions that require cross-frame reasoning. Consider "Video B" in Fig. 3 as an example. All captions contain the same object entities present in the video, *i.e.*, "a man", "the field", and "a football", but differ in the action specified. We cannot determine the most accurate caption because all suggested actions, "kick", "catch", and "throw", are plausible given the video's context. To retrieve the corresponding caption, it is crucial to grasp the action's semantics by inspecting the interactions between the man and the football across a series of frames. This underscores the necessity of understanding the video semantics over time in order to succeed in our task, rather than relying shortcuts or biases.

3.2   Feint6K: Data Collection

We utilize a human-in-the-loop system [9] to counterfactually manipulate the positive captions in existing video-text datasets [22,28]. The generated counterfactually augmented data, when paired with the original videos, forms negative pairs to be used in our proposed task. We recruit 40 annotators to manually manipulate the actions in existing captions. Specifically, the new actions introduced must be plausible within the context in the caption but are not occurring in the corresponding

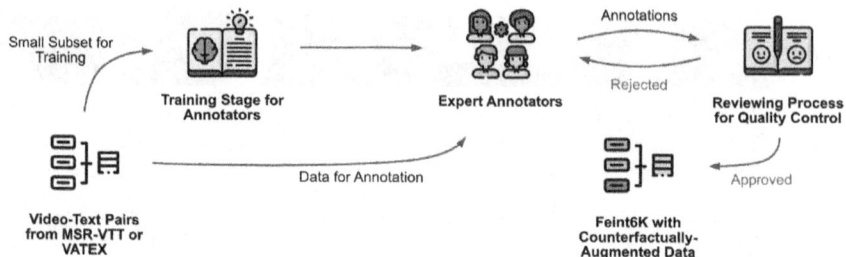

Fig. 4. Overview of our data collection pipeline for Feint6K dataset featuring a human-in-the-loop system.

video. In summary, a total of 6,243 videos from the validation set of the MSR-VTT dataset [28] and the test set of the VATEX dataset [22]. For detailed statistics of our Feint6K dataset, please refer to our supplementary materials.

The overview of our human-in-the-loop system is demonstrated in Fig. 4. We start by sampling a small subset of video-text pairs and manually annotate the counterfactually augmented captions for the purpose of training and demonstration. Specific guidance and feedback are given to the annotators on the practice questions during the training stage. Moreover, we adopt a reviewing process during the annotation stage to further ensure the quality of our annotated data. Each annotation is reviewed before acceptance, where rejected annotations are marked and sent back for refinement. We refer the readers to the supplementary materials regarding the annotation interface and annotator guidance.

3.3   Human Performance

We employ the same group of annotators to establish a human-level baseline on our Feint6K dataset. Analyzing the human performance serves two goals: (i) verifying the legitimacy of our Feint6K dataset – if each question is answerable and has one unique answer, and (ii) comparing the state-of-the-art video-text models with human-level performance.

Each annotator is first presented with the full video and then asked to select the one out of six captions that best matches the semantics in the video. To avoid leakage, we make sure the annotators assigned to the testing questions are not the same annotators annotating this video. We also randomly shuffle the order of the candidate captions to remove any biases. In Sect. 5.2 we present the quantitative comparisons between human-level performance and current state-of-the-art video-text models on our Feint6K dataset.

4   LLM-Teacher

Contrastive learning are widely adopted in previous video-text representation learning works [16,23,26]. The contrastive models learn to distinguish between similar and dissimilar pairs of video-text data by contrasting their feature representations. Despite the simple formulation and great generalization ability, our results on Feint6K show that current video-text models built heavily on contrastive objectives have very limited understanding of action semantics in natural videos and can be easily fooled by counterfactually augmented data. Therefore we aim to develop a more effective contrastive approach and learns more discriminative action representations from existing video-text data.

Section 4.1 discusses the limitations of current contrastive approaches due to numerous shortcuts in video-text dataset. Inspired by this finding, we present LLM-teacher in Sect. 4.2, a simple but effective approach to learn more discriminative action embeddings by incorporating knowledge from pretrained large language models.

4.1 Shortcuts in Video-Text Data

The CLIP model [18] trained on image-text data with a contrastive objective demonstrated strong zero-shot capabilities in many vision tasks [13,30] and was widely adopted as the vision encoder in many large vision-language models [14]. However, directly generalizing this idea to video-text data may face unprecedented challenges due to shortcuts in the video-text data.

As we proceed from image-text to video-text representation learning, we aim to learn a powerful embedding for not only the object entities, but also the actions specified by texts or sequence of frames. Naturally we hope to achieve this by contrasting videos with different actions. However, as we are optimizing the contrastive objectives in a mini-batch of videos, object entities become the shortcuts that contrastive models exploit to saturate the contrastive objective, leading to an ineffective action embeddings. Consider the example in Fig. 2a, the contrastive loss would be small as long as the model learns discriminative embeddings for shortcut objects such as "cymbal" and "football", even without any understanding of the actions. Since many video-text models start with pretrained CLIP encoders [23,26], they often posses strong embeddings for shortcut objects from the beginning of the video-text pretraining, which hinders the model from further learning effective action representations.

In Fig. 5 we present quantitative results of changes in cosine similarities when the video is unchanged but the object or action in the caption is manipulated. We find that state-of-the-art video-text models learn less discriminative embeddings for actions as compared to objects, which also explains the small gap between image-text and video-text models on standard video-text tasks.

4.2 LLM-Teacher

We introduce LLM-teacher, an LLM-powered approach to learn better action representations with contrastive objectives. Specifically, LLM serves as the "teacher" and provides synthesized captions as extra knowledge for the video-text model to learn from. These new captions serve as negative captions of the original video, removing possible shortcuts and biases and enabling a more effective learning of action semantics.

We start with positive captions from common video-text datasets and run an abstract meaning representation (AMR) parser, resulting in a list of action

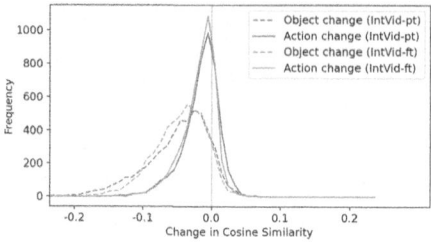 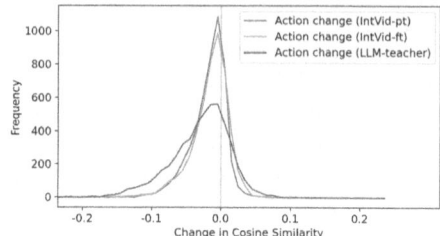

(a) Comparison between Δs w.r.t. object or action changes.

(b) Comparison between Δs w.r.t. action using InternVideo and our LLM-teacher.

Fig. 5. Change of cosine similarity w.r.t. objects or actions. (a): Comparison between the change in cosine similarity when the action or object is swapped. Results show that current video-text models learn a more effective embedding for objects than for actions. (b): Comparison between the change in cosine similarity using InternVideo or our LLM-teacher. This demonstrate that LLM-teacher learns a more discriminative embedding for actions by enabling a more effective contrastive learning using knowledge from LLMs. Refer to Sect. 5.2 for more details.

or object tokens. To generate the hard negative captions, two methods are considered to find novel actions or objects given the original caption and the substitution token w.

Method I: Mask Filling. Masked language modeling (MLM) [5] is a widely adopted self-supervised task for LLMs. In the pretraining stage, a certain percentage of word tokens are substituted with a special token [mask] and the LLM is optimized to recover the masked tokens. To find negative captions with novel actions given the original caption c_0 and substitution token a, we first substitute the token a with [mask] and use a LLM pretrained with MLM to predict a list of k possible tokens $\{\hat{w}_i\}_{i=1}^k$ such that $\hat{w}_i \neq w$ and $\hat{w}_i$ is an action token.

Method II: LLM-Powered Chatbot. Although mask filling can find appropriate words to substitute w, it is limited by one single token. In many cases, we want to update the prepositions following the change of verbs. For example, when "install" is substituted by "uninstall", we should also update the preposition from "install ... to" to "uninstall ... from". Here we consider a more flexible approach that builds on LLM-powered chatbots. Besides a prompt describing the text substitution, we leverage the in-context learning ability of LLMs [2,7] and provide multiple in-context examples to obtain the desirable negative captions.

LLM-Teacher. With the procedure above, we obtain k synthesized negative captions $\{n_i\}_{i=1}^k$ besides the original positive caption p for each video v in existing datasets [22,28]. For each of the original and generated captions, LLM serves as a "teacher" and provide binary pseudo-labels – if the caption have matched

("1") or unmatched ("0") semantics with the video. Now we may optimize the video-text model with the normalized temperature-scaled cross entropy loss

$$l = -\log \frac{\exp(\text{sim}(f_v, f_p)/\tau)}{\exp(\text{sim}(f_v, f_p)/\tau) + \sum_{i=1}^{k} \exp(\text{sim}(f_v, f_{n_i})/\tau)} \tag{1}$$

where f_v, f_p, f_{n_i} are the visual and textual embeddings.

In practice, we find that certain negative captions have unmatched but similar semantics to the video. It is undesirable to regard them as a strictly negative pair in the contrastive loss. Therefore, we extend the binary pseudo-labels to soft logits by computing caption similarities with a pretrained LLM and then optimize the video-text model to match model outputs with the soft logits from the LLM teacher model [8].

$$l = \mathcal{L}_{\text{KL}}(z_{\text{video-text}}, z_{\text{LLM}}) \tag{2}$$

$$z_{\text{video-text,t}} = \frac{\exp(\text{sim}(f_v, f_t)/\tau)}{\exp(\text{sim}(f_v, f_p)/\tau) + \sum_{i=1}^{k} \exp(\text{sim}(f_v, f_{n_i})/\tau)}$$

$$z_{\text{LLM,t}} = \frac{\exp(\text{sim}(e_p, e_t)/\tau)}{\exp(\text{sim}(e_p, e_p)/\tau) + \sum_{i=1}^{k} \exp(\text{sim}(e_p, e_{n_i})/\tau)}$$

where $t \in \{n_i\}_{i=1}^{k} \cup \{p\}$, f's are textual and visual embeddings of the video-text model, and e's are textual embeddings of a pretrained LLM.

5   Experiments

5.1   Experimental Setup

Datasets. We use MSR-VTT dataset [28] and VATEX dataset [22] for standard video-to-text retrieval and our Feint6K dataset for retrieval from counterfactually augmented data. Note that Feint6K dataset contains the same video sequences for evaluation as in MSR-VTT dataset and VATEX dataset.

Evaluation Metrics. For standard video-to-text retrieval we report rank-1 accuracy (R@1) as this is the most challenging metric and distinguish the performance of previous video-text models. For retrieval from counterfactually-augmented data we use rank-1 accuracy (R@1), rank-2 accuracy (R@2), and mean rank (MeanR).

Video-Text Models. We consider multiple state-of-the-art public models in this work. CLIP4Clip [15] extends CLIP model [18] to the video domain using a mean pooling mechanism for zero-shot video-text retrieval and a contrastive loss for finetuning. SimVTP [16] is pretrained on the WebVid2M dataset [1] with a combination of contrastive learning and masked modeling. InternVideo [23] is a video-text foundation model pretrained on a combination of 7 large-scale video-text datasets, with a supervised video post-pretraining for better video recognition. LanguageBind [31] is a language-based multi-modal pretraining model with remarkable performance on a wide range of benchmarks.

Model	MSR-VTT	Feint6K (MSR-VTT)			VATEX	Feint6K (VATEX)		
	R@1	R@1	R@2	MeanR	R@1	R@1	R@2	MeanR
Human		95.2				96.8		
Random	<1e-3	16.7			<1e-3	16.7		
Zero-Shot								
CLIP [18]	26.3	37.3	55.3	2.6	38.8	34.8	54.3	2.7
VideoCLIP [27]	14.5	35.1	71.3	2.6	13.7	33.0	70.7	2.7
InternVideo [23]	37.5	45.8	63.6	2.3	76.9	44.1	63.9	2.3
LanguageBind [31]	42.8	41.3	77.0	2.4		42.3	77.0	2.4
Finetuned								
LanguageBind [31]								
CLIP4Clip [15]	43.1	50.8	72.4	2.0				
VindLU [3]	46.6	53.4	70.9	2.0				
SimVTP [16]	**50.2**	35.7	70.8	2.6	76.6	33.6	68.4	2.6
w/ LLM-teacher-lbl	49.0 ↓1.2	40.0 ↑4.3	73.1 ↑2.3	2.3 ↓0.3	74.8 ↓1.8	37.3 ↑3.7	72.2 ↑3.8	2.4 ↓0.2
w/ LLM-teacher-lgt	<u>49.5</u> ↓0.7	43.5 ↑7.8	75.0 ↑4.2	2.2 ↓0.4	75.3 ↓1.3	40.1 ↑6.5	73.5 ↑5.1	2.3 ↓0.3
InternVideo [23]	49.1	58.6	80.2	1.8	**87.9**	58.2	76.9	1.9
w/ LLM-teacher-lbl	48.2 ↓0.9	64.2 ↑5.6	82.5 ↑2.3	1.7 ↓0.1	85.2 ↓2.7	63.8 ↑5.6	80.5 ↑3.6	1.7 ↓0.2
w/ LLM-teacher-lgt	48.9 ↓0.2	**65.8** ↑7.2	**83.8** ↑3.6	**1.7** ↓0.1	<u>87.3</u> ↓0.6	**65.6** ↑7.4	**81.7** ↑4.8	**1.7** ↓0.2

Table 1. Performance of standard video-to-text retrieval on MSR-VTT dataset [28] and VATEX dataset [22], as compared to performance of retrieval from counterfactually-augmented data on our Feint6K dataset. In the "zero-shot" setting, models are evaluated directly after the pretraining stage, while in the "finetuned" setting, models are finetuned on the training set of MSR-VTT or VATEX. With LLM-teacher, we enable a more efficient learning of action semantics and effectively improves R@1 accuracy on Feint6K. Here "LLM-teacher-lbl" stands for our approach with binary pseudo-labels and "LLM-teacher-lgt" uses soft logits.

To demonstrate the efficacy of our LLM-teacher, we adopt this method on two state-of-the-art video-text models, SimVTP and InternVideo. In the default setting we generate 10 action-based captions with LLM for each training video. For text generation we use a pretrained XLM-RoBERTa [4] for mask filling. To measure text similarities and compute soft logits, we utilize a pretrained Sentence-BERT model [19]. In Sect. 5.3 we ablate on the choices of parameters and caption generation.

5.2   Main Results

Performance of Previous State-of-the-Art. In Table 1 we report the quantitative results of standard video-text retrieval on MSR-VTT and VATEX, and of retrieval from counterfactually augmented data on our Feint6K. We also report a "Random" performance where a model predicts random guesses. We make the following observations: (i) Results show that previous state-of-the-art video-text models demonstrate limited understanding of the action semantics in a video with less than 60% R@1 accuracy on RCAD, given a 16.7% R@1 accuracy when taking random guesses. (ii) Although it is hard to compare the quantitative results between two tasks, we note that our RCAD is much more challenging given the performance gap between "Random" and previous state-of-the-arts. This supports our previous arguments that our RCAD removes shortcuts from standard video-text datasets and focuses on harder questions in the video domain that require cross-frame reasoning.

Performance of LLM-Teacher. We apply our LLM-teacher to two pretrained video-text models, SimVTP [16] and InternVideo [23]. Specifically, we consider the two objectives in Eq. (1) and Eq. (2), labeled as "LLM-teacher-lbl" and "LLM-teacher-lgt" respectively. Results in Table 1 show that by exploiting knowledge from pretrained LLMs, LLM-teacher enables a more efficient leaning of action semantics and achieved improved results on various metrics of RCAD, increasing the R@1 accuracy on RCAD by 7.2% and 7.4%. We also note that with binary pseudo-labels, results demonstrate a trade-off between the performance on standard video-to-text retrieval and RCAD; and in comparison, LLM-teacher with soft logits achieves the highest performance on RCAD (65.8% and 65.6%) with negligible drops (0.2% and 0.6%) on standard retrieval. This is because soft logits account for the text similarities between different actions and avoid overfitting on binary pseudo-labels.

Comparison with Human-Level Performance. The human-level performance are reported in Table 1. We see that human annotators achieved an almost perfect performance on Feint6K dataset, with a 95.2% accuracy on MSR-VTT videos and a 96.8% accuracy on VATEX videos. These results show that: (i) most questions for retrieval from counterfactually-augmented data are answerable with an unique answer, and (ii) this task is fairly simple for human. As a comparison, all previous video-text models with large-scale pretraining fall behind by a wide margin. Specifically, the state-of-the-art video-text model InternVideo [23] features a heavy two-stage pretraining on a collection of 7 large-scale video-text datasets but falls behind by 36.6% and 38.6% even with downstream finetuning.

Analyses of Failure Cases. In Fig. 1b we present two failure cases by the InternVideo [23] model on our Feint6K dataset. Both examples are trivial for human, yet challenging from video-text models due to incapability of cross-frame reasoning. For the first example, a model must analyze the interactions between the person and the rock over a sequence of frames to predict the correct answer. And in the second example, the model must observe the change of appearances over time. As we can see, with the counterfactually augmented data in our Feint6K, models cannot "guess" the correct answer by exploiting shortcuts extracted from a single frame. The findings on Feint6K highlight significant weaknesses in current video-text models, offering valuable insights for future researches in this area. Additional qualitative examples of video-text models on Feint6K dataset are provided in the supplementary materials for readers' reference.

Sensitivity to the Change of Object or Action. With the counterfactually augmented data collected, we investigate how the cosine similarities between the video and text change when the object or action in the text are swapped with an object or action not present in the image. Given the original video-text pair (v, t) and a counterfactually-augmented text $\hat{t}$, the *change in cosine similarity* is given by $\Delta s = s(v, \hat{t}) - s(v, t)$. As $\hat{t}$ contains objects or actions not present in the video, Δs should be negative for an ideal video-text model. Moreover, larger $|\Delta s|$ implies the model being more sensitive to the changes. We compute Δs using the InternVideo with pretraining only ("IntVid-pt"), InternVideo with downstream finetuning ("IntVid-ft"), and our LLM-teacher ("LLM-teacher").

Results in Fig. 5a show that the Δs are almost always negative when the objects are swapped and the absolute changes are larger. In comparison, Δs are sometimes positive when actions are modified and the absolute changes are smaller. This demonstrate that InternVideo learns a more effective embedding for objects than for actions. This further highlights the necessity of our new evaluation paradigm for exploring the limitations of video-text models beyond existing tasks. Moreover, in Fig. 5b we note that with LLM-teacher the action embeddings are more effective than the ones learned by InternVideo, as shown by the more reasonable changes in cosine similarity when actions are swapped.

Please refer to the supplementary materials for a detailed discussion regarding the influence of textual encoders.

5.3 Ablation Study

We conduct ablation studies on the VATEX dataset [22] and our Feint6K dataset. We follow the same settings as above and report rank-1 (R@1) accuracy for standard video-to-text retrieval and rank-1 accuracy (R@1), rank-2 accuracy (R@2), and mean rank (MeanR) for counterfactual augmented data.

Choice of Captions in LLM-Teacher. Table 2 shows the comparison between different numbers and types of LLM captions generated. In the "default" setting we use 10 action-based LLM captions and compare to settings with 5 action-based LLM captions or 5 action-based and 5 object-based LLM captions. Results show with 5 action-based LLM captions the R@1 accuracy on Feint6K drops by 0.9% while the R@1 accuracy on VATEX increases by 0.3%. We also experiment on object-based LLM captions and find them not beneficial for both standard video-to-text retrieval or our RCAD. This is consistent with our assumption that pretrained video-text models already learns a discriminative embedding for objects and a more efficient training objective for action is needed.

Choice of Caption Generation. Besides a pretrained XLM-RoBERTa model for mask filling, we also experiment on generating captions with LLM-powered agents that are finetuned on chat datasets for dialogue applications. Specifically we use the "chat" model finetuned for dialogue applications. Empirically we find that captions generated by LLM chatbots achieves a higher overall quality,

Model	VATEX	Feint6K (VATEX)		
	R@1	R@1	R@2	MeanR
Default	87.3	65.6	81.7	1.7
(10 action captions; XLM-RoBERTa)				
5 action captions	87.6 ↑0.3	64.7 ↓0.9	81.0 ↓0.7	1.7 ↑0.0
5 object + 5 object captions	87.5 ↑0.2	64.2 ↓1.4	80.6 ↓1.1	1.7 ↑0.0
LLM Chatbot	87.0 ↓0.3	65.9 ↑0.3	81.8 ↑0.1	1.7 ↑0.0

Table 2. Ablation studies on the number of LLM captions and caption generation methods. In the default setting we use a total of 10 action-based LLM captions and utilize a pretrained XLM-RoBERTa [4] for mask filling.

exploring a more diverse caption space and better flexibility. However we don't observe significant improvements from our ablation study experiments – the R@1 accuracy on VATEX drops by 0.3% and the R@1 accuracy on Feint6K increases by 0.3%. We choose XLM-RoBERTa in our main experiments as the model runs faster and scale up easily to bigger settings. We provide qualitative comparisons between the two types of caption generation in our supplementary materials.

6   Conclusions

In this work we propose a new evaluation task, retrieval from counterfactually augmented data, and a benchmark dataset, Feint6K. The idea is to remove short-cuts in the video-text questions with a human-in-the-loop system and produce more challenging questions where the video-text model must derive a comprehensive understanding of the video with cross-frame reasoning. Quantitative and qualitative evaluation results on our Feint6K dataset show that despite the task is trivial for human, previous state-of-the-art video-text models can be easily fooled by the counterfactually augmented data. This implies that the prominent results on previous video-text benchmarks could be misleading – models may largely exploit shortcuts without genuinely understand the video contents. Moreover, we identify a key limitation of current contrastive learning on video-text data being the shortcut learning and propose LLM-teacher that enables a more effective learning of action semantics by utilizing knowledge from pretrained LLMs. Experimental results and analyses show that our method can learn a more discriminative representation for actions.

Supplementary Materials. We present the following: (1) limitations of our work, (2) more details about our Feint6K dataset, (3) ethics statements, and (4) extra quantitative and qualitative results.

Acknowledgements. We would like to thank Daniel Khashabi, Yiyan Li, and the anonymous reviewers for their helpful comments and suggestions. Wufei Ma is supported by ONR with N00014-23-1-2641.

References

1. Bain, M., Nagrani, A., Varol, G., Zisserman, A.: Frozen in time: a joint video and image encoder for end-to-end retrieval. In: IEEE International Conference on Computer Vision (2021)
2. Brown, T., et al.: Language models are few-shot learners. Adv. Neural Inf. Process. Syst. **33**, 1877–1901 (2020)
3. Cheng, F., Wang, X., Lei, J., Crandall, D., Bansal, M., Bertasius, G.: Vindlu: a recipe for effective video-and-language pretraining. arXiv preprint arXiv:2212.05051 (2022)
4. Conneau, A., et al.: Unsupervised cross-lingual representation learning at scale. arXiv preprint arXiv:1911.02116 (2019)
5. Devlin, J., Chang, M.W., Lee, K., Toutanova, K.: Bert: pre-training of deep bidirectional transformers for language understanding. arXiv preprint arXiv:1810.04805 (2018)
6. Caba Heilbron, F., Escorcia, V., Ghanem, B., Carlos Niebles, J.: Activitynet: a large-scale video benchmark for human activity understanding. In: Proceedings of the IEEE Conference on Computer Vision and Pattern Recognition, pp. 961–970 (2015)
7. Gupta, T., Kembhavi, A.: Visual programming: compositional visual reasoning without training. In: Proceedings of the IEEE/CVF Conference on Computer Vision and Pattern Recognition, pp. 14953–14962 (2023)
8. Hinton, G., Vinyals, O., Dean, J.: Distilling the knowledge in a neural network. arXiv preprint arXiv:1503.02531 (2015)
9. Kaushik, D., Hovy, E., Lipton, Z.C.: Learning the difference that makes a difference with counterfactually-augmented data. arXiv preprint arXiv:1909.12434 (2019)
10. Kay, W., et al.: The kinetics human action video dataset. arXiv preprint arXiv:1705.06950 (2017)
11. Krishna, R., Hata, K., Ren, F., Fei-Fei, L., Carlos Niebles, J.: Dense-captioning events in videos. In: Proceedings of the IEEE International Conference on Computer Vision, pp. 706–715 (2017)
12. Li, K., et al.: Unmasked teacher: towards training-efficient video foundation models. arXiv preprint arXiv:2303.16058 (2023)
13. Lin, J., Gong, S.: Gridclip: one-stage object detection by grid-level clip representation learning. arXiv preprint arXiv:2303.09252 (2023)
14. Liu, H., Li, C., Wu, Q., Lee, Y.J.: Visual instruction tuning. In: NeurIPS (2023)
15. Luo, H., et al.: CLIP4Clip: an empirical study of clip for end to end video clip retrieval and captioning. Neurocomputing **508**, 293–304 (2022)
16. Ma, Y., Yang, T., Shan, Y., Li, X.: SimVTP: simple video text pre-training with masked autoencoders. arXiv preprint arXiv:2212.03490 (2022)
17. Miech, A., Zhukov, D., Alayrac, J.B., Tapaswi, M., Laptev, I., Sivic, J.: HowTo100M: learning a text-video embedding by watching hundred million narrated video clips. In: ICCV (2019)
18. Radford, A., et al.: Learning transferable visual models from natural language supervision. In: International Conference on Machine Learning, pp. 8748–8763. PMLR (2021)
19. Reimers, N., Gurevych, I.: Sentence-bert: sentence embeddings using Siamese bert-networks. In: Proceedings of the 2019 Conference on Empirical Methods in Natural Language Processing. Association for Computational Linguistics, November 2019. https://arxiv.org/abs/1908.10084

20. Soomro, K., Zamir, A.R., Shah, M.: Ucf101: a dataset of 101 human actions classes from videos in the wild. arXiv preprint arXiv:1212.0402 (2012)
21. Tong, Z., Song, Y., Wang, J., Wang, L.: Videomae: masked autoencoders are data-efficient learners for self-supervised video pre-training. Adv. Neural Inf. Process. Syst. **35**, 10078–10093 (2022)
22. Wang, X., Wu, J., Chen, J., Li, L., Wang, Y.F., Wang, W.Y.: Vatex: a large-scale, high-quality multilingual dataset for video-and-language research. In: Proceedings of the IEEE/CVF International Conference on Computer Vision, pp. 4581–4591 (2019)
23. Wang, Y., et al.: Internvideo: general video foundation models via generative and discriminative learning. arXiv preprint arXiv:2212.03191 (2022)
24. Wu, Z., Yao, T., Fu, Y., Jiang, Y.G.: Deep learning for video classification and captioning. In: Frontiers of Multimedia Research, pp. 3–29 (2017)
25. Xu, D., et al.: Video question answering via gradually refined attention over appearance and motion. In: Proceedings of the 25th ACM International Conference on Multimedia, pp. 1645–1653 (2017)
26. Xu, H., et al.: VideoCLIP: contrastive pre-training for zero-shot video-text understanding. arXiv preprint arXiv:2109.14084 (2021)
27. Xu, H., et al.: VideoCLIP: contrastive pre-training for zero-shot video-text understanding. In: Proceedings of the 2021 Conference on Empirical Methods in Natural Language Processing (EMNLP). Association for Computational Linguistics, Online, November 2021
28. Xu, J., Mei, T., Yao, T., Rui, Y.: MSR-VTT: a large video description dataset for bridging video and language. In: Proceedings of the IEEE Conference on Computer Vision and Pattern Recognition, pp. 5288–5296 (2016)
29. Yan, S., et al.: Video-text modeling with zero-shot transfer from contrastive captioners. arXiv preprint arXiv:2212.04979 (2022)
30. Zhou, Z., Lei, Y., Zhang, B., Liu, L., Liu, Y.: ZegCLIP: towards adapting clip for zero-shot semantic segmentation. In: Proceedings of the IEEE/CVF Conference on Computer Vision and Pattern Recognition, pp. 11175–11185 (2023)
31. Zhu, B., et al.: Languagebind: extending video-language pretraining to n-modality by language-based semantic alignment. In: The Twelfth International Conference on Learning Representations (2024). https://openreview.net/forum?id=QmZKc7UZCy

Risk-Aware Self-consistent Imitation Learning for Trajectory Planning in Autonomous Driving

Yixuan Fan[1,2], Yali Li[1,2], and Shengjin Wang[1,2]([⊠])

[1] Department of Electronic Engineering, Tsinghua University, Beijing, China
fan-yx21@mails.tsinghua.edu.cn
[2] Beijing National Research Center for Information Science and Technology, Beijing, China
{liyali13,wgsgj}@tsinghua.edu.cn

Abstract. Planning for the ego vehicle is the ultimate goal of autonomous driving. Although deep learning-based methods have been widely applied to predict future trajectories of other agents in traffic scenes, directly using them to plan for the ego vehicle is often unsatisfactory. This is due to misaligned objectives during training and deployment: a planner that only aims to imitate human driver trajectories is insufficient to accomplish driving tasks well. We argue that existing training processes may not endow models with an understanding of how the physical world evolves. To address this gap, we propose **RaSc**, which stands for **Ra**isk-aware **S**elf-**c**onsistent imitation learning. RaSc not only imitates driving trajectories, but also learns the motivations behind human driver behaviors (to be risk-aware) and the consequences of its own actions (by being self-consistent). These two properties stem from our novel prediction branch and training objectives regarding Time-To-Collision (TTC). Moreover, we enable the model to better mine hard samples during training by checking its self-consistency. Our experiments on the large-scale real-world nuPlan dataset demonstrate that RaSc outperforms previous state-of-the-art learning-based methods, in both open-loop and, more importantly, closed-loop settings.

1 Introduction

Using data-driven methods to achieve driving planning is a rapidly developing field in autonomous driving. Unlike trajectory prediction, planning pursues generating feasible, safe, efficient, and comfortable motion for the ego vehicle, not just output trajectories that are close to human driver records in terms of distance-based metrics. In the context of planning, closed-loop testing serves to evaluate whether the former goal is achieved, whereas open-loop testing serves

Supplementary Information The online version contains supplementary material available at https://doi.org/10.1007/978-3-031-72624-8_16.

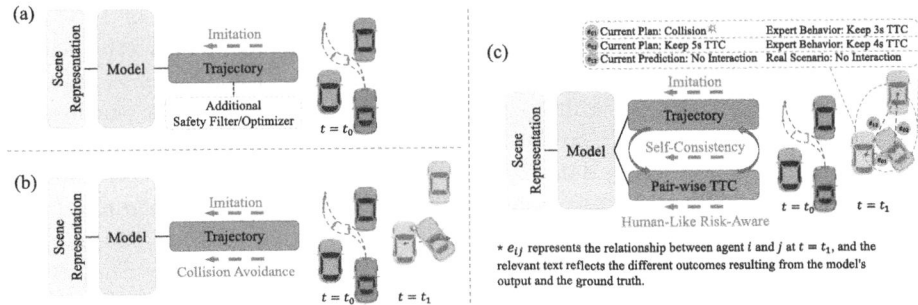

Fig. 1. Different imitation learning paradigms for safe planning. (**a**) [8,27,29,51,61] The trainable model is only supervised by trajectory imitation, while safety is ensured by an additional non-parameterized safety filter/optimizer. (**b**) [2,32,65,67] Hand-crafted collision avoidance constraints are added as training objectives. (**c**) Our method learns human awareness of risks from data and enables the model to understand the consequences of its own actions through self-consistency constraints. This is achieved through training objectives related to pair-wise Time-To-Collision.

to evaluate the latter. Notably and maybe counterintuitively, closed-loop scores and open-loop scores are often weakly correlated or even negatively correlated [11,14,59,66]. One reason that leads to this phenomenon is that open-loop testing metrics fail to reflect real driving proficiency: when model outputs deviate only slightly from ground truth, collisions may actually occur; while when deviations are large, the model may have provided another viable path.

Analogous to the little correlation between open-loop and closed-loop performance in evaluation, existing training processes of imitation learning models suffer from misaligned training objectives and actual deployment needs since supervision signals almost solely come from human driver trajectories, as shown in Fig. 1(a–b). To improve closed-loop performance, current methods adopted remedial approaches that explicitly avoid collisions, by incorporating a non-parameterized safety filter/optimizer or penalizing unsafe behaviors in the network output. However, we argue that these approaches are insufficient for guiding the network to learn the motivations behind human driver behaviors.

In this paper, we adopt Time-To-Collision (TTC) [23] as the medium to capture human comprehension of dynamic information in traffic scenes and as a cue for planning. TTC is defined as the time for two objects to collide if they continue at their present speed and heading. Compared to distance, TTC incorporates both position and velocity information, making it more flexible, as safe distances obviously differ at different speeds. A larger TTC is not always better; while an autonomous driving system maintaining a sufficiently large TTC ensures safety, it can be overly cautious and affect efficiency. Reviewing the literature on transportation, TTC has long been a vital consideration in evaluating the safety of traffic environments [56], and the use of TTC as a cue for decision-making in traffic scenarios is well-established [26]. For human drivers, information about TTC is one of the most critical factors affecting their visual control of braking [35]. To the best of our knowledge, We are the first to integrate the concept of TTC into pure learning-based trajectory planning, and to demonstrate its effectiveness in aiding neural networks to comprehend decision-making mechanisms.

Instead of manually designing safety constraints for different scenarios, we adopt pair-wise future TTC prediction to learn human drivers' risk awareness and management skills from data. Such data includes not only the ego vehicle's experiences, but all its historical observations, as similar intelligence manifests in all traffic participants. Furthermore, the TTCs computed from the model's output trajectories should be consistent with the TTCs predicted by the model, establishing self-consistency. This enables the model to understand the consequences of its own behaviors. The vision achieved by these two designs is illustrated in Fig. 1(c), where the model understands the interactive relationships between agents and better comprehends the ego vehicle's capability of seeking overtaking opportunities in the depicted road scene.

It is noteworthy that supervision from imitating human TTCs and the self-consistent constraint can produce divergent gradients. To address this, we designed the weighting for the self-consistency constraint at different training stages. Moreover, by examining the network's self-consistency, we obtain superior online hard example mining (OHEM). Relevant designs enable the model to find the right learning direction in most cases.

In summary, with the proposed RaSc model, our main contribution is three-fold: (1) We introduce risk-aware imitation, which leverages TTC data from human driving records to guide our model in learning human drivers' motivations and risk awareness skills; (2) We propose self-consistent planning, by constraint our model to perform consistent trajectory planning and TTC estimation, thereby teaching it to interpret its own actions and comprehend the consequences of its behavior; (3) We present a superior OHEM strategy, which can find more accurate hard samples by employing a self-consistency check.

We conduct experiments on the nuPlan dataset, and attain state-of-the-art performance. We believe our method serves as a pioneering work in addressing the fundamental mismatch between open-loop training and closed-loop deployment of imitation learning-based planning models.

2   Related Work

Ego Forecasting. Deep learning methods have shown adeptness at understanding complex driving scenarios and mapping them to potential trajectories of human drivers, where they significantly outperform non-parametric methods represented by the Intelligent Driver Model (IDM) [53]. Ego forecasting models can process raw sensor signals from cameras, lidars, etc. [5,8,9,43,47,60,62], or utilize perception results as inputs, projecting scenes in a BEV space [2,6,13,25,34,36] or analyzing vectorized representations [18,33,37,40,49,54]. Driven by datasets like nuScenes [3], Argoverse [7,58], Waymo Open Motion Dataset [16], etc., and open-loop metrics, related methods have made remarkable progress in aspects including scene representation [15,19,20,31,63,64], model architecture with attention mechanisms [39,42,55,68], among others. Moreover, the future trajectories of different agents influence each other. Therefore, joint prediction across multiple targets has been shown beneficial for both performance and speed [17,20,30,41]. These studies form the foundation for the method designed in this paper.

From Ego Forecasting to Safe Planning. If our prediction of trajectories were perfect, we would only need to control the vehicle to move along the predicted trajectory to realize driving like human drivers. However, this ideal situation has not materialized with the advancement of prediction algorithms. This stems from the inherent limitations of behavior cloning [11,12,14,48,57,59,66], in that they lack lucid explanations of the decision-making process.

To achieve better closed-loop performance, previous studies incorporated manually designed prior knowledge through additional modules. The designs of these modules differ across works, but all make important contributions to the results of corresponding studies. In works adopting the paradigm in Fig. 1(a), [8] designed a two-layer safety mechanism, first checking for collisions based on trajectory prediction results, applying a hard brake if a collision is detected; then using another neural network specialized for predicting brake values as a safety redundancy. [61] performs safety checks on planned trajectories in order of confidence. [27,29] preset cost functions related to safety, comfort, etc., and optimize trajectories through the Gauss-Newton method and CasADi [1] ipopt solver respectively. [51] solves its proposed optimization problem using linear programming. In works adopting the paradigm in Fig. 1(b), [2,67] generate loss when collisions or driving off-road occur under the rasterized representation, and the loss is proportional to the overlapping area with corresponding objects. [32] applies three manually designed safety constraints under the vectorized representation. [28] utilizes both paradigms (a) and (b). In addition, sampling-based planning methods such as [5,44,65] also incorporate the summarized safety mechanisms. Compared to these previous studies, our method aims to systematically enhance the neural network's intrinsic understanding of driving motivations.

3   Methodology

3.1   Problem Formulation and Approach Overview

We denote the ego vehicle as A_0 and the context agents as $A_{1:N}$. Each agent including the ego vehicle has a semantic class (i.e., vehicle, bicycle, or pedestrian), and its state at time t is denoted as s_i^t, where i is the agent index. We also introduce a vectorized high-definition map (including map elements denoted as $M_{1:M}$) and traffic light signals. For the ego vehicle, we have route roadblocks identifying its target road. Assuming the current time point is $t = 0$, given the states of all agents in the previous H time steps and the current time step $s_{0:N}^{-H:0}$, the model needs to make motion decision for the ego vehicle in the F future time steps $s_0^{1:F}$. In practice, we provide L possible plans in parallel along with corresponding confidence scores.

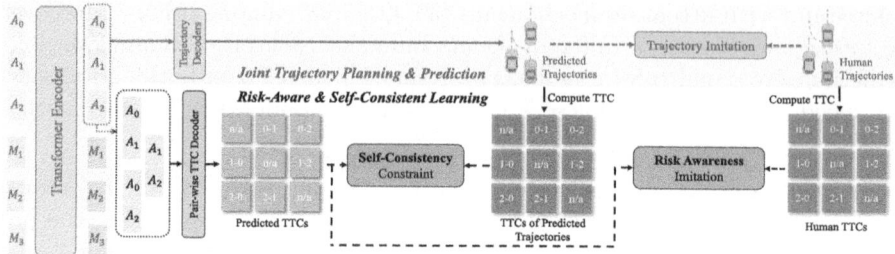

Fig. 2. Overview of our proposed RaSc framework. It expands our joint trajectory planning and prediction baseline (the gray part) by adding a pair-wise future TTC decoder (depicted in Fig. 3). The risk awareness imitation loss is computed by comparing the TTCs predicted by the model with the TTCs derived from human drivers' trajectories, and the self-consistency constraint is computed by comparing the predicted TTCs with the TTCs derived from the model's planned and predicted trajectories.

We also predict K possible future trajectories for all context agents across the F future time steps $s_{1:N}^{k,1:F}$, $k = 1, \cdots, K$. This is because predicting the future trajectories of other agents in the scene holds great importance for the ego vehicle's planning and is also an important basis for the methodology devised in this paper. Section 3.2 describes how we represent the scene and achieve joint trajectory planning and prediction. Notably, the method proposed in this paper is generic, and this part may be flexibly adapted.

The module we propose in this paper extends the conventional joint planning and prediction pipeline. In Sect. 3.3, we introduce our pair-wise TTC decoder, whose function is to predict the TTCs between any two considered agents at P future time points. Building on this, Sect. 3.4 presents how we enable the model to learn human drivers' risk control abilities from data, while achieving self-consistent outputs to explain its own behaviors. We refer the reader to Fig. 2 for a visualized understanding of the model framework. To improve training efficiency and performance, Sect. 3.5 describes how we realize precise online hard example mining by leveraging the introduced concept of self-consistency.

3.2   Joint Trajectory Planning and Prediction

We employ a vectorized representation to encode information from traffic participants' historical states and the high-definition map. Each scene element is encoded in its local coordinate system. For agents, the coordinates' origin and main axis orientation are their position and heading at the last observable timestamp. For each lane center line, the coordinate origin is the average coordinate of all points on the segment, and the main axis orientation aligns with the direction from the first point to the last along the driving direction. For each crosswalk or stop line, the origin is the average coordinate of vertices of each element's polygon, and the main axis orientation is randomly selected.

For the ego vehicle, we encode its state at the last observable timestamp using an MLP, following [10, 21, 27, 61]. For each of the other agents, we use an LSTM [24] to encode the historical states. The state of any agent at each timestamp comprises position, velocity, acceleration, and length-width dimensions. As for map elements, we employ PointNets [45] for encoding.

To embed global position information, the ego vehicle's pose at the last observable timestamp is referenced for the origin and orientation of the global coordinate system. Then the transformation between each element's local coordinate and the global coordinate is utilized to compute a position embedding for each scene element:

$$p_i = \mathrm{MLP}([\Delta x_i, \Delta y_i, \sin \Delta \theta_i, \cos \Delta \theta_i]) \tag{1}$$

where Δx_i, Δy_i, and $\Delta \theta_i$ represent the displacement in x, y coordinates, and orientation between the local coordinate system of scene element i and the global coordinate system, respectively.

Attributes including scene element types, traffic light states for each lane, and binary variables indicating whether each lane belongs to the route for the ego vehicle, are represented by learnable embeddings, respectively. The agent and map embeddings are summed with their corresponding attribute embeddings and position embeddings, and then concatenated into a combined sequence. We employ Transformer [55] Encoder layers to achieve information passing. Finally, MLPs are applied to tokens of the ego vehicle and other agents to obtain the planned and predicted trajectories. The output size of the planning head and prediction head are $L \times F \times 3$ and $K \times F \times 3$, respectively. 3 for the lateral position, longitudinal position, and orientation. There are also output heads for estimating confidence scores. Predicting multiple possible future trajectories for context agents accounts for the multi-modality of intelligent agents' behaviors. Predicting multiple ego vehicle trajectories retains backup plans during deployment. By default, we use the trajectory with the highest confidence.

3.3   TTC Prediction

Estimating future TTCs between traffic participants with neural networks is one of the central pieces of our method. Future TTCs refer to TTCs at a future time point, it is calculated by agents' position, velocity, and heading at that time point. The hyperparameter introduced here is how far into the future we want to predict TTCs. We can also estimate TTCs at multiple future

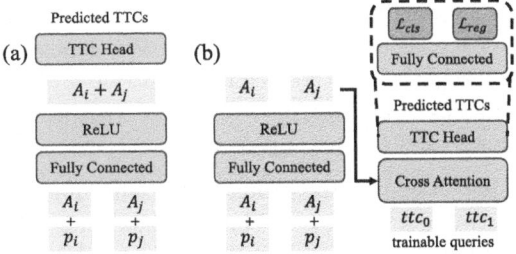

Fig. 3. Diagram of the two pair-wise TTC decoder designs. $\mathcal{L}_{cls}$ and $\mathcal{L}_{reg}$ are introduced in Sect. 3.4.

time points simultaneously. The contextualized tokens output by the Transformer Encoder are utilized by the trajectory decoder to plan or predict trajectories in the local coordinate systems of agents. However, computing TTCs requires the relative positional relationship between two agents. Therefore, we re-add the global position embeddings to these output tokens to form the input for the pair-wise TTC decoder.

For the input tokens constructed for the TTC decoder, we first apply a transformation to the feature space using a fully connected linear layer, followed by a ReLU activation. The basic requirement the TTC decoder needs to satisfy is permutation invariance. Accordingly, we design two optional structures, as shown in Fig. 3. In structure (a), we simply add up the input tokens representing the two agents; in structure (b), we use a learnable embedding as the query for each future time point we estimate TTCs, and inquire about the two agents through a multi-head cross attention layer. Finally, a fully connected layer is utilized as the head to output prediction results. Regarding structure (a), the output head dimension is $P * 2$, where P is the number of future time points we estimate TTCs; for structure (b), the output head dimension is 2. The specifics of what the outputs stand for will be described in Sect. 3.4. We use structure (b) by default because it has better performance.

3.4   Training Objective

Trajectory Imitation. For trajectory planning and prediction, we adopt the Winner-Takes-All principle for loss computing, which applies loss to the trajectory deemed best for each agent. Here, best refers to the trajectories with the minimum final displacement error compared to the trajectories recorded in the dataset. We use the Smooth L1 loss for trajectory regression and the Cross Entropy loss for confidence estimation, the trajectory imitation loss for each agent is a weighted sum of them. Recognizing the importance of trajectories at earlier timestamps, we apply an exponentially decaying weight to the loss at each timestamp. Considering that context agents closer to the ego vehicle typically have a greater impact on its planning and also more accurate position and size annotations (since they are closer to the ego vehicle's sensing system), we weight the loss for each context agent based on its distance from the ego vehicle, using an exponentially decaying law. We denote the trajectory imitation loss for the ego vehicle and the weighted sum of the losses for all context agents are $\mathcal{L}_{plan}$ and $\mathcal{L}_{pred}$, respectively, the total trajectory imitation loss for each scenario will be $\mathcal{L}_{traj} = \mathcal{L}_{plan} + \mathcal{L}_{pred}$.

Risk Awareness Imitation. Our model learns human risk control abilities in traffic scenarios by encouraging the future TTCs predicted by our model to match the corresponding TTCs of human traffic participants. For the $N + 1$ agents in a scenario, their pairwise combinations at each future time point where TTC needs to be estimated result in $N(N+1)/2$, while the majority of them have infinite TTCs (means collision will never happen at their status). To enhance efficiency, we pre-label the training set by identifying all agent pairs with a

TTC of less than 30 s at all considered future time points. During training, we focus on these pairs and a randomly selected subset of negative samples, thereby constructing the set of agent pairs to be considered, termed $\mathbb{Q}$, for each scenario, keeping the number of agent pairs to be computed at an $\mathcal{O}(N)$ level.

Our pair-wise TTC decoder outputs two values for each pair of agents at each future time point where TTC is estimated, representing whether the TTC to be predicted is below a predefined threshold and the numerical value of the predicted TTC, respectively. The training objective for risk awareness imitation for each training scenario is defined as:

$$\mathcal{L}_{ra} = \mathcal{L}_{ttc}\left(\{p_{t,i,j}\}, \{ttc_{t,i,j}\} ; \{ttc^*_{t,i,j}\}\right) = \frac{1}{|\mathbb{Q}|} \sum_{(t,i,j) \in \mathbb{Q}} (w_i + w_j)$$
$$\cdot \left(\lambda \mathcal{L}_{cls}\left(p_{t,i,j}, \sigma\left(ttc_{th} - ttc^*_{t,i,j}\right)\right) + p^*_{t,i,j}\mathcal{L}_{reg}\left(ttc_{t,i,j}, ttc^*_{t,i,j}\right)\right) \tag{2}$$

Here, t, i, j represent the indices for a specific future time point and an agent pair within the set $\mathbb{Q}$, which denotes an element referring to agents i and j at future time t. The term $p_{t,i,j}$ is the predicted probability that this agent pair will have a TTC less than a threshold value ttc_{th} at the future time point t, ttc_{th} is set to 10 s. $ttc_{t,i,j}$ is the TTC predicted by the decoder. $ttc^*_{t,i,j}$ is the TTC we calculate from trajectories in the dataset. The ground truth label $p^*_{t,i,j}$ is set to 1 if $ttc^*_{t,i,j} < ttc_{th}$, and to 0 otherwise. $\mathcal{L}_{cls}$ and $\mathcal{L}_{reg}$ are the focal loss [38] and the Smooth L1 loss, used to compute classification and regression loss respectively. λ is the weight of the classification loss. σ is the Sigmoid function, we use soft labels to enable the model to leverage more information beyond binary classifications. The term $p^*_{t,i,j}\mathcal{L}_{reg}$ means the regression loss is activated only for interacting agent pairs ($p^*_{t,i,j} = 1$) and is disabled otherwise ($p^*_{t,i,j} = 0$). w_i, w_j are weights corresponding to the two agents. As aforementioned, we assign higher weights to agents closer to the ego vehicle.

Self-consistency Constraint. To enable the model to account for its own behaviors, we apply a self-consistency constraint by encouraging the two outputs of the model to agree with each other. Continuing with the previously mentioned set of agent pairs $\mathbb{Q}$ for each scenario, we now replace the supervisory signal for TTC estimation from the TTCs calculated from trajectories in the dataset for risk awareness imitation ($\{ttc^*_{t,i,j}\}$), with $\{ttc^{traj}_{t,i,j}\}$, the TTCs calculated from trajectories planned and predicted by the model. For the multiple output trajectories, we use the one with the highest confidence. This allows us to compute the self-consistency constraint following Eq. (2):

$$\mathcal{L}_{sc} = \mathcal{L}_{ttc}\left(\{p_{t,i,j}\}, \{ttc_{t,i,j}\} ; \left\{ttc^{traj}_{t,i,j}\right\}\right) \tag{3}$$

Overall Training Objective. The total loss for each scenario during training is a weighted sum of the three aforementioned training objectives:

$$\mathcal{L} = \lambda_{traj}\mathcal{L}_{traj} + \lambda_{ra}\mathcal{L}_{ra} + \lambda_{sc}\mathcal{L}_{sc} \tag{4}$$

Here, $\lambda_{traj}, \lambda_{ra}, \lambda_{sc}$ represent the weights assigned to the trajectory imitation loss, risk awareness imitation loss, and self-consistency constraint, respectively. Notably, Since the model has poor performance in predicting trajectories during the early stages of training, λ_{sc} is designed to gradually increase according to a sine law as training progresses: $\lambda_{sc} = \lambda_{sc}^0 \sin(\frac{step}{total\_steps} \cdot \frac{\pi}{2})$, where λ_{sc}^0 is the maximum value of λ_{sc}, $step$ is the current training step and $total\_steps$ is the total training steps. In the later stages of training, λ_{sc} significantly surpasses λ_{ra} to enable the OHEM described below to work and obtain more closed-loop performance gains.

3.5 OHEM Induced by Self-consistency

Traditional Online Hard Example Mining (OHEM) [52] accelerates convergence by sorting sample losses to identify hard samples. For current decision models based on imitation learning, the loss value indicates the discrepancy between the model's output trajectories and those in the dataset. Given the multi-modal nature of human traffic participants' behaviors, a model with certain performance might generate high-confidence trajectories that significantly differ from the ground truth but are still plausible for some samples. Training on these samples could negatively impact the model, and directly applying OHEM might exacerbate this issue. On the other hand, a model may exhibit poor closed-loop performance on some samples with low loss values, indicating that

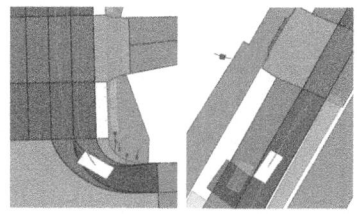

Fig. 4. Two common situations for a learning-based planner. In the left image, the highest confidence trajectory planned for the ego vehicle (the blue line) is perfectly reasonable but distant from the human record (the orange line). In the right image, the planned trajectory is close to the human record, but a small gap causes the ego vehicle to sideswipe a parked car. (Color figure online)

despite the output trajectories being spatially close to the ground truth, minor discrepancies could lead to severe outcomes such as collisions. Current methods might mistakenly classify these samples as easy due to their low loss values. The above phenomenon is illustrated in Fig. 4.

We address this challenge by identifying hard samples through an examination of the model's self-consistency. For each training sample, our first step is to check the prediction loss of the context agents who have future TTCs with the ego vehicle within 10 s in the human records. If the prediction loss of any of these agents is higher than the average of the training batch, this sample is categorized as a hard sample and is kept for training. Otherwise, we proceed to the next step. In the second step, we categorize the samples into the following four types, with corresponding treatments:

- Both $\mathcal{L}_{plan}$ and $\mathcal{L}_{sc}$ are low (lower than the average of the training batch). These samples are considered easy for the model. To prevent the model from forgetting these samples and to maintain a large enough batch size, we include the loss from these samples in the backward propagation with a certain probability.

- High $\mathcal{L}_{plan}$ but low $\mathcal{L}_{sc}$: These samples may be the aforementioned wrong hard samples. To prevent model collapse, we include the loss from these samples in the backward propagation with a certain probability.
- Low $\mathcal{L}_{plan}$ but high $\mathcal{L}_{sc}$: Some of these samples may be challenging for the TTC decoder: despite the model producing trajectories similar to those of human drivers, it fails to adequately explain its own actions. Some other of these may be the aforementioned wrong easy samples: despite the planned trajectories close to those of human drivers, the results are quite different. We retain these samples for training.
- Both $\mathcal{L}_{plan}$ and $\mathcal{L}_{sc}$ are high: These samples are hard and are kept.

The treatments outlined above, especially the treatment of the second type of samples, are only meaningful when the model has achieved sufficient performance levels. Therefore, we only apply OHEM in the later stages of training. Moreover, before deciding to discard the loss from any sample, we perform an additional check on the TTC calculated from the model's highest confidence planned and predicted trajectories, between the ego vehicle and other agents. If the ego vehicle has any TTC less than one second, we do not discard the loss from this sample.

4 Experiments

Benchmarks. We use the nuPlan [4] dataset to evaluate our method. It provides 1,200 h of diverse real-world driving data and a closed-loop simulator. Evaluations are conducted in three different settings: (1) In open-loop evaluation, the model is scored by comparing its predicted trajectory against human driver records, using open-loop metrics including average/final displacement error, average/final heading error, and miss rate. (2) In closed-loop evaluation with non-reactive agents, the planned trajectory controls the ego vehicle in the simulator using an LQR controller, with replayed non-reactive agents, and performance is measured by closed-loop metrics including those related to safety, efficiency, comfort, etc. (3) In closed-loop evaluation with reactive agents, other vehicles react to the ego vehicle using an IDM [53] policy and closed-loop metrics are used to measure performance. More details about experiment settings are illustrated in [4]. We benchmarked our model on three validation sets, namely Val14 from [14], Test14-random and Test14-hard from [10]. Test14-hard consists of specially selected hard scenarios.

4.1 Comparison with SOTA

Table 1 compares the performance of RaSc with others. Our pure learning-based model demonstrates strengths especially in open-loop and non-reactive closed-loop evaluations when compared to other learning-based approaches. In reactive closed-loop evaluations, our method shows a slight underperformance, which can

Table 1. Comparison with state-of-the-arts. RaSc* refers to our model with post-processing, as described in Sect. 4.1. OL means open-loop evaluation, NR-CL means nonreactive closed-loop evaluation and R-CL means reactive closed-loop evaluation. All metrics range from 0 to 100, with higher scores indicating better performance. The runtime includes feature extraction and model inference.

| Planners | | Val14 | | | Test14-random | | | Test14-hard | | | Time (ms) |
Type	Method	OL	NR-CL	R-CL	OL	NR-CL	R-CL	OL	NR-CL	R-CL	
Human	Log-Replay	100	94	80	100	94	76	100	86	69	–
Rule-based	IDM [53]	38	76	77	34	70	72	20	56	62	30
Learning-based	PlanCNN [46]	64	73	72	63	70	68	52	49	52	71
	UrbanDriver [50]	82	53	50	82	63	61	77	52	49	106
	GC-PGP [22]	83	59	55	77	56	51	74	43	40	113
	PDM-Open [14]	86	50	54	84	53	57	79	34	36	**28**
	PlanTF [10]	89	85	**77**	87	**86**	**81**	83	73	62	140
	RaSc (Ours)	**90**	**86**	75	**91**	**86**	76	**86**	**75**	**63**	75
w/ post-processing	GameFormer [29]	–	–	–	79	81	79	75	67	69	395
	PDM-Closed [14]	42	**93**	**92**	46	**90**	**92**	26	65	**75**	127
	PDM-Hybrid [14]	84	**93**	**92**	82	**90**	**92**	74	65	**75**	139
	RaSc* (Ours)	**87**	91	86	**88**	**91**	82	**82**	**77**	65	92

be attributed to its reliance on predicting the behavior of context agents. The IDM-driven agents exhibit significantly different behavior patterns from those observed in the dataset.

To further enhance the model's reliability, we can also incorporate a post-processing module into our method, combining paradigms (a) and (c) in Fig. 1. This module conducts two checks on the planned trajectories: (1) the TTCs between the ego vehicle and other agents, calculated from the model's planned trajectories and the highest confidence predicted trajectories. Planned trajectories resulting in any TTC less than one second are flagged as dangerous; (2) the presence of off-road situations. Should these occur, the trajectories are flagged as dangerous. In instances where high-confidence trajectories are deemed dangerous, we consider other candidate trajectories. If all candidate trajectories are evaluated as dangerous, we resort to initiating a hard brake. Comparison of our model with post-processing between others is in the latter part of Table 1.

To improve the speed of training and inference, we implemented a two-dimensional TTC calculation method capable of efficient batch processing on GPUs. We also use a lightweight structural design. These ensure that our approach has a significant efficiency advantage over other methods with comparable performance.

4.2   Ablation Study

Effect of Each Design. We evaluate the contribution of each design in our model. As our approach is specifically aimed at enhancing the model's closed-loop performance, we report our results in the non-reactive closed-loop evalua-

Table 2. Ablation for designs of our method. RA means applying the risk awareness imitation, SC means applying the self-consistency constraint, OHEM means applying the OHEM we propose. Coll means the metric "no ego at fault collisions", TTC means the metric "time to collision within bound" (not same as in the rest of the paper).

RA	SC	OHEM	Val14			Test14-random			Test14-hard		
			NR-CL	Coll	TTC	NR-CL	Coll	TTC	NR-CL	Coll	TTC
			79	87	82	78	88	81	66	82	77
✓			83	93	88	82	94	89	72	85	80
✓	✓		85	95	89	85	**95**	91	73	88	81
✓	✓	✓	**86**	**97**	**92**	**86**	**95**	**92**	**75**	**90**	**85**

Table 3. Comparison between different pair-wise TTC decoder designs on the Val14 set. P means precision, R means recall, RMSE means root mean square error.

TTC decoder	Driving Score			Risk Awareness			Self-Consistency		
	OL	NR-CL	R-CL	P↑	R↑	RMSE↓	P↑	R↑	RMSE↓
(a) Add up	89	82	72	0.61	0.55	0.79	0.69	0.75	0.53
(b) Cross attention	**90**	**86**	**75**	**0.65**	**0.71**	**0.52**	**0.72**	**0.89**	**0.14**

Table 4. Comparison between different time point choices for TTC estimation on the Val14 set. P means precision, R means recall, RMSE means root mean square error.

Time Point	Driving Score			Risk Awareness			Self-Consistency		
	OL	NR-CL	R-CL	P↑	R↑	RMSE↓	P↑	R↑	RMSE↓
0.5 s	90	84	75	0.70	0.86	0.40	0.75	0.93	0.10
1.0 s	90	86	75	0.65	0.71	0.52	0.72	0.89	0.14
1.5 s	90	85	73	0.62	0.68	0.61	0.68	0.85	0.29
2.0 s	89	82	69	0.59	0.64	0.65	0.66	0.79	0.42
0.5 s & 1.0 s	90	85	74	0.66	0.78	0.44	0.73	0.92	0.12
1.0 s & 1.5 s	90	86	73	0.63	0.70	0.55	0.70	0.86	0.24

tion. Additionally, we report on two critical safety-related metrics: no ego at fault collisions, indicating whether the ego vehicle has not been at fault in any possible collisions, and time to collision within bound (0.95 s), reflecting whether the ego vehicle maintains a sufficient TTC from other agents. The results, as illustrated in Table 2, demonstrate that all our design components contribute to performance gains in our model.

Comparison of TTC Decoder Designs. In Sect. 3.3 we introduce two structures for predicting future TTCs that meet our design criteria. Their perfor-

Table 5. Comparison between different OHEM strategies. Coll means the metric "no ego at fault collisions", TTC means the metric "time to collision within bound" (not same as in the rest of the paper).

OHEM strategy	Val14			Test14-random			Test14-hard		
	NR-CL	Coll	TTC	NR-CL	Coll	TTC	NR-CL	Coll	TTC
w/o OHEM	85	95	89	85	**95**	91	73	88	81
Trajectory loss based	83	91	87	84	92	89	70	85	80
Self-consistency based	**86**	**97**	**92**	**86**	**95**	**92**	**75**	**90**	**85**

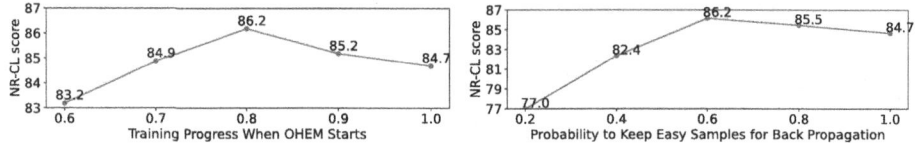

Fig. 5. Comparison between different OHEM configurations on the Val14 set. Starting OHEM at 1.0 training progress or keeping easy samples at 1.0 probability means not applying OHEM.

mance is compared in Table 3, focusing on their influence on the model's final driving capabilities and their abilities in predicting pair-wise TTCs. The latter is evaluated through the precision and recall for the classification task and the root mean square error for the regression task, for predicting the ego vehicle's future TTCs with all context agents. The cross-attention-based model, with its higher capacity, can capture the interrelations between two agents more flexibly, thereby significantly outperforming the alternative.

Choice of the Time Point for TTC Estimation. The choice of how far into the future to predict TTCs is an important hyperparameter for our method. In Table 4, we present the impact of this choice on both the model's final driving capability and its performance in predicting pair-wise TTCs for the ego vehicle. Predicting TTCs at more distant future time points is inherently more challenging, yet an appropriately challenging TTC estimation can better facilitate improvements in driving capability, particularly in closed-loop evaluations. We find that predicting TTCs at 1.0 s into the future offers the most significant benefit to driving capability.

Comparison Between OHEM Strategies. In Table 5, we compare our proposed self-consistency-induced OHEM against the traditional OHEM based on trajectory imitation loss, with a particular focus on their impact on closed-loop performance. The experimental results indicate that selecting hard samples based on trajectory imitation loss leads to performance degradation, and our method demonstrates a clear advantage.

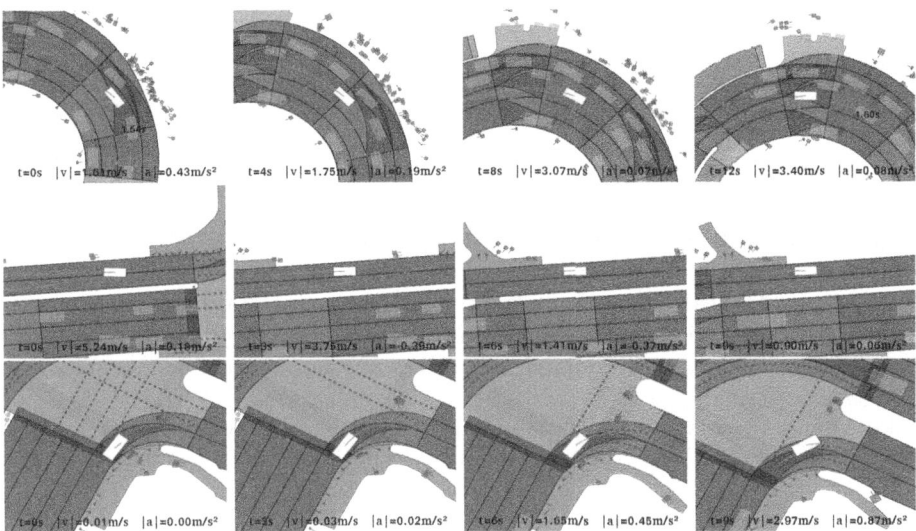

Fig. 6. Visualized results. Each row presents four sequential snapshots of one scenario, progressing from left to right. The current time step, ego vehicle's speed, and acceleration are documented at the bottom of each image. The bounding box in white indicates the ego vehicle, the orange line depicts a 15-s human driving trajectory from the dataset, and the blue line represents the trajectory planned by our model for the next 8 s. In the first scenario, the first and fourth images each highlight a context agent with the model's prediction of its TTC with the ego vehicle in one second. From top to bottom, the scenarios illustrate: (1) the ego vehicle accelerates and safely navigates through a complex pickup/dropoff area; (2) the ego vehicle notices pedestrians crossing the road and brakes to wait; (3) the ego vehicle waits for pedestrians to cross and then accelerates to initiate a right turn.

Tuning OHEM Configuration. As discussed in Sect. 3.5, the effectiveness of our proposed OHEM strategy is contingent on the model already having attained a certain level of performance; thus, it is only feasible to apply our OHEM in the later stages of training. It is also necessary to retain the two types of easy samples with a certain probability. Figure 5 illustrates the impact of these two hyperparameters on the model's performance in the non-reactive closed-loop evaluation. For optimal performance, we begin OHEM after 80% of the training process and retain easy samples with a 0.6 probability.

4.3   Visualized Results

In Fig. 6, we showcase some visualized experimental results of our method in the non-reactive closed-loop simulation.

5   Conclusion

We present RaSc, an efficient model for trajectory planning. By incorporating Time-To-Collision as a basis for driving decisions into training, we develop an approach that enables an imitation learning-based model to (1) learn human drivers' risk awareness abilities and the motivations behind their actions; (2) comprehend the consequences of its own actions; (3) identify hard samples during training. RaSc achieves significant closed-loop performance gains and surpasses previous state-of-the-arts. By RaSc, we demonstrate a way of combining traditional safety metrics with scalable learning algorithms, to bridge the prominent mismatch between open-loop training and closed-loop deployment of planning models, thus creating more reliable and aware autonomous driving systems.

Acknowledgements. This study is supported by Tsinghua University-Toyota Joint Research Center for AI Technology of Automated Vehicle (TTAD 2023-07).

References

1. Andersson, J.A., Gillis, J., Horn, G., Rawlings, J.B., Diehl, M.: CasADI: a software framework for nonlinear optimization and optimal control. Math. Program. Comput. **11**, 1–36 (2019)
2. Bansal, M., Krizhevsky, A., Ogale, A.: ChauffeurNet: learning to drive by imitating the best and synthesizing the worst. arXiv preprint arXiv:1812.03079 (2018)
3. Caesar, H., et al.: nuScenes: a multimodal dataset for autonomous driving. In: CVPR, pp. 11621–11631 (2020)
4. Caesar, H., et al.: nuPlan: a closed-loop ml-based planning benchmark for autonomous vehicles. arXiv preprint arXiv:2106.11810 (2021)
5. Casas, S., Sadat, A., Urtasun, R.: MP3: a unified model to map, perceive, predict and plan. In: CVPR, pp. 14403–14412 (2021)
6. Chai, Y., Sapp, B., Bansal, M., Anguelov, D.: Multipath: multiple probabilistifc anchor trajectory hypotheses for behavior prediction. arXiv preprint arXiv:1910.05449 (2019)
7. Chang, M.F., et al.: Argoverse: 3D tracking and forecasting with rich maps. In: CVPR, pp. 8748–8757 (2019)
8. Chen, D., Krähenbühl, P.: Learning from all vehicles. In: CVPR, pp. 17222–17231 (2022)
9. Chen, D., Zhou, B., Koltun, V., Krähenbühl, P.: Learning by cheating. In: CoRL, pp. 66–75. PMLR (2020)
10. Cheng, J., Chen, Y., Mei, X., Yang, B., Li, B., Liu, M.: Rethinking imitation-based planner for autonomous driving. arXiv preprint arXiv:2309.10443 (2023)
11. Codevilla, F., Lopez, A.M., Koltun, V., Dosovitskiy, A.: On offline evaluation of vision-based driving models. In: ECCV, pp. 236–251 (2018)
12. Codevilla, F., Santana, E., López, A.M., Gaidon, A.: Exploring the limitations of behavior cloning for autonomous driving. In: ICCV, pp. 9329–9338 (2019)
13. Cui, H., et al.: Multimodal trajectory predictions for autonomous driving using deep convolutional networks. In: ICRA, pp. 2090–2096. IEEE (2019)
14. Dauner, D., Hallgarten, M., Geiger, A., Chitta, K.: Parting with misconceptions about learning-based vehicle motion planning. arXiv preprint arXiv:2306.07962 (2023)

15. Deo, N., Wolff, E., Beijbom, O.: Multimodal trajectory prediction conditioned on lane-graph traversals. In: CoRL, pp. 203–212. PMLR (2022)
16. Ettinger, S., et al.: Large scale interactive motion forecasting for autonomous driving: the Waymo open motion dataset. In: ICCV, pp. 9710–9719 (2021)
17. Fan, Y., Liu, X., Li, Y., Wang, S.: Look before you drive: boosting trajectory forecasting via imagining future. In: IROS, pp. 5551–5558. IEEE (2023)
18. Gao, J., et al.: VectorNet: encoding HD maps and agent dynamics from vectorized representation. In: CVPR, pp. 11525–11533 (2020)
19. Gilles, T., Sabatini, S., Tsishkou, D., Stanciulescu, B., Moutarde, F.: GOHOME: graph-oriented heatmap output for future motion estimation. In: ICRA, pp. 9107–9114. IEEE (2022)
20. Gilles, T., Sabatini, S., Tsishkou, D., Stanciulescu, B., Moutarde, F.: THOMAS: trajectory heatmap output with learned multi-agent sampling. In: ICLR (2022)
21. Guo, K., Jing, W., Chen, J., Pan, J.: CCIL: context-conditioned imitation learning for urban driving. arXiv preprint arXiv:2305.02649 (2023)
22. Hallgarten, M., Stoll, M., Zell, A.: From prediction to planning with goal conditioned lane graph traversals. arXiv preprint arXiv:2302.07753 (2023)
23. Hayward, J.C.: Near miss determination through use of a scale of danger (1972)
24. Hochreiter, S., Schmidhuber, J.: Long short-term memory. Neural Comput. 9(8), 1735–1780 (1997)
25. Hong, J., Sapp, B., Philbin, J.: Rules of the road: predicting driving behavior with a convolutional model of semantic interactions. In: CVPR, pp. 8454–8462 (2019)
26. Van der Horst, A., Hogema, J.: Time-to-collision and collision avoidance systems. Verkeersgedrag in Onderzoek (1994)
27. Hu, Y., et al.: Imitation with spatial-temporal heatmap: 2nd place solution for nuPlan challenge. In: CVPRW (2023)
28. Hu, Y., et al.: Planning-oriented autonomous driving. In: CVPR, pp. 17853–17862 (2023)
29. Huang, Z., Liu, H., Mo, X., Lyu, C.: Gameformer planner: a learning-enabled interactive prediction and planning framework for autonomous vehicles. In: CVPRW (2023)
30. Jia, X., Sun, L., Zhao, H., Tomizuka, M., Zhan, W.: Multi-agent trajectory prediction by combining egocentric and allocentric views. In: CoRL, pp. 1434–1443. PMLR (2022)
31. Jia, X., Wu, P., Chen, L., Liu, Y., Li, H., Yan, J.: HDGT: heterogeneous driving graph transformer for multi-agent trajectory prediction via scene encoding. IEEE TPAMI (2023)
32. Jiang, B., et al.: VAD: vectorized scene representation for efficient autonomous driving. In: ICCV (2023)
33. Khandelwal, S., Qi, W., Singh, J., Hartnett, A., Ramanan, D.: What-if motion prediction for autonomous driving. arXiv preprint arXiv:2008.10587 (2020)
34. Konev, S., Brodt, K., Sanakoyeu, A.: MotionCNN: a strong baseline for motion prediction in autonomous driving. arXiv preprint arXiv:2206.02163 (2022)
35. Lee, D.N.: A theory of visual control of braking based on information about time-to-collision. Perception 5(4), 437–459 (1976)
36. Lee, N., Choi, W., Vernaza, P., Choy, C.B., Torr, P.H., Chandraker, M.: DESIRE: distant future prediction in dynamic scenes with interacting agents. In: CVPR, pp. 336–345 (2017)
37. Liang, M., et al.: Learning lane graph representations for motion forecasting. In: Vedaldi, A., Bischof, H., Brox, T., Frahm, J.-M. (eds.) ECCV 2020. LNCS, vol.

12347, pp. 541–556. Springer, Cham (2020). https://doi.org/10.1007/978-3-030-58536-5_32

38. Lin, T.Y., Goyal, P., Girshick, R., He, K., Dollár, P.: Focal loss for dense object detection. In: ICCV, pp. 2980–2988 (2017)

39. Liu, Y., Zhang, J., Fang, L., Jiang, Q., Zhou, B.: Multimodal motion prediction with stacked transformers. In: CVPR, pp. 7577–7586 (2021)

40. Mercat, J., Gilles, T., El Zoghby, N., Sandou, G., Beauvois, D., Gil, G.P.: Multi-head attention for multi-modal joint vehicle motion forecasting. In: ICRA, pp. 9638–9644. IEEE (2020)

41. Mo, X., Huang, Z., Xing, Y., Lv, C.: Multi-agent trajectory prediction with heterogeneous edge-enhanced graph attention network. IEEE TITS **23**(7), 9554–9567 (2022)

42. Ngiam, J., et al.: Scene transformer: a unified architecture for predicting future trajectories of multiple agents. In: ICLR (2022)

43. Ohn-Bar, E., Prakash, A., Behl, A., Chitta, K., Geiger, A.: Learning situational driving. In: CVPR, pp. 11296–11305 (2020)

44. Phan-Minh, T., et al.: Driving in real life with inverse reinforcement learning. arXiv preprint arXiv:2206.03004 (2022)

45. Qi, C.R., Su, H., Mo, K., Guibas, L.J.: PointNet: Deep learning on point sets for 3D classification and segmentation. In: CVPR, pp. 652–660 (2017)

46. Renz, K., Chitta, K., Mercea, O.B., Koepke, A., Akata, Z., Geiger, A.: PlanT: explainable planning transformers via object-level representations. arXiv preprint arXiv:2210.14222 (2022)

47. Rhinehart, N., McAllister, R., Levine, S.: Deep imitative models for flexible inference, planning, and control. arXiv preprint arXiv:1810.06544 (2018)

48. Ross, S., Gordon, G., Bagnell, D.: A reduction of imitation learning and structured prediction to no-regret online learning. In: AISTATS, pp. 627–635. JMLR Workshop and Conference Proceedings (2011)

49. Salzmann, T., Ivanovic, B., Chakravarty, P., Pavone, M.: Trajectron++: dynamically-feasible trajectory forecasting with heterogeneous data. In: Vedaldi, A., Bischof, H., Brox, T., Frahm, J.-M. (eds.) ECCV 2020. LNCS, vol. 12363, pp. 683–700. Springer, Cham (2020). https://doi.org/10.1007/978-3-030-58523-5_40

50. Scheel, O., Bergamini, L., Wolczyk, M., Osiński, B., Ondruska, P.: Urban driver: learning to drive from real-world demonstrations using policy gradients. In: CoRL, pp. 718–728. PMLR (2022)

51. Shao, H., Wang, L., Chen, R., Li, H., Liu, Y.: Safety-enhanced autonomous driving using interpretable sensor fusion transformer. In: CoRL, pp. 726–737. PMLR (2023)

52. Shrivastava, A., Gupta, A., Girshick, R.: Training region-based object detectors with online hard example mining. In: CVPR, pp. 761–769 (2016)

53. Treiber, M., Hennecke, A., Helbing, D.: Congested traffic states in empirical observations and microscopic simulations. Phys. Rev. E **62**(2), 1805 (2000)

54. Varadarajan, B., et al.: MultiPath++: efficient information fusion and trajectory aggregation for behavior prediction. In: ICRA, pp. 7814–7821. IEEE (2022)

55. Vaswani, A., et al.: Attention is all you need. In: NeurIPS, vol. 30 (2017)

56. Vogel, K.: A comparison of headway and time to collision as safety indicators. Accid. Anal. Prevent. **35**(3), 427–433 (2003)

57. Wen, C., Lin, J., Darrell, T., Jayaraman, D., Gao, Y.: Fighting copycat agents in behavioral cloning from observation histories. In: NeurIPS, vol. 33, pp. 2564–2575 (2020)

58. Wilson, B., et al.: Argoverse 2: next generation datasets for self-driving perception and forecasting. arXiv preprint arXiv:2301.00493 (2023)

59. Wu, H., Phong, T., Yu, C., Cai, P., Zheng, S., Hsu, D.: What truly matters in trajectory prediction for autonomous driving? arXiv preprint arXiv:2306.15136 (2023)
60. Wu, P., Jia, X., Chen, L., Yan, J., Li, H., Qiao, Y.: Trajectory-guided control prediction for end-to-end autonomous driving: a simple yet strong baseline. In: NeurIPS, vol. 35, pp. 6119–6132 (2022)
61. Xi, W., Shi, L., Cao, G.: An imitation learning method with data augmentation and post processing for planning in autonomous driving. In: CVPRW (2023)
62. Xu, H., Gao, Y., Yu, F., Darrell, T.: End-to-end learning of driving models from large-scale video datasets. In: CVPR, pp. 2174–2182 (2017)
63. Ye, M., Cao, T., Chen, Q.: TPCN: temporal point cloud networks for motion forecasting. In: CVPR, pp. 11318–11327 (2021)
64. Zeng, W., Liang, M., Liao, R., Urtasun, R.: LanerCNN: distributed representations for graph-centric motion forecasting. In: IROS, pp. 532–539. IEEE (2021)
65. Zeng, W., Wang, S., Liao, R., Chen, Y., Yang, B., Urtasun, R.: DSDNet: deep structured self-driving network. In: Vedaldi, A., Bischof, H., Brox, T., Frahm, J.-M. (eds.) ECCV 2020. LNCS, vol. 12366, pp. 156–172. Springer, Cham (2020). https://doi.org/10.1007/978-3-030-58589-1_10
66. Zhai, J.T., et al.: Rethinking the open-loop evaluation of end-to-end autonomous driving in nuScenes. arXiv preprint arXiv:2305.10430 (2023)
67. Zhou, J., et al.: Exploring imitation learning for autonomous driving with feedback synthesizer and differentiable rasterization. In: IROS, pp. 1450–1457. IEEE (2021)
68. Zhou, Z., Ye, L., Wang, J., Wu, K., Lu, K.: HiVT: hierarchical vector transformer for multi-agent motion prediction. In: CVPR, pp. 8823–8833 (2022)

Dual-Level Adaptive Self-labeling for Novel Class Discovery in Point Cloud Segmentation

Ruijie Xu[1]([✉]), Chuyu Zhang[1], Hui Ren[1], and Xuming He[1,2]

[1] ShanghaiTech University, Shanghai, China
{xurj2022,zhangchy2,renhui,hexm}@shanghaitech.edu.cn
[2] Shanghai Engineering Research Center of Intelligent Vision and Imaging, Shanghai, China

Abstract. We tackle the novel class discovery in point cloud segmentation, which discovers novel classes based on the semantic knowledge of seen classes. Existing work proposes an online point-wise clustering method with a simplified equal class-size constraint on the novel classes to avoid degenerate solutions. However, the inherent imbalanced distribution of novel classes in point clouds typically violates the equal class-size constraint. Moreover, point-wise clustering ignores the rich spatial context information of objects, which results in less expressive representation for semantic segmentation. To address the above challenges, we propose a novel self-labeling strategy that adaptively generates high-quality pseudo-labels for imbalanced classes during model training. In addition, we develop a dual-level representation that incorporates regional consistency into the point-level classifier learning, reducing noise in generated segmentation. Finally, we conduct extensive experiments on two widely used datasets, SemanticKITTI and SemanticPOSS, and the results show our method outperforms the state of the art by a large margin.

Keywords: Novel class discovery · Point clouds semantic segmentation · Long-tailed learning

1 Introduction

Point cloud segmentation is a core problem in 3D perception [19] and potentially useful for a wide range of applications, such as autonomous driving and intelligent robotics [21,29]. Recently, there has been tremendous progress in semantic segmentation of point clouds due to the utilization of deep learning

---

R. Xu and C. Zhang—Contributed equally. Code is available at Github.

Supplementary Information The online version contains supplementary material available at https://doi.org/10.1007/978-3-031-72624-8_17.

techniques [16,17]. However, current segmentation methods primarily focus on a closed-world setting where all the semantic classes are known beforehand. As such it has difficulty in coping with open-world scenarios where both known and novel classes coexist, which are commonly seen in real-world applications.

For open-world perception, a desirable capability is to automatically acquire new concepts based on existing knowledge [15]. While there has been much effort into addressing the problem of novel class discovery for 2D or RGBD images [10,14,25,41], few works have explored the corresponding task for 3D point clouds. Only recently, Riz et al. [28] propose an online point-wise clustering method for discovering novel classes in 3D point cloud segmentation. To avoid degenerate solutions, their method relies on an equal class-size constraint on the novel classes. Despite its promising results, such a simplified assumption faces two key challenges: First, the distribution of novel classes in point clouds is inherently imbalanced due to the different physical sizes of objects and the density of points. Imposing the equal-size constraint can be restrictive, causing the splitting of large classes or the merging of smaller ones. In addition, point-wise clustering tends to ignore the rich spatial context information of objects, which leads to less expressive representation for semantic segmentation.

To tackle the above challenges, we propose a dual-level adaptive self-labeling framework for novel class discovery in point cloud segmentation. The key idea of our approach is two-fold: 1) We design a novel self-labeling strategy that adaptively generates high-quality imbalanced pseudo-labels for model training, which facilitates clustering novel classes of varying sizes; 2) To incorporate semantic context, we develop a dual-level representation of 3D points by grouping points into regions and jointly learns the representations of novel classes at both the point and region levels. Such a dual-level representation imposes additional constraints on grouping the points likely belonging to the same category. This helps in mitigating the noise in the generated segmentation.

Specifically, our framework employs an encoder to extract point features for the input point cloud and average pooling to compute representations of pre-computed regions. Both types of features are fed into a prototype-based classifier to generate predictions across both known and novel categories for each point and region. To learn the feature encoder and class prototypes, we introduce a self-labeling-based learning procedure that iterates between pseudo-label generation for the novel classes and the full model training with cross-entropy losses on points and regions. Here the key step is to generate imbalanced pseudo labels, which is formulated as a semi-relaxed Optimal Transport (OT) problem with adaptive regularization on class distribution. Along with the training, we employ a data-dependent annealing scheme to adjust the regularization strength. Such a design prevents discovering degenerate solutions and meanwhile enhances the model flexibility in learning the imbalanced data distributions.

To demonstrate the effectiveness of our approach, we conduct extensive experiments on two widely-used datasets: SemanticKITTI [3] and SemanticPOSS [26]. The experimental results show that our method outperforms the state-of-the-art approaches by a large margin. Additionally, we conduct comprehensive ablation studies to evaluate the significance of the different components of our method. The contributions of our method are summarized as follows:

1. We propose a novel adaptive self-labeling framework for novel class discovery in point cloud segmentation, better modeling imbalanced novel classes.
2. We develop a dual-level representation for learning novel classes in point cloud data, which incorporates semantic context via augmenting the point prediction with regional consistency.
3. Our method achieves significant performance improvement on the Semantic-POSS and SemanticKITTI datasets across nearly all the experiment settings.

2   Related Work

Point Cloud Semantic Segmentation. Point cloud semantic segmentation has attracted much attention in recent years [7,22,38,42]. While previous methods have made significant progress, their primary focus is on closed-world scenarios that heavily rely on annotations for each class and cannot address open-world challenges. In contrast, we aim to develop a model to discover novel classes in 3D open-world scenarios. In the context of point cloud representation learning, incorporating spatial context is pivotal for enhancing representation learning. Several works [24,39] introduce a hierarchical representation learning strategy that leverages regions as intermediaries to connect points and semantic clusters. Unlike them, we develop a dual-level learning strategy that concurrently learns to map points and regions to semantic classes. Thanks to the learning of region-level representation, our method is less sensitive to the local noises in point clouds. Moreover, we cluster regions into semantic classes by an imbalance-aware self-labeling algorithm instead of simple K-Means.

Novel Class Discovery. The majority of research on Novel Class Discovery (NCD) has focused on learning novel visual concepts in the 2D image domain via designing a variety of unsupervised losses on novel class data or regularization strategies [10,13,15,34,37,40]. Among them, EUMS [41] addresses novel class discovery in semantic segmentation, employing a saliency model for clustering novel classes, along with entropy ranking and dynamic reassignment for clean pseudo labels. More relevantly, Zhang et al. [36] consider the NCD task in long-tailed classification scenarios, and develop a bi-level optimization strategy for model learning. It adopts a fixed regularization to prevent degeneracy, imposing strong restrictions on learned representations, and a complex dual-loop iterative optimization procedure. In contrast, we propose an adaptive regularization strategy, which is critical for the success of our self-labeling algorithm. Moreover, our formulation leads to a convex pseudo-label generation problem, efficiently solvable by a fast scaling algorithm [6,8] (see Appendix A for detailed comparisons). Perhaps most closely related to our work is [28], which explored the NCD problem for the task of point cloud semantic segmentation. Assuming a uniform distribution of novel classes, they develop an optimal-transport-based self-labeling algorithm to cluster novel classes. However, the method neglects intrinsic class imbalance and spatial context in point cloud data, often leading to sub-optimal clustering results.

Optimal Transport for Pseudo Labeling. Unlike naive pseudo labeling [20], Optimal Transport (OT) [27,33]-based methods allow us to incorporate prior class distribution into pseudo-labels generation. Therefore, it has been used as a pseudo-labels generation strategy for a wide range of machine learning tasks, including semi-supervised learning [18,30,31], clustering [1,4,35], and domain adaptation [5,11,23]. However, most of these works assume the prior class distribution is either known or simply the uniform distribution, which is restrictive for NCD. By contrast, we consider a more practical scenario, where the novel class distribution is unknown and imbalanced, and design a semi-relaxed OT formulation with a novel adaptive regularization.

3   Method

In this section, we first introduce the problem setup of novel class discovery for point cloud segmentation and an overview of our method in Sect. 3.1. We then describe our network architecture, including dual-level representation of point clouds in Sect. 3.2. Subsequently, we present in detail our adaptive self-labeling framework for model learning that discovers the novel classes in Sect. 3.3. Finally, we introduce our strategy to estimating the number of novel classes in Sect. 3.4.

3.1   Problem Setup and Overview

For the task of point cloud segmentation, the novel class discovery problem aims to learn to classify 3D points of a scene into known and novel semantic classes from a dataset consisting of annotated points from the known classes and unlabeled points from novel ones.

Formally, we consider a training set of 3D scenes, where each scene comprises two parts: 1) an annotated part of the scene $\{(x_n^s, y_n^s)\}_{n=1}^N$, which belongs to the known classes C^s and consists of original point clouds along with the corresponding labels for each point; 2) an unknown part of the scene $\{(x_m^u)\}_{m=1}^M$, which belongs to the novel classes C^u and does not contain any label information. These two sets C^s and C^u are mutually exclusive, i.e., $C^s \cap C^u = \emptyset$. Our goal is to learn a point cloud segmentation network that can accurately segment new scenes in a test set, each of which includes both known and novel classes.

To tackle the challenge of discovering novel classes in point clouds, we introduce a dual-level adaptive self-labeling framework to learn a segmentation network for both known and novel classes. The key idea of our method includes two aspects: 1) utilizing the spatial smooth prior of point clouds to generate regions and developing a dual-level representation that incorporates regional consistency into the point-level classifier learning; 2) generating imbalance pseudo-labels with a novel adaptive regularization. An overview of our framework is depicted in Fig. 1.

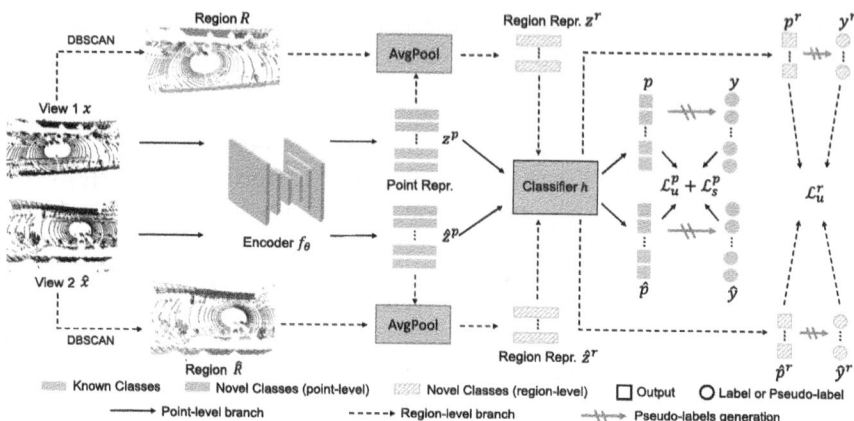

Fig. 1. Our method starts with two views of the same point cloud (x and $\hat{x}$) and clusters the points into corresponding regions. Then, we extract individual point features via a forward pass and calculate regional representations by averaging the point features within each region. Next, we make predictions p by the classifier h, and generate pseudo labels y for unlabeled points and regions using our novel adaptive self-labeling algorithm (generating pseudo-labels does not involve gradients). Lastly, we exchange the pseudo labels between the two views and update the model accordingly.

3.2   Model Architecture

We adopt a generic segmentation model architecture consisting of a feature encoder for the input point cloud and a classifier head to generate the point-wise class label prediction. Note that to capture both known and novel classes, our feature encoder is shared by all the classes $C^s \cup C^u$ and the output space of our classifier head also includes known and novel classes. Below we first introduce our feature representation and encoder, followed by the classifier head.

Dual-Level Representation. Instead of treating each point independently, we exploit the spatial smoothness prior to 3D objects in our representation learning. To this end, we adopt a dual-level representation of point clouds that describes the input scene at different granularity. Specifically, given an input point cloud $\mathbf{X}$, we first use a backbone network f_θ to compute a point-wise feature $\mathbf{Z}^p = \{\mathbf{z}_i^p\}$, where $\mathbf{z}_i^p \in \mathbb{R}^{D \times 1}$. In this work, we employ MinkowskiUNet [7] for the backbone. In addition, we cluster points into regions based on their coordinates and then compute regional features by average pooling of point features. Concretely, during training, we first utilize DBSCAN [9] to generate K_i regions, $\mathcal{R} = \{r_k\}_{k=1}^{K_i}$, for unlabeled point in sample i, and calculate the regional features as follows,

$$\{r_k\}_{k=1}^{K_i} \leftarrow \text{DBSCAN}(\{x_i^u\}_{i=1}^M), \quad \mathbf{z}_k^r = \text{AvgPool}\{\mathbf{z}_i^p | \mathbf{z}_i^p = f_\theta(x_i^u), \ x_i^u \in r_k\}, \tag{1}$$

where $\mathbf{z}_k^r$ is the feature of region r_k. Such a dual-level representation allows us to enforce regional consistency in representation learning.

Algorithm 1: Semi-relaxed Optimal Transport Algorithm

Function Self-Labeling$(-\log \mathbf{P}, \gamma, \epsilon)$

$\quad \mathbf{K} = \exp(\log \mathbf{P}/\epsilon), \quad f \leftarrow \frac{\gamma}{\gamma+\epsilon}$

$\quad \boldsymbol{\mu}, \boldsymbol{\nu} \leftarrow \frac{1}{M}\mathbf{1}_M, \frac{1}{|C^u|}\mathbf{1}_{|C^u|}$ //Marginal distribution

$\quad \mathbf{b}_0 \leftarrow \mathbf{1}_{|C^u|}$ //Initialize b

$\quad$ **while** $\|\mathbf{b}^{t+1} - \mathbf{b}^t\| < 1e-4$ **do**

$\quad\quad \mathbf{a} \leftarrow \frac{\boldsymbol{\mu}}{\mathbf{K}\mathbf{b}_t}$

$\quad\quad \mathbf{b}_{t+1} \leftarrow (\frac{\boldsymbol{\nu}}{\mathbf{K}^\top \mathbf{a}})^f$

$\quad$ **end**

$\quad \mathbf{Q} \leftarrow M\mathrm{diag}(\mathbf{a})\mathbf{K}\mathrm{diag}(\mathbf{b})$

$\quad$ **return** $\mathbf{Q}$

end

Prototype-Based Classifier. We adopt a prototype-based classifier design for generating the point-wise predictions. Specifically, we introduce a set of prototypes for known and novel classes, denoted as $h = [h^s, h^u] \in \mathbb{R}^{D \times (|C^s|+|C^u|)}$, and D denotes the dimension of the last-layer feature. For each point or region, we compute the cosine similarity between its feature and the prototypes, followed by Softmax to predict the class probabilities. Here we use the same set of prototypes for the points and regions, which enforces a consistency constraint within each region and results in a more compact representation for each class.

3.3 Adaptive Self-labeling Framework

To handle class-imbalanced data, we propose an adaptive self-labeling framework that dynamically generates imbalanced pseudo-labels. To this end, we adopt the following loss function for the known and novel classes,

$$\mathcal{L} = \mathcal{L}_s + \alpha \mathcal{L}_u^p + \beta \mathcal{L}_u^r, \tag{2}$$

where $\mathcal{L}_s$ is the cross-entropy loss for known classes, $\mathcal{L}_u^p$ is point-level loss and $\mathcal{L}_u^r$ is region-level loss for novel classes. α and β are weight parameters. For the novel classes, we first generate pseudo-labels for points and regions by solving a semi-relaxed Optimal Transport problem and then adopt the cross-entropy loss with the generated labels. The pseudo-code of our algorithm is shown in Appendix B and below we will focus on our novel pseudo-label generation process.

Imbalanced Pseudo Label Generation. The pseudo-labels generation for balanced classes can be formulated as an optimal transport problem as follows [1,36]:

$$\min_{\mathbf{Q}} \frac{1}{M} \langle \mathbf{Q}, -\log \mathbf{P}^u \rangle_F, \quad \text{s.t. } \mathbf{Q}\mathbf{1}_{|C^u|} = \mathbf{1}_M, \mathbf{Q}^\top \mathbf{1}_M = \frac{M}{|C^u|}\mathbf{1}_{|C^u|}, \tag{3}$$

where $\mathbf{Q} \in \mathbb{R}^{M \times |C^u|}$ are the pseudo labels of unlabeled data, $<,>_F$ is Frobenius inner product and $\mathbf{P}^u$ are the output probabilities of the model. For imbalanced point cloud data, we relax the second constraint on the class sizes in Eq. (3), which leads to a parameterized semi-relaxed optimal transport problem as below:

$$\min_{\mathbf{Q}} \mathcal{F}_u(\mathbf{Q}, \gamma) = \frac{1}{M} \langle \mathbf{Q}, -\log \mathbf{P}^u \rangle_F + \gamma KL(\frac{1}{M} \mathbf{Q}^\top \mathbf{1}_M, \frac{1}{|C^u|} \mathbf{1}_{|C^u|})$$

$$\text{s.t. } \mathbf{Q} \in \{\mathbf{Q} \in \mathbb{R}^{M \times |C^u|} | \mathbf{Q} \mathbf{1}_{|C^u|} = \mathbf{1}_M\}, \tag{4}$$

where γ is a weight coefficient for balancing the constraint on cluster size distribution in the second term. We further add an entropy term $-\epsilon \mathcal{H}(\frac{1}{M} \mathbf{Q})$ to Eq. (4) and for any given γ, this entropic semi-relaxed OT problem can be efficiently solved by fast scaling algorithms [6,8]. Algorithm 1 outlines the optimization process, and further details are provided in Appendix A.

In this work, we propose a novel adaptive regularization strategy that adjusts the weight γ according to the progress of model learning, significantly improving pseudo-label quality. Details of our strategy will be illustrated subsequently.

Adaptive Regularization Strategy. The objective Eq. (4) aims to strike a balance between the distribution represented by model prediction $\mathbf{P}^u$ and the uniform prior distribution. A large γ tends to prevent the model from learning a degenerate solution, e.g. assigning all the samples into a single novel class, but it also restricts the model's capacity to learn the imbalanced data. One of our key insights is that the imbalanced NCD learning requires an adaptive strategy for setting the value of γ during the training. Intuitively, in the early training stage where the model performance is relatively poor, a larger constraint on $\mathbf{Q}^\top \mathbf{1}_M$ is needed to prevent degenerate solutions. As the training progresses, the model gradually learns meaningful clusters for novel classes, and the constraint should be relaxed to increase the flexibility of pseudo-label generation.

To achieve that, we develop an annealing-like strategy for adjusting γ, inspired by the ReduceLROnPlateau method that reduces the learning rate when the loss does not decrease. Here we employ the KL term in Eq. (4) as a guide for decreasing γ, as the value of the KL term reflects the relationship between the distribution of pseudo labels and the uniform distribution. Specifically, our formulation for the adaptive regularization factor is as follows:

$$\gamma_{t+1} = \lambda \gamma_t, \text{ if } KL(\frac{1}{M} \mathbf{Q}^\top \mathbf{1}_M, \frac{1}{|C^u|} \mathbf{1}_{|C^u|}) \leq \rho \text{ consecutively for } T \text{ iter.} \tag{5}$$

where ρ, λ, T and γ_0 are hyperparameters. Compared to typical step decay and cosine decay strategies, our adaptive strategy is aware of the model learning process and allows for more flexible control of γ based on the characteristics of the input itself.

Hyperparameter Search. To search the values of our hyperparameters, we design an indicator score that can be computed on the training dataset. Specifically, our indicator regularizes the total loss in Eq. (2) with a KL term that measures the distance between the distribution of novel classes and the uniform distribution. Formally, the indicator is defined as follows:

$$\mathcal{I} = \mathcal{L} + \gamma KL(\frac{1}{M}\mathbf{Q}^\top \mathbf{1}_M, \frac{1}{|C^u|}\mathbf{1}_{|C^u|}), \tag{6}$$

where γ is obtained by Eq. (5). Empirically, this indicator score provides a balanced evaluation of the model's performance in the known and novel classes.

3.4   Estimate the Number of Novel Classes

To deal with realistic scenarios, where the number of novel classes (C^u) is unknown, we extend the classic estimation method [32] in NCD to point clouds semantic segmentation for estimating C^u. Specifically, we extract representation from a known-class pre-trained model for training data, define a range of possible total class counts ($|C^s|<|C_{all}|<$ max classes), and apply Kmeans to cluster the labeled and unlabeled point clouds across different $|C_{all}|$. Then, we evaluate the clustering performance of known classes under different $|C_{all}|$, and select $|C_{all}|$ with the highest clustering performance as the estimated $|C_{all}|$.

4   Experiments

4.1   Experimental Setup

Dataset. We perform evaluation on the widely-used SemanticKITTI [2,3,12] and SemanticPOSS [26] datasets. The SemanticKITTI dataset consists of 19 semantic classes, while the SemanticPOSS dataset contains 13 semantic classes. Both datasets have intrinsic class imbalances. For a fair comparison with existing works [28], we divide the dataset into 4 splits and select one split as novel classes, while treating the remaining splits as the known classes. Additionally, to assess the effectiveness of our method under more challenging conditions, we further split the SemanticPOSS dataset into two parts, selecting one part as novel classes. The dataset details are provided in Appendix C.

Evaluation Metric. Following the official guidelines in SemanticKITTI and SemanticPOSS, we conduct evaluations on sequences 08 and 03, respectively. These sequences contain both known and novel classes. For the known classes, we report the IoU for each class. Regarding the novel classes, we employ the Hungarian algorithm to initially match cluster labels with their corresponding ground truth labels. Subsequently, we present the IoU values for each of these novel classes. Additionally, we calculate the mean of columns across all known and novel classes.

Implementation Details. We follow [28] to adopt the MinkowskiUNet-34C [7] network as our backbone. For the parameters in DBSCAN, we set the min_samples to a reasonable value of 2, and select an epsilon value of 0.5, ensuring that 95% of the point clouds are included in the region branch learning process. A detailed

Table 1. The novel class discovery results on SemanticPOSS dataset. 'Number' denotes the number of points. 'Full' denotes the results obtained by supervised learning. The gray values are the novel classes in each split.

Split	Method	bike	build.	car	cone.	fence	grou.	pers.	plants	pole	rider	traf.	trashc.	trunk	Novel	Known	All
	Full	45.0	83.3	52.0	36.5	46.7	77.6	68.2	77.7	36.0	58.9	30.3	4.2	14.4	-	-	48.5
	EUMS	25.7	4.0	0.6	16.4	29.4	36.8	43.8	28.5	13.1	26.8	18.2	3.3	16.9	17.4	21.5	20.3
0	NOPS	35.5	30.4	1.2	13.5	24.1	69.1	44.7	42.1	19.2	47.7	24.4	8.2	21.8	35.7	26.6	29.4
	Ours	46.3	**51.5**	**6.0**	35.7	48.5	**83.0**	67.9	**53.1**	35.5	59.3	31.0	2.8	15.5	**48.4**	38.0	41.2
	EUMS	15.2	68.0	28.0	24.0	11.9	75.1	36.0	74.5	26.9	48.6	26.0	5.6	23.1	21.0	40.0	35.6
1	NOPS	29.4	71.4	28.7	12.2	3.9	78.2	**56.8**	74.2	18.3	38.9	23.3	13.7	23.5	30.0	38.2	36.4
	Ours	**31.5**	83.2	48.7	25.4	**23.9**	77.3	53.1	77.1	32.5	57.3	35.0	9.3	18.0	**36.2**	46.4	44.0
	EUMS	40.1	69.5	27.7	13.5	34.9	76.0	54.7	75.6	5.3	39.2	7.8	8.5	11.9	8.3	44.0	35.7
2	NOPS	37.2	71.8	29.7	14.6	28.4	77.5	52.1	73.0	**11.5**	47.1	0.5	10.2	**14.8**	9.0	44.2	36.0
	Ours	45.3	82.8	49.8	28.4	46.3	76.7	66.2	77.2	10.9	58.4	**18.6**	7.3	8.2	**12.6**	53.8	44.3
	EUMS	41.2	70.7	28.1	**4.3**	38.3	76.7	38.3	75.4	25.8	34.3	28.3	0.4	24.4	13.0	44.7	37.4
3	NOPS	38.6	70.4	30.9	0.0	29.4	76.5	56.0	71.8	17.0	31.9	26.2	1.0	22.6	10.9	43.9	36.3
	Ours	45.5	82.9	47.7	0.0	45.1	77.8	66.3	77.7	34.3	**49.1**	35.6	**4.0**	15.3	**17.7**	52.8	44.7

Table 2. Results on splits of SemanticPOSS dataset with more severe imbalance. The gray values are the novel classes in each split. NOPS is based on its released code.

Split	Method	bike	build.	car	cone.	fence	grou.	pers.	plants	pole	rider	traf.	trashc.	trunk	Novel	Known	All
	Full	45.0	83.3	52.0	36.5	46.7	77.6	68.2	77.7	36.0	58.9	30.3	4.2	14.4	-	-	48.5
0	NOPS	37.4	22.9	8.1	0.0	30.3	78.9	4.8	72.9	**1.0**	42.9	25.8	9.2	**9.6**	7.7	42.5	26.5
	Ours	46.0	**26.1**	**27.5**	**2.8**	46.9	77.6	**35.0**	77.8	0.2	58.6	30.5	3.2	0.0	**15.3**	48.7	33.2
1	NOPS	6.1	71.3	35.6	21.2	3.1	**42.9**	44.5	26.0	24.4	**0.7**	0.6	**0.1**	24.8	11.4	37.0	23.2
	Ours	**26.3**	82.0	51.4	18.0	**10.4**	40.0	67.5	**32.5**	31.2	0.0	**6.3**	0.0	11.7	**16.5**	44.5	29.4

analysis of DBSCAN is included in Appendix J. For the input point clouds, we set the voxel size as 0.05 and utilize the scale and rotation augmentation to generate two views. The scale range is from 0.95 to 1.05, and the rotation range is from $-\pi/20$ to $\pi/20$ for three axes. We train 10 epochs and set batch size as 4 for all experiments. The optimizer is Adamw, and the initial learning rate is 1e−3, which decreases to 1e−5 by a cosine schedule. For the hyperparameters, we set $\alpha = \beta = 1$ and fix λ at 0.5. We choose $T = 10$ and $\rho = 0.005$ based on the indicator mentioned in Sect. 3.3 and analyze them in the ablation study. Both the point- and region-level self-labeling algorithms employ the same parameters. All experiments are conducted on a single NVIDIA A100.

4.2   Results

SemanticPOSS Dataset. As presented in Table. 1, our approach exhibits significant improvements in novel classes over the previous method across all four splits. Specifically, we achieve an increase of **12.7%** and **6.2%** in split 0 and 1, respectively. It is worth noting that the fully supervised upper bounds for novel classes in split 0 and 1 are 72.7% and 53.3%, respectively, and the performance gaps have been significantly reduced. In the more challenging split 2 and split 3, we observe gains of **3.6%** and **4.7%**, respectively. The corresponding upper bounds for these splits are 26.9% and 33.2%, indicating their increased difficulty compared to splits 0 and 1. On average, we achieve an IoU of 30.2% for novel classes across all four splits, outperforming NOPS (22.5%) by **7.7%**. In addition, we provide a detailed comparison with NOPS on head, medium, and tail classes in Appendix D, as well as under a more comparable setting that applies our training strategy to NOPS in Appendix E.

To further verify that our method can alleviate the imbalanced problem, we divided the SemanticPOSS dataset into two splits, creating a more severe imbalance scenario that poses a greater challenge for clustering novel classes. As shown in Table 2, on novel classes, our method outperforms NOPS significantly on both splits, with a margin of **7.6%** on split 0 and **5.1%** on split 1. In particular, for the novel classes, we observe that our improvement mainly stems from the medium classes, such as person and bike. It is worth noting that NOPS employs extra training techniques, such as multihead and overclustering, whereas we use a simpler pipeline without needing them, further demonstrating our effectiveness.

SemanticKITTI Dataset. The results in Table 3 demonstrate our superior performance compared to previous methods on different splits. Specifically, we achieve significant improvements of **8.6%**, **3.3%**, and **3.6%** on splits 0, 1, and 2, respectively, for novel classes. The supervised upper bounds for these splits are 82.0%, 42.4%, and 39.6%, respectively. In split 3, our results are slightly higher than NOPS by 0.2%, possibly due to the scarce presence of these novel classes in split 3. On average across all four splits, our approach achieves an IoU of 27.5%, surpassing NOPS (23.4%) by **4.1%** on novel classes.

Visualization Analysis. Additionally, in Fig. 2, we perform visual comparisons on the results between NOPS and our method, and it is evident that our method shows significant improvements compared to NOPS. Specifically, as shown in the first row of Fig. 2, NOPS produces noisy predictions due to uniform constraints, mixing medium classes (e.g., building) and tail classes (e.g., car). In the second and third rows of Fig. 2, NOPS often confuses between medium and head classes, such as building and plants, as well as parking and car. In contrast, our method achieves better results for both datasets due to adaptive regularization and dual-level representation learning, generating high-quality imbalanced pseudo labels. More visual comparisons for additional splits are provided in the Appendix K.

4.3   Ablation Study

Component Analysis. To analyze the effectiveness of each component, we conduct extensive experiments on split 0 of the SemanticPOSS dataset. Here we provide ablation on three components, including Imbalanced Self-Labeling (ISL), Adaptive Regularization (AR), and Region-Level Branch (Region). As shown in

Table 3. The novel class discovery results on the SemanticKITTI dataset. 'Full' denotes the results obtained by supervised learning. The four groups represent the four splits in turn, and the gray values are the novel classes in each split.

Method	bi.cle	b.clst	build.	car	fence	mt.cle	m.clst	oth-g.	oth-v.	park.	pers.	pole	road	side2.	terra.	traff.	truck	trunk	veget.	Novel	Known	All
Full	2.9	55.4	89.5	93.5	27.9	27.4	0.0	0.9	19.9	35.8	31.2	60.0	93.5	77.8	62.0	39.8	50.8	53.9	87.0	-	-	47.9
EUMS	5.3	40.0	15.8	79.2	9.0	16.9	2.5	0.1	11.4	14.4	12.7	29.2	42.6	26.1	0.1	10.3	47.4	37.9	38.4	24.6	21.1	23.1
NOPS	5.6	47.8	52.7	82.6	13.8	25.6	1.4	1.7	14.5	19.8	25.9	32.1	56.7	8.1	23.8	14.3	49.4	36.2	44.2	37.1	26.5	29.3
Ours	5.5	51.1	74.6	92.3	29.8	22.8	0.0	0.0	23.3	24.8	27.7	59.7	41.4	22.5	23.6	39.3	43.6	51.1	66.4	45.7	33.7	36.8
EUMS	7.5	42.4	80.0	76.8	8.6	19.6	1.4	0.6	12.0	14.1	14.0	40.7	86.3	66.5	56.3	12.0	44.8	20.9	72.4	24.2	37.1	35.6
NOPS	7.4	51.2	84.5	50.9	7.3	28.9	1.8	0.0	22.2	19.4	30.4	37.6	90.1	72.2	60.8	16.8	57.3	49.3	85.1	25.4	46.2	40.7
Ours	3.7	57.4	89.2	56.5	17.3	20.3	0.0	0.0	20.0	30.6	34.8	60.6	93.2	77.6	62.0	38.7	56.9	39.2	86.7	28.7	50.1	44.5
EUMS	8.3	50.8	83.0	88.1	17.9	2.8	2.3	0.2	3.2	25.4	25.0	20.2	88.3	71.0	57.9	8.6	27.2	38.4	77.0	12.4	42.2	36.6
NOPS	6.7	49.2	86.4	90.8	23.7	2.7	0.6	1.9	15.5	29.5	27.9	36.4	90.3	73.4	61.2	17.8	10.3	46.2	84.3	16.5	48.0	39.7
Ours	3.6	54.2	88.9	93.3	28.4	10.2	0.0	0.9	9.6	33.4	32.2	36.1	92.7	77.4	62.2	10.7	34.2	51.7	86.9	20.1	50.4	42.5
EUMS	4.0	2.5	80.1	87.2	16.8	14.0	15.0	0.3	14.1	20.8	6.8	37.6	86.8	66.5	55.3	16.2	40.6	38.4	76.2	7.1	43.4	35.7
NOPS	2.3	27.8	86.0	89.9	23.1	24.5	2.9	3.1	18.2	30.1	16.3	39.9	90.7	73.5	61.0	17.4	49.8	44.0	83.2	12.4	49.0	41.2
Ours	2.6	32.5	88.7	93.3	28.1	24	0.1	1.0	23.7	35.6	15.3	59.8	93.2	77.6	61.4	37.8	56.6	52.1	86.7	12.6	54.6	45.8

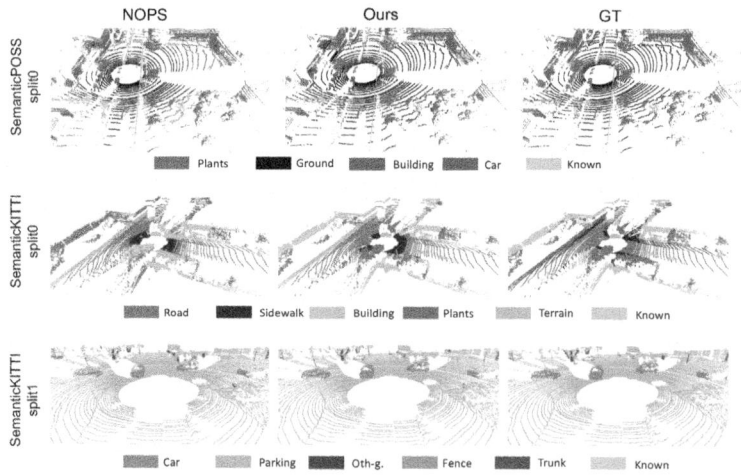

Fig. 2. Visualization comparison between Our method and NOPS on the SemanticPOSS and SemanticKITTI datasets. In the first and second rows, compared to NOPS, our method achieves much better segmentation for the 'Building', significantly reducing confusion with tail classes (such as 'Car') or medium classes (like 'Plants'). In the third row, our approach generates better segmentation for 'Parking' and 'Car'.

Table 4. Ablation study on SemanticPOSS, focusing on novel classes. **ISL** and **AR** denote imbalanced self-labeling and regularization. **Region** denotes region-level learning. The last two columns represent the average mIoU for split 0 and across all splits.

ISL	AR	Region	Split0					Overall
			Building	Car	Ground	Plants	Avg	Avg
			21.6	2.7	76.6	26.1	31.8	20.9
✓			27.6	3.1	81.2	32.1	36.0	23.9
✓	✓		53.1	5.3	81.1	37.4	44.2	28.4
✓		✓	41.9	9.3	83.6	45.6	45.1	26.9
✓	✓	✓	51.5	6.0	83.0	53.1	48.4	30.2

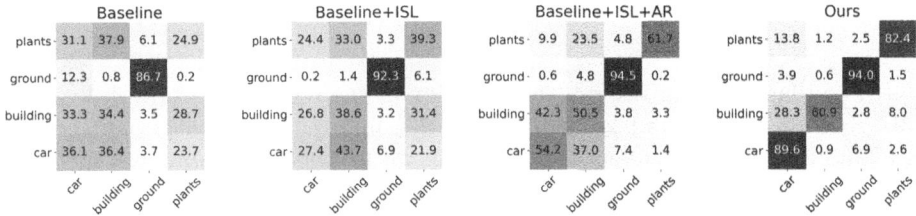

Fig. 3. Confusion Matrix, GT on the y-axis, Pseudo Label on the x-axis. (i, j) represents the % of GT in class j assigned pseudo label i. We categorize 'plants' and 'ground' as head classes, 'building' as medium, and 'car' as tail classes.

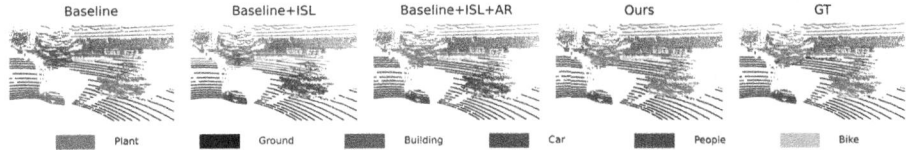

Fig. 4. Visualization analysis. The introduction of ISL notably reduces the misclassification between 'Plant' and 'Car'. Then, the integration of AR further mitigates the confusion between 'Plant' and 'Building'. Ultimately, the incorporation of Region component (Ours) effectively minimizes the mix-up between 'Plant', 'Car', and 'Building'.

Table 4, compared to baseline which employs equal-size constraints, imbalanced self-labeling improves performance by **4.2%**. The confusion matrix in Fig. 3 indicates that except for the highly-accurate class "ground", there is a significant improvement in the head and medium classes. This phenomenon is clearly depicted in Fig. 4, where the predictions of the baseline exhibit noticeable noise. From the second and third rows of Table 4, the adaptive regularization leads to a significant improvement of **8.2%** in split0 and **4.5%** in overall splits. As shown in Fig. 3, adaptive regularization enhances the quality of pseudo-labels for each class, especially for the head class (plants). We also visualize the class distribution of pseudo-labels in Appendix F, which shows adaptive regularization provides greater flexibility than fixed regularization term. According to the third,

Table 5. Analysis of adaptive regularization on SemanticPOSS dataset. GT denotes we directly assign the ground truth distribution of cluster size.

γ	0.01	0.05	0.1	0.5	1	5	$+\infty$	GT	Adaptive
Split0	10.1	33.8	33.3	36.0	32.2	33.8	31.8	32.5	44.2

Table 6. Comparison between Ours and NOPS with the estimated number of novel classes in Split 0 of SemanticPOSS. The estimated C^u is 3, and the ground truth is 4.

Method	Building	Car	Ground	Plants	Avg
NOPS	25.54	0.00	68.15	34.12	31.95
Ours	64.05	0.00	82.22	67.63	**53.47**

fourth and last rows of Table 4, the inclusion of the region-level branch leads to a **9.1%** improvement and an additional **4.2%** improvement built upon the AR. In addition, more experiments and analysis on prototype learning are included in Appendix G. In Fig. 3, there's a significant improvement in pseudo-labels for each category, particularly for the tail class (car) and the head class (plants). From Fig. 4, it is evident that the region-level branch can correct cases where a single object is mistakenly labeled as multiple categories. Due to the utilization of spatial priors, where closely-located points are highly likely to belong to the same category, our region-level branch can correct misclassifications by considering context from neighboring points, preventing splitting a single object into multiple entities. Those experiments validate the effectiveness of each component in our method.

Estimate the Number of Novel Classes. For computational simplicity, we conduct experiments on splits 0 of the SemanticPOSS dataset and randomly sample 800,000 points from all scenes to estimate $|C^u|$. We set max classes to 50, which is an estimate of the maximum number of new classes that might appear in a typical scene. The estimated $|C^u|$ is 3, which is close to the ground truth value (GT is 4). Finally, we conduct experiments with $|C^u|$ as 3. As Table 6 illustrated our method still outperforms NOPS by a large margin.

Adaptive Regularization and Hyperparameters Selection. To analyze the impact of adaptive regularization, we compare it with various fixed regularization factors, as illustrated in Table 5. We notice that employing a very small fixed γ, such as 0.05 as indicated in the table, results in a weak prior constraint, and the model tends to learn a degenerate solution where all samples are assigned to a single cluster. When the γ increases to 0.5, the model achieves optimal results, but the increment decreases when the γ further increases. Compared with adaptive γ, the optimal results of fixed γ is nearly **8.2%** lower, demonstrating that the adoption of an adaptive γ not only enhances the model's flexibility but also prevents any performance degradation. Furthermore, we experiment with the setup adopting the GT class distribution and substituting the KL constraint in

Table 7. Step decay results

λ	0.1	0.3	0.5	0.7	0.9
Step decay	34.0	34.2	32.9	34.6	33.3

Table 8. Cosine annealing results

min γ	0.1	0.05	0.01	0.005	0.001
Cosine annealing	32.2	32.5	35.8	36.1	32.0

Table 9. The results for different ρ

ρ	0.05	0.01	0.005	0.001
Split0	10.08	45.84	44.21	33.11

Table 10. The results for different T

T	5	10	20	30
Split0	44.46	44.21	44.12	32.48

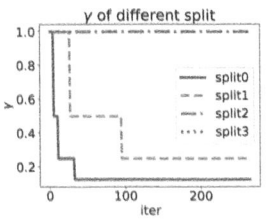

Fig. 5. γ variation

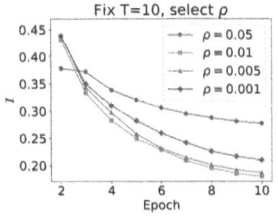

Fig. 6. Selecting ρ

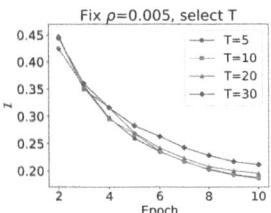

Fig. 7. Selecting T

Eq. (4) with an equality constraint. Surprisingly, the results indicate that the GT class distribution constraint is not the optimal solution for clustering imbalanced novel classes. At last, in Fig. 5, we visualize the γ curves for SemanticPOSS in four splits. Split 0 exhibits the highest rate of change, followed by Split 1, while Splits 2 and 3 remain constant, indicating that our strategy is adaptive to each dataset.

To further validate the effectiveness of adjusting γ based on KL divergence, we also compare it with typical step decay and cosine annealing strategies. For the step decay, we set the initial γ to 1 and decay it by multiplying it with λ every epoch. For the cosine annealing approach, we also set the initial γ to 1 and reduce it to the minimum value (min γ). From the Table 7 and Table 8, we observe that the results of simple step decay and cosine annealing are nearly **10%** worse than adaptive γ (which is 44.2). We believe that these two typical strategies lack flexibility compared to adaptive γ. They might not facilitate the adaptive control of the γ decay process based on the model learning process.

To choose the hyperparameters ρ and T according to the indicator outlined in Sect. 3.3, we conduct experiments for various values of ρ and T. The results are displayed in Table 9 and Table 10. Additionally, we plot the indicator's curve for each experiment in Fig. 6 and 7. The plots reveal that when ρ falls within the range of 0.01 to 0.005, and T is set between 5 and 20, the indicator value remains low while achieving a high novel IoU. Those results demonstrate the efficiency of our hyperparameters selection strategy and the robustness of our method.

Limitations. One limitation is our problem setup which follows [28] and only addresses scenarios where unlabelled data constitutes novel classes. In contrast, a more realistic open-world setting necessitates handling situations where both known classes and novel classes lack labels. Nevertheless, we anticipate that our method will establish a robust baseline and stimulate further research aimed at addressing the challenges presented by practical open-world situations.

5   Conclusion

In this paper, we propose a novel dual-level adaptive self-labeling framework for novel class discovery in point cloud segmentation. Our framework formulates the pseudo label generation process as a Semi-relaxed Optimal Transport problem and incorporates a novel data-dependent adaptive regularization factor to gradually relax the constraint of the uniform prior based on the distribution of pseudo labels, thereby generating higher-quality imbalanced pseudo labels for model learning. In addition, we develop a dual-level representation that leverages the spatial prior to generate region representation, which reduces the noise in generated segmentation and enhances point-level classifier learning. Furthermore, we propose a hyperparameters search strategy based on training sets. Extensive experiments on two widely used datasets, SemanticKITTI and Semantic-POSS, demonstrate the effectiveness of each component and the superiority of our method.

Acknowledgments. This work was supported by National Science Foundation of China under grant 62350610269, Shanghai Frontiers Science Center of Human-centered Artificial Intelligence, and MoE Key Lab of Intelligent Perception and Human-Machine Collaboration (ShanghaiTech University).

References

1. Asano, Y.M., Rupprecht, C., Vedaldi, A.: Self-labelling via simultaneous clustering and representation learning. In: International Conference on Learning Representations (ICLR) (2020)
2. Behley, J., Garbade, M., Milioto, A., Quenzel, J., Behnke, S., Gall, J., Stachniss, C.: Towards 3d lidar-based semantic scene understanding of 3d point cloud sequences: the semantickitti dataset. Int. J. Robot. Res. **40**(8–9), 959–967 (2021)
3. Behley, J., et al.: Semantickitti: a dataset for semantic scene understanding of lidar sequences. In: Proceedings of the IEEE/CVF International Conference on Computer Vision, pp. 9297–9307 (2019)
4. Caron, M., Misra, I., Mairal, J., Goyal, P., Bojanowski, P., Joulin, A.: Unsupervised learning of visual features by contrasting cluster assignments. Adv. Neural. Inf. Process. Syst. **33**, 9912–9924 (2020)

5. Chang, W., Shi, Y., Tuan, H., Wang, J.: Unified optimal transport framework for universal domain adaptation. Adv. Neural. Inf. Process. Syst. **35**, 29512–29524 (2022)

6. Chizat, L., Peyré, G., Schmitzer, B., Vialard, F.X.: Scaling algorithms for unbalanced optimal transport problems. Math. Comput. **87**(314), 2563–2609 (2018)

7. Choy, C., Gwak, J., Savarese, S.: 4d spatio-temporal convnets: minkowski convolutional neural networks. In: Proceedings of the IEEE/CVF Conference on Computer Vision and Pattern Recognition, pp. 3075–3084 (2019)

8. Cuturi, M.: Sinkhorn distances: Lightspeed computation of optimal transport. Advances in Neural Information Processing Systems **26** (2013)

9. Ester, M., Kriegel, H.P., Sander, J., Xu, X., et al.: A density-based algorithm for discovering clusters in large spatial databases with noise. In: KDD. vol. 96, pp. 226–231 (1996)

10. Fini, E., Sangineto, E., Lathuilière, S., Zhong, Z., Nabi, M., Ricci, E.: A unified objective for novel class discovery. In: Proceedings of the IEEE/CVF International Conference on Computer Vision, pp. 9284–9292 (2021)

11. Flamary, R., Courty, N., Tuia, D., Rakotomamonjy, A.: Optimal transport for domain adaptation. IEEE Trans. Pattern Anal. Mach. Intell. **1**, 1–40 (2016)

12. Geiger, A., Lenz, P., Urtasun, R.: Are we ready for autonomous driving? the kitti vision benchmark suite. In: 2012 IEEE Conference on Computer Vision and Pattern Recognition, pp. 3354–3361. IEEE (2012)

13. Gu, P., Zhang, C., Xu, R., He, X.: Class-relation knowledge distillation for novel class discovery. In: 2023 IEEE/CVF International Conference on Computer Vision (ICCV), pp. 16428–16437. IEEE Computer Society (2023)

14. Han, K., Rebuffi, S.A., Ehrhardt, S., Vedaldi, A., Zisserman, A.: Autonovel: automatically discovering and learning novel visual categories. IEEE Transactions on Pattern Analysis and Machine Intelligence (2021)

15. Han, K., Vedaldi, A., Zisserman, A.: Learning to discover novel visual categories via deep transfer clustering. In: Proceedings of the IEEE/CVF International Conference on Computer Vision, pp. 8401–8409 (2019)

16. Hu, Q., et al.: Randla-net: Efficient semantic segmentation of large-scale point clouds. In: Proceedings of the IEEE/CVF Conference on Computer Vision and Pattern Recognition, pp. 11108–11117 (2020)

17. Lai, X., et al.: Stratified transformer for 3d point cloud segmentation. In: Proceedings of the IEEE/CVF Conference on Computer Vision and Pattern Recognition, pp. 8500–8509 (2022)

18. Lai, Z., Wang, C., Cheung, S.c., Chuah, C.N.: Sar: self-adaptive refinement on pseudo labels for multiclass-imbalanced semi-supervised learning. In: Proceedings of the IEEE/CVF Conference on Computer Vision and Pattern Recognition, pp. 4091–4100 (2022)

19. Landrieu, L., Simonovsky, M.: Large-scale point cloud semantic segmentation with superpoint graphs. In: Proceedings of the IEEE Conference on Computer Vision and Pattern Recognition, pp. 4558–4567 (2018)

20. Lee, D.H., et al.: Pseudo-label: the simple and efficient semi-supervised learning method for deep neural networks. In: Workshop on challenges in representation learning, ICML. vol. 3, p. 896. Atlanta (2013)

21. Li, Y., et al.: Deep learning for lidar point clouds in autonomous driving: a review. IEEE Trans. Neural Netw. Learn. Syst. **32**(8), 3412–3432 (2020)

22. Li, Z., et al.: Bevformer: learning bird's-eye-view representation from multi-camera images via spatiotemporal transformers. In: European Conference on Computer Vision, pp. 1–18. Springer (2022). https://doi.org/10.1007/978-3-031-20077-9_1

23. Liu, Y., Zhou, Z., Sun, B.: Cot: Unsupervised domain adaptation with clustering and optimal transport. In: Proceedings of the IEEE/CVF Conference on Computer Vision and Pattern Recognition (CVPR), pp. 19998–20007 (June 2023)
24. Long, F., Yao, Ting abd Qiu, Z., Li, L., Mei, T.: Pointclustering: unsupervised point cloud pre-training using transformation invariance in clustering. In: CVPR (2023)
25. Nakajima, Y., Kang, B., Saito, H., Kitani, K.: Incremental class discovery for semantic segmentation with rgbd sensing. In: Proceedings of the IEEE/CVF International Conference on Computer Vision, pp. 972–981 (2019)
26. Pan, Y., Gao, B., Mei, J., Geng, S., Li, C., Zhao, H.: Semanticposs: a point cloud dataset with large quantity of dynamic instances. In: 2020 IEEE Intelligent Vehicles Symposium (IV), pp. 687–693. IEEE (2020)
27. Phatak, A., Raghvendra, S., Tripathy, C., Zhang, K.: Computing all optimal partial transports. In: International Conference on Learning Representations (2023)
28. Riz, L., Saltori, C., Ricci, E., Poiesi, F.: Novel class discovery for 3d point cloud semantic segmentation. In: Proceedings of the IEEE/CVF Conference on Computer Vision and Pattern Recognition, pp. 9393–9402 (2023)
29. Roriz, R., Cabral, J., Gomes, T.: Automotive lidar technology: a survey. IEEE Trans. Intell. Transp. Syst. **23**(7), 6282–6297 (2021)
30. Taherkhani, F., Dabouei, A., Soleymani, S., Dawson, J., Nasrabadi, N.M.: Transporting labels via hierarchical optimal transport for semi-supervised learning. In: Vedaldi, A., Bischof, H., Brox, T., Frahm, J.-M. (eds.) ECCV 2020. LNCS, vol. 12349, pp. 509–526. Springer, Cham (2020). https://doi.org/10.1007/978-3-030-58548-8_30
31. Tai, K.S., Bailis, P.D., Valiant, G.: Sinkhorn label allocation: semi-supervised classification via annealed self-training. In: International Conference on Machine Learning, pp. 10065–10075. PMLR (2021)
32. Vaze, S., Han, K., Vedaldi, A., Zisserman, A.: Generalized category discovery. In: Proceedings of the IEEE/CVF Conference on Computer Vision and Pattern Recognition, pp. 7492–7501 (2022)
33. Villani, C., et al.: Optimal transport: old and new, vol. 338. Springer (2009). https://doi.org/10.1007/978-3-540-71050-9
34. Yang, M., Zhu, Y., Yu, J., Wu, A., Deng, C.: Divide and conquer: compositional experts for generalized novel class discovery. In: Proceedings of the IEEE/CVF Conference on Computer Vision and Pattern Recognition, pp. 14268–14277 (2022)
35. Zhang, C., Ren, H., He, X.: P^2ot: Progressive partial optimal transport for deep imbalanced clustering. In: The Twelfth International Conference on Learning Representations (2023)
36. Zhang, C., Xu, R., He, X.: Novel class discovery for long-tailed recognition. Transactions on Machine Learning Research (2023)
37. Zhang, S., Khan, S., Shen, Z., Naseer, M., Chen, G., Khan, F.S.: Promptcal: contrastive affinity learning via auxiliary prompts for generalized novel category discovery. In: Proceedings of the IEEE/CVF Conference on Computer Vision and Pattern Recognition, pp. 3479–3488 (2023)
38. Zhang, Y., et al.: Polarnet: an improved grid representation for online lidar point clouds semantic segmentation. In: Proceedings of the IEEE/CVF Conference on Computer Vision and Pattern Recognition, pp. 9601–9610 (2020)
39. Zhang, Z., Yang, B., Wang, B., Li, B.: Growsp: unsupervised semantic segmentation of 3d point clouds. In: Proceedings of the IEEE/CVF Conference on Computer Vision and Pattern Recognition, pp. 17619–17629 (2023)

40. Zhao, B., Han, K.: Novel visual category discovery with dual ranking statistics and mutual knowledge distillation. In: Advances in Neural Information Processing Systems, vol. 34 (2021)
41. Zhao, Y., Zhong, Z., Sebe, N., Lee, G.H.: Novel class discovery in semantic segmentation. In: Proceedings of the IEEE/CVF Conference on Computer Vision and Pattern Recognition, pp. 4340–4349 (2022)
42. Zhu, X., et al.: Cylindrical and asymmetrical 3d convolution networks for lidar segmentation. In: Proceedings of the IEEE/CVF Conference on Computer Vision and Pattern Recognition, pp. 9939–9948 (2021)

EBDM: Exemplar-Guided Image Translation with Brownian-Bridge Diffusion Models

Eungbean Lee[1]🆔, Somi Jeong[2]🆔, and Kwanghoon Sohn[1,3](✉)

[1] Yonsei University, Seoul, Korea
{eungbean,khsohn}@yonsei.ac.kr
[2] NAVER LABS, Seoul, South Korea
somi.jeong@naverlabs.com
[3] Korea Institute of Science and Technology (KIST), Seoul, South Korea

Abstract. Exemplar-guided image translation, synthesizing photo-realistic images that conform to both structural control and style exemplars, is attracting attention due to its ability to enhance user control over style manipulation. Previous methodologies have predominantly depended on establishing dense correspondences across cross-domain inputs. Despite these efforts, they incur quadratic memory and computational costs for establishing dense correspondence, resulting in limited versatility and performance degradation. In this paper, we propose a novel approach termed Exemplar-guided Image Translation with Brownian-Bridge Diffusion Models (EBDM). Our method formulates the task as a stochastic Brownian bridge process, a diffusion process with a fixed initial point as structure control and translates into the corresponding photo-realistic image while being conditioned solely on the given exemplar image. To efficiently guide the diffusion process toward the style of exemplar, we delineate three pivotal components: the Global Encoder, the Exemplar Network, and the Exemplar Attention Module to incorporate global and detailed texture information from exemplar images. Leveraging Bridge diffusion, the network can translate images from structure control while exclusively conditioned on the exemplar style, leading to more robust training and inference processes. We illustrate the superiority of our method over competing approaches through comprehensive benchmark evaluations and visual results.

Keywords: Generative model · Image Translation · Diffusion Models · Image Synthesis

1 Introduction

The rising interest in applications of image synthesis has led to a notable surge in demand for image generation capabilities that extend beyond text prompts,

Supplementary Information The online version contains supplementary material available at https://doi.org/10.1007/978-3-031-72624-8_18.

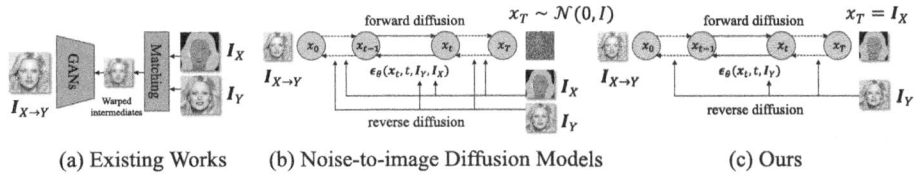

(a) Existing Works   (b) Noise-to-image Diffusion Models   (c) Ours

Fig. 1. Motivation. (a) Existing methods with matching-than-generation framework, (b) Widely used framework based on conditional noise-to-image diffusion model, and (c) our framework based on Brownian bridge diffusion models.

emphasizing control through exemplar images or structured inputs. Exemplar-guided image translation task aims to generate photo-realistic images conditioned on both a style exemplar image and specific structural controls, such as segmentation masks, edge maps, or pose keypoints (Fig. 1).

To synthesize images guided by the style of an exemplar and structure controls, pioneer works [13,31,42,43,48] have emerged. Formulated as an ill-posed problem, these methods globally leverage the style of an exemplar. Despite promising results, these methods overlook local details, which leads to compromised generation quality.

To enhance the capture of local styles from exemplar, significant efforts [20, 39,54–56,58,61] have been explored to establish cross-domain correspondences between input control and exemplar images, thereby imposing local style through a matching process. Zhang *et al.* [58] have explored the construction of cross-domain correspondences using cosine similarity for exemplar-based image translation and subsequent works [54–56,61] introduced various techniques to reduce computational complexity and address many-to-one matching problems. Moreover, they predominantly capture style information at a coarse scale, which leads to performance degradation, because their methods are significantly influenced by the quality of warped intermediate features derived from sparse correspondences between two domains, often failing to accurately reflect the dynamic nature of matching. This failure leads to local distortion, blurred details, and semantic inconsistency. Furthermore, models leveraging Generative Adversarial Networks (GANs) face the intrinsic limitations of GANs, such as mode collapse, limited diversity, and the out-of-range problem [15,53].

Recently, diffusion models [8,34,40,41], which generate high-quality images through iterative denoising processes, have attained significant success in the field of image synthesis for their several advantages, including broader distribution coverage, more stable training, and enhanced scalability compared to GANs [2,8]. As the surge in demand for customized image generation has advanced, Text-to-image (T2I) synthesis, conditioned on text prompts, has been extensively explored in works such as [4,29,33]. Beyond mere text prompts, numerous studies have sought solutions to address precise style through model fine-tuning [10,35], prompt engineering [5,49]. Additionally, efforts have been made to incorporate structure controls (*i.e.*, edge, depth, mask, pose, *etc..*) [11,28,47,57,60] as generative guidance.

Although diffusion models have demonstrated impressive performance, exemplar-based image translation remains largely unexplored. First, it is challenging to find an accurate prompt that conveys every desired aspect of an image. Second, it is hard to address exemplar style because fine-tuning offers quality at a high cost, prompt engineering is more affordable but less detailed, and CLIP representations are not sufficient to address all details in visual cues. Lastly, achieving simultaneous conditioning on both style exemplars and structured controls is challenging, particularly because the diffusion process used in such tasks can be highly sensitive to hyperparameters, including the guidance scales of structure control and embeddings.

To solve the above issues, we introduce a sophisticated technique called EBDMs (Exemplar-based image translation with Brownian bridge Diffusion Models) that fully leverages diffusion models. Our method leverages a stochastic Brownian bridge process [19] that directly learns translation between two domains, thus generating images from structure controls without any conditioning mechanism. To explore desired style control, we propose a Global Encoder and Exemplar Network to leverage coarse and fine details from exemplar images. Moreover, the Exemplar Attention Module effectively consilates the texture information from the exemplar into the denoising process. Our method can generate images by conditioning on structure control and style exemplars with single conditioning simultaneously. We conduct extensive experiments on various datasets, including mask-to-image, edge-to-image [23], and keypoint-to-image [22]. The experimental results demonstrate the superiority of our approach not only performance but also computational efficiency. The contribution of this work can be summarized as follows:

- We introduce the EBDMs, a novel framework leveraging the stochastic Brownian Bridge diffusion process that translates from structure control to a photo-realistic image while effectively exploiting style from exemplars.
- The proposed method formulated the problem into a single-conditioned bridge diffusion process that ensures the training and inference more robust.
- We propose Global Encoder, Exemplar Network, and Exemplar Attention Module to address both global style and detailed texture of exemplar image.
- Extensive experiments demonstrate that our approach achieves favorable performance on various exemplar-guided image translation tasks.

2   Related Works

2.1   Controllable Diffusion Models

Diffusion models [2,8,34,40,41] aim to synthesize images from random Gaussian noise via an iterative denoising process. For customized image generation, recent methods have explored text-guided image generation (T2I) [4,29,33,34,37] and demonstrated extraordinary generative capabilities in modeling the intricacies of complex images. GLIDE [29] aggregated the CLIP texture representations utilizing classifier-free guidance [9]. DALLE-2 [33] proposed a cascade model using the CLIP latent. VQ-Diffusion [4] proposed to learn the diffusion process on the discrete latent space of VQ-VAE [46].

To address the structure controls (such as mask, edge, pose, *etc.*.), several works have proposed fine-tuning approach [47] or adaptive models [28,57,60] in addition to text prompts. ControlNet [57] proposed an adaptive network to provide structure guidance to T2I models followed by Uni-ControlNet [60] which expands to a unified framework to accept diverse control signals at once. Concurrently, T2I-Adapter [28] introduces a more simple and lightweight adapter. Such methods enable to provide the structural guidance to existing T2I diffusion models thus providing more precise spatial control.

On the other hand, to accurately reflect the style of an exemplar, numerous studies have been conducted such as model fine-tuning [3,10,16,35,52], prompt engineering [5,21,49]. DreamBooth [35] proposed to fine-tune the T2I models with exemplar image and LoRA [10] proposed a more effective tuning method. IP-Adapter [52] proposed decoupled cross-attention to effectively inject exemplar image features into the denoising network. Moreover, Guo *et al.* [5] proposed the image-specific prompt learning method to learn domain-specific prompt vectors. While other methods [6,26,27,36] enable zero-shot editing of an exemplar image based on a target caption. Despite these method capabilities, it is challenging to find the prompt that accurately generates the image a user envisions, mainly because effectively reflects all desired aspects of an image through text, especially those that are difficult or impossible to describe precisely.

2.2   Exemplar-Guided Image Translation

The exemplar-guided image translation task involves generating an image based on an input exemplar and structure controls such as an edge, pose, or mask. A major challenge lies in effectively guiding the context within exemplars relative to the input controls. The SPADE [31] framework proposed spatially-adaptive normalization to generate an image from the semantic mask followed by class-adaptive [43] and instance-adaptive [42]. While these approaches have shown promising in global-style translation, they overlooked local details compromising generation quality.

To address the local details, significant efforts have been focused on building dense correspondence. Zhang *et al.* [58] proposed building dense correspondence between input semantic and exemplar image. Although their method has shown promising results, their method is limited by many-to-one matching issues and the quadratic computational and memory complexities of dense matching operations, restricting it to capturing only coarse-scale warped features. To alleviate these issues, recent works introduced effective correspondence learning such as GRU-assisted Patch-Match [61], unbalanced optimal transport [54], bi-level feature alignment strategy [55], multi-scale dynamic sparse attention [20], Cross-domain Feature Fusion Transformer [25] and Masked Adaptive Transformer [14]. Although they have demonstrated promising results, their matching-based framework still suffers from inherent problems such as sparse matching.

Meanwhile, recent progress [11,39,50,51] has leveraged diffusion models to bridge the gap between style exemplars and structural controls. Seo *et al.* [39]

proposed a two-staged framework, which is a matching module followed by a diffusion module. Although they have successfully applied diffusion models, they still heavily rely on a matching-based framework that does not fully utilize the diffusion models. Paint-by-example [50] proposed self-supervised training for image disentanglement and reorganization, while Composer [11] conceptualized an image as a composition of several representations, suggesting a decompose-then-recompose approach and ImageBrush [51] learns visual instructions. Although they have demonstrated promising results, they offer limited control, typically constrained to structure-preserving appearance changes or uncontrolled image-to-image translation.

3 Preliminaries

3.1 Diffusion Models

The general idea of Denoising Diffusion Probabilistic Model (DDPM) [8] is to generate images from Gaussian noise via T steps of an iterative denoising process. It consists of two processes: the forward process and the reverse process. Given the original data $\boldsymbol{x}_0 \sim q_{data}(\boldsymbol{x}_0)$, the forward diffusion process maps $\boldsymbol{x}_0$ into noisy latent variables $\{\boldsymbol{x}_t\}_{t=0}^T$ can be obtained as: $\boldsymbol{x}_t = \sqrt{\alpha_t}\boldsymbol{x}_0 + \sqrt{1-\alpha_t}\boldsymbol{\epsilon}$ where $\boldsymbol{\epsilon}$ is the Gaussian noise and $\{\alpha_t\}_{t=0}^T$ is pre-defined schedule. On the other hand, the corresponding reverse process aims to predict the original data $\boldsymbol{x}_0$ starting from the pure Gaussian noise $\boldsymbol{x}_T \sim \mathcal{N}(\boldsymbol{0}, \boldsymbol{I})$ through iterative denoising processes with pre-defined time steps. It is formulated as another Markov chain as $p_\theta(\boldsymbol{x}_{t-1} \mid \boldsymbol{x}_t) := \mathcal{N}(\boldsymbol{x}_{t-1}; \boldsymbol{\mu}_\theta(\boldsymbol{x}_t, t), \sigma_t^2 \boldsymbol{I})$ with learned mean and fixed variance. The denoising network $\boldsymbol{\epsilon}_\theta$ is trained to predict the noise by minimizing a weighted mean squared error loss, defined as:

$$L(\theta) = \mathbb{E}_{t,\boldsymbol{x}_0,\boldsymbol{\epsilon}}[\|\boldsymbol{\epsilon} - \boldsymbol{\epsilon}_\theta(\boldsymbol{x}_t, t)\|_2^2]. \tag{1}$$

Similarly, the conditional diffusion models [36,38] directly inject the condition $\boldsymbol{y}$ into the training objective (Eq. 1), such as $L(\theta) = \mathbb{E}_{t,\boldsymbol{x}_0,\boldsymbol{\epsilon}} \|\boldsymbol{\epsilon} - \boldsymbol{\epsilon}_\theta(\boldsymbol{x}_t, \boldsymbol{y}, t)\|_2^2$.

3.2 Brownian Bridge Diffusion Models

A Brownian Bridge Diffusion Model (BBDM) [19] is an image-to-image translation framework based on a stochastic Brownian Bridge process. Unlike DDPM that conclude at Gaussian noise $\boldsymbol{x}_T \sim \mathcal{N}(0, \boldsymbol{I})$, BBDM assumes that both endpoints of the diffusion process as fixed data points from an arbitrary joint distribution, i.e.$(\boldsymbol{x}_T, \boldsymbol{x}_0) \sim q_{data}(\mathcal{X}, \mathcal{Y})$. The BBDM directly learns image-to-image translation $q(\boldsymbol{x}_0|\boldsymbol{x}_T)$ with boundary distribution $q_{data}(\boldsymbol{x}_0, \boldsymbol{x}_T)$ independent of any conditional process, that enhances the fidelity and diversity of the generated samples. The forward process of the Brownian Bridge forms a bridge between two fixed endpoints at $t = 0$ and T:

$$q(\boldsymbol{x}_t \mid \boldsymbol{x}_0, \boldsymbol{y}) = \mathcal{N}(\boldsymbol{x}_t; (1 - m_t)\boldsymbol{x}_0 + m_t\boldsymbol{y}, \delta_t\boldsymbol{I}), \quad \text{where} \quad \boldsymbol{y} = \boldsymbol{x}_T \tag{2}$$

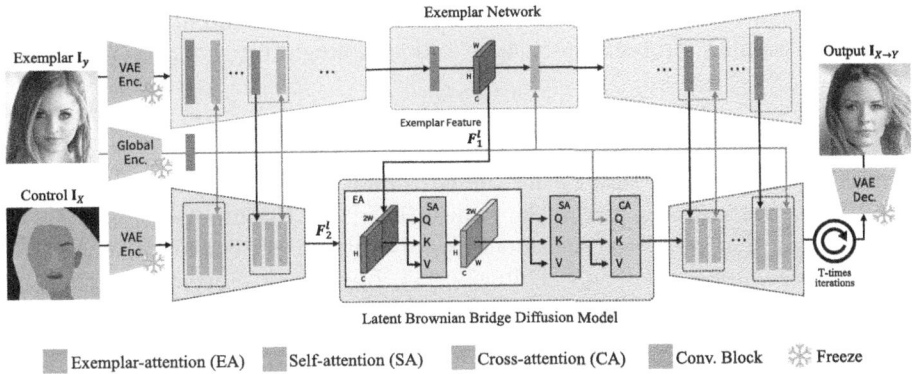

Fig. 2. Framework overview. The proposed EBDM framework is a based on (a) Brownian Bridge Diffusion Model and composed of (b) Exemplar Network and a (c) Global Encoder. Global Encoder encodes global style information and Exemplar Network extracts texture information from exemplar image. Extracted texture and global information is then used to guide the diffusion process via Exemplar Attention Module and cross-attention, respectively.

where $m_t = t/T$ and variance term $\delta_t = 2(m_t - m_t^2)$.

The reverse process of BBDM aims to predict $\boldsymbol{x}_{t-1}$ given $\boldsymbol{x}_t$:

$$p_\theta(\boldsymbol{x}_{t-1} \mid \boldsymbol{x}_t, \boldsymbol{y}) = \mathcal{N}(\boldsymbol{x}_{t-1}; \boldsymbol{\mu}_\theta(\boldsymbol{x}_t, t), \tilde{\delta}_t \mathbf{I}), \tag{3}$$

where $\tilde{\delta}_t$ is the variance of Gaussian noise at step t and $\boldsymbol{\mu}_\theta(\boldsymbol{x}_t, t)$ is the predicted mean value of the noise, which is network to be learned. The training objective of BBDM is optimizing the Evidence Lower Bound (ELBO), simplified as:

$$\mathbb{E}_{\boldsymbol{x}_0, \boldsymbol{y}, \boldsymbol{\epsilon}} \left[c_{\epsilon t} \| m_t (\boldsymbol{y} - \boldsymbol{x}_0) + \sqrt{\delta_t} \boldsymbol{\epsilon} - \boldsymbol{\epsilon}_\theta(\boldsymbol{x}_t, t) \|^2 \right]. \tag{4}$$

where $c_{\epsilon t}$ is the coefficient term of estimated noise $\boldsymbol{\epsilon}_\theta$ in mean value term, $\tilde{\boldsymbol{\mu}}_t$.

4 Methodology

In this section, we delineate our framework based upon discrete-time stochastic Brownian Bridge diffusion process [19] for Exemplar-guided image translation (Fig. 2). Given a control $\boldsymbol{I}_\mathcal{X}$ sampled from domain $\mathcal{X}$ alongside an exemplar image $\boldsymbol{I}_\mathcal{Y}$ from domain $\mathcal{Y}$, the primary objective is to generate a target image $\boldsymbol{I}_{\mathcal{X} \to \mathcal{Y}}$ that retains the structure of $\boldsymbol{I}_\mathcal{X}$ embodying the style of $\boldsymbol{I}_\mathcal{Y}$. The key to our method is the infusion of style information from $\boldsymbol{I}_\mathcal{Y}$ to guide the diffusion trajectory of the target image. To facilitate this, our method integrates three components: a denoising network equipped with an Exemplar Attention Module, a Global Encoder, and Exemplar Network. The Global Encoder extracts global style information of $\boldsymbol{I}_\mathcal{Y}$ and the Exemplar Network captures the appearance features of $\boldsymbol{I}_\mathcal{Y}$. The Exemplar Attention Module selectively incorporates appearance information into the denoising process.

In the following sections, we present a detailed explanation of the framework (Sect. 4.1), training strategy (Sect. 4.2), and sampling strategy (Sect. 4.3).

4.1 Exemplar-Guided Brownian Bridge Diffusion Models

Denoising Network. Employing the Brownian Bridge diffusion process, our denoising U-Net directly learns the translation from the input controls $I_\mathcal{X}$ to images $I_{\mathcal{X} \rightarrow y}$ that preserves the structures of controls. For efficient training and inference, we employ the Stable Diffusion [34] framework. Specifically, given an image I, the encoder $\mathcal{E}$ maps it into a latent space $z = \mathcal{E}(I)$, subsequently reconstructed by the decoder $\hat{I} = \mathcal{D}(z)$. The denoising U-Net ϵ_θ learns to establish the bridge from fixed initial point $x_T = z_\mathcal{X}$ to the target, $x_0 = z_{\mathcal{X} \rightarrow y}$.

Unlike existing noise-to-image diffusion frameworks [28,57] embed the structural information through intricate frameworks, our approach translates from structural control to images without explicit conditional operation. Consequently, our framework is able to solely focus on exemplar information that fosters enhanced stable training and inference performance.

Global Encoder. The Global Encoder, utilizing DINOv2 [30], captures the global style information from the exemplar image I_y. Specifically, exemplar image I_y is processed through the Global Encoder, subsequently, [CLS] token is extracted and passed through a linear layer to encapsulate global style attributes:

$$\tau_\theta(I_y) = \text{Linear}\left(\text{DINO}(I_y)_{[CLS]}\right) \in \mathbb{R}^c, \tag{5}$$

where c denotes the dimension of [CLS] token. The global features are utilized as global style information through a cross-attention mechanism, ensuring that the synthesized output accurately reflects the exemplar's global style.

In the context of text-to-image synthesis [8,34], prior works have extensively leveraged the CLIP image encoder to convey high-level semantic prompts via cross-attention. This approach, however, primarily focuses on the semantic alignment of prompts and images, thereby overlooking the representation of detailed textures. Furthermore, our method does not need textual prompt alignment. Motivated by recent studies [17,45] that have demonstrated the superior proficiency of DINO [1,30] over CLIP [32] in encapsulating a broader capability of semantic features in images, attributed to its self-supervised learning strategy, our method incorporates the use of a pre-trained DINOv2 encoder to enhance the semantic fidelity of generated images.

Exemplar Network. Notwithstanding the capability of Global Encoder in capturing overarching style information, it is limited to the retention of fine-grained details because it encodes exemplar in low resolution (224^2). In contrast, the exemplar-guided image translation tasks require higher fidelity to detail. To this end, we introduce Exemplar Network, referred to ψ_θ, of which the objective is to capture the detailed texture information from the exemplar image, thereby compensating for the global information.

The Exemplar Network adopts a siamese configuration akin to a denoising U-Net, streamlined by omitting the redundant layers for enhanced efficiency during training and inference. It encodes the exemplar z_y into a feature maps $\{F_1^l\}_{l=0}^N$ across N blocks. Additionally, it processes the global information through cross-attention mechanisms in each block. The exemplar features $\{F_1^l\}_{l=0}^N$ are then integrated into the noise prediction branch via Exemplar Attention Module.

Exemplar Attention Module. The straightforward approaches to integrate additional features into the denoising network are concatenation [57,60] or addition [28]/However, in contrast to existing works in that control features are spatially aligned with the target image, this approach is not suitable for our task because the exemplar image and target control are not spatially aligned. Therefore, we propose an Exemplar Attention Module to integrate the exemplar features from the Exemplar Network, $F_1^l \in \mathbb{R}^{C \times H \times W}$, into noise prediction features, $F_2^l \in \mathbb{R}^{C \times H \times W}$ for each l block. First, these features are concatenated into a spatial-wise: $F_{in}^l = \mathtt{concat}(F_1^l, F_2^l) \in \mathbb{R}^{C \times H \times 2W}$. Following this, self-attention is applied to compute the spatial attention across the features:

$$
Q = \phi_q^l(F_{in}^l), \quad K = \phi_k^l(F_{in}^l), \quad V = \phi_v^l(F_{in}^l)
$$
$$
F_{att}^l = \frac{QK^T}{\sqrt{V}}, \quad F_{EA}^l = W^l \mathrm{Softmax}(F_{att}^l)V + F_{in}^l
\tag{6}
$$

where Q, K and V represents query, key and value, respectively, $\phi(\cdot)$ is layer-specific 1×1 convolution operation, and W^l is trainable parameter. Subsequently, exemplar-attended feature $F_{EA}^l \in \mathbb{R}^{C \times H \times 2W}$ is segmented, with portions corresponding to the denoising features are extracted and forwarded toward the output, $F_{out}^l = \mathrm{Chunk}(F_{EA}^l, 2, \dim=0) \in \mathbb{R}^{C \times H \times W}$. The Exemplar Attention Module computes the region of interest for each query position, a crucial step in effectively directing the denoising steps towards the target exemplar style. This approach enables the denoising process to selectively assimilate features from the Exemplar Network, enhancing the fidelity of the output to the desired stylistic attributes.

Training Objectives. The training process is performed by optimizing the Evidence Lower Bound (ELBO), following BBDM [19], where the marginal distribution is conditioned on x_T. Thus, the training objective ELBO in Eq. 4 can be simplified as:

$$
\mathbb{E}_{x_0, y, I_y, \epsilon}\left[c_{\epsilon t} \left\| m_t(x_T - x_0) + \sqrt{\delta_t}\epsilon - \epsilon_\theta(x_t, t, \tau_\theta(I_y), \psi_\theta(z_y, \tau_\theta(I_y))) \right\|^2 \right],
\tag{7}
$$

where $c_{\epsilon t}$ is the loss weighting function that develops into $1/t$, and δ_t denotes the preserved variance schedule, $\delta_t = 2(m_t - m_t^2)$.

Algorithm 1. Training

1: **repeat**
2:     $(\boldsymbol{x}_T, \boldsymbol{x}_0) \sim q_{\text{data}}(\mathcal{X}, \mathcal{Y})$ ▷ Sample paired data
3:     $\boldsymbol{I}_\mathcal{Y} \sim q_{\text{data}}(\mathcal{Y})$ ▷ Sample exemplar
4:     $t \sim \text{Uniform}(1, \ldots, T)$ ▷ diffusion timesteps
5:     $t_{ref} \leftarrow 0$ ▷ reference timestep
6:     $\epsilon \sim \mathcal{N}(\boldsymbol{0}, \boldsymbol{I})$ ▷ sample Gaussian noise
7:     $\boldsymbol{G} \leftarrow \tau_\theta(\boldsymbol{I}_\mathcal{Y})$ ▷ Forward pass through Global Encoder
8:     $\boldsymbol{F} \leftarrow \psi_\theta(\boldsymbol{x}_\mathcal{Y}, t_{ref}, \boldsymbol{G})$ ▷ Forward pass through Exemplar Network
9:     $\boldsymbol{x}_t \leftarrow (1 - m_t)\boldsymbol{x}_0 + m_t \boldsymbol{y} + \sqrt{\delta_t}\epsilon$ ▷ Forward bridge diffusion process
10:     $\nabla_\theta \left\| m_t(\boldsymbol{y} - \boldsymbol{x}_0) + \sqrt{\delta_t}\epsilon - \epsilon_\theta(\boldsymbol{x}_t, \boldsymbol{G}, \boldsymbol{F}, t) \right\|^2$ ▷ Gradient descent step
11: **until** converged

Algorithm 2. Sampling

1: $\boldsymbol{x}_T \sim q_{\text{data}}(\mathcal{X})$ ▷ Sample control input
2: $\boldsymbol{I}_\mathcal{Y} \sim q_{\text{data}}(\mathcal{Y})$ ▷ Sample exemplar input
3: $t'_s \leftarrow \{t'_S, \cdots t'_1\} \sim \{t_T, \cdots, t_1\}$ ▷ S Inference timesteps
4: $\boldsymbol{G} \leftarrow \tau_\theta(\boldsymbol{I}_\mathcal{Y})$ ▷ Forward pass through Global Encoder
5: $\boldsymbol{F} \leftarrow \psi_\theta(\boldsymbol{x}_\mathcal{Y}, \boldsymbol{G})$ ▷ Forward pass through Exemplar Network
6: **for** $s = S, \ldots, 1$ **do**
7:     $\epsilon \sim \mathcal{N}(\boldsymbol{0}, \boldsymbol{I})$ if $s > 1$, else $\epsilon = 0$
8:     $\boldsymbol{x}_{t'_{s-1}} = c_{xt'_s}\boldsymbol{x}_{t'_s} + c_{yt'_s}\boldsymbol{x}_T - c_{\epsilon t'_s}\epsilon_\theta\left(\boldsymbol{x}_{t'_s}, \boldsymbol{G}, \boldsymbol{F}, t'_s\right) + \sqrt{\tilde{\delta}_{t'_s}}\epsilon$ ▷ Take sampling step
    return $\boldsymbol{x}_0$

4.2   Training Strategy

The training process is unfolded in two stages. In the first stage, denoising U-Net, which utilizes the Global Encoder and cross-attention mechanism, is trained to integrate the global style cues from the exemplar image. Throughout this phase, the Exemplar Network is not engaged, and pre-trained parameters of VAE and Global Encoder are kept frozen. The primary goal of this stage is to learn the model to translate from the control into high-quality images that simultaneously preserve the structure of the target control and embody the coarse style of the exemplar. This is achieved through a reconstruction manner, wherein the target image is synthesized using its control and the target image itself as the exemplar.

In the second stage, the Exemplar Network and Exemplar Attention Module are incorporated into previously trained denoising U-Net. It enables focused training of the Exemplar Network and the Exempler Attention Module within the denoising U-Net, while the other parameters of the network are kept frozen. The overall training is conducted following the strategy outlined in [58], which employs the predefined exemplar and target pairs. This strategy facilitates a concentrated learning process while emphasizing the detailed integration of the exemplar style and specific characteristics of the target.

Table 1. Quantitative Results in image quality. Comparing our methods with state-of-the-art exemplar-guided image translation methods.

Method	DeepFashion			CelebA-HQ (Edge)			CelebA-HQ (Mask)		
	FID↓	SWD↓	LPIPS↑	FID↓	SWD↓	LPIPS↑	FID↓	SWD↓	LPIPS↑
Pix2PixHD [48]	25.20	16.40	N/A	42.70	33.30	N/A	43.69	34.82	N/A
SPADE [31]	36.20	27.80	0.231	31.50	26.90	0.187	39.17	29.78	0.254
SelectionGAN [44]	38.31	28.21	0.223	34.67	27.34	0.191	42.41	30.32	0.277
SMIS [63]	22.23	23.73	0.240	23.71	22.23	0.201	28.21	24.65	0.301
SEAN [62]	16.28	17.52	0.251	18.88	19.94	0.203	17.66	14.13	0.285
CoCosNet [58]	14.40	17.20	0.272	14.30	15.30	0.208	21.83	12.13	0.292
CoCosNetv2 [61]	12.81	16.53	0.283	12.85	14.62	0.218	20.64	11.21	0.303
UNITE [54]	13.08	16.65	0.278	13.15	14.91	0.213	N/A	N/A	N/A
RABIT [55]	12.58	16.03	<u>0.284</u>	**11.67**	14.22	0.219	20.44	**11.18**	0.307
MCL-Net [56]	12.89	16.24	**0.286**	12.52	14.21	0.216	N/A	N/A	N/A
MIDMs [39]	<u>10.89</u>	**10.10**	0.279	15.67	**12.34**	<u>0.224</u>	N/A	N/A	N/A
Ours	**10.62**	<u>12.40</u>	0.255	<u>11.84</u>	<u>12.10</u>	**0.227**	**12.21**	<u>11.34</u>	**0.215**

4.3 Sampling Strategy

The inference process is similar to BBDM [19] that employs the deterministic ODE sampler [40]. Given a inference timesteps $\{t'_s\}_{s=1}^{S} \sim [1:T]$, the sampling process is formulated as:

$$\boldsymbol{x}_{t'_{s-1}} = c_{xt'_s}\boldsymbol{x}_{t'_s} + c_{yt'_s}\boldsymbol{x}_T - c_{\epsilon t'_s}\epsilon_\theta\left(\boldsymbol{x}_{t'_s}, \tau_\theta(\boldsymbol{I}_y), \psi_\theta(\boldsymbol{x}_y, \tau_\theta(\boldsymbol{z}_y)), t'_s\right) + \sqrt{\tilde{\delta}_{t'_s}}\epsilon \quad (8)$$

where $c_{\epsilon xt}, c_{\epsilon yt}, c_{\epsilon t}$ are weighting coefficients for each terms. The whole training process and sampling process are summarized in Algorithm 1 and 2.

5 Experiments

In this section, we present the experimental results of the proposed method. We conduct three tasks to evaluate our model: Edge-to-photo, mask-to-photo, and pose-to-photo. We perform extensive ablation studies to analyze the effect of each essential component of the proposed method. Also, we provide qualitative and quantitative comparisons with state-of-the-art methods. Implementation details and detailed architecture are described in supplementary material. **Datasets.** We conduct three tasks to evaluate our model: Edge-to-photo, mask-to-photo, and pose-to-photo. For mask-guided and edge-guided image generation tasks, the CelebA-HQ [23] dataset is used and we construct the edge maps using the Canny edge detector following [58,61]. For the pose-guided image generation task, we use deepfashion [22] dataset that consists of $52,712$ images with a keypoints annotation. For all tasks, the split of train and validation pairs is consistent with CoCosNet [58] policies.

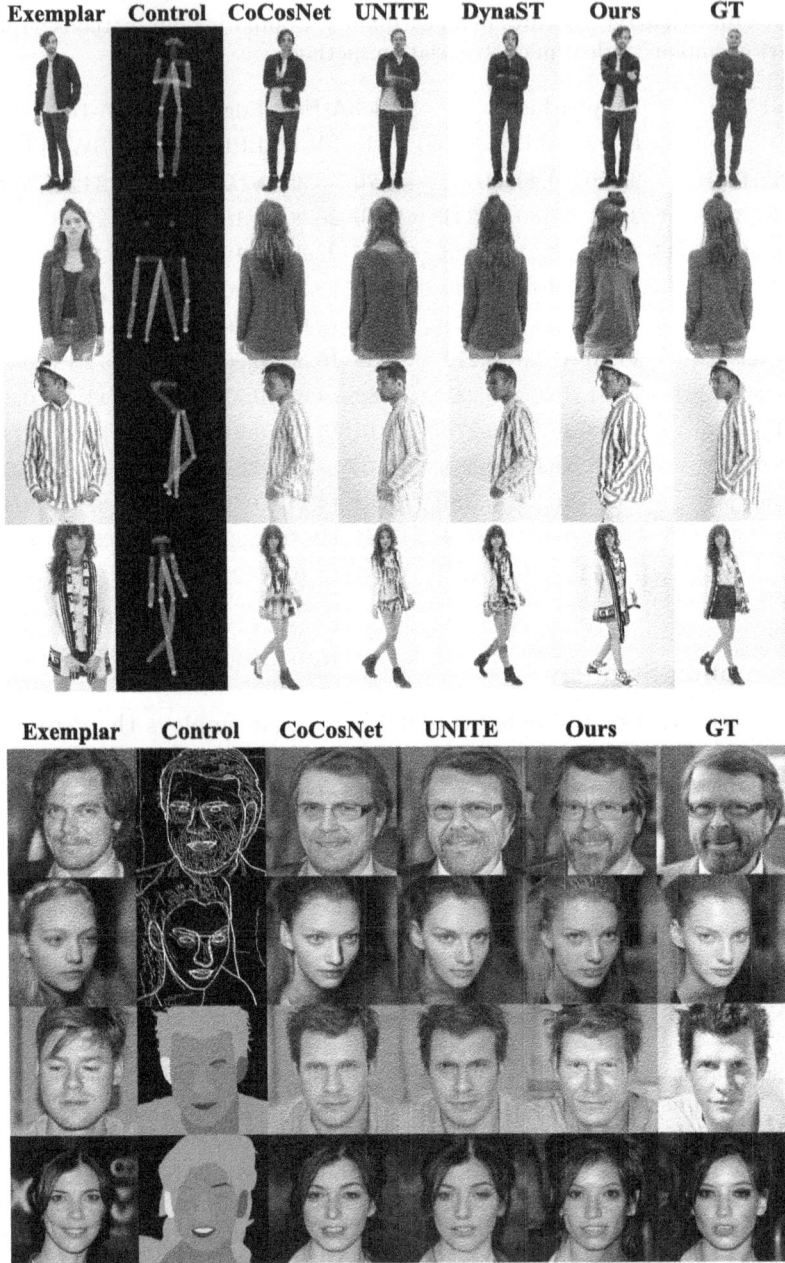

Fig. 3. Qualitative Results. Visual comparisons of the proposed EBDM and state-of-the-art methods over three types of exemplar-guided image translation tasks.

Table 2. Quantitative metrics of semantic (Sem.), color (Col.), and texture (Tex.) consistency on two datasets compared with state-of-the-art image synthesis methods.

Method	DeepFashion			CelebA-HQ (Edge)		
	Sem. ↑	Col. ↑	Tex. ↑	Sem. ↑	Col. ↑	Tex. ↑
Pix2PixHD [48]	0.943	N/A	N/A	0.914	N/A	N/A
SPADE [31]	0.936	0.943	0.904	0.922	0.955	0.927
MUNIT [12]	0.910	0.893	0.861	0.848	0.939	0.884
EGSC-IT [24]	0.942	0.945	0.916	0.915	0.965	0.942
CoCosNet [58]	0.968	0.982	**0.958**	0.949	0.977	0.958
CoCosNet-v2 [61]	0.969	0.974	0.925	0.948	0.975	0.954
UNITE [54]	0.957	0.973	0.930	**0.952**	0.966	0.950
DynaST [20]	**0.975**	0.974	0.937	**0.952**	0.980	**0.969**
MIDMs [39]	N/A	N/A	N/A	0.915	0.982	0.962
MATEBIT [14]	N/A	N/A	N/A	0.949	**0.986**	0.966
Ours	0.932	**0.982**	0.939	0.920	0.984	0.968

5.1 Qualitative Evaluation

We present a comparison of qualitative results (Fig. 3) with existing methods [20,54,58] at three tasks. The results demonstrate that our method effectively transfers the detailed texture from the exemplar to the target, concurrently preserving the structure of controls. Notably, in pose-to-photo, our approach exhibits superiority in capturing detailed patterns and minor objects, such as a cap, which other methods often overlook due to the limitation of matching frameworks. These advantages show the capability of our proposed method that fully leverages the diffusion framework that ensures a more holistic and precise depiction. On the other hand, in edge-to-photo and mask-to-photo tasks, while existing methods also achieve photo-realism, they often tend to overfit to the ground truth (*e.g.* UNITE [54]), thereby constraining its generality. However, our method not only accurately transposes the texture of the exemplar but also adeptly conserves the structure. Moreover, the images synthesized through our method demonstrably excel in photo-realistic attributes against other methods.

5.2 Quantitative Evaluation

Evaluation Metrics. We report the Fréchet Inception Distance (FID) [7] and Sliced Wasserstein Distance (SWD) [18] metrics to evaluate the image perceptual quality by reflecting the distance of feature distributions between real images and generated samples. And we also measure LPIPS [59] to evaluate the diversity of translated images. On the other hand, we show the semantic, color, and texture consistency in Table 2, also under the same setting as [58].

Image Quality. Table 1 presents a quantitative evaluation against state-of-the-art matching-based methods [20,39,54,58,61], showing that our method is

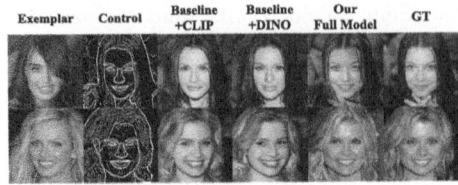

Fig. 4. Visual Comparison based on the choice of Exemplar Encoder.

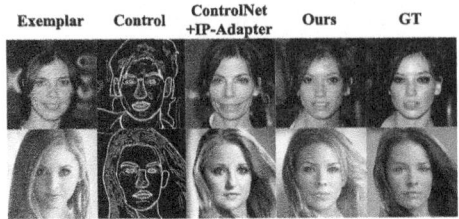

Fig. 5. Qualitative Comparison to Stable Diffusion based model.

Table 3. Quantitative Results from the Ablation Study.

Method	SSIM↑	FID↓	Sem.↑
Baseline	0.831	16.31	0.531
Baseline+CLIP	0.632	23.42	0.752
Baseline+DINO	0.754	21.32	0.786
Ours	**0.901**	**11.84**	**0.920**

Table 4. Quantitative comparison to Stable diffusion based model.

Method	SSIM↑	PSNR↓
SD+ControlNet	0.882	35.30
SD+ControlNet+CLIP	0.894	35.94
Ours	**0.901**	**36.40**

competitive both on image quality and diversity across various tasks. Additionally, in the mask-to-photo task, our method demonstrates superior performance, whereas matching-based methods struggle due to their reliance on cross-domain matching-a notably arduous endeavor when masks offer scant correspondence cues. Conversely, by leveraging diffusion models, our method iteratively translates images from masks via noise prediction. This enables our approach to excel in scenarios with limited direct correspondences, showcasing its robustness and adaptability.

Consistency. The semantic and style consistency analysis (Table 2) evidences that our method either leads or remains competitive in style relevance scores, encompassing color and texture dimensions. In the pose-to-photo domain, despite achieving scores comparable to other methods [20,58], a visual assessment (Fig. 3) reveals our method's distinct proficiency in retaining intricate details such as patterns or textures. This achievement is attributable to our integrated framework, which combines the Exemplar Network and Global Encoder within a Brownian bridge diffusion model construct. As a result, our methodology not only yields photo-realistic images but also ensures the preservation of texture and style congruence with the exemplar input, underscoring its effectiveness in generating visually coherent outputs.

5.3   Comparison to State-of-the-Arts Diffusion Methods

We compare our framework against prevalent state-of-the-art (SOTA) diffusion-based techniques, as shown in Fig. 5 and Table 4. Based on the Stable Diffusion framework [34], we incorporate the ControlNet [57] and IP-Adapter [52] to facilitate structured and stylistic control, respectively. While the existing SOTA method adeptly captures the control structure and generates photo-realistic images, our method more accurately reflects the style of the exemplar. Notably, diffusion-based approaches, conditioned on multiple information including those derived from ControlNet, textual prompts, and image embeddings, tend to be overly sensitive to hyperparameters such as control and embedding guidance scales. Conversely, our model, predicated on a Brownian Bridge diffusion process and exclusively conditioned on the exemplar, assures a more effective generation process. Moreover, the capacity of the existing methods for transferring the finer details in exemplar is somewhat constrained by their reliance on CLIP embeddings which often overlook small details. In contrast, our framework, underpinned by the Exemplar Network and Exemplar Attention Module, demonstrates superior adeptness in transposing textures from the exemplar.

5.4 Ablation Study

To validate the efficacy of our proposed architecture, we conduct ablation studies focusing on the following configurations: (1) omitting the Global Encoder, (2) utilizing the baseline model [19] integrated with CLIP, (3) implementing DINOv2, and (4) employing our complete architecture on the edge-to-photo translation task. As illustrated in Fig. 4, our findings reveal that the DINOv2-based Global Encoder surpasses the CLIP in generating images with higher detail fidelity. While CLIP effectively captures the general characteristics of the reference image, ensuring a level of resemblance, it does not fully encapsulate the intricacies of the details. Additionally, with our Exemplar Network, inputs with spatial misaligned control and exemplar often result in the generation of "blurry" images when relying exclusively on features of Global Encoder. In contrast, our complete framework demonstrates superior performance across all assessed dimensions, highlighting its architectural advantage. Quantitative assessments further underscore the importance of our design choices, as detailed in Table 3.

6   Conclusion

In this study, we presented EBDM, a novel stochastic Brownian bridge diffusion-based approach for exemplar-guided image translation. By leveraging the Brownian Bridge framework, which translates from fixed data points as structural control to photo-realistic images, our method is exclusively conditioned to the style information, thereby the framework more robust and stable. Additionally, we propose the Exemplar Network and Exemplar Attention Module to selectively incorporate the style information from exemplar images into the denoising process. Our method not only stands competitive or surpasses existing methods across the three distinct tasks. Furthermore, our methods also achieve a significant improvement in visual results not only in photorealism but also in the precise transfer of fine details such as patterns and accessories present in the exemplar images.

Acknowledgements. This research was supported by the National Research Foundation of Korea (NRF) grant funded by the Korea government (MSIP) (NRF2021R1A2C2006703) and the Yonsei Signature Research Cluster Program of 2024 (2024-22-0161).

References

1. Caron, M., et al.: Emerging properties in self-supervised vision transformers. In: ICCV (2021)
2. Dhariwal, P., Nichol, A.: Diffusion models beat GANs on image synthesis. In: NeurIPS (2021)
3. Gal, R., et al.: An image is worth one word: personalizing text-to-image generation using textual inversion. arXiv preprint arXiv:2208.01618 (2022)
4. Gu, S., et al.: Vector quantized diffusion model for text-to-image synthesis. In: CVPR (2022)
5. Guo, J., et al.: Zero-shot generative model adaptation via image-specific prompt learning. In: CVPR (2023)
6. Hertz, A., Mokady, R., Tenenbaum, J., Aberman, K., Pritch, Y., Cohen-Or, D.: Prompt-to-prompt image editing with cross attention control. In: ICLR (2023)
7. Heusel, M., Ramsauer, H., Unterthiner, T., Nessler, B., Hochreiter, S.: GANs trained by a two time-scale update rule converge to a local nash equilibrium. In: NeurIPS, vol. 30 (2017)
8. Ho, J., Jain, A., Abbeel, P.: Denoising diffusion probabilistic models. In: NeurIPS (2020)
9. Ho, J., Salimans, T.: Classifier-free diffusion guidance. In: NeurIPS Workshop (2022)
10. Hu, E.J., et al.: LoRA: low-rank adaptation of large language models. In: ICLR (2022)
11. Huang, L., Chen, D., Liu, Y., Shen, Y., Zhao, D., Zhou, J.: Composer: creative and controllable image synthesis with composable conditions. arXiv preprint arXiv:2302.09778 (2023)
12. Huang, X., Liu, M.Y., Belongie, S., Kautz, J.: Multimodal unsupervised image-to-image translation. In: ECCV, pp. 172–189 (2018)
13. Isola, P., Zhu, J.Y., Zhou, T., Efros, A.A.: Image-to-image translation with conditional adversarial networks. In: CVPR, pp. 1125–1134 (2017)
14. Jiang, C., Gao, F., Ma, B., Lin, Y., Wang, N., Xu, G.: Masked and adaptive transformer for exemplar based image translation. In: Proceedings of the IEEE/CVF Conference on Computer Vision and Pattern Recognition, pp. 22418–22427 (2023)
15. Kang, K., Kim, S., Cho, S.: GAN inversion for out-of-range images with geometric transformations. In: ICCV (2021)
16. Kawar, B., et al.: Imagic: text-based real image editing with diffusion models. In: CVPR (2023)
17. Kwon, G., Ye, J.C.: Diffusion-based image translation using disentangled style and content representation. In: ICLR (2023)
18. Lee, C.Y., Batra, T., Baig, M.H., Ulbricht, D.: Sliced Wasserstein discrepancy for unsupervised domain adaptation. In: CVPR, pp. 10285–10295 (2019)
19. Li, B., Xue, K., Liu, B., Lai, Y.K.: BBDM: image-to-image translation with Brownian bridge diffusion models. In: CVPR (2023)

20. Liu, S., Ye, J., Ren, S., Wang, X.: DynaST: dynamic sparse transformer for exemplar-guided image generation. In: Avidan, S., Brostow, G., Cissé, M., Farinella, G.M., Hassner, T. (eds.) ECCV 2022. LNCS, vol. 13676, pp. 72–90. Springer, Cham (2022). https://doi.org/10.1007/978-3-031-19787-1_5
21. Liu, V., Chilton, L.B.: Design guidelines for prompt engineering text-to-image generative models. In: ACM CHI (2022)
22. Liu, Z., Luo, P., Qiu, S., Wang, X., Tang, X.: DeepFashion: powering robust clothes recognition and retrieval with rich annotations. In: CVPR, pp. 1096–1104 (2016)
23. Liu, Z., Luo, P., Wang, X., Tang, X.: Large-scale celebfaces attributes (CelebA) dataset. Retrieved August 15, 11 (2018)
24. Ma, L., Jia, X., Georgoulis, S., Tuytelaars, T., Van Gool, L.: Exemplar guided unsupervised image-to-image translation with semantic consistency. In: ICLR (2019)
25. Ma, T., Li, B., Liu, W., Hua, M., Dong, J., Tan, T.: CFFT-GAN: cross-domain feature fusion transformer for exemplar-based image translation. In: Proceedings of the AAAI Conference on Artificial Intelligence, vol. 37, pp. 1887–1895 (2023)
26. Meng, C., He, Y., Song, Y., Song, J., Wu, J., Zhu, J.Y., Ermon, S.: SDEdit: guided image synthesis and editing with stochastic differential equations. In: ICLR (2022)
27. Mokady, R., Hertz, A., Aberman, K., Pritch, Y., Cohen-Or, D.: Null-text inversion for editing real images using guided diffusion models. In: CVPR (2023)
28. Mou, C., et al.: T2I-Adapter: learning adapters to dig out more controllable ability for text-to-image diffusion models. arXiv preprint arXiv:2302.08453 (2023)
29. Nichol, A., et al.: GLIDE: towards photorealistic image generation and editing with text-guided diffusion models. In: ICML (2022)
30. Oquab, M., et al.: DINOv2: learning robust visual features without supervision. TMLR (2024)
31. Park, T., Liu, M.Y., Wang, T.C., Zhu, J.Y.: Semantic image synthesis with spatially-adaptive normalization. In: CVPR, pp. 2337–2346 (2019)
32. Radford, A., et al.: Learning transferable visual models from natural language supervision. In: ICML (2021)
33. Ramesh, A., Dhariwal, P., Nichol, A., Chu, C., Chen, M.: Hierarchical text-conditional image generation with clip latents. arXiv preprint arXiv:2204.06125 (2022)
34. Rombach, R., Blattmann, A., Lorenz, D., Esser, P., Ommer, B.: High-resolution image synthesis with latent diffusion models. In: CVPR (2022)
35. Ruiz, N., Li, Y., Jampani, V., Pritch, Y., Rubinstein, M., Aberman, K.: DreamBooth: fine tuning text-to-image diffusion models for subject-driven generation. In: CVPR (2023)
36. Saharia, C., et al.: Palette: image-to-image diffusion models. In: ACM SIGGRAPH (2022)
37. Saharia, C., et al.: Photorealistic text-to-image diffusion models with deep language understanding. In: NeurIPS (2022)
38. Sasaki, H., Willcocks, C.G., Breckon, T.P.: UNIT-DDPM: unpaired image translation with denoising diffusion probabilistic models. arXiv preprint arXiv:2104.05358 (2021)
39. Seo, J., Lee, G., Cho, S., Lee, J., Kim, S.: MIDMs: matching interleaved diffusion models for exemplar-based image translation. In: AAAI (2023)
40. Song, J., Meng, C., Ermon, S.: Denoising diffusion implicit models. In: ICLR (2021)

41. Song, Y., Sohl-Dickstein, J., Kingma, D.P., Kumar, A., Ermon, S., Poole, B.: Score-based generative modeling through stochastic differential equations. In: ICLR (2021)
42. Tan, Z., et al.: Diverse semantic image synthesis via probability distribution modeling. In: CVPR, pp. 7962–7971 (2021)
43. Tan, Z., et al.: Efficient semantic image synthesis via class-adaptive normalization. TPAMI **4** (2021)
44. Tang, H., Xu, D., Sebe, N., Wang, Y., Corso, J.J., Yan, Y.: Multi-channel attention selection GAN with cascaded semantic guidance for cross-view image translation. In: CVPR, pp. 2417–2426 (2019)
45. Tumanyan, N., Bar-Tal, O., Bagon, S., Dekel, T.: Splicing ViT features for semantic appearance transfer. In: CVPR, pp. 10748–10757 (2022)
46. Van Den Oord, A., Vinyals, O., et al.: Neural discrete representation learning. In: NeurIPS, vol. 30 (2017)
47. Wang, T., et al.: Pretraining is all you need for image-to-image translation. arXiv preprint arXiv:2205.12952 (2022)
48. Wang, T.C., Liu, M.Y., Zhu, J.Y., Tao, A., Kautz, J., Catanzaro, B.: High-resolution image synthesis and semantic manipulation with conditional GANs. In: CVPR, pp. 8798–8807 (2018)
49. Witteveen, S., Andrews, M.: Investigating prompt engineering in diffusion models. In: NeurIPS Workshop (2022)
50. Yang, B., et al.: Paint by example: exemplar-based image editing with diffusion models. In: CVPR (2023)
51. Yang, Y., et al.: ImageBrush: learning visual in-context instructions for exemplar-based image manipulation. In: Advances in Neural Information Processing Systems, vol. 36 (2024)
52. Ye, H., Zhang, J., Liu, S., Han, X., Yang, W.: IP-adapter: text compatible image prompt adapter for text-to-image diffusion models. arXiv preprint arXiv:2308.06721 (2023)
53. Yin, F., et al.: StyleHEAT: one-shot high-resolution editable talking face generation via pre-trained styleGAN. In: Avidan, S., Brostow, G., Cissé, M., Farinella, G.M., Hassner, T. (eds.) ECCV 2022. LNCS, vol. 13677, pp. 85–101. Springer, Cham (2022). https://doi.org/10.1007/978-3-031-19790-1_6
54. Zhan, F., et al.: Unbalanced feature transport for exemplar-based image translation. In: CVPR, pp. 15028–15038 (2021)
55. Zhan, F., et al.: Bi-level feature alignment for versatile image translation and manipulation. In: Avidan, S., Brostow, G., Cissé, M., Farinella, G.M., Hassner, T. (eds.) ECCV 2022. LNCS, vol. 13676, pp. 224–241. Springer, Cham (2022). https://doi.org/10.1007/978-3-031-19790-1_6
56. Zhan, F., Yu, Y., Wu, R., Zhang, J., Lu, S., Zhang, C.: Marginal contrastive correspondence for guided image generation. In: CVPR, pp. 10663–10672 (2022)
57. Zhang, L., Rao, A., Agrawala, M.: Adding conditional control to text-to-image diffusion models. In: ICCV (2023)
58. Zhang, P., Zhang, B., Chen, D., Yuan, L., Wen, F.: Cross-domain correspondence learning for exemplar-based image translation. In: CVPR, pp. 5143–5153 (2020)
59. Zhang, R., Isola, P., Efros, A.A., Shechtman, E., Wang, O.: The unreasonable effectiveness of deep features as a perceptual metric. In: CVPR, pp. 586–595 (2018)
60. Zhao, S., et al.: Uni-ControlNet: all-in-one control to text-to-image diffusion models. In: NeurIPS (2024)

61. Zhou, X., et al.: CoCosNet v2: Full-resolution correspondence learning for image translation. In: CVPR, pp. 11465–11475 (2021)
62. Zhu, P., Abdal, R., Qin, Y., Wonka, P.: SEAN: image synthesis with semantic region-adaptive normalization. In: CVPR, pp. 5104–5113 (2020)
63. Zhu, Z., Xu, Z., You, A., Bai, X.: Semantically multi-modal image synthesis. In: CVPR, pp. 5467–5476 (2020)

DreamDrone: Text-to-Image Diffusion Models Are Zero-Shot Perpetual View Generators

Hanyang Kong[1], Dongze Lian[1], Michael Bi Mi[2],
and Xinchao Wang[1]([✉])

[1] National University of Singapore, Singapore, Singapore
hanyang.k@u.nus.edu , xinchao@nus.edu.sg
[2] Huawei International Pte. Ltd., Singapore, Singapore

Abstract. We introduce *DreamDrone*, a novel zero-shot and training-free pipeline for generating unbounded flythrough scenes from textual prompts. Different from other methods that focus on warping images frame by frame, we advocate explicitly warping the intermediate latent code of the pre-trained text-to-image diffusion model for high-quality image generation and generalization ability. To further enhance the fidelity of the generated images, we also propose a feature-correspondence-guidance diffusion process and a high-pass filtering strategy to promote geometric consistency and high-frequency detail consistency, respectively. Extensive experiments reveal that *DreamDrone* significantly surpasses existing methods, delivering highly authentic scene generation with exceptional visual quality, without training or fine-tuning on datasets or reconstructing 3D point clouds in advance.

1 Introduction

Recent advances in vision and graphics have enabled the synthesis of multi-view consistent 3D scenes along extended camera trajectories [3,7,18,20]. This emerging task, termed *perpetual view generation* [20], involves synthesizing views from a flying camera along an arbitrarily long trajectory, starting from a single RGBD image.

Previous methodologies predominantly engage in warping images frame by frame with traditional 3D geometric knowledge when given RGBD images and subsequent camera extrinsic. However, this operation often leads to blurriness and distortion in images, which arises from inaccurate interpolation, the mismatch between discrete pixels and continuous transformations, and inaccurate depth data. Moreover, such blurriness and distortion tend to amplify with the accumulation of warp operations (Fig. 1).

Supplementary Information The online version contains supplementary material available at https://doi.org/10.1007/978-3-031-72624-8_19.

A. Leonardis et al. (Eds.): ECCV 2024, LNCS 15071, pp. 324–341, 2025.
https://doi.org/10.1007/978-3-031-72624-8_19

Backyards of Old Houses in Antwerp in the Snow, van Gogh.

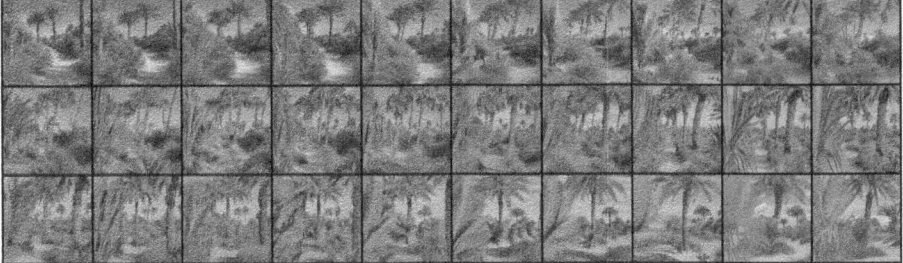

the narrow path of a lush oasis in the midst of a vast desert. Palm trees and tropical plants surround a natural spring, creating a haven for wildlife. The golden sands of the desert stretch out in every direction, meeting the clear blue sky at the horizon.

Fig. 1. Visualization results of DreamDrone. Given a single scene image and the textual description, our approach generates novel views corresponding to user-defined camera trajectory, without fine-tuning on any dataset or reconstructing the 3D point cloud in advance.

To further alleviate the errors caused by frame-by-frame warp operations, two primary paths have been proposed. i) Some methods [3,18,20] try to train a refiner on natural scene datasets. The advantage of this frame-by-frame approach is that it allows for arbitrary changes in camera trajectory during the scene generation process, offering users a higher degree of freedom and enabling infinite generation. However, this training-based method can only be used in natural scenes and cannot be generalized to arbitrary indoor/outdoor scenes or scenes of various styles. ii) Another solution is to first reconstruct the 3D scene model using text prompts, then render 2D RGB images according to the camera trajectory [7,10,52]. Although this solution yields more coherent 2D image sequences, the quality of the rendered images highly depends on the quality of the 3D scene model. This method cannot guarantee good rendering effects from every viewpoint. Additionally, since this method requires the reconstruction of 3D point clouds, it cannot achieve "infinite" scene generation in the same way as the frame-by-frame strategy.

In this paper, we advocate that a more general and flexible perpetual view generation pipeline should possess the following capabilities:

i) are versatile across diverse scenes, including indoor and outdoor scenes, as well as scenes depicted in various styles; ii) allow users to interactively control the camera trajectory during the process of scene generation, while ensuring the high quality of the generated images and the semantic consistency between adjacent frames; and iii) enable seamless transitions from one scene to another.

To this end, we introduce DreamDrone, a novel zero-shot, training-free, infinite scene generation pipeline from text prompts, which does not require any optimization or fine-tuning on any dataset. A core principle of our approach is to warp the latent code of a pre-trained text-to-image diffusion model rather than the frames, enriching it with temporal and geometric consistency. To be specific, given RGBD image I of the current view and camera rotation $\mathcal{R}$ and translation $\mathcal{T}$ for the next view (which is interactively defined by users), we first obtain the latent code x_{t_1} of the diffusion model at timestep t_1, warp it to latent code of the next view x'_{t_1} based on $\mathcal{R}$ and $\mathcal{T}$, and denoise x'_{t_1} to the image I' of the next view. To ensure geometry consistency across adjacent views, we propose a novel feature-correspondence-guidance diffusion process when denoising from x'_{t_1} to image at the next view I'. Moreover, we propose a novel high-pass filter mechanism when warping the latent code x_{t_1}, for preserving high-frequency details across adjacent views.

Our experiments demonstrate that the proposed DreamDrone effectively leads to high-quality and geometry-consistent scene generation. Quantitative and qualitative results demonstrate our comparable, even superior performance compared with other training-based and training-free methods from the aspects of temporal consistency and image quality. Moreover, the significant advantage of DreamDrone is its versatility: it is adept not only at generating real-world scenarios but also shows promising capabilities in creating imaginative scenes. Additionally, users can interactively control the camera trajectory (Fig. 5) and shuttle from one scene to another (Fig. 4). Our contributions are summarized as follows:

- To our best knowledge, we are the first attempt to generate novel views by explicitly warping the latent code of the pre-trained diffusion model.
- A novel feature-correspondence-guidance diffusion process is proposed to enforce geometry consistency across adjacent views. Moreover, a high-pass filtering strategy is introduced to preserve high-frequency details for novel views.
- Extensive experiments demonstrate that our method generates high-quality and geometry-consistent novel views for any scene, from realistic to fantastical. More interestingly, our method realizes the scene shuttle, *i.e.*, travels from one scene to another when the user controls the camera trajectory.

2   Related Works

Perpetual View Generation. Perpetual view generation extrapolates unseen content outside a single image. InfNat [20], InfNat-0 [18], and DiffDreamer [3] use iterative training for long-trajectory perpetual view extrapolation. InfNat [20] pioneered the *perpetual view generation* task with a database for infinite 2D landscapes. InfNat-0 [18] adapted this to 3D, introducing a render-refine-repeat phase for novel views. DiffDreamer [3] improved consistency with image-conditioned diffusion models. However, these methods lack robustness in complex and urban environments. In very recent concurrent work, SceneScape [7] and WonderJourney [52] firstly generate 3D point cloud for scene by zoom-out and inpainting strategy. 2D image sequences are further rendered based on the reconstructed 3D point cloud. However, the accuracy of the 3D model critically impacts performance, particularly with novel camera trajectories.

Text-to-3D Generation. Several text-to-3D generation methods [1, 4, 15, 28, 30, 54] apply text-3D pair databases to learning a mapping function. However, supervised strategies remain challenging due to the lack of large-scale aligned text-3D pairs. CLIP-based [33] 3D generation methods [12, 13, 16, 29, 56] apply pre-trained CLIP model to create 3D objects by formulating the generation as an optimization problem in the image domain. Recent text-to-3D methods like [19, 25, 26, 31, 38, 47, 50] blend text-to-image diffusion models [36] with neural radiance fields [27] for training-free 3D object generation. Other approaches [21, 22, 35, 41] focus on novel view synthesis from a single image, often limited to single objects or small camera motion ranges. Text2room [10] generates 3D indoor scenes from text prompts, but is confined to room meshes.

Text-to-Video Generation. Generating videos from textual descriptions [2, 9, 11, 24, 39, 40, 46, 55] poses significant challenges, primarily due to the scarcity of high-quality, large-scale text-video datasets and the inherent complexity in modeling temporal consistency and coherence. CogVideo [11] addresses this by incorporating temporal attention modules into the pre-trained text-to-image model CogView2 [6]. The video diffusion model [9] employs a space-time factorized U-Net, utilizing combined image and video data for training. Video LDM [2] adopts a latent diffusion approach for generating high-resolution videos. However, these methods typically do not account for the underlying 3D scene geometry in scene-related video generation, nor do they offer explicit control over camera movement. Additionally, their reliance on extensive training with large datasets can be prohibitively costly. While T2V-0 [14] introduced the concept of zero-shot text-to-video generation, its capability is limited to generating a small number of novel frames, with diminished quality in longer video sequences.

3 Method

We formulate the task of perpetual view generation as follows: given a starting image I, we generate the next view image I' corresponding to an arbitrary camera pose $\{\mathcal{R}, \mathcal{T}\}$, where the camera pose can be specified or via user's control.

3.1 Preliminaries

We implement our method based on the recent state-of-the-art text-to-image diffusion model (*i.e.* Stable Diffusion [36]). Stable diffusion is a latent diffusion model (LDM), which contains an autoencoder $\mathcal{D}(\mathcal{E}(\cdot))$ and a U-Net [37] denoiser. Diffusion models are founded on two complementary random processes. The *DDPM forward* process, in which Gaussian noise is progressively added to the latent code of a clean image: x_0:

$$x_t = \sqrt{\alpha_t}x_0 + \sqrt{1 - \alpha_t}z, \tag{1}$$

where $z \sim \mathcal{N}(0, \mathbf{I})$ and $\{\alpha_t\}$ are the noise schedule.

The *backward* process is aimed at gradually denoising x_T, where at each step a cleaner image is obtained. This process is achieved by a U-Net ϵ_θ that predicts the added noise z. Each step of the backward process consists of applying ϵ_θ to the current x_t, and adding a Gaussian noise perturbation to obtain a cleaner x_{t-1}.

Classifier-guided DDIM sampling [5] aims to generate images from noise conditioned on the class label. Given the diffusion model ϵ_θ, the latent code x_t at timestep t, the classifier $p_\theta(y|x_t)$, and the gradient scale s, the sampling process for obtaining x_{t-1} is formulated as:

$$\hat{\epsilon} = \epsilon_\theta(x_t) - \sqrt{\bar{\alpha}_{t-1}}\nabla_{x_t} \log p_\phi(y|x_t), \tag{2}$$

and

$$x_{t-1} = \sqrt{\bar{\alpha}_{t-1}} \cdot \frac{x_t - \sqrt{1 - \bar{\alpha}_t}\hat{\epsilon}}{\sqrt{\bar{\alpha}_t}} + \sqrt{1 - \bar{\alpha}_{t-1}}\hat{\epsilon}, \tag{3}$$

where α is the denoise schedule.

In the self-attention block of the U-Net, features are projected into queries $\mathbf{Q}$, keys $\mathbf{K}$, and values $\mathbf{V}$. The output of the block o is obtained by:

$$o = \mathbf{A}\mathbf{V}, \quad \text{where } \mathbf{A} = \text{Softmax}(\mathbf{Q}\mathbf{K}^\top) \tag{4}$$

The self-attention operation allows for long-range interactions between image tokens.

3.2   Overview

Perpetual view generation as the camera moves presents a complex challenge. This process involves seamlessly filling in unseen regions caused by image warping, adding details to objects as they come closer, while ensuring the imagery remains realistic and diverse. Prior works [3,18,20] have focused on training a refiner to enhance details and create new content for areas requiring inpainting or outpainting. These efforts have shown promising outcomes, yet the effectiveness of the refiner is generally limited to scenarios that align with the training dataset.

Since diffusion models can generate high-quality large-variety images from random latent code, a direct solution arises: can we modify the powerful pretrained text-to-image diffusion model as a refiner? Empirically, DDIM inversion strategy [14,43] can obtain the intermediate latent code at each timestep and the image can be reconstructed by those latent codes. To this end, we attempt

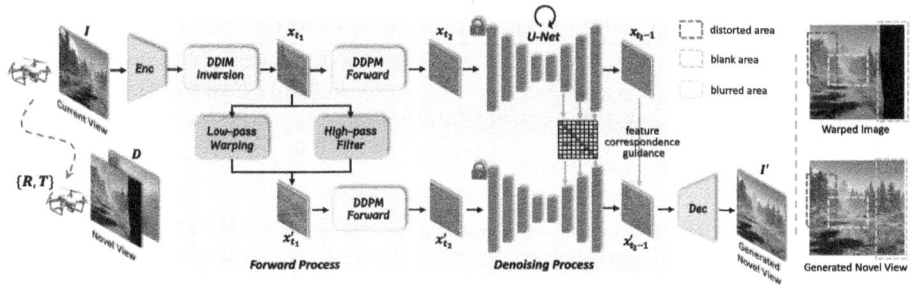

Fig. 2. Overview of our proposed pipeline. Starting from a real or generated RGBD (I, D) image at the current view, we apply DDIM inversion to obtain intermediate latent code x_{t_1} at timestep t_1 using a pre-trained U-Net model. A warping with the high-pass filter strategy is applied to generate latent code for the next novel view. A few more DDPM forward steps from timestep t_1 to t_2 are applied for enlarging the degree of freedom $w.r.t.$ the warped latent code. In the denoising process, we apply pre-trained U-Net to generate the novel view from x'_{t_2}. The cross-view self-attention module and feature-correspondence guidance are applied to maintain the geometry correspondence between x_{t_2} and x'_{t_2}. The right side shows the warped image and our generated novel view I'. Our method greatly alleviates blurring, inconsistency, and distortion. The overall pipeline is zero-shot and training-free.

to explicitly warp the latent code of the current view and generate the novel view by the pre-trained text-to-image diffusion model.

Our overall pipeline is illustrated in Fig. 2. Initially, we obtain the latent code x_{t_1} of the current view's RGB image I at timestep t_1 through the DDIM inversion process. We then warp the current frame's latent code x_{t_1} to the next view x'_{t_1} using depth information and camera extrinsic parameters. However, directly denoising from x'_{t_1} to the image also suffers from blurry, which results in the non-integer pixel coordinates and the interpolation operation. More noise is added from x'_{t_1} at timestep t_1 to x'_{t_2} at timestep t_2 by DDPM forward operation, for generating high-quality images. The side effect of DDPM is the geometry inconsistency between adjacent views. To this end, we propose a feature-correspondence-guidance denoising strategy to enforce geometry consistency. Moreover, a high-pass filtering strategy is proposed to maintain the consistency of high-frequency details between adjacent views. Please refer to Fig. 3 for the motivation of our proposed modules. Our overall pipeline requires only a pre-trained text-to-image diffusion model and a depth estimation model, eliminating the need for any additional training or fine-tuning.

3.3 Warping Latent Codes

The results in the right side of Fig. 2 reveal that directly warping images based on camera intrinsics K, extrinsics $\{\mathcal{R}, \mathcal{T}\}$, and depth information leads to regions of distortion in the images. Additionally, the use of inpainting [23,36,53] and outpainting [17,49,51] models to fill these gaps does not achieve satisfactory

---

Algorithm 1. Warping latent code with high-pass filter

Require: x_t ▷ latent code at timestep t of current view c

1: $F(x_t) \leftarrow FFT(x_t)$ ▷ Apply Fast Fourier Transform

2: Split $F(x_t)$ into F_{low} and F_{high} using threshold σ

3: $x_t^{low} \leftarrow IFFT(F_{low})$ ▷ Inverse FFT on low-frequency component

4: $x_t^{low-warped} \leftarrow warp(x_t^{low})$ ▷ warp the low-frequency content

5: $F_{warped} \leftarrow FFT(x_t^{low-warped})$ ▷ FFT on warped content

6: $F' \leftarrow F_{warped} + F_{high}$ ▷ Combine low-frequency warped content with high-frequency of original content

7: $x_t' \leftarrow IFFT(F')$ ▷ Inverse FFT to get latent code for next view c'

     return x_t' ▷ warped latent code at timestep t for next view c'

---

outcomes. In pursuit of photo-realistic images, we opt to edit the latent code corresponding to timestep t. PnP [44] and DIFT [42] have shown that the features of diffusion possess strong semantic information, with semantic parts being shared across images at each step. The simplest method for warping the latent code follows the same approach as warping the image. The only difference between warping the latent code and warping the image is a slight modification in the camera intrinsics; this entails scaling the camera intrinsics proportionally based on the different resolutions of the image and latent code.

The overall procedure for warping the latent code is illustrated in Alg. 1. Initially, a latent code x_t is obtained and transformed via Fast Fourier Transform (FFT) to $F(x_t)$. This is divided into low-frequency F_{low} and high-frequency F_{high} components, segregated at threshold σ. The key step involves warping the Inverse FFT (IFFT) processed low-frequency component $x_t^{\text{low}} = \text{IFFT}(F_{\text{low}})$, warping to the next view $x_t^{low-warped}$. Merging $\text{FFT}(x_t^{low-warped})$ with F_{high}, we obtain F', from which the final latent code $x_t' = \text{IFFT}(F')$ is reconstructed. This approach efficiently preserves high-frequency details, enabling high-fidelity scene generation aligned with text prompts.

3.4 Feature-Correspondence-Guidance Design

After obtaining the latent code x_t' corresponding to the next frame, we employ the DDPM (Denoising Diffusion Probabilistic Models) method to increase the degrees of freedom of the latent code, enabling the generation of richer image details. However, increasing freedom introduces a challenge: the correlation between frames. An unconstrained diffusion denoising process can result in poor semantic correlation between adjacent frames. To address this, we propose a feature-correspondence guidance strategy with a cross-view self-attention mechanism. We introduce these approaches in detail below.

Cross-view Self-attention. To maintain consistency between the generated result and the original image, inspired by recent image and video editing works [8, 44,45,48], we modify the process of the self-attention module of U-Net when denoising the latent code x_t'. Specifically, we denoise the views for the current and next view together. The key and value of the self-attention modules from the next view are replaced by that of the current view. To be specific, for obtaining the original view, the self-attention module is defined the same as Eq. (4). The modified cross-view self-attention for generating a novel view is defined as:

$$o' = \mathbf{A}'\mathbf{V}, \text{ where } \mathbf{A}' = \text{Softmax}(\mathbf{Q}'\mathbf{K}^\top), \qquad (5)$$

where $\mathbf{Q}'$, $\mathbf{A}'$, and o' are query, attention matrix, and output features for the novel views. $\mathbf{K}$ and $\mathbf{V}$ are injected keys and values obtained from the self-attention module for generating the original view. Please note that the $\mathbf{K}$ and $\mathbf{V}$ are also warped before injection.

Feature-Correspondence Guidance. Maintaining geometry consistency between adjacent views using the cross-view self-attention mechanism presents challenges, especially in preserving high-frequency details as the camera moves forward. The recent DIFT [42] highlights the potential of using intermediate features of diffusion models for accurate point-to-point image matching [32]. Additionally, the concept of vanilla classifier guidance [5] steers the diffusion sampling process using pre-trained classifier gradients towards specific class labels. Building on these ideas, we integrate feature correspondence guidance into the DDIM sampling process to enhance consistency between adjacent views, addressing the challenge of detail preservation in dynamic scenes.

Specifically, we obtain the features of the current and next view at each timestep t of the DDIM process and calculate the cosine distance between the warped original features and features from the next novel views:

$$\mathcal{L}_{sim}^t = \frac{1 - \cos\left[\text{warp}(f_t), f_t'\right]}{2}, \qquad (6)$$

where f_t and f_t' are intermediate features extracted from pre-trained U-Net ϵ_θ at timestep t and warp is the warping functions. The lower $\mathcal{L}_{sim}^t$, the higher the similarity.

We further introduce the similarity score $\mathcal{L}_{sim}^t$ to the DDIM sampling process, for generating novel views with geometry consistency. The predicted noise $\hat{\epsilon}$ is formulated as:

$$\hat{\epsilon} = \epsilon_\theta(\boldsymbol{x}_t) - \lambda\sqrt{\bar{\alpha}_{t-1}}\nabla_{\boldsymbol{x}_t}\mathcal{L}_{sim}^t, \qquad (7)$$

where λ is the constant hyper-parameter and latent code $\boldsymbol{x}_{t-1}$ is calculated by Eq. (3)

4   Experiments

4.1   Implementation Details

We take Stable Diffusion [36] with the pre-trained weights from version 2.1[1] as the basic text-to-image diffusion and MiDas [34] with weights dpt_beit_large_512[2].

---

[1] https://huggingface.co/stabilityai/stable-diffusion-2-1-base.
[2] https://github.com/isl-org/MiDaS.

The overall diffusion timesteps is 1000. We warp the latent code at timestep $t_1=21$ and add more degrees of noise to timestep $t_2=441$. The threshold σ for the high-pass filter is 20 and the hyper-parameter λ for feature-correspondence guidance is 300. Due to the page limit, please refer to the supplementary material (*supp.*) for details.

4.2   Baselines

We compare against 1) two supervised methods for perpetual view generation: *InfNat* [20] and *InfNat-0* [18]. 2) one text-conditioned 3D point cloud-based scene generation: *SceneScape* [7]. 3) two supervised methods for text-to-video generation: *CogVideo* [11] and *VideoFusion* [24]. 4) one method for zero-shot text-to-video generation: *T2V-0* [14].

4.3   Evaluation Metrics

We evaluate our zero-shot perpetual scene generation into two aspects: 1) the quality of generated images and text-image alignment, and 2) the temporal consistency of generated image sequences.

Image Quality and Text-Image Alignment. We evaluate CLIP score [33], which indicates text-scene alignment for quantitative comparisons. A high average CLIP score indicates not only that the generated images are more aligned with the corresponding prompts but also that they consistently maintain high quality [14]. CogVideo [11], VideoFusion [24], SceneScape [7], and T2V-0 [14] are all engaged in text-conditioned generation tasks. We generated 50 scene-related text prompts using GPT-4[3] and then created videos using each of the three methods. For the InfNat [20] and InfNat-0 [18] methods, we used Stable Diffusion to generate the initial frame, followed by subsequent frame generation based on this initial frame. We calculated the distance between each generated frame and the text embedding, known as the CLIP score. Considering that the InfNat [20] and InfNat-0 [18] methods trained on natural scene datasets, we further provided 10 very general prompts such as 'an image of the landscape' and 'an image of the mountain' for these methods, and then selected the highest CLIP score as the CLIP score for the current frame.

Temporal Consistency of Generated Image Sequences. We demonstrate our advancements in temporal consistency against other SOTA methods by calculating average PSNR and SSIM scores across adjacent frames for generated videos with different lengths. The higher scores demonstrate the superiority in terms of cross-view consistency.

---

[3] https://openai.com/gpt-4.

Table 1. Ablations of image quality and temporal coherence of generated image sequences with various lengths. Please refer to Fig. 3 for quality comparisons.

Methods	PSNR ↑			SSIM ↑			CLIP ↑		
	8 frames	16 frames	32 frames	8 frames	16 frames	32 frames	8 frames	16 frames	32 frames
warp image	26.90	22.46	21.62	0.25	0.23	0.24	0.138	0.112	0.106
warp latent	28.35	28.57	28.75	0.27	0.28	0.24	0.135	0.122	0.125
warp latent+DDPM	24.67	23.04	22.59	0.12	0.10	0.06	0.302	0.297	0.308
warp latent+DDPM+guidance	28.27	28.21	28.10	0.34	0.30	0.26	0.317	0.316	0.313
warp latent+DDPM+guidance+cross-view attn.	28.89	28.83	28.75	0.32	0.31	0.27	0.318	0.315	0.315
warp latent+DDPM+guidance+cross-view attn.+high pass filter	29.91	29.86	29.79	0.39	0.38	0.35	0.320	0.318	0.319

4.4   Ablation Studies

We perform ablation studies on our three proposed modules: 1) warping latent with high-pass filter, 2) cross-view self-attention module, and 3) feature-correspondence guidance. The quantitative ablation results are shown in Table 1 and we visualize the ablation samples in Fig. 3.

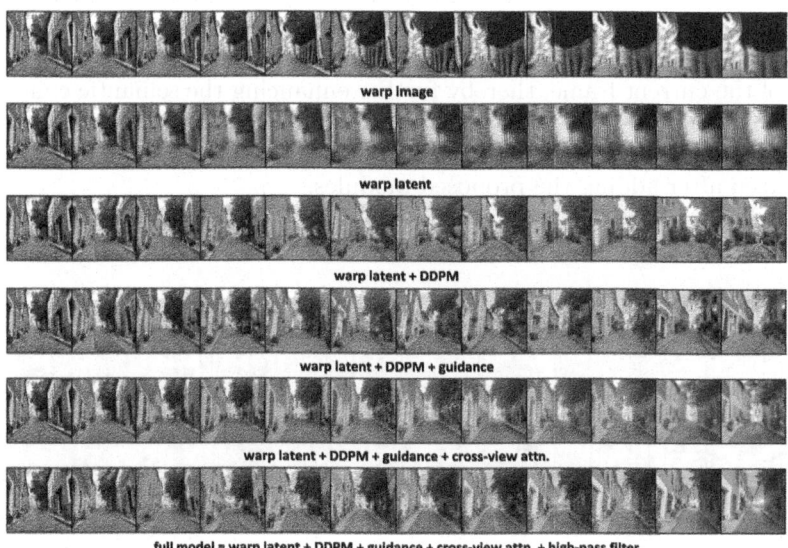

Fig. 3. Ablation results for the key components. We perform ablation studies by disabling the key components of our method. We illustrate every five frames for each ablation experiment. Please zoom in for better comparisons.

The simplest method for infinite scene generation tasks is frame-by-frame image warping, but this approach is unfeasible, as is directly warping the latent code. Warping images leads to non-integer pixel coordinates, resulting in interpolation-induced blurring and distortion. Moreover, these errors accumulate with each frame generated, leading to a collapse in quality. The first two rows of Table 1 show that directly warping images or warping latent codes (i.e., removing DDPM) results in very low CLIP scores, indicating poor quality of the generated images. The generated images becoming progressively blurred can also be observed in the first two rows of Fig. 3. We introduce DDPM to increase the degrees of freedom of the diffusion model, thereby generating high-quality images. However, the introduction of DDPM has the side effect of worsening the semantic consistency between adjacent frames (3rd row in Tab. 1 and Fig. 3). With the help of DDPM, the CLIP score increases from 0.125 to 0.308 when generating 32 images. Please refer to the *supp.* for the generated results with different scales of the DDPM forward process.

To ensure the quality of image generation while also maintaining consistency with adjacent views, we propose a feature-correspondence guidance strategy. Comparing the third and fourth rows of Fig. 3, it is evident that the semantic consistency between adjacent frames is significantly enhanced after adding guidance, with noticeable improvements in both PSNR and SSIM scores in Tab. 1. To further enhance cross-view consistency, we adopted cross-view attention modules and high-pass filtering. From the visualized results at the 5th and 6th rows in Fig. 3, it is clear that the semantic consistency of adjacent camera perspectives is further strengthened after incorporating the cross-view attention module. The operation of the high-pass filter further preserves the high-frequency details of the current frame, thereby further enhancing the semantic consistency of high-frequency details between adjacent frames. For instance, comparing the left side house at the 4th, 5th, and 6th rows in Fig. 3, the cross-view consistency is enhanced after adding the proposed modules.

In addition to conducting ablation experiments on the modules we propose, we continue to explore two more questions:

A serene of *realistic → Lego* suburban street, with rows of *realistic → LEGO* colorful houses, 4K.

The vibrant and electric streets of Inazuma City from 'Genshin Impact'. → An urban street known for its vibrant graffiti and street art.

Fig. 4. Ablation study for scene travel. We visualize two image sequences and change the prompt when generating novel views. We illustrate every five images and the prompts are changed when generating 31th image (7th image shown in each row).

Q1: Can DreamDrone Shuttle from One Scene to Another by Changing Text Prompts? During the frame-by-frame process, we changed the textual prompts, with the generated results shown in Fig. 4. The visualized results demonstrate that DreamDrone can smoothly complete the scene travel (from streets in

Inazuma City to urban art street) or the transition of scene styles (from realistic to Lego style) while ensuring the semantic consistency of adjacent views, according to the changes in textual prompts.

Q2: Can Explicitly Warping the Latent Code Vontrol the Trajectory of Camera Perspective Movement? Since our method generates image sequences frame by frame, we can freely adjust the camera's flight angle by altering the camera's extrinsic parameters. In Fig. 5, we provide sequences of images generated under different camera trajectories. The results show that our method possesses a high degree of freedom, allowing for the free customization of the camera's trajectory. Other state-of-the-art methods cannot achieve this functionality.

4.5   Qualitative Comparison

Hyper-realistic Eiffel Tower, with the intricate iron lattice work.

camera
trajectory   A scene of a city, Lego style.

Fig. 5. Ablation study on customized camera trajectory. We generate images with different camera directions. For the sample of the Eiffel Tower, our camera perspective continuously ascends. For the 2nd scene of the Lego city, our camera not only moves forward but also shifts upwards and to the right.

In our comparison with InfNat-0 [18] (Fig. 6), focusing on various scenes including coastlines, rivers, Van Gogh-style landscapes, and city streetscapes, we identified four main differences: Firstly, InfNat-0 shows proficiency in coastline scenes, a reflection of its training data, but our training-free *DreamDrone* surpasses it in later frames due to InfNat-0's cumulative errors over time. Secondly, in natural scenes with closer objects, InfNat-0's flawed generation becomes more apparent, whereas our method maintains consistency. Thirdly, InfNat-0's limited approach to gap filling leads to poor performance in stylized scenes, in contrast to *DreamDrone* which preserves high-frequency details and frame correspondence. Finally, in urban environments, InfNat-0 struggles significantly, while *DreamDrone* achieves realistic and geometry-consistent views, demonstrating its versatility across varied scenarios.

T2V-0 [14] introduces unsupervised text-conditioned video generation using stable diffusion. SceneScape [7] focuses on 'zoom out' effects during backward

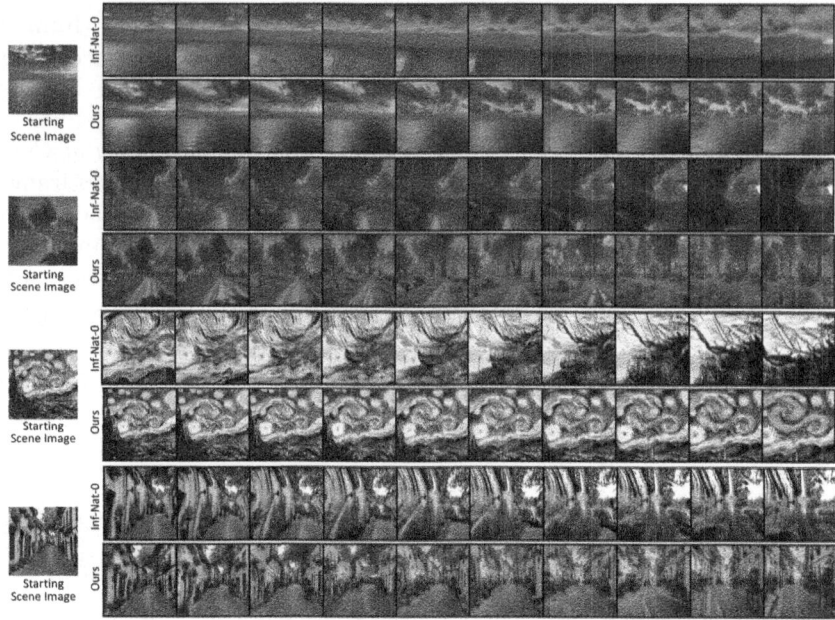

Fig. 6. Qualitative comparisons of InfNat-0 [18] **and ours.** We provide four starting scene images with various styles and categories as start points and ask models to fly through the images. 50 frames are generated and we illustrate every five frames for each starting scene image.

camera movement. However, as seen in Fig. 7, both methods have limitations. SceneScape struggles with outdoor scenes and forward camera movement, leading to blurred and distorted results after 8 steps due to its reliance on a pretrained inpainting model. T2V-0 displays a drop in quality beyond the third frame in complex environments like Lego-style cities, likely from its latent code editing approach that compromises frame continuity and geometric consistency. Conversely, our *DreamDrone* excels across various scenes. It maintains detail, continuity, and quality in advancing camera scenarios, evident in even simpler landscapes like mountains where T2V-0 and SceneScape cannot effectively portray dynamic elements like cloud movement. Our approach ensures the preservation of fine details such as shadows and sunlight, creating a more dynamic and realistic video experience. Please refer to *supp.* for more comparisons.

As our task bears similarities to text-to-video generation, we further provide qualitative comparisons with VideoFusion [24]. Due to the page limit, please refer to *supp.* for detailed comparisons.

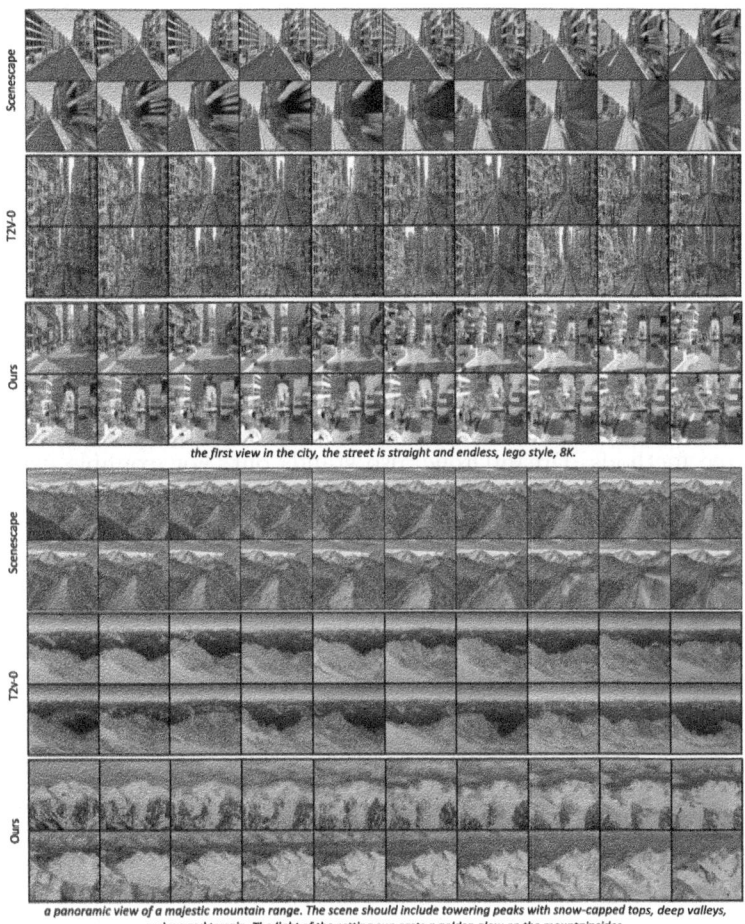

the first view in the city, the street is straight and endless, lego style, 8K.

a panoramic view of a majestic mountain range. The scene should include towering peaks with snow-capped tops, deep valleys, and rugged terrain. The light of the setting sun casts a golden glow on the mountainsides.

Fig. 7. Qualitative comparisons of SceneScape [7], **T2V-0** [14], **and our DreamDrone.** We visualize 20 continuous frames for each textual prompt. As the camera flies, our method generates geometry-consistent scene sequences.

4.6   Quantitative Comparison

Table 2 offers a detailed comparison of various SOTA methods for generating image sequences, including our method, DreamDrone. When compared to other training-based methods, DreamDrone, despite being training-free, consistently achieves higher CLIP scores across all frame lengths (0.320, 0.318, 0.319 for 8, 16, and 32 frames respectively). This is particularly noteworthy as the CLIP scores for training-based methods generally degrade as the number of generated frames increases. For instance, VideoFusion's [24] CLIP scores decrease from 0.281 for 8 frames to 0.272 for 32 frames. This trend suggests a decline in the quality of generated images with an increase in sequence length for training-based methods.

Table 2. Qualitative comparisons with other SOTA methods. We evaluate the quality and temporal coherence of the generated image sequences with various lengths.

	Methods	PSNR ↑			SSIM ↑			CLIP ↑		
		8 frames	16 frames	32 frames	8 frames	16 frames	32 frames	8 frames	16 frames	32 frames
training-based	InfNat [20]	28.75	28.67	28.65	0.32	0.30	0.30	0.125	0.123	0.118
	InfNat-0 [18]	28.92	28.89	28.87	0.37	0.35	0.34	0.128	0.125	0.122
	CogVideo [11]	31.03	30.08	29.32	0.45	0.39	0.31	0.255	0.249	0.241
	VideoFusion [24]	29.89	28.36	28.78	0.41	0.37	0.31	0.281	0.283	0.272
training-free	T2V-0 [14]	27.25	26.17	26.03	0.27	0.24	0.23	0.312	0.305	0.287
	Scenescape [7]	29.87	29.75	29.66	0.41	0.38	0.34	0.318	0.282	0.279
	DreamDrone (Ours)	29.91	29.86	29.79	0.39	0.38	0.35	0.320	0.318	0.319

In contrast, DreamDrone maintains high CLIP scores even as the sequence length increases, indicating superior image quality. When compared to other training-free methods, DreamDrone also stands out. For example, while T2V-0's [14] CLIP scores decrease from 0.312 for 8 frames to 0.287 for 32 frames, DreamDrone's CLIP scores remain relatively stable, further demonstrating its robustness in maintaining image quality across varying sequence lengths. This analysis underscores the effectiveness of DreamDrone in generating high-quality, temporally coherent image sequences without the need for training.

5   Conclusion

In this work, we propose *DreamDrone*, a novel approach for generating flythrough scenes from textual prompts without the need for training or fine-tuning. Our method explicitly warps the intermediate latent code of a pre-trained text-to-image diffusion model, enhancing the quality of the generated images and the generalization ability. We propose a feature-correspondence-guidance diffusion process and a high-pass filtering strategy to ensure geometric and high-frequency detail consistency. Experimental results indicate that *DreamDrone* surpasses current methods in terms of visual quality and authenticity of the generated scenes.

Acknowledgement. This project is supported by the Ministry of Education, Singapore, under its Academic Research Fund Tier 2 (Award Number: MOE-T2EP20122-0006).

References

1. Bautista, M.A., et al.: Gaudi: a neural architect for immersive 3d scene generation. Adv. Neural. Inf. Process. Syst. **35**, 25102–25116 (2022)
2. Blattmann, A., et al.: Align your latents: high-resolution video synthesis with latent diffusion models. In: Proceedings of the IEEE/CVF Conference on Computer Vision and Pattern Recognition, pp. 22563–22575 (2023)
3. Cai, S., et al.: Diffdreamer: towards consistent unsupervised single-view scene extrapolation with conditional diffusion models. In: Proceedings of the IEEE/CVF International Conference on Computer Vision, pp. 2139–2150 (2023)

4. Chen, K., Choy, C.B., Savva, M., Chang, A.X., Funkhouser, T., Savarese, S.: Text2Shape: generating shapes from natural language by learning joint embeddings. In: Jawahar, C.V., Li, H., Mori, G., Schindler, K. (eds.) ACCV 2018. LNCS, vol. 11363, pp. 100–116. Springer, Cham (2019). https://doi.org/10.1007/978-3-030-20893-6_7

5. Dhariwal, P., Nichol, A.: Diffusion models beat gans on image synthesis. Adv. Neural. Inf. Process. Syst. **34**, 8780–8794 (2021)

6. Ding, M., Zheng, W., Hong, W., Tang, J.: Cogview2: faster and better text-to-image generation via hierarchical transformers. Adv. Neural. Inf. Process. Syst. **35**, 16890–16902 (2022)

7. Fridman, R., Abecasis, A., Kasten, Y., Dekel, T.: Scenescape: text-driven consistent scene generation. arXiv preprint arXiv:2302.01133 (2023)

8. Geyer, M., Bar-Tal, O., Bagon, S., Dekel, T.: Tokenflow: consistent diffusion features for consistent video editing. arXiv preprint arXiv:2307.10373 (2023)

9. Ho, J., Salimans, T., Gritsenko, A., Chan, W., Norouzi, M., Fleet, D.J.: Video diffusion models. arXiv:2204.03458 (2022)

10. Höllein, L., Cao, A., Owens, A., Johnson, J., Nießner, M.: Text2room: extracting textured 3d meshes from 2d text-to-image models. arXiv preprint arXiv:2303.11989 (2023)

11. Hong, W., Ding, M., Zheng, W., Liu, X., Tang, J.: Cogvideo: large-scale pretraining for text-to-video generation via transformers. arXiv preprint arXiv:2205.15868 (2022)

12. Jain, A., Mildenhall, B., Barron, J.T., Abbeel, P., Poole, B.: Zero-shot text-guided object generation with dream fields. In: Proceedings of the IEEE/CVF Conference on Computer Vision and Pattern Recognition, pp. 867–876 (2022)

13. Jiang, Z., et al.: 3d-togo: towards text-guided cross-category 3d object generation. In: Proceedings of the AAAI Conference on Artificial Intelligence, vol. 37, pp. 1051–1059 (2023)

14. Khachatryan, L., et al.: Text2video-zero: text-to-image diffusion models are zero-shot video generators. arXiv preprint arXiv:2303.13439 (2023)

15. Kong, H., Gong, K., Lian, D., Mi, M.B., Wang, X.: Priority-centric human motion generation in discrete latent space. In: Proceedings of the IEEE/CVF International Conference on Computer Vision, pp. 14806–14816 (2023)

16. Lee, H.H., Chang, A.X.: Understanding pure clip guidance for voxel grid nerf models. arXiv preprint arXiv:2209.15172 (2022)

17. Li, J., Bansal, M.: Panogen: text-conditioned panoramic environment generation for vision-and-language navigation. arXiv preprint arXiv:2305.19195 (2023)

18. Li, Z., Wang, Q., Snavely, N., Kanazawa, A.: Infinitenature-zero: learning perpetual view generation of natural scenes from single images. In: Avidan, S., Brostow, G., Cisse, M., Farinella, G.M., Hassner, T. (eds.) European Conference on Computer Vision, vol. 13661, pp. 515–534. Springer, Heidelberg (2022). https://doi.org/10.1007/978-3-031-19769-7_30

19. Lin, C.H., et al.: Magic3d: high-resolution text-to-3d content creation. In: Proceedings of the IEEE/CVF Conference on Computer Vision and Pattern Recognition, pp. 300–309 (2023)

20. Liu, A., Tucker, R., Jampani, V., Makadia, A., Snavely, N., Kanazawa, A.: Infinite nature: perpetual view generation of natural scenes from a single image. In: Proceedings of the IEEE/CVF International Conference on Computer Vision, pp. 14458–14467 (2021)

21. Liu, M., et al.: One-2-3-45: any single image to 3d mesh in 45 seconds without per-shape optimization. arXiv preprint arXiv:2306.16928 (2023)

22. Liu, R., Wu, R., Van Hoorick, B., Tokmakov, P., Zakharov, S., Vondrick, C.: Zero-1-to-3: zero-shot one image to 3d object. In: Proceedings of the IEEE/CVF International Conference on Computer Vision, pp. 9298–9309 (2023)

23. Lugmayr, A., Danelljan, M., Romero, A., Yu, F., Timofte, R., Van Gool, L.: Repaint: inpainting using denoising diffusion probabilistic models. In: Proceedings of the IEEE/CVF Conference on Computer Vision and Pattern Recognition, pp. 11461–11471 (2022)

24. Luo, Z., et al.: Videofusion: decomposed diffusion models for high-quality video generation. In: Proceedings of the IEEE/CVF Conference on Computer Vision and Pattern Recognition, pp. 10209–10218 (2023)

25. Melas-Kyriazi, L., Laina, I., Rupprecht, C., Vedaldi, A.: Realfusion: 360deg reconstruction of any object from a single image. In: Proceedings of the IEEE/CVF Conference on Computer Vision and Pattern Recognition, pp. 8446–8455 (2023)

26. Metzer, G., Richardson, E., Patashnik, O., Giryes, R., Cohen-Or, D.: Latent-nerf for shape-guided generation of 3d shapes and textures. In: Proceedings of the IEEE/CVF Conference on Computer Vision and Pattern Recognition, pp. 12663–12673 (2023)

27. Mildenhall, B., Srinivasan, P.P., Tancik, M., Barron, J.T., Ramamoorthi, R., Ng, R.: Nerf: representing scenes as neural radiance fields for view synthesis. Commun. ACM **65**(1), 99–106 (2021)

28. Mo, S., et al.: Dit-3d: exploring plain diffusion transformers for 3d shape generation. arXiv preprint arXiv:2307.01831 (2023)

29. Mohammad Khalid, N., Xie, T., Belilovsky, E., Popa, T.: Clip-mesh: generating textured meshes from text using pretrained image-text models. In: SIGGRAPH Asia 2022 Conference Papers, pp. 1–8 (2022)

30. Nichol, A., Jun, H., Dhariwal, P., Mishkin, P., Chen, M.: Point-e: a system for generating 3d point clouds from complex prompts. arXiv preprint arXiv:2212.08751 (2022)

31. Poole, B., Jain, A., Barron, J.T., Mildenhall, B.: Dreamfusion: text-to-3d using 2d diffusion. arXiv preprint arXiv:2209.14988 (2022)

32. Qiu, J., Wang, X., Fua, P., Tao, D.: Matching seqlets: an unsupervised approach for locality preserving sequence matching. IEEE Trans. Pattern Anal. Mach. Intell. **43**(2), 745–752 (2019)

33. Radford, A., et al.: Learning transferable visual models from natural language supervision. In: International Conference on Machine Learning, pp. 8748–8763. PMLR (2021)

34. Ranftl, R., Lasinger, K., Hafner, D., Schindler, K., Koltun, V.: Towards robust monocular depth estimation: Mixing datasets for zero-shot cross-dataset transfer. IEEE Trans. Pattern Anal. Mach. Intell. **44**(3), 1623–1637 (2020)

35. Rockwell, C., Fouhey, D.F., Johnson, J.: Pixelsynth: generating a 3d-consistent experience from a single image. In: Proceedings of the IEEE/CVF International Conference on Computer Vision, pp. 14104–14113 (2021)

36. Rombach, R., Blattmann, A., Lorenz, D., Esser, P., Ommer, B.: High-resolution image synthesis with latent diffusion models. In: Proceedings of the IEEE/CVF Conference on Computer Vision and Pattern Recognition, pp. 10684–10695 (2022)

37. Ronneberger, O., Fischer, P., Brox, T.: U-net: convolutional networks for biomedical image segmentation. In: Navab, N., Hornegger, J., Wells, W.M., Frangi, A.F. (eds.) MICCAI 2015. LNCS, vol. 9351, pp. 234–241. Springer, Cham (2015). https://doi.org/10.1007/978-3-319-24574-4_28

38. Shen, Q., Yang, X., Wang, X.: Anything-3d: towards single-view anything reconstruction in the wild. arXiv preprint arXiv:2304.10261 (2023)

39. Singer, U., et al.: Make-a-video: text-to-video generation without text-video data. arXiv preprint arXiv:2209.14792 (2022)

40. Tan, Z., Yang, X., Liu, S., Wang, X.: Video-infinity: distributed long video generation. arXiv preprint arXiv:2406.16260 (2024)

41. Tang, J., et al.: Make-it-3d: high-fidelity 3d creation from a single image with diffusion prior. arXiv preprint arXiv:2303.14184 (2023)

42. Tang, L., Jia, M., Wang, Q., Phoo, C.P., Hariharan, B.: Emergent correspondence from image diffusion. arXiv preprint arXiv:2306.03881 (2023)

43. Tumanyan, N., Geyer, M., Bagon, S., Dekel, T.: Plug-and-play diffusion features for text-driven image-to-image translation. In: Proceedings of the IEEE/CVF Conference on Computer Vision and Pattern Recognition (CVPR), pp. 1921–1930 (2023)

44. Tumanyan, N., Geyer, M., Bagon, S., Dekel, T.: Plug-and-play diffusion features for text-driven image-to-image translation. In: Proceedings of the IEEE/CVF Conference on Computer Vision and Pattern Recognition, pp. 1921–1930 (2023)

45. Wang, W., et al.: Zero-shot video editing using off-the-shelf image diffusion models. arXiv preprint arXiv:2303.17599 (2023)

46. Wang, Y., et al.: Lavie: high-quality video generation with cascaded latent diffusion models. arXiv preprint arXiv:2309.15103 (2023)

47. Wang, Z., et al.: Prolificdreamer: high-fidelity and diverse text-to-3d generation with variational score distillation. arXiv preprint arXiv:2305.16213 (2023)

48. Wu, J.Z., et al.: Tune-a-video: one-shot tuning of image diffusion models for text-to-video generation. In: Proceedings of the IEEE/CVF International Conference on Computer Vision, pp. 7623–7633 (2023)

49. Yang, C.A., et al.: Scene graph expansion for semantics-guided image outpainting. In: Proceedings of the IEEE/CVF Conference on Computer Vision and Pattern Recognition, pp. 15617–15626 (2022)

50. Yang, X., Wang, X.: Hash3d: training-free acceleration for 3d generation. arXiv preprint arXiv:2404.06091 (2024)

51. Yu, H., Li, R., Xie, S., Qiu, J.: Shadow-enlightened image outpainting. In: Proceedings of the IEEE/CVF Conference on Computer Vision and Pattern Recognition, pp. 7850–7860 (2024)

52. Yu, H.X., et al.: Wonderjourney: going from anywhere to everywhere. arXiv preprint arXiv:2312.03884 (2023)

53. Yu, T., et al.: Inpaint anything: segment anything meets image inpainting. arXiv preprint arXiv:2304.06790 (2023)

54. Zeng, X., et al.: Lion: latent point diffusion models for 3d shape generation. arXiv preprint arXiv:2210.06978 (2022)

55. Zhang, D.J., et al.: Show-1: marrying pixel and latent diffusion models for text-to-video generation. arXiv preprint arXiv:2309.15818 (2023)

56. Zhang, R., et al.: Pointclip: point cloud understanding by clip. In: Proceedings of the IEEE/CVF Conference on Computer Vision and Pattern Recognition, pp. 8552–8562 (2022)

Harnessing Text-to-Image Diffusion Models for Category-Agnostic Pose Estimation

Duo Peng[1], Zhengbo Zhang[1], Ping Hu[2], Qiuhong Ke[3],
David K. Y. Yau[1], and Jun Liu[1,4(✉)]

[1] Singapore University of Technology and Design, Singapore, Singapore
{duo_peng,Zhengbo_Zhang}@mymail.sutd.edu.sg
[2] University of Electronic Science and Technology of China, Chengdu, China
[3] Monash University, Melbourne, Australia
Qiuhong.Ke@monash.edu
[4] Lancaster University, Lancaster, UK
j.liu81@lancaster.ac.uk

Abstract. Category-Agnostic Pose Estimation (CAPE) aims to detect keypoints of an arbitrary unseen category in images, based on several provided examples of that category. This is a challenging task, as the limited data of unseen categories makes it difficult for models to generalize effectively. To address this challenge, previous methods typically train models on a set of predefined base categories with extensive annotations. In this work, we propose to harness rich knowledge in the off-the-shelf text-to-image diffusion model to effectively address CAPE, without training on carefully prepared base categories. To this end, we propose a Prompt Pose Matching (PPM) framework, which learns pseudo prompts corresponding to the keypoints in the provided few-shot examples via the text-to-image diffusion model. These learned pseudo prompts capture semantic information of keypoints, which can then be used to locate the same type of keypoints from images. We also design a Category-shared Prompt Training (CPT) scheme, to further boost our PPM's performance. Extensive experiments demonstrate the efficacy of our approach.

Keywords: Category-Agnostic Pose Estimation · Diffusion Model

1 Introduction

Pose estimation [2,25,66] is one of the fundamental tasks in computer vision and has a wide range of real-world applications including AR/VR [3,13,23], autonomous driving [10,17,63], human care [14,34,58], scene understanding [15, 32,56], etc. It aims to estimate the locations of keypoints that are pre-defined by

Supplementary Information The online version contains supplementary material available at https://doi.org/10.1007/978-3-031-72624-8_20.

humans. In the field of pose estimation, there exist diverse categories such as human body, human face, furniture, vehicles, etc. While existing methods [47,65] have achieved notably good performance on these categories, they typically focus on training deep neural networks to handle a single category only, which hinders their generalization to real-world scenarios that may contain diverse unseen categories.

To address the generalization problem, recent research [57] has proposed the task of Category-Agnostic Pose Estimation (CAPE). In CAPE, before testing, the model is given one or a few examples (i.e., few-shot examples) of an arbitrary unseen category. After learning on the given few-shot examples, the model is expected to perform inferences on test images of this unseen category, aiming to find pixel locations in test images that share the same semantic meanings and structural characteristics as the keypoints in the provided few-shot example images. In this way, the model can achieve category-agnostic pose estimation with only few-shot examples.

Considering the limited information provided by few-shot examples, existing CAPE methods generally [5,28,45,46,57] resort to external knowledge to compensate for the scarcity of pose data. They typically train the model on a set of predefined base categories, and then adapt the model to unseen categories based on the given few-shot examples. While being effective, the construction of base categories requires collecting and labeling data of as many categories as possible, which can be labor-intensive and time-consuming. Also, relying solely on one or a few examples to learn for unseen categories, the model trained on base categories still struggles to generalize effectively. Existing efforts in this area [45,57] have shown that even with large-scale training, these methods may still yield unsatisfactory pose estimation results when tested on unseen categories (e.g., *furniture* and *vehicle*) that differ significantly from the base categories (e.g., *face* and *human body*). These limitations spark the need for a data-efficient and generalization-effective method for CAPE.

Recently, text-to-image diffusion models [31,36,43], such as Stable Diffusion [40], have shown impressive performance in generating photo-realistic images in response to user-defined prompts. Given that existing text-to-image diffusion models can generate high-fidelity images that are visually reasonable, we believe that they contain a wealth of knowledge about the object's semantics, structures, compositions, and spatial relations. In other words, for a text-to-image diffusion model to successfully create realistic images of an object, it must know what components make up this object and it also must understand the correct positions of these components. This concept is similarly reflected in [20]. Inspired by this insight, in this work, we propose to harness the power of text-to-image diffusion models to address the CAPE task that involves spatial composition reasoning, by only learning from few-shot examples of unseen categories. In this way, we may not need to carefully prepare labeled data of base categories but can still achieve good results. However, it is a non-trivial problem, as the text-to-image diffusion model focuses on generating images from texts, lacking a mechanism to estimate keypoint locations for images, thus leading to the challenge in utilizing the potential benefits of the diffusion model.

To tackle the above challenge, in this paper, we consider the CAPE task from the perspective of regarding this task as establishing spatial correspon-

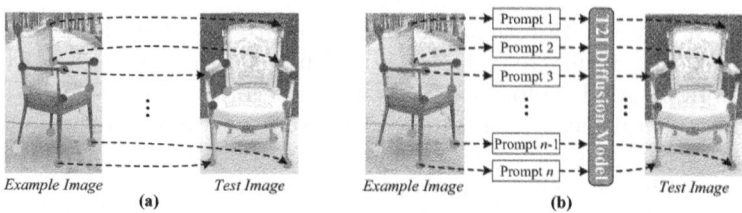

Fig. 1. **(a)** The CAPE task can be regarded as establishing spatial (keypoint) correspondences between example images and test images. **(b)** To address CAPE, our PPM learns prompts that serve as bridges to build these correspondences.

dences (mappings) between example images and test images, which is shown in Fig. 1(a). This perspective inspires us a feasible way to leverage text-to-image diffusion models to address CAPE, raising a question: *Given that the text-to-image diffusion model essentially builds the correspondences (mappings) between texts and images, can we use such text-image correspondences in diffusion models to establish image-image spatial correspondences for CAPE?* To address this question, we delve deeper into the architecture of text-to-image diffusion models. In particular, we focus on the commonly-used cross-attention mechanism in text-to-image diffusion models, which facilities the visual-textual interactions in the text-to-image generation process. [20] has revealed that the cross-attention maps extracted from text-to-image diffusion models help to highlight the text-related regions of the images. As illustrated in the Fig. 3(a), given an image of a bicycle and the text prompt "saddle", the cross-attention map highlights the saddle area. Similarly, if the text prompt is changed to "front wheel" or "back wheel", the corresponding regions are highlighted (see Fig. 3b), showcasing the model's capability to perceive different parts of the object based on the given text input. This is because the diffusion model is trained on massive text-image pairs. After its training, the cross-attention mechanism in the diffusion model is able to understand the structural and spatial information of object compositions in a joint visual-textual manner.

Motivated by this observation, we propose to extend the capabilities of text-to-image diffusion models beyond image generation to address the practical CAPE problem through the following approach: *In CAPE, before testing on images of an unseen category, we are given only one or a few labeled examples of this category. If we can find out what text 'prompt' corresponds to the particular keypoint in the given example image, then the found 'prompt' can be used to drive the text-to-image diffusion model to locate (highlight) a semantically corresponding keypoint in the test image. In this way, the found 'prompt' can serve as a bridge that builds image-to-image correspondences (see Fig. 1b), thereby harnessing the knowledge in the diffusion model to address the CAPE task.* Following this idea, an intuitive solution is to find an actual text prompt for each keypoint in the given examples. However, the text descriptions of image keypoints are usually not easy to obtain [39], since many keypoints (e.g., keypoints in Figs. 1, 6 and 7) can be difficult to describe in texts even for human beings.

To overcome these challenges, in this paper, we propose a novel framework, which we call Prompt Pose Matching (PPM), to find suitable prompt representations by learning pseudo prompts. Specifically, given few-shot examples of an unseen category, PPM uses the labeled examples to learn (optimize) pseudo prompts, to render them to be semantically aligned with the corresponding keypoints of this unseen category. Learning from the few-shot examples, the pseudo prompts can serve as bridges to find the corresponding keypoints in the test images via cross-attention maps of the diffusion model, thereby establishing the required keypoint correspondences between the example images and the test images. In this way, PPM can work solely based on the few-shot examples of unseen categories, even without training on base categories. Experiments on MP-100 [57] benchmark, which contains over 20K instances covering 100 categories, demonstrate that our PPM solely learning on few-shot examples, performs better than previous methods that require training on extensive data of base categories.

Beyond this core technical contribution, we also design a scheme, which we call Category-shared Prompt Training (CPT) for improved keypoint identification of the PPM. Given that pseudo prompts learned from few-shot examples are category-specific (i.e., customized to the unseen category), they may overlook common pose knowledge shared across different categories, such as the common positional characteristics of keypoints (e.g., keypoints tend to be located at important positions, like corners, edges, or other semantically meaningful regions, regardless of the object's category). Therefore, we further introduce a CPT scheme to acquire a category-shared pseudo prompt to enrich the prompt representation for better keypoint localization. Using CPT, the proposed PPM yields further improved performance, significantly outperforming previous methods.

2   Related Work

Pose Estimation endeavors to localize semantic or interest keypoints, such as human body parts [1,25,50], facial landmarks [2,27,48], hand keypoints [7,8,66], and vehicle poses [38,44,64] from the input data. It has evolved into a prominent task in the realm of computer vision. Existing methods can be categorized into two main types: regression-based methods [4,11,51,60] and heatmap-based methods [29,35,49,61]. Regression-based methods exhibit high efficiency for implementation on applications. However, these approaches inherently output singular 2D coordinates for each keypoint, neglecting to consider the surrounding area of the keypoint. To address this limitation, heatmap-based approaches have been introduced to localize keypoints through probabilistic heatmaps rather than fixed coordinates. In this paper, our method is based on heatmaps.

Category-Agnostic Pose Estimation aims to develop a pose estimation model that can detect the pose of various object categories, given few-shot examples as reference. Previous methods [5,28,45,46,57] typically train the model on a set of predefined base categories, and then adapt it on the given few-shot

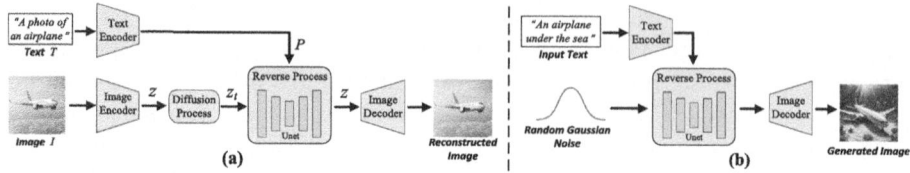

Fig. 2. (a) Overall architecture of Stable Diffusion. (b) The architecture of Stable Diffusion for text-to-image generation.

examples of unseen categories. In this work, we handle this task from a novel perspective that harnesses the knowledge within the off-the-shelf Stable Diffusion by leveraging the cross-attention mechanism to find correspondences between example images and test images, without requiring training on base categories.

Text-to-Image Diffusion Models [31,36,40,43] are generation models that employ diffusion techniques. They are primarily utilized for generating detailed images based on textual descriptions. Recently, text-to-image diffusion models have made striking advancements across a spectrum of domains such as image inpainting [55], image editing [18], 3D scene understanding [26], semantic segmentation [20], visual tracking [62], object counting [16], and various interdisciplinary applications [30]. Their versatility and efficacy are reshaping the landscape of these fields. In this paper, we are the first to study the harness of text-to-image diffusion models for category-agnostic pose estimation.

3   Preliminary

In this paper, we propose a novel framework PPM to harness the knowledge in text-to-image diffusion models for CAPE. Our PPM has good extensibility, which can be applied to various text-to-image diffusion models. For clarity, in the main paper, we use Stable Diffusion [40] as an example to illustrate our framework. Below, we briefly review Stable Diffusion to facilitate a better understanding of our method.

Architecture. As shown in Fig. 2(a), Stable Diffusion mainly consists of five parts: image encoder, image decoder, text encoder, diffusion process, and reverse process. Given an image I, the image is first mapped into a latent representation Z through an image encoder. Simultaneously, a text prompt T is transformed into an embedding P using a text encoder. The diffusion process is then applied to Z, which introduces noise over t steps to produce a noisy representation Z_t. Then, a reverse process incorporating a U-Net [41] takes in Z_t and P to denoise Z_t back to the clean Z. Finally, the image decoder reconstructs the image I from Z. After training on a vast amount of image-text pairs, the image encoder and the diffusion process are discarded, and the remaining parts can be regarded as a text-to-image generation model, as illustrated in Fig. 2(b). During inference, we can input a random Gaussian noise into the generation model to generate a text-relevant image. This works because during training, the noisy Z_t can be very close to the pure Gaussian noise after many steps of the diffusion process.

Diffusion Training. As for training the diffusion model, the image encoder, the image decoder, and the text encoder are off-the-shelf frozen modules. The diffusion process does not involve learnable parameters. Therefore, it solely focuses on training the UNet (in the reverse process) to generate text-relevant visual content by denoising. In the diffusion process, noise ϵ is added to the latent image representation Z, resulting in Z_t. The training objective is to predict the added noise ϵ from Z_t, conditioned on P. This is formulated as:

$$L_{\mathrm{dm}} = \mathbb{E}_{\mathcal{E}_i(I), \mathcal{E}_t(T), \epsilon \sim \mathcal{N}(0,1), t} \left[\|\epsilon - \epsilon_\theta (Z_t, P, t)\|_2^2 \right], \tag{1}$$

where $\mathcal{E}_i$ is the image encoder, $\mathcal{E}_t$ represents the text encoder, t denotes the time step in the diffusion process, and ϵ_θ is the UNet in the reverse process. Through this training approach, the connection between textual information P and image information Z_t is established within the UNet.

4   Method

Problem Definition. This paper aims to address the task of Class-Agnostic Pose Estimation (CAPE). Unlike most pose estimation tasks that predict keypoints for a single known (seen) category, CAPE requires the model to efficiently generalize to novel categories, which have not only different appearances, but also varying numbers of keypoints. In CAPE, during training, we are given extensively labeled data of some predefined categories (called base categories). Here, we denote the data of base categories as D_{base}. In D_{base}, we use I_{base} and M_{base} to represent the image and the corresponding keypoint localization label, respectively. Following [57], we use the keypoint labels in heatmap format. For each image I_{base}, its corresponding heatmap label M_{base} contains multiple sub-labels, i.e., $M_{\mathrm{base}} = \{M_{\mathrm{base}}^1, M_{\mathrm{base}}^2, ..., M_{\mathrm{base}}^n, ..., M_{\mathrm{base}}^N\}$ where each sub-label represents the heatmap of one keypoint, and N denotes the number of keypoints in the image. Previous methods [5,28,46,57] generally train the model on D_{base} and then test the model on novel categories to validate its generalization capacity. Before testing, for each unseen category, the model is provided with one or a few labeled examples, denoted as D_{exm}, for few-shot learning. In D_{exm}, we use I_{exm} and M_{exm} to denote the example image and its corresponding label. Similarly, M_{exm} also contain multiple sub-labels for all keypoints, i,e, $M_{\mathrm{exm}} = \{M_{\mathrm{exm}}^1, M_{\mathrm{exm}}^2, ..., M_{\mathrm{exm}}^n, ..., M_{\mathrm{exm}}^N\}$. During testing, based on the given few-shot examples D_{exm}, the model is required to detect the corresponding keypoints of the same category for the unlabeled test images I_{test}. Notably, since different object categories have varying numbers of keypoints, N may change accordingly based on the category. However, within the same category, all samples including example images and test images share the same number of keypoints.

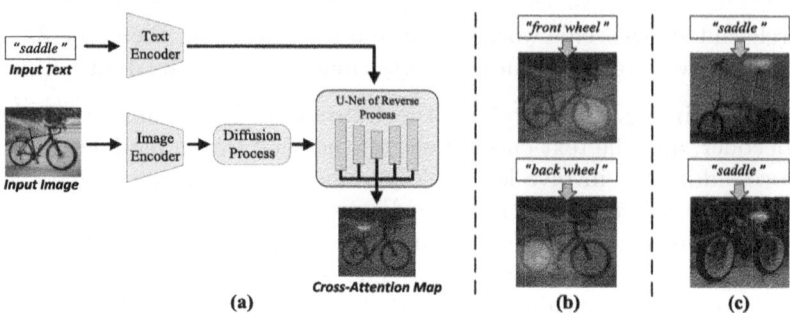

Fig. 3. **(a)** The architecture of Stable Diffusion for extracting cross-attention maps, which highlight the area that corresponds semantically to the input text. **(b)** With the same image, using different texts describing parts of the object, the attention maps highlight corresponding different areas. **(c)** With the same text "saddle", the cross-attention highlights relevant regions in different images.

In this paper, we propose a novel PPM framework that leverages the off-the-shelf Stable Diffusion to build the spatial (keypoint) correspondences between example images I_{exm} and text images I_{test}, thus addressing the CAPE task without the need of training on carefully prepared base categories D_{base}. Below, we present our PPM in detail.

4.1   Prompt Pose Matching (PPM)

Here, we propose a novel framework, Prompt Pose Matching (PPM), to effectively handle the CAPE task by learning only on the given few-shot examples D_{exm}. After learning on D_{exm}, our PPM can be directly used for inference on I_{test}, without training on D_{base}.

Inspired by the observation [20] that the off-the-shelf Stable Diffusion encompasses a wealth of knowledge that should be useful for understanding the object semantics in images, our PPM is designed to address CAPE by leveraging such Stable Diffusion embedded knowledge. To do so, we explore the cross-attention map in Stable Diffusion as a handle. We observe that the cross-attention map extracted from the U-Net of Stable Diffusion effectively captures the correspondence between the image and text. For example, as shown in Fig. 3(a), when feeding an image of a bicycle into Stable Diffusion and providing the input text "saddle", the cross-attention map extracted from the U-Net highlights the corresponding saddle region, which occurs in a semantically consistent manner. Even when different images are fed into Stable Diffusion, if we use the same text prompt (e.g., "saddle"), the resulting cross-attention maps will consistently highlight corresponding regions, as shown in Fig. 3(c). In light of this, given few-shot examples D_{exm} of an unseen category, if we can identify the text prompt T corresponding to a particular keypoint of this category, the diffusion model could also use this text prompt T to find the corresponding keypoints in test images I_{test} via cross-attention maps. However, since the actual text prompts of keypoints can be hard to obtain, we turn to learn a pseudo prompt P in the embedding space, aiming to properly represent the semantics of the particular keypoint. Based on this idea, our PPM framework is designed to handle the CAPE task via prompt optimization.

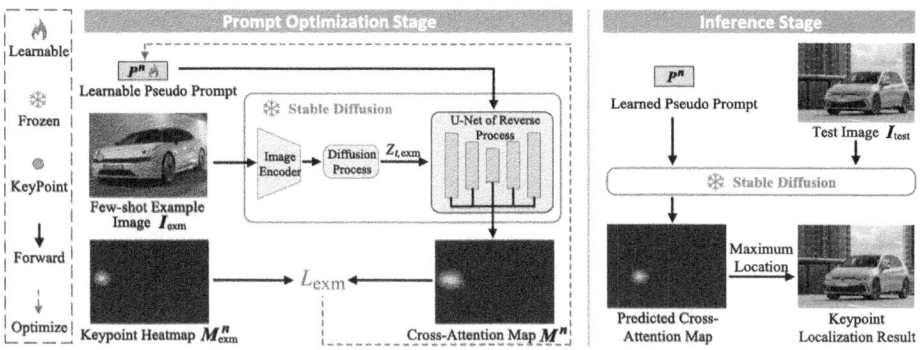

Fig. 4. Overview of our Prompt Pose Matching (PPM) framework.

Overview. As illustrated in Fig. 4, our PPM involves two stages. In the first stage (prompt optimization), for the n-th keypoint in the given example image I_{exm}, we seek to learn a pseudo prompt P^n that properly represents the semantics of this keypoint, by using the keypoint heatmap label M_{exm}^n to constrain the cross-attention map extracted from the U-Net of Stable Diffusion. For image I_{exm} with N keypoints, we learn N pseudo prompts $\{P^1, P^2, ..., P^n, ..., P^N\}$ respectively. In the second stage (inference), the N learned pseudo prompts from the first stage are kept fixed. With the help of these N learned prompts, N cross-attention maps are computed for the test image I_{test}. In each computed cross-attention map, the location with the highest attention value indicates the corresponding keypoint location of the test image I_{test}. In this way, we use the learned pseudo prompts to bridge the keypoint correspondences between example images I_{exm} and text images I_{test}, which subtly harnesses Stable Diffusion to address CAPE. Next, we introduce the two stages of our PPM.

Prompt Optimization Stage. In this stage, we aim to learn pseudo prompts to represent the semantics of the keypoints from few-shot examples D_{exm}. Taking the learning for the n-th keypoint as an example, at the beginning, the pseudo prompt $P^n \in \mathbb{R}^{1 \times 1 \times c}$ is randomly initialized, where c is the channel number of the embedding space. Also, we are given a test-time example image $I_{\text{exm}} \in \mathbb{R}^{H \times W \times 3}$ and the n-th keypoint heatmap label $M_{\text{exm}}^n \in \mathbb{R}^{H \times W \times 1}$. As shown in Fig. 4 (left), we input the prompt P^n along with the example image I_{exm} into the stable diffusion to obtain the cross-attention map M^n. Specifically, the image encoder and the diffusion process of Stable Diffusion convert the image I_{exm} into the noisy latent representation $Z_{t,\text{exm}}$. Then, $Z_{t,\text{exm}}$ and P^n are fed into the

U-Net of the reverse process. At each layer of the U-Net, a cross-attention map can be obtained, which calculates the similarity (correlation) between $Z_{t,\text{exm}}$ and P^n. More specifically, for the n-th keypoint, the cross-attention map at the l-th layer in the U-Net can be obtained as:

$$M_l^n = Sigmoid\left(Avg\left(\frac{F_{Z_{t,\text{exm}}} * F_{P^n}^\top}{\sqrt{d_l}}\right)\right), \tag{2}$$

where $F_{Z_{t,\text{exm}}} \in \mathbb{R}^{H_l \times W_l \times d_l}$ is the visual feature map, and $F_{P^n} \in \mathbb{R}^{1 \times 1 \times d_l}$ is the prompt feature; $F_{Z_{t,\text{exm}}}$ and F_{P^n} are obtained from $Z_{t,\text{exm}}$ and P^n, respectively, via $F_{Z_{t,\text{exm}}} = \phi_l(Z_{t,\text{exm}})$ and $F_{P^n} = \psi_l(P^n)$, where ϕ_l and ψ_l denote the linear mappings in the attention module at the l-th layer of the U-Net; H_l, W_l, and d_l respectively represent the height, width, and channel number of the visual feature map at the l-th layer; Avg denotes the average along the channel dimension; $Sigmiod$ is the normalization that transforms the values into $[0, 1]$. In Eq. 2, the prompt feature F_{P^n} is multiplied with each pixel-level visual feature of the feature map $F_{Z_{t,\text{exm}}}$, and thus if the prompt feature and visual feature are semantically similar (correlated), the attention value at the corresponding pixel will be high, thereby highlighting the prompt-related regions. We denote the l-th layer cross-attention map as M_l, where $M_l \in \mathbb{R}^{H_l \times W_l \times 1}$.

To incorporate cross-attention maps from different layers, as shown in Fig. 4 (left), we average cross-attention maps across the U-Net layers, to obtain:

$$M^n = \frac{1}{L} \sum_{l=1}^{L} M_l^n, \tag{3}$$

where L denotes the number of layers in U-Net. Note that when averaging, we first use the bilinear interpolation to upsample all the cross-attention maps into the original image scale $H \times W$. Thus, we can obtain $M^n \in \mathbb{R}^{H \times W \times 1}$.

After obtaining the cross-attention map M^n for the n-th keypoint of the example image, given the keypoint heatmap label M_{exm}^n of this image, we calculate the L2 loss between M^n and M_{exm}^n, and use the loss to optimize the pseudo prompt P^n, ensuring that P^n can drive Stable Diffusion to highlight the specific keypoint region. The optimization loss can be formulated as:

$$L_{\text{exm}} = \|M^n - M_{\text{exm}}^n\|^2 + L_{\text{dm}}, \tag{4}$$

where $\|\cdot\|^2$ denotes the L_2 distance. The second items L_{dm} is the diffusion model's loss function (Eq. 1). We use this loss to keep the learned prompt understandable by Stable Diffusion. After optimization, we can obtain a learned pseudo prompt P^n that represents the semantics of the n-th keypoint. By learning from $n = 1$ to $n = N$, we can obtain N learned pseudo-prompts. Not that the loss is calculated across all the given few-shot examples by averaging. That is, for a novel category (N keypoints) with multiple labeled examples (i.e., more than 1-shot), we still obtain N learned pseudo-prompts by averaging across the examples.

Next, we describe how to use the learned prompt P^n to conduct keypoint localization for the test image I_{test} (i.e., the inference stage).

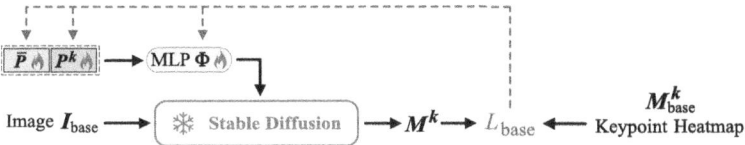

Fig. 5. Illustration of our PPM with Category-shared Prompt Training (CPT). $\bar{P}$ and P^k denote the category-shared prompt and category-specific prompt, respectively. Φ denotes a light-weight MLP.

Inference Stage. To build the image-to-image correspondences for CAPE, we use the prompt P^n learned from example images, to find the corresponding keypoint in each test image. As shown in Fig. 4 (right), given a test image I_{test}, we feed P^n and I_{test} into Stable Diffusion to compute the cross-attention map. The highlighted region in the output cross-attention map denotes the detected corresponding keypoint region. As P^n is learned from the keypoint regions in the example images using the knowledge of Stable Diffusion, the P^n-activated keypoint regions in the test images share the same semantic meaning and structural characteristics as those in the example images. Finally, we locate the keypoint by finding the position with maximum value in the cross-attention map, as shown in Fig. 4 (right). In this way, based on all learned prompts $\{P^1, P^2, ..., P^n, ..., P^N\}$, we can estimate the location of all keypoints for test images of the same category.

4.2   Category-Shared Prompt Training (CPT)

Though keypoints of different categories show various characteristics, there can be some common knowledge shared by keypoints of different categories, which could help for keypoint localization (e.g., keypoints tend to be located at "important" locations, like center points, corners, edges, etc.). In light of this, we propose to capture common knowledge by additionally learning a category-shared prompt. Specifically, we propose a Category-shared Prompt Training (CPT) scheme, to enable our framework to learn one category-shared prompt $\bar{P}$ from data of base categories D_{base} provided in the CAPE task. Our goal is to train this category-shared prompt $\bar{P}$ on each keypoint of every category from base categories, thus enabling $\bar{P}$ to acquire category-agnostic knowledge. This knowledge can equally benefit pose estimation for novel categories.

To disentangle the shared knowledge and the specific knowledge during the training on base categories, we slightly modify the PPM framework. We assume that across all categories in the base-category data, there are a total of K keypoints. To represent these K keypoints precisely, we require K category-specific prompts, denoted as $\{P^1, P^2, ..., P^k, ..., P^K\}$. As shown in Fig. 5, before sending it into the Stable Diffusion, we concatenate the learnable category-shared prompt $\bar{P}$ with the learnable category-specific prompt P^k, and then pass the combined prompt through a light-weight MLP [6] (two layers) to reduce the channel dimension, aiming to ensure compatibility with the Stable Diffusion. Specifically, given

I_{base} (the image from base categories) and M_{base}^k (the corresponding heatmap label of the k-th keypoint), the training loss L can be calculated as:

$$M^k = f_{SD}\left(I_{\text{base}}, \Phi([\bar{P}, P^k])\right), \tag{5}$$

$$L_{\text{base}} = \left\| M^k - M_{\text{base}}^k \right\|^2 + L_{\text{dm}}, \tag{6}$$

where M^k is the output cross-attention map; f_{SD} denotes the Stable Diffusion which treats I_{base} and $\Phi([\bar{P}, P^k])$ as two inputs; Φ denotes the MLP; $\bar{P}$ and P^k represent the category-shared prompt and the category-specific prompt respectively. As shown in Fig. 5, we use the loss in Eq. 6 to optimize the category-shared prompt $\bar{P}$, category-specific prompt P^k, and MLP Φ, where the Stable Diffusion is kept frozen. During this training process, the (one) category-shared prompt $\bar{P}$ and the MLP Φ are trained on all keypoints of the base categories, while the category-specific prompt P^k is independently trained for each keypoint of each base category. Thus, after training on K base-category keypoints, we will get K learned category-specific prompts $\{P^1, P^2, ..., P^K\}$ and only one learned category-shared prompt $\bar{P}$. In this disentangled learning way, $\bar{P}$ learns the common knowledge of keypoint localization, P^k learns the specific knowledge, and Φ learns how to merge the common knowledge and specific knowledge. As each learned P^k is specific to the keypoint of base categories, we discard all the learned P^k, but introduce the learned $\bar{P}$ and the learned Φ into our PPM framework to help to enhance the category-specific pseudo prompt P^n learned from few-show examples of unseen categories.

4.3   Training and Testing

As for PPM *without* CPT, we do not use data of based categories. Given few-shot examples of an unseen category, for each keypoint of this category, we randomly initialize a pseudo prompt P^n and use the labeled examples to optimize P^n. Then during testing, we use the learned P^n to locate the corresponding keypoint for test images.

As for PPM *with* CPT, it involves training on base categories. During training on base categories, we use labeled data of base categories to train a category-shared prompt $\bar{P}$ and a MLP Φ. Then, we initialize the pseudo prompt P^n for learning on few-shot examples. We concatenate the learnable P^n with the learned (frozen) category-shared $\bar{P}$ using the learned (frozen) MLP Φ, to enhance the prompt representation. We follow the same procedures of PPM *without* CPT to optimize P^n on few-shot examples. Finally, during testing, we still concatenate the learned P^n with the learned $\bar{P}$ using the learned MLP Φ, to produce the enhanced prompt for inference on test images.

5   Experiments

Dataset and Metric. The majority of pose estimation datasets are not suitable for the CAPE task, since they predominantly consist of objects belonging to a single category. Following prior CAPE research [57], we evaluate our approach on the benchmark MP-100 [57], which is composed of many popular 2D pose estimation datasets, including COCO [25], 300W [42], AFLW [22], OneHand10K [52], DeepFasion2 [9], MacaquePose [24], Vinegar Fly [33], Desert Locust [12], CUB-200 [53], CarFusion [37], AnimalWeb [19], and Keypoint-5 [54]. In total, MP-100 covers eight super-categories (i.e., *human hand, human face, human body, animal face, animal body, clothes, furniture,* and *vehicle*) and 100 sub-categories (e.g., *bus, sofa, bed,* and *skirt*, etc.), providing a very comprehensive evaluation platform. The MP-100 benchmark is organized into training, validation, and testing sets, comprising categories without overlapping, thus facilitating the evaluation of models for CAPE. On this benchmark, we use the standard metric of PCK (Probability of Correct Keypoint) [59] with a threshold of 0.2 to assess the algorithm's performance, following previous work [45,57].

Parameter Setting. The training is conducted on two NVIDIA RTX 3090 GPUs with a batch size of four in each GPU. During training in CPT, we adopt the Adam optimizer [21] with a learning rate of $5e - 4$. During the test-time optimization on few-shot examples, the learning rate is set to $2e - 3$. In the diffusion process of Stable Diffusion, we use $t = 10$ to obtain z^t.

Experiment Settings. In CAPE, there are two generalization settings: (1) Cross Super-Category Generalization and (2) Cross Sub-Category Generalization. In Cross Super-Category Generalization, the generalization performance is evaluated based on 8 super-categories, where 6 for train, 1 for val, and 1 for test. In Cross Sub-Category Generalization, the evaluation is conducted based on 100 sub-categories, where 70 for train, 10 for val, and 20 for test. Besides, there are also two few-shot settings: (1) 1-shot and (2) 5-shot, which represent the number of few-shot examples. As existing methods [5,28,45,46,57] conduct experiments under three scenarios: Cross Super-Category Generalization (1-shot), Cross Sub-Category Generalization (1-shot), and Cross Sub-Category Generalization (5-shot), for fair comparisons, we follow them to conduct experiments under the same 3 scenarios.

Table 1. Results in the setting of Cross Super-Category Generalization (1-shot). ✓ and ✗ denote use or not use of base categories.

Method	Base Cate.	Human Body	Human Face	Vehicle	Furniture	Mean
ProtoNet [46]	✓	37.61	57.80	28.35	42.64	41.60
MAML [5]	✓	51.93	25.72	17.68	20.09	23.08
Fine-tune [28]	✓	52.11	25.53	17.46	20.76	28.97
POMNet [57]	✓	73.82	79.63	34.92	47.27	58.91
CapeFormer [45]	✓	83.44	80.96	45.40	52.49	65.57
PPM	✗	84.03	81.52	50.59	58.61	68.68
PPM + CPT	✓	**85.13**	**82.44**	**52.08**	**60.59**	**70.06**

5.1   Quantitative Results

Cross Super-Category Generalization Results. In MP-100 dataset, there are eight super-categories. We follow [45,57] to undertake a cross super-category pose estimation evaluation on the MP-100 benchmark. Specifically, the model is trained and validated on six and one super-categories, and performance is evaluated on the remaining one super-category. For fair comparisons, we follow previous work [45,57] to evaluate the generalization to *human body*, *human face*, *vehicle*, and *furniture*, respectively.

As shown in Table 1, the performance of our proposed PPM, even with no base categories, surpasses that of previous methods trained on base categories, clearly demonstrating the effectiveness of our framework. We can see that previous methods get suboptimal performances. This could be attributed to the significant divergence between training categories and the testing one, which hinders the generalization of methods that highly rely on knowledge from the training base categories. In contrast, our approach leverages the wealth of knowledge within the Stable Diffusion model to mitigate overfitting to the training categories, thus achieving a significant performance improvement over prior methods (e.g., on *furniture*, we outperform [45] by 8.1%). It is noteworthy that even using PPM only, our method can significantly outperform existing methods, and with the addition of CPT, the performance of our method is further boosted, which demonstrates the effect of our method.

Table 2. Results in the setting of Cross Sub-Category Generalization.

Method	Base Cate.	1-shot						5-shot					
		Split1	Split2	Split3	Split4	Split5	Mean	Split1	Split2	Split3	Split4	Split5	Mean
ProtoNet [46]	✓	46.05	40.84	49.13	43.34	44.54	44.78	60.31	53.51	61.92	58.44	58.61	58.56
MAML [5]	✓	68.14	54.72	64.19	63.24	57.20	61.50	70.03	55.98	63.21	64.79	58.47	62.50
Fine-tune [28]	✓	70.60	57.04	66.06	65.00	59.20	63.58	71.67	57.84	66.76	66.53	60.24	64.61
POMNet [57]	✓	84.23	78.25	78.17	78.68	79.17	79.70	84.72	79.61	78.00	80.38	80.85	80.71
CapeFormer [45]	✓	89.45	84.88	83.59	83.53	85.09	85.31	91.94	88.92	89.40	88.01	88.25	89.30
PPM	✗	89.82	85.63	83.72	83.90	86.03	85.82	92.31	89.43	89.84	89.97	89.30	90.17
PPM+CPT	✓	91.03	88.06	84.48	86.73	87.40	87.54	93.64	92.71	91.76	92.85	91.94	92.58

Cross Sub-category Generalization Results. To further assess generalization capability to unseen sub-categories, following [45,57], we evaluate our framework against previous methods under two few-shot settings: 1-shot and 5-shot, which are shown in Table 2. The evaluation is conducted on five different train/val/test category splits for a comprehensive evaluation. In both 1-shot and 5-shot settings, we show the results of each split and the mean results averaged over all the five splits. From Table 2, we can see: (1) in the absence of CPT, our PPM, even using no training data of base categories, can outperform previous methods trained on base categories across all splits and few-shot settings. (2)

When using PPM equipped with CPT, our method achieves further enhanced performance and significantly surpasses existing methods, which demonstrates the effectiveness of CPT.

5.2   Qualitative Results

In Fig. 6, we qualitatively compare the generalization ability of our method with existing approaches. We can observe that when the few-shot example largely differs from the test images in terms of viewpoint or appearance, our method still performs robustly and generates predictions very close to the ground-truth labels. To demonstrate our CPT can effectively enhance our PPM's keypoint identification capacity, we further visualize cross-attention maps of the test set. As shown in Fig. 7, CPT can drive Stable Diffusion to generate cross-attention maps with finer and preciser keypoint regions. This indicates that, compared to only use category-specific prompt, the addition of category-shared prompt provided by CPT facilitates the Stable Diffusion to better locate the keypoints.

Fig. 6. Visual comparison with state-of-the-art methods. Red circles show failure keypoints.   **Fig. 7.** Visualization of attention maps in our method.

5.3   Ablation Study

Effect of Incorporating Multi-layer Attention Maps. In our PPM framework, we integrate cross-attention maps across different layers. In this study, we explore the impact of incorporating multi-layer cross-attention maps on the overall performance of the framework. For comparison, we evaluate the performance of cross-attention maps at each single layer. Because of the limited space, in Table 3, we only present the best result among all single-layer cross-attention maps. We can see that the integration of cross-attention maps across different layers shows an obvious performance advantage, even compared to the best-performing single-layer cross-attention map. More ablation studies can be seen in *Supplementary*.

Table 3. Ablation on incorporation of multi-layer attention maps.

Method	Cross Super-Cate. (1-shot)	Cross Sub-Cate. (1-shot)	Cross Sub-Cate. (5-shot)
Ours (Single Layer)	68.81	86.49	91.37
Ours (Multi Layer)	**70.06**	**87.54**	**92.58**

6 Conclusion

In this paper, we introduce a novel PPM framework to address CAPE by harnessing the knowledge in text-to-image diffusion models, without requiring base categories. We further propose a CPT scheme to enhance the keypoint identification capability of PPM. Experiments show that our method significantly outperforms existing state-of-the-art methods.

Acknowledgements. This research/project is supported by the National Research Foundation, Singapore under its AI Singapore Programme (AISG Award No: AISG2-PhD-2022-01-027[T]) and the Ministry of Education, Singapore, under the AcRF Tier 2 Projects (MOE-T2EP20222-0009 and MOE-T2EP20123-0014).

References

1. Bulat, A., Tzimiropoulos, G.: Human pose estimation via convolutional part heatmap regression. In: Leibe, B., Matas, J., Sebe, N., Welling, M. (eds.) ECCV 2016. LNCS, vol. 9911, pp. 717–732. Springer, Cham (2016). https://doi.org/10.1007/978-3-319-46478-7_44
2. Bulat, A., Tzimiropoulos, G.: How far are we from solving the 2D & 3D face alignment problem?(and a dataset of 230,000 3D facial landmarks). In: Proceedings of the IEEE International Conference on Computer Vision, pp. 1021–1030 (2017)
3. Cao, Z., Simon, T., Wei, S.E., Sheikh, Y.: Realtime multi-person 2D pose estimation using part affinity fields. In: Proceedings of the IEEE Conference on Computer Vision and Pattern Recognition, pp. 7291–7299 (2017)
4. Chan, C., Ginosar, S., Zhou, T., Efros, A.A.: Everybody dance now. In: Proceedings of the IEEE/CVF International Conference on Computer Vision, pp. 5933–5942 (2019)
5. Finn, C., Abbeel, P., Levine, S.: Model-agnostic meta-learning for fast adaptation of deep networks. In: International Conference on Machine Learning, pp. 1126–1135. PMLR (2017)
6. Gardner, M.W., Dorling, S.: Artificial neural networks (the multilayer perceptron)-a review of applications in the atmospheric sciences. Atmos. Environ. **32**(14–15), 2627–2636 (1998)
7. Ge, L., Cai, Y., Weng, J., Yuan, J.: Hand PointNet: 3D hand pose estimation using point sets. In: Proceedings of the IEEE Conference on Computer Vision and Pattern Recognition, pp. 8417–8426 (2018)
8. Ge, L., et al.: 3D hand shape and pose estimation from a single RGB image. In: Proceedings of the IEEE/CVF Conference on Computer Vision and Pattern Recognition, pp. 10833–10842 (2019)

9. Ge, Y., Zhang, R., Wang, X., Tang, X., Luo, P.: DeepFashion2: a versatile benchmark for detection, pose estimation, segmentation and re-identification of clothing images. In: Proceedings of the IEEE/CVF Conference on Computer Vision and Pattern Recognition, pp. 5337–5345 (2019)

10. Gilroy, S., Glavin, M., Jones, E., Mullins, D.: Pedestrian occlusion level classification using keypoint detection and 2d body surface area estimation. In: Proceedings of the IEEE/CVF International Conference on Computer Vision, pp. 3833–3839 (2021)

11. Gong, J., Fan, Z., Ke, Q., Rahmani, H., Liu, J.: Meta agent teaming active learning for pose estimation. In: Proceedings of the IEEE/CVF Conference on Computer Vision and Pattern Recognition, pp. 11079–11089 (2022)

12. Graving, J.M., et al.: DeepPoseKit, a software toolkit for fast and robust animal pose estimation using deep learning. Elife **8**, e47994 (2019)

13. Guleryuz, O.G., Kaeser-Chen, C.: Fast lifting for 3D hand pose estimation in AR/VR applications. In: 2018 25th IEEE International Conference on Image Processing (ICIP), pp. 106–110 (2018)

14. Huang, M.H., Foo, L.G., Liu, J.: Learning to unlearn for robust machine unlearning. In: European Conference on Computer Vision. Springer, Heidelberg (2024)

15. Huang, S., Qi, S., Xiao, Y., Zhu, Y., Wu, Y.N., Zhu, S.C.: Cooperative holistic scene understanding: unifying 3D object, layout, and camera pose estimation. In: Advances in Neural Information Processing Systems, vol. 31 (2018)

16. Hui, X., Wu, Q., Rahmani, H., Liu, J.: Class-agnostic object counting with text-to-image diffusion model. In: European Conference on Computer Vision. Springer, Heidelberg (2024)

17. Iftikhar, S., Zhang, Z., Asim, M., Muthanna, A., Koucheryavy, A., Abd El-Latif, A.A.: Deep learning-based pedestrian detection in autonomous vehicles: substantial issues and challenges. Electronics **11**(21), 3551 (2022)

18. Kawar, B., et al.: Imagic: text-based real image editing with diffusion models. In: Proceedings of the IEEE/CVF Conference on Computer Vision and Pattern Recognition, pp. 6007–6017 (2023)

19. Khan, M.H., et al.: AnimalWeb: a large-scale hierarchical dataset of annotated animal faces. In: Proceedings of the IEEE/CVF Conference on Computer Vision and Pattern Recognition, pp. 6939–6948 (2020)

20. Khani, A., Taghanaki, S.A., Sanghi, A., Amiri, A.M., Hamarneh, G.: SLiME: segment like me. arXiv preprint arXiv:2309.03179 (2023)

21. Kingma, D.P., Ba, J.: Adam: a method for stochastic optimization. arXiv preprint arXiv:1412.6980 (2014)

22. Koestinger, M., Wohlhart, P., Roth, P.M., Bischof, H.: Annotated facial landmarks in the wild: A large-scale, real-world database for facial landmark localization. In: 2011 IEEE International Conference on Computer Vision Workshops (ICCV Workshops), pp. 2144–2151. IEEE (2011)

23. Krauß, V., Boden, A., Oppermann, L., Reiners, R.: Current practices, challenges, and design implications for collaborative AR/VR application development. In: Proceedings of the 2021 CHI Conference on Human Factors in Computing Systems, pp. 1–15 (2021)

24. Labuguen, R., et al.: MacaquePose: a novel "in the wild" macaque monkey pose dataset for markerless motion capture. Front. Behav. Neurosci. **14**, 581154 (2021)

25. Lin, T.-Y., et al.: Microsoft COCO: common objects in context. In: Fleet, D., Pajdla, T., Schiele, B., Tuytelaars, T. (eds.) ECCV 2014. LNCS, vol. 8693, pp. 740–755. Springer, Cham (2014). https://doi.org/10.1007/978-3-319-10602-1_48

26. Liu, R., Wu, R., Van Hoorick, B., Tokmakov, P., Zakharov, S., Vondrick, C.: Zero-1-to-3: zero-shot one image to 3d object. In: Proceedings of the IEEE/CVF International Conference on Computer Vision, pp. 9298–9309 (2023)
27. Liu, Z., Chen, Z., Bai, J., Li, S., Lian, S.: Facial pose estimation by deep learning from label distributions. In: Proceedings of the IEEE/CVF International Conference on Computer Vision Workshops (2019)
28. Nakamura, A., Harada, T.: Revisiting fine-tuning for few-shot learning. arXiv preprint arXiv:1910.00216 (2019)
29. Newell, A., Yang, K., Deng, J.: Stacked hourglass networks for human pose estimation. In: Leibe, B., Matas, J., Sebe, N., Welling, M. (eds.) ECCV 2016. LNCS, vol. 9912, pp. 483–499. Springer, Cham (2016). https://doi.org/10.1007/978-3-319-46484-8_29
30. Nguyen, L.X., Aung, P.S., Le, H.Q., Park, S.B., Hong, C.S.: A new chapter for medical image generation: the stable diffusion method. In: 2023 International Conference on Information Networking (ICOIN), pp. 483–486. IEEE (2023)
31. Nichol, A.Q., et al.: GLIDE: towards photorealistic image generation and editing with text-guided diffusion models. In: Proceedings of the 39th International Conference on Machine Learning, vol. 162, pp. 16784–16804. PMLR (2022)
32. Peng, H., et al.: The multi-modal video reasoning and analyzing competition. In: Proceedings of the IEEE/CVF International Conference on Computer Vision, pp. 806–813 (2021)
33. Pereira, T.D., et al.: Fast animal pose estimation using deep neural networks. Nat. Methods 16(1), 117–125 (2019)
34. Probst, T., Fossati, A., Van Gool, L.: Combining human body shape and pose estimation for robust upper body tracking using a depth sensor. In: Hua, G., Jégou, H. (eds.) ECCV 2016. LNCS, vol. 9914, pp. 285–301. Springer, Cham (2016). https://doi.org/10.1007/978-3-319-48881-3_20
35. Ramakrishna, V., Munoz, D., Hebert, M., Andrew Bagnell, J., Sheikh, Y.: Pose machines: articulated pose estimation via inference machines. In: Fleet, D., Pajdla, T., Schiele, B., Tuytelaars, T. (eds.) ECCV 2014. LNCS, vol. 8690, pp. 33–47. Springer, Cham (2014). https://doi.org/10.1007/978-3-319-10605-2_3
36. Ramesh, A., Dhariwal, P., Nichol, A., Chu, C., Chen, M.: Hierarchical text-conditional image generation with clip latents. arXiv preprint arXiv:2204.06125 1(2), 3 (2022)
37. Reddy, N.D., Vo, M., Narasimhan, S.G.: CarFusion: combining point tracking and part detection for dynamic 3D reconstruction of vehicles. In: Proceedings of the IEEE Conference on Computer Vision and Pattern Recognition, pp. 1906–1915 (2018)
38. Reddy, N.D., Vo, M., Narasimhan, S.G.: Occlusion-net: 2D/3D occluded keypoint localization using graph networks. In: Proceedings of the IEEE/CVF Conference on Computer Vision and Pattern Recognition, pp. 7326–7335 (2019)
39. Reed, S., Akata, Z., Lee, H., Schiele, B.: Learning deep representations of fine-grained visual descriptions. In: Proceedings of the IEEE Conference on Computer Vision and Pattern Recognition (CVPR) (2016)
40. Rombach, R., Blattmann, A., Lorenz, D., Esser, P., Ommer, B.: High-resolution image synthesis with latent diffusion models. In: Proceedings of the IEEE/CVF Conference on Computer Vision and Pattern Recognition, pp. 10684–10695 (2022)
41. Ronneberger, O., Fischer, P., Brox, T.: U-net: convolutional networks for biomedical image segmentation. In: Navab, N., Hornegger, J., Wells, W.M., Frangi, A.F. (eds.) MICCAI 2015, Part III. LNCS, vol. 9351, pp. 234–241. Springer, Cham (2015). https://doi.org/10.1007/978-3-319-24574-4_28

42. Sagonas, C., Antonakos, E., Tzimiropoulos, G., Zafeiriou, S., Pantic, M.: 300 faces in-the-wild challenge: database and results. Image Vis. Comput. **47**, 3–18 (2016)
43. Saharia, C., et al.: Photorealistic text-to-image diffusion models with deep language understanding. In: Advances in Neural Information Processing Systems, vol. 35, pp. 36479–36494 (2022)
44. Sánchez, H.C., Martínez, A.H., Gonzalo, R.I., Parra, N.H., Alonso, I.P., Fernandez-Llorca, D.: Simple baseline for vehicle pose estimation: Experimental validation. IEEE Access **8**, 132539–132550 (2020)
45. Shi, M., Huang, Z., Ma, X., Hu, X., Cao, Z.: Matching is not enough: a two-stage framework for category-agnostic pose estimation. In: Proceedings of the IEEE/CVF Conference on Computer Vision and Pattern Recognition, pp. 7308–7317 (2023)
46. Snell, J., Swersky, K., Zemel, R.: Prototypical networks for few-shot learning. In: Advances in Neural Information Processing Systems, vol. 30 (2017)
47. Sun, K., Xiao, B., Liu, D., Wang, J.: Deep high-resolution representation learning for human pose estimation. In: Proceedings of the IEEE/CVF conference on computer vision and pattern recognition, pp. 5693–5703 (2019)
48. Sun, Y., Wang, X., Tang, X.: Deep convolutional network cascade for facial point detection. In: Proceedings of the IEEE Conference on Computer Vision and Pattern Recognition, pp. 3476–3483 (2013)
49. Tang, W., Wu, Y.: Does learning specific features for related parts help human pose estimation? In: Proceedings of the IEEE Conference on Computer Vision and Pattern Recognition, pp. 1107–1116 (2019)
50. Toshev, A., Szegedy, C.: DeepPose: human pose estimation via deep neural networks. In: Proceedings of the IEEE Conference on Computer Vision and Pattern Recognition, pp. 1653–1660 (2014)
51. Wang, J., Long, X., Gao, Y., Ding, E., Wen, S.: Graph-PCNN: two stage human pose estimation with graph pose refinement. In: Vedaldi, A., Bischof, H., Brox, T., Frahm, J.-M. (eds.) ECCV 2020. LNCS, vol. 12356, pp. 492–508. Springer, Cham (2020). https://doi.org/10.1007/978-3-030-58621-8_29
52. Wang, Y., Peng, C., Liu, Y.: Mask-pose cascaded CNN for 2D hand pose estimation from single color image. IEEE Trans. Circuits Syst. Video Technol. **29**(11), 3258–3268 (2018)
53. Welinder, P., et al.: Caltech-UCSD birds 200 (2010)
54. Wu, J., et al.: Single image 3D interpreter network. In: Leibe, B., Matas, J., Sebe, N., Welling, M. (eds.) ECCV 2016. LNCS, vol. 9910, pp. 365–382. Springer, Cham (2016). https://doi.org/10.1007/978-3-319-46466-4_22
55. Xie, S., Zhang, Z., Lin, Z., Hinz, T., Zhang, K.: SmartBrush: text and shape guided object inpainting with diffusion model. In: Proceedings of the IEEE/CVF Conference on Computer Vision and Pattern Recognition, pp. 22428–22437 (2023)
56. Xu, L., Huang, M.H., Shang, X., Yuan, Z., Sun, Y., Liu, J.: Meta compositional referring expression segmentation. In: Proceedings of the IEEE/CVF Conference on Computer Vision and Pattern Recognition, pp. 19478–19487 (2023)
57. Xu, L., et al.: Pose for everything: towards category-agnostic pose estimation. In: Avidan, S., Brostow, G., Cissé, M., Farinella, G.M., Hassner, T. (eds.) ECCV 2022. LNCS, vol. 13666, pp. 398–416. Springer, Cham (2022). https://doi.org/10.1007/978-3-031-20068-7_23
58. Xu, W., Su, P.c., Sen-ching, S.C.: Human pose estimation using two RGB-D sensors. In: 2016 IEEE International Conference on Image Processing (ICIP), pp. 1279–1283. IEEE (2016)

59. Yang, Y., Ramanan, D.: Articulated human detection with flexible mixtures of parts. IEEE Trans. Pattern Anal. Mach. Intell. **35**(12), 2878–2890 (2012)
60. Zhang, D., Guo, G., Huang, D., Han, J.: PoseFlow: a deep motion representation for understanding human behaviors in videos. In: Proceedings of the IEEE Conference on Computer Vision and Pattern Recognition, pp. 6762–6770 (2018)
61. Zhang, J., Cai, Y., Yan, S., Feng, J., et al.: Direct multi-view multi-person 3D pose estimation. In: Advances in Neural Information Processing Systems, vol. 34, pp. 13153–13164 (2021)
62. Zhang, Z., Xu, L., Peng, D., Rahmani, H., Liu, J.: Diff-tracker: text-to-image diffusion models are unsupervised trackers. In: European Conference on Computer Vision. Springer, Heidelberg (2024)
63. Zhang, Z., Zhou, C., Tu, Z.: Distilling inter-class distance for semantic segmentation. arXiv preprint arXiv:2205.03650 (2022)
64. Zhao, C., Fu, C., Dolan, J.M., Wang, J.: L-shape fitting-based vehicle pose estimation and tracking using 3D-LiDAR. IEEE Trans. Intell. Veh. **6**(4), 787–798 (2021)
65. Zheng, C., Zhu, S., Mendieta, M., Yang, T., Chen, C., Ding, Z.: 3D human pose estimation with spatial and temporal transformers. In: Proceedings of the IEEE/CVF International Conference on Computer Vision, pp. 11656–11665 (2021)
66. Zimmermann, C., Brox, T.: Learning to estimate 3D hand pose from single RGB images. In: Proceedings of the IEEE International Conference on Computer Vision, pp. 4903–4911 (2017)

SC4D: Sparse-Controlled Video-to-4D Generation and Motion Transfer

Zijie Wu[1,2], Chaohui Yu[2], Yanqin Jiang[2], Chenjie Cao[2], Fan Wang[2],
and Xiang Bai[1(✉)]

[1] Huazhong University of Science and Technology, Wuhan, China
{xbai,zjw1031}@hust.edu.cn
[2] DAMO Academy, Alibaba Group, Hangzhou, China
{huakun.ych,jiangyanqin.jyq,caochenjie.ccj,fan.w}@alibaba-inc.com
https://sc4d.github.io/

Abstract. Recent advances in 2D/3D generative models enable the generation of dynamic 3D objects from a single-view video. Existing approaches utilize score distillation sampling to form the dynamic scene as dynamic NeRF or dense 3D Gaussians. However, these methods struggle to strike a balance among reference view alignment, spatio-temporal consistency, and motion fidelity under single-view conditions due to the implicit nature of NeRF or the intricate dense Gaussian motion prediction. To address these issues, this paper proposes an efficient, sparse-controlled video-to-4D framework named SC4D, that decouples motion and appearance to achieve superior video-to-4D generation. Moreover, we introduce Adaptive Gaussian (AG) initialization and Gaussian Alignment (GA) loss to mitigate shape degeneration issue, ensuring the fidelity of the learned motion and shape. Comprehensive experimental results demonstrate that our method surpasses existing methods in both quality and efficiency. In addition, facilitated by the disentangled modeling of motion and appearance of SC4D, we devise a novel application that seamlessly transfers the learned motion onto a diverse array of 4D entities according to textual descriptions.

Keywords: Video-to-4D generation · Dynamic Gaussian splatting · Motion transfer

1 Introduction

In recent years, with the advancement of generative AI, we have witnessed significant progress in 3D generation techniques, which include the generation of static objects' shape, texture, and even an entire scene from a text prompt or a single image. Compared to static 3D assets, dynamic 3D (4D) content offers

Supplementary Information The online version contains supplementary material available at https://doi.org/10.1007/978-3-031-72624-8_21.

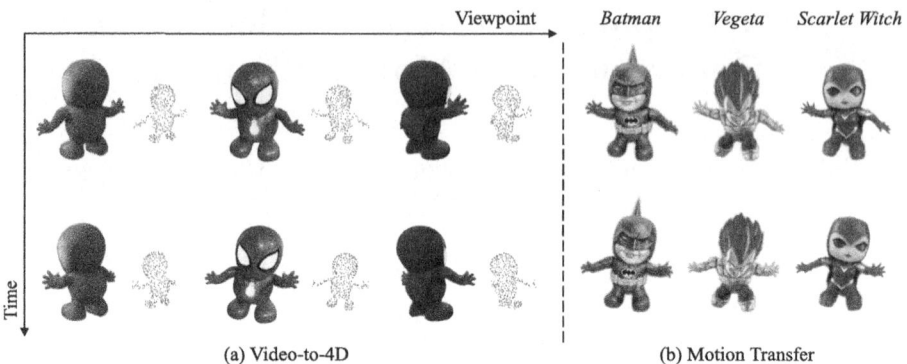

Viewpoint   *Batman*   *Vegeta*   *Scarlet Witch*

(a) Video-to-4D   (b) Motion Transfer

Fig. 1. We illustrate: (a) video-to-4D results of SC4D and corresponding control points visualizations, and (b) examples of our motion transfer applications in the figure.

greater spatio-temporal flexibility and thus harbors more substantial potential for applications in AR/VR, filming, animation, simulation, and other domains. However, generating 4D objects from text descriptions or video references is rather formidable due to the difficulties of maintaining spatio-temporal consistency and ensuring motion fidelity. Nevertheless, as humans, we are adept at resolving the above challenging tasks, a capability attributable to our possession of extensive prior knowledge of the real world (Fig. 1).

Very recently, building upon the foundations laid by extant text-to-3D [25,38, 57,60,69] and image-to-3D [26,28,41,52,54] pipelines, several methods [16,36,68] distill prior knowledge from novel view synthesis models [29,30] to generate the target 4D object as dynamic NeRF [16] or dynamic 3D Gaussians [36,68] and impose constraints to ensure the temporal consistency among frames. Despite the commendable progress achieved by these methods, they still struggle to strike a balance among reference view alignment, spatio-temporal consistency, and motion fidelity. We argue that the representation is critical for video-to-4D generation. On the one hand, dynamic NeRF [2,10,33,39] cannot sustain temporal coherence under novel views without additional restrictions due to its implicit nature and stochasticity inherent in Score Distillation Sampling (SDS) [38]. On the other hand, it is intricate to predict accurate trajectories and rotations for dense Gaussians [17,31,62,66] with only single-view conditions.

To tackle the aforementioned problems, inspired by a recent dynamic scene reconstruction method [14], we decouple the appearance and motion of the dynamic 3D object into dense Gaussians (∼50k) and sparse control points (∼512) and design an efficient two-stage video-to-4D generation framework, named **SC4D**. Specifically, in the coarse stage, we initialize a set of sparse control points as spherical Gaussians and a Multilayer Perceptron (MLP) conditioned on time and location to predict the motion of these sparse Gaussians. Then, we optimize the parameters of these Gaussians and the MLP under the guidance of reference view reconstruction and novel view score distillation. In the fine stage, we utilize the sparse control Gaussians as implicit control points and perform Linear Binding Skinning (LBS) [51] to drive the dense Gaussians. In this stage, we jointly optimize the parameters of control points, dense Gaussians, and deformation MLP to obtain the final results. Notably, since only a single-view ground truth video is provided, we empirically find that there is a proclivity for shape degeneration issues in the fine stage, which results in thickening, position displacements, and texture blur of the dynamic 3D object. To address these challenges, we devise Adaptive Gaussian (AG) initialization that inherits the shape and motion of the control Gaussians in the coarse stage with a random amount of Gaussians. Moreover, we present Gaussian Alignment (GA) loss to ensure shape and motion fidelity in the fine stage.

Benefiting from the disentangled modeling of appearance and motion within our method, SC4D effectively mitigates the ambiguity of these two attributes during optimization. Additionally, since the motion of the dynamic object is governed by a set of sparse control points rather than dense Gaussians, our approach simplifies the learning of motion and naturally exhibits enhanced local rigidity. Comprehensive evaluations reveal that our method surpasses existing video-to-4D generation methods [16,68] in both quality and efficiency. Moreover, after obtaining the motions of implicit control points, we introduce a novel application that transfers the learned motion onto other entities under the guidance of text-to-image models [44,70] and dense Gaussians' depth. In conclusion, our contributions can be summarized as follows:

1. We propose SC4D, a sophisticated video-to-4D generation framework based on sparse points control, which generates dynamic 3D objects with superior quality and efficiency compared to existing methods.
2. We devise an Adaptive Gaussian (AG) initialization approach and Gaussian Alignment (GA) loss that effectively mitigate the shape degeneration issue, ensuring accurate motion and shape learning.
3. We propose a novel application based on the control points' motion and design a motion transfer pipeline that maps the learned motion onto distinct entities, as directed by textual descriptions.

2   Related Works

Optimization-Based 3D Generation. Optimization-based 3D generation methods typically optimize a NeRF [33] or 3D Gaussians [17] utilizing prior knowledge from image-text matching model [42] or diffusion-based generative models [13,29,44,45]. CLIP-based text-to-3D methods [15,19,34,65] generally employ CLIP [42] to align each viewpoint of the target 3D scene with the given text description for optimization. DreamFusion [38] substitutes the guidance model from CLIP to a 2D diffusion model [45], and introduces Score Distillation Sampling (SDS) to distill prior knowledge from text-to-image models. As a concurrent work, SJC [57] shares a similar idea, which distills scores in a Perturb-and-Average manner. Following the paradigm of SDS, a series of methods aim at bringing finer texture details [4,25,60] or alleviating the Janus problem [46,69] utilizing DMTet [47] representation, Variational Score Distillation [60], Point-E [35] condition, etc. Recently, a succession of methods further enhanced the view quality [21,74] and multi-view consistency [48]. There are also several 3D Gaussian-based methods [5,53,67] that achieve comparable results. As for image-to-3D, RealFusion [32] performs textual inversion [11] before 3D generation to match the intended concept. Make-it-3D [54] improves the view quality and consistency in an inpainting manner. Zero123 [29] utilizes large-scale multi-view data from Objaverse dataset [7,8] to turn Stable Diffusion (SD) [44] into a novel-view generator. Based on Zero123, a bunch of methods [28,41,52] achieves high-fidelity image-to-3D generation. There are also several approaches [20,26,30,58] acquire notable progress in enhancing multi-view consistency.

4D Representation. Current 4D representations predominantly bifurcate into two principal categories: dynamic NeRF [33] and dynamic 3D Gaussians [17]. Dynamic NeRF-based methods can be further divided into deformable NeRF [37,39,56,63] and time-varying NeRF [2,9,10,12,23]. Recently, a variety of methods predicated on dynamic 3D Gaussians [3,22,24,31,62,66,66] have emerged, leveraging Gaussians' explicit nature and real-time rendering capabilities. There are also some methods that ameliorate the dense motion prediction by learning a set of sparse trajectories [18], control points [14], or basis vectors [6].

4D Generation. Compared to 3D generation, high-quality 4D generation is even more challenging since the temporal dimension is involved. Existing text-to-4D methods [1,27,43,50,59,72,73] distill geometry and temporal information from diffusion-based text-to-image models [44] and text-to-video models [49] with SDS [38]. In recent developments, several video-to-4D frameworks [16,36,68] have been introduced. These pipelines endeavor to recover the dynamic 3D objects from single-view video inputs, facilitated by Zero123 [29]. However, these video-to-4D methods struggle to strike a balance among reference view alignment, spatio-temporal consistency, and motion fidelity.

3   Method

Given a single-view reference video, video-to-4D methods [16,36,68] aim to recover a plausible dynamic 3D object that aligns with the video source. Inspired by [14], we propose an efficient video-to-4D framework based on sparse control points (shown in Fig. 2), named **SC4D**, which utilizes separated modeling of appearance and motion to yield superior outcomes. To ensure the fidelity of learned shape and motion, we introduce Adaptive Gaussian (AG) initialization based on control points, and Gaussian Alignment (GA) loss as an additional constraint. Furthermore, we devise a novel application that enables motion transfer with text descriptions after acquiring the control point motions.

3.1   Preliminaries

Score Distilltion Sampling. Score Distillation Sampling (SDS) [38] is widely adopted to distill prior knowledge from image generation models [29,44]. In this work, we utilize Zero123 [29] as the source of novel view information. Given a reference image I_r, a relative camera extrinsic (R, T) between the queried and input views, the 3D model as θ, Zero123 as ϕ, then the SDS loss is as follows:

$$\bigtriangledown_\theta L_{SDS}(\phi, x) = \mathbb{E}_{t, \epsilon}[\omega(t)(\hat{\epsilon}_\phi(z_t; I_r, R, T, t) - \epsilon)\frac{\partial x}{\partial \theta}], \tag{1}$$

where t is the randomly sampled timestep in the diffusion process, x is the rendered image, and $\omega(t)$ is a weighting function depending on the timestep t.

3D Gaussian Splatting. 3D Gaussian Splatting (3DGS) [17] represents a scene as a set of explicit 3D Gaussians. Each Gaussian G has a center position μ, a rotation quaternion q, a scaling parameter s, opacity σ and sphere harmonic (SH) coefficients sh. It can be defined as $G(x) = e^{-\frac{1}{2}(x-\mu)^T \Sigma^{-1}(x-\mu)}$, where Σ is the 3D covariance matrix, calculated by $\Sigma = RSS^T R^T$ (R, S is equivalent to q, s). The color of a pixel u is rendered using α-blending:

$$Color(u) = \sum_i SH(sh_i, v_i)\alpha_i \prod_{j=1}^{i-1}(1 - \alpha_j), \tag{2}$$

where $\alpha_i = \sigma_i G(u)$, v_i is the view direction, and SH is the spherical harmonic function. To enhance the accuracy of fitting across diverse scenes, densification and pruning are adopted based on gradient accumulation during optimization.

Sparse-Controlled Gaussian Splatting (SC-GS). SC-GS [14] is an effective dynamic scene reconstruction method, which decouples the appearance and motion of a 4D scene as 3D Gaussians and control points. SC-GS utilizes a time-condition MLP Ψ to predict the translation T_i^t and rotation R_i^t for each control point. Then, 3D Gaussians are driven by control points following LBS [51]. For each Gaussian G_j, the warped location μ_j^t and rotation q_j^t can be computed as a weighted sum of its KNN control points M_j:

$$\mu_j^t = \sum_{k \in M_j} \mathrm{w}_{jk}(R_k^t(\mu_j - p_k) + p_k + T_k^t), \tag{3}$$

$$q_j^t = (\sum_{k \in M_j} \mathrm{w}_{jk} r_k^t) \otimes q_j, \tag{4}$$

where w_{jk} is a weighting ratio depending on the distance d_{jk} between Gaussian G_j center and its neighboring control point p_k, and a learned control radius o_k. p_k, r_k^t denote the position and rotation quaternion for the control point.

3.2   Coarse Stage: Sparse Control Points Initialization

As illustrated in SC-GS [14], sparse control points initialization is critical for decoupling motion and appearance of the dynamic object/scene. Joint optimization of sparse control points and dense Gaussians directly can result in uneven distribution of control points and may even lead to training collapse. In this

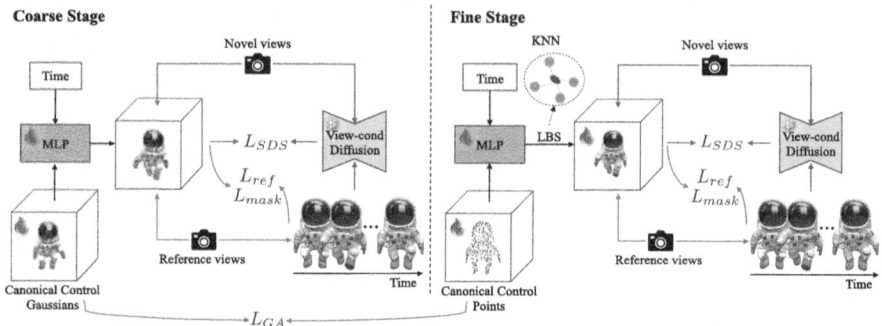

Fig. 2. Overall pipeline of the proposed SC4D. In the coarse stage, SC4D learns a proper shape and motion initialization with a set of sparse control Gaussians. Then, in the fine stage, we propose Adaptive Gaussian (AG) initialization, and Gaussian Alignment (GA) loss to prevent shape and motion degeneration, and jointly optimize control points, dense Gaussians, and deformation MLP for the final results.

stage, we aim to obtain a good initialization for sparse control points' locations and motion that align with the reference video.

As shown in Fig. 2, since no ground truth 3D data is available, we initialize M control points (denoted as control Gaussians in this stage) as 3D Gaussians with randomly sampled positions inside a sphere. In order to obtain control Gaussians that are more evenly distributed, we constrain them as spheres with the same scaling parameter (s). We denote the control Gaussians as $C_i : (p_i, r_i, s, \sigma_i, sh_i)$, and the reference image sequence as $\{I_r^f, f = 1, 2, \cdots, F\}$, where F stands for the number of frames of the reference video. Then, for a randomly sampled timestep $t = \frac{f-1}{F}$, we predict the control Gaussians' movement using MLP $\mathbf{\Psi}$. To be noted, since the control Gaussians are spherical, we only need to compute their new position as follows:

$$p_i^t = p_i + T_i^t, \tag{5}$$

where p_i is the position of C_i in the canonical space. After obtaining the deformed object at timestep t, we project it from the reference view to get $\hat{I}^t$ following Eq. (2), and compute the reconstruction loss as:

$$L_{ref} = \left\| \hat{I}^t - I_r^f \right\|_2^2. \tag{6}$$

To better leverage the information from the reference images, we additionally introduce a mask loss term:

$$L_{mask} = \left\| \alpha^t - \mathbf{M}_r^f \right\|_2^2, \tag{7}$$

where $\mathbf{M}_r$ is the foreground mask of the reference image, and α^t is the accumulated opacity obtained during rendering. As for novel view optimization, we sample B viewpoints randomly within the pitch angle range of $-ver$ to ver

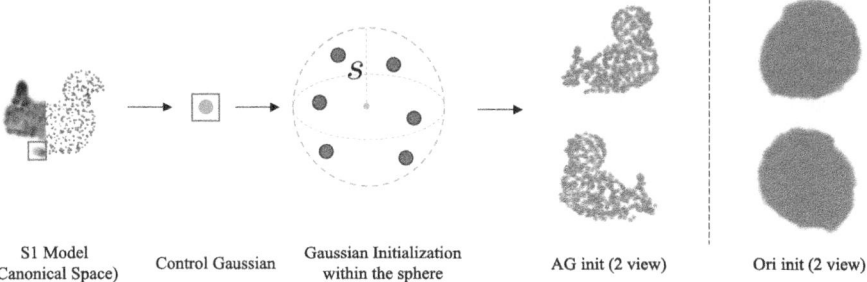

S1 Model   Control Gaussian   Gaussian Initialization   AG init (2 view)   Ori init (2 view)
(Canonical Space)   within the sphere

Fig. 3. Illustration of Adaptive Gaussian (AG) initialization. s is the scaling parameter of control Gaussians in the coarse stage. *Ori init* represents randomly initializing all the dense Gaussians within a sphere in the canonical space.

degrees and the yaw angle range of -180 to $180°$, and compute the average SDS loss following Eq. (1). The overall objective in this stage is a weighted combination of the above three losses:

$$L_{total} = \lambda_{ref}L_{ref} + \lambda_{mask}L_{mask} + \lambda_{SDS}L_{SDS}. \qquad (8)$$

In this stage, we perform densification and pruning following 3DGS [17] in the first n iterations. Then we sample M (same amount as initialization) control Gaussians utilizing Farthest Point Sampling (FPS) [40] and continue training without densification in the remaining procedure.

3.3 Fine Stage: Dense Gaussians Optimization

After the coarse stage, we get a reasonable estimation for the dynamic 3D object's motion and shape. Then in the fine stage, we aim to optimize the texture details to match the source video and to further refine the motion and shape for better fidelity. To be noted, the explicit control Gaussians in the coarse stage have transitioned to implicit control points in this stage, denoted as $C_i : (p_i, o_i)$, where o_i is the control radius of each control point, initialized with the scaling parameter s of control Gaussians in the coarse stage.

Adaptive Gaussian Initialization. In this stage, dense Gaussians are driven by neighboring sparse control points, as illustrated in Eq. (3), (4). We empirically find that the dense Gaussian initialization significantly impacts the quality of the final result. One straightforward initialization approach is to initialize the dense Gaussians uniformly within a sphere in the canonical space as in the coarse stage. However, we observe that this initialization fails to generate promising results. The target object is prone to increased thickness, and issues such as diminished texture and positional drift may arise (as shown in Fig. 7).

Instead, we propose Adaptive Gaussian (AG) initialization based on the learned control Gaussians. As shown in Fig. 3, we have learned M control Gaussians in the coarse stage with the same scaling parameter s. Then, for each control Gaussian, we consider it as a sphere with a radius of s, and randomly initialize K Gaussians within it following DreamGaussian [53]. In total, we get $N = M \times K$ Gaussians as initialization. As shown in Fig. 3, our designed initialization approach successfully inherits the shape and motion modeled in the coarse stage. The dense Gaussians are distributed near the surface of the object, which facilitates the subsequent optimization. In comparison, if directly optimizing all the dense Gaussians uniformly within a sphere in the canonical space, the deformed shape misaligns with that in the coarse stage.

Gaussian Alignment Loss. Even with a good dense Gaussian initialization, the shape and motion of the dynamic 3D object are still prone to degeneration in the later phases of training. The main reason is that: when employing Score Distillation Sampling (SDS) [38] to distill prior knowledge of novel views from Zero123 [29], a larger noise timestep biases SDS towards ensuring the plausibility of the overall shape. Conversely, with a smaller timestep, SDS tends to focus more on optimizing textures, while its capability to preserve shape diminishes.

To allow refining texture without degrading motion and shape, we propose the Gaussian Alignment (GA) loss as an additional constraint. At the beginning of this stage, we preserve the control Gaussians' parameters and the deformation MLP to query the initial positions (denoted as $\overline{p}^t$) of those control points at random timestep t. Then, we compute the Gaussian Alignment loss as:

$$L_{GA} = \left\| p^t - \overline{p}^t \right\|_2^2, \tag{9}$$

where p^t denotes the position of the current control point at timestep t. Although the proposed GA loss is quite simple, it can effectively mitigate the shape degeneration issue encountered during the latter training procedure. The Chamfer loss is another commonly used metric for constraining the distances between point clouds. However, compared to GA loss, we observe that the Chamfer loss can sometimes result in the current control points aggregating towards certain target points, thereby compromising the uniform distribution of the control points (See Sect. 4.4 of *Supp.*[1] for the comparison of GA loss and the Chamfer loss).

In this stage, the overall training objective is formulated as follows:

$$L_{total} = \lambda_{ref} L_{ref} + \lambda_{mask} L_{mask} + \lambda_{SDS} L_{SDS} + \lambda_{GA} L_{GA}. \tag{10}$$

3.4   Motion Transfer Application

Utilizing our video-to-4D framework, we can successfully extract the resultant dynamic 3D object as well as its motion represented by a set of moving control points. Leveraging this capability, we devise an application tailored for motion transfer predicated on the trajectories of these sparse control points. This application aims to synthesize dynamic objects of distinct entities that exhibit identical motion patterns, all instantiated through text descriptions.

---

[1] *Supp.*: supplementary file.

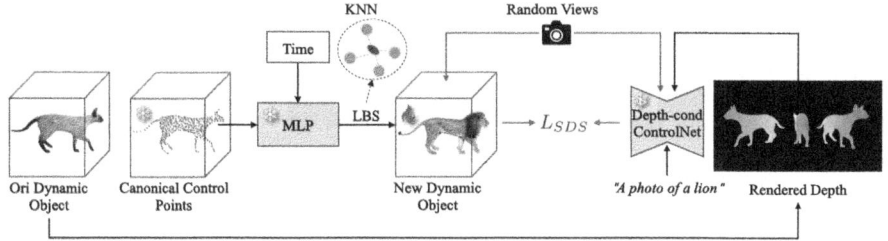

Fig. 4. Illustration of the pipeline for our motion transfer application.

As shown in Fig. 4, we employ the same initialization method as outlined in Sect. 3.3. During training, we fix the parameters of the learned control points as well as those of the deformation MLP, enabling us to preserve the motion of the target dynamic object and acquire its appearance from any viewpoint at any given time. To prevent the degeneration of motion and shape due to the displacement of dense Gaussians away from the control points during optimization, we elect to use a depth-condition ControlNet [70] as supervision, and query the depth from the dynamic object resultant from the video-to-4D pipeline. Since no ground truth reference is involved, we only utilize SDS [38] loss for optimization. We find that a small guidance scale can lead to plausible shape changes without degenerating the learned motion. Therefore, we first utilize a reduced guidance scale during the initial phase of the training regimen to capture the shape of the new entity that is congruent with the textual input, and its blurring appearance. Then, we save the intermediate dense Gaussians and employ the updated depth information as new conditions. In the remaining training period, we increase the guidance scale and decrease the noise scale of SDS to refine the texture, culminating in the final outcome. Please refer to Sect. 4.1 for more details.

4 Experiments

4.1 Experimental Settings

Implementation Details. We employ an MLP with a similar structure utilized in SC-GS [14] to predict the motion of control points. The MLP receives the positional embeddings [33] of time and control points' positions as input, with frequencies of 6 and 10. In the coarse stage, we initialize $M = 512$ control Gaussians uniformly within a sphere with the same scaling parameter. We perform densification and pruning following 3DGS [17] for the first 1,000 iterations with an interval of 100. Then we perform Farthest Point Sampling (FPS) [40] to sample M control Gaussians and train for another 500 iterations without densification. In the fine stage, we optimize all the parameters of control points, MLP, and dense Gaussians together. During training, we sample 4 random camera poses and the fixed pose of time t at a fixed radius of 2, with the azimuth in $[-180, 180]$ degrees and elevation in $[-ver, ver]$ degrees, where $ver = 30$.

We set the loss weight of $\lambda_{ref}, \lambda_{mask}, \lambda_{SDS}, \lambda_{GA}$ to $5000.0, 500.0, 1.0, 10000.0$ by default. As for our motion transfer application, during the initial 1,000 iterations, it is optimized with a guidance scale of 7.5 using the dynamic object's depth obtained from the video-to-4D pipeline as a condition. Subsequently, the depth from the intermediate result is utilized as the new condition, and we continue to optimize the remaining 1,000 iterations with a guidance scale of 30.0. We will release the source code later. Please refer to Sect. 1 of *Supp.* for more details.

Dataset. For fair comparisons, we utilize the dataset from Consistent4D [16] for evaluations. The dataset consists of 12 synthetic and 12 in-the-wild videos, all captured with a stationary camera oriented perpendicularly to the dynamic objects. Each video has 32 frames and spans around 2 s.

Evaluation Metrics. We have summarized the criteria for evaluating the quality of video-to-4D generation into three main aspects: alignment with the reference video, spatio-temporal consistency, and motion fidelity. Following NeRF [33], we utilize PSNR and SSIM [61] as measurements of the reference view alignment. We also add LPIPS [71] as a view quality metric, which is akin to Consistent4D [16]. As for multi-view (spatial) consistency, we adopt the commonly used CLIP [42] score to measure the visual similarity of two different renderings. To evaluate the temporal consistency and motion fidelity of the generated dynamic object, we follow [64] to utilize RAFT [55] to compute the optical flows of the reference image sequence, and warp the reference-view projections to calculate the temporal error (Temp), which can effectively reflect the temporal consistency under the reference view and motion accuracy (check Sect. 1 of *Supp.* for details of the temporal error metric). We also include human evaluation, which is more representative in generation tasks. To do so, we randomly choose 10 videos to train the compared methods and render the dynamic objects from the reference view and a random novel view. Then, we invite 20 participants to select their preferred reference and novel view videos based on the reference view alignment, spatio-temporal consistency, and motion fidelity. Overall, we get $10 \times 20 = 200$ votes for reference view and novel view, respectively. Then, we calculate the percentage of votes as a measurement for user preference, as shown in Fig. 6.

4.2   Comparisons

To evaluate the effectiveness of the proposed video-to-4D framework, we compare our method with the only two open-source methods: Consistent4D [16] and 4DGen [68]. We train the dataset provided by Consistent4D with these methods and conduct comprehensive evaluations of the outcomes. All the results of Consistent4D and 4DGen are obtained using their official code and settings. Qualitative and quantitative comparisons are shown in Fig. 5 and Table 5.

Qualitative Comparison. As demonstrated in Fig. 5, we show four instances (*skull, triceratops, elephant, egret*) and the results generated by the compared

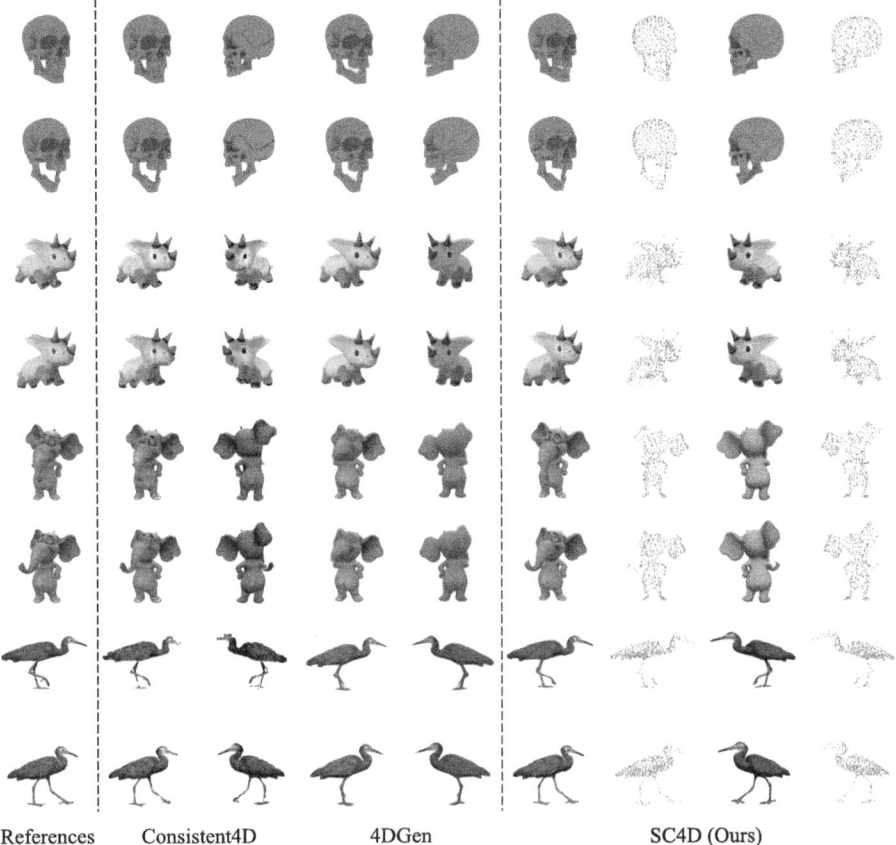

Fig. 5. Qualitative Comparisons. We compare our method with Consistent4D [16] and 4DGen [68]. For each instance, we render two viewpoints at two timesteps. We also visualize the sparse control points to show their correspondence with dense Gaussians.

methods. Consistent4D [16] generates results with artifacts and color distortions in some cases (*triceratops* and *egret*), and there is also a noticeable temporal discontinuity between frames of the same viewpoint, which is attributable to the limited capacity of low-resolution NeRF [33] and the diversity of time-varying NeRF under single view conditions. As for 4DGen [68], the generated dynamic objects often exhibit minor motion and present discernible discrepancies from the reference video (*skull* and *triceratops*). Regions with large motion are subject to blurring or may even experience loss of detail (nose of *elephant*, claw of *egret*). In comparison, the results generated by our method surpass the compared methods in terms of alignment with the reference video, spatio-temporal consistency, and motion fidelity. This is also evidenced by the visualized sparse control points, from which it can be discerned that our method has learned a set of evenly

Table 1. Quantitative comparison of different methods. Temp represents the temporal error introduced in Sect. 4.1.

Methods	PSNR ↑	SSIM ↑	LPIPS ↓	CLIP ↑	Temp ↓	Training time ↓
Consistent4D [16]	23.97	0.91	0.09	0.89	0.0089	1.9 h
4DGen [68]	21.80	0.90	0.10	0.87	0.0089	3.0 h
SC4D (Ours)	**29.50**	**0.95**	**0.08**	**0.90**	**0.0081**	**1.0 h**

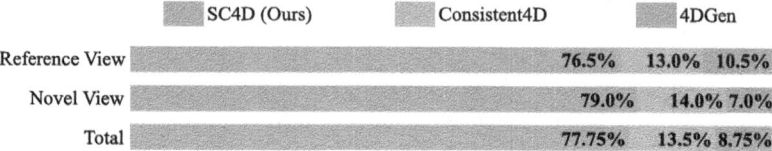

SC4D (Ours)   Consistent4D   4DGen

Reference View   76.5%   13.0%   10.5%
Novel View   79.0%   14.0% 7.0%
Total   77.75%   13.5% 8.75%

Fig. 6. User preference for video-to-4D generation methods.

distributed control points that accurately capture the dynamics and shape of the subject. Please check Sect. 5 of *Supp.* for more qualitative comparisons.

Quantitative Comparison. To quantitatively evaluate the performance of the compared methods, we randomly choose 10 reference videos from the dataset provided by Consistent4D [16], and optimize the dynamic objects using the compared methods, respectively. After optimization, we render one input view and four testing views (azimuth degrees: [72, 144, 216, 288], elevation degree: [0]) for every timestep of the reference video to calculate the metrics mentioned in Sect. 4.1. As reported in Table 1, our method outperforms Consistent4D [16] and 4DGen [68] in reference view alignment (PSNR, SSIM, LPIPS), multi-view consistency (CLIP), temporal consistency and motion fidelity (Temp), revealing the effectiveness of SC4D. As depicted in Fig. 6, user study also reveals that our proposed SC4D is more favored in human evaluations, in which the novel view preference metric also reveals the remarkable spatio-temporal consistency and motion fidelity of SC4D. Besides, our method exhibits a significant advantage in terms of training duration. All experiments are conducted on a single Tesla V100 GPU with 32 GB of graphics memory. We also utilize the test set and metrics in Consistent4D [16] to evaluate the compared methods. The results align with the conclusion drawn above. We attach them in Sect. 3 of *Supp.*.

4.3   Ablation Studies

As mentioned in Sect. 1, we propose the Adaptive Gaussian (AG) initialization and Gaussian Alignment (GA) loss to alleviate the shape degeneration issue in the fine stage. As shown in Fig. 7, without GA loss and AG initialization, the generated object is prone to suffer from over-thickness and texture blurring problems. Besides, the control points scatter, which indicates that the learned shape and motion in the coarse stage also degenerate significantly. When training

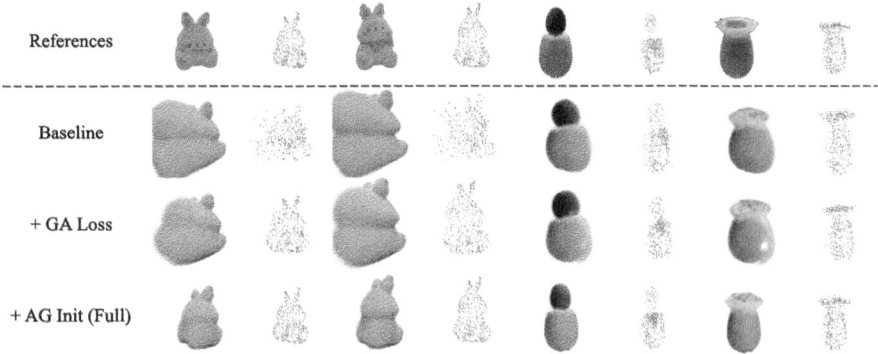

Fig. 7. Ablation studies of the proposed GA loss and AG initialization. In the first row, we show the source frames and control points from the coarse stage as references.

with GA loss, the general shape of control points can be preserved, yet there can be observable instances of increased thickness. Additionally, when a viewpoint is excessively close to the camera, a corresponding attenuation in texture detail is also noticeable. Upon utilizing AG initialization, our method has experienced a pronounced enhancement in its final output. The dynamic objects generated exhibit tangible advancements in both motion fidelity and shape plausibility. Moreover, the textural details have also undergone considerable refinement.

Table 2. Quantitative evaluations of the proposed GA loss and AG initialization.

Methods	PSNR ↑	SSIM ↑	LPIPS ↓	CLIP ↑	Temp ↓
Baseline	29.81	0.95	0.10	0.82	**0.0016**
+ GA Loss	30.19	0.96	0.09	0.83	**0.0016**
+ AG Init (Full)	**31.35**	**0.96**	**0.08**	**0.89**	**0.0016**

Table 2 also illustrates the effectiveness of our proposed techniques. The most noticeable improvement in metrics is observed in the CLIP score, indicating a substantial enhancement in the quality of the generated novel views. This further reveals the effectiveness of the proposed GA loss and AG initialization in mitigating the challenges associated with shape degeneration. Please refer to Sect. 4 of *Supp.* for more ablation studies.

4.4   Application Results

As mentioned in Sect. 3.4, we design an innovative application that is capable of flexibly generating diverse 4D entities with the same motion, based on control points learned during our video-to-4D pipeline, in conjunction with a depth-condition ControlNet [70] that operates upon textual descriptions.

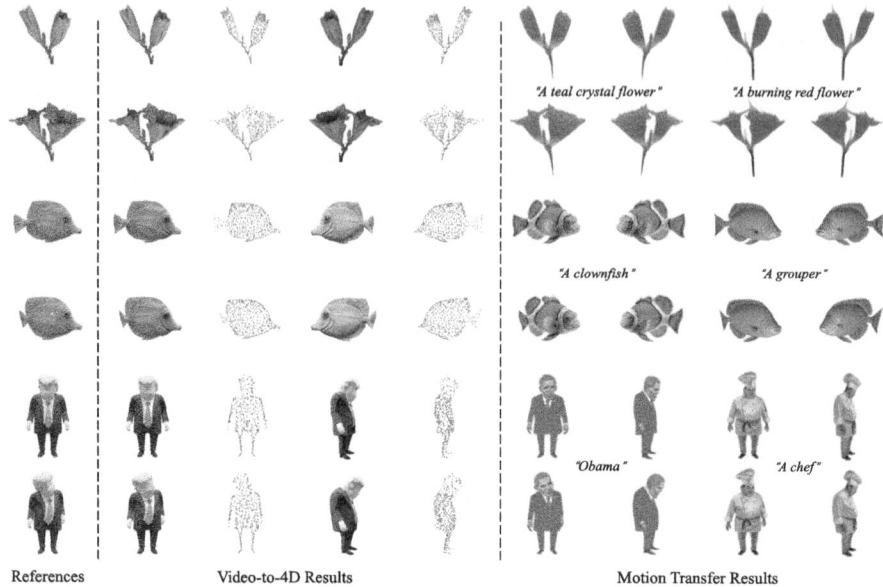

References   Video-to-4D Results   Motion Transfer Results

Fig. 8. Motion transfer application of SC4D. The prompt is attached between lines.

In Fig. 8, we sequentially present: the reference frames, video-to-4D results generated using SC4D and corresponding control points visualizations, and the new entities with identical motion synthesized based on text descriptions. From the figure, it is observable that the newly generated 4D objects possess vivid textures and align well with the motion patterns of the reference video. Moreover, the synthesized dynamic objects exhibit appreciable variations in shape, yet the motion remains coherent, which further illustrates the robustness and flexibility of our method. Please refer to Sect. 6 of *Supp.* for more application examples.

5   Limitations

The main limitations of our method are twofold: First, our approach relies on models such as Zero123 [29] to provide novel view information, and the capability of these viewpoint synthesis models is still limited, often underperforming on many complex objects. Second, similar to existing video-to-4D methods [16, 36, 68], we have not taken into account the 4D generation under moving camera scenarios, which will be one of the directions for our future research.

6   Conclusion

In this work, we propose a sophisticated video-to-4D pipeline, named SC4D, which decouples the motion and appearance of dynamic objects as sparse control points and dense 3D Gaussians. SC4D excels over contemporary approaches in generating dynamic objects with better reference view alignment, spatio-temporal consistency, and motion fidelity. Moreover, we craft an innovative application utilizing the control points learned by SC4D, which allows seamless motion transfer onto new entities, as directed by textual descriptions.

Acknowledgements. This work was supported by the National Science Fund for Distinguished Young Scholars of China (Grant No. 62225603).

References

1. Bahmani, S., et al.: 4D-fy: text-to-4D generation using hybrid score distillation sampling. arXiv preprint arXiv:2311.17984 (2023)
2. Cao, A., Johnson, J.: HexPlane: a fast representation for dynamic scenes. In: Proceedings of the IEEE/CVF Conference on Computer Vision and Pattern Recognition, pp. 130–141 (2023)
3. Chen, G., Wang, W.: A survey on 3D gaussian splatting. arXiv preprint arXiv:2401.03890 (2024)
4. Chen, R., Chen, Y., Jiao, N., Jia, K.: Fantasia3D: disentangling geometry and appearance for high-quality text-to-3D content creation. arXiv preprint arXiv:2303.13873 (2023)
5. Chen, Z., Wang, F., Liu, H.: Text-to-3D using gaussian splatting. arXiv preprint arXiv:2309.16585 (2023)
6. Das, D., Wewer, C., Yunus, R., Ilg, E., Lenssen, J.E.: Neural parametric Gaussians for monocular non-rigid object reconstruction. arXiv preprint arXiv:2312.01196 (2023)
7. Deitke, M., et al.: Objaverse-XL: a universe of 10M+ 3D objects. In: Advances in Neural Information Processing Systems, vol. 36 (2024)
8. Deitke, M., et al.: Objaverse: a universe of annotated 3D objects. In: Proceedings of the IEEE/CVF Conference on Computer Vision and Pattern Recognition, pp. 13142–13153 (2023)
9. Fang, J., et al.: Fast dynamic radiance fields with time-aware neural voxels. In: SIGGRAPH Asia 2022 Conference Papers, pp. 1–9 (2022)
10. Fridovich-Keil, S., Meanti, G., Warburg, F.R., Recht, B., Kanazawa, A.: K-planes: explicit radiance fields in space, time, and appearance. In: Proceedings of the IEEE/CVF Conference on Computer Vision and Pattern Recognition, pp. 12479–12488 (2023)
11. Gal, R., et al.: An image is worth one word: personalizing text-to-image generation using textual inversion. arXiv preprint arXiv:2208.01618 (2022)
12. Gao, C., Saraf, A., Kopf, J., Huang, J.B.: Dynamic view synthesis from dynamic monocular video. In: Proceedings of the IEEE/CVF International Conference on Computer Vision, pp. 5712–5721 (2021)
13. Ho, J., Jain, A., Abbeel, P.: Denoising diffusion probabilistic models. In: Advances in Neural Information Processing Systems, vol. 33, pp. 6840–6851 (2020)
14. Huang, Y.H., Sun, Y.T., Yang, Z., Lyu, X., Cao, Y.P., Qi, X.: SC-GS: sparse-controlled gaussian splatting for editable dynamic scenes. arXiv preprint arXiv:2312.14937 (2023)
15. Jain, A., Mildenhall, B., Barron, J.T., Abbeel, P., Poole, B.: Zero-shot text-guided object generation with dream fields. In: Proceedings of the IEEE/CVF Conference on Computer Vision and Pattern Recognition, pp. 867–876 (2022)

16. Jiang, Y., Zhang, L., Gao, J., Hu, W., Yao, Y.: Consistent4D: D360 {\deg} dynamic object generation from monocular video. arXiv preprint arXiv:2311.02848 (2023)
17. Kerbl, B., Kopanas, G., Leimkühler, T., Drettakis, G.: 3D gaussian splatting for real-time radiance field rendering. ACM Trans. Graph. **42**(4) (2023)
18. Kratimenos, A., Lei, J., Daniilidis, K.: DynMF: neural motion factorization for real-time dynamic view synthesis with 3D gaussian splatting. arXiv preprint arXiv:2312.00112 (2023)
19. Lee, H.H., Chang, A.X.: Understanding pure CLIP guidance for voxel grid nerf models. arXiv preprint arXiv:2209.15172 (2022)
20. Li, J., et al.: Instant3D: fast text-to-3D with sparse-view generation and large reconstruction model. arXiv preprint arXiv:2311.06214 (2023)
21. Li, W., Chen, R., Chen, X., Tan, P.: SweetDreamer: aligning geometric priors in 2D diffusion for consistent text-to-3D. arXiv preprint arXiv:2310.02596 (2023)
22. Li, Z., Chen, Z., Li, Z., Xu, Y.: Spacetime gaussian feature splatting for real-time dynamic view synthesis. arXiv preprint arXiv:2312.16812 (2023)
23. Li, Z., Niklaus, S., Snavely, N., Wang, O.: Neural scene flow fields for space-time view synthesis of dynamic scenes. In: Proceedings of the IEEE/CVF Conference on Computer Vision and Pattern Recognition, pp. 6498–6508 (2021)
24. Liang, Y., et al.: GauFRe: Gaussian deformation fields for real-time dynamic novel view synthesis. arXiv preprint arXiv:2312.11458 (2023)
25. Lin, C.H., et al.: Magic3D: high-resolution text-to-3D content creation. In: Proceedings of the IEEE/CVF Conference on Computer Vision and Pattern Recognition, pp. 300–309 (2023)
26. Lin, Y., Han, H., Gong, C., Xu, Z., Zhang, Y., Li, X.: Consistent123: one image to highly consistent 3D asset using case-aware diffusion priors. arXiv preprint arXiv:2309.17261 (2023)
27. Ling, H., Kim, S.W., Torralba, A., Fidler, S., Kreis, K.: Align your gaussians: text-to-4D with dynamic 3D gaussians and composed diffusion models. arXiv preprint arXiv:2312.13763 (2023)
28. Liu, M., et al.: One-2-3-45: any single image to 3D mesh in 45 seconds without per-shape optimization. In: Advances in Neural Information Processing Systems, vol. 36 (2024)
29. Liu, R., Wu, R., Van Hoorick, B., Tokmakov, P., Zakharov, S., Vondrick, C.: Zero-1-to-3: zero-shot one image to 3D object. In: Proceedings of the IEEE/CVF International Conference on Computer Vision, pp. 9298–9309 (2023)
30. Liu, Y., et al.: SyncDreamer: generating multiview-consistent images from a single-view image. arXiv preprint arXiv:2309.03453 (2023)
31. Luiten, J., Kopanas, G., Leibe, B., Ramanan, D.: Dynamic 3D gaussians: tracking by persistent dynamic view synthesis. arXiv preprint arXiv:2308.09713 (2023)
32. Melas-Kyriazi, L., Laina, I., Rupprecht, C., Vedaldi, A.: RealFusion: 360deg reconstruction of any object from a single image. In: Proceedings of the IEEE/CVF Conference on Computer Vision and Pattern Recognition, pp. 8446–8455 (2023)
33. Mildenhall, B., Srinivasan, P.P., Tancik, M., Barron, J.T., Ramamoorthi, R., Ng, R.: NeRF: representing scenes as neural radiance fields for view synthesis. Commun. ACM **65**(1), 99–106 (2021)
34. Mohammad Khalid, N., Xie, T., Belilovsky, E., Popa, T.: CLIP-mesh: generating textured meshes from text using pretrained image-text models. In: SIGGRAPH Asia 2022 Conference Papers, pp. 1–8 (2022)
35. Nichol, A., Jun, H., Dhariwal, P., Mishkin, P., Chen, M.: Point-E: a system for generating 3D point clouds from complex prompts. arXiv preprint arXiv:2212.08751 (2022)

36. Pan, Z., Yang, Z., Zhu, X., Zhang, L.: Fast dynamic 3D object generation from a single-view video. arXiv preprint arXiv:2401.08742 (2024)
37. Park, K., et al.: NeRFies: deformable neural radiance fields. In: Proceedings of the IEEE/CVF International Conference on Computer Vision, pp. 5865–5874 (2021)
38. Poole, B., Jain, A., Barron, J.T., Mildenhall, B.: DreamFusion: text-to-3D using 2D diffusion. arXiv preprint arXiv:2209.14988 (2022)
39. Pumarola, A., Corona, E., Pons-Moll, G., Moreno-Noguer, F.: D-NeRF: neural radiance fields for dynamic scenes. In: Proceedings of the IEEE/CVF Conference on Computer Vision and Pattern Recognition, pp. 10318–10327 (2021)
40. Qi, C.R., Yi, L., Su, H., Guibas, L.J.: PointNet++: deep hierarchical feature learning on point sets in a metric space. In: Advances in Neural Information Processing Systems, vol. 30 (2017)
41. Qian, G., et al.: Magic123: one image to high-quality 3D object generation using both 2D and 3D diffusion priors. arXiv preprint arXiv:2306.17843 (2023)
42. Radford, A., et al.: Learning transferable visual models from natural language supervision. In: International Conference on Machine Learning, pp. 8748–8763. PMLR (2021)
43. Ren, J., et al.: DreamGaussian4D: generative 4D gaussian splatting. arXiv preprint arXiv:2312.17142 (2023)
44. Rombach, R., Blattmann, A., Lorenz, D., Esser, P., Ommer, B.: High-resolution image synthesis with latent diffusion models. In: Proceedings of the IEEE/CVF Conference on Computer Vision and Pattern Recognition, pp. 10684–10695 (2022)
45. Saharia, C., et al.: Photorealistic text-to-image diffusion models with deep language understanding. In: Advances in Neural Information Processing Systems, vol. 35, pp. 36479–36494 (2022)
46. Seo, J., et al.: Let 2D diffusion model know 3D-consistency for robust text-to-3D generation. arXiv preprint arXiv:2303.07937 (2023)
47. Shen, T., Gao, J., Yin, K., Liu, M.Y., Fidler, S.: Deep marching tetrahedra: a hybrid representation for high-resolution 3d shape synthesis. In: Advances in Neural Information Processing Systems, vol. 34, pp. 6087–6101 (2021)
48. Shi, Y., Wang, P., Ye, J., Long, M., Li, K., Yang, X.: MVDream: multi-view diffusion for 3D generation. arXiv preprint arXiv:2308.16512 (2023)
49. Singer, U., et al.: Make-a-video: text-to-video generation without text-video data. arXiv preprint arXiv:2209.14792 (2022)
50. Singer, U., et al.: Text-to-4D dynamic scene generation. arXiv preprint arXiv:2301.11280 (2023)
51. Sumner, R.W., Schmid, J., Pauly, M.: Embedded deformation for shape manipulation. In: ACM SIGGRAPH 2007 Papers, pp. 80–es (2007)
52. Sun, J., et al.: DreamCraft3D: hierarchical 3D generation with bootstrapped diffusion prior. arXiv preprint arXiv:2310.16818 (2023)
53. Tang, J., Ren, J., Zhou, H., Liu, Z., Zeng, G.: DreamGaussian: generative gaussian splatting for efficient 3D content creation. arXiv preprint arXiv:2309.16653 (2023)
54. Tang, J., et al.: Make-it-3D: high-fidelity 3D creation from a single image with diffusion prior. arXiv preprint arXiv:2303.14184 (2023)
55. Teed, Z., Deng, J.: RAFT: recurrent all-pairs field transforms for optical flow. In: Vedaldi, A., Bischof, H., Brox, T., Frahm, J.-M. (eds.) ECCV 2020, Part II. LNCS, vol. 12347, pp. 402–419. Springer, Cham (2020). https://doi.org/10.1007/978-3-030-58536-5_24
56. Tretschk, E., Tewari, A., Golyanik, V., Zollhöfer, M., Lassner, C., Theobalt, C.: Non-rigid neural radiance fields: reconstruction and novel view synthesis of a

dynamic scene from monocular video. In: Proceedings of the IEEE/CVF International Conference on Computer Vision, pp. 12959–12970 (2021)

57. Wang, H., Du, X., Li, J., Yeh, R.A., Shakhnarovich, G.: Score Jacobian chaining: lifting pretrained 2D diffusion models for 3D generation. In: Proceedings of the IEEE/CVF Conference on Computer Vision and Pattern Recognition, pp. 12619–12629 (2023)

58. Wang, P., Shi, Y.: ImageDream: image-prompt multi-view diffusion for 3D generation. arXiv preprint arXiv:2312.02201 (2023)

59. Wang, X., et al.: AnimatableDreamer: text-guided non-rigid 3D model generation and reconstruction with canonical score distillation. arXiv preprint arXiv:2312.03795 (2023)

60. Wang, Z., et al.: ProlificDreamer: high-fidelity and diverse text-to-3D generation with variational score distillation. In: Advances in Neural Information Processing Systems, vol. 36 (2024)

61. Wang, Z., Bovik, A.C., Sheikh, H.R., Simoncelli, E.P.: Image quality assessment: from error visibility to structural similarity. IEEE Trans. Image Process. **13**(4), 600–612 (2004)

62. Wu, G., et al.: 4D gaussian splatting for real-time dynamic scene rendering. arXiv preprint arXiv:2310.08528 (2023)

63. Wu, T., Zhong, F., Tagliasacchi, A., Cole, F., Oztireli, C.: D^2NeRF: self-supervised decoupling of dynamic and static objects from a monocular video. In: Advances in Neural Information Processing Systems, vol. 35, pp. 32653–32666 (2022)

64. Wu, Z., Zhu, Z., Du, J., Bai, X.: CCPL: contrastive coherence preserving loss for versatile style transfer. In: Avidan, S., Brostow, G., Cissé, M., Farinella, G.M., Hassner, T. (eds.) ECCV 2022. LNCS, vol. 13676, pp. 189–206. Springer, Cham (2022). https://doi.org/10.1007/978-3-031-19787-1_11

65. Xu, J., et al.: Dream3D: zero-shot text-to-3D synthesis using 3D shape prior and text-to-image diffusion models. In: Proceedings of the IEEE/CVF Conference on Computer Vision and Pattern Recognition, pp. 20908–20918 (2023)

66. Yang, Z., Yang, H., Pan, Z., Zhu, X., Zhang, L.: Real-time photorealistic dynamic scene representation and rendering with 4D gaussian splatting. arXiv preprint arXiv:2310.10642 (2023)

67. Yi, T., et al.: GaussianDreamer: fast generation from text to 3D gaussian splatting with point cloud priors. arXiv preprint arXiv:2310.08529 (2023)

68. Yin, Y., Xu, D., Wang, Z., Zhao, Y., Wei, Y.: 4DGen: grounded 4D content generation with spatial-temporal consistency. arXiv preprint arXiv:2312.17225 (2023)

69. Yu, C., Zhou, Q., Li, J., Zhang, Z., Wang, Z., Wang, F.: Points-to-3D: bridging the gap between sparse points and shape-controllable text-to-3D generation. In: Proceedings of the 31st ACM International Conference on Multimedia, pp. 6841–6850 (2023)

70. Zhang, L., Rao, A., Agrawala, M.: Adding conditional control to text-to-image diffusion models. In: Proceedings of the IEEE/CVF International Conference on Computer Vision, pp. 3836–3847 (2023)

71. Zhang, R., Isola, P., Efros, A.A., Shechtman, E., Wang, O.: The unreasonable effectiveness of deep features as a perceptual metric. In: Proceedings of the IEEE Conference on Computer Vision and Pattern Recognition, pp. 586–595 (2018)

72. Zhao, Y., Yan, Z., Xie, E., Hong, L., Li, Z., Lee, G.H.: Animate124: animating one image to 4D dynamic scene. arXiv preprint arXiv:2311.14603 (2023)

73. Zheng, Y., Li, X., Nagano, K., Liu, S., Hilliges, O., De Mello, S.: A unified approach for text-and image-guided 4D scene generation. arXiv preprint arXiv:2311.16854 (2023)
74. Zhu, J., Zhuang, P.: HiFA: high-fidelity text-to-3D with advanced diffusion guidance. arXiv preprint arXiv:2305.18766 (2023)

Overcoming Distribution Mismatch
in Quantizing Image Super-Resolution Networks

Cheeun Hong[1] and Kyoung Mu Lee[1,2](✉)

[1] Department of ECE & ASRI, Seoul, South Korea
{cheeun914,kyoungmu}@snu.ac.kr
[2] IPAI, Seoul National University, Seoul, South Korea

Abstract. Although quantization has emerged as a promising approach to reducing computational complexity across various high-level vision tasks, it inevitably leads to accuracy loss in image super-resolution (SR) networks. This is due to the significantly divergent feature distributions across different channels and input images of the SR networks, which complicates the selection of a fixed quantization range. Existing works address this distribution mismatch problem by dynamically adapting quantization ranges to the varying distributions during test time. However, such a dynamic adaptation incurs additional computational costs during inference. In contrast, we propose a new quantization-aware training scheme that effectively **O**vercomes the **D**istribution **M**ismatch problem in SR networks without the need for dynamic adaptation. Intuitively, this mismatch can be mitigated by regularizing the distance between the feature and a fixed quantization range. However, we observe that such regularization can conflict with the reconstruction loss during training, negatively impacting SR accuracy. Therefore, we opt to regularize the mismatch only when the gradients of the regularization are aligned with those of the reconstruction loss. Additionally, we introduce a layer-wise weight clipping correction scheme to determine a more suitable quantization range for layer-wise weights. Experimental results demonstrate that our framework effectively reduces the distribution mismatch and achieves state-of-the-art performance with minimal computational overhead. Codes are available at https://github.com/Cheeun/ODM.

Keywords: Image super-resolution · Network quantization · Quantization-aware training

1 Introduction

Image super-resolution (SR) is a core low-level vision task aimed at reconstructing high-resolution (HR) images from their corresponding low-resolution (LR) counterparts. Recent advances in deep learning [7,11,31,36,37,43,45,54,55] have led to astonishing achievements in producing high-fidelity images. However, the remarkable

---

Supplementary Information The online version contains supplementary material available at https://doi.org/10.1007/978-3-031-72624-8_22.

Table 1. Quantization methods on SR. For performance, existing methods rely on channel-wise quantization or input-adaptive module, which incur computational overhead. Our method achieves high accuracy without utilizing channel-wise quantization or input-adaptive modules.

Method	Channel-wise Q	Input-adaptive Modules	PSNR	SSIM
PAMS [33]	✗	✗	29.51	0.835
DDTB [57]	✗	✓	30.97	0.876
DAQ [18]	✓	✓	31.01	0.871
Ours	✗	✗	**31.50**	**0.882**

performance is based on heavy network architectures that incur significant computational costs, limiting practical viability, such as mobile deployment.

To mitigate the computational complexity of neural networks, quantization has emerged as a promising avenue. Network quantization has proven effective in reducing computational costs without significant accuracy loss, particularly in high-level vision tasks, such as image classification [8,20,58]. However, when it comes to quantizing SR networks to lower bit-widths, substantial performance degradation [24] often occurs, posing a persistent and challenging problem to be addressed.

This degradation can be attributed to the significant variance in the feature (activation) distributions of SR networks. The feature distribution of a layer exhibits substantial discrepancies across different channels and images, which makes it difficult to determine a single quantization range for a layer. Early approaches to SR quantization [33] adopt a training scheme to learn the quantization range of each layer. However, despite careful selection, the quantization ranges do not align with the varied values within the channel and image dimensions, which we refer to as *distribution mismatch* in features.

Recent approaches aim to address this issue by incorporating dynamic adaptation modules that accommodate the varying distributions. For example, the quantization range is dynamically adjusted by directly leveraging the distribution mean and variance at test time [18] or by employing input-adaptive dynamic modules [57]. Although adapting the quantization function to each image during inference might handle variable distributions, the dynamic modules introduce considerable computational overhead, potentially compromising the computational benefits of quantization.

In this study, we propose a novel quantization-aware training framework that addresses the distribution mismatch problem with a loss term that regulates the mismatch in distributions. Although directly minimizing the feature mismatch presents the potential for quantization-friendliness, whether it preserves reconstruction accuracy is questionable. We observe that concurrent optimization with the mismatch regularization and the original reconstruction loss can disrupt the image reconstruction process. Therefore, we introduce a cooperative mismatch regularization strategy, where the mismatch is regulated only when it collaborates harmoniously with the reconstruction loss. To determine the cooperative behavior, we assess the cosine similarity of the gradients from each loss, then we weigh the gradients of mismatch regularization based on this similarity. Consequently, we effectively update the SR network to hold both quantization-friendliness and reconstruction accuracy.

Furthermore, we identify the distribution mismatch among the weights of different layers. We discover that employing a fixed policy to determine the layer-wise weight quantization range [17, 18, 33, 47] can be suboptimal and that a further precise range can be obtained by considering both the current distribution of weights and the distinct tendencies of each layer. Therefore, we additionally incorporate layer-specific variations using a correction parameter for each layer. This strategy allows us to accurately find the quantization range for weights while incurring only a minimal overhead (0.01% additional storage size and no additional bitOPs)–a significantly smaller impact compared to methods that use dynamic modules. Overall, the contributions of our work include the following:

- We introduce the first quantization framework to address the distribution mismatch problem in SR networks without dynamic modules, as compared in Table 1. Our framework updates the SR network to be quantization-friendly and accurate simultaneously.
- Based on the observations of the distribution mismatch in SR networks, we effectively reduce the mismatch by introducing a cooperative mismatch regularization term and a weight clipping correction term.
- Compared to existing approaches to SR quantization, ours achieves state-of-the-art performance with similar or fewer computations.

2   Related Works

2.1   Image Super-Resolution

Convolutional Neural Network (CNN)-based approaches have demonstrated remarkable advancements in the image super-resolution (SR) task [32, 37], but at the cost of substantial computational resources. The intensive computations required by SR networks have spurred interest in developing lightweight SR architectures [10, 22, 23, 28, 54]. Furthermore, various strategies for lightweight SR networks have been explored, including neural architecture search [9, 30, 34, 35, 46], knowledge distillation [22, 23, 53], and pruning [26, 42, 56]. While these methods predominantly focus on reducing network depth or the number of channels, our work specifically aims to lower the precision of floating-point operations through network quantization.

2.2   Network Quantization

By mapping 32-bit floating point values of features and weights in convolutional layers to lower-bit representations, network quantization provides a dramatic reduction in computational resources [5, 8, 14, 29, 58, 59]. Recent works have successfully quantized various networks to low bit-widths with minimal compromise in accuracy [6, 12, 16, 27, 39, 49, 52]. However, these efforts primarily target high-level vision tasks, whereas networks for low-level vision tasks remain vulnerable to low-bit quantization.

2.3   Quantized Super-Resolution Networks

In contrast to high-level vision tasks, SR poses different challenges due to its inherent high sensitivity to quantization [24, 40, 48, 51]. Some works have attempted to recover accuracy by modifying the network architecture [2, 25, 51] or by assigning different bits for each image [17, 19, 47] or network stage [38]. However, the primary challenge in quantizing SR networks lies in the vastly distinct feature distributions. To address this, Li *et al.* [33] adopted a learnable quantization range for different layers. Subsequently, recognizing that the distributions vary not only by layer, but also by channel and input, Hong *et al.* [18] introduced a channel-wise dynamic quantization function. Additionally, Zhong *et al.* [57] utilized an input-adaptive dynamic module to tailor quantization ranges for each specific input image. However, such dynamic adaptations of quantization functions during test-time incur non-negligible computational overheads. Instead of relying on test-time adaptive modules, our approach focuses on mitigating the feature mismatch before quantization. Our framework reduces the inherent distribution mismatch in SR networks with minimal overhead, accurately quantizing SR networks without dynamic modules. More recently, Qin *et al.* [44] introduced additional transformation functions in both the forward and backward processes. However, performance degradation is still evident in ultra-low bit (*e.g.*, 2-bit) scenarios.

3 Proposed Method

In this section, after a brief introduction to network quantization (Sect. 3.1), we first analyze the mismatch in the features of SR networks (Sect. 3.2). Subsequently, we propose a solution to reduce this feature mismatch during training (Sect. 3.3). Additionally, we examine the mismatch in weights across SR networks (Sect. 3.4) and introduce a quantization range selection scheme that addresses the weight mismatch (Sect. 3.5). The overall training process is summarized in Algorithm 1.

3.1 Preliminaries

To reduce the heavy computations of convolutional and linear layers in neural networks, the input features (activations) and weights of each convolution/linear layer are quantized to low-bit values [5,8,15,29]. Given the input feature of the i-th convolution/linear layer $\boldsymbol{X}_i \in \mathbb{R}^{B \times C \times H \times W}$, where $B, C, H,$ and W represent the dimensions of the input batch, channel, height, and width, respectively, a quantization operator $q(\cdot)$ quantizes the feature $\boldsymbol{X}_i$ with bit-width b:

$$q(\boldsymbol{X}_i; l, u) = \text{Int}(\frac{\text{clip}(\boldsymbol{X}_i, l, u) - l}{s}) \cdot s + l, \qquad (1)$$

where $\text{clip}(\cdot, l, u)$ truncates the input into the quantization range of $[l, u]$ and $s = \frac{u-l}{2^b-1}$. After truncation, the truncated feature is scaled to $[0, 2^b - 1]$, then rounded to integer values with $\text{Int}(\cdot)$, and rescaled to range $[l, u]$.

For activation quantization, the quantization range is defined by $[l_a, u_a]$. To obtain better quantization ranges in SR networks, the clipping parameters l_a, u_a for each layer are typically learned through quantization-aware training [33,57]. Since the rounding function is not differentiable, a straight-through estimator (STE) [3] is used to train the clipping parameters in an end-to-end manner. Following [57], we initialize l_a and u_a as the $(100-j)$-th and j-th percentile values of the feature, averaged among the training data, where j is set to 99 in our experiments to avoid outliers that corrupt the quantization range. For weights, as their distributions tend to be symmetric and the mean approximates zero, we utilize a symmetric quantization function. Thus, for the weight $\boldsymbol{W}$ of each convolution/linear layer, the quantization range is defined by $[-u_w, u_w]$.

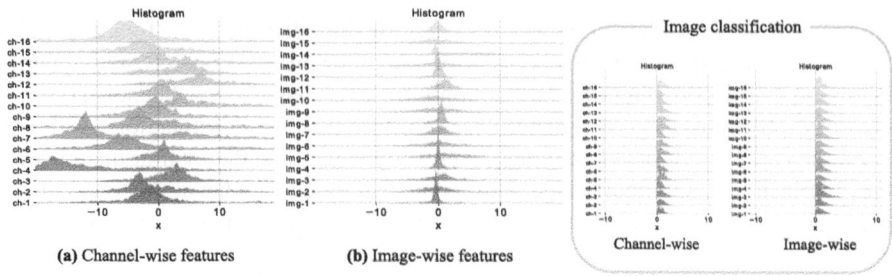

Fig. 1. Distribution mismatch in SR networks. Compared to a classification network (*e.g.*, ResNet-18), an SR network (*e.g.*, EDSR) exhibits significant mismatches within the feature distributions across channel and image dimensions. The large distribution mismatch complicates the selection of an appropriate quantization range. Channels and images from the 2nd layer are randomly selected for visualization. Additional results are available in the supplementary material.

3.2   Distribution Mismatch in SR Networks

The unfriendliness to quantization in SR networks arises from diverse feature (activation) distributions, as reported in previous studies [18,33,57], primarily due to the absence of batch normalization layers in SR networks. As illustrated in Fig. 1, where there are notable discrepancies between the channel and image distributions, quantization grids are unnecessarily allocated to regions with minimal feature density. Early SR quantization methods tackled this issue by employing learnable quantization range parameters [33] for each feature. However, even though the quantization-aware training process strives to find the optimal range for each feature, it fails to account for the channel-wise and input-wise variance in distributions. This mismatch results in a large quantization error that can impair SR performance. To deal with these distribution mismatches, existing methods adopt different quantization ranges for each channel [18] or input image [18,57]. Nevertheless, the test-time adaptation modules used to determine these ranges introduce unwanted computational overhead during inference. Thus, our straightforward solution is to pre-adjust the distributions to be quantization-friendly, thereby eliminating the need for additional adaptations at inference. The following sections will introduce a new quantization-aware training scheme designed to resolve the distribution mismatch problem.

3.3   Cooperative Mismatch Regularization

Instead of trying to identify a quantization range capable of accommodating diverse feature distributions, our approach aims to regularize the distribution mismatch beforehand. Obtaining an appropriate quantization range for a feature with high image- and channel-wise variance is difficult, as a certain number of channels or images will invariably be distant from the selected quantization range. In this work, we refer to the total distance of each feature from the selected quantization grid as the mismatch,

$$M(\boldsymbol{X}_i) = ||\boldsymbol{X}_i - q(\boldsymbol{X}_i; l_a, u_a)||_2, \tag{2}$$

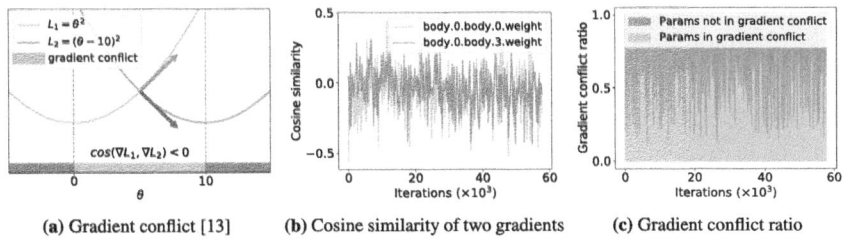

(a) Gradient conflict [13]   (b) Cosine similarity of two gradients   (c) Gradient conflict ratio

Fig. 2. Conflict between mismatch regularization and reconstruction loss. Mismatch regularization updates a number of parameters in the *contradictory* direction to the reconstruction loss, which we refer to as gradient conflict. (b) When the two losses are jointly used, gradient conflict consistently occurs during training, outputting a negative cosine similarity value. (c) We plot the ratio of conflicted gradients during training. Nearly half of the parameters undergo gradient conflict, which indicates that merely combining mismatch regularization with the reconstruction loss can impair SR accuracy. Visualizations are done on EDSR.

where $||\cdot||_2$ calculates the Frobenius norm. Further analyses of the definition of mismatch are provided in the supplementary material. We can reduce the overall mismatch by directly regularizing the mismatch of each feature to be quantized. The mismatch regularization loss is obtained by summing the mismatch over all quantized features:

$$\mathcal{L}_M = \sum_i^{\#\text{layers}} M(\boldsymbol{X}_i). \tag{3}$$

The mismatch regularization loss can be used in line with the original reconstruction loss typically used in the general quantization-aware training pipeline for SR networks:

$$\mathcal{L}_R = \mathcal{L}_1(\mathcal{Q}(\boldsymbol{I}_{LR}), \boldsymbol{I}_{HR}), \tag{4}$$

where $\mathcal{L}_1$ loss indicates the l_1 distance between the reconstructed image using the quantized network $\mathcal{Q}$ and the ground-truth HR image $\boldsymbol{I}_{HR}$. Then, the optimization of the parameter θ^t is formulated as:

$$\theta^{t+1} = \theta^t - \beta^t \cdot (\nabla_\theta \mathcal{L}_R(\theta^t) + \nabla_\theta \mathcal{L}_M(\theta^t)), \tag{5}$$

where $\nabla_\theta \mathcal{L}_R(\theta^t)$ denotes the gradient from the original reconstruction loss and $\nabla_\theta \mathcal{L}_M(\theta^t)$ is the gradient from mismatch regularization loss, and β^t refers to the learning rate. Updating the network to minimize the mismatch regularization loss will reduce the overall error from feature quantization.

However, then a question arises: *does reducing the quantization error of each feature lead to improved reconstruction accuracy?* The answer is, according to our observation in Fig. 2, not necessarily. During the training process, the mismatch regularization loss can collide with the original reconstruction loss. That is, for some parameters, the direction of the gradient from reconstruction loss and that of the mismatch regularization are opposing, referred to as gradient conflict [13]. As in Fig. 2b, the cosine similarity of two gradients oscillates between positive and negative values during training,

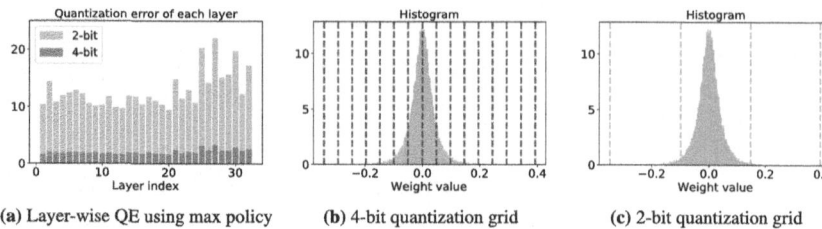

(a) Layer-wise QE using max policy   **(b)** 4-bit quantization grid   **(c)** 2-bit quantization grid

Fig. 3. Layer-wise variation in error from weight quantization. (a) Quantization error (QE) varies across different layers when a fixed global policy (*i.e.*, max) is used to determine the quantization range, particularly for low bits. For some layers, using max does not effectively serve as a proper policy for quantization range selection. (b) Outliers often dominate the quantization range, leading to quantization grids being wasted on low-density areas. (c) At low bits, quantization grids fail to cover high-density regions adequately. Therefore, the quantization range should be adjusted for certain layers.

indicating that the directions of two gradients do not converge and the gradient occasionally conflicts. Furthermore, as shown in Fig. 2c, the proportion of parameters undergoing gradient conflict is not minor, implying that the regularization loss can severely hinder the reconstruction loss.

We aim to avoid the conflict between these two losses, in other words, to minimize the mismatch as long as it does not hinder the reconstruction loss. Thus, we dismiss the mismatch regularization term when it is not cooperative with reconstruction loss and make more use of it when it is cooperative. Specifically, we determine whether the two losses are cooperative by examining the cosine similarity of the gradients of each loss and then simply weigh the gradient of mismatch regularization by the gradient similarity. Our cooperative mismatch regularization can be formulated as follows:

$$\theta^{t+1} = \theta^t - \beta^t \cdot (\lambda_R \cdot \nabla_\theta \mathcal{L}_R(\theta^t) + \lambda_M \cdot \underline{sim(\nabla_\theta \mathcal{L}_R(\theta^t), \nabla_\theta \mathcal{L}_M(\theta^t))} \cdot \nabla_\theta \mathcal{L}_M(\theta^t)), \quad (6)$$

where the underlined term, gradient similarity, is defined as $sim(v_a, v_b) = \frac{cos(v_a, v_b)+1}{2}$ and $cos(\cdot, \cdot) \in [-1, 1]$ calculates the cosine similarity between two vectors. λ_R, λ_M are hyper-parameters to balance the two gradients. If the directions of two gradients are similar (*i.e.*, smaller than 90° and closer to 0°), the gradient similarity is a large value, then the parameter is updated in the direction where mismatch regularization is also substantially considered. On the contrary, if the two gradients point in the other direction (*i.e.*, larger than 90° and closer to 180°), the two losses restrain each other, and the gradient similarity is close to 0. In this case, we follow the gradient of reconstruction loss. This allows the network to reduce the quantization error cooperatively with the reconstruction error. Details on gradient similarity are in the supplementary material.

3.4   Distribution Mismatch in Weights

The weight distributions of SR networks have remained relatively unexplored in previous literature. This is because the weights in SR networks typically exhibit bell-shaped

---

Algorithm 1. Quantization-aware training process of ODM

Input: Pre-trained 32-bit network $\mathcal{P}$.
Output: Quantized network $\mathcal{Q}$.

   for $t = 1, \cdots, \#$ iters **do**
      for $i = 1, \cdots, \#$ layers **do**
         if $t = 1$ **then**
             Initialize activation quantization range $[l_a, u_a]$
             Initialize weight quantization range $[-u_w, u_w]$
         Given quantization range, obtain $q(\boldsymbol{X}_i; l_a, u_a)$ using Eq. (1)
         Given $\boldsymbol{X}_i$ and $q(\boldsymbol{X}_i; l_a, u_a)$, obtain mismatch using Eq. (2)
         Adjust weight quantization range $[-u_w, u_w]$ using Eq. (9)
         Given quantization range, obtain $q(\boldsymbol{W}_i; -u_w, u_w)$ using Eq. (1)
         Replace $\boldsymbol{X}_i, \boldsymbol{W}_i$ in $\mathcal{P}$ with $q(\boldsymbol{X}_i), q(\boldsymbol{W}_i)$
      Calculate mismatch regularization loss (Eq. (3)) and reconstruction loss (Eq. (4))
      Update parameters of $\mathcal{Q}$ with two losses cooperatively using Eq. (6)

---

distributions, which are considered easier to quantize compared to the long-tailed, input-wise and channel-wise distinct activation distributions. Consequently, many studies [17,33,47] simply adopt max quantization for weights, setting the quantization range with the maximum value of the current weight distribution. However, we notice that this is suboptimal and may contribute to low performance when SR networks are quantized to ultra-low bits (*e.g.*, 2-bit). The issue arises because, in some layers, the outliers are far from the distribution mean, as visualized in Figs. 3b and 3c. Thus, when the maximum value is used to determine the quantization range for such distributions, the quantization range is dominated by outliers, with quantization grids not allocated to high-density regions (*e.g.*, near 0). This leads to substantial quantization errors that can accumulate and degrade restoration performance. In the case of 4-bit quantization (Fig. 3b), although grids still cover high-density regions to an extent, a number of grids are wasted on low-density areas. The problem intensifies with low-bit (2-bit) quantization (Fig. 3c), where quantization errors become significantly larger. This observation underscores the need for careful selection of the quantization range for layer-wise weights, particularly in low-bit quantization scenarios.

3.5 Weight Clipping Correction

Also, we notice that the error from weight quantization varies across different layers, as shown in Fig. 3a. For instance, employing the maximum value as the quantization range can be an effective policy for certain layers, yet this policy proves inadequate for others (*e.g.*, the layer shown in Fig. 3c). Given the unique tendency of each layer, applying a uniform global policy for selecting the quantization range across all layers is suboptimal. Existing methods [17,33,47] utilize a fixed global policy throughout training to set the quantization range clipping parameter u_w as follows:

$$u_w^t = f(\boldsymbol{W}^t), \tag{7}$$

Table 2. Quantitative comparisons on EDSR of scale 4

Model	Bit	Set5		Set14		B100		Urban100	
		PSNR	SSIM	PSNR	SSIM	PSNR	SSIM	PSNR	SSIM
EDSR	32	32.10	0.894	28.58	0.781	27.56	0.736	26.04	0.785
EDSR-PAMS	4	31.59	0.885	28.20	0.773	27.32	0.728	25.32	0.762
EDSR-DAQ	4	31.85	0.887	28.38	0.776	27.42	0.732	25.73	0.772
EDSR-DDTB	4	31.85	0.889	28.39	0.777	27.44	0.732	25.69	0.774
EDSR-ODM (Ours)	4	**32.00**	**0.891**	**28.47**	**0.779**	**27.51**	**0.735**	**25.80**	**0.778**
EDSR-PAMS	3	27.25	0.780	25.24	0.673	25.38	0.644	22.76	0.641
EDSR-DAQ	3	31.66	0.884	28.19	0.771	27.28	0.728	25.40	0.762
EDSR-DDTB	3	31.52	0.883	28.18	0.771	27.30	0.727	25.33	0.761
EDSR-ODM (Ours)	3	**31.85**	**0.888**	**28.38**	**0.776**	**27.43**	**0.732**	**25.59**	**0.771**
EDSR-PAMS	2	29.51	0.835	26.79	0.734	26.45	0.696	23.72	0.688
EDSR-DAQ [†]	2	31.01	0.871	27.89	0.762	27.09	0.719	24.88	0.740
EDSR-DDTB	2	30.97	0.876	27.87	0.764	27.09	0.719	24.82	0.742
EDSR-ODM (Ours)	2	**31.50**	**0.882**	**28.14**	**0.770**	**27.27**	**0.726**	**25.17**	**0.755**

† We note that the reported results of DAQ [18] are obtained using EDSR of 32 residual blocks. For a fair comparison, we reproduce EDSR-DAQ using EDSR of 16 residual blocks (*i.e.*, EDSR-baseline).

where the global policy $f(\cdot)$ is the same function for all layers (*e.g.*, $max(\cdot)$). A simple solution to accommodate layer-specific variations is to make the clipping parameter u_w for each layer a learnable parameter [14]:

$$u_w^{t+1} = u_w^t - \beta^t \cdot \nabla_{u_w} \mathcal{L}_R(u_w^t), \qquad (8)$$

where β^t denotes the learning rate. This process determines the clipping parameter u_w^t to quantize W^t based on the weight of the previous iteration, W^{t-1}. However, since the weight is also updated at iteration step t ($W^{t-1} \rightarrow W^t$), a mismatch occurs between the current weight and the weight quantization range derived from the previous weight. To address this, we first obtain the quantization range by applying the global policy to the current weight, then adjust the range with a learnable parameter that accounts for layer-specific tendencies. Our clipping parameter is formulated as follows:

$$u_w^{t+1} = f(W^{t+1}) \cdot (\gamma_w^t - \beta^t \cdot \nabla_{\gamma_w} \mathcal{L}_R(\gamma_w^t)), \qquad (9)$$

where γ_w is the learnable parameter for each layer representing the layer-wise adjustment. Each γ_w is initially set to 1. To prevent outliers from dominating the initial quantization range, we set the global policy $f(\cdot)$ as the j-th percentile function. Our clipping correction scheme introduces only one additional parameter per layer; thus, the overall computational overhead is minimal. For further details, please refer to Sect. 4.3.

4　Experiments

The efficacy and adaptability of the proposed quantization framework, ODM, are assessed through its application across several SR networks. The experimental settings are described (Sect. 4.1), and quantitative (Sect. 4.2), qualitative (Sect. 4.4), and complexity (Sect. 4.3) evaluations are conducted on various SR networks. Ablation studies are conducted to examine each component of the framework (Sect. 4.5).

Table 3. Quantitative comparisons on RDN of scale 4

Model	Bit	Set5		Set14		B100		Urban100	
		PSNR	SSIM	PSNR	SSIM	PSNR	SSIM	PSNR	SSIM
RDN	32	32.24	0.896	28.67	0.784	27.63	0.738	26.29	0.792
RDN-PAMS	4	30.44	0.862	27.54	0.753	26.87	0.710	24.52	0.726
RDN-DAQ	4	31.91	0.889	28.38	0.775	27.38	0.733	25.81	0.779
RDN-DDTB	4	31.97	0.891	28.49	0.780	27.49	0.735	25.90	0.783
RDN-ODM (Ours)	4	**32.06**	**0.892**	**28.49**	**0.780**	**27.52**	**0.736**	25.88	0.783
RDN-PAMS	3	29.54	0.838	26.82	0.734	26.47	0.696	23.83	0.692
RDN-DAQ	3	31.57	0.883	28.18	0.771	27.27	0.728	25.47	0.765
RDN-DDTB	3	31.49	0.883	28.17	0.772	27.30	0.728	25.35	0.764
RDN-ODM (Ours)	3	**31.79**	**0.887**	**28.33**	**0.776**	**27.42**	**0.732**	**25.51**	**0.770**
RDN-PAMS	2	29.73	0.843	26.96	0.739	26.57	0.700	23.87	0.696
RDN-DAQ	2	30.71	0.866	27.61	0.755	26.93	0.715	24.71	0.731
RDN-DDTB	2	30.57	0.867	27.56	0.757	26.91	0.714	24.50	0.728
RDN-ODM (Ours)	2	**31.37**	**0.880**	**28.08**	**0.770**	**27.24**	**0.727**	**25.09**	**0.755**

4.1 Implementation Details

Models and Training. The proposed framework is applied directly to existing representative SR networks that produce satisfactory SR results, but involve heavy computations: EDSR (baseline) [37] and RDN [55]. Furthermore, we apply our method to the Transformer-based SR model, SwinIR-S [36]. Following prior works on SR quantization [17,18,33,40,51,57], weights and activations of the high-level feature extraction module are quantized, which is the most computationally demanding. Training and validation are conducted using the DIV2K [1] dataset. ODM trains the network for 60K iterations with a batch size of 8. The weights are updated with an initial learning rate of $\beta^0 = 10^{-4}$. For cooperative update, we update clipping parameters with $10 \cdot \beta^0$ initial learning rate. The learning rates are halved every 15K iteration. The hyperparameter for the percentile is set to $j = 99$, and to balance the gradient of the loss terms, we set $\lambda_R = 1$ and $\lambda_M = 10^{-5}$. Specially, we set $\lambda_M = 10^{-6}$ for RDN whose overall mismatch is large. Ablation studies on the hyperparameters are in the supplementary material. All our experiments are implemented using PyTorch and run on an RTX 2080Ti GPU.

Evaluation. We evaluate our framework on the standard benchmark (Set5 [4], Set14 [32], BSD100 [41], and Urban100 [21]) by measuring the peak signal-to-noise ratio (PSNR) and the structural similarity index (SSIM [50]). To assess the computational complexity of our framework, we measure bitOPs and storage size. BitOPs refers to the number of operations weighted by the bit-widths of the two operands. Storage size is calculated as the number of stored parameters weighted by the precision of each parameter value.

Table 4. Quantitative comparisons on **SwinIR** of scale 4

Model	Bit	Set5		Set14		B100		Urban100	
		PSNR	SSIM	PSNR	SSIM	PSNR	SSIM	PSNR	SSIM
SwinIR	32	32.44	0.898	28.77	0.786	27.69	0.741	26.47	0.798
SwinIR-PAMS	4	31.99	0.890	28.50	0.779	27.49	0.735	25.86	0.780
SwinIR-DAQ	4	31.82	0.887	28.34	0.775	27.37	0.730	25.68	0.772
SwinIR-DDTB	4	32.09	0.891	28.55	0.780	27.54	0.735	26.01	0.783
SwinIR-ODM (Ours)	4	**32.17**	**0.892**	**28.59**	**0.781**	**27.56**	**0.736**	**26.06**	**0.785**
SwinIR-PAMS	3	31.62	0.884	28.23	0.771	27.31	0.728	25.38	0.762
SwinIR-DAQ	3	31.50	0.882	27.99	0.770	27.12	0.727	25.29	0.761
SwinIR-DDTB	3	31.80	0.887	28.34	0.775	27.40	0.731	25.63	0.771
SwinIR-ODM (Ours)	3	**31.94**	**0.889**	**28.39**	**0.777**	**27.45**	**0.733**	**25.72**	**0.775**
SwinIR-PAMS	2	29.48	0.834	26.79	0.733	26.46	0.698	23.72	0.688
SwinIR-DAQ	2	29.10	0.824	26.55	0.725	26.30	0.691	23.51	0.678
SwinIR-DDTB	2	31.01	0.873	27.80	0.762	27.04	0.719	24.79	0.739
SwinIR-ODM (Ours)	2	**31.44**	**0.880**	**28.06**	**0.769**	**27.23**	**0.725**	**25.14**	**0.754**

4.2   Quantitative Results

To evaluate the effectiveness of our proposed scheme, we compare the results with existing SR quantization works using their official codes: PAMS [33], DAQ [18], and DDTB [57]. For a fair comparison, we reproduce other methods using the same training iterations, 60K iterations. The supplementary materials provide additional experiments that further demonstrate the applicability of ODM, including results on 300K iterations, scale 2, and fully quantized settings.

EDSR. As shown in Table 2, ODM outperforms other methods in the 4, 3, and 2-bit settings, and notably, the improvement is significant for 2-bit, achieving a gain of more than 0.49 dB for Set5. We notice that 4-bit EDSR-ODM achieves closer accuracy to 32-bit EDSR, with a marginal difference of 0.1 dB for Set5. This indicates that ODM can effectively bridge the gap between the quantized network and the 32-bit network.

RDN. Similarly, Table 3 compares the results on RDN, whose computational complexity is more burdensome than EDSR. The results show that ODM consistently achieves superior performance on 4, 3, and 2-bit quantization. The gain over existing methods is especially large for the 2-bit setting, where it exceeds 0.66 dB for Set5.

SwinIR. Furthermore, we evaluate our framework on the Transformer-based architecture, SwinIR. The linear and convolutional layers of SwinIR are quantized. According to Table 4, ODM is also proven effective in quantizing SwinIR across all bit settings, where the improvement is most notable in the 2-bit setting (0.43 dB).

Comparison with QuantSR. We also compare our method with the concurrent work, QuantSR [44]. As the training code of QuantSR has not been released, we base our comparison on the reported performance. For a fair comparison with QuantSR's reported performance, we also train our model for 300K iterations on SRResNet [32] and SwinIR [36]. In Table 5, the results demonstrate that our method achieves better results

Table 5. Quantitative comparisons with QuantSR on SRResNet and SwinIR of scale 4. For a fair comparison, our model (ODM*) is trained for 300K iterations following QuantSR.

Model	Bit	Set5		Set14		B100		Urban100	
		PSNR	SSIM	PSNR	SSIM	PSNR	SSIM	PSNR	SSIM
SRResNet	32	32.07	0.893	28.50	0.780	27.52	0.735	25.86	0.779
SRResNet-QuantSR	2	31.30	0.882	28.08	0.769	27.23	0.725	25.13	0.754
SRResNet-ODM*(Ours)	2	**31.81**	**0.888**	**28.32**	**0.774**	**27.38**	**0.730**	**25.54**	**0.767**
SwinIR	32	32.44	0.898	28.77	0.786	27.69	0.741	26.47	0.798
SwinIR-QuantSR	2	31.53	0.885	28.16	0.772	27.28	0.727	25.26	0.761
SwinIR-ODM*(Ours)	2	**31.67**	**0.885**	**28.23**	**0.772**	**27.33**	**0.728**	**25.36**	**0.762**

Table 6. Computational complexity comparison with SR quantization methods

Model	Bit	Storage size	BitOPs	PSNR	SSIM
EDSR	32	1517.6K	527.1T	32.10	0.894
EDSR-PAMS	2	411.7K	215.1T	29.51	0.835
EDSR-DAQ	2	411.7K	215.6T	31.01	0.871
EDSR-DDTB	2	413.6K	215.1T	30.97	0.876
EDSR-ODM (Ours)	2	**411.8K**	215.1T	**31.50**	**0.882**

(a) EDSR

Model	Bit	Storage size	BitOPs	PSNR	SSIM
RDN	32	22271.1K	6032.9T	32.24	0.896
RDN-PAMS	2	1715.9K	236.6T	29.73	0.843
RDN-DAQ	2	1715.9K	287.7T	30.71	0.866
RDN-DDTB	2	1761.6K	236.6T	30.57	0.867
RDN-ODM (Ours)	2	**1716.1K**	236.6T	**31.37**	**0.880**

(b) RDN

than QuantSR; compared to QuantSR, ours shows a gain of 0.51 dB on 2-bit SRResNet and a gain of 0.14 dB on 2-bit SwinIR for Set5.

4.3 Complexity Analysis

Along with the accuracy of SR, we also evaluate the computational complexity of our framework in Table 6. We measure the storage size for the model weights and the bitOPs required for generating a 1920×1080 image with ×4 SR network. Overall, our framework, ODM, achieves higher restoration accuracy with similar or fewer computational resources. Specifically, our weight-clipping correction involves an additional storage size of 0.06K/0.15K for EDSR/RDN. Since the correction process can be pre-determined before test time, no extra bitOPs are required. Compared to existing methods, our method achieves ×31.7 smaller storage size overhead than DDTB and 0.5T fewer bitOPs than DAQ on EDSR. For RDN, the gap is larger; our method's overhead is ×304.7 smaller in storage size than DDTB and 51.1T fewer in bitOPs than DAQ. Moreover, we note that DAQ adopts a channel-wise dynamic quantization function and DDTB an asymmetric weight quantization function, which are not favorably supported by hardware. Despite incurring a minor additional storage size of 0.06K/0.15K over PAMS, the accuracy improvement over PAMS is significant (+1.99 dB/+1.64 dB). The results prove that our method achieves significant accuracy gain with minimal or no computational overhead.

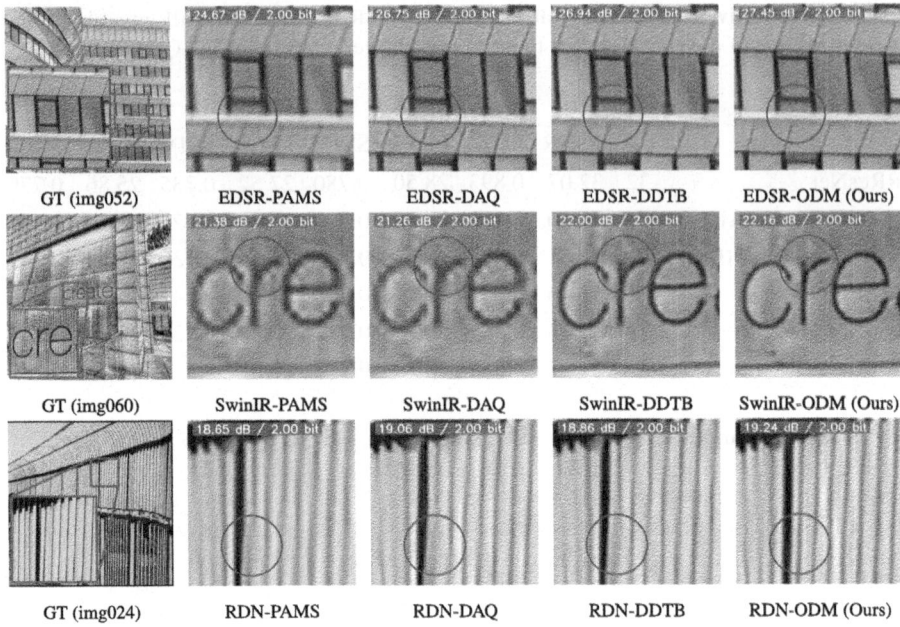

GT (img052)	EDSR-PAMS	EDSR-DAQ	EDSR-DDTB	EDSR-ODM (Ours)
GT (img060)	SwinIR-PAMS	SwinIR-DAQ	SwinIR-DDTB	SwinIR-ODM (Ours)
GT (img024)	RDN-PAMS	RDN-DAQ	RDN-DDTB	RDN-ODM (Ours)

Fig. 4. Qualitative results on Urban100 with EDSR, SwinIR, and RDN-based models

4.4 Qualitative Results

Figure 4 provides qualitative results and comparisons with the output images of quantized EDSR, RDN, and SwinIR. Our method, ODM, produces visually cleaner output images compared to existing methods. In contrast, existing methods, especially PAMS, suffer from blurred lines or artifacts. These qualitative results stress the importance of alleviating the distribution mismatch problem in SR networks.

4.5 Ablation Study

In Table 7, we verify the importance of each attribute of our framework: cooperative mismatch regularization and weight clipping correction. According to the results, each attribute individually improves baseline accuracy. Weight clipping correction improves the baseline by +1.18 dB for Set5, indicating that considering both the layer-wise trend and the current weight distribution is important for weight quantization. Also, while directly integrating mismatch regularization with the reconstruction loss (Model (**d**)) rather degrades performance by −0.13 dB, our cooperative scheme (Model (**e**)) improves the SR accuracy by +0.51 dB. This shows that reducing the mismatch in both activations and weights is important for accurately quantizing SR networks.

Furthermore, we visualize the feature distributions in Fig. 5 to validate the importance of our cooperative regularization term. After training only with the reconstruction loss, the outliers remain far from the quantization grid. When mismatch regularization loss is simply added to the original reconstruction loss, the activation distribution falls

Table 7. Ablation study on each attribute of our framework. *WCC* refers to weight clipping correction and *MR* refers to mismatch regularization. *Cooperative* denotes whether the mismatch regularization is cooperatively used with the reconstruction loss. Non-*cooperative MR* denotes that the two losses are simply used together.

Model	WCC	Cooperative	MR	Set5 (PSNR/SSIM)	Urban100 (PSNR/SSIM)
(a)	–	–	–	29.94/0.848	23.99/0.703
(b)	-	✓	✓	30.34/0.859	24.27/0.715
(c)	✓	–	–	31.12/0.876	24.91/0.746
(d)	✓	–	✓	30.99/0.871	24.79/0.735
(e)	✓	✓	✓	**31.50/0.882**	**25.17/0.755**

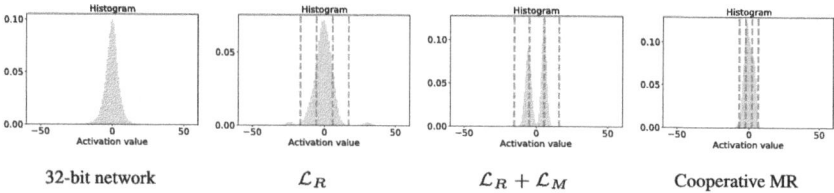

32-bit network $\mathcal{L}_R$ $\mathcal{L}_R + \mathcal{L}_M$ Cooperative MR

Fig. 5. Distribution of activations before quantization. Using our cooperative mismatch regularization results in distributions more robust to low-bit quantization. 8th conv layer of EDSR-ODM (2-bit) on 'baby' (Set5) are visualized.

into a narrower range and resembles a multi-modal distribution near the quantization grids. Although this distribution is quantization-friendly, it significantly deviates from the original activation of the 32-bit network and removes originally dense values (near 0). This can lead to an accuracy drop, as demonstrated in Table 7(**d**). In contrast, our cooperative mismatch regularization results in quantization-friendly distribution while largely preserving the activations in the dense region of the original 32-bit network.

5   Conclusion

SR networks suffer accuracy loss from quantization due to the inherent mismatch in feature distributions. Instead of employing resource-demanding dynamic modules to handle distinct distributions during test time, we introduce a new quantization-aware training technique that alleviates this mismatch through distribution optimization. We utilize cooperative mismatch regularization to update the SR network to be quantization-friendly and accurate. Additionally, to address the mismatch in layer-wise weights, we propose a weight-clipping correction strategy. These straightforward solutions effectively reduce the distribution mismatch with minimal computational overhead.

Acknowledgment. This work was supported in part by the IITP grants [No. 2021-0-01343, Artificial Intelligence Graduate School Program (Seoul National University), No. 2021-0-02068, and No. 2023-0-00156], the NRF grant [No. 2021M3A9E4080782] funded by the Korean government (MSIT).

References

1. Agustsson, E., Timofte, R.: NTIRE 2017 challenge on single image super-resolution: dataset and study. In: CVPR Workshops (2017)
2. Ayazoglu, M.: Extremely lightweight quantization robust real-time single-image super resolution for mobile devices. In: CVPR Workshops (2021)
3. Bengio, Y., Léonard, N., Courville, A.: Estimating or propagating gradients through stochastic neurons for conditional computation. arXiv preprint arXiv:1308.3432 (2013)
4. Bevilacqua, M., Roumy, A., Guillemot, C., Alberi-Morel, M.L.: Low-complexity single-image super-resolution based on nonnegative neighbor embedding. In: BMVC (2012)
5. Cai, Z., He, X., Sun, J., Vasconcelos, N.: Deep learning with low precision by half-wave gaussian quantization. In: CVPR (2017)
6. Cai, Z., Vasconcelos, N.: Rethinking differentiable search for mixed-precision neural networks. In: CVPR (2020)
7. Chen, X., Wang, X., Zhou, J., Qiao, Y., Dong, C.: Activating more pixels in image super-resolution transformer. In: CVPR (2023)
8. Choi, J., Wang, Z., Venkataramani, S., Chuang, P.I.J., Srinivasan, V., Gopalakrishnan, K.: PACT: parameterized clipping activation for quantized neural networks. arXiv preprint arXiv:1805.06085 (2018)
9. Chu, X., Zhang, B., Ma, H., Xu, R., Li, Q.: Fast, accurate and lightweight super-resolution with neural architecture search. In: ICPR (2021)
10. Dong, C., Loy, C.C., He, K., Tang, X.: Learning a deep convolutional network for image super-resolution. In: Fleet, D., Pajdla, T., Schiele, B., Tuytelaars, T. (eds.) ECCV 2014. LNCS, vol. 8692, pp. 184–199. Springer, Cham (2014). https://doi.org/10.1007/978-3-319-10593-2_13
11. Dong, C., Loy, C.C., He, K., Tang, X.: Image super-resolution using deep convolutional networks. IEEE TPAMI **38**(2), 295–307 (2015)
12. Dong, Z., Yao, Z., Gholami, A., Mahoney, M.W., Keutzer, K.: HAWQ: hessian aware quantization of neural networks with mixed-precision. In: ICCV (2019)
13. Du, Y., Czarnecki, W.M., Jayakumar, S.M., Farajtabar, M., Pascanu, R., Lakshminarayanan, B.: Adapting auxiliary losses using gradient similarity. arXiv preprint arXiv:1812.02224 (2018)
14. Esser, S.K., McKinstry, J.L., Bablani, D., Appuswamy, R., Modha, D.S.: Learned step size quantization. In: ICLR (2020)
15. Gholami, A., Kim, S., Dong, Z., Yao, Z., Mahoney, M.W., Keutzer, K.: A survey of quantization methods for efficient neural network inference. In: Low-Power Computer Vision, pp. 291–326. Chapman and Hall/CRC (2022)
16. Habi, H.V., Jennings, R.H., Netzer, A.: HMQ: hardware friendly mixed precision quantization block for CNNs. In: Vedaldi, A., Bischof, H., Brox, T., Frahm, J.-M. (eds.) ECCV 2020. LNCS, vol. 12371, pp. 448–463. Springer, Cham (2020). https://doi.org/10.1007/978-3-030-58574-7_27
17. Hong, C., Baik, S., Kim, H., Nah, S., Lee, K.M.: CADyQ: content-aware dynamic quantization for image super-resolution. In: Avidan, S., Brostow, G., Cissé, M., Farinella, G.M., Hassner, T. (eds.) ECCV 2022. LNCS, vol. 13667, pp. 367–383. Springer, Cham (2022)
18. Hong, C., Kim, H., Baik, S., Oh, J., Lee, K.M.: DAQ: channel-wise distribution-aware quantization for deep image super-resolution networks. In: WACV (2022)
19. Hong, C., Lee, K.M.: AdaBM: on-the-fly adaptive bit mapping for image super-resolution. In: CVPR (2024)
20. Hou, L., Kwok, J.T.: Loss-aware weight quantization of deep networks. In: ICLR (2018)

21. Huang, J.B., Singh, A., Ahuja, N.: Single image super-resolution from transformed self-exemplars. In: CVPR (2015)
22. Hui, Z., Gao, X., Yang, Y., Wang, X.: Lightweight image super-resolution with information multi-distillation network. In: ACMMM (2019)
23. Hui, Z., Wang, X., Gao, X.: Fast and accurate single image super-resolution via information distillation network. In: CVPR (2018)
24. Ignatov, A., Timofte, R., Denna, M., Younes, A.: Real-time quantized image super-resolution on mobile NPUs, mobile AI 2021 challenge: Report. In: CVPR Workshops (2021)
25. Jiang, X., Wang, N., Xin, J., Li, K., Yang, X., Gao, X.: Training binary neural network without batch normalization for image super-resolution. In: AAAI (2021)
26. Jiang, X., Wang, N., Xin, J., Xia, X., Yang, X., Gao, X.: Learning lightweight super-resolution networks with weight pruning. Neural Netw. **144**, 21–32 (2021)
27. Jin, Q., Yang, L., Liao, Z.: AdaBits: neural network quantization with adaptive bit-widths. In: CVPR (2020)
28. Jo, Y., Kim, S.J.: Practical single-image super-resolution using look-up table. In: CVPR (2021)
29. Jung, S., et al.: Learning to quantize deep networks by optimizing quantization intervals with task loss. In: CVPR (2019)
30. Kim, H., Hong, S., Han, B., Myeong, H., Lee, K.M.: Fine-grained neural architecture search for image super-resolution. J. Vis. Commun. Image Represent. **89**, 103654 (2022)
31. Kim, J., Lee, J., Lee, K.M.: Accurate image super-resolution using very deep convolutional networks. In: CVPR (2016)
32. Ledig, C., et al.: Photo-realistic single image super-resolution using a generative adversarial network. In: CVPR (2017)
33. Li, H., et al.: PAMS: quantized super-resolution via parameterized max scale. In: Vedaldi, A., Bischof, H., Brox, T., Frahm, J.-M. (eds.) ECCV 2020. LNCS, vol. 12370, pp. 564–580. Springer, Cham (2020). https://doi.org/10.1007/978-3-030-58595-2_34
34. Li, Y., Gu, S., Zhang, K., Van Gool, L., Timofte, R.: DHP: differentiable meta pruning via hypernetworks. In: Vedaldi, A., Bischof, H., Brox, T., Frahm, J.-M. (eds.) ECCV 2020. LNCS, vol. 12353, pp. 608–624. Springer, Cham (2020). https://doi.org/10.1007/978-3-030-58598-3_36
35. Li, Y., et al.: The heterogeneity hypothesis: Finding layer-wise differentiated network architectures. In: CVPR (2021)
36. Liang, J., Cao, J., Sun, G., Zhang, K., Van Gool, L., Timofte, R.: SwinIR: image restoration using swin transformer. In: ICCV (2021)
37. Lim, B., Son, S., Kim, H., Nah, S., Lee, K.M.: Enhanced deep residual networks for single image super-resolution. In: CVPR Workshops (2017)
38. Liu, J., Wang, Q., Zhang, D., Shen, L.: Super-resolution model quantized in multi-precision. Electronics **10**(17), 2176 (2021)
39. Lou, Q., Guo, F., Liu, L., Kim, M., Jiang, L.: AutoQ: automated kernel-wise neural network quantization. In: ICLR (2020)
40. Ma, Y., Xiong, H., Hu, Z., Ma, L.: Efficient super resolution using binarized neural network. In: CVPR Workshops (2019)
41. Martin, D., Fowlkes, C., Tal, D., Malik, J.: A database of human segmented natural images and its application to evaluating segmentation algorithms and measuring ecological statistics. In: ICCV (2001)
42. Oh, J., Kim, H., Nah, S., Hong, C., Choi, J., Lee, K.M.: Attentive fine-grained structured sparsity for image restoration. In: CVPR (2022)
43. Park, J., Son, S., Lee, K.M.: Content-aware local GAN for photo-realistic super-resolution. In: ICCV (2023)

44. Qin, H., Zhang, Y., Ding, Y., Liu, X., Danelljan, M., Yu, F.: QuantSR: accurate low-bit quantization for efficient image super-resolution. In: NeurIPS (2024)
45. Son, S., Kim, J., Lai, W.S., Yang, M.H., Lee, K.M.: Toward real-world super-resolution via adaptive downsampling models. IEEE TPAMI **44**(11), 8657–8670 (2021)
46. Song, D., Xu, C., Jia, X., Chen, Y., Xu, C., Wang, Y.: Efficient residual dense block search for image super-resolution. In: AAAI (2020)
47. Tian, S., Lu, M., Liu, J., Guo, Y., Chen, Y., Zhang, S.: CABM: content-aware bit mapping for single image super-resolution network with large input. In: CVPR (2023)
48. Wang, H., Chen, P., Zhuang, B., Shen, C.: Fully quantized image super-resolution networks. In: ACMMM (2021)
49. Wang, K., Liu, Z., Lin, Y., Lin, J., Han, S.: HAQ: hardware-aware automated quantization with mixed precision. In: CVPR (2019)
50. Wang, Z., Bovik, A.C., Sheikh, H.R., Simoncelli, E.P., et al.: Image quality assessment: from error visibility to structural similarity. IEEE TIP **13**(4), 600–612 (2004)
51. Xin, J., Wang, N., Jiang, X., Li, J., Huang, H., Gao, X.: Binarized neural network for single image super resolution. In: Vedaldi, A., Bischof, H., Brox, T., Frahm, J.-M. (eds.) ECCV 2020. LNCS, vol. 12349, pp. 91–107. Springer, Cham (2020). https://doi.org/10.1007/978-3-030-58548-8_6
52. Yang, L., Jin, Q.: FracBits: mixed precision quantization via fractional bit-widths. In: AAAI (2021)
53. Zhang, Y., Chen, H., Chen, X., Deng, Y., Xu, C., Wang, Y.: Data-free knowledge distillation for image super-resolution. In: CVPR (2021)
54. Zhang, Y., Li, K., Li, K., Wang, L., Zhong, B., Fu, Y.: Image super-resolution using very deep residual channel attention networks. In: ECCV (2018)
55. Zhang, Y., Tian, Y., Kong, Y., Zhong, B., Fu, Y.: Residual dense network for image super-resolution. In: CVPR (2018)
56. Zhang, Y., Wang, H., Qin, C., Fu, Y.: Learning efficient image super-resolution networks via structure-regularized pruning. In: ICLR (2022)
57. Zhong, Y., et al.: Dynamic dual trainable bounds for ultra-low precision super-resolution networks. In: Avidan, S., Brostow, G., Cissé, M., Farinella, G.M., Hassner, T. (eds.) ECCV 2022. LNCS, vol. 13678, pp. 1–18. Springer, Cham (2022). https://doi.org/10.1007/978-3-031-19797-0_1
58. Zhou, S., Wu, Y., Ni, Z., Zhou, X., Wen, H., Zou, Y.: DoReFa-Net: training low bitwidth convolutional neural networks with low bitwidth gradients. arXiv preprint arXiv:1606.06160 (2016)
59. Zhuang, B., Shen, C., Tan, M., Liu, L., Reid, I.: Towards effective low-bitwidth convolutional neural networks. In: CVPR (2018)

Large Motion Model for Unified Multi-modal Motion Generation

Mingyuan Zhang[1]([✉])[iD], Daisheng Jin[1][iD], Chenyang Gu[1][iD], Fangzhou Hong[1][iD], Zhongang Cai[1,2][iD], Jingfang Huang[1], Chongzhi Zhang[1][iD], Xinying Guo[1][iD], Lei Yang[2][iD], Ying He[1][iD], and Ziwei Liu[1]([✉])[iD]

[1] S-Lab, Nanyang Technological University, Singapore, Singapore
mingyuan001@e.ntu.edu.sg
[2] SenseTime Research, Shanghai, China
https://mingyuan-zhang.github.io/projects/LMM.html

Abstract. Human motion generation, a cornerstone technique in animation and video production, has widespread applications in various tasks like text-to-motion and music-to-dance. Previous works focus on developing specialist models tailored for each task without scalability. In this work, we present **Large Motion Model (LMM)**, a motion-centric, multi-modal framework that unifies mainstream motion generation tasks into a generalist model. A unified motion model is appealing since it can leverage a wide range of motion data to achieve broad generalization beyond a single task. However, it is also challenging due to the heterogeneous nature of substantially different motion data and tasks. LMM tackles these challenges from three principled aspects: 1) *Data:* We consolidate datasets with different modalities, formats and tasks into a comprehensive yet unified motion generation dataset, **Motion-Verse**, comprising 10 tasks, 16 datasets, a total of 320k sequences, and 100 million frames. 2) *Architecture:* We design an articulated attention mechanism **ArtAttention** that incorporates body part-aware modeling into Diffusion Transformer backbone. 3) *Pre-Training:* We propose a novel pre-training strategy for LMM, which employs variable frame rates and masking forms, to better exploit knowledge from diverse training data. Extensive experiments demonstrate that our generalist LMM achieves competitive performance across various standard motion generation tasks over state-of-the-art specialist models. Notably, LMM exhibits strong generalization capabilities and emerging properties across many unseen tasks. Additionally, our ablation studies reveal valuable insights about training and scaling up large motion models for future research.

Keywords: Motion Generation · Unified Model · Multi-Modality

1 Introduction

Humans perform a variety of motions in response to environmental changes, personal thoughts, and emotions to achieve their goals. These intricate motions often

M. Zhang, D. Jin and C. Gu—co-first authors.

Supplementary Information The online version contains supplementary material available at https://doi.org/10.1007/978-3-031-72624-8_23.

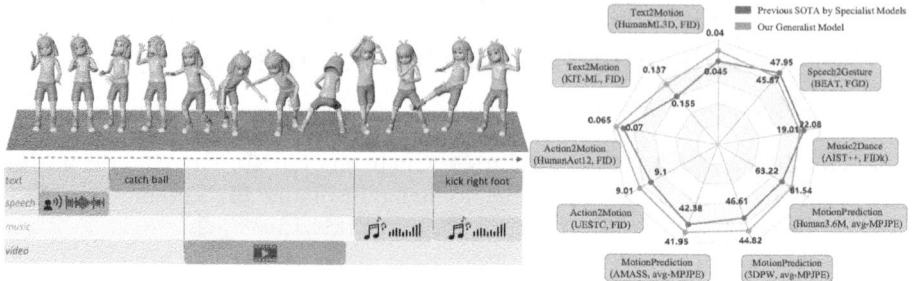

Fig. 1. We present Large Motion Model (LMM), the first generalist multi-modal motion generation model, that can perform multiple motion generation tasks simultaneously and achieve competitive performance across nine widely used benchmarks.

serve as the most information-rich element within videos and animations that feature characters, making motion generation a critical component of Generative AI, significantly influencing visual experiences and content quality. Automated human motion generation, aimed at producing continuous, natural, and logical human movements based on specific commands and control conditions, has attracted considerable attention in the computer vision field.

These specific sub-tasks are characterized by their defined inputs and outputs, with clear objectives. For instance, action-to-motion generates movements based on the category of the motion [34,102]; text-to-motion takes textual descriptions and produces corresponding movements [32,155]; music-to-dance creates dance moves in tune with the style and beat of the music input [29,121]. Dividing tasks allows for a more focused approach to each sub-task: constructing dedicated datasets and devising methods tailored to the task at hand. However, these approaches, when designed for a singular sub-task or modality, face challenges due to limited data quantity and a narrow data domain, which in turn can lead to models with restricted capabilities and poor generalization performance. In contrast, the objective of this work is to build a unified yet versatile foundation model for human motion generation, leveraging resources from a wide range of applications and achieving strong performance across the board.

Leveraging multi-modal and multi-task motion generation datasets presents significant challenges. First, disparate datasets feature varying motion formats and evaluation metrics, such as keypoint-based or rotation-based formats, and metrics assessing realism or diversity. Consequently, employing a single model to tackle multiple tasks and perform evaluations across different datasets is exceedingly difficult. Furthermore, transferring motion knowledge across tasks within these datasets is challenging, complicating the model's ability to integrate useful knowledge from various data sources to enhance its capabilities. For instance, differences in frame rates and the number of keypoints (sometimes even missing parts of the body) make it hard to unify the learned knowledge. Although some studies attempt to address multiple tasks simultaneously [79,166], they often utilize only two or three datasets with the same motion format, which limits their ability to achieve enhanced controllability and generalizability. In summary, integrated motion generation models for multi-modal and multi-task applications encounter the following problems: 1) Non-uniformity of motion data formats; 2) Different evaluation metrics due to varying task objectives; 3) Difficulty in transferring action knowledge across multiple tasks.

To deal with these challenges, we first amass multiple cross-modal motion datasets, encompassing 16 datasets with a total of 320k sequences and 100 million frames. These datasets span seven standard tasks: text-to-motion, action-to-motion, motion prediction, speech-to-gesture, music-to-dance, motion imitation, and motion in-betweening. Additionally, based on the standard tasks and multi-modal control signals, we introduce three new tasks: conditional motion prediction, conditional motion in-betweening, and multi-condition motion generation. Together, these datasets and tasks form our cross-modal motion benchmark, **MotionVerse**. To align the diverse formats of motion data, we employ a two-step approach: 1) We use the TOMATO representation [91] as a unified intermediary format, and then divide the entire representation into 10 parts. All kinds of motion representations are aligned to this format, with annotations indicating which body parts are present in each sequence; 2) We train a series of representation translators to convert the unified motion representation into the specific representations of each dataset during the testing phase. With Motion-Verse, we can smoothly use training data from various tasks and modalities, and conduct tests across different datasets.

Building on MotionVerse, we introduce the multi-modal Large Motion Model (**LMM**), which is built on a transformer-based diffusion model. Addressing the motion format inconsistency, we developed a body part-aware motion generation model. This model divides the human body into 10 segments and employs a specialized attention mechanism **ArtAttention**, featuring multi-conditioning, spatial-temporal independence, and mask injection, allowing for distinct control over different body parts. Furthermore, body part-aware modeling decomposes motion data from various datasets into relatively independent segments, thereby enabling the model to more effectively leverage knowledge learned across different datasets. Lastly, we adopt learning strategies from large language models (LLM), proposing a training method for LMM that combines unsupervised and supervised learning. In unsupervised learning, we enhance model robustness to frame rates through random frame rate augmentation and improve control over the continuity of body part movements by applying random masks to sequences and body parts in various ways. This training approach significantly leverages large amounts of multi-modal data to bolster LMM's capabilities. In supervised learning, we refine the capabilities of models to enhance their performance on specific tasks. Experimental results show that LMM achieves state-of-the-art results across various tasks, demonstrating its exceptional generalization performance, as shown in Fig. 1. Furthermore, LMM has the ability to process multi-modal inputs simultaneously, enabling it to accomplish unseen tasks.

In summary, our core contributions are as follows:

1. We present MotionVerse, a mega-scale, multi-modal, multi-task motion generation dataset that features a unified motion representation across a wide range of tasks and motion formats.
2. We introduce a Large Motion Model (LMM) that incorporates an advanced attention mechanism ArtAttention, allowing for precise and robust control, achieving finer results.
3. We devise a pre-training strategy for the LMM, including random frame rates and various masking techniques to fully leverage extensive motion datasets and enhance the model's capabilities. Additionally, through ablation studies on our training approach, we explored certain characteristics inherent in LMM's training process, laying a foundation for future research on the LMM.

2   Related Work

2.1   Motion Synthesis

Subtasks of motion generation are differentiated based on the types of control signals they utilize. Some tasks require the algorithm to synthesize motion sequences based on the given uncompleted motion sequences. For example, in the motion prediction task [1,2,8,9,14,20,30,35,48,94,123,124,133,142], the control condition is the first several poses, and the generated motion must logically follow the preceding sequence to ensure the extended sequence appears natural and reasonable. There are also works focusing on recovering the whole body motion with the given sparse upper-body tracking signals [12,21].

Action-to-motion [13,34,51,59,85,102,136,161] is an inverse task derived from the motion recognition task. While the latter identifies the action category from a motion sequence, the former involves generating a corresponding action sequence from an input action category. In the music-to-dance task [22,43,61,62,65,66,99,109,121,122,129,144,147,167], the control signal is music, and the motion sequence (dance) should be in accordance with the style and rhythm of the music. Bailando [121] addresses crucial spatial constraints and temporal coherence in dance generation by utilizing a codebook to store standardized dancing units, confining spatial constraints. It achieves temporal fluidity through a GPT designed to detect music beats. Another task based on audio is speech-to-gesture [4,23,57,84,141,149,150,163], where the input is the speaker's audio and the output is the corresponding gestures of the speaker, taking into account the speech's pauses, emotional fluctuations, etc. GestureDiffuCLIP [4] utilizes CLIP to extract style information from the input and then employs a diffusion model to generate gestures. Text-to-motion is one of the most attracting topic in conditional motion generation [3,5–7,15,18,24,25,28,29,31,33,36,38,40,42,47,49,50,52,55,56,67,74,76, 77,79,80,82,90,91,103,104,106,110–114,116,118,127,128,130,132,134,137,139, 140,145,146,148,151–157,159,160,164–166], where motions are generated based on textual descriptions. This requires the model to comprehend the meaning of the text and produce a corresponding sequence of motions. Previous works attempted to apply advanced generative model [128,155,159] to improve performance. While some other works focused on enhancing controllability [6,110,157]. Physical reality [113,151] and out-of-domain performance [41,82,127] are also vital topics in this field.

In addition, some works focus on human-scene interaction generation [44,73, 86,138], human-object interaction generation [19,37,60,63,64,78,101,105,119, 126,135,158] and human-human interaction generation [11,16,26,71,87,120,125, 162]. These tasks greatly expanded the scope of motion generation applications.

However, action generation targeting a single task often struggles with limited data volume and a singular data domain, leading to models with restricted capabilities and poor generalization. Our paper integrates various motion generation tasks, designing the Large Motion Model (LMM) to utilize multi-modality data from different tasks for model training. This enables the model to learn from various domains, enhancing its generalization capabilities.

2.2   Diffusion Model for Multiple Tasks

As diffusion models have made remarkable strides in image generation tasks, researchers have extended their application to a broader array of fields, including video generation [54,131,143], image editing [10,17,39,53,98], and motion generation [18,128,155,157,165], etc. Moreover, given the limited functionality and control over generation provided by single-modal conditions, multi-modal inputs have been introduced into diffusion models to enhance their versatility and control capabilities. In the realm of image generation, UNIMO-G [70] takes both images and text as inputs, utilizing the subjects in pictures and textual prompts to generate realistic images that match complex semantics. For video generation, MM-Diffusion [115] incorporates audio and video in a multi-modal manner, enabling the harmonious adaptation of audio and visuals to produce realistic videos with sound. In image editing tasks, the integration of multi-modal inputs with diffusion models has made image editing more flexible and convenient. Control-color [72] combines text, strokes, exemplars, and other conditions to achieve interactive, multi-modal controlled image coloring. InstructAny2pix [68] uses similar conditions for multi-modal control over inpainting. Likewise, in the field of motion generation, works like MotionDiffuse [155] and MDM [128] have tackled text-to-motion and action-to-motion. MCM [79], UDE [166] has addressed text-to-motion and music-to-dance. MoFusion [92] attempted to learn a better motion prior. In summary, across different domains, multi-modal diffusion models enrich the content and enable more precise control of generation tasks. Previous dual-modal motion generation methods often utilize similar annotation formats, such as SMPL [88], facilitating easy data alignment. However, the multi-modal datasets we collected cover a broader domain span, and the formats for motion annotation are extremely diverse. Thus we propose a comprehensive benchmark, MotionVerse, to unify the motion representation, ultimately leading to the development of LMM.

3   MotionVerse: Unified Motion Generation Datasets

3.1   Motivation

Large models have been extensively studied in fields such as language, image, and video. These models, by absorbing common knowledge from vast amounts of data and leveraging a unified task format, demonstrate outstanding performance across multiple tasks. However, on the path towards large motion models, a significant challenge lies ahead: the inconsistent motion formats across datasets. Specifically, there are three types of inconsistencies:

1. **Inconsistent pose representations:** For instance, the UESTC [46] benchmark adopts a 6D rotation representation based on SMPL [89], while the Human3.6M [45] motion prediction benchmark uses keypoint coordinates.
2. **Inconsistent number of keypoints:** For example, TED-Gesture++ [150] only includes upper body keypoints, while NTU-RGBD 120 [83,117] lacks fine-grained keypoints for the hands.
3. **Inconsistent frame rates:** KIT-ML [107] operates at 12.5 fps, whereas Motion-X [75] runs at 30 fps.

Such differences not only demand models capable of handling diverse data formats but also pose significant challenges in acquiring common knowledge.

To address this challenge, we introduce the first unified and comprehensive motion-centric benchmark **MotionVerse**, the workflow of which is shown in Fig. 2. MotionVerse possesses three advantages:

Table 1. Task definitions. x and k_* are the frame indices, and k_* represents the boundary of the mask.

Task	Mask			Condition
T2M	None			Text
A2M	None			Action
M2D	None			Music
S2G	None			Speech
MIm	None			Video
MP	$m =$	$\begin{cases} 0 & \text{if } x > k \\ 1 & \text{Otherwise} \end{cases}$		None
MIn	$m =$	$\begin{cases} 0 & \text{if } k_1 < x \le k_2 \\ 1 & \text{Otherwise} \end{cases}$		None
CMP	$m =$	$\begin{cases} 0 & \text{if } x > k \\ 1 & \text{Otherwise} \end{cases}$		Any single modal
CMI	$m =$	$\begin{cases} 0 & \text{if } k_1 < x \le k_2 \\ 1 & \text{Otherwise} \end{cases}$		Any single modal
MMG	None			Multi-modal

Table 2. Dataset information. We collect 16 widely used dataset, process all motion data into our intermediate format.

Dataset	#Seq	#Frames	Repr	Condition
HumanML3D [32]	14614	2M	H3D	Text
KIT-ML [107]	2485	245K	H3D	Text
Motion-X [75]	50863	9M	SMPLX	Text
BABEL [108]	5123	7M	SMPLX	Text
UESTC [46]	25600	10M	SMPL	Action
HumanAct12 [34]	1191	90K	Kpt3D	Action
NTU-RGB-D 120 [83,117]	139656	10M	Kpt3D	Action
AMASS [93]	14244	20M	SMPLX	–
3DPW [96]	81	140K	SMPL	Video
Human3.6M [45]	210	530K	Kpt3D	Video
TED-Gesture++ [150]	34491	10M	Kpt3D	Speech
TED-Expressive [84]	27221	8M	Kpt3D	Speech
Speech2Gesture-3D [58]	1047	1M	Kpt3D	Speech
BEAT [81]	1639	18M	Kpt3D	Speech
AIST++ [66]	1408	1M	SMPL	Music
MPI-INF-3DHP [97]	16	1M	Kpt3D	Video
Total	320K	100M	–	–

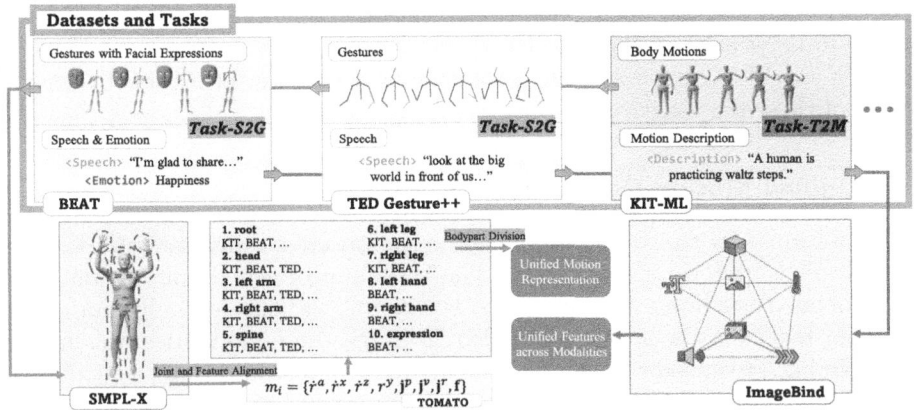

Fig. 2. MotionVerse. We preprocess distinct motion-centric datasets into a unified format. As for motion sequences, we initially convert them to the TOMATO [91] representation and then further divide them into 10 independent body parts, serving as our unified motion representation. To tackle multi-modal condition signals, we employ ImageBind [27] to transform them into unified features across modalities.

1. **Unified Problem Formulation:** we describe mainstream tasks within a unified framework, reducing the need to consider task-specific properties during model design.
2. **Unified Motion Representation:** we convert the motion formats of various datasets into a unified intermediate representation, enabling the model to acquire common knowledge from the originally diverse data formats and to be evaluated on different datasets smoothly.
3. **Systematicness and Comprehensiveness:** we encompass 10 tasks across 16 datasets, comprising 320K sequences and nearly 100M frames of motion data, as shown in Table 2, which enables us to explore large motion models.

3.2 Unified Problem Formulation

Motion generation encompasses a variety of sub-tasks with differing objectives. To standardize the data for these tasks, we first formalize the problem for motion generation tasks as:

$$\Theta = M(\mathbf{x}, \mathrm{m}, \mathrm{c}), \tag{1}$$

where $\mathbf{x} \in \mathbb{R}^{F \times D}$ represents the motion data, $\mathrm{m} \in \{0,1\}^{F \times D}$ defines the model's visibility scope, which is used in motion completion tasks, such as motion prediction and motion in-betweening. Here F is the number of frames of the target motion sequence. D is the dimensionality of each pose state. c is a set of condition control signals, including text, speech, music, and video.

We have included seven standard tasks: action-to-motion (A2M), text-to-motion (T2M), music-to-dance (M2D), speech-to-gesture (S2G), motion prediction (MP), motion in-betweening (MIn) and motion imitation (MIm), along with three multi-modal tasks: conditional motion prediction (CMP), conditional motion in-betweening (CMI), and multi-condition motion generation (MMG). For each specific task, we can align their inputs using Eq. (1). Table 1 lists the details of these ten tasks.

3.3   Unified Motion Representation

For data formats with varying inputs and outputs across different tasks, we pre-process to ensure format consistency. The basic unit of general motion datasets can be defined as <input-output> pairs. For inputs, we consider multi-modalities including text, speech, music, and video. To align these modalities, we employ Imagebind [27] to encode different inputs into unified features of the same dimension, which ensures semantic consistency across modalities.

As for the output, it encompasses motion sequences in various formats, such as keypoints and SMPL [89]. To standardize motion representation, we define a unified format similar to TOMATO [91], where we remove foot contact from TOMATO. Our representation is described as:

$$m_i = \{\dot{r}^a, \dot{r}^x, \dot{r}^z, r^y, \mathbf{j}^p, \mathbf{j}^v, \mathbf{j}^r, \mathbf{f}\}, \tag{2}$$

where r denotes information related to the root. Specifically, $\dot{r}^a \in \mathbb{R}$ is the angular velocity along the Y-axis, $(\dot{r}^x, \dot{r}^z \in \mathbb{R})$ represent linear velocities on the XZ-plane, and r^y indicates the root's height. $\mathbf{j}^p \in \mathbb{R}^{3(J-1)}$, $\mathbf{j}^v \in \mathbb{R}^{3J}$, and $\mathbf{j}^r \in \mathbb{R}^{6(J-1)}$ correspond to the position, velocity, and rotation of local keypoints relative to the root, with J denoting the number of joints ($J-1$ means all joints without the root joint). Here we follow SMPL-X [100] and consider 22 main body joints and 30 hand joints. Lastly, $\mathbf{f}$ denotes facial expression [69].

We further divide this representation into ten independent parts: global orientation and trajectory, face expression, head, spine, left arm, right arm, left

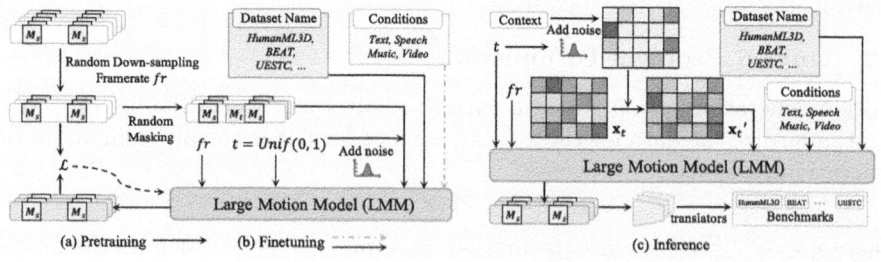

(a) Pretraining ——→   (b) Finetuning ‑‑‑‑→   (c) Inference

Fig. 3. Overall pipeline of LMM. Left: Our two-stage training procedure, including unsupervised pretraining and supervised fine-tuning. Random down-sampling and random mask strategies are applied to enhance knowledge absorption. **Right:** The generic inference process of LMM. The noised motion sequence and the given context are initially merged before being input into the network. LMM will then synthesize motion sequences, consistent with the provided multi-modal condition signals.

leg, right leg, left hand, and right hand. When processing raw data, we allow for missing body parts and annotate them in the metadata. For missing keypoints, we manually estimate the missing points from known ones for completion, while extra keypoints are discarded in this process. We then train a series of motion translators to map our unified motion representation to each dataset's specific one. Thus, in the testing process of different tasks, once the model outputs in our unified format, we can map the output to the corresponding motion format through the translator, facilitating smooth metric evaluation.

4 Large Motion Model

Our model architecture closely follows the literature [128, 155], built upon a transformer-based diffusion model. We primarily reference the FineMoGen [157] as our baseline and extend it to support various condition signals, multi-tasking, multiple frame rates, and various mask forms. The overall workflow is illustrated in Fig. 3 while the detailed architecture is shown in Fig. 4.

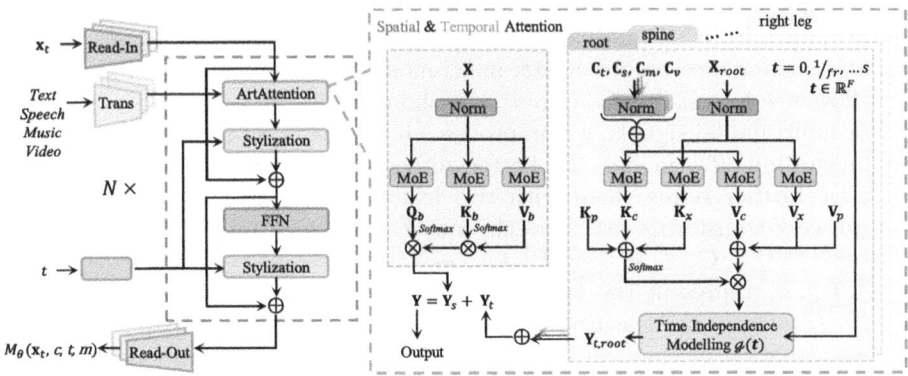

Fig. 4. Architecture of LMM. LMM is a transformer-based diffusion model. Dataset-dependent Read-In layers and Read-Out layers facilitate the conversion of the motion sequence between our intermediate representation and the latent feature space. In the stem of LMM, ArtAttention refines the feature representations through the spatial and temporal attention branches.

4.1 Transformer-Based Diffusion Model

Diffusion models are powerful generative models, capable of producing high-quality, diverse outcomes, garnering widespread attention, and demonstrating formidable generative capabilities in many fields. Its essence lies in two intertwined processes: the forward diffusion process and the reverse diffusion process. More details are provided in the supplementary materials.

4.2   Read-In Layer and Read-Out Layer

The read-in layer is responsible for transforming noised motion data into feature representations of the form $F \times H \times D$, while the read-out layer generates clean motion data from the feature space. Here, the first dimension represents the number of frames in the motion sequence, and the second dimension represents the number of body parts. Although we standardize all motion data into a unified motion format, the differences in distribution among various datasets cannot be completely ignored. Therefore, in the backbone network's read-in and read-out stages, we employ dataset-dependent motion encoders and decoders. Additionally, to obtain more comprehensive knowledge for practical applications, during training, there is a 10% probability of replacing the original dataset name with "all". Consequently, the corresponding read-in and read-out layers can be better applied in real-world application scenarios.

4.3   ArtAttention

To achieve outstanding zero-shot continuous generation capability, our model builds upon the SAMI module from FineMoGen [157], incorporating upgrades to address three new requirements: multi-modal condition, various frame rates, and allowance for missing body parts, as shown in Fig. 4.

For multi-modal signals, we preprocess all signals into token sequences using the ImageBind [27] model. To better integrate these features into our network, we further refine them with two learnable transformer encoder layers. This process transforms text, speech, music, and video into feature sequences $\mathbf{C}_t \in \mathbb{R}^{L_t \times (H \cdot D)}, \mathbf{C}_s \in \mathbb{R}^{L_s \times (H \cdot D)}, \mathbf{C}_m \in \mathbb{R}^{L_m \times (H \cdot D)}, \mathbf{C}_v \in \mathbb{R}^{L_v \times (H \cdot D)}$, where L_t, L_s, L_m, L_v represent the lengths of the corresponding condition sequences, and $H \cdot D$ denotes the feature length of each element.

Our attention mechanism can be divided into two main components: body-part attention(spatial attention) and temporal attention. Assuming the feature representation of our motion sequence is $\mathbf{X} \in \mathbb{R}^{F \times H \times D}$. In the body-part attention section, due to the presence of inherent missing body parts in our data and the masked body parts introduced artificially during pre-training, we cannot utilize a fixed set of coefficients to determine the mutual contributions among body parts. Therefore, unlike the design in FineMoGen, for each frame, we opt to use an attention structure to obtain a set of refined features $\mathbf{Y_s} \in \mathbb{R}^{F \times H \times D}$.

In the temporal attention section, we aim to leverage the self-correlation inherent in the motion features $\mathbf{X}$ and guidance obtained from the condition signals $\mathbf{C}_t, \mathbf{C}_s, \mathbf{C}_m, \mathbf{C}_v$ to generate higher-quality features for each body part in every frame. Here, we employ a Multi-head Attention mechanism, with each head focusing on a specific body part, emphasizing the utilization of temporal correlations to optimize features. We begin by utilizing Mixture-of-Expert to obtain a set of features $\mathbf{K} \in \mathbb{R}^{L \times D}$ from these condition features, where L represents the sequence length of the corresponding feature source.

Empirically, we found that directly concatenating motion sequences and condition sequences like FineMoGen and then applying Softmax processing hinders the modeling of self-correlation in multi-condition scenarios, resulting in lower motion quality, especially when the condition feature sequence is much longer than the motion sequence. Therefore, we independently normalize the motion feature $\mathbf{K}_x \in \mathbb{R}^{F \times D}$ obtained and then normalize condition features. To support unconditional generation, we introduce 64 learnable tokens $\mathbf{K}_p$ and $\mathbf{V}_p$ as placeholders. These tokens are concatenated with all condition signals for normalization, resulting in $\mathbf{K}_c \in \mathbb{R}^{(64+L_t+L_s+L_m+L_v) \times D}$. This approach also facilitates better blending of the model across different conditions. The remaining processing is similar to FineMoGen. After obtaining a series of time-varying signals, for each frame's pose, we use time as the sole query feature to calculate the correlation between each frame and each time-varying signal, as well as the values at each signal point. Unlike FineMoGen, which uses frame indices to represent time, we use real time to support different frame rates. More details are introduced in the supplementary material. Suppose the output of the temporal attention section is $\mathbf{Y}_t \in \mathbb{R}^{F \times H \times D}$, then the output of the entire ArtAttention is $\mathbf{Y} = \mathbf{Y}_s + \mathbf{Y}_t$.

4.4 Pre-training and Fine-Tuning

Leveraging MotionVerse, we have collected a vast array of <input-motion> data pairs. Disregarding the inputs, the abundant motion data inherently contains valuable information, which can enable the model to comprehend numerous characteristics of human motion, such as coherence, balance, and joint-based rotations, among others. To enable the model to better acquire common knowledge across datasets, we divide the entire training process into two main parts: **Unsupervised Pre-Training** and **Supervised Fine-Tuning**. In the first stage, we require our model to learn motion priors independent of conditions. While in the second stage, the model is supposed to learn the correlation between condition signals and motion sequences.

As illustrated in Fig. 3, in the unsupervised pretraining phase, to enrich the model's prior knowledge from these motion sequences, we employ random downsampling and random masking strategies for data augmentation. Since our representation includes terms related to velocity, when downsampling, we need to recalculate the velocity values to match the downsampling rate, while the terms related to states remain unchanged. This approach enables the model to better learn from data with different original frame rates. As mentioned earlier in the data preprocessing part, some body parts in the sequences are masked out, such as the detailed keypoints of the left and right hands in the KIT dataset. We denote this original mask as $\mathbf{M}_s \in \{0,1\}^{F \times H}$. Based on this, we additionally apply masking with a certain probability to obtain a new mask $\mathbf{M}_t \in \{0,1\}^{F \times H}$. After the model performs the read-in operation, we replace the body parts marked as 1 in $\mathbf{M}_t$ with learnable empty tokens. When calculating the loss, we only ignore the parts marked by $\mathbf{M}_s$. This means that the model not only needs to restore the noised sequence to its clean parts but also utilize the visible part to infer the rest. With this modeling, the knowledge embedded in the data with missing body parts can be better absorbed by the model.

In the supervised fine-tuning phase, our goal is to enable the model to learn the relationship between condition signals and motion sequences. Here, we pass the preprocessed condition token sequences as additional inputs to the model. To support classifier-free guidance, during training, we randomly mask out the condition signals with a probability of 10%.

5   Experiments

5.1   Implementation Details

We designed four variants: LMM-Tiny, LMM-Small, LMM-Base, and LMM-Large, which have 90M, 160M, 410M, 760M parameters respectively. We use all data in MotionVerse for both pretraining and finetuning, except for the sequences used in evaluation. We maintained a fixed total batch size of 512. For the Tiny model, we conducted training directly on 8 NVIDIA V100 GPUs with 32 GB memory each, with a batch size of 64 per GPU. For larger models, we use FP16 and gradient accumulation to achieve training effects equivalent to a batch size of 512 without exceeding 32 V100 GPUs. During the pre-training phase, we employed the Adam optimizer with a fixed learning rate of 2×10^{-4} for 80K iterations. In the fine-tuning phase, we used the same optimizer, initially iterating for 20K steps with a learning rate of 2×10^{-4}, followed by 20K steps with a learning rate of 2×10^{-5}. For more details, please refer to the supplementary material.

Table 3. Quantitative results of text-to-motion generation on the HumanML3D test set. '↑'('↓') indicates that the values are better if the metric is larger (smaller). We run all the evaluations 20 times and report the average metric and 95% confidence interval. "MM" is MultiModality. The best scores are bold, and the second-best results are underlined.

Methods	R Precision↑			FID↓	MM Dist↓	Diversity↑	MM↑
	Top 1	Top 2	Top 3				
Real motions	$0.511^{\pm.003}$	$0.703^{\pm.003}$	$0.797^{\pm.002}$	$0.002^{\pm.000}$	$2.974^{\pm.008}$	$9.503^{\pm.065}$	–
T2M-GPT [153]	$0.491^{\pm.003}$	$0.680^{\pm.003}$	$0.775^{\pm.002}$	$0.116^{\pm.004}$	$3.118^{\pm.011}$	$\underline{9.761}^{\pm.081}$	$1.856^{\pm.011}$
MDM [128]	–	–	$0.611^{\pm.007}$	$0.544^{\pm.044}$	$5.566^{\pm.027}$	$9.559^{\pm.086}$	$\mathbf{2.799}^{\pm.072}$
FineMoGen [157]	$0.504^{\pm.002}$	$0.690^{\pm.002}$	$0.784^{\pm.002}$	$0.151^{\pm.008}$	$2.998^{\pm.008}$	$9.263^{\pm.094}$	$\underline{2.696}^{\pm.079}$
MoMask [31]	$\underline{0.521}^{\pm.002}$	$\underline{0.713}^{\pm.002}$	$\underline{0.807}^{\pm.002}$	$\underline{0.045}^{\pm.002}$	$2.958^{\pm.008}$	–	$1.241^{\pm.040}$
LMM-Tiny	$0.496^{\pm.002}$	$0.685^{\pm.002}$	$0.785^{\pm.002}$	$0.415^{\pm.002}$	$3.087^{\pm.012}$	$9.176^{\pm.074}$	$1.465^{\pm.048}$
LMM-Small	$0.505^{\pm.002}$	$0.693^{\pm.002}$	$0.789^{\pm.002}$	$0.227^{\pm.002}$	$3.051^{\pm.012}$	$9.295^{\pm.076}$	$1.761^{\pm.049}$
LMM-Base	$0.511^{\pm.002}$	$0.710^{\pm.002}$	$0.802^{\pm.002}$	$0.138^{\pm.002}$	$2.971^{\pm.012}$	$9.573^{\pm.076}$	$2.426^{\pm.054}$
LMM-Large	$\mathbf{0.525}^{\pm.002}$	$\mathbf{0.719}^{\pm.002}$	$\mathbf{0.811}^{\pm.002}$	$\mathbf{0.040}^{\pm.002}$	$\mathbf{2.943}^{\pm.012}$	$\mathbf{9.814}^{\pm.076}$	$2.683^{\pm.054}$

Table 4. Quantitative results of motion prediction on the AMASS and 3DPW test set for different time steps (ms). We report the MPJPE error in *mm*.

Method	AMASS-BMLrub								3DPW							
	80	160	320	400	560	720	880	1000	80	160	320	400	560	720	880	1000
LTD-10-10 [95]	<u>10.3</u>	**19.3**	36.6	44.6	61.5	75.9	86.2	91.2	<u>12.0</u>	**22.0**	38.9	46.2	59.1	69.1	76.5	81.1
SIMLPE [35]	10.8	<u>19.6</u>	<u>34.3</u>	40.5	<u>50.5</u>	<u>57.3</u>	62.4	65.7	12.1	<u>22.1</u>	38.1	44.5	54.9	62.4	68.2	72.2
GCNext [133]	**10.2**	**19.3**	**34.1**	<u>40.3</u>	50.6	<u>57.3</u>	<u>62.0</u>	<u>65.3</u>	**11.8**	**22.0**	<u>37.9</u>	44.2	<u>55.1</u>	<u>62.1</u>	<u>67.8</u>	<u>72.0</u>
LMM-Tiny	15.9	24.1	38.2	45.9	61.2	73.4	80.3	87.2	17.3	26.2	40.1	47.3	62.8	74.8	82.5	87.0
LMM-Small	14.7	23.2	37.5	43.8	58.3	69.2	75.2	81.9	16.2	25.7	39.4	45.9	60.7	71.5	79.5	82.8
LMM-Base	13.1	21.5	35.9	41.1	53.6	60.8	66.9	70.3	14.1	23.9	38.2	<u>44.1</u>	55.3	64.8	70.3	73.6
LMM-Large	12.8	20.9	<u>34.3</u>	**39.6**	**49.1**	**55.3**	**60.5**	**63.1**	13.1	22.6	**37.1**	**42.4**	**52.4**	**59.2**	**63.8**	**68.0**

5.2 Quantitative Results

We evaluate our LMM variants on three tasks: text-to-motion, music-to-dance, motion prediction, and four datasets: HumanML3D [32], 3DPW [96], AMASS [93], and AIST++ [66], as shown in Table 3, Table 4 and Table 5. More experimental results are reported in the supplementary material.

Text-to-Motion. Table 3 demonstrates that our LMM-Large surpasses other existing works in terms of accuracy and fidelity while maintaining comparable diversity. On the other hand, LMM-Tiny, which shares a similar structure with FineMoGen, performs worse than it. This discrepancy can be attributed to the significant challenges posed by the large amounts of diverse data and the trade-offs across different tasks during model training, especially for smaller models like LMM-Tiny.

Motion Prediction. Table 4 reports the performance on AMASS and 3DPW test splits. The superior performance of LMM-Large can be attributed to its

Table 5. Quantitative results for Music-conditioned Dance Generation. Quantitative results on AIST++ test set.

Methods	Motion Quality		Motion Diversity		Freezing		Best Align Score↑
	FID_k ↓	$FID_g^\dagger$ ↓	Div_k ↑	$FID_g^\dagger$ ↑	PFF↓	AUC_f ↓	
Ground-truth	17.10	10.60	8.19	7.45	0.00	0.00	0.2374
DanceNet [167]	69.18	25.49	2.86	2.85	**0.00**	<u>0.98</u>	0.1430
DanceRevolution [43]	73.42	25.52	3.52	4.87	<u>11.01</u>	12.22	0.1950
Bailando [121]	28.16	**9.62**	7.83	6.34	14.91	13.25	**0.2332**
TM2D [29]	**19.01**	<u>20.09</u>	<u>9.45</u>	6.36	**0.00**	**0.00**	0.2049
LMM-Tiny	37.62	28.95	6.92	5.94	**0.00**	**0.00**	0.1736
LMM-Small	34.18	27.53	7.46	6.17	**0.00**	**0.00**	0.1791
LMM-Base	25.43	24.18	9.05	<u>6.55</u>	**0.00**	**0.00**	0.2084
LMM-Large	<u>22.08</u>	21.97	**9.85**	**6.72**	**0.00**	**0.00**	<u>0.2249</u>

robust motion prior, which is obtained from the mega-scale data. It is worth noting that due to the errors introduced by the motion translation step, the accuracy of LMM-Large is still lower than other methods in short-distance prediction. However, it exhibits a significant advantage in long-distance prediction. Furthermore, we observed that the advantage of LMM-Large is more pronounced on 3DPW. This is because the 3DPW benchmark demands higher generalization ability from the model. After extensive learning of motion priors, our LMM-Large exhibits a more prominent performance on out-of-distribution tests.

Music-to-Dance. Our Large model achieves comparable performance to the current state-of-the-art, as shown in Table 5. In terms of diversity-related metrics, our approach demonstrates a significant advantage. Our performance in the metrics FID_k and FID_g did not surpass existing methods. One possible reason could be the relatively small proportion of the music2dance dataset in the current dataset composition.

5.3   Ablation Study

Table 6. Ablation of the pretraining strategy. All experiments utilized LMM-Base as the base model.

	Downsample	Random Mask	Attention	HumanML3D			3DPW		
				Top 1	FID	MModality	80	400	1000
1)	–	–	ArtAttention	$0.031^{\pm.001}$	$32.814^{\pm.176}$	$5.293^{\pm.129}$	17.3	48.4	89.3
2)	✓	–	ArtAttention	$0.028^{\pm.001}$	$31.365^{\pm.171}$	$5.714^{\pm.147}$	16.2	46.5	78.4
3)	–	✓	ArtAttention	$0.515^{\pm.002}$	$0.151^{\pm.002}$	$2.214^{\pm.051}$	14.5	45.5	76.1
4)	✓	✓	SAMI [157]	$0.400^{\pm.009}$	$1.866^{\pm.009}$	$2.983^{\pm.071}$	15.9	47.2	80.9
5)	✓	✓	ArtAttention	$0.511^{\pm.002}$	$0.138^{\pm.002}$	$2.426^{\pm.054}$	14.1	44.1	73.6

Table 6 shows the ablation results. We observed that random masking is a necessary component. When the model's expressive power is strong enough, it can directly recover the clean motion sequence from the noised motion sequence. Consequently, during the fine-tuning stage, our condition signal may not play its expected role. Introducing random masking during training will make it more difficult for the model to solely restore the original sequence from the motion sequence, leading it to rely more on the additional information provided by the condition signal. Additionally, we found that both downsampling and random masking strategies are beneficial for improving the multimodality metrics in the text-to-motion task. This implies that the model can better absorb knowledge from different datasets with the help of these two strategies. These strategies also significantly impact the effectiveness of motion prediction. Finally, we compared our proposed ArtAttention with the original SAMI and found that our proposed method is more suitable for the scenario of large motion models.

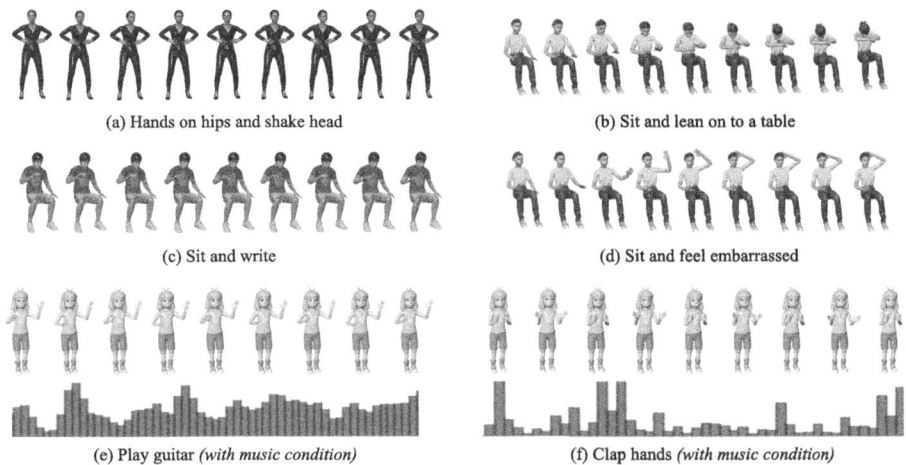

(a) Hands on hips and shake head

(b) Sit and lean on to a table

(c) Sit and write

(d) Sit and feel embarrassed

(e) Play guitar *(with music condition)*

(f) Clap hands *(with music condition)*

Fig. 5. Visualization results of LMM-Large. Figure a)–d) show examples of text-driven motion generation. Figure e) and f) show synthesized motion sequences under both textual and musical constraints.

5.4 Qualitative Results

As shown in Fig. 5(a)–(d), LMM-Large can response to diverse textual descriptions with fine-grained control, which benefits from the large-scale training data and the well-designed architecture. In addition, Fig. 5(e)–(f) provide examples for motion generation under both text description and music rhythms. Our generated motions successfully execute the given commands and follow the music beats simultaneously. For more visualization results, please kindly refer to the demo video in our homepage.

6   Conclusion and Discussion

In this paper, we establish a comprehensive motion-centric benchmark, Motion-Verse, comprising then conditional motion generation and motion completion tasks. We align all motion data to a unified intermediate format and convert all condition signals into token sequences that are closer in feature space. Building upon this foundation, we introduce the first large motion model, LMM, capable of generating high-quality actions under multi-condition guidance. We identify and address three challenges encountered in constructing large motion models through careful model structure design, especially our used novel attention module, ArtAttention. Our proposed LMM model achieves comparable performance and even surpasses existing state-of-the-art methods.

Acknowledegment. This study is supported by the Ministry of Education, Singapore, under its MOE AcRF Tier 2 (MOET2EP20221- 0012), NTU NAP, and under the RIE2020 Industry Alignment Fund - Industry Collaboration Projects (IAF-ICP)

Funding Initiative, as well as cash and in-kind contribution from the industry partner(s).

References

1. Ahn, H., Mascaro, E.V., Lee, D.: Can we use diffusion probabilistic models for 3D motion prediction? In: 2023 IEEE International Conference on Robotics and Automation (ICRA) (2023)
2. Ahn, H., Mascaro, E.V., Lee, D.: Can we use diffusion probabilistic models for 3D motion prediction? arXiv preprint arXiv:2302.14503 (2023)
3. Ahuja, C., Morency, L.P.: Language2Pose: natural language grounded pose forecasting. In: 2019 International Conference on 3D Vision (3DV), pp. 719–728. IEEE (2019)
4. Ao, T., Zhang, Z., Liu, L.: GestureDiffuCLIP: gesture diffusion model with clip latents. ACM Trans. Graph. (2023)
5. Athanasiou, N., Petrovich, M., Black, M.J., Varol, G.: TEACH: temporal action composition for 3D humans. In: 2022 International Conference on 3D Vision (3DV), pp. 414–423. IEEE (2022)
6. Athanasiou, N., Petrovich, M., Black, M.J., Varol, G.: SINC: spatial composition of 3d human motions for simultaneous action generation. In: IEEE/CVF International Conference on Computer Vision, ICCV 2023, Paris, France, 1–6 October 2023, pp. 9950–9961 (2023)
7. Azadi, S., Shah, A., Hayes, T., Parikh, D., Gupta, S.: Make-an-animation: large-scale text-conditional 3d human motion generation. In: IEEE/CVF International Conference on Computer Vision, ICCV 2023, Paris, France, 1–6 October 2023, pp. 14993–15002 (2023)
8. Barquero, G., Escalera, S., Palmero, C.: BeLFusion: latent diffusion for behavior-driven human motion prediction. In: Proceedings of the IEEE/CVF International Conference on Computer Vision, pp. 2317–2327 (2023)
9. Barsoum, E., Kender, J., Liu, Z.: HP-GAN: probabilistic 3D human motion prediction via GAN. In: Proceedings of the IEEE Conference on Computer Vision and Pattern Recognition Workshops, pp. 1418–1427 (2018)
10. Brooks, T., Holynski, A., Efros, A.A.: InstructPix2Pix: learning to follow image editing instructions. In: Proceedings of the IEEE/CVF Conference on Computer Vision and Pattern Recognition, pp. 18392–18402 (2023)
11. Cai, Z., et al.: Digital life project: autonomous 3D characters with social intelligence. arXiv preprint arXiv:2312.04547 (2023)
12. Castillo, A., et al.: BoDiffusion: diffusing sparse observations for full-body human motion synthesis (2023)
13. Cervantes, P., Sekikawa, Y., Sato, I., Shinoda, K.: Implicit neural representations for variable length human motion generation. In: Avidan, S., Brostow, G., Cissé, M., Farinella, G.M., Hassner, T. (eds.) ECCV 2022. LNCS, vol. 13677, pp. 356–372. Springer, Cham (2022). https://doi.org/10.1007/978-3-031-19790-1_22
14. Chen, L.H., Zhang, J., Li, Y., Pang, Y., Xia, X., Liu, T.: HumanMAC: masked motion completion for human motion prediction (2023)
15. Chen, X., et al.: Executing your commands via motion diffusion in latent space. In: Proceedings of the IEEE/CVF Conference on Computer Vision and Pattern Recognition, pp. 18000–18010 (2023)

16. Chopin, B., Tang, H., Daoudi, M.: Bipartite graph diffusion model for human interaction generation. In: Proceedings of the IEEE/CVF Winter Conference on Applications of Computer Vision, pp. 5333–5342 (2024)
17. Couairon, G., Verbeek, J., Schwenk, H., Cord, M.: DiffEdit: diffusion-based semantic image editing with mask guidance. In: The Eleventh International Conference on Learning Representations (2022)
18. Dabral, R., Mughal, M.H., Golyanik, V., Theobalt, C.: MoFusion: a framework for denoising-diffusion-based motion synthesis. In: Proceedings of the IEEE/CVF Conference on Computer Vision and Pattern Recognition, pp. 9760–9770 (2023)
19. Diller, C., Dai, A.: CG-HOI: contact-guided 3D human-object interaction generation. arXiv preprint arXiv:2311.16097 (2023)
20. Diller, C., Funkhouser, T., Dai, A.: Forecasting characteristic 3D poses of human actions (2022)
21. Du, Y., Kips, R., Pumarola, A., Starke, S., Thabet, A., Sanakoyeu, A.: Avatars grow legs: generating smooth human motion from sparse tracking inputs with diffusion model. In: Proceedings of the IEEE/CVF Conference on Computer Vision and Pattern Recognition, pp. 481–490 (2023)
22. Gao, X., Hu, L., Zhang, P., Zhang, B., Bo, L.: DanceMeld: unraveling dance phrases with hierarchical latent codes for music-to-dance synthesis. arXiv preprint arXiv:2401.10242 (2023)
23. Ghorbani, S., Ferstl, Y., Holden, D., Troje, N.F., Carbonneau, M.A.: ZeroEGGS: zero-shot example-based gesture generation from speech. In: Computer Graphics Forum, vol. 42, pp. 206–216. Wiley Online Library (2023)
24. Ghosh, A., Cheema, N., Oguz, C., Theobalt, C., Slusallek, P.: Synthesis of compositional animations from textual descriptions. In: Proceedings of the IEEE/CVF International Conference on Computer Vision, pp. 1396–1406 (2021)
25. Ghosh, A., Cheema, N., Oguz, C., Theobalt, C., Slusallek, P.: Text-based motion synthesis with a hierarchical two-stream RNN. In: ACM SIGGRAPH 2021 Posters, pp. 1–2 (2021)
26. Ghosh, A., Dabral, R., Golyanik, V., Theobalt, C., Slusallek, P.: ReMoS: reactive 3D motion synthesis for two-person interactions. arXiv preprint arXiv:2311.17057 (2023)
27. Girdhar, R., et al.: ImageBind: one embedding space to bind them all. In: Proceedings of the IEEE/CVF Conference on Computer Vision and Pattern Recognition (CVPR), pp. 15180–15190 (2023)
28. Goel, P., Wang, K.C., Liu, C.K., Fatahalian, K.: Iterative motion editing with natural language. arXiv preprint arXiv:2312.11538 (2023)
29. Gong, K., et al.: TM2D: bimodality driven 3D dance generation via music-text integration. In: Proceedings of the IEEE/CVF International Conference on Computer Vision, pp. 9942–9952 (2023)
30. Gopalakrishnan, A., Mali, A., Kifer, D., Giles, L., Ororbia, A.G.: A neural temporal model for human motion prediction. In: Proceedings of the IEEE/CVF Conference on Computer Vision and Pattern Recognition, pp. 12116–12125 (2019)
31. Guo, C., Mu, Y., Javed, M.G., Wang, S., Cheng, L.: MoMask: generative masked modeling of 3D human motions. arXiv preprint arXiv:2312.00063 (2023)
32. Guo, C., et al.: Generating diverse and natural 3D human motions from text. In: Proceedings of the IEEE/CVF Conference on Computer Vision and Pattern Recognition, pp. 5152–5161 (2022)
33. Guo, C., Zuo, X., Wang, S., Cheng, L.: TM2T: stochastic and tokenized modeling for the reciprocal generation of 3D human motions and texts. In: Avidan, S.,

Brostow, G., Cissé, M., Farinella, G.M., Hassner, T. (eds.) ECCV 2022. LNCS, vol. 13695, pp. 580–597. Springer, Cham (2022). https://doi.org/10.1007/978-3-031-19833-5_34

34. Guo, C., et al.: Action2Motion: conditioned generation of 3D human motions. In: Proceedings of the 28th ACM International Conference on Multimedia, pp. 2021–2029 (2020)

35. Guo, W., Du, Y., Shen, X., Lepetit, V., Alameda-Pineda, X., Moreno-Noguer, F.: Back to MLP: a simple baseline for human motion prediction. In: Proceedings of the IEEE/CVF Winter Conference on Applications of Computer Vision, pp. 4809–4819 (2023)

36. Han, B., et al.: AMD autoregressive motion diffusion. arXiv preprint arXiv:2305.09381 (2023)

37. Hao, Y., Zhang, J., Zhuo, T., Wen, F., Fan, H.: Hand-centric motion refinement for 3D hand-object interaction via hierarchical spatial-temporal modeling. arXiv preprint arXiv:2401.15987 (2024)

38. He, X., Huang, S., Zhan, X., Wen, C., Shan, Y.: SemanticBoost: elevating motion generation with augmented textual cues. arXiv preprint arXiv:2310.20323 (2023)

39. Hertz, A., Mokady, R., Tenenbaum, J., Aberman, K., Pritch, Y., Cohen-Or, D.: Prompt-to-prompt image editing with cross attention control. arXiv preprint arXiv:2208.01626 (2022)

40. Hoang, N.M., Gong, K., Guo, C., Mi, M.B.: MotionMix: weakly-supervised diffusion for controllable motion generation. arXiv preprint arXiv:2401.11115 (2024)

41. Hong, F., Zhang, M., Pan, L., Cai, Z., Yang, L., Liu, Z.: AvatarCLIP: zero-shot text-driven generation and animation of 3D avatars. ACM Trans. Graph. (TOG) **41**(4), 1–19 (2022)

42. Hu, V.T., et al.: Motion flow matching for human motion synthesis and editing. arXiv preprint arXiv:2312.08895 (2023)

43. Huang, R., Hu, H., Wu, W., Sawada, K., Zhang, M., Jiang, D.: Dance revolution: long-term dance generation with music via curriculum learning. arXiv preprint arXiv:2006.06119 (2020)

44. Huang, S., Wang, Z., Li, P., Jia, B., Liu, T., Zhu, Y., Liang, W., Zhu, S.C.: Diffusion-based generation, optimization, and planning in 3D scenes. In: Proceedings of the IEEE/CVF Conference on Computer Vision and Pattern Recognition, pp. 16750–16761 (2023)

45. Ionescu, C., Papava, D., Olaru, V., Sminchisescu, C.: Human3.6M: large scale datasets and predictive methods for 3d human sensing in natural environments. IEEE Trans. Pattern Anal. Mach. Intell. **36**(7), 1325–1339 (2013)

46. Ji, Y., Xu, F., Yang, Y., Shen, F., Shen, H.T., Zheng, W.S.: A large-scale RGB-D database for arbitrary-view human action recognition. In: Proceedings of the 26th ACM International Conference on Multimedia, pp. 1510–1518 (2018)

47. Jiang, B., Chen, X., Liu, W., Yu, J., Yu, G., Chen, T.: MotionGPT: human motion as a foreign language. In: Advances in Neural Information Processing Systems, vol. 36 (2024)

48. Jiang, C., et al.: MotionDiffuser: controllable multi-agent motion prediction using diffusion. In: Proceedings of the IEEE/CVF Conference on Computer Vision and Pattern Recognition, pp. 9644–9653 (2023)

49. Jin, P., Wu, Y., Fan, Y., Sun, Z., Wei, Y., Yuan, L.: Act as you wish: fine-grained control of motion diffusion model with hierarchical semantic graphs. In: NeurIPS (2023)

50. Jing, B., Zhang, Y., Song, Z., Yu, J., Yang, W.: AMD: anatomical motion diffusion with interpretable motion decomposition and fusion. arXiv preprint arXiv:2312.12763 (2023)
51. Kalakonda, S.S., Maheshwari, S., Sarvadevabhatla, R.K.: Action-GPT: leveraging large-scale language models for improved and generalized action generation. In: 2023 IEEE International Conference on Multimedia and Expo (ICME), pp. 31–36. IEEE (2023)
52. Karunratanakul, K., Preechakul, K., Suwajanakorn, S., Tang, S.: Guided motion diffusion for controllable human motion synthesis. In: Proceedings of the IEEE/CVF International Conference on Computer Vision, pp. 2151–2162 (2023)
53. Kawar, B., et al.: Imagic: text-based real image editing with diffusion models. In: Conference on Computer Vision and Pattern Recognition 2023 (2023)
54. Kim, G., Shim, H., Kim, H., Choi, Y., Kim, J., Yang, E.: Diffusion video autoencoders: toward temporally consistent face video editing via disentangled video encoding. In: Proceedings of the IEEE/CVF Conference on Computer Vision and Pattern Recognition, pp. 6091–6100 (2023)
55. Kim, J., Kim, J., Choi, S.: FLAME: free-form language-based motion synthesis & editing. In: Proceedings of the AAAI Conference on Artificial Intelligence, vol. 37, pp. 8255–8263 (2023)
56. Kong, H., Gong, K., Lian, D., Mi, M.B., Wang, X.: Priority-centric human motion generation in discrete latent space. In: Proceedings of the IEEE/CVF International Conference on Computer Vision, pp. 14806–14816 (2023)
57. Kucherenko, T., Hasegawa, D., Henter, G.E., Kaneko, N., Kjellström, H.: Analyzing input and output representations for speech-driven gesture generation. In: Proceedings of the 19th ACM International Conference on Intelligent Virtual Agents, pp. 97–104 (2019)
58. Kucherenko, T., Hasegawa, D., Kaneko, N., Henter, G.E., Kjellström, H.: Moving fast and slow: analysis of representations and post-processing in speech-driven automatic gesture generation. Int. J. Hum.-Comput. Interact. (2021). https://doi.org/10.1080/10447318.2021.1883883
59. Kulal, S., Mao, J., Aiken, A., Wu, J.: Programmatic concept learning for human motion description and synthesis. In: Proceedings of the IEEE/CVF Conference on Computer Vision and Pattern Recognition, pp. 13843–13852 (2022)
60. Kulkarni, N., et al.: NIFTY: neural object interaction fields for guided human motion synthesis. arXiv preprint arXiv:2307.07511 (2023)
61. Lee, H.Y., et al.: Dancing to music. In: Advances in Neural Information Processing Systems, vol. 32 (2019)
62. Li, B., Zhao, Y., Shi, Z., Sheng, L.: DanceFormer: music conditioned 3D dance generation with parametric motion transformer. In: AAAI (2022)
63. Li, J., Clegg, A., Mottaghi, R., Wu, J., Puig, X., Liu, C.K.: Controllable human-object interaction synthesis. arXiv preprint arXiv:2312.03913 (2023)
64. Li, J., Wu, J., Liu, C.K.: Object motion guided human motion synthesis. ACM Trans. Graph. (TOG) **42**(6), 1–11 (2023)
65. Li, J., et al.: Learning to generate diverse dance motions with transformer. arXiv preprint arXiv:2008.08171 (2020)
66. Li, R., Yang, S., Ross, D.A., Kanazawa, A.: AI choreographer: music conditioned 3D dance generation with AIST++. In: Proceedings of the IEEE/CVF International Conference on Computer Vision, pp. 13401–13412 (2021)
67. Li, S., Zhuang, S., Song, W., Zhang, X., Chen, H., Hao, A.: Sequential texts driven cohesive motions synthesis with natural transitions. In: Proceedings of the IEEE/CVF International Conference on Computer Vision, pp. 9498–9508 (2023)

68. Li, S., Singh, H., Grover, A.: Instructany2Pix: flexible visual editing via multi-modal instruction following. arXiv preprint arXiv:2312.06738 (2023)
69. Li, T., Bolkart, T., Black, M.J., Li, H., Romero, J.: Learning a model of facial shape and expression from 4D scans. ACM Trans. Graph. **36**(6), 194-1 (2017)
70. Li, W., Xu, X., Liu, J., Xiao, X.: UNIMO-G: Unified image generation through multimodal conditional diffusion. arXiv preprint arXiv:2401.13388 (2024)
71. Liang, H., Zhang, W., Li, W., Yu, J., Xu, L.: InterGen: diffusion-based multi-human motion generation under complex interactions. arXiv preprint arXiv:2304.05684 (2023)
72. Liang, Z., Li, Z., Zhou, S., Li, C., Loy, C.C.: Control color: multimodal diffusion-based interactive image colorization. arXiv preprint arXiv:2402.10855 (2024)
73. Lim, D., Jeong, C., Kim, Y.M.: MAMMOS: mapping multiple human motion with scene understanding and natural interactions. In: Proceedings of the IEEE/CVF International Conference on Computer Vision, pp. 4278–4287 (2023)
74. Lin, A.S., Wu, L., Corona, R., Tai, K., Huang, Q., Mooney, R.J.: Generating animated videos of human activities from natural language descriptions. In: NeurIPS Workshop (2018)
75. Lin, J., et al.: Motion-X: a large-scale 3D expressive whole-body human motion dataset. In: Advances in Neural Information Processing Systems (2023)
76. Lin, J., et al.: OHMG: zero-shot open-vocabulary human motion generation. arXiv preprint arXiv:2210.15929 (2022)
77. Lin, J., et al.: Being comes from not-being: Open-vocabulary text-to-motion generation with wordless training. In: Proceedings of the IEEE/CVF Conference on Computer Vision and Pattern Recognition, pp. 23222–23231 (2023)
78. Lin, P., et al.: HandDiffuse: generative controllers for two-hand interactions via diffusion models. arXiv preprint arXiv:2312.04867 (2023)
79. Ling, Z., Han, B., Wong, Y., Kangkanhalli, M., Geng, W.: MCM: multi-condition motion synthesis framework for multi-scenario. arXiv preprint arXiv:2309.03031 (2023)
80. Liu, C., Zhao, M., Ren, B., Liu, M., Sebe, N., et al.: Spatio-temporal graph diffusion for text-driven human motion generation. In: British Machine Vision Conference (2023)
81. Liu, H., et al.: BEAT: a large-scale semantic and emotional multi-modal dataset for conversational gestures synthesis. In: Avidan, S., Brostow, G., Cissé, M., Farinella, G.M., Hassner, T. (eds.) ECCV 2022. LNCS, vol. 13667, pp. 612–630. Springer, Cham (2022). https://doi.org/10.1007/978-3-031-20071-7_36
82. Liu, J., Dai, W., Wang, C., Cheng, Y., Tang, Y., Tong, X.: Plan, posture and go: towards open-world text-to-motion generation. arXiv preprint arXiv:2312.14828 (2023)
83. Liu, J., Shahroudy, A., Perez, M., Wang, G., Duan, L.Y., Kot, A.C.: NTU RGB+D 120: a large-scale benchmark for 3D human activity understanding. IEEE Trans. Pattern Anal. Mach. Intell. **42**(10), 2684–2701 (2019)
84. Liu, X., et al.: Learning hierarchical cross-modal association for co-speech gesture generation. In: Proceedings of the IEEE/CVF Conference on Computer Vision and Pattern Recognition, pp. 10462–10472 (2022)
85. Liu, X., Chen, G., Tang, Y., Wang, G., Lim, S.N.: Language-free compositional action generation via decoupling refinement. arXiv preprint arXiv:2307.03538 (2023)
86. Liu, X., Hou, H., Yang, Y., Li, Y.L., Lu, C.: Revisit human-scene interaction via space occupancy. arXiv preprint arXiv:2312.02700 (2023)

87. Liu, Y., Chen, C., Yi, L.: Interactive humanoid: online full-body motion reaction synthesis with social affordance canonicalization and forecasting. arXiv preprint arXiv:2312.08983 (2023)
88. Loper, M., Mahmood, N., Romero, J., Pons-Moll, G., Black, M.J.: SMPL: a skinned multi-person linear model. ACM trans. graph. (TOG) **34**(6), 1–16 (2015)
89. Loper, M., Mahmood, N., Romero, J., Pons-Moll, G., Black, M.J.: SMPL: a skinned multi-person linear model. ACM Trans. Graph. (Proc. SIGGRAPH Asia) **34**(6), 248:1–248:16 (2015)
90. Lou, Y., Zhu, L., Wang, Y., Wang, X., Yang, Y.: DiverseMotion: towards diverse human motion generation via discrete diffusion. arXiv preprint arXiv:2309.01372 (2023)
91. Lu, S., et al.: HumanTOMATO: text-aligned whole-body motion generation. arXiv preprint arXiv:2310.12978 (2023)
92. Ma, J., Bai, S., Zhou, C.: Pretrained diffusion models for unified human motion synthesis. arXiv preprint arXiv:2212.02837 (2022)
93. Mahmood, N., Ghorbani, N., Troje, N.F., Pons-Moll, G., Black, M.J.: AMASS: archive of motion capture as surface shapes. In: International Conference on Computer Vision, pp. 5442–5451 (2019)
94. Mao, W., Liu, M., Salzmann, M.: History repeats itself: human motion prediction via motion attention. In: Vedaldi, A., Bischof, H., Brox, T., Frahm, J.-M. (eds.) ECCV 2020, Part XIV. LNCS, vol. 12359, pp. 474–489. Springer, Cham (2020). https://doi.org/10.1007/978-3-030-58568-6_28
95. Mao, W., Liu, M., Salzmann, M., Li, H.: Learning trajectory dependencies for human motion prediction. In: Proceedings of the IEEE/CVF International Conference on Computer Vision, pp. 9489–9497 (2019)
96. von Marcard, T., Henschel, R., Black, M., Rosenhahn, B., Pons-Moll, G.: Recovering accurate 3D human pose in the wild using IMUs and a moving camera. In: European Conference on Computer Vision (ECCV) (sep 2018)
97. Mehta, D., et al.: Monocular 3D human pose estimation in the wild using improved cnn supervision. In: 3D Vision (3DV), 2017 Fifth International Conference on. IEEE (2017). https://doi.org/10.1109/3dv.2017.00064, http://gvv.mpi-inf.mpg.de/3dhp_dataset
98. Nguyen, T., Li, Y., Ojha, U., Lee, Y.J.: Visual instruction inversion: image editing via visual prompting. In: Thirty-seventh Conference on Neural Information Processing Systems (2023). https://openreview.net/forum?id=l9BsCh8ikK
99. Okamura, M., Kondo, N., Sakamoto, T.F.M., Ochiai, Y.: Dance generation by sound symbolic words. arXiv preprint arXiv:2306.03646 (2023)
100. Pavlakos, G., et al.: Expressive body capture: 3D hands, face, and body from a single image. In: Proceedings IEEE Conference on Computer Vision and Pattern Recognition (CVPR) (2019)
101. Peng, X., Xie, Y., Wu, Z., Jampani, V., Sun, D., Jiang, H.: HOI-Diff: text-driven synthesis of 3D human-object interactions using diffusion models. arXiv preprint arXiv:2312.06553 (2023)
102. Petrovich, M., Black, M.J., Varol, G.: Action-conditioned 3D human motion synthesis with transformer VAE. In: Proceedings of the IEEE/CVF International Conference on Computer Vision, pp. 10985–10995 (2021)
103. Petrovich, M., Black, M.J., Varol, G.: TEMOS: generating diverse human motions from textual descriptions. In: Avidan, S., Brostow, G., Cissé, M., Farinella, G.M., Hassner, T. (eds.) ECCV 2022. LNCS, vol. 13682, pp. 480–497. Springer, Cham (2022). https://doi.org/10.1007/978-3-031-20047-2_28

104. Petrovich, M., et al.: Multi-track timeline control for text-driven 3D human motion generation. arXiv preprint arXiv:2401.08559 (2024)
105. Pi, H., Peng, S., Yang, M., Zhou, X., Bao, H.: Hierarchical generation of human-object interactions with diffusion probabilistic models. In: Proceedings of the IEEE/CVF International Conference on Computer Vision, pp. 15061–15073 (2023)
106. Pinyoanuntapong, E., Wang, P., Lee, M., Chen, C.: MMM: generative masked motion model. arXiv preprint arXiv:2312.03596 (2023)
107. Plappert, M., Mandery, C., Asfour, T.: The kit motion-language dataset. Big Data 4(4), 236–252 (2016)
108. Punnakkal, A.R., Chandrasekaran, A., Athanasiou, N., Quiros-Ramirez, A., Black, M.J.: BABEL: bodies, action and behavior with English labels. In: Proceedings of the IEEE/CVF Conference on Computer Vision and Pattern Recognition, pp. 722–731 (2021)
109. Qi, Q., et al.: DiffDance: cascaded human motion diffusion model for dance generation. In: Proceedings of the 31st ACM International Conference on Multimedia, pp. 1374–1382 (2023)
110. Qian, Y., Urbanek, J., Hauptmann, A.G., Won, J.: Breaking the limits of text-conditioned 3D motion synthesis with elaborative descriptions. In: Proceedings of the IEEE/CVF International Conference on Computer Vision, pp. 2306–2316 (2023)
111. Qing, Z., Cai, Z., Yang, Z., Yang, L.: Story-to-motion: synthesizing infinite and controllable character animation from long text. In: SIGGRAPH Asia 2023 Technical Communications, SA Technical Communications 2023, Sydney, NSW, Australia, 12–15 December 2023, pp. 28:1–28:4 (2023)
112. Raab, S., Leibovitch, I., Tevet, G., Arar, M., Bermano, A.H., Cohen-Or, D.: Single motion diffusion. arXiv preprint arXiv:2302.05905 (2023)
113. Ren, J., Zhang, M., Yu, C., Ma, X., Pan, L., Liu, Z.: InsActor: instruction-driven physics-based characters. In: Advances in Neural Information Processing Systems, vol. 36 (2024)
114. Ribeiro-Gomes, J., et al.: MotionGPT: human motion synthesis with improved diversity and realism via GPT-3 prompting. In: Proceedings of the IEEE/CVF Winter Conference on Applications of Computer Vision, pp. 5070–5080 (2024)
115. Ruan, L., et al.: MM-diffusion: learning multi-modal diffusion models for joint audio and video generation. In: Proceedings of the IEEE/CVF Conference on Computer Vision and Pattern Recognition, pp. 10219–10228 (2023)
116. Shafir, Y., Tevet, G., Kapon, R., Bermano, A.H.: Human motion diffusion as a generative prior. arXiv preprint arXiv:2303.01418 (2023)
117. Shahroudy, A., Liu, J., Ng, T.T., Wang, G.: NTU RGB+ D: a large scale dataset for 3D human activity analysis. In: Proceedings of the IEEE Conference on Computer Vision and Pattern Recognition, pp. 1010–1019 (2016)
118. Shi, X., Luo, C., Peng, J., Zhang, H., Sun, Y.: Generating fine-grained human motions using ChatGPT-refined descriptions. arXiv preprint arXiv:2312.02772 (2023)
119. Shimada, S., et al.: MACS: mass conditioned 3D hand and object motion synthesis. arXiv preprint arXiv:2312.14929 (2023)
120. Siyao, L., et al.: Duolando: follower GPT with off-policy reinforcement learning for dance accompaniment. In: The Twelfth International Conference on Learning Representations (2023)

121. Siyao, L., et al.: Bailando: 3D dance generation by actor-critic GPT with choreographic memory. In: Proceedings of the IEEE/CVF Conference on Computer Vision and Pattern Recognition, pp. 11050–11059 (2022)

122. Sun, G., Wong, Y., Cheng, Z., Kankanhalli, M.S., Geng, W., Li, X.: DeepDance: music-to-dance motion choreography with adversarial learning. IEEE Trans. Multimedia **23**, 497–509 (2020)

123. Sun, J., Lin, Z., Han, X., Hu, J.F., Xu, J., Zheng, W.S.: Action-guided 3D human motion prediction. In: Advances in Neural Information Processing Systems, vol. 34, pp. 30169–30180 (2021)

124. Sun, J., Chowdhary, G.: Towards globally consistent stochastic human motion prediction via motion diffusion. arXiv preprint arXiv:2305.12554 (2023)

125. Tanaka, M., Fujiwara, K.: Role-aware interaction generation from textual description. In: Proceedings of the IEEE/CVF International Conference on Computer Vision, pp. 15999–16009 (2023)

126. Tendulkar, P., Surís, D., Vondrick, C.: FLEX: full-body grasping without full-body grasps. In: Proceedings of the IEEE/CVF Conference on Computer Vision and Pattern Recognition, pp. 21179–21189 (2023)

127. Tevet, G., Gordon, B., Hertz, A., Bermano, A.H., Cohen-Or, D.: MotionCLIP: exposing human motion generation to clip space. In: Avidan, S., Brostow, G., Cissé, M., Farinella, G.M., Hassner, T. (eds.) ECCV 2022. LNCS, vol. 13682, pp. 358–374. Springer, Cham (2022). https://doi.org/10.1007/978-3-031-20047-2_21

128. Tevet, G., Raab, S., Gordon, B., Shafir, Y., Cohen-or, D., Bermano, A.H.: Human motion diffusion model. In: The Eleventh International Conference on Learning Representations (2022)

129. Tseng, J., Castellon, R., Liu, K.: EDGE: editable dance generation from music. In: Proceedings of the IEEE/CVF Conference on Computer Vision and Pattern Recognition, pp. 448–458 (2023)

130. Voas, J., Wang, Y., Huang, Q., Mooney, R.: What is the best automated metric for text to motion generation? In: SIGGRAPH Asia 2023 Conference Papers, pp. 1–11 (2023)

131. Voleti, V., Jolicoeur-Martineau, A., Pal, C.: MCVD-masked conditional video diffusion for prediction, generation, and interpolation. In: Advances in Neural Information Processing Systems, vol. 35, pp. 23371–23385 (2022)

132. Wan, W., Dou, Z., Komura, T., Wang, W., Jayaraman, D., Liu, L.: TLControl: trajectory and language control for human motion synthesis. arXiv preprint arXiv:2311.17135 (2023)

133. Wang, X., Cui, Q., Chen, C., Liu, M.: GCNext: towards the unity of graph convolutions for human motion prediction. arXiv preprint arXiv:2312.11850 (2023)

134. Wang, Y., Leng, Z., Li, F.W., Wu, S.C., Liang, X.: Fg-T2M: fine-grained text-driven human motion generation via diffusion model. In: Proceedings of the IEEE/CVF International Conference on Computer Vision, pp. 22035–22044 (2023)

135. Wang, Y., Lin, J., Zeng, A., Luo, Z., Zhang, J., Zhang, L.: PhysHOI: physics-based imitation of dynamic human-object interaction. arXiv preprint arXiv:2312.04393 (2023)

136. Wang, Z., et al.: Learning diverse stochastic human-action generators by learning smooth latent transitions. In: Proceedings of the AAAI Conference on Artificial Intelligence, vol. 34, pp. 12281–12288 (2020)

137. Wei, D., et al.: Enhanced fine-grained motion diffusion for text-driven human motion synthesis (2023)

138. Xiao, Z., et al.: Unified human-scene interaction via prompted chain-of-contacts. arXiv preprint arXiv:2309.07918 (2023)
139. Xie, Y., Jampani, V., Zhong, L., Sun, D., Jiang, H.: OmniControl: control any joint at any time for human motion generation. arXiv preprint arXiv:2310.08580 (2023)
140. Xie, Z., Wu, Y., Gao, X., Sun, Z., Yang, W., Liang, X.: Towards detailed text-to-motion synthesis via basic-to-advanced hierarchical diffusion model. arXiv preprint arXiv:2312.10960 (2023)
141. Xu, Z., Zhang, Y., Yang, S., Li, R., Li, X.: Chain of generation: multi-modal gesture synthesis via cascaded conditional control. arXiv preprint arXiv:2312.15900 (2023)
142. Yan, H., Hu, Z., Schmitt, S., Bulling, A.: GazeMoDiff: gaze-guided diffusion model for stochastic human motion prediction. arXiv preprint arXiv:2312.12090 (2023)
143. Yang, S., Zhou, Y., Liu, Z., , Loy, C.C.: Rerender a video: zero-shot text-guided video-to-video translation. In: ACM SIGGRAPH Asia 2023 Conference Proceedings (2023)
144. Yang, S., Yang, Z., Wang, Z.: LongDanceDiff: long-term dance generation with conditional diffusion model. arXiv preprint arXiv:2308.11945 (2023)
145. Yang, Z., Su, B., Wen, J.R.: Synthesizing long-term human motions with diffusion models via coherent sampling. In: Proceedings of the 31st ACM International Conference on Multimedia, pp. 3954–3964 (2023)
146. Yao, H., Song, Z., Zhou, Y., Ao, T., Chen, B., Liu, L.: MoConVQ: unified physics-based motion control via scalable discrete representations. arXiv preprint arXiv:2310.10198 (2023)
147. Yao, S., Sun, M., Li, B., Yang, F., Wang, J., Zhang, R.: Dance with you: the diversity controllable dancer generation via diffusion models. In: Proceedings of the 31st ACM International Conference on Multimedia,D pp. 8504–8514 (2023)
148. Yazdian, P.J., Liu, E., Cheng, L., Lim, A.: MotionScript: natural language descriptions for expressive 3D human motions. arXiv preprint arXiv:2312.12634 (2023)
149. Yin, L., et al.: EMoG: synthesizing emotive co-speech 3D gesture with diffusion model. arXiv preprint arXiv:2306.11496 (2023)
150. Yoon, Y., et al.: Speech gesture generation from the trimodal context of text, audio, and speaker identity. ACM Trans. Graph. (TOG) **39**(6), 1–16 (2020)
151. Yuan, Y., Song, J., Iqbal, U., Vahdat, A., Kautz, J.: PhysDiff: Physics-guided human motion diffusion model. In: Proceedings of the IEEE/CVF International Conference on Computer Vision, pp. 16010–16021 (2023)
152. Zhai, Y., et al.: Language-guided human motion synthesis with atomic actions. In: Proceedings of the 31st ACM International Conference on Multimedia, pp. 5262–5271 (2023)
153. Zhang, J., et al.: T2M-GPT: generating human motion from textual descriptions with discrete representations. arXiv preprint arXiv:2301.06052 (2023)
154. Zhang, J., et al.: TapMo: shape-aware motion generation of skeleton-free characters. arXiv preprint arXiv:2310.12678 (2023)
155. Zhang, M., et al.: MotionDiffuse: text-driven human motion generation with diffusion model. IEEE Trans. Pattern Anal. Mach. Intell. (2024)
156. Zhang, M., et al.: ReMoDiffuse: retrieval-augmented motion diffusion model. In: IEEE/CVF International Conference on Computer Vision, ICCV 2023, Paris, France, 1–6 October 2023, pp. 364–373 (2023)
157. Zhang, M., Li, H., Cai, Z., Ren, J., Yang, L., Liu, Z.: FineMoGen: fine-grained spatio-temporal motion generation and editing. In: Advances in Neural Information Processing Systems, vol. 36 (2024)

158. Zhang, X., Bhatnagar, B.L., Starke, S., Guzov, V., Pons-Moll, G.: COUCH: towards controllable human-chair interactions. In: Avidan, S., Brostow, G., Cissé, M., Farinella, G.M., Hassner, T. (eds.) ECCV 2022. LNCS, vol. 13665, pp. 518–535. Springer, Cham (2022). https://doi.org/10.1007/978-3-031-20065-6_30

159. Zhang, Y., et al.: MotionGPT: finetuned LLMs are general-purpose motion generators. arXiv preprint arXiv:2306.10900 (2023)

160. Zhang, Y., Tsipidi, E., Schriber, S., Kapadia, M., Gross, M., Modi, A.: Generating animations from screenplays. In: Proceedings of the Eighth Joint Conference on Lexical and Computational Semantics, *SEM@NAACL-HLT 2019, Minneapolis, MN, USA, 6–7 June 2019, pp. 292–307 (2019)

161. Zhao, M., Liu, M., Ren, B., Dai, S., Sebe, N.: MoDiff: action-conditioned 3D motion generation with denoising diffusion probabilistic models. arXiv preprint arXiv:2301.03949 (2023)

162. Zhao, W., Hu, L., Zhang, S.: DiffuGesture: generating human gesture from two-person dialogue with diffusion models. In: Companion Publication of the 25th International Conference on Multimodal Interaction, pp. 179–185 (2023)

163. Zhi, Y., et al.: LivelySpeaker: towards semantic-aware co-speech gesture generation. In: Proceedings of the IEEE/CVF International Conference on Computer Vision, pp. 20807–20817 (2023)

164. Zhong, C., Hu, L., Zhang, Z., Xia, S.: AttT2M: text-driven human motion generation with multi-perspective attention mechanism. In: Proceedings of the IEEE/CVF International Conference on Computer Vision, pp. 509–519 (2023)

165. Zhou, W., et al.: EMDM: efficient motion diffusion model for fast, high-quality motion generation. arXiv preprint arXiv:2312.02256 (2023)

166. Zhou, Z., Wang, B.: UDE: A unified driving engine for human motion generation. In: Proceedings of the IEEE/CVF Conference on Computer Vision and Pattern Recognition, pp. 5632–5641 (2023)

167. Zhuang, W., Wang, C., Chai, J., Wang, Y., Shao, M., Xia, S.: Music2Dance: DanceNet for music-driven dance generation. ACM Trans. Multimed. Comput. Commun. Appl. (TOMM) **18**(2), 1–21 (2022)

FisherRF: Active View Selection and Mapping with Radiance Fields Using Fisher Information

Wen Jiang[1]([envelope]) [iD], Boshu Lei[1] [iD], and Kostas Daniilidis[1,2] [iD]

[1] University of Pennsylvania, Philadelphia, USA
wenjiang@seas.upenn.edu
[2] Archimedes, Athena RC, Marousi, Greece

Abstract. This study addresses the challenging problem of active view selection and uncertainty quantification within the domain of Radiance Fields. Neural Radiance Fields (NeRF) have greatly advanced image rendering and reconstruction, but the cost of acquiring images poses the need to select the most informative viewpoints efficiently. Existing approaches depend on modifying the model architecture or hypothetical perturbation field to indirectly approximate the model uncertainty. However, selecting views from indirect approximation does not guarantee optimal information gain for the model. By leveraging Fisher Information, we directly quantify observed information on the parameters of Radiance Fields and select candidate views by maximizing the Expected Information Gain (EIG). Our method achieves state-of-the-art results on multiple tasks, including view selection, active mapping, and uncertainty quantification, demonstrating its potential to advance the field of Radiance Fields.

1 Introduction

Neural Radiance Fields brought back image rendering and reconstruction from multiple views to the center of attention in the field of computer vision. Novel volumetric representations of radiance fields and differentiable volumetric rendering enabled unprecedented advances in image-based rendering of complex scenes both in terms of perceptual quality and speed.

Recently, 3D Gaussian Splatting [25] has demonstrated distinct advantages in real-time rendering and explicit point-based parameterizations without neural representations. However, to achieve satisfactory rendering quality, numerous viewpoints are required to train a radiance field, especially in large-scale scenarios. It is crucial to establish a criterion for selecting information-maximizing views before obtaining image observations at those locations. Quantifying the observed information of a Radiance Field model is challenging, given that Radiance Field models are typically regression-based and scene-specific. The challenge

Supplementary Information The online version contains supplementary material available at https://doi.org/10.1007/978-3-031-72624-8_24.

A. Leonardis et al. (Eds.): ECCV 2024, LNCS 15071, pp. 422–440, 2025.
https://doi.org/10.1007/978-3-031-72624-8_24

intensifies when we aim to leverage quantified observed information for active view selection and mapping, especially when the selection candidates are only $SE(3)$ camera poses for acquiring new observations, a.k.a capturing new images.

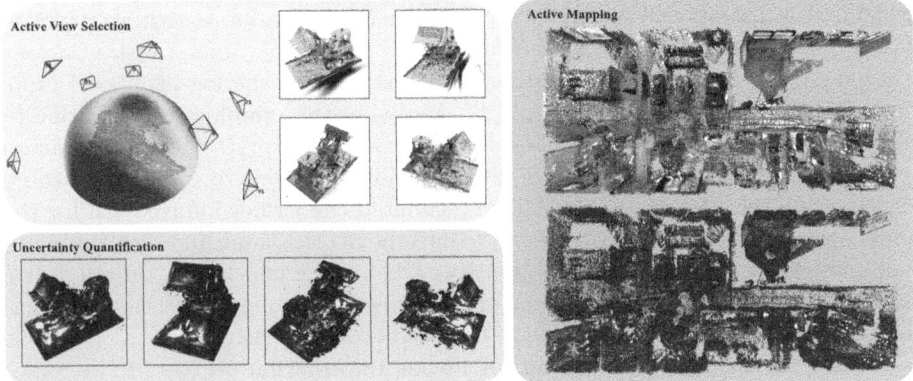

Fig. 1. A brief overview of our method. Given a Radiance Field that was trained with a limited number of views, our method could find the next best view that could maximize information gain by computing the Fisher Information of the radiance field. We illustrate the Information Gain as a heat map on the viewing sphere and show four of the candidate views. Our model can quantify pixel-wise uncertainty, visualized at the bottom left, by examining the Fisher Information on the related parameters of the ray. Our algorithm can be also adapted into an active mapping system, as showcased at the right, that could actively explore and reconstruct the environment.

Previous approaches select viewpoints by quantifying the uncertainty in Radiance Fields. They can be broadly categorized into two groups: variational white-box models and black-box models. White-box models integrate conventional NeRF architectures with Bayesian models, such as reparameterization [36,43,48] and Normalizing Flows [55]. Black-box methods, on the other hand, do not modify the existing model architecture but seek to quantify predictive uncertainty by examining the distribution of predicted outcomes [31,43,67,71]. White-box models depend on specific model architectures and are often characterized by slower training times due to the challenges associated with probabilistic learning. Conversely, existing black-box models either focus solely on studying the uncertainty on NeRF-style query points through network prediction [43] or hypothetical perturbation field [16], or rely on Monte Carlo sampling techniques [67] to quantify the uncertainty at the candidate views.

In this study, our primary objective is to quantify the observed information of a Radiance Field model and utilize it to select the optimal view with the highest information gain for downstream tasks, Fig. 1. To achieve this, we propose using Fisher Information, which represents the expectation of observation information. This quantity is directly linked to the second-order derivatives or Hessian matrix of the loss function involved in optimizing Radiance Field models. Importantly, the Hessian of the objective function in volumetric rendering is independent of ground truth data or the actual image measurement. This property allows us to compute the information gain between the training dataset and the candidate viewpoint pool using only the camera parameters of the candidate views. This capability facilitates an efficient next-best-view selection algorithm. In addition to the formulation of Information Gain, the Hessian of the loss function has an intuitive interpretation: the perturbation at flat minima of the loss function. From an intuitive standpoint, one can perceive the Fisher Information matrix as a metric of the curvature of the log-likelihood function at specific parameter instantiations denoted as $\mathbf{w}^*$. Lower Fisher Information suggests that the log-likelihood function exhibits a flatter profile around $\mathbf{w}^*$, implying that the loss is less prone to changes when $\mathbf{w}^*$ is perturbed. The flat minimum interpretation has attracted substantial attention and research within various domains of machine learning [19,20,28,34,58]. Moreover, as we have estimated the Fisher information for the model's parameters, which correspond to specific 3D locations in recent Radiance Field models such as 3D Gaussian Splatting [25] and Plenoxels [52], we can derive pixel-wise uncertainty in the model's predictions by examing the Fisher Information on parameters that contribute to the prediction for each pixel.

We implement the computation of Fisher Information on top of two types of Radiance Field models: point-based 3D Gaussian Splatting [25] and Plenoxels [52]. In 3D Gaussian Splatting, we compute the Hessians of the parameters on each 3D Gaussian with respect to the negative log-likelihood. As Fisher Information is additive, we further extend our view selection algorithm into batch selection and path selection, which are closer to real-world applications. To make the computation of information gain tractable, we use an approximation of the decrease in entropy by its upper bound, which is the trace of the product of the Hessian of the candidate view times the Hessian with respect to all previous views. These matrices are highly sparse due to the disentangling of scene parameters with respect to disparate rays. The sparsity allows us a computation of the matrix trace above that is as cost-effective as back-propagation, enabling us to evaluate views at 70 fps when we use 3D Gaussian Splatting. We carried out extensive evaluations in different benchmarks, including active view selection, active mapping, and uncertainty quantification. The quantitative and qualitative results unequivocally demonstrate that our approach surpasses previous methods and heuristic baseline by a significant margin. To summarize, our main contributions are as follows:

– A novel formulation to quantify the observed information and Information Gain for Radiance Fields.
– An effective and efficient view selection method exploiting the sparse structure of the scene rendering problem.
– An efficient pixel-wise uncertainty quantification and visualization method for 3D Gaussian Splatting.
– Extensive comparative study showing that our method outperforms existing active approaches on view selection, active mapping, and uncertainty quantification.

2   Related Works

Active Learning and Radiance Fields are prosperous research fields with various research directions. In this section, we limit our literature review to works that directly relate to our tasks. We refer the readers to literature reviews if they are interested in Radiance Fields [10,13] or Deep Active Learning [50,72]. Besides, Fisher Information has been extensively studied in deep active learning [1,2, 27,29]. Notably, Kirsch et al. [28] unified existing works in active learning for deep learning problems from the perspective of Fisher Information, which shares many insights with our method.

Uncertainty Quantifications for Radiance Fields. In the Neural Radiance Fields, Lee et al. [31], Yan et al. [66] and Zhan et al. [71] attempted to quantify the uncertainty in a scene by the distribution of densities on a casted ray. Shen et al. [55,56] designed Bayesian models by re-parametrizing the NeRF model. However, they only tackled the predictive uncertainty of the model that did not relate to the observed information of the parameters. Sunderhauf et al. [61] propose an additional uncertainty measure in uncharted regions, determined by the ray termination probability on the learned geometry. Regarding concurrent work, Goli et al. [16] introduces a hypothetical perturbation field and interpolates uncertainty over NeRF-style query points for uncertainty quantification and NeRF clean-up task. The perturbation field with respect to query points is also not compatible with recent radiance field models like 3D Gaussian Splatting [25], which do not take query points as inputs. Besides, their algorithm is not designed for an active learning problem since selecting candidates with the largest per-pixel uncertainty may not be optimal, as the information gain for the model parameters is not concerned. In contrast, we focus on active learning problems. We derive our theory from the expected information gain for the model and compute the Hessian for the model parameters, allowing us to apply our algorithm to recent advancements in radiance fields such as 3D Gaussian splatting. Our paper focuses on active view selection and mapping based on our novel expected information gain objective. Moreover, their implementation relies on multiple function calls to PyTorch backward engine whereas our efficient CUDA implementation for the diagonal Hessian computation consumes much less time (11.3 ms ± 33.3 µs v.s. 1.1 s ± 2.5 ms). To the best of our knowledge, we are the first method that quantifies the uncertainty directly on model parameters for Radience Fields, thanks to the powerful framework of 3D Gaussian and our efficient formulation.

Active View Selection and Mapping. Although the next best view selection problem was extensively studied before radiance fields gained popularity, there has not been much literature on active "training" view selection for volumetric rendering. Jin et al. [22] trains a neural network to predict per-pixel uncertainty for view selection. However, their method requires an image capture as the input of the network, which is closer to dataset subsampling rather than active learning. ActiveNeRF [43] studied the next best view selection by directly adding variances as the output of the vanilla NeRF, and it is the closest approach to the view selection problem we are working on. In a larger scope, Active Perception [3,4] has been studied along with the advancement of robotics and computer vision. Traditionally, frontier-based approach [64] and Rapidly exploring

Random Tree (RRT) [30] are the most representative and have been extended and enhanced for different data representations [12,23,57,74] and complex conditions [44,68]. For the larger scope of Active Simultaneous Localization and Mapping (SLAM), please refer to literature reviews [33,45]. In recent years, Guedon *et al.* [17,18] attempted to improve the coverage for mapping systems using point cloud representation with neural networks. Dhami *et al.* [11] studied the next best view selection in view planning by first completing the point cloud from partial observation. Georgakis *et al.* [14] measured uncertainty through deep ensemble. Chaplot *et al.* [8] and Ramakrishnan *et al.* [46] developed active exploration methods with deep reinforcement learning. The mapping system that uses implicit feild [42,51,59,75] especially 3D Gaussians [24,37,65,70] as representation has been widely studied. This presents the need for active mapping algorithms that could maximize the mapping efficiency in an embodied robot system. Similar to ActiveNeRF [43], Ran *et al.* [48] performed trajectory planning for object-centric autonomous reconstruction systems by predicting variance as an extra output of the network. Yan *et al.* [67] discussed the intuition of flat minimum and quantified the predictive uncertainty of neural mapping models through the lens of loss landscape. However, they only approximate the uncertainty of the variances from the output of neighboring points. Our method directly quantifies the observed information of the parameters by computing the Hessian matrix of the log-likelihood function of the model's objective. Besides, we introduce the Expected Information Gain for mapping systems from the first principle as a complementary method to previous heuristics.

3   Technical Approach

In this section, we first introduce how we use Fisher Information in Sect. 3.1. By leveraging Fisher Information, we demonstrate how to select the next best view and next best batch of views in Sect. 3.2 and Sect. 3.3. We then extend our method to large-scale scenes with active mapping. Finally, we showcase that our method could quantify the pixel-wise uncertainties in Sect. 3.5.

3.1   Fisher Information in Volumetric Rendering

Fisher Information is a measurement of information that an observation $(\mathbf{x}, \mathbf{y})$ carries about the unknown parameters $\mathbf{w}$ that model $p(\mathbf{y}|\mathbf{x}; \mathbf{w})$. In the problem of novel view synthesis, $(\mathbf{x}, \mathbf{y})$ are the camera pose $\mathbf{x}$ and image observation $\mathbf{y}$ at pose $\mathbf{x}$, respectively, whereas $\mathbf{w}$ are the parameters of the radiance field. The objective of neural rendering is equivalent to minimizing the negative log likelihood (NLL) between rendered images and ground truth images on the holdout set, which inherently represents the quality of scene reconstruction

$$- \log p(\mathbf{y}|\mathbf{x}, \mathbf{w}) = (\mathbf{y} - f(\mathbf{x}, \mathbf{w}))^{T} (\mathbf{y} - f(\mathbf{x}, \mathbf{w})) \tag{1}$$

where $f(\mathbf{x}, \mathbf{w})$ is our rendering model. Under regularity conditions [54], the Fisher Information of the model $\log p(\mathbf{y}|\mathbf{x}; \mathbf{w})$ is the Hessian of the log-likelihood function with respect to the model parameters $\mathbf{w}$:

$$\mathcal{I}(\mathbf{w}) = -\mathbb{E}_{\log p(\mathbf{y}|\mathbf{x},\mathbf{w})}\left[\frac{\partial^2 \log p(\mathbf{y}|\mathbf{x},\mathbf{w})}{\partial \mathbf{w}^2}|\mathbf{w}\right] = \mathbf{H}''[\mathbf{y}|\mathbf{x},\mathbf{w}] \qquad (2)$$

where $\mathbf{H}''[\mathbf{y}|\mathbf{x},\mathbf{w}]$ is the Hessian matrix of Eq. (1).

3.2   Next Best View Selection Using Fisher Information

In the active view selection problem, we start with a training set D^{train} and have an initial estimation of parameters $\mathbf{w}^*$ using D^{train}. The aim is to select the next best view that maximizes the Information Gain [21,26,32] between candidates views $\mathbf{x}_i^{acq} \in D^{pool}$ and D^{train}, where D^{pool} is the pool of candidate views:

$$\mathcal{I}[\mathbf{w}^*; \{\mathbf{y}_i^{acq}\}|\{\mathbf{x}_i^{acq}\}]$$
$$= H[\mathbf{w}^*|D^{train}] - H[\mathbf{w}^*|\{\mathbf{y}_i^{acq}\}, \{\mathbf{x}_i^{acq}\}, D^{train}] \qquad (3)$$

where $H[\cdot]$ is the entropy [28].

When the log-likelihood has the form of Eq. (1), in our case, the rendering error, the difference of the entropies in the R.H.S. of Eq. (3) can be approximated as [28]:

$$\frac{1}{2} \log \det \left(\mathbf{H}''[\{\mathbf{y}_i^{acq}\}|\{\mathbf{x}_i^{acq}\}, \mathbf{w}^*] \, \mathbf{H}''[\mathbf{w}^*|D^{train}]^{-1} + I\right)$$
$$\leq \frac{1}{2} \operatorname{tr}\left(\mathbf{H}''[\{\mathbf{y}_i^{acq}\}|\{\mathbf{x}_i^{acq}\}, \mathbf{w}^*] \, \mathbf{H}''[\mathbf{w}^*|D^{train}]^{-1}\right). \qquad (4)$$

As Fisher Information is additive, $\mathbf{H}''[\mathbf{w}^*|D^{train}]^{-1}$ can be computed by summing the Hessians of model parameters across all different views in $\{D_{train}\}$ before inverting. We can choose the next best view $\mathbf{x}_i^{acq}$ by optimizing

$$\underset{\mathbf{x}_i^{acq}}{\arg\max} \operatorname{tr}\left(\mathbf{H}''[\mathbf{y}_i^{acq}|\mathbf{x}_i^{acq}, \mathbf{w}^*] \, \mathbf{H}''[\mathbf{w}^*|D^{train}]^{-1}\right). \qquad (5)$$

The Hessian $\mathbf{H}''[\mathbf{y}|\mathbf{x},\mathbf{w}^*]$ of our model can be computed as:

$$\mathbf{H}''[\mathbf{y}|\mathbf{x},\mathbf{w}^*] = \nabla_{\mathbf{w}} f(\mathbf{x};\mathbf{w}^*)^T \nabla^2_{f(\mathbf{x};\mathbf{w}^*)} H[\mathbf{y}|f(\mathbf{x};\mathbf{w}^*)] \nabla_{\mathbf{w}} f(\mathbf{x};\mathbf{w}^*) \qquad (6)$$

where $\mathbf{H}''[\mathbf{y}|\mathbf{x},\mathbf{w}^*]$ in our case is equal to the covariance of the RGB measurement that we set equal to one. Hence, the Hessian matrix can be computed just from the Jacobian matrix of $f(\mathbf{x},\mathbf{w})$

$$\mathbf{H}''[\mathbf{y}|\mathbf{x},\mathbf{w}^*] = \nabla_{\mathbf{w}} f(\mathbf{x};\mathbf{w}^*)^T \nabla_{\mathbf{w}} f(\mathbf{x};\mathbf{w}^*). \qquad (7)$$

We can optimize the objective in Eq. (5) without knowing the ground truth of candidate views $\{\mathbf{y}_i^{acq}\}$, which was expected since the Fisher Information never depends on the observations themselves. The Hessian in (7) has only a limited number of off-diagonal elements because each pixel is considered independent in $-\log p(\mathbf{y}|\mathbf{x},\mathbf{w})$. Furthermore, recent NeRF models [40,49,52,60] typically employ structured local parameters that each parameter only contributes to the radiance and density of a limited spatial region for faster convergence and rendering. Therefore, only parameters that contribute to the color of the pixels would share non-zero values in the Hessian matrix $\mathbf{H}''[\mathbf{y}|\mathbf{x}, \mathbf{w}^*]$. However, the number of optimizable parameters is typically more than 200 million, which means it is impossible to compute without sparsification or approximation. In practice, we apply Laplace approximation [9, 35] that approximates the Hessian matrix with its diagonal values plus a log-prior regularizer λI

$$\mathbf{H}''[\mathbf{y}|\mathbf{x}, \mathbf{w}^*] \simeq \text{diag}(\nabla_{\mathbf{w}} f(\mathbf{x}, \mathbf{w}^*)^T \nabla_{\mathbf{w}} f(\mathbf{x}, \mathbf{w}^*)) + \lambda I. \tag{8}$$

3.3   Batch Active View Selection

Selecting multiple views to capture new images is useful for its possible applications, such as view planning and scene reconstruction. If we simply use Eq. (5) to select a batch of acquisition samples $\{\mathbf{x}_i^{acq}\}$, we could possibly select very similar views inside the acquisition set $\{\mathbf{x}_i^{acq}\}$ as we do not consider the mutual information between our selections. However, we would face a combinatorial explosion if we directly attempt to maximize the expected information gain between training and acquisition samples and simultaneously minimize the mutual information across acquisition samples. Therefore, we employ a greedy optimization algorithm as illustrated in Algorithm 1, which is $1/e$-approximate in Fisher Information [41]. When batch size B is 1, our algorithm is equivalent to Eq. (5). Please note that we focus on batch active view selection instead of dataset subsampling as it is more related to real-world scenarios where we wish to plan a trajectory for an agent to acquire more training views.

Algorithm 1: Batch Active Views Selection

Input: $\{\mathbf{H}''[\mathbf{w}^*|\mathbf{x}_i^{acq}]\}$, $\mathbf{H}''[\mathbf{w}^*|D^{train}]$, number of views to select B
Output: Selected Views S_B
1 $S_0 \leftarrow \emptyset$;
2 $H_0 \leftarrow \mathbf{H}''[\mathbf{w}^*|D^{train}]$;
3 **for** $b \leftarrow 1$ **to** B **do**
4   $\quad i \leftarrow \arg\max_{\mathbf{x}_i^{acq} \in D^{pool}]\backslash S_{b-1}} \text{tr}\left(\mathbf{H}''[\mathbf{y}_i^{acq}|\mathbf{x}_i^{acq}, \mathbf{w}^*] H_{b-1}^{-1}\right)$;
5   $\quad S_b \leftarrow S_{b-1} \cup \{i\}$;
6   $\quad H_b \leftarrow H_{b-1} + \mathbf{H}''[\mathbf{y}_i^{acq}|\mathbf{x}_i^{acq}, \mathbf{w}^*]$;
7 **end**

3.4   Active Mapping with 3D Gaussian Splatting

With the batch selection algorithm discussed in Sect. 3.3, our method can be extended as an active mapping algorithm. By maximizing the EIG from candi-

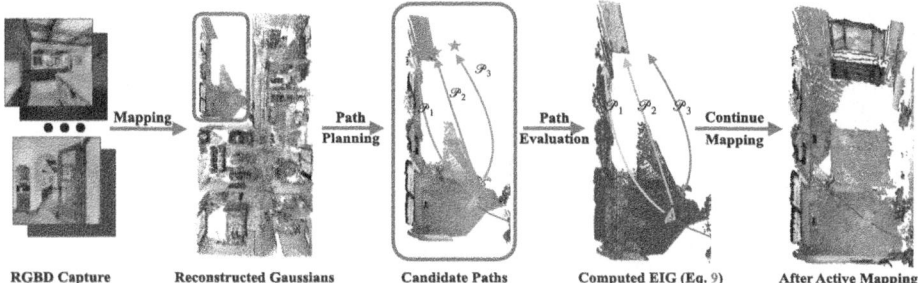

RGBD Capture   Reconstructed Gaussians   Candidate Paths   Computed EIG (Eq. 9)   After Active Mapping

Fig. 2. An Illustration of Our Active Mapping System. Given RGBD captures, we first reconstruct the environment using 3D Gaussians as representation. Afterward, we select a set of goal points from map frontiers and plan the shortest path to each goal. The EIG is computed for each path, and we choose the one with the highest EIG to continue exploration.

date trajectories $\{\mathcal{P}_j\}$, we can determine the most informative path for exploration:

$$\arg\max_{\mathcal{P}_j} \operatorname{tr} \left(\left(\sum_{\mathbf{x}_i^{acq} \in \mathcal{P}_j} \mathbf{H}''[\mathbf{y}_i^{acq}|\mathbf{x}_i^{acq}, \mathbf{w}^*] \right) \mathbf{H}''[\mathbf{w}^*|D^{train}]^{-1} \right). \tag{9}$$

We build our active mapping system on top of existing passive mapping architecture SplaTAM [24] that uses 3D Gaussians as basic data representation. We employ frontier-based exploration [64] to propose path candidates and sample cameras along the path concerning the mobility of the agent. The pipeline is illustrated in Fig. 2. Our path selection algorithm is compatible with other active mapping systems and can be used as an extension to select the most informative path.

3.5   Pixel-Wise Uncertainty with Volumetric Rendering

Previously, we discussed quantifying the Information Gain for any camera views. Our model can also be adapted to obtain pixel-wise uncertainties. As we have discussed in Sect. 3.2, we can approximate the uncertainty on each parameter with the diagonal elements of the hessian matrix $\mathbf{H}''[\mathbf{y}_i^{acq}|\mathbf{x}_i^{acq}, \mathbf{w}^*]$. In recent Radiance Field models [25,52], the parameters directly correspond to a spatial location in the scene. Therefore, we can compute the uncertainty in our rendered pixels by examining the diagonal Hessians along the casted ray for volumetric rendering

$$\mathbf{U}(\mathbf{r}) = \sum_{n=1}^{N_s} T_i \left(1 - \exp(-\sigma_n \delta_n)\right) tr(\mathbf{G}_n) \tag{10}$$

where $\mathbf{G}_n$ is the submatrix of $\mathbf{H}''[\mathbf{w}^*|D^{train}]$ containing the rows and columns that correspond to parameters at location n. Please note that this is a relative uncertainty derived from the observed information on the model parameters, which does not involve metric scales. If we wish to estimate absolute uncertainties on predictions like depth maps, we need to denormalize the uncertainty in each term by its depth d_n.

4 Experiments

In this Section, we present the empirical evaluation of our approach. The running time analysis for our customized CUDA kernel and other implementation details can be found in the supplementary. We first focus on the experiments of active view selection and compare our method quantitatively and qualitatively against the previous method (Subsect. 4.1). We further present our results on active mapping (Subsect. 4.2). Finally, we present more results of our method on uncertainty quantification results (Subsect. 4.3).

4.1 Active View Selection

We conducted extensive experiments on view selections to demonstrate that our expected information gain could help the model find the next best views. Here, we first introduce the dataset we use and detailed experimental settings. Then, we compare our method with random baselines and previous state-of-the-art ActiveNeRF [43] quantitatively and qualitatively.

Datasets. Our approach is extensively evaluated on two common benchmark datasets: Blender Dataset [38] and the challenging real-world Mip-NeRF360 dataset [5]. The Blender dataset comprises eight synthetic objects with intricate geometry and realistic non-Lambertian materials. Each scene in this dataset includes 100 training and 200 test views, all with a resolution of 800×800. Our method uses the 100 training views as a candidate pool to select training views, and we evaluate all the models on the full 200-view test set. We use the default training configuration as in 3D Gaussian and Plenoxels for this dataset. Mip-NeRF360 [5] is a real-world 360° dataset captured for nine different scenes. It has been widely used as a quality benchmark for novel view synthesis models [6,39]. We train our models at the resolution of 1066×1600 following 3D Gaussian Splatting [25].

Metrics. Our evaluations utilize image quality metrics such as peak signal-to-noise ratio (PSNR) and structural similarity index (SSIM) [62]. Additionally, we incorporate LPIPS [73], which provides a more accurate reflection of human perception.

Baselines. We quantitatively and qualitatively compare our method against the current state-the-art ActiveNeRF [43] and random selection baseline. To make a fair comparison with ActiveNeRF [43], We re-implemented a similar variance estimation algorithm in 3D Gaussian Splatting and Plenoxels in CUDA. We

Fig. 3. Qualitative Study of our method on Mip360 Dataset. From the top to bottom are results from ActiveNeRF, random baseline, our method, and the ground truth. All the models in this figure are implemented on top of 3D Gaussian Splatting [25] for better performance on this challenging dataset. We could see baseline models exhibited artifacts in some renderings due to their lack of constraints from nearby training views.

assign each 3D Gaussian (or grid cell in Plenoxels) a variance parameter and use volumetric rendering to render a variance map. The supplementary materials provide more details about our re-implementation. To show the difference between EIG and uncertainty, we adapt the uncertainty estimation method BayesRays [16] into the view selection algorithm by directly using the rendered uncertainty value on each candidate view as a metric for view selection.

Experiment Settings. We experiment with our model with 3D Gaussian Splatting backend on both the next view selection and the next batch of view selections across both the Blender and Mip360 Dataset. Each model is initialized with the same random seed and was trained on the same four uniform views. Each model is trained for 30,000 iterations following the default configurations of 3D Gaussian Splatting [25]. Similar to the training program of Gaussian Splatting, we reset the opacity every time we select new views to avoid degeneration of the training procedure. All the external settings in the experiment are kept the same except for the view selection algorithms.

Similarly, we also showcase the implementations of our active view selection algorithm on Plenoxels in the Blender Dataset. The experimental settings for initial views and view selection schedules are the same, except view selection was made every four epochs.

- Sequential Active View Selection: 1 new view is selected every 100 epochs till the model has 20 training views.
- Batch Active View Selection: 4 new views are selected every 300 epochs till the model has 20 training views.

The quantitative results of active view selections can be found in Table 1, Table 2 and Table 4. As can be seen, our method achieved better results across

20 Training Views   **10 Training Views**
ActiveNeRF Random BayesRays   Ours   ActiveNeRF Random BayesRays   Ours

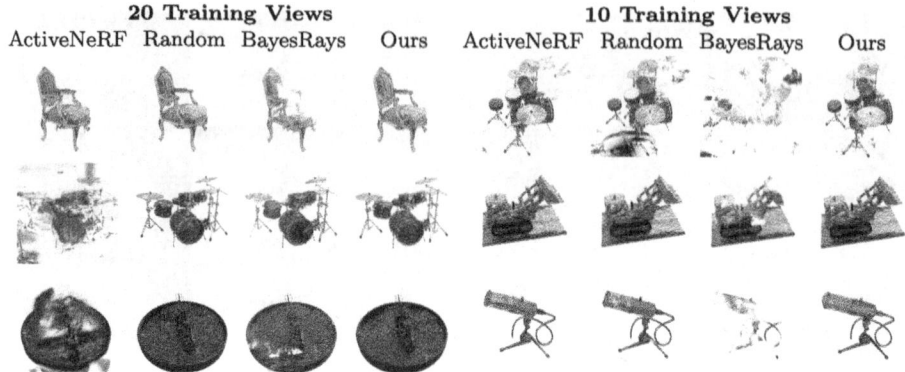

Fig. 4. Qualitative Results on Blender Dataset with 20 and 10 Training Views. All the methods are implemented on 3D Gaussian Splatting and compared in the same training configuration except for different training views selected by different methods.

different metrics and datasets. We also compare our method qualitatively on the Blender and Mip-NeRF360 datasets in Fig. 3 and Fig. 4. Our method demonstrates superior results in both single-view selection and batch-view selection. Moreover, our method achieved better performance improvements compared to the random backbone compared to BayesRays. This further renders the importance of using EIG rather than uncertainty as the metric for view selection. Furthermore, we experimented with our model with the challenging ten-view selection task on the Blender Dataset. Each model is initialized with the same random seed and two uniform initial views. Each new view is added after every 100 epochs till the model has ten training views. The quantitative and qualitative results are in Table 3 and Fig. 4. Again, our method selects necessary views given the extremely limited observations and preserves fine details of reconstructed objects.

4.2   Active Mapping

We experimented with our active mapping system on room-scale scene dataset Matterport3D [7] and Gibson [63] through the Habitat Simulator [47,53]. Our active mapping system takes posed RGB-D images as input and uses the same configurations for the simulator as previous approaches [15,67]. We turn our 3D Gaussian representation into a point cloud using only the mean $\mu \in R^3$ of each 3D Gaussian as the previous evaluation benchmark is designed for point cloud or 3D mesh. We use the following evaluation metric to evaluate the quality of our active mapping system.

Comp. The completeness metric quantifies the completeness of the active exploration algorithm in 3D space by computing the per-point nearest distance between points sampled from the ground truth mesh and the final reconstruction

Table 1. Active View Selections on Blender Dataset. The best, second, and third results are highlighted in red, orange, and yellow, respectively. *: Numbers are taken directly from ActiveNeRF [43] as reference.

Method	PSNR ↑	SSIM ↑	LPIPS ↓
ActiveNeRF*	26.240	0.856	0.124
Plenoxels + Random	23.906	0.863	0.174
Plenoxels + ActiveNeRF	23.522	0.857	0.150
Nerfacto + Random	22.614	0.890	0.111
Nerfacto + BayesRays	22.504	0.887	0.115
3D Gaussian + Random	28.732	0.939	0.053
3D Gaussian + ActiveNeRF	26.610	0.905	0.081
Plenoxel + Ours	24.513	0.876	0.157
3D Gaussian + Ours	29.525	0.944	0.043

Table 2. Active Batch View Selections on Blender Dataset. Each time, 4 views are selected based on different view selection methods. *: Numbers are taken directly from ActiveNeRF [43] as reference.

Method	PSNR ↑	SSIM ↑	LPIPS ↓
ActiveNeRF*	26.240	0.856	0.124
Plenoxels + Random	23.242	0.862	0.158
Plenoxels + ActiveNeRF	23.147	0.857	0.148
Nerfacto + Random	21.622	0.875	0.137
Nerfacto + BayesRays	22.160	0.878	0.127
3D Gaussian + Random	27.135	0.927	0.065
3D Gaussian + ActiveNeRF	27.326	0.912	0.076
Plenoxels + Ours	24.212	0.878	0.139
3D Gaussian + Ours	29.094	0.938	0.053

Table 3. Active View Selections on Blender Dataset with only 10 views. Our view selection algorithm could select necessary views when the number of training views is extremely limited.

Method	PSNR ↑	SSIM ↑	LPIPS ↓
Nerfacto + Random	16.430	0.809	0.254
Nerfacto + BayesRays	14.950	0.798	0.282
Plenoxels + Random	19.950	0.812	0.233
Plenoxels + ActiveNeRF	19.770	0.804	0.210
3D Gaussian + Random	22.493	0.873	0.112
3D Gaussian + ActiveNeRF	22.979	0.876	0.111
Plenoxels + Ours	20.670	0.824	0.205
3D Gaussian + Ours	23.681	0.883	0.102

of the scene. We calculate both the percentage of points within a 5cm threshold (Comp. (%)) and the average nearest distance (Comp. (cm)).

PSNR & Depth MAE. We uniformly sample 1000 navigable poses in the scene and render color and depth images at the sampled poses and compute average PSNR and Mean Absolute Error (MAE) for rendered color and depth images, respectively.

We compare with previous state-of-the-art in Table 5 and Fig. 5. Our active mapping system outperformed previous methods by a large margin on both geometry and rendering quality. Notably, we are the first active mapping system to capture fine-grained details and high-quality textures, thanks to our view selection objectives, which aim to improve rendering quality when measuring the EIG. In Fig. 5, we demonstrate the qualitative comparison between our method and Active-INR [67]. In Table 5, our method outperforms Active-INR by a large margin in Comp. metric, showing the superiority of the active mapping algorithm. To compare the rendering quality with previous methods, we train a 3D Gaussian Splatting model using the trajectory from UPEN [15] and Active-

Table 4. Active View Selections on Mip-NeRF360 Dataset. We compare our method on real-world dataset Mip360. †: batch active view selection setting. Our method outperformed previous state-of-the-art with a clear margin.

Method	PSNR ↑	SSIM ↑	LPIPS ↓
3D Gaussian + Random	17.914	0.564	0.430
3D Gaussian + Random†	19.542	0.568	0.376
3D Gaussian + ActiveNeRF	17.889	0.533	0.414
3D Gaussian + ActiveNeRF†	18.303	0.539	0.406
3D Gaussian + Ours	20.351	0.601	0.361
3D Gaussian + Ours†	20.568	0.608	0.365

Table 5. Scene Coverage on Gibson and MP3D dataset. Our method outperformed previous methods by a large margin when evaluating the coverage metric. Numbers for other methods are taken from Active Nerual Mapping [67].

Method	Gibson Comp. (%) ↑	Gibson Comp. (cm) ↓	MP3D Comp. (%) ↑	MP3D Comp. (cm) ↓
Random	45.80	34.48	45.67	26.53
FBE	68.91	14.42	71.18	9.78
UPEN	63.30	21.09	69.06	10.60
OccAnt	61.88	23.25	71.72	9.40
Active Neural Mapping	80.45	7.44	73.15	9.11
Ours	92.89	5.64	89.41	2.91

Ground Truth   Active-INR   Ours

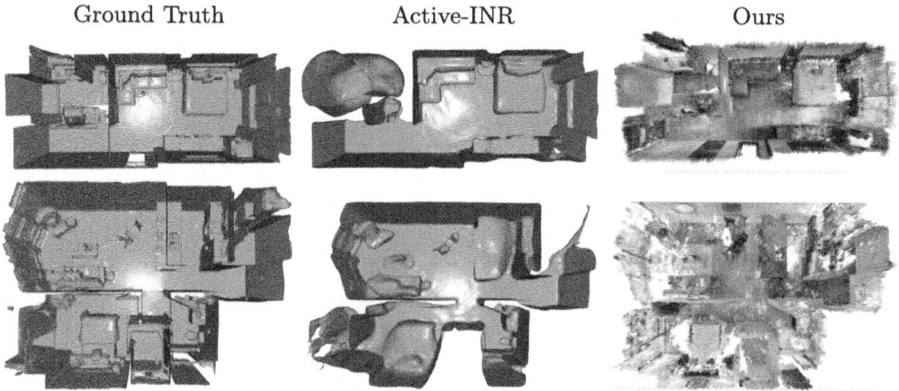

Fig. 5. Qualitative Comparisons on Active Mapping. We compare our method against the ground truth mesh and Active Neural Mapping. Our method exhibited better details and coverage. Our method can successfully produce detailed and high-fidelity reconstruction for the rooms compared to the previous method.

INR [67] and evaluate the render quality at the same holdout set. The result is summarized in Table 6. Our method efficiently selects the most informative path and outperforms other methods when using the same backbone model. We also notice that UPEN is worse than Active Neural Mapping in Table 5 but better than the other in 6 when evaluated using the 3D Gaussian Splatting model. We ascribe this gap in UPEN to the difference in the scene representations and the evaluation protocol.

4.3 Uncertainty Quantification

As discussed in Sect. 3.5, our model can be extended to compute pixel-wise uncertainties on training views. Following previous methods on uncertainty estimation [55,56], we evaluate our method on the Light Field (LF) Dataset [69]

Table 6. Evaluation for Render Quality on Gibson and MP3D. We also compare our method with previous method on the rendering quality for both color and depth reconstruction.

Method	Gibson		MP3D	
	PSNR ↑	Dpeth MAE (m) ↓	PSNR ↑	Depth MAE (m) ↓
FBE	19.35	0.1751	16.68	0.3627
UPEN	21.13	0.1893	18.40	0.2690
Active Neural Mapping	19.64	0.2943	17.16	0.4621
Ours	22.58	0.0924	19.96	0.1667

Table 7. Uncertainty Estimation on LF Dataset. Numbers are AUSE, the lower the better. The best results are highlighted in red, and the second-best results are in orange color.

Method	Statue ↓	Africa ↓	Torch ↓	Basket ↓	Average ↓
CF-NeRF	0.54	0.34	0.50	0.14	0.38
ActiveNeRF	0.40	0.39	0.21	0.34	0.33
BayesRays	0.17	0.27	0.22	0.28	0.23
Ours	0.21	0.26	0.24	0.18	0.22

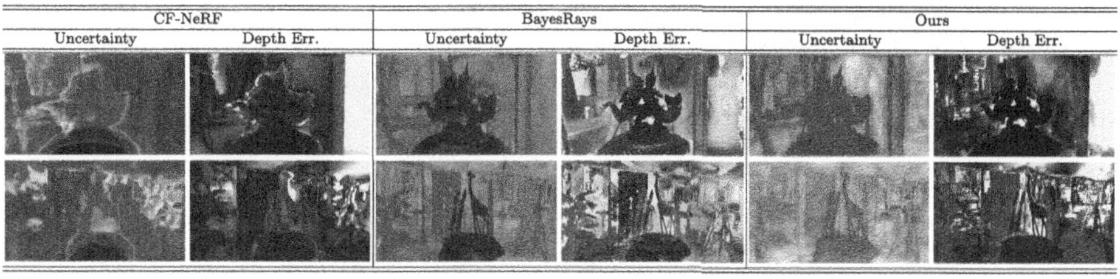

Fig. 6. Visualizations for Uncertainty on Depth Prediction the LF Dataset. We compare the per-pixel uncertainty maps from CF-NeRF [55], BayesRays [16], and ours. Our per-pixel uncertainty maps demonstrate a strong correlation with the actual depth error.

using the Area Under Sparsification Error (AUSE) metric. The pixels are filtered twice, once using the absolute error with ground truth depth and once using the uncertainty. The difference in the mean absolute error on the remaining pixels between the two sparsification processes produces two different error curvatures, where the area between those two curvatures is the AUSE, which evaluates the correlation between uncertainties and the predicted error. A low AUSE indicates our model is confident in the correctly estimated depths and could predict a high uncertainty in the regions where we are likely to have larger errors. As seen in Table 7 and Fig. 6, our model exhibited a better correlation between our uncertainty estimation and depth error than the previous state-of-the-art CF-NeRF [55]. More quantitative results and visualizations on uncertainty estimation can be found in the supplementary materials.

5   Conclusion and Limitations

We presented FisherRF, a novel method for active view selection, active mapping, and uncertainty quantification in Radiance Fields. Leveraging Fisher Information, our method provides an efficient and effective means to quantify the observed information of Radiance Field models. The flexibility of our approach allows it to be applied to various model parametrizations, including 3D Gaussian Splatting and Plenoxels. Our extensive evaluation of active view selection, active mapping, and uncertainty quantification has consistently shown superior performance compared to existing methods and heuristic baselines. These results highlight the potential of our approach to significantly enhance the quality and efficiency of image rendering and reconstruction tasks with limited viewpoints. However, our method is limited to static scenes in a confined scenario. Reconstructing dynamically changing the Radiance Field and quantifying its Fisher Information is still an open problem. More work could be done to overcome the limitations and extend the proposed method to more challenging settings.

Acknowledgements. The authors gratefully appreciate support through the following grants: NSF FRR 2220868, NSF IIS-RI 2212433, NSF TRIPODS 1934960, NSF CPS 2038873. The authors thank Pratik Chaudhari for the insightful discussion and Yinshuang Xu for proofreading the drafts.

References

1. Ash, J.T., Goel, S., Krishnamurthy, A., Kakade, S.M.: Gone fishing: neural active learning with fisher embeddings. In: NeurIPS, vol. abs/2106.09675 (2021)
2. Ash, J.T., Zhang, C., Krishnamurthy, A., Langford, J., Agarwal, A.: Deep batch active learning by diverse, uncertain gradient lower bounds. In: ICLR (2020). https://openreview.net/forum?id=ryghZJBKPS
3. Bajcsy, R.: Active perception. Proc. IEEE **76**(8), 966–1005 (1988)
4. Bajcsy, R., Aloimonos, Y., Tsotsos, J.K.: Revisiting active perception. Auton. Robot. **42**, 177–196 (2018)
5. Barron, J.T., Mildenhall, B., Verbin, D., Srinivasan, P.P., Hedman, P.: Mip-NeRF 360: unbounded anti-aliased neural radiance fields. In: CVPR (2022)
6. Barron, J.T., Mildenhall, B., Verbin, D., Srinivasan, P.P., Hedman, P.: Zip-NeRF: anti-aliased grid-based neural radiance fields. In: ICCV (2023)
7. Chang, A., et al.: Matterport3D: learning from RGB-D data in indoor environments. In: International Conference on 3D Vision (2017)
8. Chaplot, D.S., Gandhi, D., Gupta, S., Gupta, A., Salakhutdinov, R.: Learning to explore using active neural slam. arXiv preprint arXiv:2004.05155 (2020)
9. Daxberger, E., Kristiadi, A., Immer, A., Eschenhagen, R., Bauer, M., Hennig, P.: Laplace redux–effortless Bayesian deep learning. In: NeurIPS (2021)
10. Dellaert, F., Yen-Chen, L.: Neural volume rendering: NeRF and beyond (2021)
11. Dhami, H., Sharma, V.D., Tokekar, P.: Pred-NBV: prediction-guided next-best-view planning for 3D object reconstruction. In: 2023 IEEE/RSJ International Conference on Intelligent Robots and Systems (IROS), pp. 7149–7154. IEEE (2023)
12. Dornhege, C., Kleiner, A.: A frontier-void-based approach for autonomous exploration in 3D. Adv. Robot. **27**(6), 459–468 (2013)
13. Gao, K., Gao, Y., He, H., Lu, D., Xu, L., Li, J.: NeRF: neural radiance field in 3d vision, a comprehensive review (2023)

14. Georgakis, G., Bucher, B., Arapin, A., Schmeckpeper, K., Matni, N., Daniilidis, K.: Uncertainty-driven planner for exploration and navigation. In: 2022 International Conference on Robotics and Automation (ICRA), pp. 11295–11302. IEEE (2022)

15. Georgakis, G., Bucher, B., Arapin, A., Schmeckpeper, K., Matni, N., Daniilidis, K.: Uncertainty-driven planner for exploration and navigation. In: ICRA (2022)

16. Goli, L., Reading, C., Sellán, S., Jacobson, A., Tagliasacchi, A.: Bayes' rays: uncertainty quantification in neural radiance fields. arXiv (2023)

17. Guédon, A., Monasse, P., Lepetit, V.: SCONE: surface coverage optimization in unknown environments by volumetric integration. Adv. Neural. Inf. Process. Syst. **35**, 20731–20743 (2022)

18. Guédon, A., Monnier, T., Monasse, P., Lepetit, V.: MACARONS: mapping and coverage anticipation with RGB online self-supervision. In: Proceedings of the IEEE/CVF Conference on Computer Vision and Pattern Recognition, pp. 940–951 (2023)

19. Hinton, G.E., van Camp, D.: Keeping the neural networks simple by minimizing the description length of the weights. In: Proceedings of the Sixth Annual Conference on Computational Learning Theory, COLT 1993, pp. 5–13. Association for Computing Machinery, New York (1993). https://doi.org/10.1145/168304.168306

20. Hochreiter, S., Schmidhuber, J.: Simplifying neural nets by discovering flat minima. In: NeurIPS (1994)

21. Houlsby, N., Huszar, F., Ghahramani, Z., Lengyel, M.: Bayesian active learning for classification and preference learning. CoRR abs/1112.5745 (2011). http://dblp.uni-trier.de/db/journals/corr/corr1112.html#abs-1112-5745

22. Jin, L., Chen, X., Rückin, J., Popović, M.: NeU-NBV: next best view planning using uncertainty estimation in image-based neural rendering. arXiv preprint arXiv:2303.01284 (2023)

23. Karaman, S., Frazzoli, E.: Incremental sampling-based algorithms for optimal motion planning. Robot. Sci. Syst. VI **104**(2), 267–274 (2010)

24. Keetha, N., et al.: SplaTAM: splat, track & map 3D gaussians for dense RGB-D SLAM. arXiv preprint arXiv:2312.02126 (2023)

25. Kerbl, B., Kopanas, G., Leimkühler, T., Drettakis, G.: 3D Gaussian splatting for real-time radiance field rendering. ACM Trans. Graph. **42**(4) (2023). https://repo-sam.inria.fr/fungraph/3d-gaussian-splatting/

26. Kirsch, A., Amersfoort, J.v., Gal, Y.: BatchBALD: efficient and diverse batch acquisition for deep Bayesian active learning. In: NeurIPS (2019)

27. Kirsch, A., Farquhar, S., Atighehchian, P., Jesson, A., Branchaud-Charron, F., Gal, Y.: Stochastic batch acquisition for deep active learning. arXiv preprint arXiv:2106.12059 (2021)

28. Kirsch, A., Gal, Y.: Unifying approaches in active learning and active sampling via fisher information and information-theoretic quantities. Trans. Mach. Learn. Res. (2022). https://openreview.net/forum?id=UVDAKQANOW. Expert Certification

29. Kothawade, S.N., Beck, N.A., Killamsetty, K., Iyer, R.K.: SIMILAR: submodular information measures based active learning in realistic scenarios. In: Beygelzimer, A., Dauphin, Y., Liang, P., Vaughan, J.W. (eds.) Advances in Neural Information Processing Systems (2021). https://openreview.net/forum?id=VGDFaLNFFk

30. LaValle, S.: Rapidly-exploring random trees: a new tool for path planning. Research Report 9811 (1998)

31. Lee, S., Chen, L., Wang, J., Liniger, A., Kumar, S., Yu, F.: Uncertainty guided policy for active robotic 3D reconstruction using neural radiance fields. IEEE Robot. Autom. Lett. **7**(4), 12070–12077 (2022)

32. Lindley, D.V.: On a measure of the information provided by an experiment. Ann. Math. Stat. **27**(4), 986–1005 (1956). https://doi.org/10.1214/aoms/1177728069
33. Lluvia, I., Lazkano, E., Ansuategi, A.: Active mapping and robot exploration: a survey. Sensors **21**(7), 2445 (2021)
34. MacDonald, L.E., Valmadre, J., Lucey, S.: On progressive sharpening, flat minima and generalisation (2023)
35. MacKay, D.J.C.: Bayesian interpolation. Neural Comput. **4**(3), 415–447 (1992). https://doi.org/10.1162/neco.1992.4.3.415
36. Martin-Brualla, R., Radwan, N., Sajjadi, M.S.M., Barron, J.T., Dosovitskiy, A., Duckworth, D.: NeRF in the wild: neural radiance fields for unconstrained photo collections. In: CVPR (2021)
37. Matsuki, H., Murai, R., Kelly, P.H.J., Davison, A.J.: Gaussian splatting SLAM (2024)
38. Mildenhall, B., Srinivasan, P.P., Tancik, M., Barron, J.T., Ramamoorthi, R., Ng, R.: NeRF: representing scenes as neural radiance fields for view synthesis. In: ECCV (2020)
39. Mildenhall, B., Verbin, D., Srinivasan, P.P., Hedman, P., Martin-Brualla, R., Barron, J.T.: MultiNeRF: a code release for Mip-NeRF 360, ref-NeRF, and RawNeRF (2022). https://github.com/google-research/multinerf
40. Müller, T., Evans, A., Schied, C., Keller, A.: Instant neural graphics primitives with a multiresolution hash encoding. ACM Trans. Graph. **41**(4), 102:1–102:15 (2022). https://doi.org/10.1145/3528223.3530127
41. Nemhauser, G.L., Wolsey, L.A., Fisher, M.L.: An analysis of approximations for maximizing submodular set functions-I. Math. Program. **14**, 265–294 (1978)
42. Ortiz, J., et al.: iSDF: real-time neural signed distance fields for robot perception. In: Robotics: Science and Systems (2022)
43. Pan, X., Lai, Z., Song, S., Huang, G.: ActiveNeRF: learning where to see with uncertainty estimation. In: Avidan, S., Brostow, G., Cissé, M., Farinella, G.M., Hassner, T. (eds.) ECCV 2022. LNCS, vol. 13693, pp. 230–246. Springer, Cham (2022). https://doi.org/10.1007/978-3-031-19827-4_14
44. Papachristos, C., Khattak, S., Alexis, K.: Uncertainty-aware receding horizon exploration and mapping using aerial robots. In: 2017 IEEE international conference on robotics and automation (ICRA), pp. 4568–4575. IEEE (2017)
45. Placed, J.A., et al.: A survey on active simultaneous localization and mapping: state of the art and new frontiers. IEEE Trans. Robot. (2023)
46. Ramakrishnan, S.K., Al-Halah, Z., Grauman, K.: Occupancy anticipation for efficient exploration and navigation. In: Vedaldi, A., Bischof, H., Brox, T., Frahm, J.-M. (eds.) ECCV 2020. LNCS, vol. 12350, pp. 400–418. Springer, Cham (2020). https://doi.org/10.1007/978-3-030-58558-7_24
47. Ramakrishnan, S.K., et al.: Habitat-matterport 3D dataset (HM3D): 1000 large-scale 3D environments for embodied AI. In: Thirty-fifth Conference on Neural Information Processing Systems Datasets and Benchmarks Track (2021)
48. Ran, Y., et al.: NeurAR: neural uncertainty for autonomous 3D reconstruction with implicit neural representations. IEEE Robot. Autom. Lett. **8**(2), 1125–1132 (2023). https://doi.org/10.1109/lra.2023.3235686
49. Reiser, C., Peng, S., Liao, Y., Geiger, A.: KiloNeRF: speeding up neural radiance fields with thousands of tiny MLPs. In: ICCV (2021)
50. Ren, P., et al.: A survey of deep active learning. ACM Comput. Surv. (CSUR) **54**(9), 1–40 (2021)

51. Sandström, E., Li, Y., Van Gool, L., Oswald, M.R.: Point-SLAM: dense neural point cloud-based SLAM. In: Proceedings of the IEEE/CVF International Conference on Computer Vision, pp. 18433–18444 (2023)
52. Sara Fridovich-Keil and Alex Yu, Tancik, M., Chen, Q., Recht, B., Kanazawa, A.: Plenoxels: radiance fields without neural networks. In: CVPR (2022)
53. Savva, M., et al.: Habitat: a platform for embodied AI research. In: Proceedings of the IEEE/CVF International Conference on Computer Vision (ICCV) (2019)
54. Schervish, M.: Theory of Statistics. Springer Series in Statistics. Springer, New York (2012). https://books.google.com/books?id=s5LHBgAAQBAJ
55. Shen, J., Agudo, A., Moreno-Noguer, F., Ruiz, A.: Conditional-flow NeRF: accurate 3D modelling with reliable uncertainty quantification. In: Avidan, S., Brostow, G., Cissé, M., Farinella, G.M., Hassner, T. (eds.) ECCV 2022. LNCS, vol. 13663, pp. 540–557. Springer, Cham (2022). https://doi.org/10.1007/978-3-031-20062-5_31
56. Shen, J., Ruiz, A., Agudo, A., Moreno-Noguer, F.: Stochastic neural radiance fields: quantifying uncertainty in implicit 3D representations. CoRR abs/2109.02123 (2021). https://arxiv.org/abs/2109.02123
57. Shen, S., Michael, N., Kumar, V.: Autonomous indoor 3D exploration with a micro-aerial vehicle. In: 2012 IEEE International Conference on Robotics and Automation, pp. 9–15. IEEE (2012)
58. Smith, S., Le, Q.V.: A Bayesian perspective on generalization and stochastic gradient descent. In: ICLR (2018). https://openreview.net/pdf?id=BJij4yg0Z
59. Sucar, E., Liu, S., Ortiz, J., Davison, A.: iMAP: implicit mapping and positioning in real-time. In: Proceedings of the International Conference on Computer Vision (ICCV) (2021)
60. Sun, C., Sun, M., Chen, H.T.: Direct voxel grid optimization: super-fast convergence for radiance fields reconstruction. In: CVPR (2022)
61. Sünderhauf, N., Abou-Chakra, J., Miller, D.: Density-aware NeRF ensembles: quantifying predictive uncertainty in neural radiance fields. In: ICRA (2023)
62. Wang, Z., Simoncelli, E.P., Bovik, A.C.: Multiscale structural similarity for image quality assessment. In: NeurIPS, vol. 2, pp. 1398–1402. IEEE (2003)
63. Xia, F., Zamir, A.R., He, Z., Sax, A., Malik, J., Savarese, S.: Gibson ENV: real-world perception for embodied agents. In: Proceedings of the IEEE Conference on Computer Vision and Pattern Recognition, pp. 9068–9079 (2018)
64. Yamauchi, B.: A frontier-based approach for autonomous exploration. In: Proceedings 1997 IEEE International Symposium on Computational Intelligence in Robotics and Automation CIRA 1097. Towards New Computational Principles for Robotics and Automation, pp. 146–151. IEEE (1997)
65. Yan, C., et al.: GS-SLAM: dense visual slam with 3D gaussian splatting. arXiv preprint arXiv:2311.11700 (2023)
66. Yan, D., Liu, J., Quan, F., Chen, H., Fu, M.: Active implicit object reconstruction using uncertainty-guided next-best-view optimization (2023)
67. Yan, Z., Yang, H., Zha, H.: Active neural mapping. In: ICCV (2023)
68. Ye, K., et al.: Multi-robot active mapping via neural bipartite graph matching. In: Proceedings of the IEEE/CVF Conference on Computer Vision and Pattern Recognition, pp. 14839–14848 (2022)
69. Yücer, K., Sorkine-Hornung, A., Wang, O., Sorkine-Hornung, O.: Efficient 3D object segmentation from densely sampled light fields with applications to 3D reconstruction. ACM Trans. Graph. 35(3) (2016). https://doi.org/10.1145/2876504
70. Yugay, V., Li, Y., Gevers, T., Oswald, M.R.: Gaussian-SLAM: photo-realistic dense slam with gaussian splatting (2023)

71. Zhan, H., Zheng, J., Xu, Y., Reid, I., Rezatofighi, H.: ActiverMap: radiance field for active mapping and planning (2022)
72. Zhan, X., Wang, Q., Hao Huang, K., Xiong, H., Dou, D., Chan, A.B.: A comparative survey of deep active learning (2022)
73. Zhang, R., Isola, P., Efros, A.A., Shechtman, E., Wang, O.: The unreasonable effectiveness of deep features as a perceptual metric. In: CVPR, pp. 586–595 (2018)
74. Zhu, C., Ding, R., Lin, M., Wu, Y.: A 3D frontier-based exploration tool for MAVs. In: 2015 IEEE 27th International Conference on Tools with Artificial Intelligence (ICTAI), pp. 348–352. IEEE (2015)
75. Zhu, Z., et al.: Nice-SLAM: neural implicit scalable encoding for SLAM (2021)

Occlusion Handling in 3D Human Pose Estimation with Perturbed Positional Encoding

Niloofar Azizi[1][(✉)], Mohsen Fayyaz[2], and Horst Bischof[1]

[1] Graz University of Technology, Graz, Austria
{azizi,bischof}@tugraz.at
[2] Microsoft, Redmond, USA
mohsenfayyaz@microsoft.com

Abstract. Understanding human behavior fundamentally relies on accurate 3D human pose estimation. Graph Convolutional Networks (GCNs) have recently shown promising advancements, delivering state-of-the-art performance with rather lightweight architectures. In the context of graph-structured data, leveraging the eigenvectors of the graph Laplacian matrix for positional encoding is effective. Yet, the approach does not specify how to handle scenarios where edges in the input graph are missing. To this end, we propose a novel positional encoding technique, PerturbPE, that extracts consistent and regular components from the eigenbasis. Our method involves applying multiple perturbations and taking their average to extract the consistent and regular component from the eigenbasis. PerturbPE leverages the Rayleigh-Schrodinger Perturbation Theorem (RSPT) for calculating the perturbed eigenvectors. Employing this labeling technique enhances the robustness and generalizability of the model. Our results support our theoretical findings, e.g. our experimental analysis observed a performance enhancement of up to 12% on the Human3.6M dataset in instances where occlusion resulted in the absence of one edge. Furthermore, our novel approach significantly enhances performance in scenarios where two edges are missing, setting a new benchmark for state-of-the-art.

1 Introduction

Estimating the 3D pose of the human skeleton is crucial for understanding human motion and behavior, which facilitates high-level computer vision tasks such as action recognition [25] and augmented and virtual reality [13]. Nevertheless, estimating 3D human joint positions presents considerable obstacles. First, there is a scarcity of labeled datasets, as acquiring 3D annotations is costly. Additionally, challenges such as self-occlusions, complex joint inter-dependencies, and small and barely visible joints further complicate the estimation process.

To address the challenge of estimating 3D human poses, several strategies have been explored, including leveraging multi-view setups [41], utilizing synthetic data [38], or incorporating motion analysis [44]. However, these methods can be cost-prohibitive,

Supplementary Information The online version contains supplementary material available at https://doi.org/10.1007/978-3-031-72624-8_25.

with multi-view configurations being impractical for real-world applications and the analysis of temporal data requiring significant resources. A more resource-efficient approach is lifting 2D-to-3D skeletons. The 2D human skeleton is a graph-structured data and thus Graph Convolutional Networks (GCNs) achieve state-of-the-art performance for 2D-to-3D human pose estimation by reducing the number of parameters in the order of magnitude, e.g. [2] (Fig. 1).

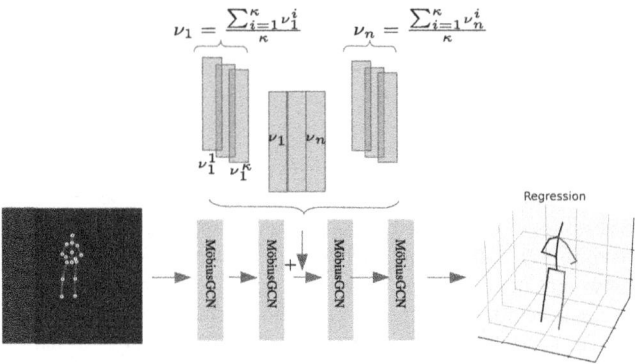

Fig. 1. Given a human skeleton graph with only blue edges provided and red edges absent, we calculate the corresponding graph Laplacian matrix. By selectively removing several edges at random (the selected elements in the graph Laplacian matrix), we then compute the perturbed eigenvectors (ν_i), which are subsequently averaged to identify consistent eigenbasis ($\frac{\sum_{i=1}^{\kappa} \nu_i}{\kappa}$). Finally, we integrate the perturbed positional features into a graph neural network architecture. The perturbed eigenvectors are computed with Rayleigh-Schrödinger Perturbation Theorem (RSPT).

Nevertheless, GCNs have limited expressivity. The seminal work [31] shows that GCNs are as expressive as 1-Weisfeiler-Lehman (1-WL) [50]. Efforts aimed at augmenting the expressiveness of GCNs have pursued four main directions: aligning with the k-WL hierarchy [27], enriching node features with identifiers, or exploiting structural information that cannot be captured by the WL test [4]. Subgraph GNNs [3] have emerged as a potential solution, enriching GNN features by encoding extracted subgraphs as novel features and incorporating them into the GNN architecture. One line of work addresses the limited expressivity of GCNs by positional encoding which utilizes the eigenbasis of graph Laplacian matrix [9,20]. However, if some edges are missing these methods are not applicable. This scenario may occur in human pose estimation when the subject is not completely visible due to obstructions, e.g., occlusion.

To address this problem, we propose a novel perturbed **P**ositional **E**ncoding (PerturbPE), where we label the nodes by considering that graph Laplacian matrix has regular and irregular parts [22]. We extract the regular part of the graph Laplacian eigenbasis for the positional encoding, by applying multiple perturbations (, i.e., removing different sets of edges each time) to the initial graph Laplacian matrix. After computing the perturbed eigenvector each time, we average over the perturbed eigenvectors to extract the consistent part of the graph Laplacian eigenbasis with missed edges. Employing this labeling technique enhances the robustness and generalizability of the model. In other words, the eigenvectors can well reflect network structural features. As proved in [22], the regularity of the network is reflected in the consistency of structural features before and after a random removal of a small set of links.

We summarize our main contributions as follows:

- We provide a novel positional encoding, PerturbPE, to extract the regular part of the graph Laplacian eigenbasis for scenarios where some edges are missing in the input graph, e.g., scenarios where occlusion happens in the input human 2D skeleton. We utilize Rayleigh-Schrödinger Perturbation Theorem (RSPT) to compute the eigenpairs. Its primary advantage is that it eliminates the need to calculate the entire eigenbasis.
- We attain state-of-the-art by while training only one neural network to adeptly manage all scenarios where specific edges (specific parts of the 2D human skeleton) are missing.
- We also provide the state-of-the-art results of positional encoding for the task of 3D human pose estimation, achieving state-of-the-art 3D human pose estimation results, despite requiring the same number of model parameters as MöbiusGCN [2] on benchmark datasets.

2   Related Work

3D Human Pose Estimation. Classical methods for 3D human pose estimation primarily rely on had-engineered features and incorporate prior knowledge (e.g., [15,40,46]). While these techniques have achieved commendable outcomes, their main limitation is their inability to generalize. Current state-of-the-art in computer vision, including 3D human pose estimation, primarily utilize Deep Neural Networks (DNNs) [18,24,26]. These models operate under the assumption that input data exhibits characteristics of locality, stationarity, and multi-scalability. While DNNs excel in areas defined by Euclidean geometry, many challenges in the real world do not conform to this Euclidean framework.

Graph Convolutional Networks (GCNs) have emerged as a powerful solution for these problem categories, encompassing two main types: spectral and spatial GCNs. Spectral GCNs operate on the principles of the Graph Fourier Transform to analyze graph signals via the graph's Laplacian matrix vector space. On the other hand, spatial GCNs focus on transforming features and aggregating neighborhood information directly on the graph, e.g., Message Passing Neural Networks [11] and Graph-SAGE [12].

GCNs stand out for delivering high-quality results with a relatively small number of parameters in the task of 3D human pose estimation. Various studies [21,52,56], have explored pose estimation using GCNs. Xu and Takano [52] introduced the Graph Stacked Hourglass Networks (GraphSH), a method that processes graph-structured features across multiple levels of human skeletal structures. Similarly, Liu et al. [21] delved into diverse strategies for feature transformation and the aggregation of neighborhood spatial features, highlighting the advantage of assigning unique weights to enhance a node's self-information. Zhao et al. [56] developed the semantic GCN (SemGCN) focusing on the concept of learning the adjacency matrix to capture semantic relationships between graph nodes. MöbiusGCN [2], dramatically reduced parameter count to just 0.042M by leveraging the Möbius transformation to explicitly define joint transformations, showcasing an even more efficient model architecture for pose estimation.

GCN Expressivity. Nonetheless, the expressiveness of GCNs is constrained, and to surmount this challenge, four principal approaches have been proposed. Evaluating the expressive capacity of GNNs requires dealing with the graph isomorphism, which has no P solution (NP-intermediate). The 1-WL test [50], effectively resolves graph isomorphism for a vast majority of graph-structured data. However, as indicated in research by Li and Leskovec [51], Morris et al. [31], Xu et al. [19], the expressivity of MPNNs is limited to that of the 1-WL test. This limitation becomes particularly significant in real-world applications involving graph structures, as the 1-WL test cannot distinguish between certain graph features. Specifically, it fails in differentiating isomorphism in attributed regular graphs, measuring distances between nodes, and counting cycles within graphs [19]. Addressing these limitations, recent studies have provided four different approaches. The first approach involves adding random attributes to nodes with identical substructures. This method aims to provide uniqueness to these nodes, enhancing their capability to be differentiated. However, this advantage comes at the cost of reduced deterministic predictability, potentially leading to difficulties during inference [42]. Furthermore, deterministic positional features (*e.g.*, [55]) argue that the incapability of the GNNs to encode the distance between nodes in the input graph raises the above issues and addresses them by injecting deterministic distance attributes. However, these methods assign different node features to isomorphic graphs and, thus, are not generalizable in inference time [19]. The third strategy involves developing higher-order GNNs [27] to surpass the 1-WL test's expressivity limits. The fourth is adopting subgraph-based methods like ESAN [3], which enhance expressivity through selected subgraphs.

Nevertheless, if some edges are missing, none of the previously mentioned methods can be applied to enhance expressivity. In the realm of 2D human pose estimation, this significant challenging scenario can happen when occlusion happens. When parts of the human body are occluded or hidden from view, accurately estimating the 3D human pose becomes more difficult. This problem is prevalent in real-world scenarios where objects may obstruct the view, or the camera angle limits visibility.

To address this problem we propose a novel positional encoding. In the seminal work [9], it was proposed that incorporating graph Laplacian eigenvectors as positional features within the Graph Neural Network (GNN) architecture could improve its effectiveness, although this approach encountered obstacles such as sign ambiguity and eigenvalue multiplicity. To address these challenges, SignNet/BasisNet [20] was developed. A Multilayer Perceptron (MLP) was employed to deal with the issue of multiplicity, while an approach using an even function—by amalgamating the function with its inverse—was implemented to tackle problems related to sign ambiguity. However, to the best of our knowledge, none of the previous works provided a solution in case some edges missing, which can be reduced to subgraph matching which is NP-complete. We propose extracting the consistent and robust part of the graph Laplacian matrix's eigenbasis by labeling the joints (graph nodes) utilizing the perturbed eigenbasis of the graph Laplacian matrix. To compute the perturbed eigenbasis, we employ the Rayleigh Schrodinger Perturbation Theory (RSPT).

Data Reduction. Semi-supervised techniques are favored due to the tedious and costly nature of data annotation. MöbiusGCN [2] stands as the most efficient framework for 3D human pose estimation tasks to date, thereby diminishing the need for extensive annotated data for training. The advantage of light architectures is their ability to be trained with less data. Our novel perturbed positional encoding architecture, which does not add additional training parameters, maintains the framework's efficiency, enabling it to achieve leading results with a minimal amount of labeled data.

3   Preliminaries

To compute the consistent part of the graph Laplacian eigenbasis for the positional encoding, we use the perturbed eigenvectors. This process involves using the Rayleigh-Schrödinger Perturbation Theorem (RSPT) to determine the perturbed eigenvectors of the graph Laplacian matrix. Hence, we provide a concise overview of the RSPT. Moreover, while PerturbPE can enhance any GNN framework, our experiments are conducted using MöbiusGCN, a model developed specifically for the task of 3D human pose estimation. Therefore, we also briefly overview the MöbiusGCN.

3.1   Rayleigh-Schrödinger Perturbation Theory

In mathematical terms, we are dealing with a discretized Laplacian-type operator represented by a real symmetric matrix that undergoes a minor symmetric linear perturbation.

$$\mathbf{A}(\epsilon) = \mathbf{A}_0 + \epsilon \mathbf{A}_1. \tag{1}$$

The Rayleigh-Schrödinger Perturbation Theory (RSPT) [29] provides estimates for the eigenvalues and eigenvectors of the matrix $\mathbf{A}$ through a series of progressively higher-order adjustments to the eigenvalues and eigenvectors of the matrix $\mathbf{A}_0$. $\mathbf{A}_0$ is likewise real and symmetric but may possess multiple eigenvalues.

An advantage of RSPT is that it doesn't necessitate the full set of $\mathbf{A}_0$ and can be approached using the Moore-Penrose Pseudoinverse. The pseudoinverse need not be explicitly calculated since only pseudoinverse vector products are required. These may be efficiently be calculated by a combination of QR-factorization and Gaussian elimination. Since we are only concerned with real-symmetric matrices, the existence of a complete set of orthonormal eigenvectors is assured.

Reconstruction of Perturbed Matrix. To compute the eigenpairs of $\mathbf{A}$ possessing respective perturbation expansions

$$\lambda_i(\epsilon) = \sum_{k=0}^{\infty} \epsilon^k \lambda_i^{(k)} \quad \mathbf{v}_i(\epsilon) = \sum_{k=0}^{\infty} \epsilon^k \mathbf{v}_i^{(k)}, \tag{2}$$

where $(i = 1, \ldots, n)$ for sufficiently small ϵ. Experimentally, we set both ϵ and k to 1.

Considering the eigenvalue problem and and taking into account the Eqs. 1 and 2, we have to solve the following recurrence relation

$$(\mathbf{A}_0 - \lambda_i^{(0)}\mathbf{I})\mathbf{x}_i^{(k)} = -(\mathbf{A}_1 - \lambda_i^{(1)}\mathbf{I})\mathbf{x}_i^{(k-1)} + \sum_{j=0}^{k-2}\lambda_i^{k-j}\mathbf{x}_i^{(j)} \tag{3}$$

for $(k = 1, \ldots, \infty; i = 1, \ldots, n)$.

The solution is either degenerate or non-degenrate.

Nondegenerate Case. By assuming all the eigenvalues are distinct the eigenpairs are computed as follows.

If j is odd, then

$$\lambda_i^{2j+1} = \langle \mathbf{x}_i^{(j)}, \mathbf{A}_1\mathbf{x}_i^{(j)}\rangle - \sum_{\mu=0}^{j}\sum_{\nu=1}^{j}\lambda_i^{(2j+1-\mu-\nu))}\langle \mathbf{x}_i^{(\nu)}, \mathbf{x}_i^{(\mu)}\rangle. \tag{4}$$

If j is even, then

$$\lambda_i^{2j} = \langle \mathbf{x}_i^{(j-1)}, \mathbf{A}_1\mathbf{x}_i^{(j)}\rangle - \sum_{\mu=0}^{j}\sum_{\nu=1}^{j}\lambda_i^{(2j-\mu-\nu))}\langle \mathbf{x}_i^{(\nu)}, \mathbf{x}_i^{(\mu)}\rangle. \tag{5}$$

The corresponding eigenvector is computed as follows

$$\mathbf{x}_i^{(k)} = (\mathbf{A}_0 - \lambda_i^{(0)}\mathbf{I})^\dagger[-(\mathbf{A}_1 - \lambda_i^{(1)}\mathbf{I})\mathbf{x}_i^{(k-1)} + \sum_{j=0}^{k-2}\lambda_i^{(k-j)}\mathbf{x}_i^{(j)}] \tag{6}$$

The unperturbed eigenvectors are assumed to have been normalized to unity so that $\lambda_i^{(0)} = \langle \mathbf{x}_i^{(0)}, \mathbf{A}_0\mathbf{x}_i^0\rangle$.

Degenerate Case. For the calculation of perturbed eigenpairs in scenarios involving multiplicity, $\lambda_1^{(0)} = \lambda_2^{(0)} = \cdots = \lambda_m^{(0)} = \lambda^{(0)}$, accompanied by m known orthonormal eigenvectors $\mathbf{v}_1^{(0)}, \ldots, \mathbf{v}_m^{(0)}$ and the assumption of first-order degeneracy which ensures the uniqueness of the first-order eigenvalues, the calculation of these perturbed eigenvalues and their corresponding degenerate eigenvectors are achieved by determining appropriate linear combinations of

$$\mathbf{y}_i^{(0)} = a_1^{(i)}\mathbf{v}_1^{(0)} + a_2^{(i)}\mathbf{v}_2^{(0)} + a_3^{(i)}\mathbf{v}_3^{(0)} + \cdots + a_m^{(i)}\mathbf{v}_m^{(0)}, \tag{7}$$

To have a solution for Eq. (3) in this scenario, it is necessary and sufficient that for each fixed i,

$$\langle x_\mu^{(0)}, (\mathbf{A}_1 - \lambda_i^{(1)}\mathbf{I})\mathbf{y}_i^{(0)}\rangle = 0 \quad (\mu = 1, \ldots, m) \tag{8}$$

By replacing Eq. (7),

$$
\begin{bmatrix}
\langle \mathbf{x}_1^{(0)}, \mathbf{A}_1 \mathbf{x}_1^{(0)} \rangle & \cdots & \langle \mathbf{x}_1^{(0)}, \mathbf{A}_1 \mathbf{x}_m^{(0)} \rangle \\
\vdots & \ddots & \vdots \\
\langle \mathbf{x}_m^{(0)}, \mathbf{A}_1 \mathbf{x}_1^{(0)} \rangle & \cdots & \langle \mathbf{x}_m^{(0)}, \mathbf{A}_1 \mathbf{x}_m^{(0)} \rangle
\end{bmatrix}
\begin{bmatrix}
a_1^{(i)} \\
\vdots \\
a_m^{(i)}
\end{bmatrix}
= \lambda_i^{(1)}
\begin{bmatrix}
a_1^{(i)} \\
\vdots \\
a_m^{(i)}
\end{bmatrix}
$$

the corresponding eigenvector of each $\lambda_i^{(1)}$ becomes $[a_1^{(i)}, \ldots, a_m^{(i)}]^\top$.

3.2 Spectral Graph Convolutional Network

Graph Definitions. Given a graph $\mathcal{G}(V, E)$ with vertices $V = \{v_1, \ldots, v_N\}$ and edges $E = \{e_1, \ldots, e_M\}$, where $e_j = (v_i, v_k)$ with $v_i, v_k \in V$. The adjacency matrix $\mathbf{A}$ marks 1 for connected vertices and 0 otherwise. The degree matrix $\mathbf{D}$ is diagonal, listing vertex degrees $\mathbf{D}ii$ for vi. Graph $\mathbf{A}$ is symmetric for undirected graphs. The graph Laplacian $\mathbf{L} = \mathbf{D} - \mathbf{A}$, and its normalized form $\bar{\mathbf{L}} = \mathbf{I} - \mathbf{D}^{-\frac{1}{2}}\mathbf{A}\mathbf{D}^{-\frac{1}{2}}$, with $\mathbf{I}$ the identity matrix, are key in analyzing graph structure. $\bar{\mathbf{L}}$ is symmetric and positive semi-definite with ordered, real, non-negative eigenvalues λ_i and orthonormal eigenvectors $\mathbf{u}_i$. A graph signal $\mathbf{x} \in \mathbb{R}^N$ assigns values to vertices, and $\mathbf{X} \in \mathbb{R}^{N \times d}$ represents a d-dimensional signal on $\mathcal{G}$ [45].

Graph Fourier Transform. Graph signals $\mathbf{x} \in \mathbb{R}^N$ admit a graph Fourier expansion $\mathbf{x} = \sum_{i=1}^{N} \langle \mathbf{u}_i, \mathbf{x} \rangle \mathbf{u}_i$, where $\mathbf{u}_i, i = 1, \ldots, N$ are the eigenvectors of the graph Laplacian [45]. Eigenvalues and eigenvectors of the graph Laplacian matrix are analogous to frequencies and sinusoidal basis functions in the classical Fourier series expansion.

Spectral Graph Convolutional Network. Spectral GCNs [5] build upon the graph Fourier transform. Let $\mathbf{x}$ be the graph signal and $\mathbf{y}$ be the graph filter on graph $\mathcal{G}$. The graph convolution $*_{\mathcal{G}}$ can be defined as:

$$
\mathbf{x} *_{\mathcal{G}} \mathbf{y} = \mathbf{U} \operatorname{diag}(\mathbf{U}^\top \mathbf{y}) \mathbf{U}^\top \mathbf{x}, \tag{9}
$$

where the matrix $\mathbf{U}$ contains the eigenvectors of the normalized graph Laplacian and $\odot$ is the Hadamard product. This can also be written as

$$
\mathbf{x} *_{\mathcal{G}} g_\theta = \mathbf{U} g_\theta(\mathbf{\Lambda}) \mathbf{U}^\top \mathbf{x}, \tag{10}
$$

where $g_\theta(\mathbf{\Lambda})$ is a diagonal matrix consisting of the learnable parameters, and is a function of the eigenvalues $\mathbf{\Lambda}$. We utilize MöbiusGCN [2], which defines the function g_θ to be Möbius transformation.

Thus a MöbiusGCN block is

$$
\mathbf{Z} = \sigma(2\Re\{\mathbf{U}\text{Möbius}(\mathbf{\Lambda})\mathbf{U}^\top \mathbf{X}\mathbf{W}\} + \mathbf{b}), \tag{11}
$$

where $\mathbf{Z} \in \mathbb{R}^{N \times F}$ is the convolved signal matrix, σ is a nonlinearity (*e.g.* ReLU [32]), and $\mathbf{b}$ is a bias term and the graph signal matrix $\mathbf{X} \in \mathbb{C}^{N \times d}$ with d input channels (*i.e.* a d-dimensional feature vector for every node) and $\mathbf{W} \in \mathbb{C}^{d \times F}$ feature maps.

4   Method

Eigenvector Positional Encoding. We aim to compute the consistent part of graph Laplacian eigenbasis to enhance generalizability, specifically in cases where the occlusions occur and the complete 2D human skeleton graph is not provided as an input to the architecture. To compute the positional encoding in such scenarios, we apply the Rayleigh-Schrödinger Perturbation Theory (RSPT) multiple times (κ-times) independently. During each iteration, we randomly eliminate a number of edges from the graph's Laplacian matrix (denoted as $\mathbf{A}_0$ in Eq. (1)), and then compute the perturbation with respect to it (we update $\mathbf{A}_1$ with eliminated edges). By executing the RSPT for each of these iterations and averaging over them, the regular part of the κ eigenvectors can be extracted with

$$\mathbf{p} = \frac{\sum_{i=1}^{\kappa} \mathbf{v}_i}{\kappa}, \tag{12}$$

where $\mathbf{v}_i$ is the ith perturbed eigenvector.

After executing the algorithm κ times and averaging the outcomes to isolate the stable elements of the graph Laplacian matrix's eigenbasis, we employ the following positional encoding technique to incorporate the features into the architecture.

Positional Features. We incorporate positional features, similar to [9], into the architecture through the subsequent method.

$$\mathbf{X}^\ell = \sigma(f(\mathbf{Z}^\ell + \mathbf{P})) \tag{13}$$

where $\mathbf{P} \in \mathbb{R}^{N \times N}$ is the PerturbPE positional encoding computed with the RSPT. For each vector, we define the positional features, denoted as $\mathbf{p}$, as the mean of the perturbed eigenvectors where it contains the consistent regular part of the graph Laplacian eigenbasis. The function f represents a Multilayer Perceptron (MLP) that is applied to both node and positional features. Therefore, Eq. 11 becomes

$$\mathbf{Z}^{\ell+1} = \sigma(2\Re\{\mathbf{U}\,\mathrm{M\ddot{o}bius}(\mathbf{\Lambda})\mathbf{U}^\top \sigma(f(\mathbf{Z}^\ell + \mathbf{P}))\mathbf{W}^{\ell+1}\} + \mathbf{b}). \tag{14}$$

Masked Condition Strategy. In our experiments, we operate under the assumption that each sample may lack certain components of the human skeleton, specifically that up to two edges between joints might be missing randomly during both testing and training phases. However, we assume that although some edges are missed the total number of joints is known.

5   Experimental Results

5.1   Datasets and Evaluation Protocols

We employ the widely recognized Human3.6M motion capture dataset for our study [15]. This extensive dataset encompasses over 3.6 million images, collected from

11 participants engaging in 15 distinct activities, captured through four calibrated RGB cameras. This setup was meticulously designed to ensure a comprehensive capture of each subject's movements, both during the training and testing phases. In alignment with previous works (*e.g.,* [2,28,36,43,47,48,52,56]), our experimental framework utilizes the data from five subjects (S1, S5, S6, S7, S8) for model training purposes, while reserving two subjects (S9 and S11) for the testing phase. Each sample is independently analyzed, reflecting the unique viewpoints provided by each camera.

To evaluate our model's ability to generalize, we employ the MPI-INF-3DHP dataset [30]. This dataset features six subjects tested across three distinct settings: a studio with a green screen (GS), a studio lacking a green screen (noGS), and an outdoor environment (Outdoor). It's important to mention that for the experiments conducted using the MPI-INF-3DHP dataset, training was only performed on the Human3.6M dataset.

Following the methodology of prior research [28,47,48,52,56], we adopt the MPJPE protocol, Protocol #1. The MPJPE is the mean per joint position error in millimeters between predicted joint positions and ground truth joint positions after aligning the pre-defined root joints (*i.e.,* the pelvis joint). Note that some works (*e.g.,* [21,37]) use the P-MPJPE metric, which reports the error after a rigid transformation to align the predictions with the ground truth joints. However, we deliberately chose the standard MPJPE metric as it is more challenging and the more equitable basis it provides for comparing our work with previous research.

In assessing performance on the MPI-INF-3DHP test set, in alignment with prior research [23,52], we adopt the 3D Percentage of Correct Keypoints (3D PCK) with a 150 mm threshold [30]. This measure allows us to accurately gauge the accuracy of our 3D joint predictions within a specified error margin, offering a comprehensive view of our model's performance in comparison to the benchmarks set by preceding studies.

5.2   Implementation Details

2D Pose Estimation. PerturbPE receives 2D joint positions as inputs, which are independently estimated from the RGB images captured by all four cameras. PerturbPE operates independently from any off-the-shelf architecture employed for the estimation of 2D joint positions. Although CPN [6] provides better 2D human skeleton estimation, similar to previous works [28,56], we use the stacked hourglass architecture [33] to estimate the 2D joint positions. The Hourglass architecture is a type of autoencoder architecture that incorporates multiple skip connections at various intervals. In line with [56], the stacked hourglass network undergoes initial pre-training on the MPII dataset [1], followed by subsequent fine-tuning using the Human3.6M dataset [15]. As detailed by Pavllo et al. [37], the input joints are adjusted to fit within image coordinates and are normalized to the range of $[-1, 1]$.

3D Pose Estimation. The Human3.6M dataset [15] provides ground truth 3D joint positions in world coordinates. To align with previous works [2,56], we utilize camera calibration parameters to transform these joint positions into camera space. Furthermore, when training the pipeline, akin to previous studies [2,28]a predefined joint (the pelvis joint) as the center of the coordinate system is selected.

We trained PerturbPE using Adam optimizer [16] with an initial learning rate of 0.001 and mini-batches of size 64. Our neural network pipeline, which operates in the complex-valued domain, is built upon the PyTorch framework [34], which leverages Wirtinger calculus [17] to enable backpropagation within the complex-valued domain.

We adopt MöbiusGCN as our baseline architecture due to its lightweight nature and impressive accuracy. In our experiments, we utilize eight MöbiusGCN blocks. Each block, excluding the first and last blocks with input and output channels set to 2 and 3 respectively, consists of either 128 channels (yielding 0.16 million parameters) or 192 channels (resulting in 0.66 million parameters). Furthermore, we incorporate positional encoding features in the BLA block, followed by a subsequent linear layer.

Same as [28,56], we predict the normalized locations [21,28,35,56] of 16 joints (*i.e.*, without the 'Neck/Nose' joint) in 3D and use the mean squared error (MSE) loss between the 3D ground truth joint locations $\mathcal{Y}$ and our predictions $\hat{\mathcal{Y}}$

$$\mathcal{L}(\mathcal{Y}, \hat{\mathcal{Y}}) = \sum_{i=1}^{k}(\mathcal{Y}_i - \hat{\mathcal{Y}}_i)^2, \tag{15}$$

where k is the number of joints [28,37]. Furthermore, similar to [2,39], to let the architecture differentiate between different 3D poses with the same 2D pose, the center of mass of the subject is provided as an additional input. Please note that during inference the scale of the outputs is calibrated by forcing the sum of the length of all 3D bones to be equal to a canonical skeleton [2,58,59].

All of our experiments were conducted using a PyTorch framework [34] on an NVIDIA GeForce RTX 2080 GPU.

Table 1. Quantitative Evaluation Using MPJPE (mm) on the Human3.6M [15] Dataset under Protocol #1, Highlighting Leading Performances. In the upper section, methods utilize stacked hourglass (HG) 2D estimates [33], with the exception of one approach using CPN [6] (denoted by *). The lower section compares methods based on 2D ground truth (GT) inputs. Best results are highlighted in **bold**, and the second-best are underlined. Lower is better.

Protocol #1	# Param.	Dir.	Disc.	Eat	Greet	Phone	Photo	Pose	Purch.	Sit	SitD.	Smoke	Wait	WalkD.	Walk	WalkT.	Average
Martinez et al. [28]	4M	51.8	56.2	58.1	59.0	69.5	78.4	55.2	58.1	74.0	94.6	62.3	59.1	65.1	49.5	52.4	62.9
Tekin et al. [48]	n/a	54.2	61.4	60.2	61.2	79.4	78.3	63.1	81.6	70.1	107.3	69.3	70.3	74.3	51.8	63.2	69.7
Sun et al. [47]	n/a	52.8	54.8	54.2	54.3	61.8	67.2	53.1	53.6	71.7	86.7	61.5	53.4	61.6	47.1	53.4	59.1
Yang et al. [54]	n/a	51.5	58.9	50.4	57.0	62.1	65.4	49.8	52.7	69.2	85.2	57.4	58.4	**43.6**	60.1	47.7	58.6
Hossain and Little [14]	16.96M	48.4	50.7	57.2	55.2	63.1	72.6	53.0	51.7	66.1	80.9	59.0	57.3	62.4	46.6	49.6	58.3
Fang et al. [10]	n/a	50.1	54.3	57.0	57.1	66.6	73.3	53.4	55.7	72.8	88.6	60.3	57.7	62.7	47.5	50.6	60.4
Pavlakos et al. [36]	n/a	48.5	54.4	54.5	52.0	59.4	65.3	49.9	52.9	65.8	71.1	56.6	52.9	60.9	44.7	47.8	56.2
SemGCN Zhao et al. [56]	0.43M	48.2	60.8	51.8	64.0	64.6	**53.6**	51.1	67.4	88.7	**57.7**	73.2	65.6	48.9	64.8	51.9	60.8
Sharma et al. [43]	n/a	48.6	54.5	54.2	55.7	62.2	72.0	50.5	54.3	70.0	78.3	58.1	55.4	61.4	45.2	49.7	58.0
GraphSH [52] *	3.7M	**45.2**	**49.9**	47.5	50.9	54.9	66.1	48.5	46.3	59.7	71.5	51.4	48.6	53.9	39.9	44.1	51.9
MöbiusGCN [2] (HG)	**0.16M**	46.7	60.7	47.3	50.7	64.1	61.5	46.2	45.3	67.1	80.4	54.6	51.4	55.4	43.2	48.6	52.6
Ours (HG)	0.66M	45.9	50.1	41.2	43.2	52.7	57.4	43.0	38.4	55.4	61.8	45.8	46.8	48.5	38.9	42.8	50.8
Liu et al. [21] (GT)	4.2M	36.8	40.3	33.0	36.3	37.5	45.0	39.7	34.9	40.3	47.7	37.4	38.5	38.6	29.6	32.0	37.8
GraphSH [52] (GT)	3.7M	35.8	38.1	31.0	35.3	35.8	43.2	37.3	31.7	38.4	45.5	35.4	36.7	36.8	27.9	30.7	35.8
SemGCN [56] (GT)	0.43M	37.8	49.4	37.6	40.9	45.1	41.4	40.1	48.3	50.1	42.2	53.5	44.3	40.5	47.3	39.0	43.8
MöbiusGCN [2] (GT)	0.16M	31.2	46.9	32.5	31.7	41.4	44.9	33.9	30.9	49.2	55.7	35.9	36.1	37.5	29.07	33.1	36.2
Ours (GT)	0.66M	30.1	35.3	30.6	27.6	36.2	38.4	30.7	30.3	35.9	40.7	32.9	34.9	35.2	27.2	32.0	32.7

RSPT Perturbed Eigenvectors. In our experiments we consider computing the perturbed eigenvectors with different scenarios. We consider scenarios where zero, one, or two random edges are removed for computing the perturbed eigenvectors from the input occluded 2D human skeleton graph. Specifically, in our experiments, we consider graph Laplacian eigenvectors [9] positional encoding, RSPT eigenbasis with zero edge is missed for perturbation, one edge is missed for perturbation, or two edges are missed for computing the perturbation with RSPT.

5.3 Complete 2D Human Skeleton

In this study, we present a comparative analysis of PerturbPE's performance using a complete 2D human skeleton against the former leading techniques in 3D human pose estimation on the Human3.6M and MPI-INF-3DHP datasets. Our comparison utilizes two types of inputs: a) 2D poses estimated through the stacked hourglass architecture (HG) [33] and b) the 2D ground truth (GT).

Comparison on Human3.6M. Table 1 presents a comparison between our PerturbPE method and the leading techniques as per Protocol #1 in the Human3.6M dataset. Utilizing the eigenvector of the graph Laplacian matrix for positional encoding leads to enhanced performance, reducing the error rate from 34.1 mm to 33.4 mm without an increase in model complexity. By introducing perturbed eigenvectors to tackle the multiplicity issue, we improved further, lowering the error rate to 32.7 mm. Notably, these advancements in positional encoding are achieved without the addition of extra parameters, ensuring the model remains efficient while benefiting from these enhancements. To evaluate the efficacy of our PerturbPE, we conducted experiments using SemGCN [56]. Please refer to the supplementary material in Table A.1 for more details.

Comparison on MPI-INF-3DHP. To assess the adaptability and robustness of our method, we conducted evaluations using the MPI-INF-3DHP dataset, despite our

Table 2. Results on the MPI-INF-3DHP test set [30]. Best in bold, second-best underlined. All methods use 2D ground truth as input. Lower is better.

Method	# Parameters	GS	noGS	Outdoor	All(PCK)
Martinez et al. [28]	4.2M	49.8	42.5	31.2	42.5
Mehta et al. [30]	n/a	70.8	62.3	58.8	64.7
Luo et al. [23]	n/a	71.3	59.4	65.7	65.6
Yang et al. [54]	n/a	–	–	–	69.0
Zhou et al. [59]	n/a	71.1	64.7	72.7	69.2
Ci et al. [7]	n/a	74.8	70.8	77.3	74.0
Zhou et al. [57]	n/a	75.6	71.3	80.3	75.3
GraphSH [52]	3.7M	**81.5**	**81.7**	75.2	<u>80.1</u>
MöbiusGCN [2]	**0.16M**	79.2	77.3	<u>83.1</u>	80.0
Ours	**0.16M**	<u>80.0</u>	<u>79.0</u>	**84.0**	**82.0**

model, PerturbPE, being exclusively trained on the Human3.6M dataset. This choice allowed us to test the generalizability of our approach beyond the conditions for which it was directly trained. In a comparative analysis with MöbiusGCN, PerturbPE exhibited outstanding performance, marking a notable improvement in the key metric of evaluation, PCK. With an identical configuration in terms of the number of parameters between the two models, our method demonstrated an overall enhancement in the PCK metric across various testing scenarios, improving the initial score of 80.0 to an improved score of 82.0. The significance of this improvement was even more pronounced in the most challenging conditions presented by outdoor scenarios, where our model achieved a state-of-the-art PCK score of 84.0. This result underscores the efficacy of PerturbPE in handling complex real-world situations. The results are in Table 2.

Comparison to Previous GCNs. Table 3 showcases how our approach stands up against previously established GCN models. By augmenting the channel count in each block of the MöbiusGCN architecture from 128 to 192, we observed an enhancement in the MPJPE metric by 2.1 Mm. Building on this improvement, our novel PerturbPE technique further advances MPJPE performance by an additional 1.4 mm, all while maintaining an equivalent number of parameters to MöbiusGCN. This strategic enhancement effectively reduces the MPJPE from 34.1 mm down to 32.7 mm, illustrating the efficacy of our modifications in refining pose estimation accuracy.

Table 3. Supervised quantitative comparison between GCN architectures on Human3.6M [15] under Protocol #1. Best in bold, second-best underlined. All methods use 2D ground truth as input. Lower is better.

Method	# Parameters	MPJPE
Liu et al. [21]	4.20M	37.8
GraphSH [52]	3.70M	35.8
Liu et al. [21]	1.05M	40.1
GraphSH [52]	0.44M	39.2
SemGCN [56]	0.43M	43.8
Yan et al. [53]	0.27M	57.4
Veličković et al. [49]	**0.16M**	82.9
MöbiusGCN [2]	**0.16M**	36.2
MöbiusGCN [2]	0.66M	<u>34.1</u>
Ours	0.66M	**32.7**

PerturbPE with Reduced Dataset. PerturbPE adeptly mirrors the parameter efficiency of the lightweight MöbiusGCN model, reaping the advantages of a compact design that necessitates a smaller volume of training data. By incorporating our novel positional encoding technique, PerturbPE dramatically refines its performance from 44.7 $textmm$ to 42.9 mm with MPJPE metric, achieved by analyzing data from just three subjects. This is particularly advantageous given the significant expense involved in acquiring 3D

Table 4. Evaluating the effects of using fewer training subjects on Human3.6M [15] under Protocol #1 (given 2D GT inputs). Lower is better.

Subject	# Parameters	MöbiusGCN [2]	PerturbPE
S1	0.15M	44.7	**42.9**
S1 S5	0.15M	50.9	**48.9**
S1 S5 S6	0.15M	67.4	**66.4**

ground truth annotations. Moreover, we demonstrate that by further reducing the subject count to two and then to one, the results still show an improvement from 50.9 mm to 48.9 mm and from 67.4 mm to 66.4 mm, respectively. Results are detailed in Table 4. All experiments were conducted using ground truth 2D human skeleton data as the input.

5.4   Recover 3D Pose from Partial 2D Observation

This section explores the impact of applying positional encoding through PerturbPE in scenarios where different numbers of edges are missing (*i.e.,* the problem reduced to subgraph matching). Initially, we examine scenarios with the absence of a single edge. Subsequently, we delve into the more challenging scenarios involving the omission of two edges. In the real world, these scenarios frequently arise since incomplete 2D pose estimates frequently occur in 2D human pose estimation. This typically happens when body parts are outside of the camera view or obscured by objects within the scene.

2D Human Skeleton with One Random Edge Missing. In this first experiment, we eliminate one random edge from the input 2D human skeleton. We experiment the PerturbPE positional encoding under three conditions: no edge, one edge, and two edges missing for computing the perturbed eigenvectors.

Initially, when we compute the perturbed eigenvectors with no edge missing, which leads to addressing multiplicity, the results enhance accuracy from 55.0 mm to 51.4 mm. Further improvements achieved by averaging results from two applications of one-time perturbations, reducing the error to 49.0 mm. Doubling the perturbation with two missing edges further refined accuracy to 48.0 mm, validating our theoretical approach.

This experiment demonstrates that increasing the frequency of perturbations and removing a greater number of edges leads to the extraction of the most robust components of the graph Laplacian eigenbasis. Consequently, leading to more consistent labeling, which in turn improves the outcomes during inference. However this improvement comes with the computational expense.

The results are demonstrated in Table 5, showcasing the effect of PerturbPE positional encoding in partially observed 2D human skeleton input graphs when one edge is missing, also compared to eigenvector labeling [9].

2D Human Skeleton with Two Random Edges Missing. In this experiment, we increase the difficulty of the task by removing two edges from the input 2D human

Table 5. Evaluating the effects of positional encoding with one edge missing on Human3.6M [15] under Protocol #1 (given 2D GT inputs). Lower is better.

Method	MPJPE
PerturbPE Label w. Eigenvector	55.0
PerturbPE Label w. Perturbed Eigenvector w. multiplicity	51.4
PerturbPE Label w. Perturbed Eigenvector w. 1-edge perturb	49.0
PerturbPE Label w. Perturbed Eigenvector w. 2-edge perturb	**48.0**

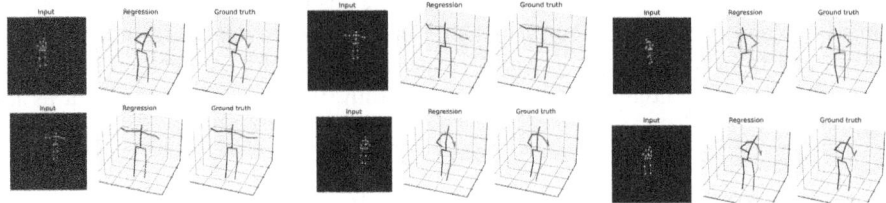

Fig. 2. Qualitative self-occlusion results of PerturbPE on Human3.6M [15]. This figure illustrates the PerturbPE's performance when trained to cope with the absence of any two arbitrary edges in the 2D human skeleton input. The network is trained for any two random edges missing in the input 2D human skeleton. A human skeleton graph with only blue edges provided and red edges absent.

skeleton. We experiment the PerturbPE positional encoding mainly under one condition: no edge is missing for computing the perturbed eigenvectors. In this scenario, we observe that when assigning labels by addressing multiplicity issues—the performance notably improves by approximately 10% (decreasing the MPJPE from 60.00 mm to 54.00 mm). The outcomes of this study are depicted in Table 6.

Further, we compare our results with GFPose [8]. While GFPose adopts a strategy of training distinct networks tailored to specific instances of missing body parts, our approach differs by training a single network that can handle any combination of missing edges. Despite this, our novel PerturbPE architecture significantly surpasses GFPose's performance, demonstrating the superior efficacy of our approach. Quantitative and qualitative results are shown in Table 7 and Fig. 2, respectively. In experiments with a 2D human skeleton missing arms, accuracy improved from 60.0 mm to 58.6 mm. The absence of both legs further demonstrated the effectiveness of our PerturbPE method, enhancing performance to 52.4 mm. PerturbPE outperforms GFPose when either the right or left leg and arm are missing and shows better or similar results when any two edges are absent.

Time Complexity. The computational complexity of the RSPT algorithm is $\mathcal{O}(n^3)$. Nevertheless, this level of complexity becomes acceptable for our purposes by taking into account the number of nodes present in a human skeleton and limiting ourselves to one order of perturbation (MöbiusGCN has an inference time of 0.009 s per sample, while our method takes 0.010 s, demonstrating similar performance).

Table 6. Assessment of PerturbPE positional encoding impact on Human3.6M [15] dataset performance with two edges removed, utilizing Protocol #1 with given 2D ground truth inputs. Lower is better.

Method	MPJPE
MöbiusGCN	60.0
PerturbPE w. Perturbed Eigenvector w. 0-edge missed	54.0

Table 7. Recover 3D pose from partial 2D observation: We train one model for two random missing edges with a masking strategy on Human3.6M dataset [15] under Protocol #1 (given 2D GT inputs). Lower is better.

Occ. Body Parts	Ours	GFPose [8]
2 Legs	**52.4**	53.5
2 Arms	**58.6**	60.0
Left Leg + Left Arm	**48.8**	54.6
Right Leg + Right Arm	**44.6**	53.1
PerturbPE(Any Two Edges Missed)	**54.0**	

6 Conclusions

In this paper, we introduced the PerturbPE technique, a novel positional encoding method that leverages the Rayleigh-Schrodinger Perturbation Theorem (RSPT) to compute perturbed eigenvectors. This technique enables the extraction of consistent and regular components from the eigenbasis in cases where the input graph has missing edges, thereby enhancing the model's robustness and generalizability. Our empirical evidence strongly supports our theoretical claims. Notably, we witnessed an improved performance of up to 12% on the Human3.6M dataset when occlusion led to the absence of an edge. The performance improvement was even more significant in scenarios where two edges were missing, setting a new state-of-the-art benchmark. While the initial results are promising, potential future work could involve refining the PerturbPE technique, investigating other potential applications of the RSPT in graph-structured data analysis, or exploring the scalability of our proposed method for larger datasets. Ultimately, this paper presents a novel encoding approach that boosts the capabilities of Graph Convolutional Networks (GCNs) in handling missing edges in the input graph-structured data, marking a significant stride in the field of 3D human pose estimation.

References

1. Andriluka, M., Pishchulin, L., Gehler, P., Schiele, B.: 2D human pose estimation: new benchmark and state of the art analysis. In: CVPR (2014)
2. Azizi, N., Possegger, H., Rodolà, E., Bischof, H.: 3D human pose estimation using Möbius graph convolutional networks. In: Avidan, S., Brostow, G., Cissé, M., Farinella, G.M., Hassner, T. (eds.) ECCV 2022. LNCS, vol. 13661, pp. 160–178. Springer, Cham (2022). https://doi.org/10.1007/978-3-031-19769-7_10

3. Bevilacqua, B., et al.: Equivariant subgraph aggregation networks. In: ICLR (2022)
4. Bodnar, C., et al.: Weisfeiler and Lehman go topological: message passing simplicial networks. In: ICML (2021)
5. Bruna, J., Zaremba, W., Szlam, A., LeCun, Y.: Spectral networks and locally connected networks on graphs. In: ICLR (2014)
6. Chen, Y., Wang, Z., Peng, Y., Zhang, Z., Yu, G., Sun, J.: Cascaded pyramid network for multi-person pose estimation. In: CVPR (2018)
7. Ci, H., Wang, C., Ma, X., Wang, Y.: Optimizing network structure for 3D human pose estimation. In: ICCV (2019)
8. Ci, H., et al.: GFPose: learning 3D human pose prior with gradient fields. In: CVPR (2023)
9. Dwivedi, V.P., Bresson, X.: A generalization of transformer networks to graphs. In: AAAI Workshop (2020)
10. Fang, H.-S., Xu, Y., Wang, W., Liu, X., Zhu, S.-C.: Learning pose grammar to encode human body configuration for 3D pose estimation. In: AAAI (2018)
11. Gilmer, J., Schoenholz, S.S., Riley, P.F., Vinyals, O., Dahl, G.E.: Neural message passing for quantum chemistry. In: ICML (2017)
12. Hamilton, W., Ying, Z., Leskovec, J.: Inductive representation learning on large graphs. In: NeurIPS (2017
13. Han, X., Zuxuan, W., Zhe, W., Yu, R., Davis, L.S.: VITON: an image-based virtual try-on network. In CVPR (2018)
14. Hossain, M.R.I., Little, J.J.: Exploiting temporal information for 3D human pose estimation. In: ECCV (2018)
15. Ionescu, C., Papava, D., Olaru, V., Sminchisescu, C.: Human3.6M: large scale datasets and predictive methods for 3D human sensing in natural environments. IEEE TPAMI **36**(7), 1325–1339 (2014)
16. Kingma, D.P., Ba, J.: Adam: a method for stochastic optimization. In: ICLR (2015)
17. Kreutz-Delgado, K.: The complex gradient operator and the CR-calculus. arXiv preprint arXiv:0906.4835 (2009)
18. Li, C., Lee, G.H.: Generating multiple hypotheses for 3D human pose estimation with mixture density network. In: CVPR (2019)
19. Li, P., Leskovec, J.: The expressive power of graph neural networks. Graph Neural Netw.: Found. Front. Appl. (2022)
20. Lim, D., et al.: Sign and basis invariant networks for spectral graph representation learning. In: ICLR (2023)
21. Liu, K., Ding, R., Zou, Z., Wang, L., Tang, W.: A comprehensive study of weight sharing in graph networks for 3D human pose estimation. In: Vedaldi, A., Bischof, H., Brox, T., Frahm, J.-M. (eds.) ECCV 2020. LNCS, vol. 12355, pp. 318–334. Springer, Cham (2020). https://doi.org/10.1007/978-3-030-58607-2_19
22. Lü, L., Pan, L., Zhou, T., Zhang, Y.-C., Stanley, H.E.: Toward link predictability of complex networks. Natl. Acad. Sci. (2015)
23. Luo, C., Chu, X., Yuille, A.: A fully convolutional network for 3D human pose estimation. In: BMVC, 2018
24. Luo, D., Songlin, D., Ikenaga, T.: Multi-task neural network with physical constraint for real-time multi-person 3D pose estimation from monocular camera. Multimed. Tools. Appl. **80**, 27223–27244 (2021)
25. Luvizon, D.C., Picard, D., Tabia, H.: 2D/3D pose estimation and action recognition using multitask deep learning. In: CVPR (2018)
26. Ma, X., Su, J., Wang, C., Ci, H., Wang, Y.: Context modeling in 3D human pose estimation: a unified perspective. In: CVPR (2021)
27. Maron, H., Ben-Hamu, H., Shamir, N., Lipman, Y.: Invariant and equivariant graph networks. In: ICLR (2019)

28. Martinez, J., Hossain, R., Romero, J., Little, J.J.: A simple yet effective baseline for 3D human pose estimation. In: ICCV (2017)
29. McCartin, B.J.: Rayleigh-Schrödinger Perturbation Theory: Pseudoinverse Formulation. Springer, Cham (2009)
30. Mehta, D., et al.: Monocular 3D human pose estimation in the wild using improved CNN supervision. In: 3DV (2017)
31. Morris, C., et al.: Weisfeiler and Leman go neural: higher-order graph neural networks. In: AAAI (2019)
32. Nair, V., Hinton, G.E.: Rectified linear units improve restricted Boltzmann machines. In: Proceedings of the ICML (2010)
33. Newell, A., Yang, K., Deng, J.: Stacked hourglass networks for human pose estimation. In: ECCV (2016)
34. Paszke, A., et al.: PyTorch: an imperative style, high-performance deep learning library. In: NeurIPS (2019)
35. Pavlakos, G., Zhou, X., Derpanis, K.G., Daniilidis, K.: Coarse-to-fine volumetric prediction for single-image 3D human pose. In: CVPR (2017)
36. Pavlakos, G., Zhou, X., Daniilidis, K.: Ordinal depth supervision for 3D human pose estimation. In: CVPR (2018)
37. Pavllo, D., Feichtenhofer, C., Grangier, D., Auli, M.: 3D human pose estimation in video with temporal convolutions and semi-supervised training. In: CVPR (2019)
38. Peng, X., Tang, Z., Yang, F., Feris, R.S., Metaxas, D.: Jointly optimize data augmentation and network training: adversarial data augmentation in human pose estimation. In: CVPR (2018)
39. Poier, G., Schinagl, D., Bischof, H.: Learning pose specific representations by predicting different views. In: CVPR (2018)
40. Ramakrishna, V., Kanade, T., Sheikh, Y.: Reconstructing 3D human pose from 2D image landmarks. In: Fitzgibbon, A., Lazebnik, S., Perona, P., Sato, Y., Schmid, C. (eds.) ECCV 2012. LNCS, vol. 7575, pp. 573–586. Springer, Heidelberg (2012). https://doi.org/10.1007/978-3-642-33765-9_41
41. Rhodin, H., et al.: Learning monocular 3D human pose estimation from multi-view images. In: CVPR (2018)
42. Sato, R., Yamada, M., Kashima, H.: Random features strengthen graph neural networks. In: SDM (2021)
43. Sharma, S., Varigonda, P.T., Bindal, P., Sharma, A., Jain, A.: Monocular 3D human pose estimation by generation and ordinal ranking. In: ICCV (2019)
44. Shere, M., Kim, H., Hilton, A.: Temporally consistent 3D human pose estimation using dual 360deg cameras. In: ICCV (2021)
45. Shuman, D.I., Narang, S.K., Frossard, P., Ortega, A., Vandergheynst, P.: The emerging field of signal processing on graphs: extending high-dimensional data analysis to networks and other irregular domains. IEEE Signal Process. Mag. **30**(3), 83–98 (2013)
46. Sminchisescu, C.: 3D human motion analysis in monocular video techniques and challenges. In: AVSS (2006)
47. Sun, X., Shang, J., Liang, S., Wei, Y.: Compositional human pose regression. In: ICCV (2017)
48. Tekin, B., Marquez-Neila, P., Salzmann, M., Fua, P.: Learning to fuse 2D and 3D image cues for monocular body pose estimation. In: ICCV (2017)
49. Veličković, P., Cucurull, G., Casanova, A., Romero, A., Liò, P., Bengio, Y.: Graph attention networks. In ICLR (2018)
50. Weisfeiler, B., Leman, A.: The reduction of a graph to canonical form and the algebra which appears therein. NTI, Series (1968)

51. Xu, K., Hu, W., Leskovec, J., Jegelka, S.: How powerful are graph neural networks? In: ICLR (2019)
52. Xu, T., Takano, W.: Graph stacked hourglass networks for 3D human pose estimation. In: CVPR (2021)
53. Yan, S., Xiong, Y., Lin, D.: Spatial temporal graph convolutional networks for skeleton-based action recognition. In: AAAI (2018)
54. Yang, W., Ouyang, W., Wang, X., Ren, J., Li, H., Wang, X.: 3D human pose estimation in the wild by adversarial learning. In: CVPR (2018)
55. Zhang, M., Chen, Y.: Link prediction based on graph neural networks. In: NeurIPS (2018)
56. Zhao, L., Peng, X., Tian, Y., Kapadia, M., Metaxas, D.N.: Semantic graph convolutional networks for 3D human pose regression. In: CVPR (2019)
57. Zhou, K., Han, X., Jiang, N., Jia, K., Lu, J.: Hemlets pose: learning part-centric heatmap triplets for accurate 3D human pose estimation. In: ICCV (2019)
58. Zhou, X., Zhu, M., Pavlakos, G., Leonardos, S., Derpanis, K.G., Daniilidis, K.: MonoCap: monocular Human Motion Capture using a CNN Coupled with a Geometric Prior. IEEE TPAMI 41(4), 901–914 (2018)
59. Zhou, X., Huang, Q., Sun, X., Xue, X., Wei, Y.: Towards 3D human pose estimation in the wild: a weakly-supervised approach. In: ICCV (2017)

Gradient-Regularized Out-of-Distribution Detection

Sina Sharifi$^{(\boxtimes)}$, Taha Entesari, Bardia Safaei, Vishal M. Patel,
and Mahyar Fazlyab

Johns Hopkins University, Baltimore, MD 21218, USA
{sshari12,tentesa1,bsafaei1,vpatel36,mahyarfazlyab}@jhu.edu

Abstract. One of the challenges for neural networks in real-life applications is the overconfident errors these models make when the data is not from the original training distribution. Addressing this issue is known as Out-of-Distribution (OOD) detection. Many state-of-the-art OOD methods employ an auxiliary dataset as a surrogate for OOD data during training to achieve improved performance. However, these methods fail to fully exploit the local information embedded in the auxiliary dataset. In this work, we propose the idea of leveraging the information embedded in the gradient of the loss function during training to enable the network to not only learn a desired OOD score for each sample but also to exhibit similar behavior in a local neighborhood around each sample. We also develop a novel energy-based sampling method to allow the network to be exposed to more informative OOD samples during the training phase. This is especially important when the auxiliary dataset is large. We demonstrate the effectiveness of our method through extensive experiments on several OOD benchmarks, improving the existing state-of-the-art FPR95 by 4% on our ImageNet experiment. We further provide a theoretical analysis through the lens of certified robustness and Lipschitz analysis to showcase the theoretical foundation of our work. Our code is available at https://github.com/o4lc/Greg-OOD.

Keywords: Out-of-Distribution Detection · Gradient Regularization · Energy-based Sampling

1 Introduction

Neural networks are increasingly being utilized across a wide range of applications and fields, achieving unprecedented performance levels and surpassing traditional state-of-the-art approaches. However, concerns about robustness and

---

S. Sharifi, T. Entesari and B. Safaei—Co-first authors, equally contributed to the work.

Supplementary Information The online version contains supplementary material available at https://doi.org/10.1007/978-3-031-72624-8_26.

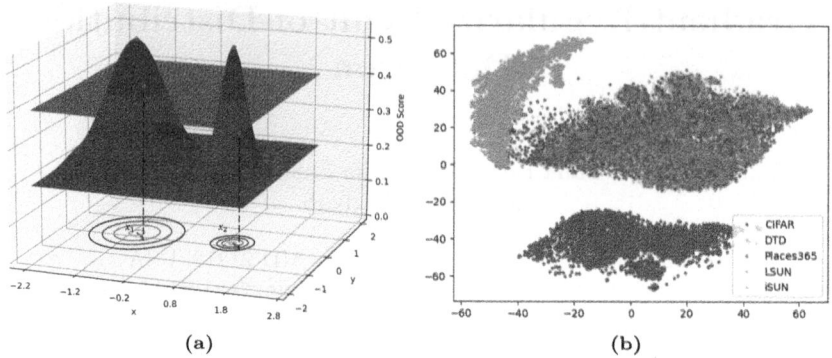

Fig. 1. *Left:* Effect of the local structure of the score manifold on a two-dimensional toy example for OOD detection. The grey plane depicts the score function's decision threshold. An equal amount of perturbation in these two scenarios results in different OOD detections, highlighting the importance of the local structure of the score manifold. *Right:* t-SNE plot showing the representation of ID and OOD datasets for CIFAR experiments.

safety, coupled with significant challenges in verifying robustness and ensuring safe performance impede their use in more sensitive applications [2,32]. One concern that arises when deep models are deployed in the real world is their tendency to produce over-confident predictions upon encountering unfamiliar samples that are distant from the space in which they were trained [47,51]. As a result, the field of Out-of-Distribution (OOD) detection has emerged to study and address this phenomenon.

Many works have been presented in recent years, trying different methods and ideas to regulate the network's output to OOD samples in terms of softmax probability [22,48,76] or energy metrics [36]. One family of methods, known as post-hoc approaches [1,34,37,41,49,53,60,66,72], operate on a given trained network and employ certain statistics, such as the confidence level of the network, the number of active neurons, the contribution of important pathways in the network, etc., to perform OOD detection without using any samples from an auxiliary distribution. Other methods [7,23,27,36,43] use OOD data for training or finetuning a network. The latter family usually tends to achieve better performances in comparison to post-hoc methods, as their access to the auxiliary dataset provides them with additional information regarding the outlier distribution. OOD detection is closely related to anomaly [54,69] and novelty detection [10,39], each with slightly different problem formulations. Our work focuses on the general OOD detection problem.

In this paper, we propose *GReg*, a novel **G**radient **Reg**ularized OOD detection method, which aims to learn the local information of the score function to improve the OOD detection performance. Using information from the gradient of the score function, our method builds on top of current state-of-the-art OOD detection methods (that require training) to incorporate local information into the training process and lean the network towards a smoother manifold.

To motivate this idea, consider Fig. 1a. This figure shows a two-dimensional example of a possible scenario in OOD detection. In this example, x_1 and x_2 are two OOD samples and the network has learned to assign the same OOD scores (denoted by red dots in the figure) to both samples and detect them correctly as OOD. However, the difference in the local behavior of the network at these points, leaves samples in the vicinity of x_2 susceptible to misdetection, whereas, samples close to x_1 are always detected as OOD. To see this, consider the perturbations indicated by the red arrows, where the resulting points in each case represent two possible samples with equal distance from points x_1 and x_2, respectively. However, as can be seen in the figure, the OOD detection algorithm cannot detect one of the samples correctly. This emphasizes the importance of regulating the local behavior of deep learning models, especially when OOD robustness is crucial. Furthermore, Fig. 1b shows the t-SNE plot of the CIFAR and some of the well-known OOD benchmarks. This further confirms that even in high dimensions the ID data (CIFAR) are well-separated from the OOD data.

On top of that, several recent works have shown the effectiveness of some form of informative sampling or clustering on OOD detection [7,27,43]. Sampling gains more importance specifically when a large dataset of outlier samples is available and the network cannot be exposed to all of the data during training. Inspired by such works, we present *GReg+*, the coupling of *GReg* with an energy-based clustering method aimed at making better use of the auxiliary data during training. This is in line with our intuition of utilizing local information as clustering enables the use of samples from diverse regions of the feature space. The diversity put forward by the clustering mechanism allows *GReg+* to achieve its state-of-the-art performance in more regions of the space rather than only performing well on regions well-represented in the dataset.

In summary, our key contributions are:

- We propose the idea of regularizing the gradient of the OOD score function during the training (or fine-tuning). This allows the network to better learn the local information embedded in ID and OOD samples.
- We propose a novel energy-based sampling method that chooses samples – based on their energy levels– from the OOD dataset that represent more vulnerable regions of the space using clustering techniques.
- We provide empirical results and ablation studies on a wide range of architectures and different datasets to showcase the effectiveness of our method, along with an extensive comparison to the state-of-the-art. In addition, we provide a detailed theoretical analysis to justify our results.

To the best of our knowledge, this is the first work that utilizes the norm of the gradient of the score function during the training phase to learn the local behavior of the ID/OOD data aiming to improve the OOD detection performance.

2   Related Work

Several works have attempted to enable deep neural networks with OOD detection capabilities in recent years [4,6,16,31,42,50,55,64,75,77]. These works can be divided into two main groups: methods that do not use auxiliary data (post-hoc) and those that utilize auxiliary data.

2.1 Post-hoc OOD Detection

In one of the earliest attempts to remedy the OOD detection problem, [22] uses the maximum softmax probability (MSP) score to recognize OOD samples during inference. ODIN [23] improves the MSP score's separability by adding small noise to the input and utilizing temperature scaling. [25,35–37] aim to formulate enhanced scoring functions to better distinguish OOD samples from ID samples. For example in [36], the authors propose using energy as a powerful OOD score. GradNorm [25] uses the vector norm of the gradient of the KL divergence between the softmax output and a uniform probability distribution to devise an OOD score. It was later observed by [58], that OOD data tend to have different activation patterns. As a solution, they propose activation truncation to improve OOD detection. Building on this observation, DICE [59] ranks weights based on the measure of contribution, and selectively uses the most contributing weights. LINe [1] further improves on DICE by using Shapley values [56] to more accurately detect important and contributing neurons.

2.2 OOD Detection with Auxiliary Outlier Dataset

When an auxiliary dataset is available, this group of methods [13,14,22,36,63, 68,71,74,80] use this data to train the model to improve the network's robustness against OOD data. OE [22] and Energy [8,36] propose their respective loss functions to be used during the training of the network. Recently, [9] improved [36] by proposing balanced Energy loss that balances the number of samples of the auxiliary OOD data across classes. Furthermore, [52] increases the expected Frobenius Jacobian norm difference between ID and OOD. This is in sharp contrast with our intuition as we aim to decrease the norm of the score function in all regions. Another line of work [15,19,29,44,45] focuses on generating artificial samples that resemble the real OOD data, using various generative models such as GANs. Most recently, OpenMix [79] was proposed as a misclassification detection method, teaching the model to reject pseudo-samples generated from the available outlier samples.

Outlier Sampling: When large amounts of auxiliary data are available, choosing an informative and diverse set of samples to perform training becomes a priority. Many of the state-of-the-art OOD methods use their custom sampling techniques. NTOM [7] uses greedy outlier sampling where outliers are selected based on the estimated confidence. POEM [43] proposes a posterior sampling-based outlier mining framework for OOD detection. Most recently DOS [27] used K-Means [38] to perform clustering on the auxiliary dataset to provide the network with diverse informative OOD samples.

3 Preliminaries

3.1 Notation

We consider a classification setup where $\mathcal{X} \in \mathbb{R}^d$ denotes the input space and $\mathcal{Y} \in \{1, 2, ..., K\}$ denotes the labels. We define a neural network $f \colon \mathbb{R}^d \to \mathbb{R}^K$

where $f_i(x)$ denotes the i-th logit of the neural network to an input x. We denote the feature extractor of the neural network as h, which is followed by a classification layer. We use $f(x) = Wh(x) + b$ for all the models in this paper. We denote the norm of a vector x as $\|x\|$. The log-sum-exp function is defined as $\mathrm{LSE}(x) = \log \sum_{i=1}^{n} \exp(x_i)$. For simplicity, for a given score function S, we use $S(x)$ as a shorthand for $S(f(x))$.

The In-Distribution (ID), Out-of-Distribution (OOD) and Auxiliary distribution over $\mathcal{X}$ are denoted by $\mathcal{D}_{\mathrm{in}}$, $\mathcal{D}_{\mathrm{out}}$ and $\mathcal{D}_{\mathrm{aux}}$, respectively. The corresponding datasets are assumed to be sampled i.i.d from their respective distributions. We denote the indicator function of an event $e \leq 0$ as $\mathbb{I}_{e \leq 0}$, i.e., if $e \leq 0$ then $\mathbb{I}_{e \leq 0} = 1$, and otherwise, $\mathbb{I}_{e \leq 0} = 0$. We denote $\mathcal{B}(x, \varepsilon) = \{y \mid \|y - x\| \leq \varepsilon\}$. Finally, a function f is said to be locally Lipschitz continuous on $\mathcal{X}$ if there exists a positive constant L such that $\|f(x) - f(y)\| \leq L\|x - y\| \ \forall x, y \in \mathcal{X}$. The smallest such constant is called the local Lipschitz constant of f.

3.2   Out-of-Distribution Detection

It has been observed that when neural networks are faced with samples from a different distribution compared to their training distribution, they produce over-confident errors [21]. The goal of OOD detection is to differentiate the ID data from the OOD data. To achieve this goal, the main approach in the literature is to define a scoring function S and use a threshold γ to distinguish different samples. That is, for a sample x, if $S(x) \leq \gamma$, we label it as ID, and label it as OOD otherwise.

Most relevant to our setup, [36] uses the scoring function $S_{\mathrm{En}}(x) = -\mathrm{LSE}(f(x))$, and defines the following loss function to train the network to better differentiate the ID and OOD samples,

$$\mathcal{L}_{S_{\mathrm{En}}} = \mathbb{E}_{(x_{\mathrm{in}}, y_{\mathrm{in}}) \sim \mathcal{D}_{\mathrm{in}}}[\mathbb{I}_{S_{\mathrm{En}}(x_{\mathrm{in}}) \geq m_{\mathrm{in}}}(S_{\mathrm{En}}(x_{\mathrm{in}}) - m_{\mathrm{in}})^2] \tag{1}$$
$$+ \mathbb{E}_{(x_{\mathrm{aux}}, y_{\mathrm{aux}}) \sim \mathcal{D}_{\mathrm{aux}}}[\mathbb{I}_{S_{\mathrm{En}}(x_{\mathrm{aux}}) \leq m_{\mathrm{aux}}}(m_{\mathrm{aux}} - S_{\mathrm{En}}(x_{\mathrm{aux}}))^2],$$

where m_{in} and m_{aux} define two thresholds to filter out points that already have an acceptable level of energy, i.e., if the energy score of an ID (OOD) sample is low (high) enough, it will be excluded from the energy loss.

4   Method

In this section, we propose our method which consists of regularizing the gradient of the score function by adding a new term to the loss, coupled with a novel energy-based sampling method to allow us to choose more informative samples during training. Figure 2 provides an overview of our method $GReg+$ with a simple illustration of how the sampling algorithm utilizes clustering to provide more informative samples.

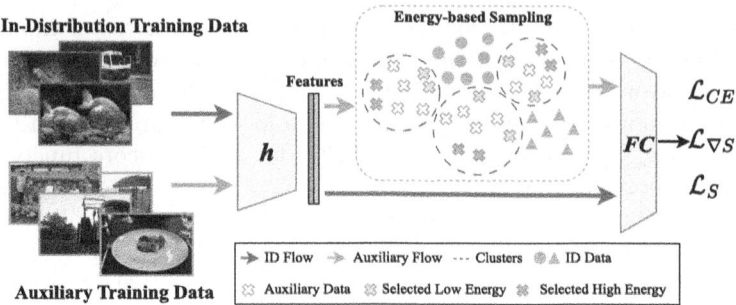

In-Distribution Training Data

Energy-based Sampling

Features

h

FC

$\mathcal{L}_{CE}$

$\mathcal{L}_{\nabla S}$

$\mathcal{L}_S$

Auxiliary Training Data

→ ID Flow   ⇢ Auxiliary Flow   --- Clusters   ● ▲ ID Data

⊗ Auxiliary Data   ⊗ Selected Low Energy   ⊗ Selected High Energy

Fig. 2. Overview of *GReg+*. The gradient loss allows the method to obtain more local information from the training data. The sampling algorithm uses the normalized features of the OOD samples to perform clustering and choose the sampled data based on the energy score (see Algorithm 1) which will be used to calculate the gradient loss to expose the network to samples that can improve the performance of the model.

4.1   Gradient Regularization

Many of the current state-of-the-art methods focus only on the value of their scoring function S and aim at minimizing or maximizing it on different regions of the input space [9, 23, 27, 36, 43, 79]. We argue that by optimizing only based on the value of the score function, we are not utilizing all of the information that is embedded in the data. As such, the generalization of the model requires the auxiliary dataset to be a good representation of the actual OOD distribution and its corresponding support in the image space (which is extremely high-dimensional). In response, we aim to use the local information embedded in the data and propose to do so by leveraging the gradient of the score function S. In other words, we focus not only on the value of the score function S on the auxiliary samples but also on the local behavior of the score function around these data points. The rationale behind this approach is that, based on the definition of the problem, the distribution of ID and the OOD tends to be well separated. Therefore, the area close to an ID sample most likely does not include any OOD samples, and vice versa. Driven by this insight, we propose to add a regularization term $\mathcal{L}_{\nabla S}$ that promotes the smoothness of the score function around the training samples by penalizing the norm of its gradient.

Suppose x is a training sample. Using the first-order Taylor approximation of the score function around x, we have

$$S(x') \simeq S(x) + \nabla S(x)^{\top}(x' - x), \tag{2}$$

for x' sufficiently close to x. As a result, if $\|\nabla S(x)\|$ is small, then equation (2) hints that the difference $|S(x') - S(x)|$ would likely remain small, implying stability of the score manifold around the point x. To this end, we propose the following regularizer. The specific definition of $\mathcal{L}_{\nabla S}$ is a design choice, however, for the case of the energy loss, which is the main focus of this paper, we use the following formulation.

$$\mathcal{L}_{\nabla S_{\mathrm{En}}} = \mathbb{E}_{(x_{\mathrm{in}}, y_{\mathrm{in}}) \sim \mathcal{D}_{\mathrm{in}}} \| \mathbb{I}_{S_{\mathrm{En}}(x_{\mathrm{in}}) \leq m_{\mathrm{in}}} \nabla_{x_{\mathrm{in}}} S_{\mathrm{En}}(x_{\mathrm{in}}) \|$$
$$+ \mathbb{E}_{(x_{\mathrm{aux}}, y_{\mathrm{aux}}) \sim \mathcal{D}_{\mathrm{aux}}} \| \mathbb{I}_{S_{\mathrm{En}}(x_{\mathrm{aux}}) \geq m_{\mathrm{aux}}} \nabla_{x_{\mathrm{aux}}} S_{\mathrm{En}}(x_{\mathrm{aux}}) \| \tag{3}$$

To elaborate on the intuition behind this choice, consider the following cases:

1. If an ID (OOD) sample is correctly detected, we would want the local area around it to be detected the same.
2. If an ID (OOD) sample is mis-detected as OOD (ID), we do not want its local area to behave the same.

Hence, we select thresholds such that the loss only penalizes the gradient of the correctly detected samples. This is in sharp contrast to (1), where the aim is to penalize the energy score of the samples that have undesired energy levels. Therefore, by designing the gradient loss as (3), we allow the energy loss to train the samples that are not yet well learned and only use the samples whose score values are properly learned. Having specified $\mathcal{L}_{\nabla S}$, we train our model using the following loss $\mathcal{L} = \mathcal{L}_{\mathrm{CE}} + \lambda_S \mathcal{L}_S + \lambda_{\nabla S} \mathcal{L}_{\nabla S}$, where λ_S and $\lambda_{\nabla S}$ are regularization constants and $\mathcal{L}_{\mathrm{CE}}$ is the well-known cross-entropy loss.

4.2   OOD Sampling

One concern faced by OOD training schemes is that the OOD data may outnumber the ID data by orders of magnitude. Having access to such a large auxiliary dataset begs the question of which samples should the network see during the training phase. This creates two possible scenarios. Either the training algorithm exposes the model to an equal number of both ID and OOD data points, which means that the model cannot fully utilize the OOD dataset as most of the OOD dataset remains unseen. Alternatively, the algorithm may provide disproportionately more OOD samples, leading to bias in the model due to the imbalance in the samples [62]. These observations hint towards employing a more informative sampling strategy [7,33] that utilizes as much of the dataset as possible while avoiding the pitfalls caused by imbalanced datasets. Whilst the first option that comes to mind is to perform a greedy sampling based on the score function S, it has been observed [27] that such greedy samples could bias the model to some specific regions of the image space, resulting in a model unable to generalize well.

To address these issues, we present Algorithm 1, which uses clustering (to discourage greedy behavior) alongside energy-based scoring to sample more informative images. Next, we explain the sampling algorithm in more detail.

Clustering: Given samples $\{x_i\}_{i=1}^{n_{\mathrm{OOD}}}$, we first calculate the features $z_i = h(x_i)$ and the scores $s_i = -\mathrm{LSE}(f(x_i))$. The next step is to perform clustering in the *feature space*. To perform clustering, we use K-Means with a fixed number of clusters. As K-Means assigns clusters based on Euclidean distances, we normalize the features to $\hat{z}_i = \frac{z_i}{\|z_i\|_2}$ to avoid skewed clusters towards samples with disproportionate feature values.

As for the number of clusters, we use the number of samples in each mini-batch of training, i.e., if in a training iteration the model is presented with n_{ID} ID samples, we cluster the n_{OOD} OOD samples into $k = n_{\text{ID}}$ clusters. We study the choice of the number of clusters in the Supplementary Material. Finally, given a clustering $\{C_j\}_{j=1}^k$, we can acquire cluster labels l_i for each data point x_i, $i = 1, \cdots, n_{\text{OOD}}$.

Energy-Based Sampling: The final step in our sampling algorithm is to choose the *best* samples from each cluster with respect to some criterion. As we use the energy scores s_i to perform OOD detection, we use the same scores for sample selection. Given our choice of loss functions in (1) and (3), we use the samples with the smallest energy scores s_i of each cluster to provide samples useful for the loss term (1), and we use the samples with the largest energy scores to provide samples

---

Algorithm 1 Energy-Based Sampling

Input: Auxiliary dataset $\mathcal{X} \sim \mathcal{D}_{\text{aux}}$ with n_{OOD} samples, number of clusters k, feature extractor h, score function S

Output: Selected   samples: $\bar{\mathcal{X}} \subset \mathcal{X}$;

1: $z_i \leftarrow h(x_i)$, $s_i \leftarrow S(x_i)$, $i = 1, \cdots, n_{\text{OOD}}$.
2: $(\hat{z}_1, \cdots, \hat{z}_{n_{\text{OOD}}}) \leftarrow h_N(z_1, \cdots, z_{n_{\text{OOD}}})$
3: Labels $l_i \leftarrow \text{K-Means}(\{\hat{z}_i\}_{i=1}^{n_{\text{OOD}}}, k)$.
4: **for** $j = 1 \cdots k$ **do**
5:     $\mathcal{I} \leftarrow \{i : l_i = j\}$
6:     $\bar{\mathcal{X}}_j \leftarrow$ Sample $\{x_i\}_{i \in I}$ based on $\{s_i\}_{i \in I}$
7: **end for**
8: return $\bar{\mathcal{X}} \leftarrow \cup_{j=1}^k \bar{\mathcal{X}}_j$

---

that are useful for the loss term (3). Intuitively, $\mathcal{L}_{\mathcal{S}_{\text{En}}}$ aims to increase the energy score of the OOD samples, and so, we need to provide samples whose energy scores are small. On the other hand, $\mathcal{L}_{\nabla \mathcal{S}_{\text{En}}}$ focuses on samples that already have high energy scores to make the model locally behave similarly, and so, we need to also provide samples with high energy scores.

Overall, by clustering, we ensure that the model is exposed to *diverse* regions of the feature space, and by sampling using the scores we ensure that from each region more *informative* samples are used.

5   Theoretical Analysis

In this section, we focus on the theoretical implications of gradient regularization. Traditionally, during the training phase, there is no control over the smoothness of the network and the score function, which can lead to overly sensitive OOD detection. The smoothness of networks has been studied in depth in certified robustness [17,26,40,57] and has been deemed as a desired or even necessary property for the robustness of neural networks. The desirable properties sought by smoothness consequently follow over to OOD detection; simply because OOD detection can be cast as a simple classification task that the aforementioned literature studies. Intuitively, the lack of smoothness decreases the network's OOD robustness as the score values can abruptly change with small perturbations in the image space. To support our intuition, consider Fig. 3 which portrays the evolution of the energy loss and the norm of its gradient with respect to the input in two cases, with and without *GReg*. Although the energy loss decreases

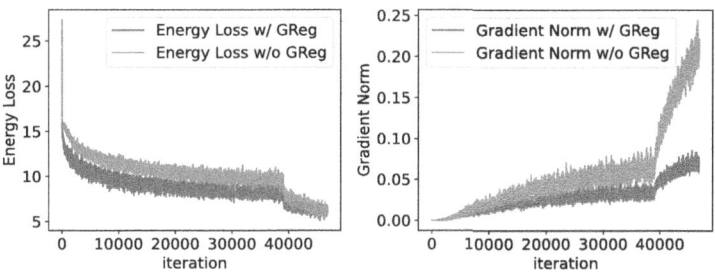

Fig. 3. The evolution of the Energy loss and the norm of the gradient with respect to the input of the score function during training with and without gradient regularization. Gradient regularization reduces the slope of the increase of the gradient norm, without negatively affecting the reduction in energy loss. The sudden change in the behavior of the plots towards the final iterations is due to learning rate scheduling.

in a similar fashion in both scenarios, the norm of the gradient increases much more rapidly in the absence of gradient regularization. The excess increase in the average norm of the gradient potentially points to a more sensitive score manifold. Note that the general increasing trend of the norm of the gradient in Fig. 3 is not unexpected as the network is initialized randomly.

We further motivate our idea using the concept of penalizing the norm of the gradient [5] or upper bounds on the norm of the gradient, like the Lipschitz constant [17,26] for achieving robustness. In other words, if the gradient of the score function S is bounded around some arbitrary point x, there exists an ε-neighborhood in which our prediction of ID/OOD would not change. This means that if the point x is detected as ID (OOD), then all the points $x' \in \mathcal{B}(x, \varepsilon)$ are also provably ID (OOD). This is desirable for us as it is usually assumed that the ID and OOD data are well separated. Formally, suppose S satisfies the following local Lipschitz property on $\mathcal{B}(x, \varepsilon)$ where $x \sim \mathcal{D}_{\text{in}}$

$$|S(x') - S(x)| \le L_S \|x' - x\|, \qquad \forall x' \in \mathcal{B}(x, \varepsilon),$$

where $L_S > 0$. Suppose $x \sim \mathcal{D}_{\text{in}}$ is correctly classified by the score function as ID, i.e., $S(x) < \gamma$. If we want x' to be also labeled as ID, it is sufficient to satisfy $S(x) + L_S \|x' - x\| \le \gamma$, which in turn gives the following certificate

$$\|x' - x\| \le \varepsilon^* = \min\{\varepsilon, \frac{\gamma - S(x)}{L_S}\}, \tag{4}$$

i.e., all points x' within ε^* distance to x will also be labeled correctly as ID. A similar argument follows for OOD samples. This bound suggests that we can increase the certified radius by controlling L_S during training, e.g., through penalizing L_S. There exists a vast literature on Lipschitz estimations for neural networks [17,18,28]. However, using these methods on deep models increases the computational load of training. Instead, we penalize the norm of the gradient, which can be considered as a proxy to the local Lipschitz constant.

Furthermore, for piecewise-linear networks like ReLU and LeakyReLU networks, the local Lipschitz constant is equal to the norm of the gradient. More specifically, for every point x^1, there exists an $\hat{\varepsilon} > 0$ such that the network remains linear in this region, i.e.,

$$f(x') = Ax' + b, \qquad \forall x' \in \mathcal{B}(x, \hat{\varepsilon}).$$

In such a region, the local Lipschitz constant is equal to the norm of the gradient of the points [3], i.e., $L_f = \|A\|$. This further motivates directly penalizing the norm of the gradient of S. Thus, for such networks, minimizing $\|\nabla S(x)\|$ decreases the local Lipschitz constant in the neighborhood of x, which in turn could increase the maximum certified perturbation radius in (4). Note that our argument for using the norm of the gradient in practice does not require the knowledge of $\hat{\varepsilon}$. We just use the fact that such an $\hat{\varepsilon}$ exists and that in the corresponding region, the network will be linear, and as a result, minimizing the norm of the gradient is equivalent to minimizing the local Lipschitz constant for this local neighborhood.

Post-hoc Methods and Gradient Regularization: So far, we have provided a certified robustness radius against OOD misdetection using the Lipschitz constant of the network. This radius (4) is proportional to the inverse of the Lipschitz constant of the score function, and by regularizing the gradient we aim to reduce the Lipschitz constant to increase this radius. Here, we note that many of the state-of-the-art OOD detection methods can also be analyzed in this framework. For example, DICE [59] and LINe [1] use activation pruning and sparsification to remove noisy neurons. It has been shown [26] that for a given matrix, removing a row or column of it results in a reduction of the Lipschitz constant. As activation pruning is essentially a form of masking the weights of the network, methods utilizing activation pruning reduce the upper bound on the Lipschitz constant, acquiring a network with a smaller Lipschitz upper bound.

6   Experiments

We perform extensive OOD detection experiments on CIFAR [30] and ImageNet [12] benchmarks. In our comparisons, we use a wide range of CNN-based architectures. We also provide ablation analysis to demonstrate the effect of each component within our framework.

6.1   Setup

Datasets. For CIFAR experiments, we use CIFAR-10 and CIFAR-100 as ID datasets. We evaluate the models on six different OOD datasets including Textures [11], SVHN [46], Places365 [78], LSUN-cropped, LSUN-resized [70], and

---

[1] We discard the negligible case of points on the boundaries of multiple piecewise-linear spaces.

Table 1. Comparison on the CIFAR-10 benchmark. Mean AUROC and FPR95 percentages on various benchmarks. The experimental results are reported over three trials. The best mean results are bolded and the runner-up is underlined. AUROC and FPR95 are percentages.

Dataset	Method	ResNet		WRN		DenseNet	
		FPR95 ↓	AUROC ↑	FPR95 ↓	AUROC ↑	FPR95 ↓	AUROC ↑
CIFAR-10	MSP [ICLR17] [22]	58.04	90.49	51.52	90.8	52.56	91.7
	ODIN [ICLR18] [34]	34.46	91.59	35.23	89.82	24.88	94.42
	Energy Score [Neurips20] [36]	39.32	92.24	33.4	91.76	28.99	94.09
	ReAct [NeurIPS21] [58]	39.44	92.30	37.54	91.49	25.83	95.27
	Dice [ECCV22] [59]	42.40	91.25	34.12	91.74	26.37	94.61
	LINe [CVPR23] [1]	45.25	90.59	36.27	90.2	14.84	96.95
	OE [ICLR18] [23]	19.92	95.71	21.12	95.55	21.76	95.8
	Energy Loss [NeurIPS20] [36]	11.14	97.53	13.11	97.14	11.26	97.43
	OpenMix [CVPR23] [79]	22.24	96.26	21.92	96	22.86	95.65
	GReg	**7.9**	**97.95**	**7.95**	**98.1**	**7.93**	**98.12**

Table 2. Comparison on the CIFAR-100 benchmark. Mean AUROC and FPR95%s on various benchmarks. The experimental results are reported over three trials. The best mean results are bolded and the runner-up is underlined.

Dataset	Method	ResNet		WRN		DenseNet	
		FPR95 ↓	AUROC ↑	FPR95 ↓	AUROC ↑	FPR95 ↓	AUROC ↑
CIFAR-100	MSP [ICLR17] [22]	74.66	80.10	80.39	75.37	85.87	68.78
	ODIN [ICLR18] [34]	64.00	84.33	67.95	79.86	69.97	81.55
	Energy Score [Neurips20] [36]	68.46	84.15	74.15	79.26	78.91	76.12
	ReAct [NeurIPS21] [58]	59.38	87.59	70.65	80.2	76.27	80.1
	Dice [ECCV22] [59]	77.29	80.35	72.32	79.62	69.80	80.50
	LINe [CVPR23] [1]	72.09	82.93	67.5	80.9	35.16	88.72
	OE [ICLR18] [23]	77.32	63.89	74.22	65.70	60.52	84.21
	Energy Loss [NeurIPS20] [36]	62.77	84.05	65.62	80.18	64.37	83.86
	POEM [ICML22] [43]	61.11	82.43	62.58	79.74	59.33	79.81
	OpenMix [CVPR23] [79]	59.64	88.09	73.12	79.23	67.29	84.47
	DOS [ICLR24] [27]	54.62	88.30	**45.26**	90.76	34.92	93.22
	GReg	59.6	82.92	58.26	86.56	56.29	87.35
	GReg+	**50.78**	**88.75**	48.12	**91.02**	**30.55**	**93.38**

iSUN [67]. Following [59], at test time, we use 10000 randomly selected samples from each of these 6 datasets. Following [79], instead of the recently retracted TinyImages, we use the 300K RandomImages [23] as the auxiliary outlier dataset.

For the ImageNet experiments, following [27], we consider a set of 10 *random* classes from ImageNet-1K as the ID dataset and the rest of the dataset as the auxiliary outlier dataset. For evaluation, we use 10000 randomly selected samples from Textures, Places, SUN [65], and iNaturalist [61] as OOD datasets.

Table 3. Comparison on the ImageNet benchmark. The experimental results are reported over three trials. The best mean results are bolded and the runner-up is underlined. AUROC and FPR95 are percentages.

Method	OOD Datasets								Average	
	iNaturalist		SUN		Places		Textures			
	FPR95 ↓	AUROC ↑	FPR95 ↓	AUROC ↑	FPR95 ↓	AUROC ↑	FPR95 ↓	AUROC ↑	FPR95 ↓	AUROC ↑
MSP [ICLR17] [22]	62.6	87.09	64.3	86.64	62.3	86.72	78.3	75.03	66.87	83.87
ODIN [ICLR18] [34]	48.5	92.07	53.9	90.77	49.3	91.06	69.6	80.77	55.32	88.66
Energy Score [Neurips20] [36]	54	90.77	52.5	90.7	46.6	91.37	73.1	77.54	56.55	87.59
ReAct [NeurIPS21] [58]	51.88	91.52	52.48	90.81	49.56	90.92	64.61	83.77	54.63	89.25
DICE [ECCV22] [59]	32.86	93.6	41.35	92.9	45.82	91.58	65.2	78.61	46.3	89.17
LINe [CVPR23] [1]	29.47	93.77	37.29	92.85	38.85	92.16	52.32	86.38	39.48	91.29
OE [ICLR18] [23]	34.56	92.75	54.63	89.75	54.86	89.05	76.2	75.55	55.06	86.78
Energy Loss [Neurips20] [36]	18.44	95.91	50.19	90.37	49.32	90.47	62.96	77.78	45.23	88.63
DOS [ICLR24] [27]	61.63	88.68	42.69	93.23	45.49	93.37	47.45	92.63	49.31	91.97
GReg	29.54	94.94	45.86	92.53	45.35	92.06	68.29	80.98	47.26	90.13
GReg+	24.11	95	33.98	92.31	32.89	92.7	49.32	88.22	**35.08**	**92.06**

Experimental Setup: For the experiments of *GReg* on CIFAR, we use a pre-trained model of the corresponding architecture and finetune for 20 epochs using SGD with cosine annealing. For the experiments of *GReg+* on CIFAR, we train a model from scratch for 50 epochs with a learning rate of 0.1 and then further train for another 10 epochs with a learning rate of 0.01, using the same optimizer. Following [36], we set $\lambda_S = 0.1$ and set $\lambda_{\nabla S} = 1$.

For the ImageNet experiment, we only perform finetuning on a pre-trained DenseNet-121. See the Supplementary Material for more details.

For the CIFAR experiments, we evaluate our results on ResNet-18 [20], WRN-40 [73], and DenseNet-101 [24] to span a wide range of architectures and parameter numbers.

Sampling Details: Following the explanations in Sect. 4.2, we cluster the auxiliary samples into n_{ID} clusters and choose the lowest and highest scoring samples of each cluster, i.e., in each iteration, we present $2n_{ID}$ auxiliary samples for the training of that iteration. For the CIFAR experiments, we use the whole auxiliary dataset in each epoch for clustering. In other words, at each epoch, the 300K images are split into mini-batches, and clustering is performed on each mini-batch at each iteration so that no auxiliary sample is missed. For the ImageNet experiment, in each epoch, we sample 600K randomly selected images from the remaining (990) classes as the auxiliary dataset and perform the clustering on each mini-batch.

Evaluation Metrics: We report the following metrics: 1) *FPR95*: the false positive rate of the OOD samples when the true positive rate of ID samples is at 95%. 2) *AUROC*: the area under the receiver operating characteristic curve.

Comparison Methods: For comparison with post-hoc OOD methods, we compare to MSP [22], ODIN [23], Energy (score) [36], ReAct [58], DICE [59] and LINe [1]. For the methods that use auxiliary data, we compare with OE [23], Energy (loss) [36], POEM [43], OpenMix [79] and DOS [27]. MSP uses the max-

Table 4. Ablation Study of GReg. AUROC and FPR95% on CIFAR benchmarks averaged over all OOD datasets and three runs on DenseNet.

Method	CIFAR-10		CIFAR-100	
	FPR95 ↓	AUROC ↑	FPR95 ↓	AUROC ↑
OE	21.76	95.8	**60.52**	84.21
OE + Grad	**18.79**	**96.32**	62.54	**84.71**
Energy	11.26	97.43	64.37	83.86
Energy + Grad	**7.93**	**98.12**	**56.29**	**87.35**
POEM	29.15	94.18	59.33	79.81
POEM + Grad	**23.87**	**95.41**	**47.94**	**86.63**

imum softmax probability to detect the OOD samples. ODIN improves upon MSP by introducing noise and temperature scaling. Energy proposes a score function for OOD detection. ReAct uses activation clipping. DICE and LINe build up on that idea by also detecting the important neurons and activation to improve OOD detection. OE and Energy train based on their respective loss functions. POEM uses posterior sampling and DOS proposes diverse sampling to improve OOD detection. OpenMix uses pseudo-samples to improve misclassification detection.

6.2   Results

Evaluation on CIFAR-10: We first evaluate the performance of our method on the CIFAR-10 dataset. Following the literature [1], Table 1 shows the average results over the six commonly used OOD datasets. The detailed results on individual OOD datasets can be found in the Supplementary Material. As the results on CIFAR-10 are saturated, in Table 1 we focus on the effectiveness of gradient regularization and report the rest of the methods in the Supplementary Material. It can be seen that GReg outperforms all the methods in both of the reported metrics. Specifically, GReg decreases the FPR95 of [36] by 3.9%.

Evaluation on CIFAR-100: For the results of this part, we use the same experimental setup and only change the ID dataset to CIFAR-100. The average results of this experiment set are presented in Table 2 and the individual results can be found in the Supplementary Material. In this setup, GReg provides competitive results on the ResNet and WRN architectures with state-of-the-art methods that do not employ sampling. Moreover, coupled with sampling, GReg+ outperforms all current state-of-the-art methods on almost all metrics of the different architectures. On average, GReg+ outperforms the FPR95 metric of DOS [27] by 1.8%, which showcases the effectiveness of energy-based sampling. Results on each OOD dataset can be found in the Supplementary Material.

Evaluation on ImageNet: For this experiment, we randomly select 10 classes of Imagenet-1K as ID and use the rest as the auxiliary data. The results of

Table 5. Ablation Study on Sampling. Comparison among sampling methods on CIFAR-100 on DenseNet averaged over six OOD datasets.

Method	CIFAR-100	
	FPR95 ↓	AUROC ↑
POEM	59.33	79.81
DOS	34.92	93.22
Energy only clustering	31.77	93.20
GReg+	**30.55**	**93.38**

Table 6. Ablation Study on ID Accuracy. Accuracy comparison on CIFAR-10 on two architectures.

Method	CIFAR-10	
	WRN	DenseNet
Clean	94.84	94.03
Energy loss	88.15	91.82
GReg	89.55	92.14
DOS	78.90	87.06
GReg+	83.06	89.89

this large-scale experiment are presented in Table 3. We can see that in terms of FPR95, *GReg+* outperforms DOS and LINe by large margins of 14.23% and 3.68%, respectively, while achieving state-of-the-art AUROC performance. The energy-based sampling approach enables our method to acquire more informative samples compared to the sampling method in [27], resulting in improved FPR95.

6.3 Ablation Study

In this section, we perform extensive experiments to show the effectiveness of components within our framework, individually. That is, we show the effect of regularizing the gradient on other methods and then compare our proposed sampling method with the state-of-the-art. We also study the effect of our OOD schemes on the ID accuracy of the models.

Gradient Regularization: Due to the generality of the idea of gradient regularization, our method can be used in conjunction with many existing methods. To leverage the gradient regularization idea, one needs to define a suitable loss function based on the norm of the gradient of the score, to promote a smoother score manifold. To show the effectiveness of *GReg* on different OOD training schemes, we choose three well-known OOD detection methods, Energy, OE, and POEM, and train a DenseNet with and without gradient regularization on both CIFAR benchmarks. Table 4 shows the overall improvement of all these methods with gradient regularization, in both the FPR95 and AUROC, supporting our claim that coupling *GReg* with other methods could improve their performance.

Sampling Methods: To study the effect of various sampling methods, we compare POEM and DOS to our method with and without gradient regularization. We denote *GReg+* without gradient regularization as Energy-only-clustering. Table 5 compares different methods that utilize some form of sampling and shows the impact of energy-based clustering. On top of that, the comparison between Energy-only-clustering and *GReg+* further encourages gradient regularization.

ID Accuracy: To study the effect of gradient regularization on accuracy, we compare the ID accuracy of our methods to the most relevant state-of-the-art on WRN and DenseNet architectures. Table 6 shows that gradient regularization (*GReg*) improves the ID accuracy of energy on both architectures and *GReg+* also has superior ID accuracy in comparison with DOS.

7   Conclusions

In this paper, we presented the idea of gradient regularization of the score function in OOD detection. We also developed an energy-based sampling algorithm to improve sampling quality when the auxiliary dataset is large. We demonstrate that *GReg* exhibits superior performance compared to methods that do not require clustering, and *GReg+* outperforms all state-of-the-art methods. Furthermore, our experiments show that regularizing the gradient could result in robust models capable of extracting local information from the ID and OOD data. We also provide detailed analytical analysis and ablation studies to support our method.

Acknowledgements. Research was sponsored by the Army Research Laboratory and was accomplished under Cooperative Agreement Number W911NF-23-2-0008. The views and conclusions contained in this document are those of the authors and should not be interpreted as representing the official policies, either expressed or implied, of the Army Research Laboratory or the U.S. Government. The U.S. Government is authorized to reproduce and distribute reprints for Government purposes notwithstanding any copyright notation herein.

References

1. Ahn, Y.H., Park, G.M., Kim, S.T.: LINe: out-of-distribution detection by leveraging important neurons. arXiv preprint arXiv:2303.13995 (2023)
2. Bafghi, R.A., Gurari, D.: A new dataset based on images taken by blind people for testing the robustness of image classification models trained for ImageNet categories. In: Proceedings of the IEEE/CVF Conference on Computer Vision and Pattern Recognition (CVPR), pp. 16261–16270 (2023)
3. Bhowmick, A., D'Souza, M., Raghavan, G.S.: LipBaB: computing exact Lipschitz constant of ReLU networks. In: Farkaš, I., Masulli, P., Otte, S., Wermter, S. (eds.) ICANN 2021, Part IV. LNCS, vol. 12894, pp. 151–162. Springer, Cham (2021). https://doi.org/10.1007/978-3-030-86380-7_13
4. Cao, C., Zhong, Z., Zhou, Z., Liu, Y., Liu, T., Han, B.: Envisioning outlier exposure by large language models for out-of-distribution detection. arXiv preprint arXiv:2406.00806 (2024)
5. Chan, A., Tay, Y., Ong, Y.S., Fu, J.: Jacobian adversarially regularized networks for robustness. arXiv preprint arXiv:1912.10185 (2019)
6. Chen, G., et al.: Learning open set network with discriminative reciprocal points. In: Vedaldi, A., Bischof, H., Brox, T., Frahm, J.-M. (eds.) ECCV 2020, Part III. LNCS, vol. 12348, pp. 507–522. Springer, Cham (2020). https://doi.org/10.1007/978-3-030-58580-8_30

7. Chen, J., Li, Y., Wu, X., Liang, Y., Jha, S.: ATOM: robustifying out-of-distribution detection using outlier mining. In: Oliver, N., Pérez-Cruz, F., Kramer, S., Read, J., Lozano, J.A. (eds.) ECML PKDD 2021, Part III. LNCS (LNAI), vol. 12977, pp. 430–445. Springer, Cham (2021). https://doi.org/10.1007/978-3-030-86523-8_26

8. Chen, S., Huang, L.K., Schwarz, J.R., Du, Y., Wei, Y.: Secure out-of-distribution task generalization with energy-based models. In: Advances in Neural Information Processing Systems, vol. 36 (2024)

9. Choi, H., Jeong, H., Choi, J.Y.: Balanced energy regularization loss for out-of-distribution detection. In: Proceedings of the IEEE/CVF Conference on Computer Vision and Pattern Recognition, pp. 15691–15700 (2023)

10. Choi, S., Lee, H., Lee, H., Lee, M.: Projection regret: reducing background bias for novelty detection via diffusion models. In: Advances in Neural Information Processing Systems, vol. 36, pp. 19230–19245 (2023)

11. Cimpoi, M., Maji, S., Kokkinos, I., Mohamed, S., Vedaldi, A.: Describing textures in the wild. In: Proceedings of the IEEE Conference on Computer Vision and Pattern Recognition, pp. 3606–3613 (2014)

12. Deng, J., Dong, W., Socher, R., Li, L.J., Li, K., Fei-Fei, L.: ImageNet: a large-scale hierarchical image database. In: 2009 IEEE Conference on Computer Vision and Pattern Recognition, pp. 248–255. IEEE (2009)

13. Dhamija, A.R., Günther, M., Boult, T.: Reducing network agnostophobia. In: Advances in Neural Information Processing Systems, vol. 31 (2018)

14. Du, X., Fang, Z., Diakonikolas, I., Li, Y.: How does unlabeled data provably help out-of-distribution detection? arXiv preprint arXiv:2402.03502 (2024)

15. Du, X., Wang, Z., Cai, M., Li, Y.: VOS: learning what you don't know by virtual outlier synthesis. arXiv preprint arXiv:2202.01197 (2022)

16. Fan, K., et al.: Test-time linear out-of-distribution detection. In: Proceedings of the IEEE/CVF Conference on Computer Vision and Pattern Recognition (CVPR), pp. 23752–23761 (2024)

17. Fazlyab, M., Entesari, T., Roy, A., Chellappa, R.: Certified robustness via dynamic margin maximization and improved lipschitz regularization (2023)

18. Fazlyab, M., Robey, A., Hassani, H., Morari, M., Pappas, G.: Efficient and accurate estimation of lipschitz constants for deep neural networks. In: Advances in Neural Information Processing Systems, pp. 11427–11438 (2019)

19. Ge, Z., Demyanov, S., Chen, Z., Garnavi, R.: Generative OpenMax for multi-class open set classification. arXiv preprint arXiv:1707.07418 (2017)

20. He, K., Zhang, X., Ren, S., Sun, J.: Deep residual learning for image recognition. In: Proceedings of the IEEE Conference on Computer Vision and Pattern Recognition, pp. 770–778 (2016)

21. Hein, M., Andriushchenko, M., Bitterwolf, J.: Why ReLU networks yield high-confidence predictions far away from the training data and how to mitigate the problem. In: Proceedings of the IEEE/CVF Conference on Computer Vision and Pattern Recognition, pp. 41–50 (2019)

22. Hendrycks, D., Gimpel, K.: A baseline for detecting misclassified and out-of-distribution examples in neural networks. In: International Conference on Learning Representations (2017). https://openreview.net/forum?id=Hkg4TI9xl

23. Hendrycks, D., Mazeika, M., Dietterich, T.: Deep anomaly detection with outlier exposure. In: International Conference on Learning Representations (2019). https://openreview.net/forum?id=HyxCxhRcY7

24. Huang, G., Liu, Z., Van Der Maaten, L., Weinberger, K.Q.: Densely connected convolutional networks. In: Proceedings of the IEEE Conference on Computer Vision and Pattern Recognition, pp. 4700–4708 (2017)

25. Huang, R., Geng, A., Li, Y.: On the importance of gradients for detecting distributional shifts in the wild. In: Advances in Neural Information Processing Systems, vol. 34, pp. 677–689 (2021)
26. Huang, Y., Zhang, H., Shi, Y., Kolter, J.Z., Anandkumar, A.: Training certifiably robust neural networks with efficient local lipschitz bounds. In: Advances in Neural Information Processing Systems, vol. 34, pp. 22745–22757 (2021)
27. Jiang, W., Cheng, H., Chen, M., Wang, C., Wei, H.: DOS: diverse outlier sampling for out-of-distribution detection. arXiv preprint arXiv:2306.02031 (2023)
28. Jordan, M., Dimakis, A.G.: Exactly computing the local lipschitz constant of ReLU networks. In: Advances in Neural Information Processing Systems, vol. 33, pp. 7344–7353 (2020)
29. Kong, S., Ramanan, D.: OpenGAN: open-set recognition via open data generation. In: Proceedings of the IEEE/CVF International Conference on Computer Vision, pp. 813–822 (2021)
30. Krizhevsky, A., Hinton, G., et al.: Learning multiple layers of features from tiny images (2009)
31. Li, J., Chen, P., He, Z., Yu, S., Liu, S., Jia, J.: Rethinking out-of-distribution (OOD) detection: masked image modeling is all you need. In: Proceedings of the IEEE/CVF Conference on Computer Vision and Pattern Recognition, pp. 11578–11589 (2023)
32. Li, T., Qiao, F., Ma, M., Peng, X.: Are data-driven explanations robust against out-of-distribution data? In: Proceedings of the IEEE/CVF Conference on Computer Vision and Pattern Recognition, pp. 3821–3831 (2023)
33. Li, Y., Vasconcelos, N.: Background data resampling for outlier-aware classification. In: Proceedings of the IEEE/CVF Conference on Computer Vision and Pattern Recognition, pp. 13218–13227 (2020)
34. Liang, S., Li, Y., Srikant, R.: Enhancing the reliability of out-of-distribution image detection in neural networks. arXiv preprint arXiv:1706.02690 (2017)
35. Lin, Z., Roy, S.D., Li, Y.: Mood: Multi-level out-of-distribution detection. In: Proceedings of the IEEE/CVF Conference on Computer Vision and Pattern Recognition, pp. 15313–15323 (2021)
36. Liu, W., Wang, X., Owens, J., Li, Y.: Energy-based out-of-distribution detection. In: Advances in Neural Information Processing Systems, vol. 33, pp. 21464–21475 (2020)
37. Liu, X., Lochman, Y., Zach, C.: GEN: pushing the limits of softmax-based out-of-distribution detection. In: Proceedings of the IEEE/CVF Conference on Computer Vision and Pattern Recognition, pp. 23946–23955 (2023)
38. Lloyd, S.: Least squares quantization in PCM. IEEE Trans. Inf. Theory $\mathbf{28}(2)$, 129–137 (1982)
39. Lo, S.Y., Oza, P., Patel, V.M.: Adversarially robust one-class novelty detection. IEEE Trans. Pattern Anal. Mach. Intell. $\mathbf{45}(4)$, 4167–4179 (2022)
40. Losch, M., Stutz, D., Schiele, B., Fritz, M.: Certified robust models with slack control and large lipschitz constants. arXiv preprint arXiv:2309.06166 (2023)
41. Lu, H., Gong, D., Wang, S., Xue, J., Yao, L., Moore, K.: Learning with mixture of prototypes for out-of-distribution detection. arXiv preprint arXiv:2402.02653 (2024)
42. Ming, Y., Cai, Z., Gu, J., Sun, Y., Li, W., Li, Y.: Delving into out-of-distribution detection with vision-language representations. In: Advances in Neural Information Processing Systems, vol. 35, pp. 35087–35102 (2022)

43. Ming, Y., Fan, Y., Li, Y.: POEM: out-of-distribution detection with posterior sampling. In: International Conference on Machine Learning, pp. 15650–15665. PMLR (2022)
44. Moon, W., Park, J., Seong, H.S., Cho, C.H., Heo, J.P.: Difficulty-aware simulator for open set recognition. In: Avidan, S., Brostow, G., Cissé, M., Farinella, G.M., Hassner, T. (eds.) ECCV 2022. LNCS, vol. 13685, pp. 365–381. Springer, Cham (2022). https://doi.org/10.1007/978-3-031-19806-9_21
45. Neal, L., Olson, M., Fern, X., Wong, W.K., Li, F.: Open set learning with counterfactual images. In: Proceedings of the European Conference on Computer Vision (ECCV), pp. 613–628 (2018)
46. Netzer, Y., Wang, T., Coates, A., Bissacco, A., Wu, B., Ng, A.Y.: Reading digits in natural images with unsupervised feature learning (2011)
47. Nguyen, A., Yosinski, J., Clune, J.: Deep neural networks are easily fooled: high confidence predictions for unrecognizable images. In: Proceedings of the IEEE Conference on Computer Vision and Pattern Recognition, pp. 427–436 (2015)
48. Noh, S., Jeong, D., Lee, J.H.: Simple and effective out-of-distribution detection via cosine-based softmax loss. In: Proceedings of the IEEE/CVF International Conference on Computer Vision, pp. 16560–16569 (2023)
49. Olber, B., Radlak, K., Popowicz, A., Szczepankiewicz, M., Chachuła, K.: Detection of out-of-distribution samples using binary neuron activation patterns. In: Proceedings of the IEEE/CVF Conference on Computer Vision and Pattern Recognition, pp. 3378–3387 (2023)
50. Oza, P., Patel, V.M.: C2AE: class conditioned auto-encoder for open-set recognition. In: Proceedings of the IEEE/CVF Conference on Computer Vision and Pattern Recognition, pp. 2307–2316 (2019)
51. Park, J., Jung, Y.G., Teoh, A.B.J.: Nearest neighbor guidance for out-of-distribution detection. In: Proceedings of the IEEE/CVF International Conference on Computer Vision, pp. 1686–1695 (2023)
52. Park, J., Park, H., Jeong, E., Teoh, A.B.J.: Understanding open-set recognition by Jacobian norm of representation. arXiv preprint arXiv:2209.11436 (2022)
53. Peng, B., Luo, Y., Zhang, Y., Li, Y., Fang, Z.: ConjNorm: tractable density estimation for out-of-distribution detection. arXiv preprint arXiv:2402.17888 (2024)
54. Perini, L., Davis, J.: Unsupervised anomaly detection with rejection. In: Advances in Neural Information Processing Systems, vol. 36 (2024)
55. Safaei, B., Vibashan, V., de Melo, C.M., Hu, S., Patel, V.M.: Open-set automatic target recognition. In: ICASSP 2023-2023 IEEE International Conference on Acoustics, Speech and Signal Processing (ICASSP), pp. 1–5. IEEE (2023)
56. Shapley, L.S., et al.: A value for n-person games (1953)
57. Srinivas, S., Matoba, K., Lakkaraju, H., Fleuret, F.: Efficient training of low-curvature neural networks. In: Advances in Neural Information Processing Systems, vol. 35, pp. 25951–25964 (2022)
58. Sun, Y., Guo, C., Li, Y.: React: Out-of-distribution detection with rectified activations. In: Advances in Neural Information Processing Systems, vol. 34, pp. 144–157 (2021)
59. Sun, Y., Li, Y.: Dice: Leveraging sparsification for out-of-distribution detection. In: Avidan, S., Brostow, G., Cissé, M., Farinella, G.M., Hassner, T. (eds.) ECCV 2022. LNCS, vol. 13684, pp. 691–708. Springer, Cham (2022). https://doi.org/10.1007/978-3-031-20053-3_40
60. Tang, K., Hou, C., Peng, W., Chen, R., Zhu, P., Wang, W., Tian, Z.: CORES: convolutional response-based score for out-of-distribution detection. In: Proceed-

ings of the IEEE/CVF Conference on Computer Vision and Pattern Recognition, pp. 10916–10925 (2024)

61. Van Horn, G., et al.: The inaturalist species classification and detection dataset. In: Proceedings of the IEEE Conference on Computer Vision and Pattern Recognition, pp. 8769–8778 (2018)

62. Wang, L., Han, M., Li, X., Zhang, N., Cheng, H.: Review of classification methods on unbalanced data sets. IEEE Access **9**, 64606–64628 (2021)

63. Wang, Q., Fang, Z., Zhang, Y., Liu, F., Li, Y., Han, B.: Learning to augment distributions for out-of-distribution detection. In: Advances in Neural Information Processing Systems, vol. 36 (2024)

64. Wang, Q., et al.: Out-of-distribution detection with implicit outlier transformation. arXiv preprint arXiv:2303.05033 (2023)

65. Xiao, J., Hays, J., Ehinger, K.A., Oliva, A., Torralba, A.: Sun database: large-scale scene recognition from abbey to zoo. In: 2010 IEEE Computer Society Conference on Computer Vision and Pattern Recognition, pp. 3485–3492. IEEE (2010)

66. Xu, M., Lian, Z., Liu, B., Tao, J.: VRA: variational rectified activation for out-of-distribution detection. In: Advances in Neural Information Processing Systems, vol. 36, pp. 28941–28959 (2023)

67. Xu, P., Ehinger, K.A., Zhang, Y., Finkelstein, A., Kulkarni, S.R., Xiao, J.: TurkerGaze: crowdsourcing saliency with webcam based eye tracking. arXiv preprint arXiv:1504.06755 (2015)

68. Yang, J., et al.: Semantically coherent out-of-distribution detection. In: Proceedings of the IEEE/CVF International Conference on Computer Vision, pp. 8301–8309 (2021)

69. Yang, Y., Lee, K., Dariush, B., Cao, Y., Lo, S.Y.: Follow the rules: reasoning for video anomaly detection with large language models. In: European Conference on Computer Vision (2024)

70. Yu, F., Seff, A., Zhang, Y., Song, S., Funkhouser, T., Xiao, J.: LSUN: construction of a large-scale image dataset using deep learning with humans in the loop. arXiv preprint arXiv:1506.03365 (2015)

71. Yu, Q., Aizawa, K.: Unsupervised out-of-distribution detection by maximum classifier discrepancy. In: Proceedings of the IEEE/CVF International Conference on Computer Vision, pp. 9518–9526 (2019)

72. Yuan, Y., He, R., Dong, Y., Han, Z., Yin, Y.: Discriminability-driven channel selection for out-of-distribution detection. In: Proceedings of the IEEE/CVF Conference on Computer Vision and Pattern Recognition, pp. 26171–26180 (2024)

73. Zagoruyko, S., Komodakis, N.: Wide residual networks. arXiv preprint arXiv:1605.07146 (2016)

74. Zhang, J., Inkawhich, N., Linderman, R., Chen, Y., Li, H.: Mixture outlier exposure: towards out-of-distribution detection in fine-grained environments. In: Proceedings of the IEEE/CVF Winter Conference on Applications of Computer Vision, pp. 5531–5540 (2023)

75. Zhang, J., et al.: OpenOOD v1.5: enhanced benchmark for out-of-distribution detection. arXiv:2306.09301 (2023)

76. Zhang, Z., Xiang, X.: Decoupling maxlogit for out-of-distribution detection. In: Proceedings of the IEEE/CVF Conference on Computer Vision and Pattern Recognition (CVPR), pp. 3388–3397 (2023)

77. Zheng, H., Wang, Q., Fang, Z., Xia, X., Liu, F., Liu, T., Han, B.: Out-of-distribution detection learning with unreliable out-of-distribution sources. In: Advances in Neural Information Processing Systems, vol. 36, pp. 72110–72123 (2023)

78. Zhou, B., Lapedriza, A., Khosla, A., Oliva, A., Torralba, A.: Places: a 10 million image database for scene recognition. IEEE Trans. Pattern Anal. Mach. Intell. **40**(6), 1452–1464 (2017)
79. Zhu, F., Cheng, Z., Zhang, X.Y., Liu, C.L.: OpenMix: exploring outlier samples for misclassification detection. In: Proceedings of the IEEE/CVF Conference on Computer Vision and Pattern Recognition (CVPR), pp. 12074–12083 (2023)
80. Zhu, J., et al.: Diversified outlier exposure for out-of-distribution detection via informative extrapolation. In: Advances in Neural Information Processing Systems, vol. 36 (2024)

Event-Based Mosaicing Bundle Adjustment

Shuang Guo[1]([✉])[iD] and Guillermo Gallego[1,2][iD]

[1] TU Berlin and Robotics Institute Germany, Berlin, Germany
shuang.guo@campus.tu-berlin.de
[2] Einstein Center Digital Future and SCIoI Excellence Cluster, Berlin, Germany

Abstract. We tackle the problem of mosaicing bundle adjustment (i.e., simultaneous refinement of camera orientations and scene map) for a purely rotating event camera. We formulate the problem as a regularized non-linear least squares optimization. The objective function is defined using the linearized event generation model in the camera orientations and the panoramic gradient map of the scene. We show that this BA optimization has an exploitable block-diagonal sparsity structure, so that the problem can be solved efficiently. To the best of our knowledge, this is the first work to leverage such sparsity to speed up the optimization in the context of event-based cameras, without the need to convert events into image-like representations. We evaluate our method, called EMBA, on both synthetic and real-world datasets to show its effectiveness (50% photometric error decrease), yielding results of unprecedented quality. In addition, we demonstrate EMBA using high spatial resolution event cameras, yielding delicate panoramas in the wild, even without an initial map. The code is available at https://github.com/tub-rip/emba.

1 Introduction

Event cameras are novel bio-inspired visual sensors that measure per-pixel brightness changes [11,28,40]. In contrast to the images/frames captured by standard cameras, the output of an event camera is a stream of asynchronous events. This unique working principle endows event cameras with great potential in the tasks of camera motion estimation and scene reconstruction, especially in scenarios of high dynamic range (HDR), low power consumption and/or fast motion [12].

Bundle Adjustment (BA) is the problem of jointly refining the camera motion and the reconstructed scene map that best fit the visual data through a given objective function [2,43] (e.g., reprojection or photometric error). It is a paramount topic in photogrammetry, computer vision and robotics, enabling accurate positioning and measurement technology for applications such as image

Supplementary Information The online version contains supplementary material available at https://doi.org/10.1007/978-3-031-72624-8_27.

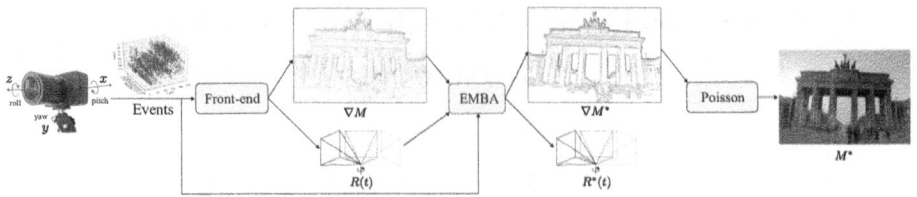

Fig. 1. Our back-end module EMBA jointly refines the camera rotations and panoramic gradient map. The intensity map can be recovered by solving Poisson's equation.

stitching [5], visual odometry (VO) [9], simultaneous localization and mapping (SLAM) [2,6] and AR/VR [10]. BA with frame-based cameras is a mature topic [2,21,26,42]. In contrast, BA with event cameras is still in its infancy, which limits the maturity of the above-mentioned applications for event cameras.

A key problem for BA and SLAM-related tasks is data association, i.e., establishing correspondences between measurements to identify which pixels observe the same scene point [17,22]. So far, event-based BA has been applied in a feature-based (i.e., indirect) manner, i.e., extracting sparse keypoints from image-like event representations and associating them over time (e.g., [7,44]). However, this discards the large amount of information contained in the events (as shown in image reconstruction [29,34,41,46]) and/or quantizes their high temporal resolution. Instead, recent development in direct methods with event cameras [20,22,35] suggest that it should be possible to achieve BA while exploiting the unique characteristics of events, namely that they are continuously (asynchronously) triggered by edges as the camera moves, and that each event is a relative brightness measurement (i.e., an increment if using logarithmic scale).

This paper proposes an event-based mosaicing bundle adjustment (EMBA) method to tackle the photometric BA problem for event cameras (Fig. 1). Rotational motion is a rich and practical scenario, as shown by previous works [3,7,8,16,20,24,25,35]. It is essential to many applications: panorama creation (e.g., in smartphones), star tracking [3], and VO/SLAM in dominantly-rotational motion cases (e.g., satellites [7]). We leverage the linearized event generation model (LEGM) to formulate the problem as a regularized non-linear least squares (NLLS) optimization in the high-dimensional space of camera motions and panoramic gradient maps. Due to the sparse property of event data, only a portion of pixels of the panoramic map are observed and are consequently refined, which naturally leads to a semi-dense gradient map. Moreover, the LEGM also yields a block-diagonal sparsity pattern within the system equations, which we leverage to design an efficient second-order solver.

Therefore, to the best of our knowledge, EMBA is novel. In the experiments, we run EMBA to refine the camera motion trajectories and maps obtained by four state-of-the-art event-based rotation estimation front-end methods [16,24,25,35], on both synthetic and real-world datasets. The results show notable improvements in terms of both camera motion and map quality, revealing previously hidden scene details. We also demonstrate the application of EMBA

Table 1. *VO/SLAM systems that use event data.* The columns indicate the number of degrees of freedom (DOFs), the type of method (**D**irect or **I**ndirect –feature-based), whether there is a refinement step (back-end [6]), and whether the method exploits the event generation model (linearized –LEGM– or not).

System	Year	DOFs	Refine	D/I	EGM	Remarks
Weikersdorfer et al. [45]	2013	3	✗	D	✗	Edge map; 2D scenario
PF-SMT [24]	2014	3	✗	D	✓	LEGM. Brightness map
RTPT [35]	2017	3	✗	D	✗	Probabilistic map
CMax-ω [16]	2017	3	✗	D	✗	The map is a local IWE
EKF-SMT [23]	2018	3	✗	D	✓	LEGM. Brightness map
Chin et al. [7]	2019	3	✓	I	✗	Converts events into frames
CMax-GAE [25]	2021	3	✗	D	✗	The map is a growing 3D-point set
CMax-SLAM [20]	2024	3	✓	D	✗	The map is a panoramic IWE
This work	2024	3	✓	D	✓	LEGM. Event-only photometric BA

to generate high-quality panoramas in outdoor scenes with high resolution event cameras, without requiring an initial map. That is, EMBA just needs a set of initial camera rotations (e.g., provided by an IMU or some front-end) to recover a delicate panorama from scratch while jointly refining the camera motion.

Our contributions can be summarized as follows:

- We propose a novel event-only mosaicing bundle adjustment method, which refines an event-camera's trajectory orientation and gradient map, producing a high quality grayscale panoramic map of the scene (Sect. 3). Its key ingredients are: formulating the BA problem as a regularized NLLS optimization and leveraging the block-diagonal sparsity pattern induced by the chosen parameterization to implement an efficient solver (Sect. 3.2).
- We conduct a comprehensive evaluation on synthetic and real-world datasets (Sect. 4) using four state-of-the-art front-ends for initialization. We demonstrate the method using high-resolution event cameras (VGA and HD), obtaining remarkable panoramas without map initialization (Sect. 4.5).
- We make the source code publicly available.

2   Related Work

Table 1 summarizes some of the VO/SLAM methods operating on event data.

Event-Based Rotation Estimation. Several works have demonstrated the capabilities of event cameras to estimate rotational motion in challenging scenarios (e.g., high speed, HDR). Kim *et al.* [24] proposed a 3-DOF simultaneous mosaicing and tracking (SMT) method consisting of two Bayesian filters operating in parallel (PF-SMT – particle filter SMT); it estimated the camera motion and a grayscale intensity map of the scene. Later, the tracker was replaced by a

Fig. 2. Initial intensity map (top), and final map M (middle), via the refined gradient map ∇M (bottom), for the *street* data from [20]. Three insets are also shown.

Kalman filter [13], yielding EKF-SMT [23]. Although EMBA uses a similar measurement model (LEGM) as SMT, the latter only performs local-time estimation since it is filter-based. Conversely, EMBA is optimization-based, so it can perform global refinement in both time and map domains. Also working in parallel, a real-time panoramic tracking and probabilistic mapping was presented in [35] (RTPT), where the panoramic map of the scene stored the spatial event rate at each point (instead of intensity). Using contrast maximization (CMax), [16] proposed to estimate the camera's angular velocity by warping events on the image plane and aligning them via a focus function [14]. The work has been extended to jointly estimate angular velocity and orientation in [25] (CMax-GAE).

Bundle Adjustment. All above-mentioned methods are short-term, i.e., front-ends of SLAM systems. They lack a BA refinement module, i.e., a SLAM back-end [6], which is desirable to improve accuracy and consistency. Surveying the literature, [7] introduced a BA approach for an event-based system; but it was feature-based and tested only on synthetic star-tracking data. Guo et al. [20] augmented [16] with a back-end, but the map was obtained as a by-product of camera trajectory refinement, resulting in a panoramic edgemap (no intensity map). Expanding the survey to 6-DOF motions, USLAM [36] fused events, frames and IMU data: keypoints were extracted from motion-compensated event-images and frames, and fed to a classical back-end [27]. Recent work [22] (EDS) proposed an event-aided direct VO system, in which event data was leveraged to track the camera motion during the blind time between frames. The system borrowed the photometric BA module from direct methods like DSO [9], which works on images. Current stereo methods are semi-dense and lack a back-end [47], or have a back-end but are feature-based (i.e., indirect) [44]. Therefore, to the best of our knowledge, event-only photometric (i.e., direct) BA is still an unexplored topic, which we address.

3   Event-Based Mosaicing Bundle Adjustment

3.1   Event Generation Model (EGM)

EGM on the Sensor. Each pixel of an event camera measures brightness changes independently, producing an event $e_k \doteq (\mathbf{x}_k, t_k, s_k)$ when the logarithmic intensity change ΔL at the pixel reaches a preset contrast threshold C [12]:

$$\Delta L \doteq L(\mathbf{x}_k, t_k) - L(\mathbf{x}_k, t_k - \Delta t_k) = s_k C, \tag{1}$$

where the event polarity $s_k \in \{+1, -1\}$ indicates the sign of the intensity change, and Δt_k is the time elapsed since the last event at the same pixel $\mathbf{x}_k = (x_k, y_k)^\top$.

Assuming brightness constancy (i.e., optical flow constraint), one can further linearize (1) to obtain the "linearized event generation model" (**LEGM**) [12]:

$$\Delta L \approx -\nabla L(\mathbf{x}_k, t_k) \cdot \mathbf{v} \Delta t_k = s_k C. \tag{2}$$

It states that the brightness change ΔL is caused by an edge ∇L moving with velocity $\mathbf{v}$ during Δt over a displacement $\Delta \mathbf{x} = \mathbf{v} \Delta t$. The dot product captures the condition that no event is triggered if the motion is parallel to the edge.

EGM on the Scene Map. Following [24], we may model the scene map using a mosaic, $M : \mathbb{R}^2 \to \mathbb{R}$ (e.g., Fig. 2), where each map point $\mathbf{p}$ holds the logarithmic intensity of the 3D world point viewed in the direction of $\mathbf{p}$. As the camera rotates, the correspondence between camera pixels $\mathbf{x}$ and map points $\mathbf{p}$ varies. This warp (i.e., geometric transformation) depends on the camera orientation $\mathsf{R}(t)$, intrinsic calibration K and type of projection model π (e.g., equirectangular) used to represent the map: $\mathbf{x} \mapsto \mathbf{p}$, i.e.,

$$\mathbf{p}(t) \doteq \mathbf{W}(\mathbf{x}; \mathsf{R}(t), \mathsf{K}, \pi). \tag{3}$$

Given this correspondence, we may reformulate (2) in terms of the map:

$$\Delta L \approx \nabla M(\mathbf{p}(t_k)) \cdot \Delta \mathbf{p}(t_k) = s_k C, \tag{4}$$

where $\Delta \mathbf{p}(t_k) \doteq \mathbf{p}(t_k) - \mathbf{p}(t_k - \Delta t_k)$ is the map displacement "traveled" by the pixel $\mathbf{x}_k$ as the camera moves during Δt_k. Hence, the LEGM (4) naturally associates each event e_k with the brightness gradient at one map point, $\nabla M(\mathbf{p}(t_k))$.

3.2   Problem Formulation

Objective or Loss Function. Stemming from (4), each event represents a brightness change of predefined size C, which can be modeled in terms of the camera motion $\mathsf{R}(t)$ and the scene texture ∇M. Hence, assuming C is known, a natural design choice consists of formulating the BA problem as finding the motion and scene parameters $\mathbf{P}$ that minimize the sum of square errors

$$g(\mathbf{P}) \doteq \sum_{k=1}^{N_e} \left(\hat{\Delta L}_k(\mathbf{P}) - \Delta L_k \right)^2 \tag{5}$$

where $\Delta L_k \doteq s_k C$ is the measurement, $\hat{\Delta L}_k(\mathbf{P}) \doteq \nabla M \cdot \Delta \mathbf{p}$ is its prediction, and N_e is the number of events considered. Stacking the per-event error terms into a vector $(\mathbf{e})_k \doteq \hat{\Delta L}_k(\mathbf{P}) - \Delta L_k$, we may rewrite the problem as:

$$\min_{\mathbf{P}} g(\mathbf{P}), \quad \text{with} \quad g = \|\mathbf{e}\|^2 = \mathbf{e}^\top \mathbf{e}, \tag{6}$$

where $\mathbf{e}(\mathbf{P}) \in \mathbb{R}^{N_e}$ is the photometric error (or "residual") vector. This is a non-linear least squares (NLLS) function of the state $\mathbf{P}$. It admits the probabilistic interpretation of maximum likelihood estimation under the assumption of zero-mean Gaussian noise in the temporal contrast ΔL, which is a sensible choice according to empirical evidence [28, Fig. 6].

Solution Approach. The standard and effective approach to minimize NLLS objectives like (6) is Gauss-Newton's (GN) method and its variations, e.g., Levenberg-Marquardt (LM) [4,21]. They linearize the errors in terms of the parameters, solve the normal equations and update the model parameters $\Delta \mathbf{P}^*$, iterating until local convergence. For GN, assuming an "operating point" $\mathbf{P}_{\mathrm{op}}$ (in a high dimensional space) and a perturbation $\Delta \mathbf{P}$ around this operating point, the errors are linearized in terms of the parameters:

$$\mathbf{e} \approx \mathbf{e}_{\mathrm{op}} + \mathsf{J}_{\mathrm{op}} \Delta \mathbf{P}, \tag{7}$$

where $\mathbf{e}_{\mathrm{op}} = \mathbf{e}(\mathbf{P}_{\mathrm{op}})$, and J_{op} is the derivative of the error with respect to $\mathbf{P}$. Inserting (7) into (6), differentiating with respect to $\Delta \mathbf{P}$ and setting the result equal to zero yields the necessary optimality condition. The optimal perturbation $\Delta \mathbf{P}^*$ satisfies the system of normal equations:

$$\mathsf{J}_{\mathrm{op}}^\top \mathsf{J}_{\mathrm{op}} \Delta \mathbf{P}^* = -\mathsf{J}_{\mathrm{op}}^\top \mathbf{e}_{\mathrm{op}} \quad \Leftrightarrow \quad \mathsf{A}\,\Delta \mathbf{P}^* = \mathbf{b} \tag{8}$$

The optimal perturbation is used to update the "operating point" and iterate.

While this approach may appear as a classic one, several challenges are involved: (i) designing a meaningful and well-behaved loss, (ii) identifying a suitable parametrization, (iii) finding efficient approximations and solvers for an actual implementation. We tackle these challenges in the upcoming paragraphs.

Parameterization. The two unknowns of the problem are the camera trajectory $\mathsf{R}(t)$ and the scene map M. The continuous-time trajectory is approximated using splines that interpolate $\mathsf{R}(t)$ linearly between two neighboring poses $\{\mathsf{R}_i, \mathsf{R}_{i+1}\} \subset \boldsymbol{\alpha}$. Thus the parameters $\boldsymbol{\alpha}$ represent the discrete "control poses" that specify the trajectory [20]. The map M is approximated by a panoramic intensity image (Fig. 2). Since the error terms $(\mathbf{e})_k$ depend directly on the intensity gradient ∇M, we use its N_p pixels $\boldsymbol{\beta}$ as the parameters to optimize in (6).

The computation of the linearized errors (7) in terms of the camera and scene parameters α and β is given in the supplementary. We use a Lie Group sensible LM approach [4] to linearize and update camera rotations. The perturbation $\Delta\mathbf{P}$ of the parameters has two parts, corresponding to the camera trajectory $\Delta\mathbf{P}_\alpha \in \mathbb{R}^{3N_{\text{poses}}}$ (with 3 DOFs per control pose), and the map pixels $\Delta\mathbf{P}_\beta \in \mathbb{R}^{2N_p}$ (with 2 values/channels per map gradient pixel).

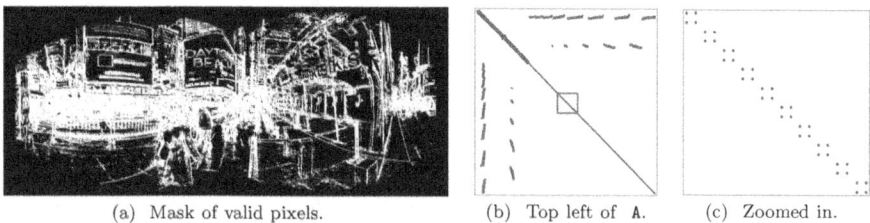

(a) Mask of valid pixels.   (b) Top left of A.   (c) Zoomed in.

Fig. 3. *Sparsity illustration.* (a) Mask of valid pixels; (b) 1000×1000 block at the top left of matrix A; (c) Zoomed-in version of A_{22} showing its block-diagonal structure.

Partitioning and Sparsity. For problems of moderate size, with millions of events (N_e), thousands of pixels (N_p), and hundreds of control poses (N_{poses}), it is intractable to store the Jacobian matrix $J_{\text{op}} \in \mathbb{R}^{N_e \times (3N_{\text{poses}}+2N_p)}$. Even in sparse format, accessing its non-zero entries is time-consuming because it does not have a simple sparsity pattern. Instead, we directly calculate the matrix in the normal equations in an efficient way (see the supplementary). This matrix only depends on the number of unknowns, which is significantly smaller than the size of J_{op}, and has a simpler sparsity pattern.

According to the parameterization, the state $\mathbf{P}$ of the BA problem has two parts: the camera rotations and the scene map. This allows us to partition the perturbation vector and the normal equations (8) in blocks:

$$\begin{pmatrix} A_{11} & A_{12} \\ A_{12}^\top & A_{22} \end{pmatrix} \begin{pmatrix} \Delta\mathbf{P}_\alpha^* \\ \Delta\mathbf{P}_\beta^* \end{pmatrix} = \begin{pmatrix} \mathbf{b}_1 \\ \mathbf{b}_2 \end{pmatrix}, \tag{9}$$

where $A_{11} \doteq J_{\text{op},\alpha}^\top J_{\text{op},\alpha}$ only depends on the derivatives w.r.t. the camera poses, $A_{22} \doteq J_{\text{op},\beta}^\top J_{\text{op},\beta}$ only depends on the derivatives w.r.t. the scene map, and $A_{12} \doteq J_{\text{op},\alpha}^\top J_{\text{op},\beta}$. There is a large size difference: the size of A_{11} (poses) is significantly smaller than that of A_{22} (map pixels), as shown in Fig. 3b. This fact can be leveraged when using well-known tools for solving block-partitioned systems.

Additionally, we can exploit sparsity to implement an efficient LM solver for this problem, as follows. Due to the sparsity of event data, only a portion of map points is observed. We select sufficiently measured map points (e.g., receiving more than five events) as "valid pixels" in the optimization, as shown in Fig. 3a. Furthermore, (6) states that each error term $(\mathbf{e})_k$ only depends on

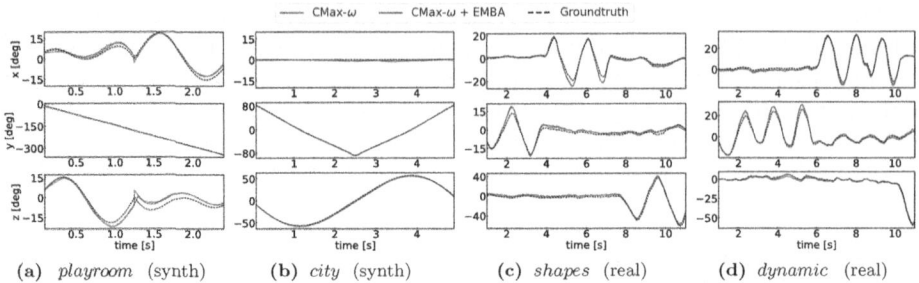

Fig. 4. Camera trajectory degrees-of-freedom (DOFs) before ("CMax-ω") and after ("CMax-ω+EMBA") refinement, for some synthetic and real sequences from [20,30].

the gradient at one map point, which leads to a block-diagonal structure of A_{22} (with blocks of size 2×2), as depicted in Fig. 3c. This makes A_{22} easy to invert. We can leverage this property, together with the block-partitioning structure of the normal equations (9) to solve them efficiently via the Schur complement [4].

Map Regularization (Loss). The fact that each error term $(\mathbf{e})_k$ only depends on the gradient at one map pixel is beneficial for speed, but it causes instabilities: during optimization the values of β at some pixels may grow rapidly, suppressing the update of other pixels. To mitigate this, we add a map prior to (5), so that map pixels evolve with regularization, yielding the objective:

$$\min_{\{\mathsf{R}_i\}, \nabla M} \|\mathbf{e}(\{\mathsf{R}_i\}, \nabla M)\|^2 + \eta \|\nabla M\|^2, \tag{10}$$

where $\{\mathsf{R}_i\} \equiv \boldsymbol{\alpha}$ are the control poses of the camera trajectory, and $\eta > 0$ is the weight of the L^2 regularizer $\|\nabla M\|^2 \equiv \|\beta\|^2$, which encourages smoothness of the estimated map. Consequently, the normal equations of (10) become:

$$\begin{pmatrix} A_{11} & A_{12} & 0 \\ A_{12}^\top & A_{22} + \eta\mathbf{1} & 0 \\ 0 & 0 & \eta\mathbf{1} \end{pmatrix} \begin{pmatrix} \Delta\mathbf{P}_\alpha^* \\ \Delta\mathbf{P}_{\beta_1}^* \\ \Delta\mathbf{P}_{\beta_2}^* \end{pmatrix} = \begin{pmatrix} \mathbf{b}_1 \\ \mathbf{b}_2 - \eta\nabla M_{\mathrm{op},\beta_1} \\ -\eta\nabla M_{\mathrm{op},\beta_2} \end{pmatrix}, \tag{11}$$

where $\mathbf{1}$ is identity matrix, and we distinguish "valid" and "invalid" pixels using β_1 and β_2, respectively. Equation (11) says that "invalid" pixels are updated only using the L^2 regularization, which just sets their gradients to zero. For valid pixels, the L^2 regularization adds a scaled identity matrix to A_{22}, which does not spoil its block-diagonal structure. Hence, it is still cheap to solve (11).

Poisson Reconstruction. Having obtained the optimized gradient map β by solving the above NLLS problem, we can recover the corresponding intensity map M by solving the well-know Poisson's equation [1,24]: $\nabla^2 M = \frac{\partial g_x}{\partial x} + \frac{\partial g_y}{\partial y}$, where g_x and g_y are the two channels of image $\beta \equiv \nabla M$.

4   Experiments

4.1   Experimental Setup

Datasets. We test EMBA on publicly available data: six synthetic sequences from [20] and four real-world sequences from [30]. All sequences contain events, frames (not used), IMU data (not used) and groundtruth (GT) poses.

The synthetic sequences were obtained with a simulator [33], with input panoramas from the Internet. The panoramas covered indoor, outdoor, daylight, night, human-made and natural scenarios, with varying resolution, from 2K (*playroom*), 4K (*bicycle*), 6K (*city* and ,), to 7K (*town* and *bay*). *playroom* was created with a DVS128 camera model (128×128 px) and a duration of 2.5 s, while the other five sequences were created with a DAVIS240C camera model (240×180 px) and a duration of 5 s.

The Event Camera Dataset (ECD) [30] contains four dominantly-rotational motion sequences (*shapes*, *poster*, *boxes* and *dynamic*) that have been commonly used for benchmarking [14–16,18,20,25,31,32,35]. They feature indoor scenes with various amounts of texture complexity and motion. A motion capture system (mocap) outputs accurate GT poses at 200 Hz. We use the ECD data from 1 to 11 s for evaluation, where the camera translation is small.

Initialization. To obtain camera rotations and gradient maps to initialize EMBA, we first run the four front-end methods (EKF-SMT [23], RTPT [35], CMax-GAE [25] and CMax-ω [16]) on all sequences. Then we feed these front-end rotations and the event data into the mapping module of EKF-SMT to obtain the initial maps (e.g., top row of Fig. 2). The front-end rotations are interpolated at 1 kHz and aligned to the GT ones at $t = t_0$ ($t_0 = 0.1$ s for synthetic data and $t_0 = 1$ s for real data) before they are used to obtain initial gradient maps and bootstrap EMBA. Unless otherwise specified, the map size is set to 1024×512 px and the control pose frequency f is set to 20 Hz.

Evaluation Metrics. We evaluate EMBA using two metrics:

Absolute Rotation Error (ARE). The ARE measures the accuracy of the estimated camera rotations. At timestamp t_k, the rotation error between the estimated rotation R_k and the corresponding GT rotation R'_k (obtained by linear interpolation), is given by the angle of their difference $\Delta \mathsf{R}_k = \mathsf{R}'^{\top}_k \mathsf{R}_k$ [4]. Since each front-end method outputs rotations at a different rate, we calculate the errors at such timestamps and aggregate them using the Root Mean Square (RMS). The refined rotations share the same control pose timestamps (regardless of the front-ends), hence errors are calculated on them.

Photometric Error (PhE). The PhE, defined by (5), measures the goodness of fit between the data and the estimated variables in the model, and is the most straightforward criterion to assess the effect of BA algorithms.

Table 2. Absolute rotation RMSE [°] on synthetic sequences. The best results per sequence are in bold. "-" means the method fails on that sequence, and "N/A" indicates that EMBA is not applicable because the corresponding front-end failed on this sequence. RTPT is not shown because it fails on all sequences.

Sequence	playroom		bicycle		city		street		town		bay	
	before	after	before	after	before	after	before	after	before	after	before	after
EKF-SMT	5.86	6.09	1.47	1.18	1.69	1.68	3.44	3.46	4.32	4.40	2.50	2.41
CMax-GAE	4.63	4.42	1.65	1.50	–	N/A	–	N/A	4.66	4.53	–	N/A
CMax-ω	3.22	2.86	1.69	0.92	1.53	0.97	0.97	0.74	1.91	0.86	1.80	1.41

Table 3. Squared photometric error [$\cdot 10^6$] on synthetic data.

Sequence	playroom		bicycle		city		street		town		bay	
	before	after	before	after	before	after	before	after	before	after	before	after
EKF-SMT	0.35	0.23	0.52	0.30	2.62	2.13	1.82	1.52	1.88	1.51	2.26	1.96
CMax-GAE	0.35	0.19	0.53	0.31	–	N/A	–	N/A	1.90	1.54	–	N/A
CMax-ω	0.33	0.15	0.55	0.30	2.71	1.98	1.90	1.34	1.92	1.43	2.30	1.83

4.2   Experiments on Synthetic Data

Figures 4a and 4b compare the initial and refined CMax-ω rotations on *playroom* and *city*. The refined orientations using EMBA agree better with the GT than the initial ones. This is further elaborated in Table 2, using all front-ends: the errors decrease on almost all synthetic sequences, which is most salient when initialized by CMax-ω. For *city*, the rotation RMSE of the CMax-ω trajectory is reduced from 1.53° to 0.97°, and that of *town* decreases from 1.91° to 0.86°.

While the plots in Fig. 4 show small differences between the DOF curves, and the numbers in Table 2 also report apparently small differences (of $\approx$ 1° RMSE), the improvement effect of EMBA is most noticeable in the photometric error PhE (Table 3) and the visual quality of the maps (Figs. 2 and 5). In all trials, the PhE values are significantly reduced (Table 3); the maximal relative decrease reaches 54.5% (when refining the CMax-ω rotations on *playroom*).

Moreover, EMBA is also able to refine higher resolution maps: Fig. 5 compares initial and refined maps of 2048 × 1024 px size. The large improvements of EMBA refinement are visually obvious: blurred regions become sharper, and subtle textures hidden at initialization are revealed, such as the wheels in *bicycle*, the billboards in *city* and the windows and tree leaves in *town*.

In short, EMBA achieves a compelling refinement on synthetic data in terms of rotation accuracy (Table 2), map quality (Fig. 5) and photometric error (Table 3).

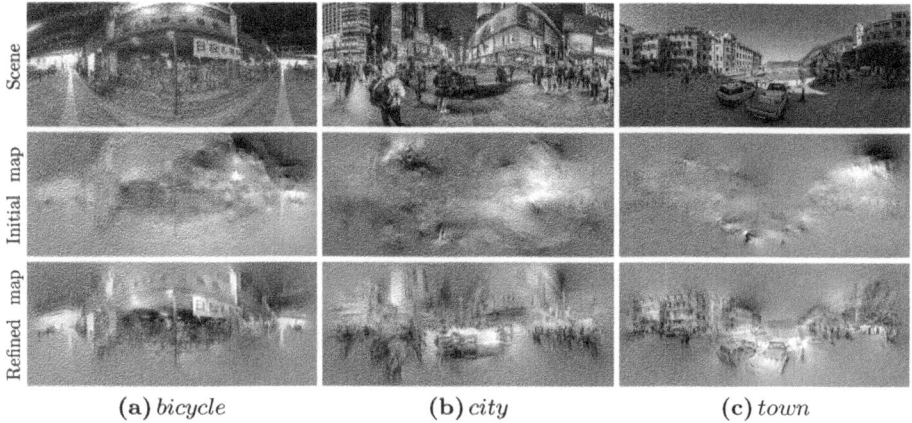

Scene

Initial map

Refined map

(a) *bicycle*   (b) *city*   (c) *town*

Fig. 5. *EMBA results on synthetic data.* Panoramic maps have 2048×1024 px. Initial camera rotations are obtained using CMax-ω [16].

4.3   Experiments on Real Data

The main difficulty of real-world evaluation lies in finding real data that is compatible with the purely rotational motion assumption of the problem [20]. Real-world sequences from established datasets [30] are recorded hand-held, hence they contain some translational motion (see top row of Fig. 6). However, such a translation cannot be removed from the input events. Hence, by design, all rotational motion estimation methods explain the translation in the data using only rotational DOFs. If the translation is non-negligible, comparing the rotations that explain additional DOFs to the rotational component of the GT provided by a 6-DOF motion-capture system [30] can be misleading. Therefore, in the context of photometric BA, the PhE becomes a sensible figure of merit.

Plots 4c and 4d compare the CMax-ω rotations before and after refinement on *shapes* and *dynamic*; the differences are small at this scale. The top part

Table 4. Results on real data. Top: absolute rotation error (ARE) [°], in RMSE form. Bottom: squared PhE [$\cdot 10^5$]. EKF-SMT is not shown since it fails on all sequences.

	Sequence	shapes		poster		boxes		dynamic	
		before	after	before	after	before	after	before	after
ARE	RTPT	2.19	2.85	3.80	3.96	1.74	2.32	2.00	2.29
	CMax-GAE	2.51	2.69	3.63	4.09	2.02	2.40	1.70	2.00
	CMax-ω	4.11	4.44	4.07	4.20	3.22	2.87	3.13	2.79
PhE	RTPT	0.68	0.37	4.69	2.58	4.46	2.30	3.29	2.24
	CMax-GAE	0.61	0.38	5.03	3.07	4.52	2.93	3.16	2.39
	CMax-ω	0.58	0.36	4.37	2.58	3.92	2.25	3.05	2.13

of Table 4 reports the errors using all front-ends. The ARE slightly decreases in some trials while it increases in others; there are no big differences between initial and refined trajectory errors because the estimated camera rotation contains compensation for the translational component. The benefits of EMBA are demonstrated in terms of the PhE (bottom part of Table 4) and the maps (Fig. 6). EMBA considerably reduces the PhE on real-world data (Table 4), around 30% to 50% reduction. Visually, Fig. 6 shows that a remarkable improvement is attained after refinement. Just for comparison, we fed the GT rotations from the mocap into the mapping part of EKF-SMT, and displayed the reconstructed maps in the top row in Fig. 6, which are blurred due to the presence of translation. In the EMBA-refined maps (bottom row), the fine textures on the stones in *poster* are revealed, and the HDR lights in the roof in *dynamic* are also recovered clearly.

In summary, although the real-world evaluation presents difficulties, the effectiveness of EMBA is still proved by the PhE criterion and the map quality.

4.4   Relationship with CMax-SLAM

On the topic of event-based rotational bundle adjustment, the closest relevant work is CMax-SLAM [20]. This section clarifies the differences between EMBA and CMax-SLAM, and demonstrates the potential of their combination.

First of all, they have different objectives and produce different types of map. The objective of CMax-SLAM is to find the camera rotations that maximize the

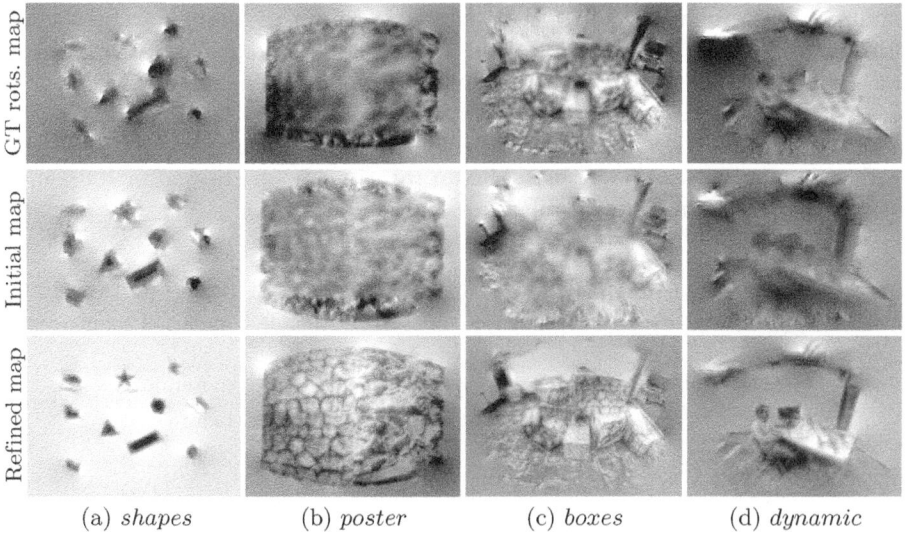

(a) *shapes*   (b) *poster*   (c) *boxes*   (d) *dynamic*

Fig. 6. *EMBA results on real-world data from* [30]. The maps in the top two rows are obtained using the mapping module of SMT [24], by feeding the GT camera rotations or the rotations estimated using CMax-ω, respectively. The refined maps are produced with our method. These are central crops from 1024×512 px panoramic maps.

contrast of the panoramic IWE. Hence, the optimization only involves the camera rotations; the scene map is obtained as a secondary result and it is an edge map (Fig. 7a). The problem is well-posed, not suffering from "event collapse" [37–39]. Conversely, EMBA aims at minimizing the event-based photometric error by refining both camera rotations and an intensity panorama. The intensity map is explicitly modeled as a problem unknown (i.e., it is not a by-product).

In terms of mode of operation, CMax-SLAM works in a sliding-window manner, whereas EMBA processes all events in batch. A sliding window means that rotations far away in time are not refined; hence if CMax-SLAM runs for a very long time interval, some events might align to wrong edges, which does not happen in EMBA. Last but not least, both methods are actually complementary: CMax-SLAM can be used to initialize EMBA and get a clean photometric map of the scene, as shown in Fig. 7. Here, the ARE decreases from 0.470° (see Table II in [20]) to 0.377°. Hence, EMBA can further refine the rotations from CMax-SLAM while jointly reconstructing a precise intensity panorama.

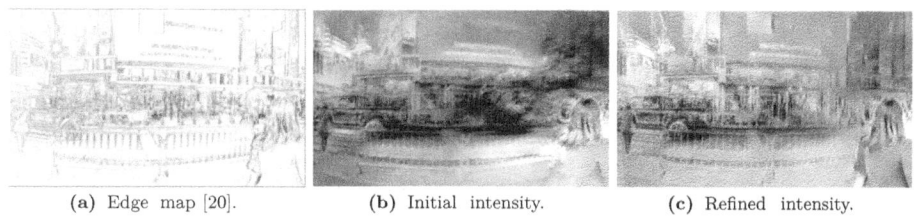

(a) Edge map [20].   (b) Initial intensity.   (c) Refined intensity.

Fig. 7. Results of initializing EMBA with CMax-SLAM [20] (*street* scene).

4.5   Experiments with VGA and HD Event Cameras

An immediate application of EMBA is panoramic imaging (mosaicing), in particular using the latest event cameras, which produce massive event rates due to their high spatial resolution. To this end, we show results on sequences with a DVXplorer (VGA, 640×480 px [40]) and a Prophesee EVK4 (HD, 1280×720 px [11]). For the DVXplorer, which is equipped with an IMU, EMBA is initialized by IMU angular velocity dead-reckoning [4]. For the Prophesee EVK4, which does not have an IMU, we feed the events into [16,20] to provide initial camera orientations. EMBA recovers the gradient maps from scratch while refining the camera motion parameters. The output panoramas in Fig. 8 have high quality, demonstrating the capabilities of EMBA to handle sequences in the wild.

(a) *bridge* .   (b) *crossroad* .

(c) *atrium* .   (d) *graffiti* .

Fig. 8. *Panoramas obtained from scratch.* (a) and (b): DVXplorer data. (c) and (d): Prophesee's EVK4 (1 Mpixel camera) data. Crops from 4K panoramic maps.

4.6   Complexity Analysis and Runtime

EMBA has three main steps. First, evaluating the objective function and its derivatives, whose complexity is $O(N_e)$. Second, forming the normal equations, whose complexity is also $O(N_e)$. Third, solving the (LM-augmented) normal equations, whose main complexity lies in working with A_{22}. Due to the block-diagonal structure, the cost of inverting A_{22} is linear with the number of blocks, i.e., $O(N_p)$, where N_p is the number of valid pixels. Overall, EMBA is lightweight and efficient, compared to the other event-based algorithms.

To support the above statements, a runtime evaluation is carried out. Table 5 reports the average runtime of each step for different scenes (e.g., texture complexity), on a standard laptop (Intel Core i7-1165G7 CPU @ 2.80 GHz). The most expensive step is evaluating the objective function and its derivatives. Obviously, the runtime of EMBA increases with N_e. Leveraging the block-diagonal sparsity, the cost of solving the normal equations using the Schur complement

Table 5. Runtime evaluation of the three main steps of EMBA [s].

ECD sequence	shapes	poster	boxes	dynamic
Obj. func. evaluation	1.114	8.873	7.436	5.837
Forming Normal Eqs.	0.300	2.366	2.106	1.574
Solving Normal Eqs. (Schur)	0.429	2.013	2.006	1.656
Solving Normal Eqs. (CG)	0.267	3.127	3.561	2.056
N_p (valid pixels)	6913	50738	49357	41313
N_e (number of events)	1.78M	12.59M	10.76M	8.80M

increases slowly as the texture complexity grows (order: *shapes* < *dynamic* < *boxes* < *poster*). For comparison, we also implement EMBA with Eigen's [19] built-in conjugate gradient (CG) solver, which does not directly exploit sparsity. We find that the Schur solver is even faster than the CG solver when N_p is large.

4.7 Limitations

Event cameras rely on scene texture to produce data. Too little texture usually leads to tracking failure (EMBA initialization failure), while high texture triggers too many events, which slows down the algorithm in spite of the linear complexity of EMBA. This limitation, shared by most event-based algorithms, might be overcome by adapting the camera's C value and/or downsampling events.

All surveyed event-based rotational SLAM methods assume static scenarios and brightness constancy. Events triggered by moving objects and flickering lights may cause inaccuracy or failure if they are plentiful. Also, the linearization in (2) due to Taylor's approximation in the LEGM model [12] is another source of inaccuracies. However, this was chosen because it endowed matrix A_{22} in the normal equations with a highly beneficial block-diagonal sparsity pattern.

The Levenberg-Marquardt method has its limitations, e.g., local convergence. EMBA may get stuck in local minima of the very large search space if the initialization is not good. This is also a problem in BA for frame-based cameras.

5 Conclusion

We have introduced EMBA, an event-only mosaicing bundle adjustment approach to jointly refine the orientations of a rotating camera and the panoramic gradient map of the scene. We have leveraged the LEGM to formulate the BA problem as a regularized NLLS optimization with a beneficial sparsity pattern so that it can be efficiently solved by exploiting well-developed tools in BA, such as the LM method. To the best of our knowledge, no previous work has constructed and utilized such a useful sparsity for event-based BA without converting events into image-like representations. Through a comprehensive evaluation, the proposed method achieves remarkable results in terms of photometric error (50% decrease), camera poses and map quality. In addition, we have demonstrated the application of EMBA to mosaicing with high-resolution event cameras, of relevance for smartphone applications, even without map initialization. We release the code and hope that our work helps bring maturity to event-based SLAM and related applications.

Acknowledgments. Funded by the Deutsche Forschungsgemeinschaft (DFG, German Research Foundation) under Germany's Excellence Strategy - EXC 2002/1 "Science of Intelligence" - project number 390523135.

References

1. Agrawal, A., Raskar, R., Chellappa, R.: What is the range of surface reconstructions from a gradient field? In: Leonardis, A., Bischof, H., Pinz, A. (eds.) ECCV 2006. LNCS, vol. 3951, pp. 578–591. Springer, Heidelberg (2006). https://doi.org/10.1007/11744023_45

2. Alismail, H., Browning, B., Lucey, S.: Photometric bundle adjustment for vision-based SLAM. In: Lai, S.-H., Lepetit, V., Nishino, K., Sato, Y. (eds.) ACCV 2016. LNCS, vol. 10114, pp. 324–341. Springer, Cham (2017). https://doi.org/10.1007/978-3-319-54190-7_20

3. Bagchi, S., Chin, T.J.: Event-based star tracking via multiresolution progressive hough transforms. In: IEEE Winter Conference on Applications of Computer Vision (WACV), pp. 2132–2141 (2020).https://doi.org/10.1109/WACV45572.2020.9093309

4. Barfoot, T.D.: State Estimation for Robotics - A Matrix Lie Group Approach. Cambridge University Press (2015). https://doi.org/10.1017/9781316671528

5. Brown, M., Lowe, D.G.: Automatic panoramic image stitching using invariant features. Int. J. Comput. Vis. **74**(1), 59–73 (2006). https://doi.org/10.1007/s11263-006-0002-3

6. Cadena, C., et al.: Past, present, and future of simultaneous localization and mapping: toward the robust-perception age. IEEE Trans. Robot. **32**(6), 1309–1332 (2016). https://doi.org/10.1109/TRO.2016.2624754

7. Chin, T.J., Bagchi, S., Eriksson, A.P., van Schaik, A.: Star tracking using an event camera. In: IEEE Conference on Computer Vision and Pattern Recognition Workshops (CVPRW), pp. 1646–1655 (2019). https://doi.org/10.1109/CVPRW.2019.00208

8. Cook, M., Gugelmann, L., Jug, F., Krautz, C., Steger, A.: Interacting maps for fast visual interpretation. In: International Joint Conference on Neural Networks (IJCNN), pp. 770–776 (2011). https://doi.org/10.1109/IJCNN.2011.6033299

9. Engel, J., Koltun, V., Cremers, D.: Direct sparse odometry. IEEE Trans. Pattern Anal. Mach. Intell. **40**(3), 611–625 (2018). https://doi.org/10.1109/TPAMI.2017.2658577

10. Engel, J., et al.: Project Aria: a new tool for egocentric multi-modal AI research (2023)

11. Finateu, T., Niwa, A., et al.: A 1280 × 720 back-illuminated stacked temporal contrast event-based vision sensor with 4.86 μm pixels, 1.066Geps readout, programmable event-rate controller and compressive data-formatting pipeline. In: IEEE International Solid-State Circuits Conference (ISSCC), pp. 112–114 (2020). https://doi.org/10.1109/ISSCC19947.2020.9063149

12. Gallego, G., et al.: Event-based vision: a survey. IEEE Trans. Pattern Anal. Mach. Intell. **44**(1), 154–180 (2022). https://doi.org/10.1109/TPAMI.2020.3008413

13. Gallego, G., Forster, C., Mueggler, E., Scaramuzza, D.: Event-based camera pose tracking using a generative event model (2015). arXiv:1510.01972

14. Gallego, G., Gehrig, M., Scaramuzza, D.: Focus is all you need: Loss functions for event-based vision. In: IEEE Conference on Computer Vision and Pattern Recognition (CVPR), pp. 12272–12281 (2019). https://doi.org/10.1109/CVPR.2019.01256

15. Gallego, G., Rebecq, H., Scaramuzza, D.: A unifying contrast maximization framework for event cameras, with applications to motion, depth, and optical flow estimation. In: IEEE Conference on Computer Vision and Pattern Recognition (CVPR), pp. 3867–3876 (2018). https://doi.org/10.1109/CVPR.2018.00407

16. Gallego, G., Scaramuzza, D.: Accurate angular velocity estimation with an event camera. IEEE Robot. Autom. Lett. **2**(2), 632–639 (2017). https://doi.org/10.1109/LRA.2016.2647639

17. Gehrig, D., Rebecq, H., Gallego, G., Scaramuzza, D.: EKLT: asynchronous photometric feature tracking using events and frames. Int. J. Comput. Vis. **128**, 601–618 (2020). https://doi.org/10.1007/s11263-019-01209-w

18. Gu, C., Learned-Miller, E., Sheldon, D., Gallego, G., Bideau, P.: The spatio-temporal Poisson point process: a simple model for the alignment of event camera data. In: International Conference on Computer Vision (ICCV), pp. 13495–13504 (2021). https://doi.org/10.1109/ICCV48922.2021.01324

19. Guennebaud, G., Jacob, B., et al.: Eigen v3 (2010). http://eigen.tuxfamily.org

20. Guo, S., Gallego, G.: CMax-SLAM: event-based rotational-motion bundle adjustment and SLAM system using contrast maximization. IEEE Trans. Robot. **40**, 2442–2461 (2024). https://doi.org/10.1109/TRO.2024.3378443

21. Hartley, R., Zisserman, A.: Multiple View Geometry in Computer Vision, 2nd edn. Cambridge University Press (2003). https://doi.org/10.1017/CBO9780511811685

22. Hidalgo-Carrió, J., Gallego, G., Scaramuzza, D.: Event-aided direct sparse odometry. In: IEEE Conference on Computer Vision and Pattern Recognition (CVPR), pp. 5781–5790 (2022). https://doi.org/10.1109/CVPR52688.2022.00569

23. Kim, H.: Real-time visual SLAM with an event camera. Ph.D. thesis, Imperial College London (2018)

24. Kim, H., Handa, A., Benosman, R., Ieng, S.H., Davison, A.J.: Simultaneous mosaicing and tracking with an event camera. In: British Machine Vision Conference (BMVC) (2014). https://doi.org/10.5244/C.28.26

25. Kim, H., Kim, H.J.: Real-time rotational motion estimation with contrast maximization over globally aligned events. IEEE Robot. Autom. Lett. **6**(3), 6016–6023 (2021). https://doi.org/10.1109/LRA.2021.3088793

26. Klenk, S.: Photometric bundle adjustment for globally consistent mapping. Master's thesis, TU Munich (2020)

27. Leutenegger, S., Lynen, S., Bosse, M., Siegwart, R., Furgale, P.: Keyframe-based visual-inertial odometry using nonlinear optimization. Int. J. Robot. Res. **34**(3), 314–334 (2015). https://doi.org/10.1177/0278364914554813

28. Lichtsteiner, P., Posch, C., Delbruck, T.: A 128 × 128 120 dB 15 μs latency asynchronous temporal contrast vision sensor. IEEE J. Solid-State Circuits **43**(2), 566–576 (2008). https://doi.org/10.1109/JSSC.2007.914337

29. Mostafavi, I.S., Wang, L., Yoon, K.J.Y.: Learning to reconstruct HDR images from events, with applications to depth and flow prediction. Int. J. Comput. Vis. **129**(4), 900–920 (2021). https://doi.org/10.1007/s11263-020-01410-2

30. Mueggler, E., Rebecq, H., Gallego, G., Delbruck, T., Scaramuzza, D.: The event-camera dataset and simulator: event-based data for pose estimation, visual odometry, and SLAM. Int. J. Robot. Res. **36**(2), 142–149 (2017). https://doi.org/10.1177/0278364917691115

31. Nunes, U.M., Demiris, Y.: Robust event-based vision model estimation by dispersion minimisation. IEEE Trans. Pattern Anal. Mach. Intell. **44**(12), 9561–9573 (2022). https://doi.org/10.1109/TPAMI.2021.3130049

32. Peng, X., Gao, L., Wang, Y., Kneip, L.: Globally-optimal contrast maximisation for event cameras. IEEE Trans. Pattern Anal. Mach. Intell. **44**(7), 3479–3495 (2022). https://doi.org/10.1109/TPAMI.2021.3053243

33. Rebecq, H., Gehrig, D., Scaramuzza, D.: ESIM: an open event camera simulator. In: Conference on Robotics Learning (CoRL). Proceedings of the Machine Learning Research, vol. 87, pp. 969–982. PMLR (2018)

34. Rebecq, H., Ranftl, R., Koltun, V., Scaramuzza, D.: High speed and high dynamic range video with an event camera. IEEE Trans. Pattern Anal. Mach. Intell. **43**(6), 1964–1980 (2021). https://doi.org/10.1109/TPAMI.2019.2963386

35. Reinbacher, C., Munda, G., Pock, T.: Real-time panoramic tracking for event cameras. In: IEEE International Conference on Computational Photography (ICCP), pp. 1–9 (2017). https://doi.org/10.1109/ICCPHOT.2017.7951488

36. Rosinol Vidal, A., Rebecq, H., Horstschaefer, T., Scaramuzza, D.: Ultimate SLAM? Combining events, images, and IMU for robust visual SLAM in HDR and high speed scenarios. IEEE Robot. Autom. Lett. **3**(2), 994–1001 (2018). https://doi.org/10.1109/LRA.2018.2793357

37. Shiba, S., Aoki, Y., Gallego, G.: Event collapse in contrast maximization frameworks. Sensors **22**(14), 1–20 (2022). https://doi.org/10.3390/s22145190

38. Shiba, S., Aoki, Y., Gallego, G.: A fast geometric regularizer to mitigate event collapse in the contrast maximization framework. Adv. Intell. Syst. 2200251 (2022). https://doi.org/10.1002/aisy.202200251

39. Shiba, S., Klose, Y., Aoki, Y., Gallego, G.: Secrets of event-based optical flow, depth, and ego-motion by contrast maximization. IEEE Trans. Pattern Anal. Mach. Intell. 1–18 (2024). https://doi.org/10.1109/TPAMI.2024.3396116

40. Son, B., et al.: A 640 × 480 dynamic vision sensor with a 9μm pixel and 300Meps address-event representation. In: IEEE International Solid-State Circuits Conference (ISSCC), pp. 66–67 (2017). https://doi.org/10.1109/ISSCC.2017.7870263

41. Stoffregen, T., et al.: Reducing the sim-to-real gap for event cameras. In: Vedaldi, A., Bischof, H., Brox, T., Frahm, J.-M. (eds.) ECCV 2020. LNCS, vol. 12372, pp. 534–549. Springer, Cham (2020). https://doi.org/10.1007/978-3-030-58583-9_32

42. Szeliski, R.: Computer Vision: Algorithms and Applications. Texts in Computer Science. Springer, Cham (2010)

43. Triggs, B., McLauchlan, P.F., Hartley, R.I., Fitzgibbon, A.W.: Bundle adjustment—a modern synthesis. In: Triggs, B., Zisserman, A., Szeliski, R. (eds.) IWVA 1999. LNCS, vol. 1883, pp. 298–372. Springer, Heidelberg (2000). https://doi.org/10.1007/3-540-44480-7_21

44. Wang, J., Gammell, J.D.: Event-based stereo visual odometry with native temporal resolution via continuous-time Gaussian process regression. IEEE Robot. Autom. Lett. **8**(10), 6707–6714 (2023). https://doi.org/10.1109/LRA.2023.3311374

45. Weikersdorfer, D., Hoffmann, R., Conradt, J.: Simultaneous localization and mapping for event-based vision systems. In: Chen, M., Leibe, B., Neumann, B. (eds.) ICVS 2013. LNCS, vol. 7963, pp. 133–142. Springer, Heidelberg (2013). https://doi.org/10.1007/978-3-642-39402-7_14

46. Zhang, Z., Yezzi, A., Gallego, G.: Formulating event-based image reconstruction as a linear inverse problem with deep regularization using optical flow. IEEE Trans. Pattern Anal. Mach. Intell. (2022). https://doi.org/10.1109/TPAMI.2022.3230727

47. Zhou, Y., Gallego, G., Shen, S.: Event-based stereo visual odometry. IEEE Trans. Robot. **37**(5), 1433–1450 (2021). https://doi.org/10.1109/TRO.2021.3062252

Author Index

GPSR Compliance

The European Union's (EU) General Product Safety Regulation (GPSR) is a set of rules that requires consumer products to be safe and our obligations to ensure this.

If you have any concerns about our products, you can contact us on ProductSafety@springernature.com

In case Publisher is established outside the EU, the EU authorized representative is:

Springer Nature Customer Service Center GmbH
Europaplatz 3
69115 Heidelberg, Germany

The manufacturer's authorised representative in the EU is Springer
Nature Customer Service Centre GmbH, Europaplatz 3, 69115 Heidelberg,
Germany. If you have any concerns regarding our products, please
contact ProductSafety@springernature.com

Printed and bound by CPI Group (UK) Ltd, Croydon, CR0 4YY

24/04/2026

02096358-0018